W0255268

Handbuch der Pflanzenphysiologie · Encyclopedia of plant physiology

Gesamtdisposition:

Teil I. Allgemeine Grundlagen.

Band I. Genetische Grundlagen physiologischer Vorgänge. Konstitution der Pflanzenzelle.
„ II. Allgemeine Physiologie der Pflanzenzelle.

Teil II. Stoff- und Energiewechsel.

Band III. Pflanze und Wasser.
„ IV. Die mineralische Ernährung der
„ Pflanze.
„ V. Die CO_2-Assimilation.
„ VI. Aufbau, Speicherung, Mobilisierung und Umbildung der Kohlenhydrate.
„ VII. Stoffwechselphysiologie der Fette und fettähnlicher Stoffe.
„ VIII. Der Stickstoff-Umsatz.
„ IX. Der Stoffwechsel der schwefel- und phosphorhaltigen Verbindungen.
„ X. Der Stoffwechsel sekundärer Pflanzenstoffe.
„ XI. Die Heterotrophie.
„ XII. Die Pflanzenatmung, einschließlich Gärungen und Säurestoffwechsel.
„ XIII. Der Stofftransport in der Pflanze.

Teil III. Wachstum, Entwicklung, Bewegungen.

Band XIV. Wachstum und Wuchsstoffe.
„ XV. Differenzierung und Entwicklung.
„ XVI. Außenfaktoren in Wachstum und Entwicklung.
„ XVII. Physiologie der Bewegungen.
1. Bewegungen durch Einflüsse mechanischer und elektrischer Natur sowie durch Strahlungen.
2. Bewegungen durch Einflüsse der Temperatur, Schwerkraft, chemischer Faktoren und aus inneren Ursachen.
„ XVIII. Sexualität, Fortpflanzung, Generationswechsel.

Synopsis of the contents:

Part I. General foundations.

Vol. I. Genetic control of physiological processes. The constitution of the plant cell.
„ II. General physiology of the plant cell.

Part II. Metabolism.

Vol. III. Water relations of plants.
„ IV. Mineral nutrition of plants.
„ V. The assimilation of carbon dioxide.
„ VI. Formation, storage, mobilization, and transformation of carbohydrates.
„ VII. The metabolism of fats and related compounds.
„ VIII. Nitrogen metabolism.
„ IX. The metabolism of sulfur- and phosphorus-containing compounds.
„ X. The metabolism of secondary plant products.
„ XI. Heterotrophy and heterotrophic plants.
„ XII. Plant respiration, incl. fermentations and acid metabolism in plants.
„ XIII. Translocation in plants.

Part III. Growth, development, movements.

Vol. XIV. Growth and growth substances.
„ XV. Differentiation and development.
„ XVI. External factors affecting growth and development.
„ XVII. Physiology of movements.
1. Movements due to mechanical and electrical stimuli and to radiations.
2. Movements due to the effects of temperature, gravity, chemical factors and internal factors.
„ XVIII. Sexuality, reproduction, alternation of generations.

ENCYCLOPEDIA OF PLANT PHYSIOLOGY

EDITED BY

W. RUHLAND

COEDITORS

E. ASHBY · J. BONNER · M. GEIGER-HUBER · W. O. JAMES
A. LANG · D. MÜLLER · M. G. STÅLFELT

VOLUME VII

THE METABOLISM OF FATS AND RELATED COMPOUNDS

CONTRIBUTORS

J. ASSELINEAU · E. BAMANN · W. FRANKE · H. FREHSE
J. A. LOVERN · M. L. MEARA · K. SCHMALFUSS · M. STEINER
P. K. STUMPF · E. ULLMANN · H. v. WITSCH · A. ZELLER

SUBEDITOR

M. STEINER

WITH 59 FIGURES

SPRINGER-VERLAG
BERLIN · GÖTTINGEN · HEIDELBERG
1957

HANDBUCH DER PFLANZENPHYSIOLOGIE

HERAUSGEGEBEN VON

W. RUHLAND

IN GEMEINSCHAFT MIT

E. ASHBY · J. BONNER · M. GEIGER-HUBER · W. O. JAMES
A. LANG · D. MÜLLER · M. G. STÅLFELT

BAND VII

STOFFWECHSELPHYSIOLOGIE DER FETTE UND FETTÄHNLICHER STOFFE

BEARBEITET VON

J. ASSELINEAU · E. BAMANN · W. FRANKE · H. FREHSE
J. A. LOVERN · M. L. MEARA · K. SCHMALFUSS · M. STEINER
P. K. STUMPF · E. ULLMANN · H. v. WITSCH · A. ZELLER

REDIGIERT VON

M. STEINER

MIT 59 ABBILDUNGEN

SPRINGER-VERLAG
BERLIN · GÖTTINGEN · HEIDELBERG
1957

ISBN-13: 978-3-642-94705-6 e-ISBN-13: 978-3-642-94704-9
DOI: 10.1007/978-3-642-94704-9

Softcover reprint of the hardcover 1st edition 1957

Druck der Universitätsdruckerei H. Stürtz AG., Würzburg

Inhaltsverzeichnis. — Contents.

I. Einführung und Übersicht.

II. Die Pflanzenfette.

(Chemie und sonstige Eigenschaften, Methoden, Verbreitung in Pflanzenarten und -organen).

III. Fermente des Fettstoffwechsels.

IV. Biochemistry of fat formation.

V. Physiologie der Fettbildung und Fettspeicherung.

Mitarbeiter von Band VII. — Contributors of volume VII.

JEAN ASSELINEAU, Ingénieur-chimiste, Docteur-es-Sciences, Chargé de Recherches au C.N.R.S., Institut de Biologie physico-chimique, 13 rue Pierre Curie, Paris 5e (France).

Professor Dr. EUGEN BAMANN, Direktor des Instituts für Pharmazie und Lebensmittelchemie der Universität München, München 15, Pettenkoferstr. 14a.

Dr. WILHELM FRANKE, a. o. Professor, Botanisches Institut der Universität Köln, Gyrhofstr. 15.

Dr. HELMUT FREHSE, Deutsches Institut für Fettforschung, Münster (Westf.), Piusallee 76.

Dr. JOHN ARNOLD LOVERN, Senior Principal Scientific Officer, Department of Scientific and Industrial Research, Food Investigation Organization, Torry Research Station, Aberdeen, Scotland (Great Britain).

M. L. MEARA, Ph. D., D. Sc., F.R.I.C., The Aliphatic Research Co. Ltd, Rhodes, Middleton, Manchester (Great Britain).

Professor Dr. KARL SCHMALFUSS, Direktor des Instituts für Pflanzenernährung und Bodenkunde der Martin Luther-Universität Halle-Wittenberg, Halle (Saale), Sophienstr. 17b.

Professor Dr. phil. MAXIMILIAN STEINER, Pharmakognostisches Institut der Universität Bonn, Meckenheimer Allee 170a.

Dr. P. K. STUMPF, Associate Professor of Plant Biochemistry, University of California, Berkeley 4, California (USA).

Privatdozentin Dr. ELSA ULLMANN, Abteilungsvorstand am Institut für Pharmazie und Lebensmittelchemie der Universität München, München 15, Pettenkoferstr. 14a.

Dr. phil. HANS v. WITSCH, o. Professor für Botanik, Direktor des Botanischen Instituts, Botanisches Institut der Technischen Hochschule München, Freising-Weihenstephan bei München.

Dr. ALFRED ZELLER, tit. a. o. Univ.-Professor (Wien), Chesterford Park Research Station Saffron Walden, Essex, (Great Britain); [Privat: 84 Hills Avenue, Cambridge (Great Britain) und Wien II, Konradgasse 3, Österreich].

Mitarbeiter von Band VII — Contributors of volume VII

Einführung und Übersicht.

Von

Maximilian Steiner.

I. Geschichtliche Bemerkungen.

Noch im Jahre 1924 haben E. F. ARMSTRONG und J. ALLEN einen Sammelbericht über die Biochemie der Fette "a neglected chapter in chemistry, the fats" betitelt. Auf der anderen Seite hat H. B. BULL (1937) auf die ständige Zunahme der Fettliteratur hingewiesen. Eine Kurve im Einleitungskapitel seines Buches "The biochemistry of the lipids" zeigt, daß die Zahl der Arbeiten auf dem Fettgebiet (nach den Referaten in den Chemical Abstracts) von etwa 60 im Jahre 1907 auf rund 800 im Jahre 1932 angestiegen war. Ohne daß eine präzise Statistik vorliegt, kann angenommen werden, daß die Kurve der jährlichen Veröffentlichungen auf dem Berichtsgebiete weiterhin bis heute stetig angestiegen ist. Es darf aber kaum übersehen werden, daß ein sehr großer Teil der Arbeiten sich mit der chemischen Technologie der Fette befaßt, mit ihrer Gewinnung, Reinigung, Veredlung, Konservierung. Das ist bei der großen Bedeutung, die den Fetten als Nahrungsmitteln und Rohstoffen der Technik zukommt, nicht weiter verwunderlich. Ein Großteil der übrigen Arbeiten ist rein deskriptiver Natur: alljährlich wird eine beträchtliche Anzahl von Fetten des Pflanzen- und Tierreiches neu beschrieben. Erst ein relativ kleiner Rest an Veröffentlichungen schließlich betrifft Fragen des Fettstoffwechsels; davon stammt wiederum der weitaus größere Teil aus tierphysiologischen (bzw. physiologisch-chemischen) Laboratorien, nur der kleinere beschäftigt sich mit dem Fettstoffwechsel der pflanzlichen Lebewesen.

1. Die Geschichte der deskriptiven *Biochemie* der Fette begann mit der Entdeckung des Glycerins im Olivenöl durch K. W. SCHEELE (1783). M. E. CHEVREUIL (1823) gelang es als erstem, Fette in Glycerin und Fettsäuren zu zerlegen und eine Reihe der letzteren (Stearin-, „Margarin"-, Öl-, Butter-, Capron-, Caprinsäure) zu beschreiben.

Damit war die eigentliche Fettforschung eingeleitet. Ihr weiterer Ausbau brachte zunächst eine starke Erweiterung der Kenntnisse der natürlich vorkommenden Fettsäuren und ihrer qualitativen und quantitativen Verteilung in Fetten beider Naturreiche. Soweit diese Forschungsergebnisse Pflanzenfette betreffen, sind sie in F. CZAPEKs Biochemie der Pflanzen (1913, insbesondere Bd. I, S. 109ff.) zusammengefaßt.

Einen wesentlichen Schritt nach vorwärts brachten die ab 1927 durchgeführten Untersuchungen von T. P. HILDITCH und seiner Schule (s. T. P. HILDITCH 1938, 1949) über die *Konstitution der Glyceride* der natürlichen Fette. Hierüber lagen bis dahin nur spärliche und fragmentarische Angaben vor. Das wichtige Ergebnis dieser Arbeiten ist die — wenigstens für die meisten Pflanzenfette gültige — „Regel der gleichmäßigen Verteilung" ("Rule of even distribution"). Danach werden von der Zelle bevorzugt gemischte Glyceride vom Typus G $(R_1R_2R_3)$ bzw. G $(R_1R_1R_2)$ oder G $(R_1R_2R_2)$ gebildet, und nicht, wie ehedem meist angenommen wurde, einfache Triglyceride [G ($R_1R_1R_1$), G ($R_2R_2R_2$) usw.][1]. Die Verbindung zwischen verschiedenen Fettsäuren und Glycerin geschieht also nicht nach den Regeln der statistischen Wahrscheinlichkeit ("random distribution"), wie es der Fall ist, wenn in vitro Glycerin mit einem Gemisch von Fettsäuren verestert wird. Trotz dieser bemerkenswerten Verteilungsregel bleibt die „Artspezifität" der Pflanzenfette von rein „quantitativer" Art. Es handelt sich um eine „Spezifität der *Bausteine*" (nämlich der Fettsäuren) und nicht um eine solche der „*Bausteine und der Anordnung*", wie wir sie für die Aminosäuren der Zellproteine annehmen müssen.

Die Entdeckung und erste Erforschung der Phosphatide ist durch die Namen L. N. VAUQUELIN (1812), M. GOBLEY (1845—1858), C. DIACONOW (1867, 1868), J. L. W. THUDICHUM (1901) gekennzeichnet. Im übrigen sei auf die kurze geschichtliche Darstellung von E. KLENK (1951) verwiesen.

Im Gegensatz zu den Fetten pflanzlicher Reserveorgane (Perikarpien, Samen) sind die Fette vegetativer Pflanzenorgane sehr wenig untersucht worden. Das hängt zweifellos mit den großen Schwierigkeiten zusammen, die sich im letzteren Falle der Beschaffung ausreichenden Analysenmaterials entgegenstellen. Aus ähnlichen Gründen ist unsere Kenntnis der Fette pflanzlicher Mikroorganismen recht gering. Noch am besten bekannt sind die „Fette" der Bakterien, die durch ihre abweichende chemische Beschaffenheit und zum Teil auch wegen ihrer Beziehungen zu pathogenen Prozessen das Interesse der Biochemiker gefunden haben. Das gilt insbesondere für die Lipoide des Tuberkelbacillus und anderer säurefester Bakterien, deren intensives Studium zuerst R. J. ANDERSON und seine Schule (erste Zusammenfassung bei R. J. ANDERSON 1939) in Angriff nahm. Unter den Pilzfetten ist eigentlich nur das Fett des pharmazeutisch wichtigen Mutterkornes von mehreren Forschern gründlich bearbeitet worden. Sehr viel geringer sind unsere Kenntnisse von den Fetten der Mycelpilze und der Sproßpilze. Noch viel seltener wurden die Fette von Algen analysiert. Da in den Kapiteln dieses Handbuchbandes, die den Mikroorganismen-Fetten gewidmet sind, das wesentliche Tatsachenmaterial ziemlich erschöpfend dargestellt wird und die einschlägigen Originalarbeiten zitiert werden, erübrigt sich hier ein Eingehen auf die geschichtliche Entwicklung dieser Spezialgebiete.

2. Schon F. J. P. MEYEN (1838) war es bekannt, daß in unreifen Ölsamen reichlich Stärke vorhanden ist, deren Menge bei zunehmender Reife zugunsten der Fette abnimmt. J. SACHS (1859) untersuchte als erster den umgekehrten Vorgang, die Umwandlung von Fett in Kohlenhydrate (Stärke) bei der Keimung von Fettsamen. J. B. LAWES (1853) und J. B. LAWES und J. H. GILBERT (1860) lieferten den Beweis, daß im Tierkörper die Nahrungskohlenhydrate zu Fetten umgewandelt werden. Daß Zucker als Ausgangsmaterial für die Fettbildung durch heterotrophe Mikroorganismen dienen kann, hat zuerst C. v. NAEGELI (1879) bei *Penicillium* gezeigt.

Den Weg der Umwandlung von Kohlenhydraten zu Fetten im lebenden Organismus hat als erster E. FISCHER (1894) theoretisch formuliert. Er nahm an, daß Zuckermoleküle ohne vorangehenden Abbau verkoppelt und durch Wegnahme bzw. Verschiebung von Sauerstoff in Fettsäuremoleküle umgewandelt werden. Demgegenüber vertrat A. MAGNUS-LEVY (1902) die Meinung, daß die Fettsäuren aus C_2-Verbindungen synthetisiert werden. Auch er konnte seine Theorie zunächst freilich nicht durch direkte experimentelle Beweise stützen. Erst 1911 berichtete F. STOCKHAUSEN über Versuche von P. LINDNER (s. auch P. LINDNER und ST. SZIFER 1912), daß Äthanol bei vielen Sproßpilzen als einzige

[1] G: Glycerin; R_1, R_2, R_3: Fettsäurereste.

C-Quelle zum Aufbau aller Zellbestandteile brauchbar ist. Später fand LINDNER (1919), daß gerade Äthanol (bei guter Durchlüftung der Kulturen) zu starker Zellverfettung führt. I. SMEDLEY-MACLEAN und D. HOFFERT (1923, 1924) zeigten an Hand sorgfältig bilanzierter Experimente, daß ruhende Hefezellen in durchlüfteten Acetatlösungen eine starke Lipoidanreicherung erfahren. Solche Ergebnisse vertrugen sich freilich auch mit der Annahme, daß *indirekte* Wege von der C_2-Verbindung zur Fettsäure führen. Und die eben genannten Forscherinnen vermuteten selbst eine Zeit lang (I. SMEDLEY-MACLEAN und D. HOFFERT 1926) eine Fettsäurebildung aus Acetat über eine Resynthese von Kohlenhydraten. Die endgültige Entscheidung im Sinne einer direkten Synthese von Fettsäuren aus C_2-Körpern wurde erst viel später durch Isotopenversuche gebracht (D_2-Acetat: R. SONDERHOFF und H. THOMAS 1937, C^{13}-Acetat: A. G. C. WHITE und C. H. WERKMAN 1947, 1948).

Die *hydrolytische Spaltung* der Glyceride als einleitender Schritt des physiologischen Fettabbaus wird durch die *Lipase* katalysiert. Dieses Ferment wurde 1846 von CLAUDE BERNARD im Pankreassekret entdeckt. 1891 konnte J. R. GREEN die erste pflanzliche Lipase in keimenden *Ricinus*samen überzeugend nachweisen. Sechs Jahre später folgte die Entdeckung eines fettspaltenden Enzyms in Schimmelpilzen (*Penicillium, Aspergillus*) durch L. CAMUS (1897a, b).

Den ersten wesentlichen Schritt zur Aufklärung des *Fettsäurenabbaues* im Organismus brachte die Entdeckung der β-Oxydation durch F. KNOOP 1904 und der Ausbau und die Erweiterung seiner Vorstellungen durch G. EMBDEN und Mitarbeiter (G. EMBDEN, H. S. SALOMON und F. SCHMIDT 1906, G. EMBDEN und A. MARX 1908) und H. D. DAKIN (1909). Erst rund 40 Jahre später konnte durch die Entdeckung der zellfreien Oxydation von Fettsäuren, der Fettsäuren-„Aktivierung“ durch Sulfhydrylbindung an Coenzym-A, die Verknüpfung des Fettsäurenabbaues mit dem Tricarbonsäurecyclus usw. der Abbauweg so festgelegt werden, wie er nunmehr im „Fettsäurecyclus“ seinen Ausdruck gefunden hat. Zu dieser raschen Entwicklung innerhalb des letzten Jahrzehnts haben zahlreiche Forscher wesentliche Beiträge geliefert; es seien hier nur — repräsentativ, aber unvollständig — die Namen F. L. BREUSCH, D. E. GREEN und Mitarbeiter, A. L. LEHNINGER, L. F. TELOIR, J. M. MUÑOZ, F. LYNEN und Mitarbeiter, S. OCHOA, E. R. STADTMAN und H. A. BARKER, H. WIELAND und Mitarbeiter genannt.[1] Der Nachweis der Reversibilität aller Teilschritte im Fettsäurecyclus legt den Weg des Aufbaus der Fettsäuren als eine Folge reduktiver Kondensationen von C_2-Bausteinen (aktivierte Essigsäure) fest. Daß in höheren Pflanzen die Synthese der Fettsäuren auf dem gleichen Wege verläuft wie bei Tieren und Mikroorganismen, haben zuerst E. H. NEWCOMB und P. K. STUMPF (1953) an Schnitten von Erdnußkeimblättern demonstriert.

II. Zur Nomenklatur und Methodik der Fettchemie; Fettkennzahlen.

Der Begriff der Fette im engeren und eigentlichen Sinne ist eindeutig definiert. Die organische Chemie versteht darunter die Glycerinester höherer Fettsäuren. Sie sind durch ihre Unlöslichkeit in Wasser und ihre Löslichkeit in polaren Solventien (Äther, Petroläther, Benzol und Homologen, Chloroform, Trichloräthylen, Tetrachlorkohlenstoff usw.) gekennzeichnet. Diese Eigenschaft teilen die Fette s. str. mit anderen Substanzen, welche bei der Fettextraktion also notwendigerweise miterfaßt werden. Soweit man nicht die eindeutigen Ausdrücke „Äther-

[1] Wegen der Zitate der Originalarbeiten sei auf den Beitrag von W. FRANKE und H. FREHSE, dieser Handbuchband S. 137ff., insbesondere S. 176ff. verwiesen.

extrakt", „Petrolätherextrakt" oder auch „Rohfett" gebrauchen wollte, entstand das Bedürfnis nach einer terminologischen Verständigung über die Gruppe der „Fette und fettähnlichen Stoffe", wobei „fettähnlich" zunächst nur im Sinne einer ähnlichen Löslichkeit, nicht aber notwendig im Sinne chemischer Verwandtschaft zu verstehen ist.

Der Begriff der „Lipoide" wurde durch E. OVERTON (1901) in seinen „Studien über die Narkose" in den wissenschaftlichen Sprachgebrauch eingeführt[1]. Er verstand darunter die „fettähnlichen" Stoffe, die in Äther, Chloroform usw. löslich sind, mit Ausschluß der Fette im engeren Sinne. Diese wurden aber bald in den Lipoidbegriff miteinbezogen, so z. B. von I. BANG (1911) und insbesondere von F. CZAPEK (in der „Biochemie der Pflanzen" 1913, und insbesondere 1919)[2]. Es fehlte nicht an Versuchen, durch klärende Definitionen der Gefahr der Verwirrung zu begegnen, die sich daraus ergab, daß die Bezeichnung „Lipoide" bald als Oberbegriff, bald als Unterbegriff verwandt wurde (z. B. W. HALDEN 1933). Schließlich wurde vorgeschlagen, das Wort als „zu stark belastet" überhaupt fallen zu lassen und durch einen anderen, eindeutig definierten Terminus zu ersetzen.

In der angelsächsischen und französischen Literatur hat sich gegenwärtig die Bezeichnung „Lipide" fast vollständig eingebürgert. Sie wurde von der Nomenklaturkommission des Internationalen Chemikerkongresses in Cambridge (1925) auf Grund von Vorarbeiten von G. BERTRAND (1923) empfohlen. Für die heutige Anwendung des „Lipidbegriffes" wird meist W. R. BLOOR (1925) zitiert.

Nach BLOOR sind „Lipide" a) durch ihre Unlöslichkeit in Wasser und ihre Löslichkeit in „Fettlösungsmitteln", b) durch ihre aktuellen oder potentiellen Beziehungen zu Fettsäureestern und c) durch ihre Verwendung im Organismus gekennzeichnet. Er gibt folgende Einteilung:

A. *Einfache Lipide* (Ester von Fettsäuren).
- I. *Fette* (Ester von Fettsäuren und Glycerin).
- II. *Wachse* (Ester von Fettsäuren mit anderen Alkoholen als Glycerin, z. B. mit Fettalkoholen oder Sterolen).

B. *Zusammengesetzte Lipide* (Ester von Fettsäuren mit zusätzlichen Gruppen).
- I. *Phospholipide* (N- und P-haltig, z. B. Lecithin, Cephalin, Sphingomyelin).
- II. *Cerebroside* (Fettsäuren + Kohlenhydrat + N-haltige Substanz) (hierher die *Glykolipide* überhaupt).
- III. *Aminolipide*.
- IV. *Sulfolipide*.

C. *Abgeleitete Lipide* (Hydrolyseprodukte von A und B).
- I. *Fettsäuren*.
- II. *Sterole*.
- III. *Alkohole*.

Die flüchtigen „ätherischen Öle" fallen also trotz ihres identischen Löslichkeitsverhaltens nicht unter die „Lipide".

Das BLOORsche Schema ist zweifellos revisionsbedürftig. Wenn unter C II und III die Sterole von den „Alkoholen" getrennt aufgeführt werden, müßten

---

[1] A. a. O. S. 54 „. . . Gehirn-Lipoide (so mögen die lecithin-cholesterinartigen Bestandteile der Zellen der Kürze wegen bezeichnet werden . . ."). Aus einer — nicht näher belegten — Bemerkung in einer Arbeit von O. ROSENHEIM (1909) geht hervor, daß schon einige Jahrzehnte vorher KLETZINSKI den Ausdruck „Lipoide" gebraucht hat.

[2] Ausführlichere Angaben über den Lipoidbegriff und seine Geschichte bei F. WINTERSTEIN (1932).

wohl auch die *Triterpenalkohole* besonders genannt werden. Es gehören in die Gruppe C außerdem die Carotinoide, die frei oder mit Fettsäuren verestert („Farbwachse“) in die „Lipidfraktion“ eingehen. Schließlich wäre es zweckmäßig, wenn auch die als Fettbegleiter auftretenden aliphatischen und cyclischen *Kohlenwasserstoffe* im Lipidschema einen Platz fänden.

Alles in allem gesehen, werden heute die Bezeichnungen „Lipide“ und „Lipoide“ praktisch als Synonyma verwendet. Die Gefahr einer Verwirrung oder Unklarheit dürfte de facto doch wohl häufig überschätzt worden sein. Dem Verfasser ist keine einzige Arbeit auf dem Fettgebiet bekannt geworden, die als Gegenbeweis herangezogen werden könnte.

Viel wichtiger als eine bis ins letzte konsequente Nomenklatur sind bei den Veröffentlichungen über die Biochemie und Physiologie der Fette präzise Angaben über die verwandte Methode. Je nach der Vorbehandlung und der Wahl der Extraktionsmittel kann bei gleichem Ausgangsmaterial Menge und Zusammensetzung des extrahierten Fettes stark schwanken.

Vor allem die Fette von Mikroorganismen (zumal von Hefen) lassen sich durch bloße Extraktion mit neutralen Solventien nur zum kleinen Teil aus der Zelle entfernen. Es muß eine gründliche mechanische Zerstörung der Zellen oder eine chemische Behandlung durch Säuren oder Alkalien der Fettextraktion vorangehen. Im letzten Falle ist natürlich mit Veränderungen des nativen Materials zu rechnen. Bei Alkaliaufschluß können überhaupt nur die „Gesamtfettsäuren“ bestimmt werden. Die Notwendigkeit der Vorbehandlung dürfte zum Teil auf Undurchlässigkeit der Zellmembran, zum Teil auf Bindung der Fette an andere Zellbestandteile zurückzuführen sein. Die Lipoproteide der Hefe können bereits durch Kochen mit wäßrigem Äthanol oder Methanol weitgehend gespalten werden (z. B. R. Reichert 1944).

Die Bestimmung gewisser *Kennzahlen* ermöglicht es ohne langwierige präparative Arbeit, wenigstens gewisse Anhaltspunkte über die Zusammensetzung eines Fettes (oder einzelner Fettfraktionen) zu erhalten. Da diese Kennzahlen in verschiedenen Beiträgen dieses Handbuches verwendet werden, erscheint es nützlich, hier eine kurze Definition der wichtigsten unter ihnen zu geben.

Verseifungszahl (VZ): mg KOH, welche zur Absättigung der in 1 g Fett enthaltenen *freien Säuren* und zur *Verseifung* der *Ester* (und Lactone) verbraucht werden. Die VZ gibt einen Anhaltspunkt für das *mittlere Molekulargewicht* (MMG) der Fettsäuren. Es errechnet sich nach der Formel: $\text{MMG} = \frac{56110}{\text{VZ}}$.

Säurezahl (Neutralisationszahl) (SZ): mg KOH, welche zur Neutralisation der in 1 g Fett enthaltenen freien Säuren verbraucht werden. Die SZ gibt ein Maß für den Gehalt an freien Säuren. Bei *freien* Fettsäuren kann (wie oben) aus der SZ das MMG berechnet werden.

Esterzahl: mg KOH, die zur Verseifung der in 1 g Fett enthaltenen Ester (und Lactone) verbraucht werden. Die Esterzahl ist also definitionsgemäß: VZ — SZ. Sie wird meist nicht direkt bestimmt, sondern aus VZ und SZ berechnet.

Hehner-*Zahl* (HZ): Prozentanteil *wasserunlöslicher Fettsäuren und unverseifbarer Substanzen* in einem Fett.

Reichert-Meissl-*Zahl* (RMZ): ml 0,1 n KOH, welche zur Neutralisation der aus 5 g Fett erhaltenen, in *Wasserdampf flüchtigen wasserlöslichen* Säuren verbraucht werden. Die RMZ gibt einen Anhaltspunkt für den Gehalt an niedermolekularen Fettsäuren.

Polenske-*Zahl* (PZ): ml 0,1 n KOH, welche zur Neutralisation der aus 5 g Fett erhaltenen, in *Wasserdampf flüchtigen wasserunlöslichen* Säuren verbraucht werden. Die PZ gibt einen Anhaltspunkt für den Gehalt an Fettsäuren mittlerer Kettenlänge.

Acetylzahl (AZ): mg KOH, welche zur Neutralisation der in 1 g acetyliertem Fett gebundenen Essigsäure verbraucht werden. Die AZ gibt ein Maß für den Gehalt an freien Hydroxylgruppen (Oxyfettsäuren).

Jodzahl (JZ): Prozentmenge Halogen (als Jod berechnet), welche ein Fett addieren kann. Die JZ kennzeichnet den Grad der Ungesättigtheit, also die Anzahl der vorhandenen Doppel- (und Dreifach-) Bindungen.

Rhodanzahl (RhZ): Prozentmenge Rhodan (als Jod berechnet), welche ein Fett addieren kann. Unter den gegebenen Bedingungen addieren Fettsäuren mit 1 Doppelbindung $\sim$100%, Fettsäuren mit 2 (nicht konjugierten) Doppelbindungen etwas mehr als 50%, Fettsäuren mit 3 (nicht konjugierten) Doppelbindungen etwa 66% der Rhodanmenge, die nach dem Grade der Ungesättigtheit theoretisch zu erwarten wäre. Ein Vergleich zwischen JZ und RhZ gibt also einen Hinweis auf den Anteil einfach, zweifach und dreifach ungesättigter Säuren[1].

Das nachstehende Schema diene der Orientierung über die Trennung und Isolierung der wichtigsten Lipoid-(Lipid-)Fraktionen.

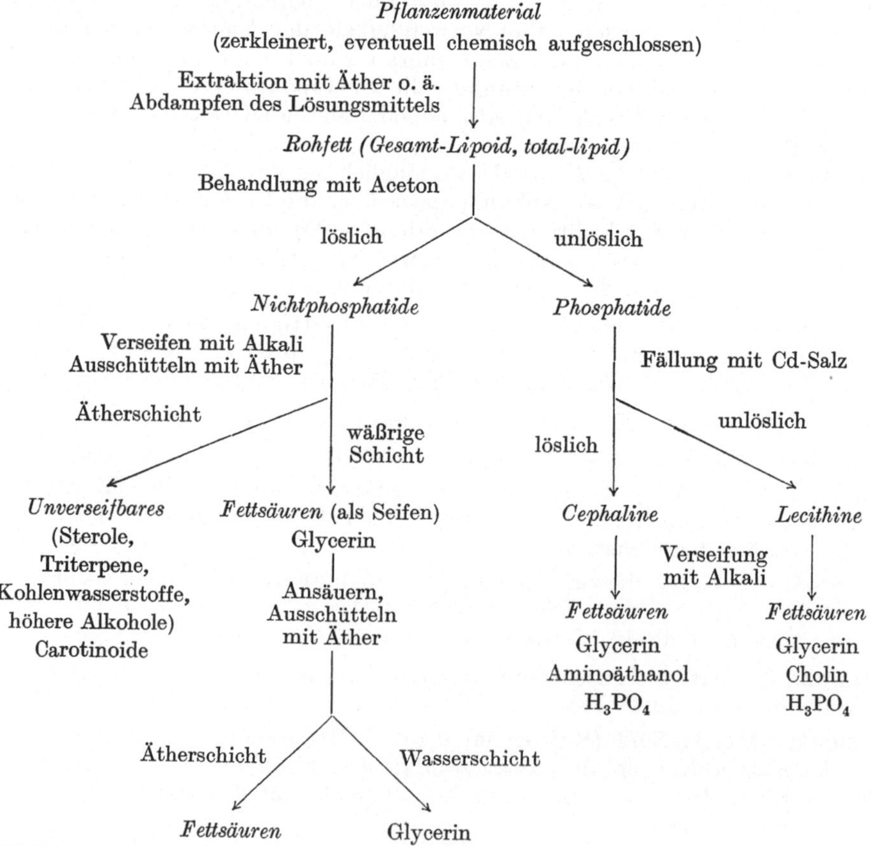

[1] Wegen weiterer Einzelheiten methodischer Art sei auf die Darstellung von M. L. MEARA (1955) verwiesen.

III. Stoffumgrenzung und Inhaltsübersicht.

Der komplexe Begriffsinhalt, der sich mit den Bezeichnungen „Fette", „Lipoide" und „Lipide" verbindet (s. oben), machte die Umgrenzung des Stoffes, der im vorliegenden Handbuchband behandelt werden sollte, etwas problematisch. Man hätte durchaus daran denken können, die genannten Begriffe in ihrer weitesten Fassung zu verwenden. Dafür ließen sich mancherlei gute Gründe anführen: Die Lipoide bilden zunächst *analytisch* eine Einheit. Sie werden bei den üblichen Bestimmungsmethoden als Gesamtheit erfaßt und in vielen Arbeiten, die sich mit der Physiologie des Fettstoffwechsels befassen, gar nicht weiter differenziert. Auch in *funktioneller* Hinsicht bilden sie eine Einheit. In der Zelle liegen wohl in der Regel Neutralfette, Phosphatide und Sterine als Gemenge vor. Die biologische Synthese der Fettsäuren und der Sterine vollzieht sich, soweit wir heute wissen, gleichzeitig nebeneinander und aus dem gleichen Ausgangsmaterial.

Nach eingehenden Überlegungen wurde schließlich einer Stoffbegrenzung unter dem Gesichtspunkt der *chemischen Verwandtschaft* der Vorzug gegeben. Den eigentlichen Gegenstand dieses Handbuches bilden also die Glyceride der Fettsäuren, wie sie in den Neutralfetten und in den Phosphatiden gegeben sind. Die ausführliche Behandlung der Sterine ist dem Band X dieses Handbuches vorbehalten. Das schließt freilich nicht aus, daß diese Stoffklasse an zahlreichen Stellen des vorliegenden Bandes mitberücksichtigt werden muß, etwa bei der Besprechung der unverseifbaren Anteile von Pflanzenfetten, bei der Darstellung der Physiologie der Fettbildung in Mikroorganismen usw.

Die Disposition des Stoffes, auf den sich die einzelnen Beiträge dieses Bandes verteilen, entspricht den wichtigsten Richtungen der biochemischen und physiologischen Forschung über die Pflanzenfette, die sich bereits in dem kurzen historischen Überblick angedeutet finden.

Die *Beschreibung der Pflanzenfette* ist eine notwendige Voraussetzung für das Verständnis der Stoffwechselvorgänge bei ihrer Bildung, ihrer Umwandlung und ihrem Abbau im Organismus. Der erste Hauptabschnitt dieses Bandes ist deshalb der deskriptiven Biochemie der Fette gewidmet. Unsere Kenntnisse von den Fetten aus verschiedenen Gruppen des Pflanzensystems sind sehr ungleichmäßig. Über die *Fette der höheren Pflanzen* liegen so zahlreiche Angaben vor, daß der Beitrag, der ihre Beschreibung zum Gegenstand hat, nicht nur eine Auswahl aus der Fülle des Materials treffen mußte, sondern auch eine Behandlung unter übergeordneten Prinzipien, insbesondere unter dem Gesichtspunkt „Pflanzenchemie und Pflanzenverwandtschaft" wählen konnte. Auf der anderen Seite schien es zweckmäßig, die *Fette der pflanzlichen Mikroorganismen (Algen, Pilze, Bakterien)* in speziellen eigenen Sonderkapiteln zu behandeln, die das spärliche Material mit einiger Vollständigkeit bringen. Es wurde also hier eine „starke Vergrößerung des Darstellungsmaßstabes" gewählt. Das dürfte schon darin genügend motiviert sein, daß Mikroorganismen bevorzugt zu Untersuchungen über den Fettstoffwechsel verwendet werden. Die monographische Darstellung der Bakterienlipoide in einem besonderen Beitrag wird auch ihrer chemischen Sonderstellung am besten gerecht.

Die *physiologische Chemie des Fettstoffwechsels* wird zunächst unter dem Aspekt der hierbei beteiligten Fermente behandelt. Es ergibt sich die naturgemäße Aufteilung in 2 Teilkapitel: Fermente der hydrolytischen Fettspaltung *(Lipasen)* und Fermente des Abbaues der Fettsäuren *(Fettsäuren-Oxydasen, -Dehydrasen* und *-Reduktasen, α- und β-Oxydation)*. Die Darstellung der β-Oxydation gibt hier bereits Anlaß, auf die gegenläufigen Vorgänge beim *Aufbau* der Fettsäuren einzugehen. Daran schließt thematisch unmittelbar der Beitrag an, der den „*Chemismus des Fettaufbaues*" mit besonderer Berücksichtigung der höheren Pflanzen zum Gegenstand hat.

Über die *Physiologie der Fettbildung und Fettspeicherung*, also die Abhängigkeit dieser Vorgänge von inneren und äußeren Bedingungen, aber mit Ausschluß der in den vorangehenden beiden Abschnitten behandelten rein chemischen Probleme, wird in zwei Beiträgen für „*höhere Pflanzen*" und „*niedere Pflanzen*" berichtet. Ein weiteres Kapitel behandelt die „*Mobilisierung der Fette*".

Bei der großen Bedeutung, welche die Pflanzenfette als Nahrungsmittel und als technische Rohstoffe besitzen, schien ein warenkundlich-technologischer Abschnitt über die *wirtschaftliche Bedeutung der Fettpflanzen und Pflanzenfette* angebracht. Daß sich dieser Beitrag im Rahmen eines „Handbuches der Pflanzenphysiologie" auf eine auswählende Übersicht ohne jede Tendenz zu erschöpfender Vollständigkeit beschränken mußte, bedarf keiner eingehenden Begründung. Auch und gerade in diesem Umfange dürfte das Kapitel vielen Benutzern des Handbuches willkommen sein.

Daß die *Phosphatide* (und Glycolipoide) in einem eigenen Beitrag behandelt werden, entspricht der Gesamtdisposition des Stoffes, die sich, wie oben gesagt, in erster Linie an chemische Gesichtspunkte hält.

Der Stand unserer Kenntnisse über die Pflanzenfette und die Physiologie ihres Stoffwechsels ist ungleichmäßig und, aufs Ganze gesehen, wenig befriedigend. An vielen Stellen sehen sich die Verfasser der einzelnen Beiträge veranlaßt, auf auffallende Lücken unserer Kenntnisse hinzuweisen. Auf der anderen Seite erscheint dieser Band gerade in einem Zeitpunkt, der auf wesentliche neue Erkenntnisse auf unserem Gebiete in naher Zukunft hoffen läßt. Schon jetzt zeichnen sich die Fortschritte ab, die die Anwendung neuer Methoden der Fettanalyse bringen wird. Die Pflanzenphysiologie beginnt soeben die Anregungen aufzugreifen und weiterzuentwickeln, die sich aus dem stürmischen Fortschreiten der Kenntnisse über den Fettstoffwechsel des Tierorganismus darbieten. Herausgeber und Mitarbeiter dieses Bandes sind sich in dem Wunsche einig, daß die hier versuchte Darstellung des gegenwärtigen Wissensstandes als Grundlage, Anregung und Hilfe für die weitere experimentelle Forschung dienen möge.

Literatur.

ANDERSON, R. J.: The chemistry of the lipoids of the tubercle bacillus and certain other microorganisms. Fortschr. Chem. organ. Naturstoffe **3**, 145—202 (1939). — ARMSTRONG, E. F., and J. ALLAN: A neglected chapter in chemistry: the fats. J. Soc. Chem. Ind. (Lond.) **43**, 207T—218T (1924).

BANG, I.: Chemie und Biochemie der Lipoide. Wiesbaden 1911. — BERNARD, C.: 1846. Siehe Handwörterbuch der Naturwissenschaften, 2. Aufl., Bd. 2, S. 800—801. Jena: Gustav Fischer 1931. — BERTRAND, G.: Projet de réforme de la nomenclature de chimie biologique. Bull. Soc. Chim. biol. Paris **5**, 96—109 (1923). — BLOOR, W. R.: Biochemistry of the fats. Chem. Rev. **2**, 243—300 (1925). — BULL, H. B.: The biochemistry of the lipids. New York: John Wiley & Sons 1937.

CAMUS, L.: Formation de lipase par le „*Penicillium glaucum*". C. r. Soc. Biol. Paris, VI. s. **4**, 192—193 (1897a). — De la lipase dans les cultures de l'*Aspergillus niger*. C. r. Soc. Biol. Paris, VI. s. **4**, 230 (1897b). — CHEVREUIL, M. E.: Recherches chimiques sur les corps gras d'origine animal. Paris 1823. Zit. nach E. KLENK 1951. — CZAPEK, F.: Biochemie der Pflanzen, 3 Bde, 2. Aufl. Jena: Gustav Fischer 1913. — Der Nachweis von Lipoiden in Pflanzenzellen. Ber. dtsch. bot. Ges. **37**, 207—216 (1919).

DAKIN, H. D.: The mode of oxidation in the animal organism of phenyl derivatives of fatty acids. IV. Further studies on the fate of phenylpropionic acid and some of its derivatives. J. of Biol. Chem. **6**, 203—219 (1909).

EMBDEN, G., u. A. MARX: Über Acetonbildung in der Leber. III. Beitr. chem. Physiol. u. Path. **11**, 318—326 (1908). — EMBDEN, G., H. SALOMON u. F. SCHMIDT: Über Acetonbildung in der Leber. II. Beitr. chem. Physiol. u. Path. **8**, 129—155 (1906).

FISCHER, E.: Die Chemie der Kohlenhydrate und ihre Bedeutung für die Physiologie. (Rede gehalten zur Feier des Stiftungstages der militärärztlichen Bildungsanstalten am 2. August 1894.) In E. FISCHER, Untersuchungen über Kohlenhydrate und Fermente, S. 96—115. Berlin: Springer 1904.

GREEN, J. R.: On the germination of the seed of the castor-oil plant *(Ricinus communis)*. Proc. Roy. Soc. Lond. **48**, 370—392 (1891).

HALDEN, W.: Zur Definition des Lipoidbegriffes. Protoplasma **20**, 209—215 (1933). — HILDITCH, T. P.: The component glycerides of vegetable fats. Fortschr. Chem. organ. Naturstoffe **1**, 24—52 (1938). — The chemical constitution of natural fats, 2. Aufl. London 1949.

KLENK, E.: Fette und Lipoide (Lipide); Allgemeines, Geschichtliches. In B. FLASCHENTRÄGER u. E. LEHNARTZ, Physiologische Chemie, Bd. 1, S. 360—263. Berlin-Göttingen-Heidelberg: Springer 1951. — KNOOP, F.: Der Abbau aromatischer Fettsäuren im Tierkörper. Beitr. chem. Physiol. u. Path. **6**, 150—162 (1904).

LAWES, J. B.: Pig feeding. J. Roy. Agricult. Soc. England **14**, 459—542 (1853). — LAWES, J. B., and J. H. GILBERT: On the composition of oxen, sheep, and pigs, and of their increase whilst fattening. J. Roy. Agricult. Soc. England **21**, 433—488 (1860). — LINDNER, P.: Zur Verflüchtigung des Biosbegriffes. Z. techn. Biol. **7**, 79—87 (1919). — LINDNER, P., u. ST. SZIFER: Der Alkohol, ein mehr oder weniger ausgezeichneter Nährstoff für verschiedene Pilze. Wschr. Brauerei **1912**, 1—6.

MAGNUS-LEVY, A.: Über den Aufbau der höheren Fettsäuren aus Zucker. Arch. f. Physiol. **1902**, 365—369. — MEARA, M. L.: Fats and other lipids. In K. PAECH u. M. V. TRACEY, Moderne Methoden der Pflanzenanalyse, Bd. 2, S. 317—402. Berlin-Göttingen-Heidelberg: Springer 1955. — MEYEN, F. J. P.: Neues System der Pflanzenphysiologie. 1838. Zit. nach F. CZAPEK 1913.

NAEGELI, C. v.: Über die Fettbildung bei niederen Pilzen. Sitzgsber. bayer. Akad. Wiss., Math.-physik. Kl. **9**, 288—316 (1879). — NEWCOMB, E. H., and P. K. STUMPF: Fat metabolism in higher plants. I. Biogenesis of higher fatty acids by slices of peanut cotyledons in vitro. J. of Biol. Chem. **200**, 233—239 (1953).

OVERTON, E.: Studien über die Narkose. Jena: Gustav Fischer 1901.

REICHERT, R.: Methode der Fettbestimmung in Hefen. Helvet. chim. Acta **27**, 961—965 (1944). — ROSENHEIM, O.: Proposals for the nomenclature of the lipoids. Biochemic. J. **4**, 331—336 (1909).

SACHS, J.: Über das Auftreten von Stärke bei der Keimung ölhaltiger Samen. Bot. Ztg. **1859**, 177—183, 185—188. — SCHEELE, K. W.: 1783. Zit. nach E. KLENK 1951. — SMEDLEY-MACLEAN, I., and D. HOFFERT: Carbohydrate and fat metabolism in yeast. Biochemic. J. **17**, 720—741 (1923). — The carbohydrate and fat metabolism of yeast. Part II. The influence of phosphate on the storage of fat and carbohydrate in the cell. Biochemic. J. **18**, 1273—1278 (1924). — The carbohydrate and fat metabolism of yeast. Part III. The nature of the intermediate stages. Biochemic. J. **20**, 343—357 (1926). — SONDERHOFF, R., u. H. THOMAS: Die enzymatische Dehydrierung der Trideuteroessigsäure. Liebigs Ann. **530**, 195—213 (1937). — STOCKHAUSEN, F.: Alkoholassimilation durch Hefe (Demonstration der Versuche von Professor P. LINDNER). Jg. Versuchs- u. Lehranstalt Brauerei, Berlin **14**, 551—556 (1911).

WHITE, A. G. C., and C. H. WERKMAN: Assimilation of acetate by yeast. Arch. of Biochem. **13**, 27—32 (1947). — Fat synthesis in yeast. Arch. of Biochem. **17**, 474—482 (1948). — WINTERSTEIN, A.: Allgemeines über „Lipoide“. In G. KLEIN, Handbuch der Pflanzenanalyse, Bd. II/1, S. 578—590. Wien: Springer 1932.

The fats of higher plants.

By

M. L. Meara.

The lipids or, as they are often more commonly called, the fats, form a group of naturally occurring substances, common both to the vegetable and animal kingdoms, and are characterized by the presence of members of the group of carboxylic acids called the higher fatty acids.

Since the lipids are in general a complex mixture of different groups of substances, each group containing fatty acids as an integral part of its constitution, it is desirable if not essential to classify them further according to their constitutions.

Two broad categories of lipids may therefore be defined.

(1) Lipids containing only carbon, hydrogen and oxygen.

(2) Lipids containing carbon, hydrogen and oxygen, and in addition certain other elements.

Both of these categories are composite, and can most conveniently be further subdivided.

Thus the chief members of group (1) comprise

(1*a*) Esters of the higher fatty acids with the trihydric alcohol glycerol, *i.e.* the triglycerides. Since the bulk of the vegetable fats consist almost entirely of triglycerides, the natural fats can be considered, from a technological standpoint at least to consist of triglycerides accompanied by minor amounts of fat-soluble impurities.

(1*b*) Esters of the higher fatty acids with alcohols other than glycerol. These alcohols are usually monohydric. This group of lipids is collectively termed the waxes, though it is now more usual to designate the esters of the higher fatty acids and higher fatty alcohols, *i.e.* the major components of the vegetable waxes, as true or ester waxes to differentiate them from the esters of the higher fatty acids and sterols or carotenols.

Similarly the more important components of group (2) fall into two further categories.

(2*a*) Esters in which only two of the three hydroxyl groups of glycerol are combined with fatty acids, the third being combined with phosphoric acid, which is itself coupled with an amino base, or an amino acid.

(2*b*) Compounds of fatty acids with carbohydrates and phosphoric acid. The constitution of components of this group of substances is not yet clear, but there now appears to be a growing amount of evidence to indicate that both glycerol, and a nitrogen base may or may not be present. It would appear therefore that when further information is available this subgroup of lipids will most conveniently be further subdivided.

The constituents of sub-groups (2a) and (2b) constitute the very highly complex category of substances termed the phosphatides, or phospholipids.

It was formerly customary to call all solid triglycerides fats, while on the other hand liquid triglycerides were called oils. This description has occasionally led to confusion in certain quarters when differentiation between these substances and the essential, and mineral oils has not been made clear. Nor must it be forgotten that the physical state of a product often depends on its geographical location. Thus the fatty material from the fruit coat of the palm is liquid in those (tropical) regions in which it is produced, but is solid in temperate climates. The calling of the triglycerides which are liquid at ordinary temperatures liquid fats, or fatty oils, obviates much of the above confusion.

I. Occurrence of fats.

Fats are distributed throughout the entire vegetable kingdom, both in the vegetative and reproductive structures, all living cells probably containing a certain amount of fatty material. The amount of fatty, or more strictly, lipid material, in the vegetative parts of plants is usually of a low order, and in these parts, *e.g.* roots, bark, stem, leaves, petals, the triglyceride material is usually accompanied by phospholipid material in amounts of approximately the same order. The seed constitutes the more usual site for fat synthesis and storage, the fat apparently forming one of the reserves of energy for the developing embryo.

The fat content of seeds is however very variable, ranging from the order of 60 per cent in the case of the seeds of some members of the families *Lauraceae* and *Myristicaceae* to 1–5 per cent in the seeds of certain members of the *Gramineae* and *Leguminosae*. In these latter cases the fat content would appear to play little role in the supply of energy to the developing embryo. Nor does the fat content of the (ripe) seed always remain of the same order in any given botanical family, for instance among the *Leguminosae*, the fat content of the common bean is of the order of 1 per cent while that of the groundnut is of the order of 50 per cent.

There are a number of plants which produce fruit, the fleshy part of which contains considerable amounts of fatty material in addition to containing fat rich seeds, the most well known examples of such plants being the palm *(Elaeis guineensis)* and the olive *(Olea europaea)*. Such fats *i.e.* palm oil and olive oil of industry are classified as fruit coat fats in this review to differentiate them from their respective seed fats. It may be noted here that only in the case of the olive does the fruit coat fat bear any resemblance in constitution to that of the seed fat.

II. Component fatty acids of the higher plant fats.

The nature of the physical and chemical properties of the fats depends in the main on the amounts and constitution of the fatty acids which, in combination with glycerol, constitute the fat.

These acids are, with a number of exceptions, monobasic acids containing an even number of carbon atoms, and belonging to the normal series, *i.e.* they are straight chain aliphatic compounds. They may be saturated as occurs when all the residual valency bonds of the carbon atoms constituting the long chain are occupied by hydrogen atoms, or unsaturated, when this condition is not fulfilled.

Unsaturated acids are classified according to the number of double (or triple) bonds in the carbon chain, polyunsaturated acids, *i.e.* acids with more than one unsaturated system in the chain, being termed non-conjugated, or conjugated, depending on whether the unsaturated systems are or are not separated by one or more methylene groups.

In addition to the occurrence of the saturated, mono, and polyethenoid acids of which the bulk of the known fats are constituted, it has been known for some considerable time that a series of acids are elaborated by the members of the family *Flacourtiaceae*, in which a *cyclo*pentenyl group occurs on the terminal carbon atom of the chain remote from the carboxyl group.

It has also been known for some considerable time that the major constituent of castor oil is the hydroxy-unsaturated acid ricinoleic acid, and further that a mono-acetylenic acid, tariric acid, occurs as a major component in the seed fat of the South American species *Picramnia*. It has only recently been shown however that long chain hydroxy-unsaturated acids other than ricinoleic acid, that acids with rings other than ω-*cyclo*pentenyl rings, and acids which contain both ethenoid and ethynoid linkages in their molecules occur in the seed fats of certain specific species.

This newer information has been due primarily to the application of newer physical and chemical techniques in the investigation of the fats.

1. The higher saturated acids.

This homologous series of acids may be represented by the general formula $C_nH_{2n}O_2$.

Formic acid, the first member of this series, has not so far been reported as a constituent of fatty material but has been variously reported to occur in the stinging nettle, certain fruits, and other plant organs.

Likewise acetic acid occurs both free and combined in the form of esters in many plants, but its occurrence as a component in a triglyceride molecule has not yet been demonstrated. The seed fat of *Schleichera trijuga* however on saponification gives rise to acetic acid in amounts of *ca* 1 per cent (by weight) of the mixed fatty acids. DHINGRA, HILDITCH and VICKERY (1929) were of the opinion that the acetic acid occurred as a component of the "unsaponifiable" fraction. WEERAKOON (1952) however as a result of a low temperature crystallization of this fat found that acetic acid was distributed fairly evenly among the fractions obtained, suggesting that the acetic acid is either attached to the glycerol molecule or that a small amount of an acetylated hydroxy acid occurs in the fat.

Neither propionic nor butyric acids, although they are produced by fermentation processes, have been shown to occur in vegetable fats, though the latter occurs to the extent of 3 per cent (wt.) or 10 per cent (mol.) in ruminant milk fats.

The odd membered branched chain isovaleric acid occurs free in valerian root, but whereas it has not been shown to occur in vegetable fats, it occurs to the extent of *ca* 3–25 per cent in dolphin and porpoise oils, depending on the site from which the oil is taken.

The lowest members of the saturated fatty acid series definitely known to occur in the mixed triglycerides of vegetable fats comprise *caproic* (*n*-hexanoic), *caprylic* (*n*-octanoic) and *capric* (*n*-decanoic) acids, all three of which occur as minor components in the seed fats of the *Palmae*.

***Lauric* (*n*-dodecanoic) acid** occurs as a major component in the seed fats of the *Lauraceae* (from which it takes its name), some tropical species of this family containing as much as 80–90 per cent of this acid. It occurs as a major constituent in the mixed fatty acids of the seed fats of the members of the *Myristicaceae* and *Palmae* families.

***Myristic* (*n*-tetradecanoic) acid** occurs chiefly in the seed fats of the members of the family *Myristicaceae* in amounts frequently attaining 75 per cent. It occurs to a somewhat lesser extent in the seed fats of the *Irvingia* sp. *(Simarubaceae)* and usually attains an amount of the order of 20 per cent in the seed fats of the *Palmae*. It occurs in somewhat smaller amounts in the seed fats of a number of plant families and it is possibly a component even though a very minor one in the seed fats of most other plant families.

***Palmitic* (*n*-hexadecanoic) acid.** Palmitic acid is the characteristic saturated acid, having been reported in practically every vegetable (and aminal) fat so far investigated. In spite of its wide distribution however it is a minor component in the fats of most plant families. It is rarely present to the extent of more than 20–25 per cent in any seed fat though its content in certain fruit coat fats, *e.g.* palm oil and stillingia tallow, attains the order of 35 per cent and 65 per cent respectively.

***Stearic* (*n*-octadecanoic) acid** resembles myristic acid in that it is fairly widely distributed though frequently occurring only in minor amounts. It appears to attain major proportions only in the seed fats of a number of tropical species, belonging primarily to the families *Guttiferae*, *Sapotaceae*, *Dipterocarpaceae* and *Sterculiaceae*.

***Arachidic* (*n*-eicosanoic) acid** is somewhat less widely distributed than stearic acid, but though it is relatively widely distributed, very rarely does it occur in more than trace or minor amounts. It is a characteristic component of the seed fats of the *Leguminosae* in which it occurs to the extent of 5 per cent. In the seed fats of certain members of the *Sapindaceae* family it occurs to the extent of 20–35 per cent.

For a long time doubt existed as to whether the arachidic acid occurring in groundnut oil possessed a straight or a branched chain, since the melting point of the acid derived from natural sources was lower than that of the synthetic product. The problem was finally resolved with the development of more efficient distillation techniques introduced by JANTZEN and TIEDCKE (1930) who showed that imperfect resolution had been obtained by former workers.

***Behenic* (*n*-docosanic) acid** does not appear to be very widely distributed. It occurs in small amounts together with arachidic acid among the higher saturated acids in the seed fats of the *Leguminosae*, and probably occurs in trace amounts in the seed fats of the *Cruciferae*. In the seed fats of two *Lophira* species *(Ochnaceae)*, *L. alata* and *L. procera*, so far examined, it has been shown (HILDITCH, MEARA and PATEL 1951) to occur to the extent of 14 and 20 per cent respectively, though stearic and arachidic acids are absent from these fats.

***Lignoceric* (*n*-tetracosanoic) acid** resembles behenic acid in being distributed in certain families, and only in trace amounts. It occurs to slightly larger amounts in the seed fats of the legumes, while in the seed fat of *Adenanthera pavonina* *(Mimosoideae)* it is reported (PARANJPE 1931) to occur to the extent of 25 per cent.

***Cerotic* (*n*-hexacosanoic) acid.** This acid is a major component of many plant waxes. While it has been reported to be present in trace amounts among the mixed fatty acids of many seed fats it has not yet been ascertained whether this small content of cerotic acid is present as mixed glyceride, or whether it is derived from wax ester which may be present on the pericarp or testa of the seed.

Among the many derivatives which have been prepared for the characterization of the saturated fatty acids may be mentioned the amides, anilides, *o*- and *p*-toluidides, *o*-bromo-*p*-toluidides, *p*-bromo-anilides, 2,4,6-tribromo-anilides, β-naphthyl-amides, α-bromo-β-naphthyl-amides, and the phenacyl, *p*-chloro, *p*-bromo, *p*-iodo and *p*-phenyl-phenacyl esters. In a number of these derivatives however the melting point of any member of the homologues series differs but little from that of the preceding or succeeding member, making their value

Table 1. *Properties of the saturated acids.*

Acid	No. of C atoms	Mol. wt. (Sap. equiv.)	M. p. °C	M. p. °C			
				amide	anilide	*p*-toluidide	*p*-bromo-phenacyl ester
Caproic	6	116.2	—3.9	101	92	73	72
Caprylic	8	144.2	16.3	105	55	70	67
Capric	10	172.3	31.3	99	70	78	67.4
Lauric	12	200.3	43.5	100	78	87	76
Myristic	14	228.4	54.4	103	84	93	81
Palmitic	16	256.4	62.9	106	89	98	86
Stearic	18	284.5	69.6	109	94	102	90
Arachidic	20	312.5	75.4	108			89
Behenic	22	340.6	80.0				90
Lignoceric	24	368.6	84.2				
Cerotic	26	396.7	87.7				

for identification purposes of restricted value. A small selection of the more frequently used derivatives are therefore included in Table 1, the melting points of other derivatives, so far as they are known, being given in other works of reference, *e.g.* MARKLEY (1947), RALSTON (1948), quoted in the bibliography, and in more recent reviews of the chemistry of the fatty acids.

2. The higher unsaturated acids.

The higher unsaturated acids are best considered in terms of their unsaturation, *i.e.* the number of double or triple bond systems present in the molecule. The simplest of the unsaturated acid series therefore consists of acids which contain one double bond per molecule, giving rise to the homologous series of monoethenoid fatty acids having the general formula $C_nH_{2n-2}O_2$.

The lowest member of this series occurring in natural oils appears to be crotonic acid $CH_3 \cdot CH{:}CH \cdot COOH$, which occurs in small amounts in the fatty matter extractable from the seeds of *Croton tiglium*. There is some evidence however that this acid is associated with the resinous rather than with the glyceride fraction of this toxic and vesicant oil.

No monoethenoid fatty acid with a chain length of less than ten carbon atoms has yet been found to occur in the glyceride fraction of vegetable oils.

Traces of dec-4-enoic acid have been reported (TOYAMA 1937) to occur in the seed fat of *Lindera obtusiloba (Lauraceae)* while traces of an isomer of this acid, dec-9-enoic acid, have been shown (GRÜN and WIRTH 1922) to occur in milk fats.

Similarly although 3 isomers of dodecenoic acid have been reported in natural fats, only dodec-4-enoic acid appears to occur in the vegetable kingdom, TOYAMA (1937) reporting its presence, with that of dec-4-enoic acid in the seed fat of *Lindera obtusiloba*. HILDITCH and LOVERN (1928) have reported the presence of *ca* 4 per cent of dodec-3-enoic acid in the waxy constituent of sperm head oil, while HILDITCH and LONGENECKER (1938) have reported 0.4 per cent of dodec-9-enoic acid in the mixed fatty acids of cow milk fat.

***Tetradecenoic* acids $C_{14}H_{26}O_2$** occur in nature in at least three isomeric forms.

ATHERTON and MEARA (1939) have shown the unique occurrence of tetradec-9-enoic acid in the seed fat of *Pycnanthus kombo,* in which fat it occurs to the extent of over 20 per cent. It is of interest to note that this acid has not so far been observed in the seed fat of any other *Myristicaceae* fat. Traces of tetradec-4-enoic acids have been reported to occur in the seed fat of *Lindera obtusiloba* (TOYAMA 1937) while HILDITCH and LOVERN (1928) have shown that the $\Delta^{5:6}$ isomer occurs to the extent of *ca* 14 per cent in sperm head oil.

***Hexadecenoic* (palmitoleic) acid.** While only one isomer of this acid has so far been reported to occur in natural fats, this acid is nevertheless widely distributed throughout both the plant and animal world, there being few fats in which it is reported to be absent. While marine animal oils usually contain 15–20 per cent of this acid, the depot fats of amphibia and reptiles 8–15 per cent, the depot fats of birds 6–8 per cent, depot fats of mammals 2–3 per cent and milk fats 3–4 per cent, its occurrence in seed fats is usually of the order of 0.5–1 per cent and in only one case, that of *Macadamia ternifolia (Proteaceae),* has it been reported (BRIDGE and HILDITCH 1950) to occur as a major component, of the order of 20 per cent. There is evidence however supporting the belief that it occurs in quantity in the fats of lower forms of both land and aquatic flora.

***Octadecenoic acids,* $C_{18}H_{34}O_2$.** Three isomeric forms of this acid occur in nature, one occurs in both vegetable and animal fats, the second in the vegetable kingdom only, the third being of animal origin.

The isomer occurring both in vegetable and animal fats, ***oleic (octadec-9-enoic)* acid** is the most widely distributed unsaturated acid of the natural fats, there being few fats either in the vegetable or animal kingdom from which it is absent. It frequently occurs to the extent of over 50 per cent of the mixed fatty acids of the fat while there are few plant families the seed fats of which contain less than 10 per cent of oleic acid.

***Petroselinic (octadec-6-enoic)* acid** has so far been found to occur only in the seed fats of two plant families, the *Umbelliferae* and the *Araliaceae,* in the seed fats of which it occurs in amounts ranging from 20–70 per cent accompanied by other acids including the more common oleic acid.

"Vaccenic" acid, which occurs in small amounts in most animal fats, is now considered to be a mixture of at least *trans* octadeca-10 and -11-enoic acids (GUPTA, HILDITCH, PAUL and SHRIVASTAVA 1950) and is believed to arise from more highly unsaturated acids through the operation of a biohydrogenation process in animal fat metabolism. There is now a certain amount of evidence to support the view that the corresponding *cis* acids occur in the liver fats of animals (BROWN 1954).

***Eicosenoic acids,* $C_{20}H_{38}O_2$.** The occurrence of eicosenoic acids in the plant world appears to be very limited though eicos-11-enoic acid occurs in the fats of most fish and marine animals. Eicos-11-enoic acid appears to be a consistent minor component (of the order of 5 per cent) in the seed fats of the *Cruciferae* (BALIGA and HILDITCH 1948, 1949), while McKINNEY and JAMIESON (1936) showed that this acid occurred to the extent of *ca* 65 per cent in the mixed fatty acids of the liquid seed wax elaborated by the shrub *Simmondsia californica (Buxaceae).*

***Docosenoic acids,* $C_{22}H_{42}O_2$.** Two isomers of this acid are known to occur, one in the plant kingdom, the other occurring in fish and marine animal oils. Docos-13-enoic (erucic) acid constitutes a major component of the seed fats of the *Cruciferae* and *Tropaeolaceae* in which it occurs to the extent of about 50 per cent and 80 per cent respectively of the total mixed fatty acids.

Docos-11-enoic acid occurs in fats of aquatic animals.

Tetracosenoic acids, $C_{24}H_{46}O_2$, have not yet been definitely proved to occur in vegetable fats though tetradec-15-enoic acid has been found to occur in the fats of certain elasmobranch fish and in the lipid material associated with brain tissue.

Hexacosenoic* (ximenic) *acid, $C_{26}H_{50}O_2$. Hexacos-17-enoic acid has been shown (Boekenoogen 1939) to occur to the extent of *ca* 20 per cent in the seed fat of the shrub *Ximenia americana (Olacaceae).*

Tricos-21-enoic* (lumequic) *acid, $C_{30}H_{58}O_2$, constitutes a minor component *ca* 5 per cent of the mixed fatty acids of *Ximenia* oil (Boekenoogen 1939).

It may be noted that low iodine values are usually recorded for vegetable (and animal) waxes, implying the presence of some unsaturated components. The presence of small amounts of the higher unsaturated (monoethenoid?) acids is therefore not precluded from the component fatty acids of the waxes.

The unsaturated acids are not so usefully characterized by the preparation of derivatives as are those of the saturated acids, since in many cases these derivatives are liquid at room temperature. They are more usually characterized therefore after separation, or concentration, by the determination of the iodine value (taken in conjunction with the saponification equivalent) followed by the preparation of the bromo addition product, which however in the case of the monoethenoid acids are too low melting to be of much significant value, and more especially by the preparation of the dihydroxy derivative either from the naturally occurring acid or from its stable geometrical isomer. There is now some indication however that the phenyl phenacyl esters, the alkyl hydroxamic acid esters and the S-benzylthiuronium salts produce crystalline derivatives which may be of use for characterization purposes.

Table 2. *Properties of monoethenoid acids.*

Acid	Mol. wt.	Iod. val.	M. p. °C		M. p. °C dihydroxy acids	
			cis	trans	*threo*	*erythro*
Dec-4-enoic	170.2	149.1				
Dodec-9-enoic	198.3	128.0				
Tetradec-9-enoic	226.4	112.2			82	123
Hexadec-9-enoic	254.0	99.8	0		87	124
Octadec-6-enoic	282.5	89.9	30	53	114	122
Octadec-9-enoic	282.5	89.9	13.4 (α) 16.3 (β)	43.7	95	132
Eicos-11-enoic	310.5	81.8				127
Docos-13-enoic	338.6	75.0	33.5	60	100	132
Tetracos-15-enoic	366.6	69.2	39	69		
Hexacos-17-enoic	394.7	64.3				
Tricos-21-enoic	450.8	56.3				

3. Diethenoid acids $C_nH_{2n-4}O_2$.

This series of acids is characterized by the presence of two double bonds in the molecule, the number of the members of this series apparently being strictly limited, only a relatively small number having so far been found to occur in nature.

Hexa-2,4-dienoic* (sorbic) *acid has been isolated from the berries of the mountain ash, but does not occur as a constituent of fatty material. On account of the presence of a conjugated diene system, and of the shortness of the chain, this acid can undergo polymerisation with great ease.

Deca-2,4-dienoic acid occurs to the extent of *ca* 4 per cent (wt.), *i.e. ca* 10 per cent (mol.) in Stillingia oil, the seed fat of *Sapium sebiferum* (CROSSLEY and HILDITCH 1950). These workers noted that earlier statements of the constitution of this oil, in terms of unsaturated acids of the C_{18} series, could not account for its very ready thermal polymerisation, more detailed investigation leading to the detection and estimation of the diethenoid acid of the C_{10} series.

Although tetradecenoic acid occurs to the extent of *ca* 25 per cent in the seed fat of *Pycnanthus kombo* (ATHERTON and MEARA 1939) no diethenoid acid containing 14 carbon atoms could be found in this fat by HAWKE (1952) nor could BRIDGE and HILDITCH (1950) detect any diethenoid acids of the C_{16} series in the seed fat of *Macadamia ternifolia*, in which hexadecenoic acid occurs to the extent of 20 per cent.

***Octadec-9,12-dienoic* (linoleic) acid.** This is the only diethenoid acid which occurs to any marked extent in the vegetable kingdom. It occurs in the seed fats of a large number of botanical families, its distribution being only slightly less wide than that of oleic, though most frequently in amounts of a lower order than that to which oleic acid occurs. In a number of families, however, the linoleic acid content assumes major proportions, attaining to 70 per cent in the mixed fatty acids of sunflower seed oil.

While diethenoid acids of the C_{20} and C_{22} series occur in the fats of aquatic animals their occurrence in seed fats appears to be limited to members of the *Cruciferae* family in which eicosa-11,13-dienoic, and docosa-13,16-dienoic acids have been found (BALIGA and HILDITCH 1949) in amounts of the order of 1–2 per cent.

It follows therefore that much of the chemistry of the fatty dienoic acids has been worked out with particular reference to linoleic acid which is readily obtained from the seed fats rich in this component.

4. Triethenoid acids $C_nH_{2n-6}O_2$.

The occurrence, in seed fats and in fruit coat fats, of acids with three double bonds appears to be limited to the C_{18} series, though a trienoic acid occurring as a minor component in the leaf fat of the rape *Brassica napus* containing 16 carbon atoms has been identified (SHORLAND 1945) as hexadec-7,10,13-trienoic acid.

Octadec-9,12,15-trienoic (linolenic) acid is by far the most commonly occurring triene acid, being found to varying amounts in the seed fats of a considerable number of botanical families, it being present in quantity in most drying oils. A non-conjugated isomer, *octadec-6,9,12-trienoic acid*, has been found to occur in the seed fat of the evening primrose *(Oenothera biennis)*, but has not yet been detected in any other seed fat, though it is possibly present in the seed fat of the nearly related species.

Octadec-9,11,13-trienoic (elaeostearic) acid occurs in the seed fats of a number of the members of the large families *Euphorbiaceae*, *Rosaceae*, and *Cucurbitaceae*. There seems to be some correlation in its occurrence in that it is absent from all members in which linolenic acid occurs.

While in the non-conjugated triene acids all three double bonds have the cis configuration it is now generally accepted, from studies of the properties of the maleic anhydride addition products, and, more recently from infra-red studies of the acid, that the double bond systems present in naturally occurring (α) elaeostearic acid are in the cis-cis-trans configuration respectively, isomerisation of α-elaeostearic acid under the influence of ultra-violet light giving rise to β-elaeostearic acid in which the configuration is considered to be all trans.

Two further conjugated triene acids, Punicic acid occurring in the fat of pomegranate seed oil, and Trichosanic acid, occurring in the seed fat of *Trichosanthes cucumeroides*, are at the moment regarded as geometric isomers of α-elaeostearic acid since although they are not identical with either α- or β-elaeostearic acid they appear to be identical with β-elaeostearic after isomerisation as a result of irradiation with ultra-violet light.

While triethenoid acids occur among the C_{16}, C_{18}, C_{20}, C_{22} and C_{24} unsaturated acids occurring in fish oils there is no evidence available so far that any of these are identical with any of the triene acids occurring in vegetable fats.

5. Tetraethenoid acids.

Only one tetraethenoid acid has so far been reported to occur in any vegetable seed fat.

Octadec-9,11,13,15-tetraenoic (parinaric) acid has been found to occur to approximately 50 per cent in the mixed fatty acids of the seed fats of *Parinarium laurinum*. Parinaric acid however is absent from the seed fat of *Parinarium macrophyllum*, which consists primarily of elaeostearic and linoleic acids (RILEY 1950).

Table 3. *Properties of the C_{18} polyunsaturated acids.*

Acid	Iod. val.	M. p. °C	M. p. °C bromo deriv.	M. p. °C phenyl phenacyl deriv.
Linoleic	181.0	— 5	116	37–38
Linolenic	273.5	—11	186	37.5–39
Elaeostearic	273.5	48	141	89–90
Parinaric		86		

6. Keto acids.

Only one higher keto acid has so far been found to occur in natural fats, and as in the case of the majority of the polyethenoid acids, contains eighteen carbon atoms in the molecule.

Octadec-4-keto-9,11,13-trienoic (licanic) acid occurs to the extent of *ca* 80 per cent amongst the mixed fatty acids of oiticica oil, the seed fat of *Licania rigida*.

For some considerable time it was believed that licanic acid was merely a geometrical isomer of elaeostearic acid, the occurrence of a keto group in the molecule first being detected by BROWN and FARMER (1935) who determined its constitution.

The seed fat of *Parinarium Sherbroense* has also been shown to contain licanic acid together with elaeostearic acid (RILEY 1950), no tetraene acid being found amongst the mixed acids of this fat as was found in the related species *Parinarium laurinum*.

7. Acetylenic acids.

Whilst the acetylenic acid **octadec-6-ynoic (tariric) acid** occurs as a major component in seed oil, the seed fat of *Picramnia Sow (Simarubaceae)* no other unsaturated acid containing only an acetylenic linkage has so far been reported.

A number of unsaturated acids however are now known to occur in some seed fats which contain one or more acetylenic linkages accompanied with one or more double bonds.

An acid of this type, **isanic acid,** is present to the extent of 20 per cent in the seed fat of *Onguekoa Gore* (ENGLER), the constitution octadec-9,11-diyne-17-enoic acid being suggested by CASTILLE (1939) this being confirmed by BLACK and WEEDON (1953) who succeeded in synthesising this acid.

A further yne-ene acid has recently been shown to exist to the extent of 1 per cent in the seed fat of *Ximenia americana.* The constitution of this acid has been reported by LIGTHELM *et al.* (1950, 1952) to be trans octadec-11-en-9-ynoic acid. GUNSTONE and MCGEE (1954) have since shown that the alleged non-conjugated octatrienoic acid occurring in the seed fat of *Santalum album* is in fact octadec-11-en-9-ynoic acid.

A number of other -yne, and yne-ene acids have been reported to occur in nature though not as components of vegetable fats.

8. Hydroxy acids.

A number of hydroxy saturated acids are known to occur in plant materials. These are believed to occur to a small extent primarily as estolides and as lactones in the coniferous waxes, sabinic acid, 12-hydroxy-decanoic, and juniperic acid, 16-hydroxy-hexadecanoic, being the most well known examples. A component of the odoriferous material extracted from musk seed oil is the lactone ambrettolide which on hydrolysis gives rise to 16-hydroxy-hexadec-7-enoic acid which is converted to juniperic acid on hydrogenation.

Ricinoleic (octadec-10-hydroxy-9-enoic) acid, is by far the most important acid of this group. Its occurrence however appears to be strictly limited to the members of the *Ricinus* species *(Euphorbiaceae)*, it occurring to the extent of *ca* 85 per cent in the seed fat of the castor seed, *Ricinus communis.* Its occurrence in seed fat of the ivory wood *Agonandra brasiliensis* has been claimed by GURGEL and DE AMORIM (1929) and in the oil of *Wrightia annamensis* by MARGAILLAN (1931). These claims would appear to require confirmation in view of the fact that an alleged isomer of this acid was claimed by VIDYARTHI and MALLYA (1939) to occur to the extent of 60 per cent in the seed fat of *Vernonia anthelmintica.* A later investigation of this fat by GUNSTONE (1954) has indicated that the oxygenated component is in fact 12:13-epoxy-octadec-9-enoic acid.

Octadec-9-hydroxy-12-enoic acid has been shown by GUNSTONE (1953) to occur in the seed fats of a number of *Strophanthus* species. CALDERWOOD and GUNSTONE (1953) have confirmed the findings of GUPTA, SHARMA and AGGARWAL (1951) that the seed fat from *Mallotus philippinensis* commonly called kamala oil contains approximately 36 per cent of 18-hydroxy-octadeca-9,11,13-trienoic acid.

Again, there is a certain amount of physical and chemical evidence to indicate the occurrence of a hydroxy-monoethenoid acid of unknown constitution in the fatty material extracted from the fungus ergot. The presence of a hydroxy acid among the mixed fatty acids of this material has been claimed by a number of workers, but BAUGHMAN and JAMIESON (1928) concluded that the high acetyl value of the oil, on which the claim was based, was due to non fatty acid hydroxylic residues, *e.g.* di- and mono-glycerides, produced by hydrolytic rancidity of the oil. DALE (1954) however has indicated both from physical (infra-red spectroscopy) and chemical evidence that hydroxy acids occur in some specimens at least of this oil, in amounts of the order of 20 per cent.

9. Dibasic acids.

The dibasic acids do not occur in nature except to a very limited extent. Apart from thapsic (tetradecanedioic) acid, which is said to occur in the extract of the dried roots of *Thapsia garganica*, the only well authenticated case so far reported is their occurrence in the fruit coat fat of the sumach berry (*Rhus* sp.). The occurrence of dibasic acids to the extent of 5–6 per cent in this seed fat is considered to be the factor contributing the specific waxy properties to the fat. It is now known that the dibasic acids present belong to the $\alpha\,\omega$ *i.e.* straight chain dicarboxylic acid series and that the molecular dimensions range from 19 to 23 carbon atoms. The arrangement of these dibasic acids in the triglyceride molecules is as yet unknown.

10. Cyclic acids.

Until comparatively recently it was believed that the only naturally occurring cyclic acids occurred in the seed fats of the *Hydnocarpus* genus of the botanical family *Flacourtiaceae*. The cyclopentenyl group common to this series of acids is attached at the end of a straight chain remote from the carboxyl group. The straight chain may or may not be saturated. It is of interest to note that in the fats containing these acids those with 18 and 16 carbon atoms predominate, lower homologues being present usually only in minor amounts, these fats therefore in this respect resembling those of the seed fats of the majority of plant families.

The acids of these fats, which are of considerable therapeutic value for the treatment of leprosy, contain an asymmetric carbon atom (at the attachment of the cyclopentenyl ring to the chain) the specific rotations of which are of a much higher order than that of ricinoleic acid in which the asymmetric carbon atom is situated in the chain.

Table 4. *Characteristics of cyclic acids.*

Acid	Systematic name	Mol. wt.	Iod. val.	M. p. °C	Rotation $[\alpha]_D$
Aleprolic . . .	$\Delta^{2,3}$ cyclopentenyl carboxylic			liq.	+120,5
Aleprestic . . .	$5\Delta^{2,3}$ cyclopentenyl-*n*-pentanoic			liq.	+100.5
Aleprylic . . .	$7\Delta^{2,3}$ cyclopentenyl-*n*-heptanoic			32	+90.8
Alepric	$9\Delta^{2,3}$ cyclopentenyl-*n*-nonanoic			48	+77.1
Hydnocarpic .	$11\Delta^{2,3}$ cyclopentenyl-*n*-undecanoic	252.4	100.6	59–60	+69.3
Chaulmoogric .	$13\Delta^{2,3}$ cyclopentenyl-*n*-tridecanoic	280.4	90.5	71	+60.3
Gorlic	$13\Delta^{2,3}$ cyclopentenyl-*n*-tridec-6-enoic	278.4	182.3	liq.	+60.7

The readily polymerisable seed oil of *Sterculia foetida* possesses an acid for which the constitution 12-methyl-octadec-9,11-enoic acid was tentatively suggested by HILDITCH, MEARA and ZAKI (1941). More recently NUNN (1952) isolated this acid by application of the urea complex technique, and suggested its structure as ω(2 *n*-octyl-*cyclo*prop-1-enyl)-octanoic acid, this structure being based on the strong band at 9.8 μ in the infra-red spectrum of dihydrosterculic acid. VERMA, NATH and AGGARWAL (1955) however have criticized this structure. They pointed out that NUNN did not take the infra-red curve of sterculic acid itself into account, and fixed the position of the double bond without any supporting evidence. VERMA *et al.* (loc. cit.) have shown that this acid gives clear evidence of a cyclopropane ring at 9.96 μ, and a double bond at 6.09 μ on the shoulder of the strong carbonyl group at 5.9 μ. Since the cyclopropane ring in sterculic acid may form a "conjugated" system with the double bond this would explain a number of indeterminacies indicated by HILDITCH, MEARA and ZAKI. In the light of the evidence now available the Indian workers suggest the constitution ω(2n-hexyl-cycloprop-1-anyl)-dec-9-enoic acid.

$$CH_3\cdot(CH_2)_5\cdot\underset{\diagdown\ CH_2\ \diagup}{CH-CH}-CH{=}CH\cdot(CH_2)_7\cdot COOH$$

Further investigation is therefore most desirable to decide conclusively the constitution of this acid.

III. Analysis.

Since in general fats consist of complex mixtures of mixed triglycerides the technical assessment of an oil or fat, *i.e.* the determination of its characteristics, gives only a limited estimate of its probable components. A much more detailed investigation is usually required for the quantitative estimation of the mixed fatty acids, though with the application of newer techniques in the field of organic chemistry more accurate information can often be obtained than was formerly possible by the older more complicated methods.

The basic principles of the methods upon which the quantitative analysis of a fat is arrived at may be summarized as follows.

The fat is saponified, and the mixed fatty acids recovered may be separated into fractions containing mainly saturated and mainly unsaturated acids. This may be accomplished by the application of the older lead salt technique, or more conveniently by the application of a low temperature crystallization technique. The latter method has the advantage that no chemical operations are involved in the resolution, eliminating any possible modification in structure which might occur, and further it has the distinct advantage that the unsaturated components of the mixed fatty acids can usually be further resolved into fractions consisting of mainly monoethenoid and mainly polyethenoid acids respectively. It has been found from experience that on account of a mutual solubility factor (FOREMAN and BROWN 1944) it is impracticable to make a complete resolution of the component acids of a fat into the above categories, nevertheless under suitable conditions high concentrates of these categories can be achieved. Further, for practical purposes, it has been found more satisfactory to start the crystallization at the lowest temperature first followed by subsequent recrystallizations from the same or different solvents respectively at successively higher temperatures. Crystallization from a 10 per cent solution of acids in acetone at —60° C. precipitates the saturated and monoethenoid acids, any di- and tri-unsaturated acids remaining in solution. Recrystallization of the mixture of mainly saturated and monoethenoid acids from ether (10 ml. per g. of acids) at —30° C. allows the saturated acids to crystallize out leaving the bulk of the monoethenoid acids in solution. It must be emphasised that the resolution obtained for each fraction consists of a concentrate of any one category of fatty acids accompanied by minor amounts of the other two categories.

The fraction consisting of the concentrate of saturated acids is methylated, the methyl esters fractionally distilled at reduced pressure. The component fatty acids of this fraction are then readily deduced from the saponification equivalents and iodine values of each ester fraction.

If it is known, as is the case with the majority of the seed fats, that the unsaturated acids present are only of the C_{18} series, the composition of the unsaturated acid fractions can readily be deduced from spectrophotometric analysis, and, knowing the proportion of each group into which the original mixed fatty acids of the oil has been resolved the mixed fatty acids of the whole fat can be computed.

When other unsaturated acids in addition to those of the C_{18} series are present it follows that the composition of these can be arrived at from a knowledge of the saponification equivalent and iodine values of each ester fraction and of the spectrophotometric data obtained from the acids recovered from selected ester fractions after alkali isomerisation under standardised conditions.

While this general technique can be applied to the majority of cases, the analysis of certain oils will require special treatment. Thus in the case of tung oil the elaeostearic acid content can be determined by direct spectrophotometric analysis of the mixed fatty acids since the conjugated triene system having an absorption band at 268 mμ is already present.

Again the hydroxy acid content of castor oil or of the oils of the *Strophanthus* species may be best determined by determining the acetyl value of the oil, either by the "double saponification equivalent method", followed by application of the ANDRE-COOK formula, or by the pyridine-acetic anhydride method as proposed by WILSON and HUGHES (1939).

Licanic acid *e.g.* in oiticica oil, may be estimated by determination of the carbonyl value, as described by KAUFMANN, FUNKE and LIU (1938). This method utilizing hydroxylamine hydrochloride suffers from the drawback that the indicator change at the end point is unsatisfactory, potentiometric titration in this instance causing no marked improvement. The spectrophotometric method developed by MENDELOWITZ and RILEY (1953) based on the determination of the optical density at 625 mμ of the colour produced by the 2,4-dinitrophenylhydrazone in alkaline solution shows considerable promise.

The epoxy acid content of the oxygenated acid now known to occur in the seed fat of *Vernonia anthelmintica*, and indeed of any epoxy acid, can readily be determined by applying the method of KING (1949, 1951) for the estimation of oxirane derivatives. This method is based on the determination of the amount of hydrochloric acid required to open the epoxy ring.

IV. Component acids of root fats.

The root is not normally a site of fat storage, the amount of lipid material extractable from the dried root tissue rarely exceeding 10 per cent, the only case so far reported in which the root can be considered to be fat rich being that of the tubers of the tropical sedge *Cyperus esculentus* in which the oil content is of the order of 20–30 per cent. BAUGHMAN and JAMIESON (1932) have found that the component fatty acids of this oil consist of myristic traces, palmitic 12, stearic 5, arachidic, lignoceric traces, oleic 76, linoleic 6, per cent. While the component fatty acids of most root fats appear to conform to the general composition, saturated acids, mainly palmitic 10–20, unsaturated acids, mainly oleic 80–90, per cent, the fat from mangel roots *(Beta vulgaris rapa)* appears to be exceptional (NEVILLE 1912) in that while the saturated acids (palmitic 14 per cent) conform to the above distribution the unsaturated acids comprise oleic 57 and erucic 29 per cent.

V. Component acids of bark fats.

Fatty material appears to occur in bark to the extent of 3 per cent. The component fatty acids of very few of these bark fats have been examined, but from what evidence is available so far it would appear that the major component in all cases is oleic acid.

Thus the component fatty acids of the bark fat of *Tilia cordata* were found by PIERAERTS (1926) to consist of oleic 94, linoleic 4, saturated acids (palmitic and stearic) 2 per cent, whilst RUCHKIN (1929) reported the presence of 37 per cent saturated acids and 63 per cent unsaturated acids (mainly oleic) in the bark fat of the sea buckthorn *Hippophaë rhamnoides*.

The fat of hawthorn *(Crataegus oxyacantha)* bark has been reported (DIETERLE and DORNER 1937) to contain palmitic, stearic and oleic acids.

The lipid material elaborated in the sugar cane is quite different from that of the bark lipids mentioned above in that it contains 44 per cent of unsaponifiable matter. The component fatty acids are present chiefly as wax esters and not as glycerides and have been shown by VIDYARTHI and NARASINGARAO (1939) to comprise palmitic 29,2, stearic 23.6, arachidic 3.5, and oleic 43.7 per cent.

VI. Component acids of leaf fats.

The composition of leaf fats has not been investigated over a wide range of plants, the chief interest to date residing in the determination of the constitution of the lipids of pasture grasses.

Dried pasture grasses usually contain 4–6 per cent of lipids of which 1.5–4 per cent occurs in the form of glycerides. An investigation of the leaf glycerides of cocksfoot *(Dactylis glomerata)* and of perennial ryegrass *(Lolium perenne)* (SMITH and CHIBNALL 1932) indicated that both oils were similar in composition, containing saturated 10,17; oleic 16,23; linoleic 30,21; and linolenic 44,39 per cent respectively. An examination by SHORLAND (1941, 1944) of the lipids of New Zealand pasture grasses led to the same general results, while a more detailed analysis of the fat obtained from English mixed pasture grasses (HILDITCH and JASPERSON 1945) was broadly in line with the findings of SHORLAND *(loc. cit.)*.

The constitutions of two *Brassica* leaf fats have so far been reported. Cabbage *(Brassica oleracea)* leaf cytoplasm was shown by CHIBNALL and CHANNON (1927) to contain ca. 1.5 per cent of ether-soluble lipids. The glycerides present in this fraction contained 10 per cent of saturated acids, mainly palmitic, the unsaturated acids comprising mainly linoleic, and a considerable amount of linolenic acids, the amount of oleic being relatively small. A more comprehensive study by SHORLAND (1945) indicated that rape *(Brassica napus)* leaf fat consists very approximately of saturated, mainly palmitic 15, tetradecenoic 1, hexadecatrienoic 14, and C_{18} acids of mean unsaturation —5.6, 70 per cent. This fat then is remarkable in that it contains C_{16} acids of triethenoid unsaturation, not observed in any other leaf fat so far reported, and in the absence of erucic acid, which occurs to the extent of 40 per cent in the seed fat of this species.

GORDON (1928) was able to show that the leaves of dried watermint *(Mentha aquatica)* contained 5 per cent of lipid material, much of which was glyceridic. The isolation of di-, tetra-, and hexahydroxy acids from the products of alkaline permanganate oxidation of the mixed acids indicated the occurrence of oleic, linoleic and linolenic acids in the fat. Similarly spinach leaves, *Spinacea oleracea*, were shown (SPEER, WISE, HART and HEYL 1929) to contain 0.4 per cent of neutral fat amongst the extracted lipids, the component fatty acids of which were saturated, small amounts, the unsaturated acids oleic, linoleic and linolenic being in the proportion of 3:5:2. Likewise the fat of the nettle *Urtica dioica* contains saturated acids to the extent of *ca.* 5 per cent, oleic 82, and linoleic 13 per cent (HILDITCH and MEARA 1944).

It appears therefore, from the small but wide range of plant families investigated, that leaf fats are broadly similar in composition. The saturated

Table 5. *Component acids (wt. per cent) of fruit coat fats.*

	Location	Saturated				Unsaturated		
		C_{12}	C_{14}	C_{16}	C_{18}	oleic	linoleic	
Palmae								
Elaeis guineensis . . .	Congo		2.4	41.6	6.3	38.0 (*a*)	9.5	HILDITCH, MEARA and ROELS (1947)
	Ivory Coast		2.3	34.3	5.6	49.5	8.3	HILDITCH and JONES (1930)
	Nigeria		1.2	39.6	5.8	42.4	11.0	DEAN and HILDITCH (1933)
	Sumatra		2.5	41.8	4.2	42.1	9.4	DEAN and HILDITCH (1933)
Oleaceae								
Olea europea	Italian		1.2	15.6	2.0	64.6	15.0	GUNDE and HILDITCH (1940)
	Californian		—	7.0	2.5	85.8	4.7	JAMIESON and BAUGHMAN (1925)
	Dodecanese		0.6	18.4	—	68.5	12.5	BRANDONISIO (1936)
	Palestinian		0.5	10.0	3.3	77.5	8.6	HILDITCH and THOMPSON (1937)
Lauraceae								
Laurus nobilis	Mediterranean	2.7	—	20.3	—	63.0	14.0	COLLIN (1931)
Neolitsea involucrata . .	India	10.2	—	28.2	3.1	48.2	10.3	GUNDE and HILDITCH (1938*b*)
Persea gratissima . . .		—	—	7	1	81	11	JAMIESON, BAUGHMAN and HANN (1928)
Sterculiaceae								
Sterculia foetida	Java			← 40 →		← 60 →		STEGER and VAN LOON (1941*a*)
Sterculia parviflora . .	Malaya		2.3	27.6	9.0 (*b*)	37.6	22.2	HILDITCH and MEARA (1944)
Sterculia tomentosa . .	Sudan			← 32 →		63	5	HENRY and GRINDLEY (1940)
Theobroma cacao . . .	*Tropics?*			← 51–56 →		33–35	16–9	BAUER and SEBER (1938)
Myricaceae								
Myrica mexicana . . .	Central America	—	61.1	37.5	—	1.4	—	JAMIESON (1943)
Euphorbiaceae								
Stillingia sebifera . . .	Hong Kong	—	0.5	63.2	7.6	27.1	1.6	GUPTA and MEARA (1950)
Anacardiaceae								
Rhus succedanea . . .		—	1.9	67.5	11.6 (*c*)	13.6	trace	SCHUETTE and CHRISTENSON (1942)
Caryocaraceae								
Caryocar villosum . . .		—	1.5	41.2	0.8	53.9	2.6	HILDITCH and RIGG (1935)
Burseraceae								
Dacryodes rostrata . . .		—	—	33.9	2.7	59.3	4.1	HILDITCH and STAINSBY (1934)
Celastraceae								
Celastrus paniculatus .	India	—	3.0	26.2	4.0	36	8 (*d*)	GUNDE and HILDITCH (1938b)
Caprifoliaceae								
Sambucus racemosa . .	Temperate zones			← 23 →		71	6	BYERS and HOPKINS (1902)

(*a*) 1.8 per cent hexadecenoic acid, 0.4 per cent linolenic; (*b*) 1.3 per cent arachidic; (*c*) 5.6 per cent dibasic acids; (*d*) 23 per cent linolenic.

acid content is usually of the order of 10 per cent, the remainder comprising unsaturated acids of the C_{18} series with linoleic, and to a lesser extent linolenic, usually predominating, oleic being a minor component, except in the case of peppermint, spinach and the nettle leaf fats, in latter case oleic being the most abundant acid in the fat.

VII. Component acids of fruit coat fats.

While the seed is the more usual site for fat storage, the fleshy part of a considerable number of fruit contains a certain amount of oil, this attaining to considerable proportions in such cases as the fruit flesh of a number of members of the *Palmae, Lauraceae* and *Oleaceae,* typical examples of such fruit coat fats of these families being palm, laurel, and olive oils.

The component fatty acids of a number of fruit coat fats have now been determined, some of the more well known examples being quoted in Table 5.

From the table it is seen that the characteristics and component fatty acids of most of the fruit coat fats, irrespective of their botanical family, are very similar. Palmitic and oleic acids appear to be the most characteristic acids though their amounts may vary over a wide range. While Chinese vegetable tallow contains 70 per cent of palmitic acid, the palmitic acid content of the commercially valuable palm oil is of the order of 40 per cent, this falling to 10 per cent in the case of olive oil. Likewise the oleic acid content falls from 80 per cent in olive oil to a value of the order of 2 per cent in *Myrica mexicana.* The linoleic acid content is on the whole less variable, usually occurring to the extent of 10–15 per cent. The fruit coat fat of *Myrica mexicana* is however an outstanding exception not only by virtue of the absence of linoleic acid and its remarkably low oleic acid content, but also by its high content, 61 per cent, of myristic acid, an acid which is present in most fruit coat fats in amounts of the order of 1–5 per cent. The fruit coat fats of a number of the members of the family *Lauraceae* are anomalous in containing appreciable quantities of lauric acid.

While there is a considerable amount of qualitative similarity, and sometimes quantitative similarity in the fruit coat fats listed in Table 5 and the seed fats of some of these species, as can be seen from Table 6, subsequent tables show no general correlation, the seed fats ranging from oils of a very high degree of unsaturation to those in which the content of unsaturated acids is of a very minor order. Within this range however, there are a number of species in which the component

Table 6. *Comparison of fruit coat and seed fats*[1].

	Fruit coat fat				Seed fat				
	C_{12}	C_{16}	oleic	linoleic	C_{12}	C_{16}	oleic	linoleic	others
Elaeis guineensis	—	35–40	40–50	5–11	47	9	18	1	14
Olea europaea	—	7–15	70–85	4–12	—	6	83	7	4
Laurus nobilis	3	20	63	14	43	6	32	18	—
Theobroma cacao	—	50[2]	35	10	—	60[2]	37	2	—
Stillingia sebifera	—	60–70	20–35	—	—	6	16	46	20
Rhus sp.	—	77	12	—	—	minor	major	major	
Caryocar villosum	—	41	54	3	—	48	46	3	
Dacryodes rostrata	—	34	59	4	—	11	44	3	40
Celastrus paniculatus	—	26	36	8	—	22	22	35	16
Sambucus racemosa	—	23	71	6	—	23	45	24	8

[1] Since minor components have been omitted, the totals do not always add up to 100 per cent.

[2] "Palmitic" here includes large quantities of stearic.

fatty acids of the seed fat, by virtue of its consisting primarily of palmitic, oleic and linoleic acids, are practically identical with those of the fruit coat fat, examples of the close similarity between fruit coat and seed fat being given in Table 6 for *Olea europaea*, *Theobroma cacao* and *Caryocar villosum*.

In spite of the qualitative similarity in constitution of the fruit coat fats, their physical properties are widely different. Thus whereas in the case of the solid seed fats, the degree of hardness, *i.e.* a rough measure of the melting point is usually due to the occurrence in quantity of solid acids other than palmitic, as in the case of the very hard Baku kernel fat, the differences in the physical properties of the fruit coat fats are controlled almost entirely by variation in the respective amounts of palmitic and oleic acids and only to a minor extent by linoleic acid. Thus fruit coat fats containing only these acids range from the hard Chinese vegetable tallow to the liquid olive oil.

Table 5 also demonstrates the degree of variation in the compositions of oils from given species. In the case of the palm oils, Dyke (1928), before the component fatty acids of the oils were known with any high degree of accuracy, had pointed out that the titre, *i.e.* the setting point of the mixed fatty acids derived from the palm oil from the more westerly regions of West Africa was of the order of 41° C., whereas that derived from the oils of the Gold Coast and Nigeria was 44° C. Since this difference is one of longitude and not latitude it would appear that the oil palm of the western areas is a different variety from that of the more easterly region. Since plantation oils from both the Belgian Congo and the East Indies, Malaya, Sumatra show close similarity amongst themselves and with the oils from the eastern region, it is probable that the seed used in these regions was originally taken from indigenous stock from the Gold Coast and Nigeria.

The component fatty acids of the oils from the easterly region, *viz.* Ivory and Gold Coasts, comprise 40 per cent palmitic, 40–45 per cent oleic, and 8–10 per cent linoleic acids, whereas the corresponding values for oils from the westerly regions are 35, 50, 10 per cent respectively, the oils from both regions containing approximately the same amounts of other minor components.

Similarly, oilve oils can be classified into two broad groups. Most of the olive oils from Asia Minor, California, Italy and Spain, *i.e.* oils from three continents, are very similar in constitution containing *ca.* 10 per cent palmitic acid, 2 per cent stearic acid, 80–85 per cent of oleic and 7 per cent linoleic acid.

In contrast to this, oils from other sources, in particular those from the Dodecanese, Tunisia, and the Tuscany region of Italy, form another broad group of olive oils the component fatty acids of which show considerable differences from those already mentioned. The palmitic acid content is somewhat higher, and of the order of 15 per cent, this also being accompanied by an increased linoleic content, the amount being of the order of 15 per cent, that is, a linoleic acid content of approximately twice that which obtains in the former class of olive oils. To counterbalance these increases in both palmitic and linoleic acids the oleic acid content decreases to a value in the region of 65–70 per cent. The suggestion of Brandonisio (1936) that, contrary to what has usually been found to obtain in the case of the seed fats, the fruit coat fat of the olive contains a higher linoleic acid content when grown in warmer climates, must be accepted with a certain amount of reserve until the component fatty acids of olive oils produced in a somewhat wider range of climatic conditions have been determined, since the differences observed so far may well be accounted for by variety differences.

It may be pointed out that, technically, this latter type of oil, by virtue of its considerably higher linoleic acid content will be much more prone to oxidative rancidity than the oils of the former type.

As has already been mentioned, the fruit coat fat of Sumach berries (*Rhus* sp.), commonly called Japan wax, is exceptional, not so much by virtue of its very high palmitic acid content (which is of the same order as that to which it occurs in stillingia tallow) but by virtue of its containing up to 6 per cent of long-chain normal dibasic acids, these acids apparently giving rise to the characteristic properties of this wax.

VIII. Seed fats.

Most seed fats contain as major components only those acids which have been found to occur in the fats from sites in the plant other than that of the seed. It follows therefore that since many seed fats contain palmitic, oleic, linoleic and linolenic acids in quantity, there is considerable scope in variation of the amounts of the major component acids from one family to another.

Thus the seed fats can be classified according to their major component acids, although a number of cases arise in which more than one of the usual distributions of the major components occurs in members of the same botanical family.

Other seed fats have, in addition to the above mentioned acids as major components one or more acid as a major component which is not usually associated (at least in quantity) with the fat from any other site in the plant. These "specific" acids may be saturated, for example as in the high stearic acid contents of the members of the botanical family *Sapindaceae,* or unsaturated, *e.g.* (erucic) as in the *Cruciferae.*

This grouping of the seed fats according to their major component acids has brought to light the observation that the seed fats of the members of any given botanical family usually belong to the same group since the members of a family frequently show a considerable amount of qualitative, and often quantitative similarities in their composition.

It has already been mentioned that in certain instances acids other than palmitic, oleic, linoleic and linolenic occur in quantity in an oil and since the occurrence of any such particular acid is limited to a given family, it is in some cases necessary to introduce a secondary classification by plant families in addition to the classification in which oils in which only palmitic, oleic, linoleic and linolenic acids predominate. A further point may be mentioned at this juncture. In certain large plant families, notably *Euphorbiaceae*, *Rosaceae*, *Cucurbitaceae*, while some members fall within the general classification indicated above, *e.g. Aleurites moluccana*, other *Aleurites* species elaborate oils in which the conjugated triethenoid elaeostearic acid is present in quantity, as in *Aleurites Fordii* and *Aleurites montana*. It follows therefore that in cases such as these classification according to plant families is more desirable.

This general classification, devised by HILDITCH, has been adopted in this review.

Of the seed fats in which only palmitic, oleic, linoleic, and linolenic acids are major components, these may be subdivided into

(*a*) those containing primarily linoleic and linolenic acids
(*b*) those containing primarily oleic and linoleic acids
(*c*) those containing primarily oleic and palmitic acids

these classes approximating to the older divisions of drying, semi-drying and non-drying oils into which oils were formerly classified. Some indication of the distribution of these fatty acids in a considerable number of plant families, together with the distribution of "specific" acids in those families which can be better classified as such, is given below.

(A) **Major component acids: linoleic, linolenic and/or oleic.** *Celastraceae, Coniferae, Elaeagnaceae, Juglandaceae, Labiatae, Linaceae, Moraceae, Oenotheraceae, Passifloraceae, Rhamnaceae, Valerianaceae.*

(B) **Major component acids: linoleic, oleic.** *Amaranthaceae, Asclepiadaceae, Betulaceae, Capparidaceae, Compositae, Dipsacaceae, Fagaceae, Hippocastanaceae, Myrtaceae, Olacaceae, Oleaceae, Papaveraceae, Pedaliaceae, Plantaginaceae, Scrophulariaceae, Staphyleaceae, Theaceae, Typhaceae, Ulmaceae, Vitaceae.*

(C) **Major component acids: linoleic, oleic, or linolenic, elaeostearic, licanic, or ricinoleic.** *Cucurbitaceae, Euphorbiaceae, Rosaceae.*

(D) **Major component acids: palmitic, oleic, linoleic.** *Acanthaceae, Anacardiaceae, Anonaceae, Apocynaceae, Berberidaceae, Bombacaceae, Caprifoliaceae, Caricaceae, Caryocaraceae, Combretaceae, Gramineae, Lecythidaceae, Magnoliaceae, Malvaceae, Martyniaceae, Menispermaceae, Rubiaceae, Rutaceae, Solanaceae, Tiliaceae.*

Families elaborating seed fats containing characteristic fatty acids.

(E) **Petroselinic, oleic and linoleic acids.** *Araliaceae, Umbelliferae.*

(F) **Acetylenic acids, tariric, octadecenynoic, acids.** *Simarubaceae (Picramnia* sp.*), Olacacea.*

(G) **Eicosenoic (oleic, linoleic).** *Olacaceae (Ximenia* sp.*), Sapindaceae, Buxaceae (Simmondsia* sp.*).*

(H) **Erucic, oleic, linoleic.** *Cruciferae, Tropaeolaceae.*

(I) **Cyclic unsaturated acids.** *Flacourtiaceae.*

(J) **Arachidic, lignoceric, oleic, linoleic.** *Leguminosae, Moringaceae, Ochnaceae, Sapindaceae.*

(K) **Stearic, palmitic, oleic.** *Gnetaceae, Burseraceae, Convolvulaceae, Dipterocarpaceae, Guttiferae, Meliaceae, Sapotaceae, Verbenaceae.*

(L) **Lauric, myristic, palmitic.** *Lauraceae, Myristicaceae, Palmae, Salvadoraceae, Simarubaceae, Ulmaceae, Vochysiaceae.*

It is of course impossible in this review to include a full list of the component fatty acids of the large number of fats of the various members of the different plant families which have now been investigated, nevertheless the component fatty acids of the more important vegetable fats, the more typical of any particular plant family, and those forming the most interesting exceptions will be quoted, given to the nearest unit per cent and in some cases, very minor components added to the nearest major homologue.

1. Seed fats in which the major components are linoleic, linolenic and/or oleic acids, *i.e.* family groups A and B.

Considering now the first of the group of families, in which linoleic, linolenic and/or oleic acids occur as major components. Comparison of Tables 7 and 8 indicates, as might have been excepted, that no marked differentiation exists between this grouping and that in which oleic and linoleic acids tend to be the major components. The subdivision into those fats in which linolenic acid is present in quantity and those in which it is only a very minor component, on the whole however serves well to differentiate these groups.

It is seen that both groups of fats are derived from the seeds of plants of all different sizes, ranging from large trees, through shrubs to herbs. Thus the seed oils of the larger trees growing in temperate climates can be considered to belong to the drying or semi-drying class. The seed oils of the conifers however show a greater variation in linolenic acid content, on the whole, the *Pinus* species conforming with the belief that where varieties can grow in both warmer and colder regions the seed oil produced in the cooler region is more unsaturated than that from warmer regions. It may be pointed out here that while the linoleic acid content of the temperate climate gymnosperms is of the order of 50 per cent, that of the tropical gymnosperm *Gnetum scandens* (see Table 15) is of the order of only 3 per cent, the latter seed fat however containing 56 per cent of stearic acid. It is not known however whether this is an extreme example of the influence of climate, or whether the tropical angiosperms produce seed fats which may be characteristic of those of the temperate regions in addition to those which are patently (at the moment) exceptional.

The seed fat of the elm constitutes a noted exception to the rule that related members of a botanical family are usually qualitatively if not quantitatively similar to each other with respect to the component acids of their seed fats. The seed fat of the American elm has been shown by ZEHNPFENNIG and SCHUETTE (1941) to consist of 11 per cent of linoleic acid, 9 per cent of oleic acid, 60 per cent of capric acid, together with small amounts of caprylic, lauric, myristic and palmitic acids. This species is therefore unique amongst the temperate trees and shrubs not only on account of the high capric acid content of its seed fat, but also by virtue of the exceptionally low content of unsaturated acids which obtains only in the *Lauraceae*, *Myristicaceae* and *Palmae* families.

The seed fats of the smaller plants, *i.e.* shrubs and herbs, are similar in general to those of the larger trees. However, in addition to the families listed in Tables 7 and 8, and to which some members of the families *Cucurbitaceae*, *Euphorbiaceae*, *Leguminosae* and *Rosaceae* conform (*cf.* Table 9), a number of plant families included in Tables 7 and 8 present some unusual features. Thus members of the *Celastraceae* contain in their seed fats in addition to the fatty acids linoleic, linolenic, by virtue of which they are classified in this group, considerable amounts of formic, acetic and benzoic acids. There is a certain amount of evidence however to show that these acids do not form part of the glyceridic portion of the fat but are combined with some other water-soluble probably cyclic alcoholic constituent.

Again the family *Olacaceae* appears to be of considerable heterogeneity. Thus while the composition of *coula* seed oil is normal for the group, the seed fat of *Onguekoa Gore* ENGLER is abnormal in that it contains as a major component the ene-diyne acid isanic acid. On the other hand the seed fat of *Ximenia americana*, in addition to containing approximately 54 per cent of oleic acid contains appreciable amounts of cerotic, ximenic, ximenynic and lumequic acids. Further the oil derived from the seed of ivory wood, *Agonandra brasiliensis*, is alleged to contain 47 per cent of ricinoleic acid. While it is clear therefore that the seed fats of the *Olacaceae* constitute a very heterogeneous group, it is highly desirable for a number of the above claims to be confirmed and for the component fatty acids of the seed fats of more members of this family to be investigated.

In a number of cases in the foregoing tables the component fatty acids of the seeds of a plant species grown in several geographical locations have been recorded. Whilst the tables in such a review are by no means complete it is seen that the oil produced by *any given species* in a colder climate tends to be more unsaturated than that which is produced in a warmer climate. Following the pioneer work of IVANOV (1929) considerable investigation of the influence of climate on the constitution of linseed oils has been carried out by WOODWARD

Table 7. *Component acids of seed fats (wt. per cent) of botanical families group A.*

		Sat.		Oleic	Lin-oleic	Lin-olenic	
		C_{16}	C_{18}				
Coniferae							
Pinus excelsa	W. Siberia	← 8 →		12	57	22	Eibner and Reitter (1926)
Pinus monophylla	U.S.A.	← 9 →		64	27	—	Adams and Holmes (1913)
Pinus pinea	Mediterranean	5	1	48	46	—	Matthes and Rossié (1918)
Pinus sylvestris		4	3	10	58	25	Eibner and Reitter (1926)
Celastraceae							
C. paniculatus	India	22.3	4.3	22	35	16	Gunde and Hilditch (1938b)
Juglandaceae							
Carya cordifolia	U.S.A.	7	5	88	—	—	Riebsomer *et al.* (1940)
Hicoria pecan	U.S.A.	3	2	79	16	—	Jamieson and Gertler (1929b)
Juglans regia	U.S.A.	4	2	35	57	3	Jamieson and McKinney (1936)
Juglans regia	European	7	1	16	72	4	Griffiths and Hilditch (1934)
Labiatae							
Perilla ocimoides	Far East	← 7 →		14	16	63	Kaufmann (1930)
Salvia hispanica	Holland	← 10 →		—	32	58	Steger and van Loon (1942a)
Linaceae							
Linum usitatissimum		7	9	13	17	54	Gunstone and Hilditch (1946)
Linum usitatissimum	Argentine	5	4	17	24	47	Griffiths and Hilditch (1934)
Moraceae							
Cannabis sativa	Temperate zones	6	2	6	70	15	Griffiths and Hilditch (1934)
Oenotheraceae							
Oenothera biennis	Temperate zones	9	2	7	72	10	Riley (1949)
Passifloraceae							
Passiflora edulis	U.S.A.	7	2	17	73	—	Jamieson and McKinney (1934)
Rhamnaceae							
Rhamnus cathartica		1	6	31	37	24	Krassowski (1906)
Valerianaceae							
Valerianella olitoria		← 12 →		18	64	9	Steger and van Loon (1937)

(1939), Rose and Jamieson (1941) and Painter and Nesbitt (1943). While a full appraisal of the import of this work can only be arrived at by consulting the original papers, it may be stated that Painter and Nesbitt point out that high temperatures are conducive to the formation of less unsaturated oils but that other factors, such as soil, moisture etc., are also operating in the determination of the final unsaturation of the oil.

A somewhat similar study of sunflower seed oils has been made by Barker and Hilditch (1950a) as a result of which Hilditch has suggested that the final unsaturation of the oil will depend on the rate of ripening of the seed

2. Seed fats in which the major components are linoleic, oleic, linolenic or other conjugated polyethenoid acids, *i.e.* family group C.

This group of seed fats can be considered to form a sub-section of group A (p. 28) since in the three families considered, *viz. Cucurbitaceae, Euphorbiaceae, Rosaceae*, while some members of each family are apparently normal and could be readily classified within the confines of group A, other members contain as major components other "specific" acids such as elaeostearic, licanic, ricinoleic

Table 8. *Component acids (wt. per cent) of seed fats of botanical families group B.*

		C_{16}	C_{18}	Oleic	Lin-oleic	Lin-olenic	
Betulaceae							
Corylus avellana . . .	Italy	3	2	91	4	—	SCHUETTE and CHANG (1933)
Compositae							
Carthamus tinctorius. .	U.S.A.	4	3	26	67	—	JAMIESON and GERTLER (1929a)
Guizotia abyssinica . .	India	8	4	15	71	—	HILDITCH, SIME *et al.* (1944)
Helianthus annuus . .	Russia	6	3	25	66	—	HILDITCH and ZAKY (1942b)
Fagaceae							
Fagus sylvetica		←12→		48	38	2	DELVAUX (1936)
Hippocastanaceae							
Aesculus hippocastanum		4	4	66	25	1	KAUFMANN and BALTES (1938)
Myrtaceae							
Psidium guyava. . . .	India	→ 16 →		56	28	—	VARMA, GODBOLE and SRIVASTAVA (1936)
Olacaceae							
Coula edulis		2		95	3	—	STEGER and VAN LOON (1935a)
Ximenia americana . .	See p. 29						
Onguekoa Gore	See p. 29						
Agonandra brasiliensis .	See p. 29						
Oleaceae							
Olea europea	Portugal	6	4	83	7	—	KLEIN (1898)
Papaveraceae							
Papaver somniferum. .	India	11	5	11	73	—	BRIDGE, CHAKRABARTY and HILDITCH (1951)
Papaver somniferum. .	England	8	3	16	73	—	BRIDGE, CHAKRABARTY and HILDITCH (1951)
Pedaliaceae							
Sesamum indicum . .	India	8	5	46	41	—	HILDITCH and RILEY (1945)
Scrophulariaceae							
Verbascum thapsus . .	Czecho-slovakia	← 6 →		28	57	8	VOTOČEK, VALENTIN and BULÍŘ (1936)
Theaceae							
Thea sinensis.	China	9	4	79	8	—	WEERAKOON (1952)
Typhaceae							
Typha latifolia	U.S.A.	←12.5→		15	12.5	—	CLOPTON and VAN KORFF (1945)
Ulmaceae							
Celtis occidentalis . . .	U.S.A.	—	6	16	78	—	SCHUETTE and ZEHNPFENNIG (1937)
Ulmus americana . . .	See p. 29						
Vitaceae							
Vitis rotundifolia . . .	U.S.A.	15		20	65	—	PICKETT (1940)
Vitis vinifera	Germany	12		17	71	—	KAUFMANN and FIEDLER (1937)

acids. Within these large families, the number of these apparently exceptional cases was originally believed to be somewhat small, but within recent years the number of "exceptions" has increased, making the classification by family desirable until further development in this field may bring to light as yet unknown relationships which may give rise to a more logical classification.

a) Cucurbitaceae seed fats.

Inspection of Table 9 shows that, as will be seen in the case of the family *Rosaceae*, the bulk of the members of this family appear to produce seed fats in which oleic and linoleic acids are the major components, though in the members of the *Cucurbitaceae* family there appears to be a definite tendency towards an increased amount of saturated (palmitic and stearic) acids, as compared

Table 9. *Component acids (wt. per cent) of seed fats of botanical families group C.*

		C_{16}	C_{18}	Oleic	Linoleic	Other unsaturated acids	
Cucurbitaceae							
Citrullus pepo	Subtropics	12	7	24	57	—	KAUFMANN and FIEDLER (1939)
Citrullus vulgaris	India	10	6	35	49	—	DHINGRA and BISARAS (1945)
Hodgsonia capniocarpa	Malaya	39	9	27	25	—	HILDITCH, MEARA and PEDELTY (1939)
Telfairia occidentalis	Africa	16	3	23	23	19 (*c*) 16 (*e*)	HILDITCH and RILEY (1946)
Telfairia pedata	Africa	33	14	14	36	3 (*a*)	HILDITCH and MEARA (1944)
Trichosanthes cucumeroides	E. Indies	9		20	42	29 (*f*)	TOYAMA and TSUCHIYA (1935)
Euphorbiaceae							
Aleurites cordata	Japan	↓ 7 →		19	—	74 (*c*)	McKINNEY and JAMIESON (1937)
Aleurites Fordii	China	4	—	9	10	77 (*c*)	HILDITCH and RILEY (1946)
Aleurites montana		← 5 →		7	10	18 (*c*)	HILDITCH and MENDELOWITZ (1951)
Aleurires moluccana	Ceylon	7	3	19	44	27 (*a*)	WEERAKOON (1952)
Aleurites trisperma	E. Indies	← 17 →		13	—	67 (*c*)	McKINNEY and JAMIESON (1937)
Cephalocroton cordofanus	Sudan	← 2 →		59	6	33 (*g*)	HENRY and GRINDLEY (1943)
Croton tiglium	Asia	1	1	56	29	13 (*h*)	FLASCHENTRÄGER and WOLFFERSDORFF (1934)
Hevea brasiliensis	Ceylon	11	13	19	35	24 (*a*)	GUNSTONE and HILDITCH (1946)
Jatropha curcas	E. Indies	17	6	36	41	—	STEGER and VAN LOON (1942c)
Ricinodendron africanum	African	10	1	10	26	11 (*a*) 53 (*c*)	HILDITCH and RILEY (1946)
Ricinus communis	India	← 3 →		5	4	86 (*g*) 2 (*j*)	SREENIVASAN, KAMATH and KANE (1956)
Sapium sebiferum		9	5	10	30	41 (*a*) 5 (*i*)	CROSSLEY and HILDITCH (1950)
Tetracarpidium conophorum	Africa	3	6	15	11	65 (*a*)	GUNSTONE, HILDITCH and RILEY (1947)
Rosaceae							
Crataegus oxyacantha	Temperate	2	1	81	11	5 (*a*)	VASTERLING (1922)
Cydonia vulgaris	Subtropics	← 9 →		43	47	1 (*a*)	STEGER and VAN LOON (1934a)
Licania arborea	Mexico	← 12 →		5	8	73 (*b*)	ROSE and JAMIESON (1943)
Licania crassifolia	E. Indies	← 8 →		5	22	65 (*b*)	SESSELER and ROWAAN (1939)
Licania rigida	Brazil						
Parinarium campestre	E. Indies	15		27	9	49 (*c*)	STEGER and VAN LOON (1942b)
Parinarium corymbosum	E. Indies	4	7	8	6	62 (*b*) 13 (*c*)	FRAHM (1941)
Parinarium laurinum	Pacific Islands	← 4 →		6	3	31 (*c*) 56 (*d*)	RILEY (1950)
Parinarium macrophyllum	W. Africa	12	2	40	15	31 (*c*)	HILDITCH and RILEY (1946)
Parinarium sherbroense	W. Africa	← 12 →		← 10 →		48 (*a*) 30 (*c*)	STEGER and VAN LOON (1938)
Prunus amygdalus	Temperate	10	2	44	44	—	MEARA (1952)
Prunus armeniaca	Temperate	3	1	64	32	—	JAMIESON and McKINNEY (1933)
Prunus cerasus	Temperate	5	3	50	42	—	JAMIESON and GERTLER (1929a)
Prunus domestica	Temperate	← 7 →		69	24	—	DELVAUX (1936)
Rosa canina	Temperate	← 5 →		9	54	32 (*a*)	STEGER and VAN LOON (1943)

(*a*) linolenic acid; (*b*) licanic acid; (*c*) elaeostearic acid; (*d*) parinaric acid; (*e*) resin acids; (*f*) trichosanic acid; (*g*) ricinoleic acid; (*h*) lower saturated acids; (*i*) decadienoic acid; (*j*) dihydroxystearic acid.

with the seed fats of the members of the *Rosaceae* and indeed of the *Euphorbiaceae*. In the *Telfairia* genus however, whilst *Telfairia pedata* conforms to the normal type, *Telfairia occidentalis* contains 19 per cent of elaeostearic acid. Again in the genus *Trichosanthes*, while *T. Kadam* is believed to consist primarily of saturated acids (20 per cent) and oleic acid (80 per cent), *T. cucumeroides* contains approximately 30 per cent of trichosanic acid, in addition to oleic 20 per cent, linoleic 42 per cent and saturated acids, 8 per cent.

b) Euphorbiaceae seed fats.

Most of the seed fats of the members of this family, so far investigated, have been of tropical origin. Of the temperate region members examined, the seed oil of the caper spurge *Euphorbia lathyrus* is remarkably simple in containing over 90 per cent of oleic acid. The seed fats of the tropical members are however much more complex though a number of them *e.g. E. calycina, E. erythraceae, Hevea brasiliensis,* are normal in that their major component acids comprise oleic, linoleic and/or linolenic acids. On the other hand, elaeostearic acid is present to the extent of 75–80 per cent in the seed fats of four members of the *Aleurites* genus, *A. Fordii, A. montana, A. cordata* and *A. trisperma,* while it is absent from *A. moluccana.*

A further genus of this family is characterized by the presence of a high content of the hydroxy acid ricinoleic acid in the seed fat of its members. This acid appears to occur in all the members of this genus so far examined and has been noted in one further member of the family, *Cephalocroton cordofanus* (Henry and Grindley 1943), and in the seed fat of *Agonandra brasiliensis (Olacaceae)* though in this latter case it would appear to require confirmation. The belief that this acid occurred in grape seed oil is now known to be unsubstantiated while its occurrence in ergot oil is yet to be finally proved. It is known however that an isomer of this acid occurs in some members of the *Strophanthus* species *(Apocynaceae).*

c) Rosaceae seed fats.

It is perhaps not surprising that within so large a botanical family some further subdivision according to the constitution of the seed fats of the members may be desirable. Although only a relatively small number of the members of this family have been investigated it appears that they can be arranged into two groups.

(a) Temperate and subtropical shrubs the seed fats of which contain mainly linoleic and oleic acids together with small amounts of linolenic and saturated acids. All members of the *Prunus, Rubus, Craetagus* and *Cydonia* species so far examined conform to this category.

(b) Certain tropical species of the genus *Licania* and *Parinarium.* In these species the amount of saturated acids is of the order of 10 per cent, the oleic and linoleic acids reduced to this order, the balance consisting of licanic (4-keto-elaeostearic) acid in the case of the members of the *Licania* genera. The seed fats of members of the genus *Parinarium* however are somewhat more complex, in some members *e.g. Parinarium macrophyllum,* elaeostearic acid being the only "specific" acid present, whereas in others both elaeostearic and licanic acids occur. *Parinarium laurinum* contains the conjugated tetraethenoid parinaric acid, which has also been reported (Tutiya 1940) to occur in balsam seed oil *(Balsaminaceae).*

Table 10. *Component acids (wt. per cent) of seed fats of botanical families group D.*

	Habitat	Sat. C_{16}	Sat. C_{18}	Oleic	Linoleic	
Acanthaceae						
Blepharis edulis	India		12	74	14	PENDSE and LAL (1938)
Anacardiaceae						
Anacardium occidentale .	Tropics	7	11	74	8	PATEL, SUDBOROUGH and W (1923)
Mangifera indica . . .	W.Indies	4	43 (*j*)	45	55	DUNN and HILDITCH (1947)
Pistacia vera	India	8	2	70	20	DHINGRA and HILDITCH (19
Anonaceae						
Anona squamosa . . .	Japan	←39→		29	32	HATA (1938)
Asimina triloba	U.S.A.	3	3	65	29	RIEBSOMER, BISHOP and R (1938)
Bombacaceae						
Bombax malabaricum .	India	11	10	46	33	JAMIESON and MCKINNEY (1936b)
Caprifoliaceae						
Sambucus racemosa . .		6	3	4	53 (*a*)	SCHUETTE *et al.* (1943)
Viburnum opulus . . .	U.S.A.	←2→		61	37	SCHUETTE and KORTH (194
Caryocaraceae						
Caryocar villosum . . .	Brazil	50	1	46	33	HILDITCH and RIGG (1935)
Gramnineae						
Andropogon sorghum . .	U.S.A.	11	4	33	52	BALDWIN and SNIEGOWSKI
Avena sativa	Temperate	10	—	59	31	AMBERGER and WHEELER-] (1927)
Hordeum vulgare seed .	Temperate	9	3	33	54	TÄUFEL and RUSCH (1929
Hordeum vulgare germ.	Temperate	10	6	21	62	TÄUFEL and RUSCH (1929
Oryza sativa meal . . .	Tropics	13	4	44	39	JAMIESON (1926)
Oryza sativa bran . . .	Tropics	←18→		47	35	MURTI and DOLLEAR (194
Panicum miliaceum . .	Subtropics	←12→		25	51 (*b*)	STEGER and VAN LOON (19
Secale cereale seed . .	Temperate	21		18	61	CROXFORD (1930)
Secale cereale germ . .	Temperate	2	9	33	53 (*c*)	STOUT and SCHUETTE (193
Triticum vulgare germ .	Temperate	16	6	12	57 (*d*)	GUNSTONE and HILDITCH (
Zea Mays germ	U.S.A.	8	3	31	58	BAUR and BROWN (1945)
Lecythidaceae						
Bertholletia excelsa . .	Brazil	16	6	48	30	SCHUETTE and ENZ (1931)
Malvaceae						
Gossypium herbaceum .	India	25	2	25	48	HILDITCH and MADDISON (
Gossypium esculentum .	Sudan	24	7	27	42	CROSSLEY and HILDITCH (1
Rubiaceae						
Coffea arabica	Brazil	32	8	9	46 (*i*)	KHAN and BROWN (1953)
Rutaceae						
Citrus Aurantium . . .	W.Indies	24	9	25	37 (*e*)	DUNN. HILDITCH and RILE (1948)
Citrus decumana . . .	W.Indies	30	2	25	37 (*f*)	DUNN, HILDITCH and RILE (1948)
Solanaceae						
Atropa belladonna . . .	Europe	6	2	25	67	HILDITCH and ICHAPORIA (
Hyoscyamus niger . . .	Europe	6	1	11	82	HILDITCH and ICHAPORIA (
Nicotiana tabacum . . .		←10–13→		8–20	68–74	CRAWFORD and HILDITCH (
Solanum esculentum . .	Temperate zone	13	6	46	35	RABAK (1917)
Tiliaceae						
Corchorus capsularis . .	India	12	5	29	41 (*g*)	MEARA and SEN (1952)
Corchorus olitorius . . .	India	17	4	9	62 (*h*)	MEARA and SEN (1952)
Tilia parvifolia		←13→		29	58	KAUFMANN and FIEDLER (

(*a*) Contains 34 per cent linolenic; (*b*) Contains 5 per cent linolenic; (*c*) Contains 3 per cent linolenic; (*d*) Contains 9 per cent linolenic; (*e*) Contains 5 per cent linolenic; (*f*) Contains 6 per cent linolenic; (*g*) Contains 4 per cent higher saturated, 5 per cent linolenic, 4 per cent C_{20} monoethenoid; (*h*) Contains 3 per cent higher saturated, 1 per cent linolenic, 4 per cent C_{20} monoethenoid; (*i*) Contains 5 per cent unsaturated acids higher than C_{18}; (*j*) Contains 3 per cent arachidic.

3. Seed fats in which palmitic, oleic and linoleic acids are major components, *i.e.* families belonging to group D.

A considerable number of plant families elaborate seed fats, the constitution of which indicates that they are best classified in this grouping of plant families. Only a small number of the more typical or more well known seed fats are recorded in Table 10, a much more complete selection being readily available from inspection of the literature and from the larger books of reference.

It is seen from Table 10 that a large number of the seed fats listed contain oleic and linoleic acids in proportions very similar to those already mentioned in Tables 7–9, but which in addition contain saturated acids, more specifically, palmitic acid, in amounts in excess of that which would be regarded as a minor component. The families which elaborate seed fats which can be classified into this category include a number of herbs and shrubs of tropical and of subtropical origin, though this is not specific of the group since a number of the *Gramineae* fall within this category though members of the *Gramineae* are indigenous to temperate and tropical regions.

While the combined content of oleic and linoleic acids often exceeds two-thirds of the mixed fatty acids, the balance is frequently made up to a large extent of palmitic acid. Sometimes linoleic acid predominates as in the case of the technically valuable semi-drying oils, *e.g.* cottonseed oil, while in other cases oleic acid predominates.

It appears that a further subdivision of this group of families may be made, depending on the palmitic acid content.

(*a*) Members of the *Malvales* contain palmitic acid usually in amounts roughly approximating to 20 per cent.

(*b*) In the seed fats of the members of the *Solanaceae* the palmitic acid content is more frequently nearer 10 per cent, leading, in the absence of major amounts of other saturated acids and of linolenic acid to the increased amounts of oleic and linoleic acids. The seed fats of the members of this family therefore are similar to a number of those of the *Cucurbitaceae* though there is no close botanical connection between the two plant families.

(*c*) The palmitic acid content is somewhat variable in the seed fats of the members of a number of families, notably *Anacardiaceae*, *Apocynaceae*, *Anonaceae*, and *Martyniaceae*. In addition, in some of the tropical families the stearic acid content is such that it is beginning to become a major component.

In the case of the *Gramineae* information is available in some cases for the constitution of the fats in both the endosperm and embryo, the latter usually containing relatively more fat than the former. It appears that while there is a marked variation in the amounts of the unsaturated acids present in the fats of the respective genera there is little difference between constitutions of the endosperm and embryo fat of a given species.

4. Seed fats containing characteristic acids.

It has already been mentioned that while many plant families elaborate seed fats containing oleic, linoleic, or linolenic and palmitic acids as major components, a considerable number of families contain one or more other acids, often in considerable proportions. In these cases there is a very marked correlation between the seed fat composition and the family to which the plant belongs. The seed fats which contain elaeostearic, licanic, parinaric, and ricinoleic acids have already been mentioned while discussing the seed fats of the families *Cucurbitaceae*, *Euphorbiaceae*, and *Rosaceae*, since for systematic purposes the fats containing these acids can best be considered as being "exceptions" to the general trend in their respective botanical families. In the following groupings it will be seen that the characteristic acid is present to a greater or lesser extent in the seed fats of all or very nearly all members of the respective family (or subfamily).

5. Seed fats containing petroselinic acid, *i.e.* family group E.

The occurrence of petroselinic acid appears to be confined to the members of the Umbelliferae and to one or two members of isolated, but related species.

It is clear from Table 11 that the members of the *Umbelliferae* are characterized by the occurrence of 20–70 per cent of petroselinic acid. While the fats of the

Table 11. *Component acids (wt. per cent) of seed fats of botanical families group E.*

	Saturated	Petroselinic	Oleic	Linoleic	
Araliaceae					
Hedera helix	5	62	20	13	Steger and van Loon (1928)
Umbelliferae					
Ammi visnaga	5	50	32	13	Grindley (1950)
Angelica sylvestris . . .	4	19	44	33	Hilditch and Jones (1928)
Anthriscus cerefolium .	5	41	1	53	Christian and Hilditch (1929)
Apium graveolens . . .	3	51	26	20	Christian and Hilditch (1929)
Carum carvi	3	26	40	31	Christian and Hilditch (1929)
Coriandrum sativum . .	8	53	32	7	Christian and Hilditch (1929)
Daucus carota.	4	58	14	24	Christian and Hilditch (1929)
Foeniculum officinale .	4	60	22	14	Christian and Hilditch (1929)
Pastinaca sativa. . . .	1	46	32	21	Christian and Hilditch (1929)
Heracleum sphondylium	4	19	52	25	Hilditch and Jones (1928)
Petroselinum sativum .	3	76	15	6	Hilditch and Jones (1927)

members of this group are further characterized by the low content of saturated acids, usually of the order of 4 per cent, consisting primarily of palmitic, but probably containing small amounts of stearic and possibly of higher saturated acids, there appears to be no marked correlation between the relative amounts of oleic and linoleic acids as the content of petroselinic acid changes. The seed fat of the ivy *Hedera helix (Araliaceae)* has been shown to be broadly similar to those of the *Umbelliferae* and contains 62 per cent of the characteristic petroselinic acid. In addition to the occurrence of petroselinic acid in the above-mentioned families, it appears in addition to be a major component of the mixed fatty acids of *Picrasma quassioides (Simarubaceae)*, see p. 42.

6. Seed fats containing tariric acid, *i.e.* family group F.

This acid has so far been reported to occur only in the fats of the members of the Central American genus *Picramnia (Simarubaceae)*. The members of other genuses of this family produce seed fats which, while not containing tariric acid, show considerable variation in their component fatty acids.

Tariric acid has so far been reported to be present in *P. tariri*, *P. Camboita*, *P. Carpinterae*, *P. Lindeniana* and *P. Sow*, the seed fat of the latter containing 95 per cent of this acid. This group of seed fats falls in a sub-group of the family *Simarubaceae*, the component fatty acids of some of the members of which are given in Table 16.

Similarly octadecen-ynoic acids have been found to occur in a number of the members of the family *Olacaceae* (*Ximenia* sp.), and an octadecen-diynoic acid in a further member *(Onguekoa Gore)* of this family. It would appear therefore that the members of the family *Olacaceae* produce seed fats which can be characterized either according to the broad grouping *B*, p. 28, or according to their characteristic acids. The seed fats of insufficient members of this interesting plant family have yet been investigated to make any general deductions as to its heterogeneity in this respect.

7. Seed fats containing erucic acid.

Inspection of Table 12 and of the fuller literature reveals that it is most probable that the seed fats of all the members of the large family *Cruciferae*, irrespective of the genera to which they belong, almost without exception contain erucic acid in greater or lesser extent. While its presence has been definitely established in all except one of the species so far examined, from the analytical constants of the oils of the species less fully investigated, it may be deduced that erucic acid is present in quantity. Its content rarely falls below 20 per cent. In the *Brassica* seed fats its content lies between 40 and 50 per cent, while it is present to over 80 per cent in the seed fats of the *Tropaeolum* species.

The sole instance in which erucic acid is reported to be absent is that of the seed fat of the Garden Rocket *Hesperis matronalis*, in which the content of linolenic acid is abnormally high.

Table 12. *Component acids (wt. per cent) of seed fats of botanical families Cruciferae and Tropaeolaceae.*

	Location	Saturated	Oleic	Linoleic	Linolenic	Erucic	
ciferae							
.lliaria officinalis . . .	Norway	1	5	8	9	78	RAMSTAD and GLOPPE (1942)
3rassica alba	England		28	15	1	52	HILDITCH, RILEY and VIDYARTHI (1927)
3rassica campestris . .	India		15	16	7	54 (*a*)	HILDITCH, LAURENT and MEARA (1947)
3rassica juncea	India		31	17	3	42	DUTT (1939)
3rassica nigra	India		25	19	2	50	HILDITCH, RILEY and VIDYARTHI (1927)
'amelina sativa			24	15	33	3 (*b*)	VON MIKUSCH (1952)
'heiranthus cheiri . . .	Temperate zone		10	32	16	39	GRIFFITHS and HILDITCH (1934)
'onringia orientalis . .	N-America	4	20–30	15–20	—	35–45 (*c*)	HOPKINS (1946)
'ruca sativa	Germany		6	28	1	59	KAUFMANN and FIEDLER (1938)
Iesperis matronalis . .	Holland		11	35	46	—	STEGER and VAN LOON (1942a)
satis tinctoria			27	19	24	21	STEGER and VAN LOON (1941b)
3isymbrium altissimum .	American	14	6	19	35	26	GOSS and RUCKMAN (1944)
3isymbrium irio . . .	India		28	35	8	18	AGGARWAL and KARIMULLAH (1947)
'hlaspe arvense			13	33	1	49	CLOPTON and TRIEBOLD (1944)
)paeolaceae							
'ropaeolum minus . .	England	1	16	1	—	82	HILDITCH and MEARA (1938)

(*a*) Contains 2 per cent docosadienoic acid; (*b*) Contains 2 per cent hexadecenoic acid, 14 per cent eicosenoic acid; (*c*) Contains 12 per cent eicosenoic acid.

A further feature of the Cruciferous seed fats appears to be the low content of saturated acids, these acids containing, in addition to palmitic acid, notable amounts of acids of higher molecular weight, in this respect the seed fats of this family resembling those of the family *Leguminosae*.

At this juncture it is appropriate to refer to the lipid material which has been found to be present in the seed of *Simmondsia californica*. The seeds of this plant are exceptional for their family *(Buxaceae)* in consisting entirely of embryo and cotyledons. Likewise the lipid material is exceptional in that it is composed entirely of esters of higher alcohols and higher acids, *i.e.* the lipid material is in fact a liquid seed wax. While McKINNEY and JAMIESON (1936) showed that the approximate compositions of alcohols and acids were respectively eicosenol 30, docosenol 70 per cent; saturated 3.5, hexadecenoic 0.5, oleic 1.4, eicosenoic 64.4, docosenoic 30.2 per cent, GREEN, HILDITCH and STAINSBY (1936) showed that the respective components were *n* eicos-11-enol, *n* docos-13-enol, *n* eicos-11-enoic, and *n* docos-13-enoic acids, *i.e.* the latter acid being erucic acid.

8. Seed fats with cyclic acids as major components.

Apart from the occurrence of the cyclic acid occurring to the extent of *ca.* 70 per cent in the seed fat of *Sterculia foetida*, and of 12:13-epoxyoctadec-9-enoic acid occurring in the seed fat of *Vernonia anthelmintica (Compositae)*, two oils

Table 13. *Component fatty acids (wt. per cent) of Flacourtiaceae seed fats.*

	Saturated (a)	Oleic	Lower cyclic homologues	Hydnocarpic	Chaulmoogric	Gorlic	
Carpotroche brasiliensis	7	6	trace	46	25	16	COLE and CARDOSO (1938a)
Hydnocarpus anthelmintica	8	13	trace	69	9	1	COLE and CARDOSO (1939a)
Hydnocarpus Kurzii	4	15	trace	35	23	23	COLE and CARDOSO (1939a)
Hydnocarpus Wightiana	2	7	3	49	27	12	COLE and CARDOSO (1939a)
Onchoba echinata	8	2	trace	—	75	15	COLE and CARDOSO (1938b)

(a) Saturated acids consist almost entirely of palmitic.

which are clearly anomalous in their composition with respect to the other members of their families so far investigated, there remains the group of cyclic acids which are characteristic of most members of the *Flacourtiaceae*.

The absence of these cyclic acids from the seed fats of *Hydnocarpus odoratus* and *Oncoba spinosa* is inferred from the low rotation of the oil since these cyclic acids have an asymmetric carbon atom with a high optical rotation.

9. Seed fats containing specific saturated acids.

Just as a number of plant families produce seeds the fat of which contains other unsaturated acids in quantity in place of or in addition to the more usual oleic, linoleic and/or linolenic acids, so the occurrence of saturated acids other than palmitic in quantity can be regarded as being very specific, acids other than palmitic occurring in quantity in a relatively few plant families only. It is abundantly clear that these families can be segregated into a number of groups, depending on the constitution of the saturated acids present. Thus while the seed fats of some families contain notable amounts of arachidic and/or higher molecular weight saturated acids, others are characterized by the appearance to an appreciable extent of stearic acid (with or without an accompanying moderately high amount of palmitic acid). Again in others myristic acid appears as the major component acid, and in others lauric acid occurs either as the sole major component acid or still in quantity but together with myristic acid and/or appreciable amounts of lower saturated acids. These respective families fall into the groupings *J–L* as is indicated on p. 28.

10. Seed fats containing arachidic, behenic or lignoceric acids.

The seed fats of the plant family *Leguminosae* form an interesting group in that the range in the amounts of the characteristic higher saturated acids varies widely. Thus while one quarter of the mixed fatty acids of the seed fat of *Adenanthera pavonina* consists of lignoceric acid, with arachidic and behenic acids absent, or present only in small amounts, the seed fat of *Parkia biglandulosa* contains behenic acid to the extent of 8 per cent but neither arachidic nor lignoceric acids. Again, while all the saturated acids from palmitic to lignoceric are present in the seed fats of the two species of *Pentaclethra*, while no great variation in the arachidic acid content occurs from one species to the other, an approximate reversal of the respective amounts of behenic and lignoceric is indicated.

Table 14. *Component acids (wt. per cent) of seed fats of botanical families group J.*

	Location	C_{16}	C_{18}	C_{20}	C_{22}	C_{24}	Oleic	Linoleic	
Leguminosae									
Acacia sp.	Sudan	←20–35→		← 1–4 →			31–45	28–44	GRINDLEY (1945)
Adenanthera pavonina	India	9	1	—	—	26	49	15	MUDBIDRI (1928)
Parkia biglandulosa	India	9	13	—	8	—	31	39	PARANJPE (1931)
Pentaclethra macrophylla	Africa	5	4	3	6	13	20	49	HILDITCH, MEARA and PATEL (1951)
Pentaclethra Eetvaldeana	Africa	4	5	5	14	3	53	16	HILDITCH, MEARA and PATEL (1951)
Arachis hypogaea	Africa	9	5	← 10 →			39	37	CRAWFORD and HILDITCH (1950b)
Arachis hypogaea	India	9	4	← 7 →			54	26	HILDITCH and RILEY (1945)
Dipteryx odorata	Malaya	6	6	← 13 →			60	15	HILDITCH and STAINSBY (1934)
Medicago sativa	Switzerland	10	5	—	—	—	11	43 (*a*)	HILDITCH and ZAKY (1944)
Soja hispida	Asian?	12	2	← 2 →			24	51 (*b*)	HILDITCH, MEARA and HOLMBERG (1947)
Moringaceae									
Moringa oleifera	Trinidad	5	8	← 9 →			77	1	DUNN and HILDITCH (1947)
Ochnaceae									
Lophira alata	Africa	29	—	—	14	2	16	33 (*c*)	HILDITCH and MEARA (1944)
Lophira procera	Africa	37	—	—	20	1	12	26 (*d*)	HILDITCH, MEARA and PATEL (1951)
Sapindaceae									
Nephelium lappaceum	Malaya	2	14	35	—	—	45 (e)	—	HILDITCH and STAINSBY (1934)
Nephelium mutabile	Malaya	3	31	22	—	—	44	—	HILDITCH and STAINSBY (1934)
Sapindus trifoliatus	India	6	9	22	—	2	61	—	PARANJPE and AYYAR (1927)
Schleichera trijuga	Ceylon	9	5	30	—	2	39	5 (*f*)	WEERAKOON (1952)

(*a*) also 32 per cent linolenic acid; (*b*) also 9 per cent linolenic acid; (*c*) also 5 per cent docosenoic acid; (*d*) also 3 per cent docosenoic acid; (*e*) also 4 per cent eicosenoic acid; (*f*) also 5 per cent eicosenoic acid.

The sub-families *Mimosoideae, Caesalpinoideae* and *Papilionatae* are characterised by the presence of lower amounts of the characteristic higher saturated acids, but though their total amounts are considerably reduced as compared with those of the *Pentaclethra* fats there is some indication that all the members are present.

By far the greatest amount of work on this family has been associated with the constitution of the oils from *Arachis hypogaea* and of *Soja hispida*, produced from various varieties of seeds and under different climatic conditions. Much of the work is of a statistical nature, but from which it may be deduced, among other things, that there is a definite tendency for a more saturated oil to be produced under the influence of a cooler climate.

While there appear to be two types of *Arachis hypogaea* with saturated acids ranging approximately from 15 to 18 and from 18 to 23 per cent, the saturated acid content of *Soja hispida* remains remarkably constant at 12–14 per cent. It has been shown that this saturated acid content remains constant irrespective of the variety of seed or geographical location in which it was grown (Dollear, Krauczunas and Markley 1940, Scholfield and Bull 1944).

While the seed fat of *Moringa oleifera* (Behen oil) contains 9 per cent of higher saturated acids, the amount of behenic acid (from the trivial name of the plant) present is believed to be very small.

The two members of the family *Ochnaceae* so far examined, *Lophira alata* and *Lophira procera*, are remarkable in that while palmitic and behenic acids are present in both cases, a very careful search failed to reveal the presence of either stearic or arachidic acids. It would therefore be of great interest to determine whether the seed fat of *Ochna pulchra* would be similar in this respect.

It is seen that the members of the *Sapindaceae* are characterised by a very high content of arachidic acid, usually of the order of 20–30 per cent. The seed fats of this family also appear to be characterized by the presence of small amounts of the corresponding unsaturated eicosenoic acid.

11. Seed fats containing stearic acid as a major constituent, *i.e.* family group K.

In many of the seed fats of families already described it can be observed that in any particular family, as the content of unsaturated acids decreases, there is some tendency for the palmitic acid to increase. In the present group of families recorded in Table 15 it is seen that while in some cases palmitic acid occurs in quantity, in nearly every case does the stearic acid content assume major proportions. The result is that many of these fats, mainly from plants of tropical origin, are very hard solids at room temperatures and provide a number of the "vegetable butters" of industry.

It is of considerable interest to note that the seed fat of the tropical gymnosperm *Gnetum scandens* elaborates a hard fat containing over 50 per cent of stearic acid in its mixed fatty acids, this being in marked contrast to that elaborated by the temperate and colder region gymnosperms (see the *Coniferae*, Table 7, p. 30) in which the stearic acid content is of a very low order, the major component being linoleic acid. The influence of climate as distinct from family difference has not as yet been investigated.

While in general the seed fats of the families *Guttiferae* and *Dipterocarpaceae* both contain large amounts of stearic acid, in the former the palmitic content is of a low order whereas in the latter it is more pronounced. This tendency appears to be somewhat more developed in the case of some of the members

Table 15. *Component acids (wt. per cent) of seed fats of botanical families group K.*

	Location	C_{16}	C_{18}	Oleic	Linoleic	
ıtaceae						
ˀnetum scandens	India	14	56	27	3	VARIER (1943)
rseraceae						
ˀanarium commune . . .	E.Indies	31	10	40	19	STEGER and VAN LOON (1940)
Ɔacryodes rostrata	Malaya	13	31 (*a*)	50	3	HILDITCH, MEARA and ZAKY (1941)
ɔterocarpaceae						
ˀhorea stenoptera.	Malaya	18	45	37	trace	BUSHELL and HILDITCH (1938b)
ˀhorea robusta.	N. India	5	44 (*b*)	42	3	HILDITCH and ZAKY (1942b)
ˀateria indica	S. India	10	41 (*c*)	42	3	BALIGA and MEARA (1949)
ıtiferae						
1llanblackia floribunda . .	Africa	3	57	40	—	MEARA and ZAKY (1940)
ˀalophyllum inophyllum .	Ceylon	15	20	36	29	WEERAKOON (1952)
ˀarcinia indica	India	3	56	39	2	HILDITCH and MURTI (1941)
Mesua ferrea	India	10	14	66	10	DHINGRA and HILDITCH (1931)
ˀentadesma butyracea. . .	Africa	5	46	49	—	HILDITCH and SALETORE (1931)
liaceae						
1zadirachta indica	India	15	15	62	8	HILDITCH and MURTI (1939)
ˀarapa guianensis	S. America	29	7 (*f*)	52	10	SRISOMBOON (1954)
ˀarapa procera	S. America	21	11	49	19	SRISOMBOON (1954)
ɔotaceae						
Butyrospermum Parkii . .	Africa	6	41	49	4	GREEN and HILDITCH (1938)
ˀalocarpum mammosum. .	Honduras	10	22	54	14	JAMIESON and MCKINNEY (1931)
Madhuca butyracea. . . .	India	57	4	36	4	BUSHELL and HILDITCH (1938a)
Madhuca latifolia	India	24	19	43	14	HILDITCH and ICHAPORIA (1938)
Mimusops Heckelii. . . .	W. Africa	4	36	60	—	ATHERTON and MEARA (1940)
ˀalaquium oblongifolium .	Malaya	6	54	40	—	HILDITCH and STAINSBY (1934)
rculiaceae						
ˀterculia foetida	see p. 20					HILDITCH, MEARA and ZAKY (1941)
ˀterculia parviflora. . . .	see below					HILDITCH and MEARA (1944)
ˀterculia tomentosa	Sudan	← 22 →		46	32	HENRY and GRINDLEY (1940)
ˀheobroma cacao.	Africa	25	35	38	2	HILDITCH and STAINSBY (1936)

(*a*) Also 3 per cent arachidic; (*b*) also 6 per cent arachidic; (*c*) also 4 per cent arachidic; (*d*) also 2 per cent arachidic; (*e*) also 2 per cent lauric, 6 per cent myristic; (*f*) also 2 per cent arachidic.

of the family *Sapotaceae*, and may be still more pronounced in some members of the family *Burseraceae*.

Further it appears that the family *Sterculiaceae* produces fats of varying types, the seed fats of *Sterculia foetida* and of *parviflora* containing a C_{19} cyclopropane ring-containing acid (see p. 20), while *S. tomentosa* contains an abnormally high content of linoleic acid for the family. Only *Theobroma cacao* shows any correlation with the members of other families of this group in containing palmitic and stearic acids in quantity.

12. Seed fats containing lauric and myristic acids as major constituents, *i.e.* family group L.

This final group of plant families (see Table 16, p. 42) is characterised by the presence in quantity of saturated acids of lower molecular weight than palmitic acid. Thus in the seed fats of the members of the family *Lauraceae* the lauric acid content frequently reaches a very high order, the only known exception being that of *Laurus nobilis* in which the lauric acid content falls to

Table 16. *Component acids (wt. per cent) of seed fats of botanical families group L.*

	Location	C_8	C_{10}	C_{12}	C_{14}	C_{16}	C_{18}	Oleic	Lin-oleic	
Lauraceae										
Cinnamomum zeylanicum	Ceylon	—	—	87	7	2	—	4	—	Weerakoon (1952)
Laurus nobilis	India	—	—	43	6	32	—	19	—	Collin (1931)
Neolitsea involucrata . .	Ceylon	—	3	86	4	—	—	4	3	Gunde and Hilditch (1938a)
Neolitsea zeylanica . .	Ceylon	—	4	78	7	2	—	7	2	Weerakoon (1952)
Myristicaceae										
Myristica fragrans . .	E. Indies	—	—	2	77	10	—	10	1	Collin and Hilditch (1930b)
Myristica malabarica . .	S. India	—	—	—	39	13	3	44	1	Collin and Hilditch (1930b)
Pycnanthus Kombo . .	S. America	—	—	10	57	4	—	6 (*a*)	—	Atherton and Mear (1939)
Virola ucuhuba	S. America	—	1	15	73	5	—	6	—	Atherton and Meara (1939)
Palmae										
Attalea cohune	Honduras	7	7	47	16	9	3	10	1	Hilditch and Vidyar (1928)
Attalea funifera	Brazil	5	7	44	15	9	3	16	1	Jackson and Longen-ecker (1944)
Cocos nucifera	Pacific Islands	8	8	46	17	9	4	6	2	Dale and Meara (195
Cocos nucifera (testa) .		2	2	28	22	12	1	23	10	Armstrong, Allan a Moore (1925)
Elaeis guineensis . . .	W. Africa	2	4	45	19	9	4	15	2	Dale and Meara (19
Elaeis guineensis (testa)	W. Africa	1	4	36	16	11	1	26	3	Carsten, Hilditch ar Meara (1945)
Salvadoraceae										
Salvadora oleoides . . .	India	—	2	21	53	19	—	5	—	Gunde and Hilditch
Salvadora persica . . .	India	—	1	20	54	20	—	5	—	Gunde and Hilditch
Simarubaceae										
Irvingia Barteri	W. Africa	—	—	39	50	—	—	11	—	Collin and Hilditc (1930a)
Irvingia gabonensis . .	W. Africa	—	3	59	33	2	1	2	—	Bushell and Hildit (1939)
Picramnia Lindeniana .	Central America	—	—	—	22	33	3	22 (*b*)	—	Grimme (1912)
Picramnia Sow	Central America	←	—	5	—	—	→	← (*c*)	→	Steger and van Loon
Picrasma quassioides .	Japan	←	—	23	—	—	→	← (*d*)	→	Tsujimoto and Koya (1933)
Ulmaceae										
Ulmus americanus . . .	U.S.A.	5	61	6	5	3	—	11	9	Zehnpfennig and Schuette (1941)
Vochysiaceae										
Erisma calcaratum . .	Brazil	—	—	24	54	19	—	3	—	Steger and van Loon (1935b)

(*a*) Also 23 per cent tetradecenoic; (*b*) also 20 per cent tariric acid; (*c*) *ca.* 95 per cent tariric acid; (*d*) *ca* 77. per cent petroselinic acid.

ca. 40 per cent while a concomitant marked increase in oleic and linoleic acids occurs.

In the *Myristicaceae* seed fats myristic acid is the dominant component though occurring in amounts somewhat less than that of lauric in the *Lauraceae*. Both lauric and palmitic acids are usually present in some quantity. The members of the *Palmae* differ from those of both the *Lauraceae* and *Myristicaceae* in that they appear to be unique in containing all the saturated acids from hexanoic acid (trace—1 per cent) to stearic acid with lauric and myristic acids comprising the chief constituents. The fats of the *Palmae* are indeed remarkable in the close similarity in the component fatty acids of its members when the large number of component fatty acids is taken into account.

Table 16 also indicates that the family *Simarubaceae* resembles the families *Euphorbiaceae* and *Burseraceae* in elaborating members the seed fats of which appear to be of a more heterogeneous character than usually holds.

Again while members of the families *Salvadoraceae* and *Vochysiaceae* conform to the general pattern of elaborating seed fats rich in lauric and myristic, these members also contain in their seed fats appreciable quantities of palmitic acid. The family *Ulmaceae* is at present unique in elaborating seed fats containing over 50 per cent of decanoic acid.

IX. Component glycerides.

For some considerable time it was considered that the natural fats consisted of mixtures of simple triglycerides. It would appear that the first reported mixed triglyceride was oleo-butyro-palmitin isolated from butter fat, by Blyth and Robertson (1889). By 1902 sufficient mixed triglycerides had been isolated to indicate that mixed triglycerides occurred very frequently in nature and Guth (1902) suggested that natural fats in general consisted of mixtures of mixed triglycerides.

With the development of methods, primarily by the Hilditch school, for the quantitative estimation of the triglycerides of a fat, it became clear that the component glycerides of the majority of vegetable fats were elaborated according to a definite pattern, which was not one of random distribution of fatty acid radicals amongst the glycerol molecules.

Hilditch and co-workers found that when a given fatty acid occurred in minor amounts *i.e.* of the order of 15 per cent (mol) or less, it is of necessity absent from many triglyceride molecules and that in those in which it occurs it is present only once in each triglyceride molecule.

When a particular fatty acid occurs to the extent of about 35 per cent (mol) it will occur once in practically every triglyceride molecule.

When the amount of the given acid present lies between 35 and 65 per cent (mol) it will occur in every triglyceride molecule, once in some and twice in the remainder, the proportions of these components naturally depending on the extent to which the given acid occurs in the mixed fatty acids of the oil.

When a given fatty acid occurs to the extent of 70 per cent (mol) or more, it is present once per molecule in only a very small number of triglyceride molecules. In the bulk of the triglyceride molecules it is present twice and only after the latter requirement has been satisfied, to give the maximum amount of the latter does the remainder of the acid appear as simple triglyceride.

These observations have for some time now come to be known as the "*Rule of Even Distribution*".

It is known that a marked tendency exists for fats to be elaborated along these lines, but whereas the Hilditch school defined this pattern of glyceride structure in general terms, and frequently pointed out that the rules were only approximately conformed to, other workers have attempted to apply a rigidly mathematical interpretation to the above rules, without much success, and have therefore criticised the principle.

A clarification of the concept of "Even distribution" was therefore put forward by Hilditch (1954), as a result of investigations by Barker and Hilditch

(1950a) and HILDITCH and SEAVELL (1950) who determined the component gylcerides of a number of oils rich in linoleic and linolenic acids.

On plotting the observed contents of mono-, di- and tri-linoleo (linoleno) glycerides against the content of the respective acid in the mixed fatty acids of the oil, smooth curves were obtained with maxima at the points where the percentages of the acid in the mixed fatty acids were 33 and 67 respectively. These maxima were found to correspond to contents of 85–90 per cent of the total glycerides of the oil.

The curves further indicate that triglycerides with two molecules of any given acid occur in small proportions before the acids form one-third of the total acids, that small amounts of triglycerides with one molecule of the given acid persist after the acid forms more than two-thirds of the total mixed acids, and that simple triglycerides begin to appear before the acid forms more than two-thirds of the mixed fatty acids.

It is now known that the component glycerides of a considerable number of oils and fats conform very closely to these graphs.

Even distribution can therefore probably be best defined through this graphical approach, an oil the component glycerides of which conform with the described graphs being considered as being elaborated according to the general principles of "Even distribution".

Literature.

ECKEY, E. W.: Vegetable fats and oils. New York: Reinhold Publishing Corp. 1954.
HILDITCH, T. P.: Chemical constitution of natural fats, 3nd Ed. London: Chapman & Hall 1956. — Industrial Chemistry of Fats and Waxes. London: Baillière, Tindall & Cox. 1949.
MARKLEY, K. S.: Fatty acids. New York: Interscience Publishers Inc. 1947.
RALSTON, A. W.: Fatty acids and their derivates. New York: J. Wiley & Sons Inc. 1948.

References.

ADAMS, M., and A. HOLMES: Pine nut oil. Industr. Engin. Chem. **5**, 285–287 (1913). — AGGARWAL, J. S., and KARIMULLAH: Chemical examination of the fatty oil from the seeds of *Sisymbrium irio*. J. Sci. Ind. Res. (India) B **5**, 57–60 (1947). — AMBERGER, K., and E. WHEELER-HILL: Über die Zusammensetzung des Haferöls. Z. Unters. Lebensmitt. **54**, 417–431 (1927). — ARMSTRONG, E. F., J. ALLAN and C. W. MOORE: The fatty acid constituents of some natural fats. I. The oils from some other members of the Palmae. J. Soc. Chem. Ind. (Lond.) **44**, 61–63T (1925). — ATHERTON, D., and M. L. MEARA: Component fatty acids and glycerides of some *Myristica* fats. J. Soc. Chem. Ind. (Lond.) **58**, 353–357 (1939). — Fatty acids and glycerides of solid seed fats. IX. *Mimusops heckelii* (Baku) kernel fat. J. Soc. Chem. Ind. (Lond.) **59**, 95–96 (1940).

BALDWIN, A. R., and M. S. SNIEGOWSKI: Fatty acid compositions of lipides from corn and grain sorghum kernels. J. Amer. Oil Chem. Soc. **28**, 24–27 (1951). — BALIGA, M. N., and T. P. HILDITCH: The component acids of rape seed oil. J. Soc. Chem. Ind. (Lond.) **67**, 258–262 (1948). — The constitution of some minor unsaturated fatty acids of rape seed oil. J. Chem. Soc. (Lond.) **1949**, 915–955. — BALIGA, M. N., and M. L. MEARA: The component fatty acids and glycerides of Dhupa fat. J. Soc. Chem. Ind. (Lond.) **68**, 52–54 (1949). — BARKER, C., and T. P. HILDITCH: The influence of environment upon the composition of sunflower-seed oils. I. Individual varieties of sunflowers grown in different parts of America. J. Sci. Food Agricult. **1**, 118–121 (1950a). — The component glycerides of drying oils. I. Linolenic rich oils. J. Oil a. Colour Chemists' Assoc. **33**, 6–23 (1950b). — BAUER, K. F., u. L. SEBER: Vergleichende Untersuchungen über die Zusammensetzung der Fette von Keimen, Samenschalen und Cotyledonen der Samen von *Theobroma cacao* LINNÉ. Fette u. Seifen **45**, 293–299 (1938). — BAUGHMAN, W. F., and G. S. JAMIESON: Chemical composition of ergot oil. Oil a. Fat Industr. **5**, 85–90 (1928). — The constituents of "Chufa" oil, a fatty oil from the tubers of *Cyperus esculentus* LINNÉ. J. Agricult. Res. **26**, 77–82 (1932). — BAUR jr., F. J., and J. B. BROWN: Fatty acids of corn oil. J. Amer. Chem. Soc. **67**, 1899–1900 (1945). — BLACK, H. K., and B. C. L. WEEDON: Synthesis of erythrogenic (isanic) acid. Chem. a. Ind. **1953**, 40–41. — BLYTH, A. W., and G. H. ROBERTSON: Notes of experiments on butter fat. Proc. Chem. Soc. **5**, 5 (1889). — BOEKENOOGEN, H. A.:

Ximenia-Öl, ein Pflanzenfett mit besonders hochmolekularen Fettsäuren. Fette u. Seifen **46**, 717–719 (1939). — BRANDONISIO, V.: Composition of olive oils from the Isles of Rhodes and Coo. Chim. e Ind. (Italy) **18**, 14–16 (1936). — BRIDGE, R. E., M. M. CHAKRABARTY and T. P. HILDITCH: Composition of poppyseed oils. J. Oil a. Colour Chemists' Assoc. **34**, 354–360 (1951). — BRIDGE, R. E., and T. P. HILDITCH: Seed fat of *Macadamia ternifolia*. J. Chem. Soc. (Lond.) **1950**, 2396–2399. — BROWN, J. B.: Private communication 1954. — BROWN, W. B., and E. H. FARMER: Unsaturated acids of natural oils. V. α- and β-licanic acids. J. Chem. Soc. (Lond.) **1935**, 1632-1633. — BUSHELL, W. J., and T. P. HILDITCH: Fatty acids and glycerides of solid seed fats. IV. Seed fat of *Madhuca butyracea* (phulwara butter). J. Soc. Chem. Ind. (Lond.) **57**, 48–49 (1938a). — Fatty acids and glycerides of solid seed fats. VI. Borneo tallow. J. Soc. Chem. Ind. (Lond.) **57**, 447–449 (1938b). — Fatty acids and glycerides of solid seed fats. VII. Dika fat. J. Soc. Chem. Ind. (Lond.) **58**, 24–26 (1939). — BYERS, H. G., and P. HOPKINS: Investigations of the red Elderberry "*Sambucus racemosa arborescens*". J. Amer. Chem. Soc. **24**, 771–774 (1902).

CALDERWOOD, R. C., and F. D. GUNSTONE: Structure of kamlolenic acid. Chem. a. Ind. **1953**, 436–437. — CARSTEN, H. A., T. P. HILDITCH and M. L. MEARA: The component acids of the testa and kernel fats of the oil palm. J. Soc. Chem. Ind. (Lond.) **64**, 207–209 (1945). — CASTILLE, A.: Zur Kenntnis des fetten Öls der Samen von *Ongokea Klaineana* PIERRE. Liebigs Ann. **543**, 104–110 (1939). — CHIBNALL, A. C., and H. J. CHANNON: The ether-soluble substances of cabbage leaf cytoplasm. III. The fatty acids. Biochemic. J. **21**, 479–483 (1927). — CHRISTIAN, B. C., and T. P. HILDITCH: Seed fats of the *Umbelliferae*. II. The seed fats of some cultivated species. Biochemic. J. **32**, 327–338 (1929). — CLOPTON, J. R., and R. W. v. KORFF: *Typha* (cattail) seed oil. Oil a. Soap **22**, 330–331 (1945). — CLOPTON, J. R., and H. O. TRIEBOLD: Fanweed seed oil. Potential substitute for rapeseed oil. Industr. Engin. Chem. **36**, 218–219 (1944). — COLE, H. I., and H. T. CARDOSO: Isolation and properties of gorlic acid, an optically active liquid fatty acid. J. Amer. Chem. Soc. **60**, 612–614 (1938a). — Analysis of Chaulmoogra oils. II. *Oncoba echinata* (Gorli) oil. J. Amer. Chem. Soc. **60**, 617–619 (1938b). — Analysis of Chaulmoogra oils. IV. *Hydnocarpus anthelmintica* oil. V. *Taraktogenos kurzii* (Chaulmoogra) oil. J. Amer. Chem. Soc. **61**, 3442–3445 (1939a). — Alepric, Aleprylic, Aleprestic and Aleprolic acids, new homologs of Chaulmoogric acid. J. Amer. Chem. Soc. **61**, 2349–2351 (1939b). — COLLIN, G.: Some peculiarities in the glyceride structure of laurel fats. Biochemic. J. **25**, 95–100 (1931). — COLLIN, G., and T. P. HILDITCH: Dika fat (*irvingia* butter). J. Soc. Chem. Ind. (Lond.) **49**, 138–139T (1930a). — Fatty acids of nutmeg (mace) butter and of expressed oil of laurel. J. Soc. Chem. Ind. (Lond.) **49**, 141–143T (1930b). — CRAWFORD, R. V., and T. P. HILDITCH: The component fatty acids of tobacco-seed oils. J. Sci. Food Agricult. **1**, 230–234 (1950a). — The component fatty acids and glycerides of groundnut oils. J. Sci. Food Agricult. **1**, 372–379 (1950b). — CROSSLEY, A., and T. P. HILDITCH: Component acids of some authentic and commercial *stillingia* oils. J. Sci. Food Agricult. **1**, 292–300 (1950). — The fatty acids and glycerides of okra seed oil. J. Sci. Food Agricult. **2**, 251–255 (1951). — CROXFORD, J. W.: Rye oil. Analyst (Lond.) **55**, 735–738 (1930).

DALE, A. P.: Private communication 1954. — DALE, A. P., and M. L. MEARA: Component fatty acids and glycerides of coconut oils. J. Sci. Food Agricult. **6**, 162–166 (1955). — DEAN, H. K., and T. P. HILDITCH: Composition of commercial palm oils. III. Some characteristic differences between the component acids of oils from Liberia or the Ivory Coast and those of native or plantation palm oils from other localities. J. Soc. Chem. Ind. (Lond.) **52**, 165–169T (1933). — DELVAUX, E.: Die Zusammensetzung des Pflaumenkern- und Bucheckernöles. Fette u. Seifen **43**, 183–184 (1936). — DHINGRA, D. R., and A. K. BISARAS: Component fat acids of oil of *Citrullus vulgaris* SCHRAD. (watermelon) seed. J. Indian Chem. Soc. **22**, 119–122 (1945). — DHINGRA, D. R., and T. P. HILDITCH: Fatty acids of some Indian seed oils: seed fats of *Mesua ferrea*, *Calophyllum inophyllum* and *Pistacia vera*. J. Soc. Chem. Ind. (Lond.) **50**, 9–12T (1931). — DHINGRA, D. R., T. P. HILDITCH and R. J. VICKERY: Fatty acids and glycerides of kusum oil. J. Soc. Chem. Ind. (Lond.) **48**, 281–286T (1929). — DIETERLE, H., u. O. DORNER: Über die Inhaltsstoffe von *Crataegus oxyacantha* L. Arch. Pharmaz. **275**, 428–437 (1937). — DOLLEAR, F. G., P. KRAUCZUNAS and K. S. MARKLEY: The chemical composition of some soybean oils of high iodine number. Oil a. Soap **17**, 120–121 (1940). — DUNN, H. C., and T. P. HILDITCH: Component acids of West Indian ben and mango seed oils. J. Soc. Chem. Ind. (Lond.) **66**, 209–211 (1947). — DUNN, H. C., T. P. HILDITCH and J. P. RILEY: Composition of seed fats of West Indian *citrus* fruits. J. Soc. Chem. Ind. (Lond.) **67**, 199–203 (1948). — DUTT, S.: Mustard oil. Indian Soap J. **5**, 279–285 (1939). — DYKE, F. M.: African World Suppl. **25** (1928).

EIBNER, A., u. F. REITTER: Zur Normung der trocknenden fetten Öle nach Verwendungseigenschaften, Gruppe der linolensäurehaltigen maltechnisch mohnölartigen Öle. Über Abietineensamenöle und die isomeren Linolensäuren. Chem. Umschau **33**, 114–124, 125–129 (1926).

Flaschenträger, B., u. R. V. Wolffersdorff: The poisons in croton-oil. I. Croton oil acids. Helvet. chim. Acta **17**, 1444–1452 (1934). — Foreman, H. D., and J. B. Brown: Solubilities of the fatty acids in organic solvents at low temperatures. Oil a. Soap **21**, 183–187 (1944). — Frahm, E. D. G.: *Parinarium* oils. Ing. Nederl.-Indie **8**, VII, 92–98 (1941).

Gordon, S. M.: Studies in the genus *Mentha*. XVI. The non-volatile constituents of *Mentha aquatica* Linné. Amer. J. Pharmacy **100**, 433–449, 509–524 (1928). — Goss, W. H., and J. E. Ruckman: Oil from tumbling mustard seed. Oil a. Soap **21**, 234—236 (1944). — Green, T. G., and T. P. Hilditch: Fat acids and glycerides of solid seed fats. V. Shea butter. J. Soc. Chem. Ind. (Lond.) **57**, 49–53T (1938). — Green, T. G., T. P. Hilditch and W. J. Stainsby: Seed wax of *Simmondsia californía*. J. Chem. Soc. (Lond.) **1936**, 1750–1755. — Griffiths, H. N., and T. P. Hilditch: Oleic-elaidic acid transformation as an acid in the analysis of mixtures of oleic, linoleic and linoleic acids. J. Soc. Chem. Ind. (Lond.) **53**, 75–81T (1934). — Grimme, C.: Über das Fett von *Picramnia Lindeniana*. Chem. Rev. Fett-Harz Ind. **19**, 51–55 (1912). — Grindley, D. N.: Seed oils of some Sudan *Mimosaceae*. J. Soc. Chem. Ind. (Lond.) **64**, 152 (1945). — *Ammi visnaga:* composition of the fatty acids present in the seed fat. J. Sci. Food Agricult. **1**, 53—56 (1950). — Grün, A., u. T. Wirth: ϑ, ι-Decylensäure, eine bisher unbekannte Säure aus der Butter. Ber. dtsch. chem. Ges. **55**, 2197–2205 (1922). — Gunde, B. G., and T. P. Hilditch: Seed and fruit-coat fats of *Neolitsea involucrata*. J. Chem. Soc. (Lond.) **1938**a, 1610–1614. — Seed and fruitcoat fats of *Celastrus paniculatus*. J. Chem. Soc. (Lond.) **1938**b, 1980–1985. — Seed fats of *Salvadora oleoides* and *Salvadora persica*. J. Chem. Soc. (Lond.) **1939**, 1015–1016. — Mixed unsaturated glycerides of liquid seed fats. I. Non-drying oils. J. Soc. Chem. Ind. (Lond.) **59**, 47–53 (1940).— Gunstone, F. D.: Vegetable oils. II. Further studies of seed oils of various *Strophanthus* species. J. Sci. Food Agricult. **4**, 129–132 (1953). — Fatty acids. II. The nature of the oxygenated acid present in *Vernonia anthelmintica* seed oil. J. Chem. Soc. (Lond.) **1954**, 1611–1616. — Gunstone, F. D., and T. P. Hilditch: The use of low-temperature crystallization in the determination of component acids of liquid fats. II. Fats which contain linolenic as well as linolenic and oleic acids. J. Soc. Chem. Ind. (Lond.) **65**, 8–13 (1946). — Gunstone, F. D., T. P. Hilditch and J. P. Riley: African drying oils. I. The seed oil of *tetracarpidium conophorum*. J. Soc. Chem. Ind. (Lond.) **66**, 293–296 (1947). — Gunstone, F. D., and M. A. McGee: Santalbic acid. Chem. a. Ind. **1954**, 1112. — Gupta, S. S., T. P. Hilditch, S. Paul and R. K. Shrivastava: Constitution of vaccenic acid. J. Chem. Soc. (Lond.) **1950**, 3484–3490. — Gupta, S. S., and M. L. Meara: Configuration of naturally occurring mixed glycerides. VI. The component fatty acids and glycerides of *stillingia* tallows. J. Chem. Soc. (Lond.) **1950**, 1337–1342. — Gupta, S. S., V. N. Sharma and J. S. Aggarwal: The chemical constitution of the hydroxy acid from the seed oil of *Mallotus philippinensis*. J. Sci. Ind. Res. (India) B **10**, 76 (1951). — Gurgel, L., e T. F. de Amorim: Oil from (seeds of) the páo marfim (ivory wood) (*Agonandra brasiliensis*, Miers.). Mem. Inst. chim. Rio de Janeiro **2**, 31–38 (1929).— Guth, F.: Über synthetisch dargestellte einfache und gemischte Glycerinester fetter Säuren. Z. biol. Chem. **44**, 78–110 (1902).

Hata, T.: Formosan plant seed oils. XV. Pineapple, tomato, *passiflora* and sugar-apple seed oils. J. Chem. Soc. Japan **59**, 1099–1103 (1938). — Hawke, F.: Private communication 1952. — Henry, A. J., and D. N. Grindley: Oil of seeds of *Cephalocroton cordofanus*. J. Soc. Chem. Ind. (Lond.) **62**, 60 (1943). — The oils of the seed of *Ocimum kilimandscharicum*, *Euphorbia calycine*, *E. erythraeae Sterculia tomentosa* and *Trichilia emetica*. J. Soc. Chem. Ind. (Lond.) **63**, 188–190 (1944). — Hilditch, T. P.: The fats: A story of natures art. J. Sci. Food Agricult. **5**, 557–566 (1954). — Hilditch, T. P., and M. B. Ichaporia: Composition of some solanaceous seed fats. J. Soc. Chem. Ind. (Lond.) **55**, 189–190T (1936). — Fatty acids and glycerides of solid seed fats. III. Seed fat of *Madhuca (Bassia) latifolia* (Mowrah fat). J. Soc. Chem. Ind. (Lond.) **57**, 44–48T (1938). — Hilditch, T. P., and H. Jasperson: The polyethenoid acids of the C_{18}-series present in milk and grass fats. J. Soc. Chem. Ind. (Lond.) **64**, 109–111 (1945). — Hilditch, T. P., and E. E. Jones: Some ill-defined acids of the oleic series. II. Petroselinic acid and the composition of English parsley-seed oil. J. Soc. Chem. Ind. (Lond.) **46**, 174–177T (1927). — Seed fats of the Umbelliferae. I. *Heracleum sphondylium* and *Angelica sylvestris*. Biochemic. J. **22**, 326–330 (1928). — Composition of commercial palm oils. I. Fatty acids and component glycerides of some palm oils of low free acidity. J. Soc. Chem. Ind. (Lond.) **49**, 363–368 369T (1930). — Hilditch, T. P., P. A. Laurent and M. L. Meara: The mixed unsaturated glycerides of liquid fats. VI. Low-temperature crystallization of rape oil. J. Soc. Chem. Ind. (Lond.) **66**, 19–22 (1947). — Hilditch, T. P., and H. E. Longenecker: Further determination and characterization of the component acids of butter fat. J. of Biol. Chem. **122**, 497–506 (1938). — Hilditch, T. P., and J. A. Lovern: Head and blubber oils of the sperm whale. I. Quantitative determinations of the mixed fatty acids present. J. Soc. Chem. Ind. (Lond.) **47**, 105–111T (1928). — Hilditch, T. P., and L. Maddison: Mixed unsaturated glycerides of liquid seed fats. II. Low-temperature crystallization of cottonseed oil. J. Soc.

Chem. Ind. (Lond.) **59**, 162–168 (1940). — HILDITCH, T. P., and M. L. MEARA: The seed fat of the annual nasturttium *(Tropaeolum var.)*. J. Chem. Soc. (Lond.) **1938**, 1608–1610. — The fat acids and glycerides of solid seed fats. XII. *Lophira alta* kernel fat (Niam fat). J. Soc. Chem. Ind. (Lond.) **63**, 114–115 (1944). — HILDITCH, T. P., M. L. MEARA and J. HOLMBERG: The component glycerides of soybean oil and of soybean oil fractions. J. Amer. Oil Chemists' Soc. **24**, 321–325 (1947). — HILDITCH, T. P., M. L. MEARA and C. B. PATEL: Component acids and glycerides of *pentaclethra (Leguminosae)* and *lophira (Ochnaceae)* seed fats. J. Sci. Food Agricult. **2**, 142–148 (1951). — HILDITCH, T. P., M. L. MEARA and W. H. PEDELTY: Fatty acids of solid seed fats. VIII. The seed fat of *Hodgsonia capniocarpa*. J. Soc. Chem. Ind. (Lond.) **58**, 27–29 (1939). — HILDITCH, T. P., M. L. MEARA and O. A. ROELS: The composition of commercial palm oils. VI. Component acids and glycerides of a Belgian Congo palm oil (studied with the acid of low-temperature crystallization). J. Soc. Chem. Ind. (Lond.) **66**, 284—288 (1947). — HILDITCH, T. P., M. L. MEARA and Y. A. K. ZAKY: The component acids of *Sterculia foetida* seed fat (*sterculia* oil): a correction of work previously reported. J. Soc. Chem. Ind. (Lond.) **60**, 198–203 (1941). — HILDITCH, T. P., and A. MENDELOWITZ: The component fatty acids and glycerides of tung oil. J. Sci. Food Agricult. **2**, 548–556 (1951). — HILDITCH, T. P., and K. S. MURTI: Fat acids and glycerides of neem *(margosa)* oil. J. Soc. Chem. Ind. (Lond.) **58**, 310–312 (1939). — Fat acids of solid seed fats. X. Seed fats of *Garcinia morella* and *Garcinia indica*. J. Soc. Chem. Ind. (Lond.) **60**, 16–18 (1941). — HILDITCH, T. P., and J. G. RIGG: Component glycerides of *piquia* fats. J. Soc. Chem. Ind. (Lond.) **54**, 109–111 T (1935). — HILDITCH, T. P., and J. P. RILEY: The use of low-temperature crystallization in the determination of component acids of liquid fats. I. Fats in which oleic and linoleic acids are major components. J. Soc. Chem. Ind. (Lond.) **64**, 204–207 (1945). — The use of low-temperature crystallization in the determination of component acids of liquid fats. III. Fats which contain eleostearic as well as linoleic and oleic acids. J. Soc. Chem. Ind. (Lond.) **65**, 74–81 (1946). — HILDITCH, T. P., T. RILEY and N. L. VIDYARTHI: The fatty acids of seed oils of *Brassica* species. The composition of rape, ravison, and mustard seed oils. J. Soc. Chem. Ind. (Lond.) **46**, 457–462 (1927). — HILDITCH, T .P., and S. A. SALETORE: Fatty acids and glycerides of solid seed fats. I. Composition of the seed fats of *Allanblackia stuhlmannii, Pentadesma butyracea, Butyrospermum parkii* (Shea) and *Vateria indica* (Dhupa). J. Soc. Chem. Ind. (Lond.) **50**, 468–472 T (1931). — HILDITCH, T. P., and A. J. SEAVELL: The component glycerides of drying oils. II. Linolenic-rich oils. J. Oil a. Colour Chemists' Assoc. **33**, 24–28 (1950). — HILDITCH, T. P., I. C. SIME, Y. A. H. ZAKY and M. L. MEARA: The component acids of various vegetable fats. J. Soc. Chem. Ind. (Lond.) **63**, 112–114 (1944). — HILDITCH, T. P., and W. J. STAINSBY: Fatty acids and glycerides of solid seed fats. II. Composition of some Malayan vegetable fats. J. Soc. Chem. Ind. (Lond.) **53**, 197–203 T (1934). — Component glycerides of cacao butter. J. Soc. Chem. Ind. (Lond.) **55**, 95–101 T (1936). — HILDITCH, T. P., and H. M. THOMPSON: Further observations on the component glycerides of olive and tea-seed oils. J. Soc. Chem. Ind. (Lond.) **56**, 434–481 T (1937). — HILDITCH, T. P., and N. L. VIDYARTHI: Fatty acids of cohune-nut fat. J. Soc. Chem. Ind. (Lond.) **47**, 35–37 T (1928). — HILDITCH, T. P., and Y. A. H. ZAKY: The component fat acids of some vegetable seed phosphatides. Biochemic. J. **36**, 815–821 (1942a). — Fat acids and glycerides of solid seed fats. XI. *Shorea robusta* kernel fat. J. Soc. Chem. Ind. (Lond.) **61**, 34–36 (1942b). — HOPKINS, C. Y.: Fat acids of hare's–ear mustard-seed oil. Canad. J. Res. B **24**, 211–220 (1946).

IVANOV, S.: Biochemistry of plant fats. Biol. generalis (Wien) **5**, 579–586 (1929).

JACKSON, F. L., and H. E. LONGENECKER: Fat acids and glycerides of babassu oil. Oil a. Soap **21**, 73–75 (1944). — JAMIESON, G. S.: Chemical composition of rice oil. J. Oil a. Fat Ind. **3**, 256–261 (1926). — Vegetable fats and oils, 2nd ed., p. 38. New York: Reinhold Publishing Corp. 1943. — JAMIESON, G. S., and W. F. BAUGHMAN: The chemical composition of California olive oil. J. Oil a. Fat Ind. **2**, 40–44 (1925). — JAMIESON, G. S., W. F. BAUGHMAN and H. HANN: Avocado oil. J. Oil a. Fat Ind. **5**, 202–207 (1928). — JAMIESON, W. F., and S. I. GERTLER: American safflower-seed oil. J. Oil a. Fat Ind. **6**, 11–13 (1929a). — Pecan oil. J. Oil a. Fat Ind. **6**, 23–24 (1929b). — American cherry kernel oil. J. Oil a. Fat Ind. **7**, 371–372 (1930). — JAMIESON, W. F., and R. S. MCKINNEY: Sapote (mammy apple) seed and oil. J. Oil a. Fat Ind. **8**, 255–256 (1931). — California apricot oil. Oil a. Soap **10**, 147–149 (1933). — Passion fruit seed oil. Oil a. Soap **11**, 193 (1934). — Composition of the oil of the American black walnut. Oil a. Soap **13**, 202 (1936a). — Expressed kapok-seed oil. Oil a. Soap **13**, 233–234 (1936b). — The composition of expressed lumbang oil. Oil a. Soap **14**, 203–205 b (1937). — JANTZEN, E., u. C. TIEDCKE: Über das Zerlegen der Säuren aus Erdnußöl unter Verwendung neuer Prinzipien bei fraktionierter Destillation. J. prakt. Chem. **127**, 277 (1930).

KAUFMANN, H. P.: Rhodanometrische Untersuchungen von *Perilla*ölen. Allg. Öl- u. Fett-Ztg **27**, 39–40 (1930). — KAUFMANN, H. P., u. J. BALTES: Beiträge zur Erschließung neuer deutscher Ölquellen. IV. Das Kastanienöl. (Studien auf dem Fettgebiet. 55. Mitt.) Fette u. Seifen **45**, 175–176 (1938). — KAUFMANN, H. P., u. H. FIEDLER: Beiträge zur Frage der

Erschließung neuer deutscher Ölquellen. 1. Traubenkernöl. (Studien auf dem Fettgebiet. 36. Mitt.) Fette u. Seifen **44**, 286–289 (1937). — Beiträge zur Erschließung neuer deutscher Ölquellen. III. Lindensamenöl. (Studien auf dem Fettgebiet. 53. Mitt.) Fette u. Seifen **45**, 149–151 (1938). — Untersuchungen über das Kürbiskernöl. (Studien auf dem Fettgebiet. 67. Mitt.) Fette u. Seifen **46**, 125–127 (1939). — Kaufmann, H. P., S. Funke u. F. Y. Liu: Beitrag zur Bestimmung der Carbonylzahl mit besonderer Berücksichtigung ihrer Anwendung auf dem Fettgebiet. I. (Studien auf dem Fettgebiet. 65. Mitt.) Fette u. Seifen **46**, 616–620 (1938). — Khan, N. A., and J. B. Brown: The composition of coffee oil and its component fatty acids. J. Amer. Oil Chemists' Soc. **30**, 606–609 (1953). — King, G.: Estimation of epoxides. Nature (Lond.) **164**, 706–707 (1949). — Determination of oxirane-oxygen, with special reference to its application to the study of autoxidation. J. Chem. Soc. (Lond.) **1951**, 1980–1984. — Klein, O.: Beiträge zur Kenntnis des Olivenkernöls. Z. angew. Chem. **1898**, 845–850. — Krassowski, N.: Über das Öl aus den Samen der Beeren von *Rhamnus carthartica* (Purgierwegdorn). J. russ. phys.-chem. Soc. **38**, 144–161 (1906).

Ligthelm, S. P., and H. M. Schwartz: Isolation of a conjugated unsaturated acid from the oil from *Ximenia caffra* kernels. J. Amer. Chem. Soc. **72**, 1868 (1950). — Ligthelm, S. P., H. M. Schwartz and M. M. v. Holdt: The chemistry of ximenynic acid. J. Chem. Soc. (Lond.) **1952**, 1088–1093.

Margaillan, L.: The oil of *Wrightia annamensis* Dubard and Eberhardt, an oil similar to castor oil. C. r. Acad. Sci. Paris **192**, 373–374 (1931). — Matthes, H., u. W. Rossié: Über Piniensamen und Piniensamenöl. Arch. Pharmaz. **256**, 289–302 (1918). — McKinney, R. S., and G. S. Jamieson: A nonfatty oil from jojoba seed. Oil and Soap **13**, 289–292 (1936). — Japanese tung oil. Oil a. Soap **14**, 2–3 (1937). — Meara, M. L.: The component acids of an English almond oil. Chem. a. Ind. **1952**, 667–668. — Meara, M. L., and N. K. Sen: Component fatty acids and glycerides of jute-seed oils. J. Sci. Food Agricult. **3**, 237–240 (1952). — Meara, M. L., and Y. A. H. Zaky: Fatty acids and glycerides of the seed fats of *Allanblackia floribunda* and *Allanblackia parviflora*. J. Soc. Chem. Ind. (Lond.) **59**, 25–26 (1940). — Mendelowitz, A., and J. P. Riley: Spectrophotometric determination of lang-chain fatty acids containing ketonic groups with particular reference to licanic acid. Analyst (Lond.) **78**, 704–709 (1953). — Mikusch, J. D. v.: Die Zusammensetzung von Leindotteröl. Farbe u. Lack **58**, 402–406 (1952). — Mudbidri, S. M., P. R. Ayyar and H. E. Watson: Oil from the seeds of *Adenanthera pavonina*. A source of lignoceric acid. J. Indian Inst. Sci. A **11**, Pt. 14, 173–180 (1928). — Murti, K. S., and F. G. Dollear: Rice-bran oil. II. Composition of oil obtained by solvent extraction. J. Amer. Oil Chemists' Soc. **25**, 211–213 (1948).

Neville, A.: The „Crude Fat" of *Beta vulgaris*. J. Chem. Soc. (Lond.) **101**, 1101–1104 (1912). — Nunn, J. R.: Structure of sterulic acid. J. Chem. Soc. (Lond.) **1952**, 313–318.

Painter, E. P., and L. L. Nesbitt: Fat acid composition of linseed oil from different varieties of flaxseed. Oil a. Soap **20**, 208–211 (1943). — Paranjpe, D. R.: Composition of *Parkia* oil. J. Indian Chem. Soc. **8**, 767–772 (1931). — Paranjpe, D. R., and P. R. Ayyar: Oil from the seeds of *Sapindus trifoliatus* (Linn.). J. Indian Inst. Sci. A **12**, Pt. 12, 179–184 (o. J.). — Patel, C. K., J. J. Sudborough and H. E. Watson: Cashew kernel oil. J. Indian Inst. Sci. **6**, 111–129 (1923). — Pendse, G. P., and J. B. Lal: Constitutents of the seeds of *Blepharis edulis*, Pers. II. Composition of the oil (Correction). J. Indian Chem. Soc. **15**, 471 (1938). — Pickett, T. A.: Note on muscadine grapeseed oil. Oil a. Soap **17**, 246 (1940). — Pieraerts, J.: Contribution to the chemical study of (oil bearing) *Malvaceae*. Mat. grasses **18**, 7611–7614, 7640–7643, 7667–7669 (1926).

Rabak, F.: The utilization of waste tomato seeds and skins. U. S. Dept. Agric. Bull. 632, 15 S., 1917. — Ramstad, E., and K. E. Gloppe: The seed fats of *Alliaria officinalis* Andrz. Medd. Norsk Farm. Selsk. **4**, 98–105 (1942). — Riebsomer, J. L., J. Bishop and C. Rector: Composition of the seeds of *Asiminia triloba*. J. Amer. Chem. Soc. **60**, 2853–2854 (1938). — Riebsomer, J. L., R. Larson and L. Bishman: Composition of the fatty oil from *Carya cordiformia* nuts. J. Amer. Chem. Soc. **62**, 3065–3066 (1940). — Riley, J. P.: Seed fat of *Oenothera biennis* L. J. Chem. Soc. (Lond.) **1949**, 2728–2731. — Seed fat of *Parinarium laurinum*. I. Component acids of the seed fat. J. Chem. Soc. (Lond.) **1950**, 12–18. — Rose, W. G., and G. S. Jamieson: The composition of seven American linseed oils. Oil a. Soap **18**, 173–174 (1941). — *Licania arborea* (Cacahuananche) seed oil. Oil a. Soap **20**, 227–231 (1943).— Ruchkin, V.: The oil in the juice of berries. Masloboino-Zhirovoe Delo **1929**, 47–48.

Scholfield, C. R., and W. C. Bull: Relation between the fat acid composition and the iodine number of soybean oil. Oil a. Soap **21**, 87–89 (1944). — Schuette, H. A., and C. Y. Chang: Hazelnut (filbert) oil. J. Amer. Chem. Soc. **55**, 3333–3335 (1933). — Schuette, H. A., and R. M. Christenson: The saturated fatty acids of Japan wax. Oil a. Soap **19**, 209–211 (1942). — Schuette, H. A., and W. W. F. Enz: Expressed Brazil-nut oil. J. Amer. Chem. Soc. **53**, 2756–2758 (1931). — Schuette, H. A., and J. A. Korth: Seed oil of the highbush cranberry. Oil a. Soap **17**, 265 (1940). — Schuette, H. A., and R. G. Zehnpfennig: The seed oil of the hackberry. Oil a. Soap **14**, 269–270 (1937). — Sesseler, W. M., and P. A. ROWAAN: Drying-oil from *Licania crassifolia* BENTH. Chem. Weekbl. **36**, 208–209 (1939). — SHORLAND, F. B.: The petroleum-ether-soluble constitutents of pasture fractions prepared for feeding experiments in facial eczema investigations. New Zealand J. Sci. Technol., Sect. A **23**, 112–120 (1941). — Leaf lipides of forage grasses and clovers. Nature (Lond.) **153**, 168 (1944). — Occurrence of hexadecatrienoic acid in the glycerides of rape (*Brassica napus* L.) leaf. Nature (Lond.) **156**, 269–270 (1945). — SMITH, J. A. B., and A. C. CHIBNALL: Glyceride fatty acids of forage grasses. I. Cocksfoot and perennial ryegrass. Biochemic. J. **26**, 218–234 (1932a). — Phosphatides of forage grasses. I. Cocksfoot. Biochemic. J. **26**, 1345–1357 (1932b). — SPEER, J. H., E. C. WISE, M. C. HART and F. W. HEYL: The composition of spinach fat. J. of Biol. Chem. **82**, 105–110 (1929). — SREENIVASAN, B., N. R. KAMATH and J. G. KANE: The fatty acid composition of castor oil. J. Amer. Oil Chem. Soc. **33**, 61 (1956). — SRISOMBOON, V.: Private communication 1954. — STEGER, A., and J. VAN LOON: The composition of ivy seed oil. Rec. Trav. chim. Pays-Bas (Amsterd.) **47**, 471–476 (1928). — Das Fett der Samen von *Picramnia sow*. Rec. Trav. chim. Pays-Bas (Amsterd.) **52**, 593–600 (1933). — Das fette Öl der Samen von *Cydonia vulgaris*. Rec. Trav. chim. Pays-Bas (Amsterd.) **53**, 24–27 (1934a). — Das fette Öl der Hirse (*Panicum miliaceum*). Rec. Trav. chim. Pays-Bas (Amsterd.) **53**, 41–44 (1934b). — Coulsamenöl. Rec. Trav. chim. Pays-Bas (Amsterd.) **54**, 502–504 (1935a). — Jaboty fat. Chem. a. Ind. **1935**b, 1095–1097. — Fatty oil from the seeds of *Valerianella olitoria*, POLL. J. Soc. Chem. Ind. (Lond.) **56**, 298–300T (1937). — Po-Yoak-Öl. Rec. Trav. chim. Pays-Bas (Amsterd.) **57**, 620–628 (1938). — Das fette Öl der Samen von *Canarium commune* L. Rev. Trac. chim. Pays-Bas (Amsterd.) **59**, 168–172 (1940). — Das Fruchtfleischöl von *Sterculia foetida* L. Rec. Trav. chim. Pays-Bas (Amsterd.) **60**, 87–90 (1941a). — Das fette Öl der Samen von *Isatis tinctoria* L. Rec. Trav. chim. Pays-Bas (Amsterd.) **60**, 947–949 (1941b). — Das fette Öl der Samen von *Hesperis matronalis* L. Rec. Trav. chim. Pays-Bas (Amsterd.) **61**, 123–126 (1942a). — Fungu- oder Behuradaöl. Fette u. Seifen **49**, 769–770 (1942b). — Das fette Öl der Samen von *Jatropha curcas* L. Fette u. Seifen **49**, 770–773 (1942c). — Hagebuttensamenöl. Fette u. Seifen **50**, 505 (1943). — STEGER, A., J. VAN LOON and IR. B. PENNEKAMP: Chiaöl. Fette u. Seifen **49**, 241–243 (1942). — STOUT, A. W., and H. A. SCHUETTE: Rye-germ oil. J. Amer. Chem. Soc. **54**, 3298–3302 (1932).

TÄUFEL, K., u. M. RUSCH: Zur Kenntnis des Fettes der Gerste und ihrer Mälzungsprodukte. Z. Unters. Lebensmitt. **57**, 422–431 (1929). — TOYAMA, Y.: Unsaturated acids of tohaku oil. J. Soc. Chem. Ind. (Japan) **40**, Suppl. 285–289 (1937). — TOYAMA, Y., and T. TSUCHIYA: Ein neues Stereoisomeres der Eläeostearinsäure im Samenöl von *Trichosanthes cucumeroides* („Karasu-Uri"). J. Soc. Chem. Ind. (Japan) **38**, 185–187B (1935). — TSUJIMOTO, M., and H. KOYANAGI: Nigaki oil. Bull. Chem. Soc. (Japan) 8, 161–167 (1933). — TUTIYA, T.: Unsaponifiable matter in the seed oil of balsam, *Impatiens balsamina* L. I. Occurrence of a new, highly unsaturated sterol, $C_{27}H_{40}O$. J. Chem. Soc. (Japan) **61**, 717–718 (1940).

VARIER, N. S.: Fixed oil from the seeds of *Gnetum scandens* (ROXB.). Proc. Indian Acad. Sci., Sect. A **17**, 195–198 (1943). — VARMA, P. S., N. N. GODBOLE u. P. D. SRIVASTAVA: Über das Samenöl von *Psidium Guyava Pyriferum* aus Indien. I. Fettchem. Umschau **43**, 8–9 (1936). — VASTERLING, P.: Untersuchungen über die Inhaltsstoffe der Hagebuttenfrüchte (*Semen Cynosbati*), insbesondere über das darin enthaltene fette Öl. Arch. Pharmaz. **260**, 27–44 (1922). — VERMA, J. P., B. NATH and J. S. AGGARWAL: Structure of sterculic acid. Nature (Lond.) **175**, 84–85 (1955). — VIDYARTHI, N. L., and M. V. MALLYA: Occurrence of an isomer of ricinoleic acid in the fatty oil from the seeds of *Vernonia anthelmintica*. J. Indian Chem. Soc. **16**, 479–480 (1939). — VIDYARTHI, N. L., and M. NARASINGARAO: Chemical examination of the wax from sugar cane. J. Indian Chem. Soc. **16**, 135–143 (1939). — VOTOČEK, E., F. VALENTIN and J. BULÍŘ: The constituents of mullen (*Verbascum*)-seed oil. Coll. českoslov. chem. Commun. 8, 455–460 (1936) (in French).

WEERAKOON, H. A.: Private communication 1952. — WILSON, H. N., and W. C. HUGHES: The pyridine acetic anhydride method for determining hydroxyl and the preparation of pyridine of suitable quality. J. Soc. Chem. Ind. (Lond.) **58**, 74–77 (1939). — WOODWARD, F. N.: Analysis of linseed oils of various origins. Analyst (Lond.) **64**, 265–269 (1939).

ZEHNPFENNIG, R. G., and H. A. SCHUETTE: Elm-seed oil. II. Studies on fat acid separation. Oil a. Soap **18**, 189–190 (1941).

Die Fette der Algen.

Von

H. v. Witsch.

Mit 1 Abbildung.

I. Vorkommen.

Fette sind als Assimilationsprodukte und Speicherstoffe bei Algen weit verbreitet; vor allem in Dauersporen und ähnlichen Stadien werden sie oft in großer Menge angehäuft. In manchen Algengruppen treten fette Öle in erwähnenswerter Menge nur gelegentlich unter besonderen Außenbedingungen oder in bestimmten Entwicklungsstadien auf. In vielen Verwandtschaftskreisen stellen Fette und fettähnliche Stoffe jedoch ein regelmäßiges, oft in großen Mengen in den Zellen gespeichertes Assimilationsprodukt dar, so (nach HARDER, Lehrbuch der Botanik für Hochschulen 1954) Lipoproteide bei *Cyanophyceen*, fette Öle bei *Chrysomonadalen*, *Heterochloridalen*, manchen *Dinoflagellaten* und *Volvocalen*, *Heteroconten*, vor allem aber bei *Diatomeen* und manchen *Chlorophyceen*. Von geringerer Bedeutung sind sie bei *Phaeophyceen* und *Rhodophyceen*.

II. Methodisches, Bedingungen für die Fettbildung.

Lange Zeit waren Algenfette nur unter allgemein physiologischen und systematischen Gesichtspunkten von Interesse und unsere Kenntnisse über Vorkommen und Natur dieser Stoffe recht spärlich. Erst in neuerer Zeit setzten eingehendere Untersuchungen ein mit dem Ziele, Algen in Massenkulturen zu züchten und als Nahrungs-, vor allem Fettquelle nutzbar zu machen. In einigen Fällen fanden, ebenfalls von praktischen Gesichtspunkten ausgehend, auch natürliche, besonders marine Algenvorkommen näheres Interesse und wurden vor allem auf Menge und Zusammensetzung ihrer Lipoide genauer untersucht (LOVERN, TAKAHASHI und Mitarbeiter, s. weiter unten). Diesen Umständen verdanken wir recht eingehende Kenntnisse der Physiologie der Fettbildung sowie der Zusammensetzung der Öle einiger für Massenkultur besonders günstiger Algenarten.

Für Massenkulturen zum Zweck der Fettproduktion eignen sich nur wenige planktontische Algen. In den zahlreichen Untersuchungen, die in den vergangenen Jahren in Deutschland, England, USA, Holland und an anderen Stellen durchgeführt worden sind, erwiesen sich vor allem die Grünalgen *Chlorella* und *Scenedesmus* und *Diatomeen* der Gattung *Nitzschia* als brauchbar.

Die Kulturen wurden in senkrecht stehenden Glaszylindern, flachen Tanks oder auch großen Rohrsystemen aus Kunststoff durchgeführt und zur Verhinderung des Absetzens und Verstärkung des Wachstums meist mit CO_2-haltiger Luft durchströmt. Bezüglich aller technischen Einzelheiten sei auf den Sammelbericht über Algenmassenkultur von BURLEW (1953) verwiesen.

Es wurde schon oben erwähnt, daß äußere und innere Faktoren großen Einfluß auf die Fettbildung von Algen besitzen. Angaben über Massenertrag und Fettgehalt von Algenkulturen können daher nur unter Berücksichtigung des Entwicklungszustandes und der Wachstumsbedingungen der Kultur beurteilt werden.

Einzelheiten über die Physiologie der Fettbildung bei Mikroorganismen werden im 5. Abschnitt des Bandes eingehend behandelt. Hier genügen daher einige orientierende Angaben, soweit sie im Zusammenhang notwendig sind.

In allen einschlägigen Untersuchungen (HARDER und v. WITSCH 1943, v. WITSCH 1946 und 1948, v. DENFFER 1948, SPOEHR und MILNER 1949, MYERS 1951, AACH 1952, Einzelheiten und weitere Literatur bei BURLEW 1953) zeigte sich, daß Fettbildung bei Algen stets dann einsetzt, wenn der Stickstoffvorrat

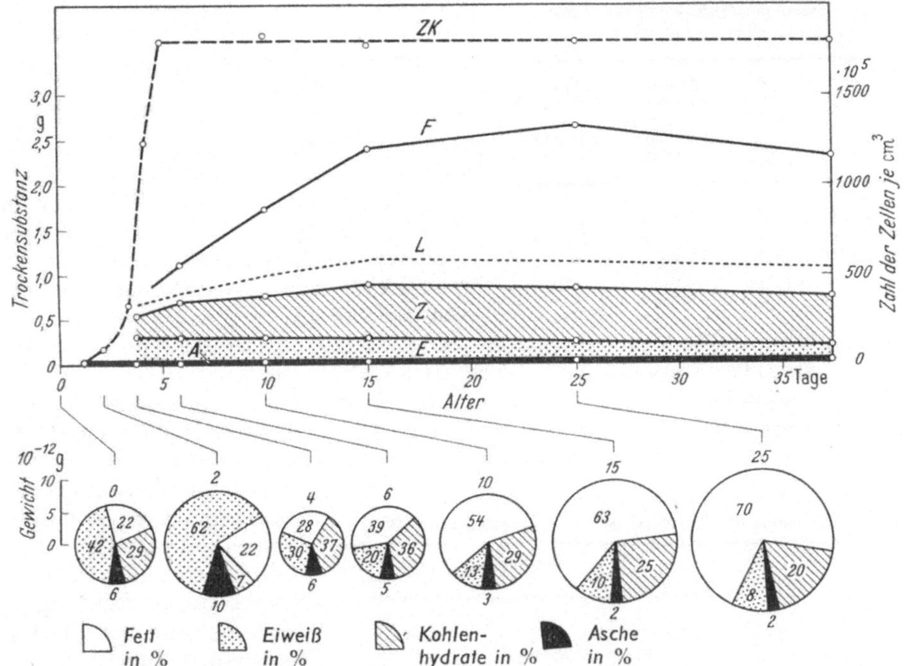

Abb. 1. Wachstum und stoffliche Zusammensetzung einer *Chlorella*-Kultur in verschiedenem Alter und Entwicklungszustand. (Nach AACH 1952, teilweise verändert.) Kurve *ZK*: Zellkonzentration der Kultur. Das Teilungswachstum ist nach 5 Tagen abgeschlossen. Übrige Kurven: Menge und Zusammensetzung der gebildeten organischen Substanz. *F:* Neutralfette und (durch Kurve L abgetrennt) sonstige Lipoide. *Z:* Cellulose, Zucker und sonstige Kohlenhydrate. *E:* Eiweiß und andere Stickstoffverbindungen. *A:* Asche. Kreisdiagramme 0—25: Trockengewicht und Zusammensetzung einer Durchschnittszelle nach 0—25tägiger Kulturdauer. Das Trockengewicht der Zelle ist dem Radius des Kreises proportional, der Prozentsatz der einzelnen Substanzen den Winkeln. Prozentzahlen in bzw. unter den Sektoren.

der Nährlösung verbraucht und damit gleichzeitig die Phase des Teilungswachstums der Kultur abgeschlossen ist. Von diesem Zeitpunkt an werden die neugebildeten Assimilate nicht mehr zum Aufbau neuer Zellen verbraucht, sondern in großen Mengen in den bereits vorhandenen, allmählich heranwachsenden Zellen als Reservestoffe gespeichert. Die stoffliche Zusammensetzung der Zellen ändert sich daher wesentlich mit dem Entwicklungszustand der Kultur und kann durch die Wahl von Nährlösung und Kulturbedingungen weitgehend beeinflußt werden. Als eindrucksvolles Beispiel hierfür seien die Untersuchungen von AACH (1952) an *Chlorella* angeführt, aus welchen einige Analysenergebnisse in Abb. 1 diagrammatisch wiedergegeben sind. Alles Nähere ist dem Abbildungstext zu entnehmen. Das abweichende Verhalten von Rotalgen und Cyanophyceen wird in anderem Zusammenhang weiter unten behandelt werden.

MILNER (1948) analysierte 4 Proben von *Chlorella pyrenoidosa* verschiedenen Alters und Entwicklungszustandes und errechnete die in Tabelle 1 verzeichneten Gehalte an Kohlenhydraten, Eiweiß und Lipoiden. Kultur 1 war in relativ stickstoffreicher Nährlösung gewachsen, Nr. 2—4 in stickstoffarmer; Kultur 2 war in jungem Zustand geerntet worden, 3 und 4 wesentlich später; Probe 4 stand im Gegensatz zu den übrigen, im natürlichen, diffusen Licht gewachsenen Kulturen in starkem künstlichem Dauerlicht. Die Bedingungen für Fettspeicherung waren demnach von Kultur 1—4 in zunehmendem Maße günstiger. Der Anteil der Lipoide an der Trockensubstanz steigt dementsprechend von Probe 1—4 von rund 20% auf 77% ständig an, während der Kohlenhydrat- und vor allem der Eiweißgehalt entsprechend sinken.

Tabelle 1. *Aus der Elementaranalyse berechnete Zusammensetzung von Chlorella in verschiedenem Entwicklungszustand.* (Nach MILNER 1948.)

Probe	Kohlenhydrate in Prozent vom Trockengewicht	Eiweiß in Prozent vom Trockengewicht	Lipoide in Prozent vom Trockengewicht
1	33,4	46,4	20,2
2	37,5	27,3	35,2
3	23,5	13,1	63,4
4	15,0	7,9	77,1

AACH (1952) untersuchte neben *Chlorella* auch durch Stickstoffmangel verfettete Kulturen von *Scenedesmus obliquus* und *Ankistrodesmus falcatus* (Tabelle 2). Im Vergleich mit *Chlorella* fallen hier die relativ große Menge von Kohlenhydraten und das starke Zurücktreten des Eiweißanteiles auf. *Chlorella*-Kulturen mit entsprechendem Stickstoffgehalt (1,8%) enthielten rund 65% Lipoide und nur 22% Kohlenhydrate, bei vergleichbarem Lipoidgehalt (39%) jedoch 3,16% Stickstoff und 36% Kohlenhydrate.

Unter optimalen Bedingungen kann der Fettgehalt der untersuchten Algen demnach erstaunlich hohe Werte erreichen. Die maximale Lipoidmenge von *Chlorella* betrug bei AACH 70% des Trockengewichtes, bei MILNER bis 77%, SPOEHR und MILNER (1949) berichten sogar von 85,6%. Kohlenhydratreiche, aber noch fettarme *Chlorella* konnte durch Erwärmung auf 37,5° und verminderte Sauerstoffzufuhr veranlaßt werden, die Kohlenhydrate kurzfristig in Öl zu verwandeln (WEISS 1953).

Tabelle 2. *Analyse der Trockensubstanz von Scenedesmus und Ankistrodesmus in N-Mangelkultur.* (Nach AACH 1952.)

		Scenedesmus obliquus	*Ankistrodesmus falcatus*
Lipoide, % der Trockensubstanz		43,3	37,7
Neutralfette	% der Lipoide	79,1	76,2
Farbstoffe	% der Lipoide	9,0	13,8
Phosphatide	% der Lipoide	2,1	1,6
Reduzierter Stickstoff, % der Trockensubstanz		1,7	1,9
Kohlenhydrate, % der Trockensubstanz		44,2	48,0
Asche, % der Trockensubstanz		1,9	2,5

Wertvolle Einblicke in die Physiologie der Fettbildung bei Algen verschiedenster systematischer Zugehörigkeit verdanken wir COLLYER und FOGG (1955). Sie untersuchten Cyanophyceen, Diatomeen, Grün- und Rotalgen, die unter optimalen Kulturbedingungen in bakterienfreien Reinkulturen gezüchtet worden waren, auf ihren Gehalt an Fetten, unverseifbaren Lipoiden, Gesamtstickstoff und hydrolysierbaren Kohlenhydraten. Nach ihren Ergebnissen sind charakteristische Unterschiede in der Zusammensetzung der Fette von Algen verschiedener Klassen kaum zu erwarten, während bekanntlich die akzessorischen Assimilationspigmente und Kohlenhydrate zum Teil wichtige systematische Merkmale darstellen. Jedoch fanden sie größere und charakteristische Unterschiede in der Quantität der unter entsprechenden Bedingungen gebildeten Fette.

Bei den untersuchten Chlorophyceen *(Chlorella vulgaris, Chlorella pyrenoidosa, Scotiella sp.)*. Xantophyceen *(Tribonema aequale, Monodus subterraneus)*, der Diatomee *Navicula pelliculosa* und bei *Euglena gracilis var. bacillaris* wurde mit zunehmendem Alter der Kulturen steigender Fett- bei sinkendem Eiweißgehalt der Zellen festgestellt. Von entscheidendem Einfluß ist auch hier wieder der Stickstoffgehalt der Nährlösung.

Die untersuchten Cyanophyceen (*Oscillatoria sp.*, nachträglich identifiziert als *Microcoleus vaginatus*, und *Anabaena cylindrica*) und die Rhodophycee *Porphyridium cruentum* bilden zwar unter Umständen ebenfalls nicht wenig Fett, doch ist bei ihnen der Fettgehalt nicht von einem Stickstoffmangel der Zellen abhängig und nimmt mit dem Alter der Kulturen sogar ab. Es kann aber bei ihnen die Fettbildung durch Wassermangel, wie zeitweises Trockenliegen oder hohe Salzkonzentration gesteigert werden. Bei *Oscillatoria* z. B. wurde bei Kultur auf Agar ein Gehalt an Fett und unverseifbaren Lipoiden von 11,7 bzw. 4,1% des Trockengewichtes gefunden, während die Maximalwerte bei Nährlösungskultur bei 6 bzw. 1,5% lagen.

Der Anteil an unverseifbaren Lipoiden wies keine klaren Beziehungen zum Fett- und Stickstoffgehalt der Kulturen auf, er war bei *Euglena* wesentlich höher, bei den *Cyanophyceen* niedriger als bei den übrigen untersuchten Arten.

Ein charakteristischer Faktor für beide Algengruppen ist der prozentuale Anteil der Fette am Gesamtreservematerial (Fette + Kohlenhydrate) in Beziehung zum Stickstoffgehalt der Algen. Die untersuchten Grünalgen, Diatomeen und *Euglena* wiesen hohe Fett:Kohlenhydratquotienten bei relativ niedrigem Stickstoffgehalt auf, *Porphyridium*, *Tribonema* und *Anabaena* niedrige Fett: Kohlenhydratwerte bei hohem Stickstoffgehalt. Außerdem zeigte die Kurve bei zunehmendem Stickstoffgehalt bei Rotalgen und Cyanophyceen steigende Tendenz, bei den übrigen Arten hingegen fallende. In diesem Zusammenhang ist eine Angabe von KANEDA und ISHII (1950) von Interesse, die in jungen Exemplaren der Phaeophycee *Sargassum Ringoldianum* mehr Öl nachweisen konnten als in alten.

CHODAT und HAAG (1940) berichten von Versuchen mit der Chlorophycee *Dictyococcus cinnabarinum*, welche zeigen, daß dieselben Bedingungen, welche die Bildung fetter Öle fördern, auch das Auftreten von Carotinoiden begünstigen. Die Volvocale *Dunaliella salina* erhält nach LERCHE (1937) durch Kultur in N- und P-armer Nährlösung durch abnorme Carotinoidbildung eine auffallend rote Färbung. Allgemein beobachtet man häufig, daß gerade Zygoten oder andere Dauerformen von Algen, in denen vielfach auch Fett gespeichert wird, durch Carotinoide rot gefärbt sind, z. B. bei *Hämatococcus* (TISCHER 1944), *Scenedesmus*, *Chara* u. v. a.

Ins Kulturmedium abgeschiedene Fettsäuren können antibakterielle Wirkungen zeigen. So enthält die Nährlösung von *Chlorella pyrenoidosa* nach SPOEHR, SMITH, MILNER und HARDIN (1949) ungesättigte 10- und 12-C-Fettsäuren, die leicht Photooxydation erleiden und dadurch antibakterielle Wirksamkeit erlangen. Über entsprechende Beobachtungen an *Stichococcus bacillaris* und *Protosiphon botryoides* (*Chlorophyceae*) berichten HARDER und OPPERMANN (1953).

III. Chemie der Algenlipoide.

Über die genauere Zusammensetzung von Algenlipoiden unterrichten uns eine Reihe neuerer Arbeiten. Eine erste Aufgliederung der Lipoidfraktionen von *Scenedesmus* und *Ankistrodesmus* ist bereits den Angaben AACHs (letzter Abschnitt, Tabelle 2) zu entnehmen. MILNER (1948 und 1953) untersuchte den Lipoidanteil der S. 51 erwähnten 4 Proben von *Chlorella pyrenoidosa* eingehend. Seine Ergebnisse sind in Tabelle 3 zusammengestellt.

Das getrocknete Algenmaterial wurde von MILNER mit der doppelten Wassermenge verrieben, mit Methanol bis zu einer Alkoholkonzentration von 95% entwässert, eine Stunde am Rückflußkühler gekocht, abfiltriert und die Extraktion hierauf 3mal mit reinem Methanol und dann abwechselnd mit Petroläther und Methanol bis zur Farblosigkeit des Rückstandes fortgesetzt. Eine letzte Extraktion erfolgte mit Äther, die Lösungsmittel wurden im Vakuum verdampft.

Dem hohen Lipoidanteil von 23,37—75,51% (Tabelle 3, Zeile 1) entspricht ein errechneter Gehalt an Triglyceriden von 6,85—68,6% (Zeile 6). Der Anteil der eigentlichen Fette an den Lipoidfraktionen steigt demnach von Probe 1 zu 4 von 29 auf 91%. *Chlorella*-Öl ist stark ungesättigt, doch zeigt sich ein deutlicher Rückgang der ungesättigten Komponenten, vor allem der C_{18}-Gruppe, mit steigendem Verfettungsgrad der Zellen (Zeile 13—15). Zu einem erheblichen Anteil

Tabelle 3. *Analyse der Chlorellalipoide.* (Nach MILNER 1948.)

Nr.		*Chlorella*-Probe 1	2	3	4
1	Gesamtlipoide, % vom Trockengewicht	23,37	33,17	62,96	75,51
2	Chlorophyll, % vom Trockengewicht	6			0,03
	Zusammensetzung der Gesamtlipoide				
3	Fettsäuren, % der Lipoide	28,0	49,5	83,0	86,8
4	Unverseifbares, % der Lipoide	12,0	7,7	3,3	3,3
5	Wasserlösliche Verseifungsprodukte, % der Lipoide	60,0	42,8	13,7	9,9
6	Errechneter Gehalt an Triglyceriden, % vom Trockengewicht	6,85	17,2	54,7	68,6
	Analyse der Fettsäuren				
7	Jodzahl	163,1	143,8	143,6	125,3
8	Äquivalentgewicht	269,5	273,6	272,7	274,1
9	Palmitinsäure, % der Gesamtfettsäuren	16,6	10,9	7,9	11,4
10	Stearinsäure, % der Gesamtfettsäuren	0,4	4,1	3,9	3,5
11	Ungesättigte C_{16}-Säuren, % der Gesamtfettsäuren	29,1	18,3	27,3	18,0
12	Ungesättigte C_{18}-Säuren, % der Gesamtfettsäuren	53,9	66,7	60,9	67,1
	Sättigungsdefizit				
13	Ungesättigte C_{16}-Säuren	—4,1 H		—4,4 H	
14	Ungesättigte C_{18}-Säuren	—4,5 H		— 3,4 H	
15	Ungesättigte C_{16}- + C_{18}-Säuren		—3,6 H		—3,2 H

handelt es sich um 3fach ungesättigte Säuren. Aus der Jodzahl der ungesättigten C_{16}-Fraktion von Probe 3 (217,2) errechnet sich z. B. ein Anteil von mindestens 17% 3fach ungesättigten Säuren.

PASCHKE und WHEELER (1954) analysierten relativ schwach verfettete Kulturen von *Chlorella pyrenoidosa* (12,4% Gesamtlipoide). Rund 50% der Lipoide bestanden aus Fettsäuren, rund 80% derselben waren ungesättigt. Der Anteil an 1-, 2-, 3- und 4fach ungesättigten C_{16}-Säuren am Gesamtfettsäuregehalt betrug 4, 6, 12 und 3%, die entsprechenden Zahlen für die C_{18}-Säuren sind 7, 11, 34 und 1%. An 2- und 3fach ungesättigten C_{18}-Säuren wurden vor allem normale Linol- und Linolensäure gefunden.

Eine vergleichende Untersuchung der Lipoide von *Pyrenoidosa chlorella*, *Scenedesmus sp.* und *Nitzschia palea* geben DEUTICKE, HARDER, KATHEN (1949) und KATHEN (1949). Die analysierten Kulturen waren relativ jung, so daß die Verfettung noch nicht sehr weit fortgeschritten war.

Das abzentrifugierte Algenmaterial wurde unter CO_2 und Evakuierung mit der Wasserstrahlpumpe im Wasserbad getrocknet, mit Quarzsand verrieben und mit Äther im Soxhlet

24 Std extrahiert. Auch hierbei wurde die Luft durch CO_2 verdrängt. Der in Petroläther-Benzol gelöste Ätherrückstand wurde durch Adsorption an einer Zucker- und einer Aluminiumoxydsäule fraktioniert.

Die Analysenergebnisse sind in Tabelle 4 zusammengestellt. Aus den gefundenen Verseifungs-, Jod- und Rhodanzahlen der Triglyceride (Tabelle 5) wird geschlossen, daß diese vor allem aus C_{18}-Fettsäuren aufgebaut sind und folgende Zusammensetzung haben:

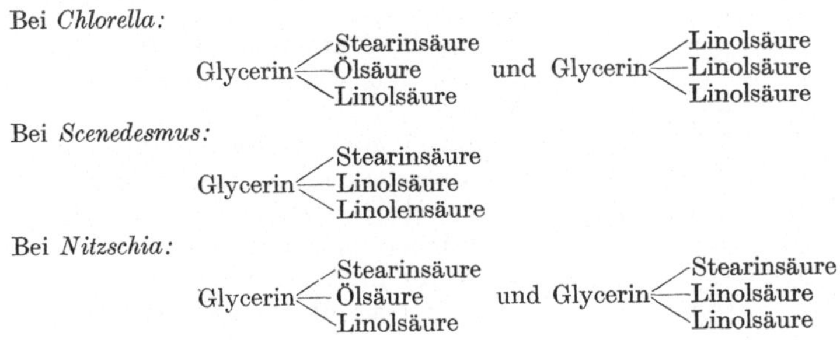

Das Vorkommen von Linolensäure im *Scenedesmus*-Öl sei besonders erwähnt, da nach PAECH (1950) diese Säure im Öl aller niederen Pflanzen nach den bisherigen Untersuchungen fehlen soll.

Diglyceride wurden im Äthereluat aller 3 Algen gefunden. Durch Wasserdampfdestillation der Lipoide ließen sich bei *Chlorella* 1,5%, *Scenedesmus* 2,0%, bei *Nitzschia* sogar 9,5% ätherische Öle abtrennen.

Fettsäuren mit 16 C-Atomen konnten von KATHEN nicht nachgewiesen werden, doch wird das Vorhandensein geringer Mengen für möglich gehalten. Freie Fettsäuren wurden im Gegensatz zu den Befunden von CLARKE und MAZUR (1941) an marinen *Diatomeen* hier nicht festgestellt.

Tabelle 4. *Prozentuale Zusammensetzung der Lipoide von Chlorella, Nitzschia und Scenedesmus.* (Nach KATHEN 1949.)

Gesamtlipoide in % der Trockensubstanz	*Chlorella* 27,6	*Nitzschia* 18,0	*Scenedesmus* 10,7
Davon: Farbstoffe	14,3%	16,0%	38,9%
Phosphatide . . .	0,9%	0,5%	0,7%
Kohlenwasserstoffe und Wachse . .	0,5%	1,4%	3,9%
Sterinester? . . .	2,4%	1,6%	0,3%
Freie Sterine . . .	0,3%	1,5%	3,6%
Triglyceride . . .	78,5%	80,5%	50,6%
Freie Fettsäuren .	—	—	—
	96,9%	101,5%	98,0%

Tabelle 5. *Charakterisierung der Triglyceride von Chlorella, Nitzschia und Scenedesmus.* (Nach DEUTICKE, KATHEN, HARDER 1949.) (Zum Vergleich die Werte für Mohnöl nach BÖMER, JUCKENACK und TILLMANS 1939.)

	Chlorella	*Nitzschia*	*Scenedesmus*	*Mohnöl*
Verseifungszahl	188 (Mittel)	192 (Mittel)	189 (Mittel)	189—194
Jodzahl	93—168	86—116	136—139	131—143
Rhodanzahl	59—94	59—64	85—90	76—79

Von ganz anderen Gesichtspunkten aus untersuchte LOVERN (1936, 1953) die Fette von verschiedenen Süßwasser- und Meeresalgen. Fischöle besitzen Fettsäuren mit 14—24 C-Atomen und fallen außerdem durch den besonders hohen Anteil an ungesättigten Fettsäuren auf. Sie sind dadurch von den Fetten der meisten Landtiere und Pflanzen scharf unterschieden. Darüber hinaus differieren Fischöle von Süß- und Salzwasserfischen in ihrem Gehalt an ungesättigten Fettsäuren. Der prozentuale Anteil an ungesättigten C_{16}- und C_{18}-Säuren ist im Öl von Süßwasserfischen nach den Untersuchungen des Autors stets höher, teilweise

Tabelle 6. *Zusammensetzung von Fettsäuregemischen von Algen in Gewichtsprozenten.* (Nach LOVERN.)

	Gesättigte Fettsäuren				Ungesättigte Fettsäuren				
	$C_{<14}$	C_{14}	C_{16}	C_{18}	C_{14}	C_{16}	C_{18}	C_{20}	C_{22}
A. *Süßwasserarten*									
Nitella opaca (Charophyceae)	—	6	18	3	3 (—2,0 H)	34 (—2,5 H)	23 (—4,5 H)	13 (—5,8 H)	—
Oedogonium sp. (Chlorophyceae)	—	2	20	1	— —	32 (—3,1 H)	35 (—4,6 H)	9 (— ? H)	1 (— ? H)
Cladophora Sauteri (Chlorophyceae)	Spuren	12	10	2	Spuren	19 (—4,7 H)	49 (—3,8 H)	8 (—7,1 H)	—
B. *Salzwasserarten*									
Nitzschia closterium (Diatomeae)	—	8	17	2	1 —	36 (—3,4 H)	20 (—5,3 H)	16 (—7,0 H)	—
Verschiedene *Chlorophyceae*	—	4	10	2	3 (—2,0 H)	39 (—3,4 H)	30 (—5,1 H)	8 (—6,5 H)	4 (— ? H)
Fucus vesiculosus (I) *(Phaeophyceae)*	Spuren	8	9	1	Spuren	6 (—2,0 H)	57 (—3,2 H)	16 (—7,3 H)	3 (— ? H)
Fucus vesiculosus (II)	Spuren	9	7	2	1 —	5 (—2,0 H)	63 (—3,0 H)	13 (—7,3 H)	—
Laminaria digitata (Phaeophyceae)	—	6	14	1	2 —	11 (—2,0 H)	42 (—4,2 H)	24 (—8,1 H)	—
Rhodymenia palmata (Rhodophyceae)	1	4	19	1	Spuren	6 (—2,9 H)	20 (—4,5 H)	36 (—9,2 H)	13 (— ? H)

sogar doppelt so hoch wie bei dem von Seefischen. Bei letzteren hingegen ist der relative Anteil von ungesättigten C_{20}- und C_{22}-Säuren größer als im Fett der Süßwasserfische. Versuche hatten ergeben, daß der Charakter der Fette weitgehend von der aufgenommenen Nahrung abhängig ist. Sie waren der Anlaß zu entsprechenden Analysen bei tierischem und pflanzlichem Plankton sowie an höheren Algen.

Soweit diese Untersuchungen in unserem Zusammenhang interessieren, sind sie in Tabelle 6 wiedergegeben. Das dem natürlichen Standort entnommene Material wurde lufttrocken pulverisiert und mit Äther extrahiert, nur *Fucus* mit Alkohol und Äther. Eine genauere Spezifikation der einzelnen Komponenten wurde nicht durchgeführt, wohl aber die in der Tabelle zusammengestellte Charakterisierung der einzelnen Fraktionen gegeben. Auffallend ist auch hier das Vorkommen von C_{14}- und C_{20}-Säuren in Mengen von zusammen 11—22%, welche z. B. in den von KATHEN (1949) analysierten Algen nicht nachgewiesen werden konnten. Bemerkenswert ist auch der hohe Gehalt an ungesättigten Säuren. Eine differente Verteilung derselben bei Süßwasser- und Meeresalgen konnte jedoch nicht beobachtet werden, doch sind die Fette systematisch verwandter Formen auch in ihrer chemischen Zusammensetzung ähnlich.

Über eine Geschmacksbeeinflussung von Süßwasserfischen durch Oscillatorienöl berichtete BANDT (1935). Nach einer starken Wasserblüte von *Oscillatoria agardhii* bzw. *Oscillatoria princeps* im Werbellinsee und in 2 Seen bei Rüdersdorf wiesen die Fische einen starken Modergeschmack auf. Ätherextraktion von lufttrockenem Material von *Oscillatoria princeps* ergab bei Frühjahrsmaterial 1,83, bei Herbstmaterial 2,86% eines zähen Öles, das erhebliche Mengen ungesättigter Säuren enthielt (Jodzahl 102, Säurezahl 55). In diesem Öl sieht BANDT die Ursache für den Modergeschmack, da Öl aus Cyanophyceen der gewöhnlichen Wasserblüte *(Anabaena, Aphanizomenon, Microcystis* und *Polycystis)*, durch welche keine Geschmacksverschlechterung verursacht wurde, wesentlich weniger ungesättigte Säuren enthielt (Jodzahl 37). Er stützt sich hierbei auf eine ältere Angabe von TSUJIMOTO, wonach der tranige Geruch mancher Algenöle durch hoch ungesättigte Säuren vom Typ $C_nH_{2n-8}O_2$, vor allem $C_{22}H_{36}O_2$ bedingt wird.

Mit den Fetten mariner Algen befaßt sich auch eine Reihe von Arbeiten von TAKAHASHI und Mitarbeitern (1933, 1935, 1936, 1938, 1939). Der Fettgehalt war im allgemeinen bei Braunalgen am höchsten, bei Rotalgen am niedrigsten. Im Fett von *Pelvetia wrightii* Yendo *(Phaeophyceae)* wurde Caprin-, Capron-, Caprylsäure, ferner Linolen-, Öl- und Myristinsäure festgestellt (1933). In einer späteren Arbeit (1935) werden als häufigste gesättigte Säuren der Algenfette vor allem Palmitin-, ferner Myristin- und Stearinsäure genannt, als wesentliche ungesättigte Komponenten Ölsäure sowie $C_{18}H_{28}O_2$ und $C_{18}H_{30}O_2$. Die Analyse des Fettes von *Turbinaria fusiformis (Phaeophyceae)* (1936) ergab 36% hochgradig ungesättigte Säuren mit 16, 18 und 20 C-Atomen, an gesättigten Säuren Palmitin- und Stearinsäure. Eine eingehendere Untersuchung liegt von der Phaeophycee *Cystophyllum hakodatense* Yendo vor (1938). Die nicht flüchtigen Fettsäuren bestanden zu 22,86% aus festen Säuren (80% Palmitin- und 20% Myristinsäure), zu 77,14% aus flüssigen und zwar hiervon: $C_{16}H_{30}O_2$:20%, $C_{16}H_{28}O_2$: 8%, $C_{18}H_{34}O_2$: 47%, $C_{18}H_{32}O_2$: 4%, $C_{18}H_{30}O_2$: 2%, $C_{18}H_{28}O_2$: 8%, $C_{22}H_{40}O_2$: 1% und $C_{22}H_{38}O_2$: 4% (total 94%).

Die Untersuchung der Phaeophycee *Alaria crassifolia* Kjellm. (1939) ergab für den Fettanteil der vegetativen Thalli 23,22% feste und 76,78% flüssige Fettsäuren. 80% der ersteren bestanden wieder aus Palmitin-, 20% aus Myristinsäure. Für die flüssigen Fettsäuren werden angegeben: $C_{16}H_{30}O_2$: 4%, $C_{16}H_{28}O_2$: 12%, $C_{18}H_{32}O_2$: 11%, $C_{18}H_{30}O_2$: 3% und $C_{18}H_{28}O_2$: 19%. Bei dieser Alge wurden die reproduktiven Thallusabschnitte gesondert analysiert. Das Verhältnis der festen zu den flüssigen Säuren war hier 51,02 zu 48,98%. Die festen zeigten die gleiche Zusammensetzung wie im vegetativen Thallus, die Angaben für die flüssigen Fettsäuren lauten hier: $C_{16}H_{30}O_2$: 5%, $C_{16}H_{28}O_2$: 12%, $C_{18}H_{32}O_2$: 17%, $C_{18}H_{30}O_2$: 3% und $C_{18}H_{28}O_2$: 5%. Allgemein wurde im vegetativen Thallus, in welchem die Assimilationsfunktion vorherrscht, ein größerer Anteil an ungesättigten Säuren festgestellt, in den reproduktiven Thallusteilen, in welchen die Speicherfunktion über die Assimilationstätigkeit überwiegt, mehr gesättigte Fettsäuren.

Literatur.

AACH, H. G.: Über Wachstum und Zusammensetzung von *Chlorella pyrenoidosa* bei unterschiedlichen Lichtstärken und Nitratmengen. Arch. Mikrobiol. **17**, 213—246 (1952).

BANDT, H. I.: Geschmacksbeeinflussung der Fische durch Oscillatorienöl. Allg. Fischereiztg. **60**, 328—330 (1935). — BÖMER, A., A. JUCKENACK, u. J. TILMANS: Handbuch der Lebensmittelchemie, Bd. IV. Berlin 1939. — BURLEW, J. S.: Algal culture from laboratory to pilot plant. Carnegie Inst. Wash. Publ. 600, **1953**.

CHODAT, F., et E. HAAG: Sur les conditions d'accumulation des caroténoides chez une algue verte. I. Accumulation concominante des caroténoides et des lipides. C. r. Soc. Physique Genève (Suppl. Arch. Sci. Phys. 22) **57**, 265—269 (1940). — CLARKE, H. T., and A. MAZUR: J. of Biol. Chem. **141**, 283—289 (1941). Zit. nach KATHEN 1949. — COLLYER, D. M., and G. E. FOGG: Studies on Fat Accumulation by Algae. J. of Exper. Bot. **6**, 256—275 (1955).

DENFFER, D. v.: Planktische Massenkultur pennater Grunddiatomeen. Arch. Mikrobiol. **14**, 159—202 (1948). — DEUTICKE, H. J., H. KATHEN u. R. HARDER: Analyse von Algenlipoiden. Naturwiss. **36**, 60 (1949).

HARDER, R.: Thallophyta. In Lehrbuch der Botanik für Hochschulen, bearbeitet von FITTING, SCHUMACHER, HARDER u. FIRBAS, 26. Aufl. Stuttgart: Gustav Fischer 1954. — HARDER, R., u. A. OPPERMANN: Über antibiotische Stoffe bei den Grünalgen *Stichococcus bacillaris* und *Protosiphon botryoides*. Arch. Mikrobiol. **19**, 398—401 (1953). — HARDER, R., u. H. v. WITSCH: Über Massenkultur von Diatomeen. Ber. dtsch. bot. Ges. **60**, (146)—(152) (1943).

Kaneda, T., u. S. Ishii: Studien über das Öl und die Sterine von Algen. I. Jahreszeitliche Veränderungen der *Sargassum*-Lipide. Bull. Jap. Soc. Fisheries **15**, 608 (1950). Zit. nach Fette u. Seifen **56**, 435 (1954). — Kathen, H.: Über die Ermittlung der chemischen Konstitution von Algenlipoiden mit Hilfe der Adsorptionsmethode. Arch. Mikrobiol. **14**, 602—634 (1949).

Lerche, W.: Untersuchungen über die Entwicklung und Fortpflanzung in der Gattung *Dunaliella*. Arch. Protistenkde 88, 236—268 (1937). Zit. nach M. Hartmann, Die Sexualität. Jena 1943. — Lovern, J. A.: Fat metabolism in fishes. IX. The fats of some aquatic plants. Biochemic. J. **30**, 387—390 (1936). — Der Ursprung der Eigenart der Fette bei Fischen. Fette u. Seifen **55**, 425—430 (1953).

Milner, H. W.: The fatty acids of *Chlorella*. J. of Biol. Chem. **176**, 813—817 (1948). — The chemical composition of algae. In Burlew, Algal culture from laboratory to pilot plant, S. 285—302. 1953. — Myers, J.: Physiology of the algae. Annual Rev. Microbiol. **5**, 157—180 (1951).

Paech, K.: Biochemie und Physiologie der sekundären Pflanzenstoffe. Berlin-Göttingen-Heidelberg: Springer 1950. — Paschke, R. F., u. D. H. Wheeler: Die ungesättigten Fettsäuren der Alge *Chlorella*. Amer. Oil Chemists Soc. **31**, 91 (1954). Zit. nach Fette u. Seifen **56**, 434 (1954).

Spoehr, H. A., u. H. W. Milner: The chemical composition of *Chlorella*; Effect of environment conditions. Plant Physiol **24**, 120—149 (1949).— Spoehr, H. A., J. H. C. Smith, H. H. Strain, H. W. Milner and G. J. Hardin: Fatty acid antibacterials from plants. Carnegie Inst. Wash. Publ. 586, **1949**.

Takahashi, E., K. Shirahama and Shun-Iti-Tase: Chemical studies on fats of algae. I. Contents of fats and their properties, especially the fatty acids of *Pelvetia wrightii Yendo*. J. Chem. Soc. Japan **54**, 619—623 (1933)[1]. — Fats of sea algae. II. J. Chem Soc. Japan **56**, 1250—1257 (1935)[1]. — Takahashi, E., and K. Shirahama: Fats of sea algae. III. Fat of *Turbinaria fusiformis*. J. Chem. Soc. Japan **57**, 411—414 (1936)[1]. — Takahashi, E., K. Shirahama and N. Ito: Fats of sea algae. IV. Fat acid of *Cystophyllum hakodatense Yendo*. J. Chem. Soc. Japan **59**, 662—667 (1938)[1]. — Takahashi, E., K. Shirahama and N. Togasawa. The fats of marine algae. V. Non volatile fat acids of *Alaria crassifolia Kjellm*. J. Chem. Soc. Japan **60**, 56—60 (1939)[1]. — Tischer, J.: Über die Carotinoide von *Hämatococcus pluvialis*. III. Z. physiol. Chem. **281**, 143—155 (1944).

Weiss, H.: Beitrag zur Frage einer Algenmassenzucht in Deutschland. Zbl. Bakter. **107**, 230—246 (1953). — Witsch, H. v.: Wachstum und Vitamin-B_1-Gehalt von Süßwasseralgen unter verschiedenen Außenbedingungen. Naturwiss. **33**, 221 (1946). — Beobachtungen zur Physiologie des Wachstums von *Chlorella* in Massenkulturen. Biol. Zbl. **67**, 95—100 (1948).

[1] Die japanisch geschriebene Arbeit lag zwar im Original vor, mußte jedoch in erster Linie nach einem Referat in den Chem. Abstr. ausgewertet werden.

Die Fette der Pilze.

Von

Maximilian Steiner.

I. Einleitung.

Unter „Pilzen" werden in diesem Beitrag alle Kryptogamen mit Ausschluß der „Algen" und der Bakterien verstanden. Unsere Kenntnisse über die Fette der Pilze sind im Vergleich zu denen der höheren Pflanzen sehr spärlich und lückenhaft. Große Gruppen des Pilzreiches — z. B. fast alle pflanzenparasitären niederen Pilze — sind in dieser Hinsicht überhaupt noch kaum erforscht. Die vorliegenden Untersuchungen beschränken sich auf einige wenige Gruppen von Pilzen: auf die *Fruchtkörper* der höheren Pilze, die als Speise- oder Giftpilze praktisches Interesse beanspruchen; auf einige pharmazeutisch verwendete Arten (insbesondere das *Mutterkorn*); auf *Hefen-* und *Schimmelpilze*, bei denen die Physiologie der Fettbildung untersucht wurde, vielfach mit dem Ziel, die Grundlagen für eine mikrobiologische großtechnische Fettsynthese zu schaffen. Einige Arbeiten beschäftigen sich mit dem Fett der *Myxomycetenplasmodien*. Spärliche Angaben liegen über die Fette der *Flechten* vor. Die eben genannten Pilzgruppen wurden in diesem Beitrag getrennt behandelt. Die Einteilung des Stoffes folgt also nicht streng taxonomischen Prinzipien.

Im Rahmen dieses Handbuches dürfte es berechtigt sein, wenn das Hauptgewicht auf die Behandlung derjenigen Organismengruppen gelegt wird, die für das Studium der *Physiologie* des Fettstoffwechsels eine gewisse Bedeutung erlangt haben: auf die „Hefen" und die „Schimmelpilze", über deren Fette übrigens auch die genauesten Angaben vorliegen.

Die ältere einschlägige Literatur ist bei W. Zopf (1890), J. Zellner (1907), F. Czapek (1913) und Ch. Pontillon (1932) zusammenfassend verarbeitet.

II. Myxomycetenplasmodien.

In ihren klassischen, viel zitierten Untersuchungen über das Protoplasma des Plasmodiums von *Fuligo varians* haben J. Reinke und H. Rodewald (1881a, b, s. auch J. Reinke 1880) auch den Gehalt an Lipoiden ermittelt. Allerdings hatte schon 70 Jahre vorher H. Braconnot (1811b) das gleiche Objekt analysiert und eine fettähnliche Substanz („matière adipeuse jaune") darin gefunden. A. Kiesel (1925, 1927, s. auch 1929) hat weitere Myxomycetenplasmodien *(Reticularia lycoperdon, Lycogala epidendron)* untersucht und dabei den Fetten große Aufmerksamkeit geschenkt. Etwa zur gleichen Zeit hat sich W. W. Lepeschkin (1923, 1926) mit dem *Fuligo*plasmodium, insbesondere mit der Frage der Lipoproteide befaßt. Tabelle 1 bringt eine Übersicht über die Analysenergebnisse der genannten Forscher.

Die Fettsäuren ihrer Plasmodien fanden REINKE und RODEWALD zum weitaus überwiegenden Teil frei und Teile auch als Ca-Salze. Im einzelnen werden angegeben: Propion-, Capron-, Caprin-, Palmitin-, Stearin- und Ölsäure.

Tabelle 1. *Lipoidgehalt und Lipoidfraktionen bei Myxomycetenplasmodien.*

	Fuligo varians		*Reticularia lycoperdon* (A. KIESEL 1925) Plasmodium % v. Trockengewicht	*Lycogala epidendron* (A. KIESEL 1927)	
	(J. REINKE u. H. RODEWALD 1881) Plasmodium % v. Trockengewicht	(W. W. LEPESCHKIN 1923) Plasmodium % v. Trockengewicht		Plasmodium % v. Trockengewicht	unreife Fruchtkörper % v. Trockengewicht
Gesamtlipoide . .	5,36—8,13	6,8—13,9	23,10[2]	40,13[3]	35,15[3]
Neutrale Fette und Fettsäuren	—	3,1—10,2	17,85	37,51	31,22
Fettsäuren	4,00				
„Lecithin"	0,20	0,7—2,0	4,67	n. b.	0,12
Sterine	1,40[1]	2,4—5,5	0,58	1,16	1,31

[1] „Cholesterin" und „Paracholesterin". Das letzte ist wahrscheinlich unreines Ergosterin (s. auch E. GÉRARD 1892). Die Zuordnung der „Cholesterine" von REINKE und RODEWALD und von KIESEL und des „Phytosterins" LEPESCHKINS zu bestimmten Stoffindividuen der Sterinreihe scheint noch nicht festzustehen.

[2] Darunter 1,20% „Öl der Lipoproteide".

[3] Darunter 0,66% bzw.2, 39% „Öl der Lipoproteide" (mit 0,7—0,8% P) und 1,20% bzw. 2,29% einer „unbekannten lipoidartigen Substanz".

Die detaillierteren Analysen von A. KIESEL lassen sich in der folgenden Tabelle 2 kurz zusammenfassen.

Tabelle 2. *Zusammensetzung und Kennzahlen der Fette einiger Myxomycetenplasmodien.* (Nach A. KIESEL 1925, 1927.)

a) *Reticularia lycoperdon, Lycogala epidendron:* Zusammensetzung der Fette.

	Reticularia lycoperdon Plasmodium	*Lycogala epidendron*	
		Plasmodien	unreife Fruchtkörper
Rohfett, % v. Trockengewicht	23,10	40,13	35,15
Fettsäuren, % des Rohfettes .	87,27	86,56	83,23
Ungesättigte Fettsäuren . .	78,73	76,02	74,44
Ölsäure	65,84		
Linolsäure	12,63		
Gesättigte Fettsäuren . . .	8,54	10,54	8,79
Palmitinsäure	8,54		

b) *Lycogala epidendron:* Fett-Kennzahlen.

Kennzahlen bestimmt in	Äthanolextrakt	Ätherextrakt	Petrolätherextrakt	
			Plasmodium	unreife Fruchtkörper
Verseifungszahl	168,5		193,2	198,4
Säurezahl	25,42		14,9	23,5
Jodzahl	84,53	95,1	102,5	112,0

III. Hefen.

1. Geschichtliches.

PAYEN (1846) scheint der erste gewesen zu sein, der den Fettgehalt der Bierhefe — 2,10% des Trockengewichtes — bestimmte. Daß der Fettgehalt dieses Pilzes sich je nach den Ernährungsbedingungen verändert und bis 5,29% ansteigen kann, zeigten C. v. NAEGELI (1878) und C. v. NAEGELI und O. LOEW (1878). Als Bestandteile des Hefefettes glaubten sie Olein, Palmitin, Stearin, „Cholesterin", aber kein „Lecithin" feststellen zu können. O. KAPPES (1890) fand in der pathogenen Soorhefe 4,25% vom Trockengewicht Ätherextrakt, P. GUICHARD (1894) konnte in Trockenhefe wiederum nur 1,4% petrolätherlösliche Substanz feststellen. W. HENNEBERG (1904) gibt für Kulturhefen einen Fettgehalt von etwa 2,1% an. Beginnend mit DUBAQUIÉ (1909) und vor allem mit I. SMEDLAY-MACLEAN (1923) wenden sich die Untersucher vor allem den Fragen über die physiologischen Bedingungen der Fettbildung bei Hefen zu. Hierüber wird an anderer Stelle dieses Handbuchbandes (S. 209ff.) berichtet. Die ersten Angaben über die qualitative Zusammensetzung eines Hefefettes bringen E. GÉRARD und P. DAREXY (1897) (s. Tabelle 3, I).

2. Die Zusammensetzung der Fette von Hefen.

In Tabelle 3 sind die Analysendaten für einige eingehender untersuchte Hefefette zusammengestellt. Tabelle 4 bringt zur Ergänzung einige Kennzahlen für die Fette von zwei Hefen, die bei Untersuchungen über die Physiologie der Fettbildung eine wichtige Rolle gespielt haben. In ihrem Fettsäurenbestand unterscheiden sich die Hefefette demnach nicht auffallend von den Reservefetten der höheren Pflanzen. Unter den gesättigten Fettsäuren überwiegt die Palmitinsäure (C_{16}); es folgen die Stearinsäure (C_{18}) und — zumeist in weitem Abstand — die Laurinsäure (C_{12}) und die höheren Fettsäuren ($C_{>20}$). Unter den einfach ungesättigten Säuren nimmt überall die Ölsäure (C_{18}) die erste Stelle ein. Die flüssige oder halbfeste Konsistenz der Hefefette findet damit eine Erklärung. Linolsäure und Linolensäure wurden bei den vollständig durchgeführten Analysen regelmäßig gefunden. Fraglich ist, ob die *flüchtigen* Säuren überhaupt zum „Fett" gehören, d.h., ob sie, wie etwa im Butterfett, in Glyceridbindung vorliegen. Es ist durchaus möglich, daß sie aus verschiedenen Stoffwechselvorgängen im Gäransatz, z. B. auch aus dem Aminosäurenabbau stammen. In einigen Fällen (z. B. Tabelle 3, VI) finden sich in den Originalarbeiten Hinweise dafür, daß die hohen Prozentzahlen für die flüchtigen Säuren verdorbene Fette kennzeichnen. J. HOLMBERG (1948) weist auf die Ähnlichkeit des von ihm untersuchten *Rhodotorula*fettes (Tabelle 3, IX) mit den Pericarp-Fetten von Samenpflanzen, z. B. dem Palmöl, hin. T. P. HILDITCH und R. K. SHRIVASTAVA (1948) betonen die Ähnlichkeit ihres Hefefettes (X) mit den Depotfetten von Landtieren. Mit Recht heben sie freilich auch hervor, daß das bisher vorliegende Material noch kaum weitergehende Schlüsse erlaubt, weil die Fettzusammensetzung wahrscheinlich sehr von der Rasse und von den Kulturbedingungen abhängt. Verschiedene Stämme zeigten z. B. unter gleichen Kulturbedingungen Jodzahlen zwischen 40 und 90; der gleiche Stamm produzierte Fette mit um so höherem Sättigungsgrad, je rascher die Fettbildung verlief (s. auch S. 240ff. dieses Bandes).

Der Gehalt an „Phosphatiden" ist in den Hefefetten, soweit geprüft, durchweg recht hoch (s. a. S. 64ff.). Besonders groß ist auch in einzelnen Fällen die Fraktion der unverseifbaren Bestandteile. Diese dürften unter dem Blickpunkt der vergleichenden Biochemie das interessanteste Kapitel bei der Behandlung der Hefelipoide darstellen (s. S. 66ff.).

Tabelle 3. *Zusammen*

	Saccharomyces cerevisiae					
	Bierhefe (E. GÉRARD u. P. DARÉXY 1897)	untergärige Bierhefe (O. HINSBERG u. E. ROOS 1903, 1904	Preßhefe (I. SMEDLEY-MACLEAN u. E. M. THOMAS 1920)	Bierhefe (J. WEICHHERZ u. R. MERLÄNDER 1931)	FLEISCHMANN-Preßhefe (M. S. NEWMAN u. R. J. ANDERSON 1933a, b)	Bierhefe (K. TÄUFEL, H. THALER u. H. SCHREYEGG 1936)
	I	II	III	IV	V	VI
Rohfett (% der Pilztrockensubstanz)	?	2,3—7,8	2,0	?	6,02	?
Verseifungszahl	—	—	187,5	—	109,6	156,6
Säurezahl	—	—	—	—	28,6	108,4
Jodzahl	—	—	121,2	—	61,3	130,4
REICHERT-MEISSL-Zahl	—	—	—	—	2,3	7,4
POLENSKE-Zahl	—	—	—	—	0,5	3,4
Acetylzahl	—	—	—	—	20,2	—
Phosphatide, % des Rohfettes	—	—	8,5	—	21,7	—
Unverseifbares, % des Rohfettes		Sterine +, „ätherisches Öl": Sp.	39,8 (17,2% Sterine, davon 12,3% verestert, 4,9% frei)	17,8	46,7	19,6 (3,3% Sterine; Ergosterin: Kryptosterin ~6:1, 16,3% Squalen)
Fettsäuren (% der gesamten Fettsäuren): Gesättigte Fettsäuren flüchtig	wenig C_4			32,5 C_5: +, C_{10}: ? C_1,C_2,C_4: —		7,3 C_5,$C_8$$C_{10}$: +
nicht flüchtig $\leq C_{14}$		„C_{15}": +	C_{12} wenig			C_{12}: —
C_{16}	~50		+		+ (~3	13,4
C_{18}	~50				+ : 1)	8,3
$\geq C_{20}$			C_{20} +			—
Einfach ungesättigte Fettsäuren $\leq C_{14}$		„C_{12}": +			Sp.	—
C_{16}					+ (~1 :	66,9
C_{18}		+	+		+ 3)	
$\geq C_{20}$						
Doppelt ungesättigte Fettsäuren C_{18}			+			11,2
Dreifach ungesättigte Fettsäuren C_{18}				+		1,0
Oxysäuren						

Erklärung I—XII s. S. 64.

setzung einiger Hefefette.

Geotrichoides sp. LANGERON et TALICE (O. TURPEINEN 1936)	*Torulopsis utilis* (R. REICHERT 1945)	*Rhodotorula gracilis* (J. HOLMBERG 1948)	*Torulopsis lipofera* (T. P. HILDITCH u. R. K. SHRIVASTAVA 1948)	*Blastomyces dermatitidis* (R. L. PECK u. C. L. HAUSER 1938)	*Monilia albicans* (R. L. PECK u. C. L. HAUSER 1939)
VII	VIII	IX	X	XI	XII
5,07—37,8	6,4	49,6%	25—33	8,5—9,0	5,3
190	180,7	—	—	191,5	183,9
78	122,4	—	33% freie Säure	45,3	96,3
104	120,5	—		106,1	81,0
—	6,1	—	—	2,7	—
—	0,5	—	—	—	—
—	19,8	—	—	—	—
36,0	37,7	—	—	8,2—34,1	3,4
5,75 (davon 39,2% Sterine, hauptsächlich Ergosterin)	12,3 (6,5% Ergosterin, wenig andere Sterine, 3,1% Squalen, 1,2% $C_{12}H_{24}O$, wenig gesättigte Kohlenwasserstoffe oder Oxykohlenwasserstoffe		2,4	8,0 (4% Sterine)	13,6 (7% Ergosterin)
~1,0	4,6	—	—	—	—
—	33	1,1	C_{12}:0,1	—	—
~6,0	7,9	29,8	27,6	11,0	19,5
~3,0	3,8	8,8	5,9	5,5	6,4
—	0,2	1,4	5,1	—	—
		C_{20}: ~0,3			
		C_{22}: ~0,4			
		C_{24}: ~0,7			
—	—	—	—	—	—
—	7,6	1,8	1,3	—	—
37,3	21,5	40,1	54,5	62,5	61,2
—	—	C_{20}:1,0	1,1	—	—
44,6	49,7	11,2	5,7	21,0	12,9
—	4,4	1,0	0,7	—	—
—	—	—	—	—	—

Erklärungen zu Tabelle 3.

I: Palmitin- und Stearinsäure (frei und als Glyceride).

II: Auch A. NEVILLE (1913) gibt eine gesättigte C_{15}-Säure an, daneben Arachinsäure (C_{20}) und ungesättigte C_{16}- und C_{18}-Säuren.

IV: Oxysäuren werden wegen der physiologischen Wirkung des Fettes im Tierversuch (Darmreizung) vermutet (vgl. Ricinolsäure). Einen recht hohen Gehalt an flüchtigen Fettsäuren (0,5%) fand auch G. WEIS (1931) bei ihren Untersuchungen eines Hefefettes. Die Anwesenheit von opt. aktiver und inaktiver Valeriansäure wird vermutet. Das Ausgangsmaterial (dunkel gefärbt, stark riechend) dürfte allerdings von nicht einwandfreier Beschaffenheit gewesen sein. In der gleichen Richtung weisen die „Anhaltspunkte" für die Anwesenheit von oxydierten Fettsäuren „höherer Oxydationsstufen".

V: Fett 2 Jahre alt (vgl. Acetylzahl). Das Unverseifbare eines Handelsmusters der Hefe enthielt cyclische Kohlenwasserstoffe ($C_{19}H_{38}$, $C_{30}H_{60}$, $C_{34}H_{66}$), die aber in einer unter kontrollierten Bedingungen gezüchteten Hefe nicht gefunden wurden. Die Mengenangaben für die Fettsäure beziehen sich auf den acetonlöslichen (Nichtphosphatid-)Anteil des Rohfettes.

VI: „Cerolin" *(C. F. Boehringer und Sohn)*: Äthanolextrakt gewaschener Bierhefe, für die Analyse durch Umlösen aus Äther gereinigt. Hoher Anteil freier und flüchtiger Säuren durch Art der Gewinnung oder Verderbnis bedingt? Hexadecensäure nicht gesondert bestimmt, in den Angaben für Ölsäure (66,9%) enthalten.

VII: Kennzahlen der acetonlöslichen (Nichtphosphatid-)Fraktion: Verseifungszahl 178, Säurezahl 105, Jodzahl 128. Auf diese Fraktion beziehen sich auch die Mengenangaben für die Fettsäuren.

VIII: Die Kennzahlen und die Mengenangaben für die Fettsäuren beziehen sich auf die acetonlösliche (Nichtphosphatid-)Fraktion.

IX: Methanolextraktion ergab 16,3%, nachfolgende Ätherextraktion 33,3% (zusammen 49,6%) vom Trockengewicht.

XI: Alle Angaben beziehen sich auf die acetonlösliche (Nichtphosphatid-)Fraktion. Die Analyse einer zweiten Fettprobe ergab ähnliche Resultate.

XII: Alle Angaben beziehen sich auf die acetonlösliche (Nichtphosphatid-)Fraktion.

Tabelle 4. *Kennzahlen einiger Hefefette (Rohfett).*

	Endomycopsis vernalis			*Candida Reukauffii* A. RIPPEL-BALDES u. Mitarb. (1950)
	nach STOCKHAUSEN (H. FINK u. Mitarb. 1937)	nach H. HAEHN u. W. KINTTOF (1925)		
		festes Fett	flüssiges Fett	
Verseifungszahl	185	238,15	228,95	179
Säurezahl	13,0	30,24%[1]	22,7%[1]	—
REICHERT-MEISSL-Zahl	2,0	—	—	—
Jodzahl	96,0	64,64	105,22	70
Acetylzahl	39,8	—	—	—
Unverseifbares	—	—	—	0,21%

[1] Als freie Ölsäure berechnet.

3. Die Phosphatide der Hefen.

F. HOPPE-SEYLER gab (1879a) für gewaschene Preßhefe Lecithin in einer Menge von 0,314% der Trockensubstanz an. Der negative Befund von C. v. NAEGELI und O. LOEW (1878, 1880) wird damit erklärt, daß diese Forscher vor der Lipoidextraktion die Bierhefe mit starker Salzsäure aufgeschlossen hatten. Das eigene Ergebnis wird aufrechterhalten (F. HOPPE-SEYLER 1879b, s. auch O. LOEW 1878) und durch den Nachweis von Glycerinphosphorsäure und Cholin unter den Spaltprodukten gestützt. Auch F. KUTSCHER und LOHMANN konnten unter den Produkten der Hefeautolyse Cholin feststellen. W. KOCH (1902) scheint der erste gewesen zu sein, der in der Hefe auch Kephalin nachweisen konnte. Seine Methode bestand in der Bestimmung des $P:CH_3$-Verhältnis, welches für Lecithin (Formel I) 1:3, für Kephalin (Formel II) 1:1 beträgt.

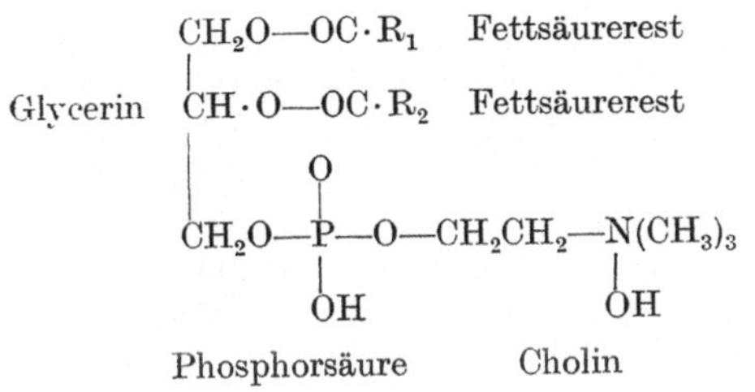

Glycerin

$CH_2O—OC \cdot R_1$ Fettsäurerest
$CH \cdot O—OC \cdot R_2$ Fettsäurerest
$CH_2O—P(=O)(OH)—O—CH_2CH_2—N(CH_3)_3(OH)$

Phosphorsäure — Cholin

I *Lecithin*

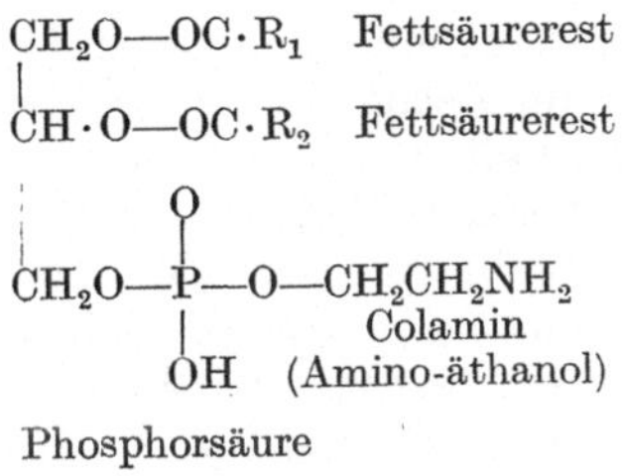

$CH_2O—OC \cdot R_1$ Fettsäurerest
$CH \cdot O—OC \cdot R_2$ Fettsäurerest
$CH_2O—P(=O)(OH)—O—CH_2CH_2NH_2$

Phosphorsäure — Colamin (Amino-äthanol)

II *Kephalin*

Aus Tabelle 3 ließ sich bereits entnehmen, daß der Phosphatidanteil in Hefefetten wechselt, im allgemeinen aber recht hoch gefunden wird. Einige nähere Angaben bringt Tabelle 5.

Tabelle 5. *Gehalt an Rohfett und an Phosphatiden bei einigen Hefen.*

Objekt	Rohfett % Trockengewicht	Phosphatide % Trockengewicht	Phosphatide % des Rohfettes	Lecithin/Kephalin	Autor
Saccharomyces cerevisiae					
„Preßhefe"	4,5	0,5	11	1:<1	W. C. AUSTIN (1924)
„Hefe der Ostwerke"	8,1	4,9	60,4	—	B. REWALD (1930)
„Braasch-Hefe"	7,25	4,5	61,9	—	B. REWALD (1930)
Torulopsis utilis					
„Bergin-Hefe"	—	—	60—80	~34:46	K. DIRR und A. RUPPERT (1948)
„Kostheim-Hefe" von Buchenholzsulfitlauge	8,05	3,40	41,7	76:24	W. DIEMAIR und W. POETSCH (1949)
„Bergin-Hefe" von Holzzucker	7,82	3,60	46,0	73:27	W. DIEMAIR und W. POETSCH (1949)
„Phrix-Hefe" von Strohvorhydrolysat	5,68	2,97	52,2	72:28	W. DIEMAIR und W. POETSCH (1949)
Oospora lactis					
„Lenzing-Hefe" von Sulfitablauge	5,13	1,72	33,6	64:36	W. DIEMAIR und W. POETSCH (1949)

Über das Verhalten der Phosphatide bei experimenteller Lipoidanreicherung in der Hefe (C. G. DAUBNEY und I. SMEDLEY-MACLEAN 1927) unterrichtet Tabelle 6. Es zeigt sich, daß sowohl die Gesamtlipoide als auch die Phosphatide unter den Versuchsbedingungen zunehmen, die letzten aber merklich weniger.

Tabelle 6. *Rohfett- und Phosphatidgehalt von Hefe (Saccharomyces cerevisiae) bei experimenteller Lipoidanreicherung.* (Nach C. G. DAUBNEY und I. SMEDLEY-MACLEAN 1927.)

	Rohfett % Trockengewicht	Phosphatide % Trockengewicht	Phosphatide % vom Rohfett
Kontrolle	1,08	0,29	26,8
Durchlüftet, 4% Glucose	2,87	0,65	22,6
Durchlüftet, 4% Glucose + Phosphat	5,34	0,97	18,2

Über die Fettsäuren der Hefephosphatide machte TH. SEDLMAYR (1903) die ersten Angaben. Er findet ein Dipalmitoylglycerylphosphorylcholin. Zu anderen Ergebnissen kommen C. G. DAUBNEY und I. SMEDLEY-MACLEAN (1927): Sowohl im Lecithin- wie im Kephalinanteil von Preßhefe ist Palmitin- *und* Ölsäure enthalten, von der letzteren im Kephalin etwas mehr als im Lecithin. Die Fettsäuren der Phosphatide sind insgesamt etwas höher gesättigt als diejenigen des Neutralfettes. Das Gesamtphosphatid von *Torulopsis utilis* liefert bei Verseifung ein Fettsäurengemisch mit einer Jodzahl von 97,96—120,93, in welchem Palmitin-, Stearin-, Öl-, Linol- und/oder Linolensäure enthalten sind (W. DIEMAIR und J. KOCH 1948). K. DIRR und A. RUPPERT (1948) geben für das gleiche Objekt 16% gesättigte feste und 84% ungesättigte flüssige Fettsäuren an. Fast genau die gleichen Verhältnisse (14—16% bzw. 84—86%) finden M. S. NEWMAN und R. J. ANDERSON (1936b) im Phosphatid von Preßhefe. In beiden Fällen enthielten die Lecithin- und Kephalinfraktion etwa die gleichen Fettsäuren. Das von R. L. PECK und C. L. HAUSER (1938) untersuchte Fett der pathogenen Hefe *Blastomyces dermatitidis* weist im Lecithin-Kephalingemisch folgende Mengenverteilung der Fettsäuren auf: Palmitinsäure 12,6%, Stearinsäure 6,4%, Ölsäure 75,6%, Linolsäure 5,4%. V. R. USDIN und R. C. BURELL (1952) fanden in einem Fett von *Rhodotorula gracilis* (31,23% vom Trockengewicht) einen Phosphatidanteil von 3,8%, der kein Kephalin enthielt. Beider Verseifung wurden 57,4% Fettsäuren mit der Neutralisationszahl 260,2 und der Jodzahl 86,1 erhalten. Vor kurzem konnten D. J. HANAHAN und M. E. JAYKO (1952) auf chromatographischem Wege aus dem Äthanolextrakt der Hefe ein reines Dipalmitoleyl-L-α-glycerylphosphorylcholin, also ein völlig ungesättigtes Lecithin isolieren, welches durch Hydrierung in das entsprechende Dipalmitoyl-Lecithin übergeführt werden konnte.

Einige Literaturangaben unterrichten über den Anteil der beiden isomeren Glycerinphosphorsäuren am Aufbau der Hefephosphatide. L. F. SALISBURY und R. J. ANDERSON (1936) finden bei Preßhefe im Lecithin nur die optisch aktive α-Glycerin-phosphorsäure, in der 4mal größeren Kephalinfraktion nur die optisch inaktive β-Verbindung. K. DIRR und A. RUPPERT (1948) geben für das *Torulopsis utilis*-Phosphatid (46% Kephalin) 79,5% α-, 20,5% β-Glycerinphosphorsäure an.

Das von V. R. USDIN und R. C. BURELL (1952) analysierte Phosphatid aus *Rhodotorula gracilis* enthält 87,6% der Glycerinphosphorsäure in der α-Form. Kurz erwähnt sei schließlich die Entdeckung von Glycerylphosphoryläthanolamin im Wasserextrakt der Bierhefe durch P. N. CAMPBELL, D. H. SIMMONDS und T. S. WORK (1951).

4. Sterine, Triterpene und andere Begleitstoffe der Hefefette.

C. v. NAEGELI und O. LOEW (1878, 1880) gebührt das Verdienst, zuerst das Vorkommen eines „*Cholesterins*" im Fett der Bierhefe nachgewiesen zu haben. Die Identität des „Hefesterins" mit dem durch C. TANRET (1889) im Mutterkorn entdeckten *Ergosterin* hat E. GÉRARD (1895) erkannt. Das wurde in der Folge mehrfach bestätigt, so von A. WINDAUS und W. GROSSKOPF (1923). I. SMEDLEY-MACLEAN und E. M. THOMAS (1928) fanden im Hefefett 20% Sterine, in der Hauptsache Ergosterin; noch im gleichen Jahr berichtet die erste Verfasserin (I. SMEDLEY-MACLEAN 1928) über ein zweites Hefesterin, das Zymosterin. Diese Angabe wird 1929 von H. PENAU und G. TANRET bestätigt. Die Erkenntnis von A. WINDAUS und R. POHL (1926), daß das Vitamin D durch UV-Bestrahlung aus dem Ergosterin hervorgeht, regte eine ganze Reihe von Arbeiten an, die sich mit dem Ergosteringehalt der Hefe (und anderer Pilze) befaßten. Einige Zahlen bringt Tabelle 7. Die Mutterlaugen der nunmehr einsetzenden großtechnischen Ergosterindarstellung aus der Hefe lieferten das Ausgangsmaterial, in dem weitere Nebensterine (und Triterpene) der Hefe gefunden wurden (F. REINDEL und Mitarbeiter 1927, 1928, 1929, 1930, H. WIELAND und M. ASANO 1929). Zusammenfassende Darstellungen bringen A. STOLL, E. JUCKER (1955) und H. LETTRÉ und R. TSCHESCHE (1954).

Danach sind bisher folgende Sterine in der Hefe gefunden worden:

Hauptsterin	Ergosterin	$C_{28}H_{44}O$
Nebensterine	Zymosterin	$C_{27}H_{44}O$
	α-Dihydroergosterin	$C_{28}H_{46}O$
	Ascosterin	$C_{28}H_{46}O$
	Faecosterin	$C_{28}H_{46}O$
	Episterin	$C_{28}H_{46}O$
	Anasterin	$C_{27}H_{44}O$(?)
	Hyposterin	$C_{27}H_{42}O$(?)
	Cerevisterin	$C_{28}H_{46}O_3$

Die meisten genannten Stoffe sind konstitutionell mit dem Ergosterin nahe verwandt. Das *Neosterin* WIELANDS hat sich als ein Gemenge von Ergosterin und α-Dihydroergosterin erwiesen (D. H. R. BARTON und J. D. COX 1948). Das „*Kryptosterin*" (H. WIELAND und M. W. STANLEY 1931) ist identisch (L. RUZICKA, R. DENSS und O. JEGER 1945) mit dem von A. WINDAUS und R. TSCHESCHE (1930) aus dem „Isocholesterin" des Wollfettes isolierten tetracyclischen Triterpenalkohol „Lanosterin" (s. unten). Wegen aller Einzelheiten und der Originalliteratur sei auf die oben genannten Zusammenfassungen verwiesen.

Tabelle 7. *Gehalt einiger Heferassen an Gesamtsterin und Ergosterin.*

	Gesamtsterin % Trockengewicht	Ergosterin % Trockengewicht	Literaturhinweis
Preßhefe I	—	1,17	A. HEIDUSCHKA und P. LINDNER (1929)
Preßhefe II	—	0,70	
Bierhefe	—	0,61	
Portweinhefe	—	0,28	
Torula pulcherrima	—	0,20	
Schizosaccharomyces mellacei	—	0,54	
Brennereihefe	0,16	0,103	A. CASTILLE und E. RUPPOL (1933)
Brauoberhefe	0,165	0,103	
Brauunterhefe	0,258	0,20	
Preßhefe	0,364	0,343—0,584	
Saccharomyces Pastorianus	0,3	0,23	
Hefe „Wildiers 100"	0,29	0,24	

Unter den *Triterpenen*, die sich im „Unverseifbaren" des Hefefettes finden, verdient vor allem das Squalen ($C_{30}H_{50}$) Interesse. Dieser hoch ungesättigte, aliphatische Triterpenkohlenwasserstoff wurde unabhängig von M. TSUJMOTO (1916, 1920) und von A. C. CHAPMAN (1917a, b) (als Spinacen) in Haifischleberölen entdeckt, von G. SANI (1930) und H. MARCELET (1936) im Olivenöl festgestellt, von K. TÄUFEL, H. THALER und H. SCHREYEGG (1936a, b) in großen Mengen (16% des Rohfettes) in der Bierhefe gefunden und sicher identifiziert. Auch im Fett von *Torulopsis utilis* stellt Squalen rund $^1/_4$ des Unverseifbaren (R. REICHERT 1945). Über die durch Isotopenversuche belegte Biosynthese des Squalens aus C_2-Bausteinen (Essigsäure) und über seine mögliche Rolle als Vor- oder Zwischenstufe der Steroid- und Triterpenbildung im Organismus vgl. z. B. R. TSCHESCHE und F. KORTE (1954), M. STEINER und H. HOLTZEM (1955, insbesondere S. 62ff.) und R. TSCHESCHE (1955).

Der tetracyclische Triterpenalkohol Lanosterin (= Lanostadienol, $C_{30}H_{50}O$) wurde von A. WIELAND und W. M. STANLEY (1931) als „Kryptosterin" aus den

Nebenprodukten der großtechnischen Darstellung des Hefeergosterins dargestellt. Wie R. TSCHESCHE und F. KORTE (1954) betonen, ist es noch unsicher, ob diese Substanz ein echtes Stoffwechselprodukt der Hefe darstellt oder ob sie eine zufällige Verunreinigung aus dem Wollfett ist, welches vielfach als schaumhemmendes Mittel den Gäransätzen zugegeben wird.

Im Rohfett der Rosahefen (*Rhodotorula*arten) sind die Carotinoidpigmente enthalten (Zusammenfassungen bei T. W. GOODWIN 1955 und bei F. T. HAXO

Tabelle 8. *Abhängigkeit des Steringehaltes der Hefe von den Züchtungsbedingungen.*

Heferasse Kulturbedingungen	Gesamtsterin % Trockengewicht	Ergosterin % Trockengewicht	Literatur
Bierhefe			
Bierwürze + Gerstenkleieextrakt	—	0,46	A. HEIDUSCHKA und P. LINDNER (1929)
Bierwürze + Gerstenmehl	—	0,66	
Bierwürze, mit O_2 belüftet	—	1,01	
Bierwürze + 5% Äthanol	—	0,73	
Preßhefe			
Getreidemaische	0,39—0,41	0,34—0,38	A. CASTILLE und E. RUPPOL (1933)
Brauwürze	0,24	0,12	
synthetisches Medium	0,20—0,24	0,12	

1955). Über den Vitamin A-Gehalt von *Rhodotorula glutinis* berichten R. NILSSON, L. ENEBO und E. BRUNIUS (1942).

Auf das *Cerebrin* der Hefe wird weiter unten (S. 81) im Zusammenhang mit den übrigen Pilzcerebrinen eingegangen.

Daß der Sterin- und Ergosteringehalt der gleichen Heferasse auch von den Kulturbedingungen abhängt, zeigen die Zahlen in Tabelle 8.

Weitere Angaben ähnlicher Art sind im Zusammenhang mit dem Fettstoffwechsel der Pilze, S. 209ff. dieses Bandes zu finden (vgl. auch in Bd. X das Kapitel „Sterine").

IV. Schimmelpilze und andere Mycelpilze.

1. Geschichtliches.

In Reinkultur-Mycelien von Schimmelpilzen wurden die ersten Fettanalysen von C. v. NAEGELI (1879) und C. v. NAEGELI und L. LOEW (1880) durchgeführt. Der Fettgehalt des *Penicillium*mycels lag je nach der Kulturbedingung zwischen 6,67 und 18,10% vom Trockengewicht. Wenig später hat N. SIEBER gemischte Schimmelpilzrassen (*Penicillium glaucum* + *Mucor mucedo)* auf verschiedenen künstlichen Nährböden untersucht und zwischen 11,19 und 18,70% vom Trockengewicht ätherlösliche Substanz festgestellt. MARSCHALL (1897) fand folgende Prozentzahlen für das Ätherextrakt: *Aspergillus glaucus* 4,7%, *Penicillium glaucum* 4,1%, *Rhizopus nigricans* (= *Mucor stolonifer*) 7,0% (Kultur auf Fleischextrakt-Peptonbrühe + 1% Weinsäure + 2% Glucose). DUBAQUIÉ (1909) fand Fettgehalte in der gleichen Größenordnung: in *Aspergillus niger* bis 12,6%, *A. glaucus* bis 3,9%, *in Penicillium glaucum* bis 4,5% je nach den Versuchsbedingungen. Daß Schimmelpilze auch zu wesentlich kräftigerer Fettbildung fähig sind, zeigte als erster J. LABORDE (1897) bei *Eurotiopsis Gayoni*, dessen Trockensubstanz bis zu 28,8%, nach A. PERRIER (1905) sogar bis zu 34,05% aus Fett bestehen kann. In den späteren Arbeiten werden die leicht kultivierbaren Schimmelpilze immer stärker zu Untersuchungen über die Physiologie der Fettbildung verwendet, insbesondere zum Studium der Fettmenge und -zusammensetzung in Abhängigkeit von den Kulturbedingungen: *Aspergillus niger* (= *Sterigmatocystes nigra*) (O. FLIEG, P. BELIN, E. F. TERROINE, R. BONNET, G. KOPP, J. VÉCHOT, P. DUQUÉNOIS, CH. PONTILLON, K. BERNHAUER, G. POSSELT); *Aspergillus fischeri* (E. B. FRED, H. J. GORCICA, H. C. GREENE,

W. H. PETERSON, P. R. WENCK); *Penicillium javanicum* (H. T. HERRICK, L. B. LOCKWOOD, O. E. MAY, *circinelloides* H. T. O'NEILL, G. E. WARD); *Mucor* (K. BERNHAUER, J. RAUCH); *Oospora laits* (H. FINK, H. HAESELER, M. SCHMIDT, H. GEFFERS usf.). Über die Ergebnisse dieser Arbeiten vgl. M. STEINER (dieser Handbuchband, S. 216ff.). Die meisten Detailanalysen über die Zusammensetzung von Schimmelpilzfetten (s. unten, insbesondere Tabelle 8 und 9) beziehen sich auf Arten, die für derartige physiologische Studien verwendet wurden.

2. Die Zusammensetzung der Fette von Schimmelpilzen und anderen Mycelpilzen.

In Tabelle 9 finden sich die Ergebnisse verschiedener Autoren zusammengefaßt, die Schimmelpilzfette eingehender untersucht haben. Tabelle 10 bringt zur Ergänzung einige Kennzahlen solcher Fette. In Tabelle 11 sind schließlich einige Daten über den Fettgehalt von Schimmelpilzconidien zusammengestellt, in denen ja Fetttropfen zumeist schon mikroskopisch festgestellt werden können. Es ist allerdings fraglich, ob durch Ätherextraktion der intakten Sporen das gesamte Rohfett erfaßt wurde.

Als übereinstimmendes Merkmal der untersuchten Fette kann der hohe Gehalt an ungesättigten Fettsäuren vermerkt werden: Ölsäure und Linolsäure machen zusammen etwa zwei Drittel des gesamten Bestandes an Fettsäuren aus. Nur das von T. P. HILDITCH und Mitarbeiter untersuchte Penicilliumfett (Tabelle 9, V) ist stärker gesättigt (53% gesättigte Fettsäuren)[1]. Regelmäßig finden sich mit kleineren Anteilen eine oder mehrere der höheren gesättigten Säuren (C_{20}—C_{26}). Fast alle neueren, mit moderneren Trennverfahren durchgeführten Analysen (V, IX) geben auch die Hexadecen-Palmitoleinsäure an. Es kann vermutet werden, daß sie allgemein verbreitet ist und in den Zahlenangaben für „Ölsäure" der alten Arbeiten miterfaßt ist.

Außerordentlich großen Schwankungen unterliegt der Anteil freier Säuren am Gesamtfett. Wie ein Vergleich der Verseifungs- und Säurezahlen zeigt, können bis drei Viertel der Fettsäuren in freier Form vorliegen. Da wohl alle Schimmelpilze kräftige Lipasebildner sind (vgl. E. BAMANN und E. ULLMANN, dieser Handbuchband, S. 110f.), wären sekundäre Spaltungen während der Aufarbeitung nicht ausgeschlossen. F. M. STRONG und W. H. PETERSON (Tabelle 9, VII) inaktivierten das Pilzmycel vor der Extraktion durch Hitzebehandlung; die rund 25% freier Fettsäuren dürften also schon im lebenden Mycel vorgelegen haben. Andererseits gibt der auffallend niedrige Gehalt freier Fettsäuren in der mit den modernsten Kautelen durchgeführten Analyse X (J. SINGH, T. K. WALKER, M. L. MEARA 1955) zu denken. Das Fett von *Penicillium javanicum* scheint immer relativ arm an freien Fettsäuren zu sein. I. T. KAIBARA (1948) gibt Konstanten an (Verseifungszahl 190,1, Säurezahl 9,02, Jodzahl 86,62), die vorzüglich mit den Werten von WARD und JAMIESON (s. Tabelle 9, IV) übereinstimmen. Auch bei den Oxysäuren des *Oospora lactis*-Fettes (XII) ist eine sekundäre Entstehung wohl denkbar.

Phosphatide und Sterine sind auch bei den Schimmelpilzen konstante Begleiter der Neutralfette und der freien Fettsäuren. Ihr prozentualer Anteil ist aber in der Regel weit geringer als bei den Hefen (vgl. Tabelle 3, 7 und 8).

[1] Dagegen ist auch das Fett von *Penicillium aurantio-brunneum*, über welches E. H. KROEKER, F. M. STRONG und W. H. PETERSON (1935) berichteten, stark ungesättigt. Unter den Fettsäuren des acetonlöslichen Glyceridanteils sind 16,4% gesättigt und 83,6% ungesättigt; im einzelnen, auf die gesamten Fettsäuren berechnet: Palmitinsäure 9,4%, Stearinsäure 6,2%, gesättigte Säuren $C_{\geq 20}$ höchstens Spur, Ölsäure 47,1%, Linolsäure 36,5%. Die Phosphatidfraktion dagegen war für ein Schimmelpilzfett ungewöhnlich hoch: etwa 20% des Äthanol-Ätherextraktes waren mit Aceton fällbar.

Tabelle 9. *Zusammensetzung der Fette einig*

	Zygomycetes		Plectascomycet		
	Rhizopus japonicus (H. LIM 1935)	*Phycomyces Blakesleeanus* (K. BERNHARD u. H. ALBRECHT 1948)	*Penicillium glaucum* aus Erdboden (M. X. SULLIVAN 1913)	*Penicillium javanicum* (G. E. WARD u. G. S. JAMIESON 1934)	*Penicillium* sp. von Soja-Lecithi (T. P. HILDITCH u. Mitarb. 1944)
	I	II	III	IV	V
Rohfett (% der Pilz-Trockensubstanz) . . .	9,7	19,5	?	11	?
Verseifungszahl	164,0	—	—	191	255,1
Säurezahl	95,4	—	—	10,6	76% freie Säure
Jodzahl	84,65	—	—	84,0	54,2
REICHERT-MEISSL-Zahl .	8,16	—	—	0,3	—
POLENSKE-Zahl	0,82	—	—	—	—
Acetylzahl	38,00	—	—	10,2	—
Phosphatide (% des Rohfettes)	1,23	9,9	—	—	—
Unverseifbares (% des Rohfettes)	3,78 (1,1% Ergosterin, 2,5% Fungisterin)	5,3 [davon 4,1% Ergosterin, Rest: Carotinoide: α und β-Carotin, Lycopin (?)]	—	2,0	—
Fettsäuren (% der gesamten Fettsäuren):					
Gesättigte Fettsäuren					
flüchtig	—	—	—	—	—
nicht flüchtig					
$\leqq C_{14}$	38,75[1] (für C_{14} bis C_{20})	—	—	—	C_{14}: 3,5
C_{16}	7,8[2]	27,7	+	23,4	40,5
C_{18}	1,2[2]	4,7	—	9,4	7,9
$\geqq C_{20}$		C_{22}: 1,7 C_{24}: 2,1 C_{26}: 1,3	+ (C_{20} ?)	C_{24}: 0,8	C_{20}: 1,1
Einfach ungesättigte Fettsäuren $\leqq C_{14}$	61,25[1] (für einfach und doppelt ungesättigte)				7,9
C_{16}	viel (für C_{16} bis C_{20})	29,6		34,6	19,0
C_{18}		C_{24}:[3] 2,9	+		
$\geqq C_{20}$		C_{26}: 4,9			
Doppelt ungesättigte Fettsäuren C_{18}	wenig	25,8	—	31,8	20,1
Dreifach ungesättigte Fettsäuren C_{18}	—	(γ) 3,4	—	—	0
Oxysäuren	—	—	—	—	—

I[1] Prozent des Gesamtfettes. [2] Prozent der Pilztrockensubstanz.
II[3] $\Delta^{17,18}$ Triacosen-carbonsäure.
V[4] Nach J. SINGH, T. K. WALKER und M. L. MEARA (1955). — "The mould (probably a common species of *Penicillium*) was unusually rich in fat." Die Fettsäuren des Substrates (Soja-Lecithin) hatten folgende Zusammensetzung (%): Myristinsäure 3,5, Palmitinsäure 40,5, Stearinsäure 7,9, Arachinsäure 1,1, Hexadecensäure (?) 7,9, Ölsäure 19,0, Linolsäure 20,0. — Wegen der Zusammensetzung des Fettes von *Penicillium lilacinum* vgl. den Nachtrag S. 89.

Schimmelpilze und anderer Mycelpilze.

Plectascomycetes					Fungi imperfecti	
Penicillium chrysogenum Q 176 (Y. ABE 1949, 1952)	*Aspergillus sydowi* (F. M. STRONG u. W. H. PETERSON 1934)	*Aspergillus niger* (K. BERNHAUER u. G. POSSELT 1937)	*Aspergillus citromyces* (E. RUPPOL 1937)	*Aspergillus nidulans* (J. SINGH, T. K. WALKER u. M. L. MEARA 1955)	*Fusarium lini* (J. V. FIORE 1948)	*Oospora lactis*, Stamm 17 (H. P. KAUFMANN u. O. SCHMIDT 1938)
VI	VII	VIII	IX	X	XI	XII
2,67	7,7—12	?	?	45	?	?
176,7	169,5	169,0	—	290,8 (Äq.)	—	191,7
58,8	43,4	71,2	—	0,8% freie Fettsäuren	—	11,2
89,0	114,4	95,1	—	73,4	—	63,1[6]
—	—	0,99	—	—	—	—
—	—	0,7	—	—	—	—
—	—	—	—	—	—	—
—	6,15—9,2[5]	—	—	—	—	—
3,0 (daraus 0,6% reines Ergosterin)	8,18 5,36% Sterine (Ergosterin), rote Farbstoffe	12,0 (1,4% Ergosterin)	10,0 (Ergosterin: + Cerylalkohol: +)	1,4	1% Ergosterin	1,25
	0,46		C_6: +			
C_{14}: 3,7			—	C_{14}: 0,7	—	—
12,0	10,9 } 28,0	10,5 } 19,2	+	20,9	—	25,7
5,7	13,6	1,3	+	15,9	+	17,1
C_{24}: 0,7	C_{24}: 0,9	C_{24}: 2,7	C_{26}: +	1,4	—	—
				1,2		
53,3	36,6 } 65,5	31,8 } 67,4	+	40,3	+	41,2
4,1	1,7			2,4		
20,8	20,2	α: 35,4	+	17,0	+	11,8
—	—	—	—	0,2	—	0,12
—	—	—	—	—	—	3,7%

VII[5] Nach D. W. WOOLLEY, F. M. STRONG, W. H. PETTERSON und E. A. PRILL (1935). Die Mengenangaben für die Fettsäuren beziehen sich auf die acetonlösliche (Nicht-Phosphatid-)Fraktion.

VIII Fett aus Mycelium extrahiert, welches von technischer Citronensäuregärung auf Melasse stammt.

IX Palmitin-, Öl- und Linolsäure zusammen ~90% der gesamten Fettsäuren.

X Herkunft des Fettes siehe H. FINK, H. HAESELER und O. SCHMIDT (1937).

XII[6] Jodzahl der durch Verseifung abgetrennten Fettsäuren; Rhodanzahl 51,6.

Tabelle 10. *Kennzahlen der Fette einiger Schimmelpilze.*

	Mucor mucedo (M. BLINC u. M. BOJEC 1942)		*Fusarium* sp. (H. DAMM 1943)
	a) in Petroläther leichtlösliche Fraktion	b) in Petroläther schwerlösliche Fraktion	
Fett, % von der Pilztrockenubstanz	6,5		
Konsistenz	halbfest	fest	Öl
Verseifungszahl	190,8	192,6	190—196
Säurezahl	107,7	96,7	0,4—4,0
Jodzahl	125,6	—	79—90
Unverseifbares	6,1%	—	Carotinoide +

Tabelle 11. *Fettgehalt (Ätherextrakt) von Schimmelpilzsporen.*

Art	Autor	Fett % v. Trockengewicht	Bemerkungen
Penicillium glaucum	E. CRAMER (1894)	3,93—13,92 Mittel 7,34	—
Aspergillus oryzae	K. ASO (1900)	0,38	—
Aspergillus oryzae	M. SUMI (1928)	0,88	„Sterin" +, „Lecithin" +

3. Die Phosphatide der Schimmelpilze.

N. SIEBER (1881) dürfte der erste gewesen sein, der aus der Anwesenheit von Lipoid-Phosphor auf die Anwesenheit von „Lecithin" in den von ihm untersuchten Schimmelpilzen schloß. R. GOUPIL (1913) untersuchte den zeitlichen Verlauf der Bildung von Fett bei *Mucor (Amylomyces) Rouxii*. Im Maximum der Entwicklung des Pilzes (nach 2 Monaten Kulturdauer) wurden 21,0% Fett/Trockengewicht, davon 2,0% „Lecithin" (aus dem Lipoid-P berechnet) gefunden. Bei jungen (5 tägigen) Kulturen entfiel sogar ein Fünftel des Gesamtfettes auf „Lecithin". RÉMOND und H. LASALLE (1925) stellten bei Brot, das mit *Penicillium glaucum* beimpft war, während 22 Tagen eine stetige Zunahme der gesamten Fettsäuren, der Sterine und des Lipoid-P (von 0 auf 7 mg/10 g) als Folge des Pilzstoffwechsels fest. A. HÉE (1930) fand in 66stündiger Kultur von *Aspergillus niger* auf CZAPEK-Lösung 73 mg-% Trockengewicht Lipoid-P.

Im Vergleich zu den Hefephosphatiden sind die Schimmelpilzphosphatide wenig erforscht. Das dürfte sicher damit zusammenhängen, daß bei dem relativ geringen Phosphatidgehalt der Mycelpilze nicht leicht Materialmengen verfügbar sind, die für eine gründliche Analyse ausreichen. D. W. WOOLLEY, F. M. STRONG, W. H. PETERSON, E. A. PRILL (1935) scheinen bisher die einzigen gewesen zu sein, die eingehendere Angaben über ein Schimmelpilzphosphatid machen. Es handelt sich übrigens um die acetonunlösliche Fraktion des gleichen Fettes von *Aspergillus sydowi*, dessen acetonlöslicher Anteil in Tabelle 9, VII berücksichtigt wurde. Aus dem Hydrolyseprodukt wurden isoliert:

Glycerinphosphorsäure	Ölsäure	
Cholin	Palmitinsäure	
Äthanolamin	Stearinsäure	kleine Mengen wahrscheinlich
	gesättigte Fettsäuren $\geq C_{20}$	

Danach lag also ein Lecithin-Kephalingemisch mit Ölsäure als vorherrschender Fettsäure vor.

4. Sterine und andere Fettbegleiter in Schimmelpilzen.

E. Gérard stellte 1892 in *Penicillium glaucum*, 1895 in *Mucor mucedo* ein „Cholesterin" (Fp. 135°) fest. Dieser Stoff ließ sich mit dem Tanretschen Ergosterin identifizieren.

Eine Reihe von Untersuchern studierte den Ergosteringehalt von leicht kultivierbaren Schimmelpilzen im Hinblick auf technische Gewinnung dieses als Provitamin D bedeutsamen Sterins. In neuerer Zeit spielt der Gedanke eine Rolle, Ergosterin als Nebenprodukt der Penicillinfabrikation zu gewinnen. Einige Ergebnisse bringt die nachfolgende Zusammenstellung (Tabelle 12).

Tabelle 12.

Pilz	Kultur	Ergosterin % Trockengewicht	Literatur
Penicillium glaucum . . .	—	0,75	A. Heiduschka und P. Lindner (1929)
Aspergillus oryzae	—	0,46	
Aspergillus oryzae	—	0,28 (kristallisiert)	R. Takata (1929)
Penicillium puberulum .	Czapek-Dox-Lösung 2—3 Wochen	0,13	J. H. Birkinshaw, R. K. Callow, C. F. Fischmann (1931)
Penicillium notatum . .	Technischer Ansatz	1,00	H. D. Zook, T. S. Oakwood, F. C. Whitmore (1944)
Penicillim notatum . . .	Oberflächenkultur;		C. J. Cavallito (1944)
	auf Lactose	1,10	
	auf Stärke	1,05	
Penicillium citrinum . .	Oberflächenkultur	1,1—1,3	
Penicillium chrysogenum	Submerskultur	0—Sp	
	Oberflächenkultur	1,0—1,1	
Penicillium notatum . .	Submerskultur	0,22	K. Savard und G. A. Grant (1946)
Penicillium notatum var. *chrysogenum* St. Q 176		2,3	G. Tappi (1948) (vgl. auch A. Saito 1949)

Das Ergosterin liegt auch bei den Schimmelpilzen zum Teil frei, zum Teil mit Fettsäuren verestert vor. So fanden A. E. Oxford und H. Raistrick (1933) in 15 von 16 Arten und Rassen der Gruppe des *Penicillium brevicompactum* Ergosteryl-Palmitat in Ausbeuten zwischen 0,02 und 0,50% des Trockengewichts. Auch H. H. Barber (1929) fand in einem *Penicillium* die Fettsäuren zum Teil frei und als Glyceride, zum Teil als Steride.

Über die *Nebensterine*, welche sicherlich auch bei den Schimmelpilzen das Hauptsterin Ergosterin begleiten (s. z. B. A. Angeletti und G. Tappi 1947), liegen bisher präzise Angaben nur spärlich vor. H. Lim (1935) fand im Mycel von *Rhizopus japonicus* neben Ergosterin das in höheren Pilzen weit verbreitete Fungisterin (1,55 bzw. 2,55 g aus 1 kg Trockenmycel). D. H. R. Barton und T. Bruun (1951) stellten im Mycel eines für technische Citronensäuregärung benutzten *Aspergillus niger*-Stammes neben Ergosterin das bis dahin unbekannte 4-Dehydro-Ergosterin fest.

Auf die „Cerebrine", die jüngst in einigen Schimmelpilzen gefunden wurden *(Aspergillus, Penicillium)*, wird weiter unten eingegangen (S. 81ff.).

An dieser Stelle sollen kurz einige höhere mehrbasische Oxysäuren Erwähnung finden, welche Raistrick und Mitarbeiter als Stoffwechselprodukte von

Schimmelpilzen aufgefunden haben. Bei der Länge ihres C-Skelets stehen sie wenigstens in formaler, vielleicht auch in biogenetischer Beziehung zu den Fettsäuren.

Penicillium spiculisporum LEHMANN (P. W. CLUTTERBUCK, H. RAISTRICK, M. L. RINTOUL 1931):

γ-Keto-Pentadecensäure.

γ-Lacton der γ-Oxy-β,δ-dicarboxypentadecansäure (Spiculisporsäure).

Penicillium crateriforme GLIMANN et ABBOT (A. E. OXFORD und H. RAISTRICK 1934):

Spiculisporsäure (s. oben).

Penicillium minioluteum DIERCK (J. H. BIRKINSHAW und H. RAISTRICK 1934):

Spiculisporsäure (s. oben).

γ-Lacton der d-α,β-dioxy-β,γ-dicarboxytetradecansäure(?) (Minioluteinsäure).

V. Mutterkorn.

1. Geschichtliches.

Das *Mutterkorn*, das Sklerotium von *Claviceps purpurea* TUL., war wegen seiner großen Bedeutung als Arzneidroge schon frühzeitig Gegenstand von chemischen Untersuchungen. Bald wurde auch dem reichlich vorhandenen Öl des Mutterkorns Beachtung geschenkt. Nach F. A. FLÜCKIGER (1891) können durch Pressen bis 13%, durch Ätherextraktion zwischen 33 und 50% vom Trockengewicht an Fett aus der Droge erhalten werden. Das Mutterkornöl hat übrigens in mehrfacher Hinsicht praktische Bedeutung: weil es beim Mahlprozeß die Walzenstühle verschmiert, so daß die Mühlen nicht nur wegen der Giftigkeit des Mutterkorns an seiner sorgfältigen Ausreuterung aus dem Roggen interessiert waren, weil eine Koppelung zwischen dem Ranzigwerden des Öls und der Alkaloidzersetzung in der Droge vermutet wurde, und schließlich, weil man im Mutterkornöl eine Zeitlang nach den Wirkstoffen der Droge fahndete (A. TSCHIRCH 1923.)

VAUQUELIN (1816) konnte durch Extraktion aus dem Mutterkorn eine „matière huileuse..., assez abondante" darstellen, wie sie nach seinen Angaben schon früher BUCQUET und CORNET durch Auspressen erhalten hatten. F. L. WINCKLER (1827) findet in den Sklerotien 23,3% Fett. PETTENKOFER (1817) erhielt aus dem Äthanolauszug eine „fettige Substanz". H. A. L. WIGGERS gibt in seiner Göttinger Preisschrift (1832) zunächst ein Sammelreferat über das Mutterkorn, das der Redakteur (J. v. LIEBIG) in seinen Annalen veröffentlicht, „weil über das Mutterkorn so vieles geschrieben wurde, wodurch die Geschichte desselben ebenso interessant wie verwickelt geworden ist". WIGGERS selbst konnte mit Äther aus dem Mutterkorn 35,006 % eines Öles ausziehen, das er „vergeblich versuchte, mit kaustischem Kali zu verbinden", das aber mit „Ätzammonflüssigkeit" ein „Liniment" lieferte. Aus dem Öl schieden sich farblose, in siedendem Äthanol leichtlösliche Blättchen mit den Eigenschaften des „Cerins" aus (Ergosterin ?). H. LUDWIG (1863) erhielt aus 16 Unzen gepulverten Mutterkorns durch Ätherextraktion etwa 4 Unzen ($\sim$ 25%) fettes Öl, das *keine* freien Säuren enthielt und das im Gegensatz zu den Angaben von WIGGERS mit NaOH eine „harte Seife" gab. Der gleiche Forscher berichtet 1869, daß er aus dem fetten Öl von 3 kg Mutterkorn durch Kristallisation mit Äther und Alkohol etwa 0,25 g „Cholesterin" in Form „perlglänzender, schuppiger Kristalle" erhalten konnte, wovon eine Portion (0,12 g) bei 141°, eine andere (0,08 g) bei 160° schmolz. Er hatte offenbar bereits unreine Präparate von Ergosterin (Fp. 148°) und Fungisterin (Fp. 165°) in Händen! (s. auch C. TANRET 1889, 1920). O. FICINUS (1873) gibt wiederum 30 Gew.-% eines zähen Öles an, welches leicht „ranzig" wird. Die Ergebnisse weiterer Untersucher bringen bereits detaillierte Daten über die Ölzusammensetzung: J. C. HERMANN (1869), J. B. GANSER (1870), H. ZEEH (1894), J. A. MJOEN (1896), A. RATHJE (1908), H. DIETERLE, H. DIESTER und T. THIMANN (1927), W. F. BAUGHMAN und G. S. JAMIESON (1928), H. VANDERMEULEN (1939).

2. Das Fett des Mutterkorns und seine Begleitstoffe.

Tabelle 12 gibt eine Übersicht über die bisher vorliegenden Analysen des Öles von *Secale cornutum*. Als übereinstimmender Zug der Angaben der verschiedenen Untersucher verdient zunächst der außerordentlich niedrige Gehalt

Tabelle 13. *Zusammensetzung des Mutternkornfettes.*

	J. C. Hermann (1869)	J. B. Ganser (1870)	H. Zeeh (1894)	J. A. Mjoen (1896)	A. Rathje (1908)	H. Dieterle, H. Diester, T. Thiemanns (1927)	W. F. Baughman u. G. S. Jamieson (1928)	H. Vandermeulen (1939)
	I	II	III	IV	V	VI	VII	VIII
Rohfett, % des Trockengewichts	33	30	23	?	?	?	?	?
Art der Gewinnung	Äther-extrakt	Äther-extrakt		Petrol-Äther-extrakt		gepreßt		
Spezifisches Gewicht	(18°) 0,92496	(15°) 0,916	—	(15°) 0,9254	0,9250	(16°) 0,9259	—	—
n_D	—	—	—	—	1,4685	1,5420	—	—
Verseifungszahl	—	—	—	178,4	179,3	193,3	—	—
Säurezahl	—	±0	—	4,94	11,38	13,19	—	—
Jodzahl	—	—	—	71,8	74,0	66,0—66,9	—	—
Hehner-Zahl	—	—	—	96,3	96,25	96,13	—	—
Reichert-Meissl-Zahl	—	—	—	0,20	0,63	0,47	—	—
Acetylzahl	—	—	—	62,9	27,44	60,4 ?	—	—
Gesamte Fettsäuren								
C_2	wenig +	—	—	—	—	—	—	0,1**
C_4	wenig +	—	+	—	—	—	—	—
C_6	—	—	—	—	—	—	—	0,1**
C_{14}	—	—	14	—	—	—	0,3*	—
C_{16}	22,70*	—	—	+	5*	+ } 20,48[2]	21,5*	30**
C_{17}	—	—	—	—	—	+ }	0,0!	—
C_{18}	—						5,3*	12**
C_{20}	—					—	0,7*	—
Ölsäure	69,21*		+ } 74*	+	68*	70,99*	62,5*	23**
Erucasäure	—	—	+ }	—	—	—	—	—
α-Linolsäure	—	—	—	—	—	4,58*	8,7*	Sp
Ricinolsäure	—	—	—	+[1]	22	0,0!	0,0!	34**
Unverseifbares	—	0,12% „Cholesterin", Harz	+ „Cholesterin"	+ „Cholesterin"	0,35% dazu 0,6% Alkaloide	2,9%, 0,25% höhere Kohlenwasserstoffe, 0,66% feste Phytosterine, 0,67% höhere Fettalkohole	1,2%	1% (Ergosterin, Sterin A und C, ätherisches Öl, Harz)

IV [1] „Noch nicht isolierte Oxyfettsäure". Fettsäuren aus der Verseifung: Fp. 39,5—42°, Mittl. M.G. 306,8, Jodzahl 75,09, Acetylzahl 75,1.
V Fettsäuren aus der Verseifung: Neutralisationszahl 182,9, Mittl. M.G. 307, Jodzahl 77,2; flüssige Säuren: Jodzahl 82,12, Acetylzahl 40,5.
VI [2] „Palmitinsäure, *Daturinsäure* u. a. feste Fettsäuren". Nachweis von Oxyfettsäuren nicht gelungen.
* Prozent der Fettsäuren.
** Prozent der Glycerid-Fettsäuren.

des Mutterkornfettes an freien Fettsäuren Beachtung. Er fällt besonders auf, wenn man die Daten für die Fette der Schimmelpilze (z. B. Tabelle 9) oder der Fruchtkörper höherer Pilze (Tabelle 14) zum Vergleich heranzieht. Bereits J. ZELLNER (1906) hat festgestellt, daß das Fett des Mutterkorns bei der Lagerung *nicht* gespalten wird. Auch J. SCHINDELMEISER (1909) fand im Pulver von *Secale cornutum* eine nur geringfügige lipatische Aktivität gegenüber Olivenöl.

Daneben verdienen aber einige merkwürdige Unstimmigkeiten zwischen den Ergebnissen der verschiedenen Analytiker unsere Aufmerksamkeit, wofür die fortschreitende Entwicklung der Untersuchungsmethodik nicht als einzige Erklärung dienen kann. Das betrifft insbesondere das Vorkommen einer Oxyfettsäure (Ricinolsäure), welches von einigen Autoren entschieden und mit quantitativen Angaben behauptet, von anderen aber ebenso entschieden in Abrede gestellt wird (vgl. z. B. VII und VIII in Tabelle 13)[1]. Es erhebt sich die Frage, ob nicht auch bei der Zusammensetzung des Fettes des Mutterkornöls rassenspezifische Unterschiede und Klimaeinflüsse eine Rolle spielen könnten (wegen der letzteren vgl. z. B. A. ZELLER, dieser Handbuchband, S. 284ff., M. L. MEARA, ebenda S. 29, M. STEINER, ebenda S. 240ff.).

Die „Daturinsäure“ von H. DIETERLE, H. DIESTER und T. THIMANN (1927) (vgl. Tabelle 13, VI) darf wohl endgültig gestrichen werden. Alle Angaben der Originalarbeit (Schmelzpunkt der Methyl- und Äthylester sowie der Hydroxamsäuren) sprechen für *Palmitin*säure.

Daß *Phosphatide* im Mutterkornfett vorhanden sind, ist durchaus anzunehmen. Direkte Angaben darüber fehlen. Nur H. ZEEH (1894) erwähnt unter den wasserlöslichen Verseifungsprodukten ausdrücklich Cholin. Es darf aber vermutet werden, daß der Phosphatidanteil gering ist.

Etwas besser sind wir über die *Sterine* des Mutterkorns unterrichtet. C. TANRET hat darin zuerst das *Ergosterin* (1889) und später, als zweites Mycosterin, das *Fungisterin* entdeckt (1920). Daß damit die Reihe der Mutterkornsterine noch nicht erschöpft war, zeigten die Untersuchungen von M. CHART und F. W. HEYL (1930), die bei 0,13% Gesamtsterin neben Ergosterin und Fungisterin noch ein drittes Sterin (Fp. 120—125°, $\alpha_{[D]} = -2°$) finden; H. VANDERMEULEN (1939) gibt neben Ergosterin 2 Sterine A und C an, deren letzteres E. RUPPOL (1949) als das Dihydroergosterin (H. D. R. BARTON und J. D. COX 1948) erkannte. H. WIELAND und G. COUTELLE (1941) gelang es, das *Cerevisterin* im Mutterkorn nachzuweisen. Gleichzeitig stellten sie die Anwesenheit des Triterpenkohlenwasserstoffs *Squalen* ($C_{30}H_{50}$) (30 mg-%) fest.

VI. Fruchtkörper höherer Pilze.

Die Fruchtkörper höherer Pilze aus den Klassen der *Ascomycetes (Plectascales, Pezizales, Tuberales)* und der *Basidiomycetes (Hymenomycetales, Gastromycetales, Tremellales, Auriculariales)* haben eine ziemlich extensive chemische Bearbeitung erfahren. Die allerersten Arbeiten französischer Forscher (BOUILLON-LAGRANGE 1801a, b, H. BRACONNOT 1811a, b, 1813, 1838, VAUQUELIN 1813) gingen von vergleichend biochemischen Fragestellungen ihrer Zeit aus; die eigentümliche Stellung der Pilze zwischen den Reichen der Tiere und der „typischen“ Pflanzen reizte zu chemischen Untersuchungen. Die Feststellung einer „matière animale“ dürfte offenbar als ein besonders wesentliches Ergebnis angesehen worden sein. Stets wurden auch lipoide Substanzen gefunden, für deren Differenzierung im einzelnen („cire“, „adipocire“, „matière grasse“, „matière huileuse“) heute kaum mehr eine Deutung möglich sein dürfte.

Zahlreiche spätere Untersuchungen gehen von lebensmittelchemischen Gesichtspunkten aus. Zur Orientierung über den Nährwert des Pilzfruchtkörpers wurden die üblichen Pausch-

[1] Hierzu auch M. L. MEARA, dieser Band S. 19.

analysen (Wassergehalt, Rohprotein, Rohfett, Rohfaser, Extraktivstoffe, Asche) ausgeführt: M. J. LEFORT (1856), M. GOBLEY (1856), O. KOHLRAUSCH (1867), O. SIEGEL (1870), A. v. LÖSECKE (1876), TH. BISSINGER (1883), J. SCHMIEDER (1886), F. STOHMER (1887), K. FRITSCH (1889), E. GÉRARD (1890), A. ZEGA (1900, 1902), J. HOFMANN (1901), A. R. CHIAPELLA (1907), auch H. KORDES (1923).

A. v. LÖSECKE (1876) untersuchte z. B. 24 Arten von Pilzen. Der Gehalt an Petrolätherlöslichem lag im Mittel bei etwa 3% Trockengewicht, mit recht beträchtlichen Schwankungen von Art zu Art, z. B. 0,81% bei *Fistulina hepatica*, 1,38% bei *Cantharellus cibarius*, aber 5,21% bei *Armillaria mellea* und sogar 9,60% bei *Polyporus ovinus*. In der gleichen Größenordnung liegen auch die Zahlenangaben der meisten anderen Untersucher. Nur die ältesten Arbeiten geben zumeist (methodisch bedingte?) niedrigere Fettgehalte an, z. B. M. GOBLEY (1856) nur 0,25% vom Trockengewicht für den Champignon. Nur in wenigen dieser Veröffentlichungen finden sich Angaben über die *Zusammensetzung* des Pilzfettes.

K. MARGEWICZ (1883) hebt hervor, daß das Hymenium von Basidiomyceten mehr Fett enthält als die sterilen Teile des Hutes: 5,81% bzw. 4,07% vom Trockengewicht bei *Boletus scaber*, 7,97 bzw. 5,82% bei *B. edulis*, 8,53 bzw. 4,79% bei *B. aurantiacus*.

Als Beiträge zur vergleichenden Pflanzenchemie hat J. ZELLNER seine Pilzanalysen durchgeführt, über welche er, zum Teil gemeinsam mit Mitarbeitern, in 27 Mitteilungen (1904—1935) berichtet hat. Die durch Extraktion gewonnenen Fette werden in vielen Fällen durch Kennzahlen charakterisiert. Häufig finden sich Angaben über die mengenmäßig hervortretenden Fettsäuren, über Sterine und andere Bestandteile des Unverseifbaren usw. Weiter unten sind einige dieser Ergebnisse zusammengestellt.

Eine halbwegs vollständige Analyse liegt bisher über kein Fett aus höheren Pilzen vor. Das hängt zweifellos mit der Schwierigkeit zusammen, ausreichende Mengen von Untersuchungsmaterial zu beschaffen. Die einzige Ausnahme bildet das Öl des Sklerotiums von *Claviceps purpurea*, des Mutterkorns, worüber im vorangehenden Abschnitt (S. 74ff.) berichtet wurde.

Mengen und Zusammensetzung der Fette von Fruchtkörpern höherer Pilze.

In Tabelle 14 finden sich einige Angaben über Menge und Art der Fette von Pilzfruchtkörpern nach den Analysen von J. ZELLNER und Mitarbeitern. Trotz der Lückenhaftigkeit des Materials scheint es möglich, einige allgemeine Gesichtspunkte herauszuheben. Der Vergleich der Verseifungs- und Säurezahlen zeigt zunächst, daß fast alle Fette höherer Pilze einen *beträchtlichen Teil der Fettsäuren in freier Form* enthalten. Ähnliches hatten auch schon frühere Untersucher festgestellt: z. B. E. OPITZ (1891) bei *Amanita pantherina* und *Boletus luridus*, 50 bzw. 62,3% freie Fettsäuren, E. GÉRARD bei *Lactariu spiperatus*, F. STOHMER (1887) bei *Boletus edulis*. Weitere Beispiele (48—78% der gesamten Fettsäuren in freier Form) bei J. ZELLNER (1906b). M. GOBLEY (1856) hatte allerdings in frischen Champignons „keine freien Fettsäuren" gefunden. J. ZELLNER hat sich die Frage gestellt, ob nicht während des Trocknens und Lagerns des Pilzmaterials eine Fettverseifung durch pilzeigene Lipasen zustande kommt, die er selbst in zahlreichen Pilzfruchtkörpern in hoher Aktivität nachweisen konnte (J. ZELLNER 1905, 1906a, b). Die nachfolgende Zahlenreihe (Tabelle 15) zeigt (nach J. ZELLNER 1905), daß sich in der Tat mit der Zeit die Säurezahl des Fliegenpilzfettes erhöht, daß aber auch in jüngeren Fruchtkörpern bereits freie Fettsäuren

Tabelle 14. *Menge, Kennzahlen und Zusammensetzung der Fette*

Art	Rohfett % Trockengewicht	Verseifungszahl	Säurezahl	Jodzahl	REICHERT-MEISSL-Zahl
Calocera viscosa	—	—	—	—	—
Scleroderma vulgare	2	144,3	68,9	—	—
Sarcodon (Hydnum) imbricatum .	—	170,07	106,51	92,52	—
Calodon (Hydnum) ferrugineum .	~5,0	188,97	104,28	—	—
Polyporus sulphureus	—	173,1	92,0	58,4	—
Polyporus betulinus	~3,5	155,0	96,3	98,6	—
Polyporus applanatus (= *Ganoderma appl.*)	—	137,5	54,58	—	—
Polyporus suaveolens (= *Trametes suav.*)	0,8	173,6	30,85	—	—
Boletinus cavipes (= *Boletus cav.*)	—	—	—	—	—
Boletus satanas	3,5	202	102	—	—
Lactarius piperatus	5,9	200,2	121,3	—	—
Lactarius scrobiculatus.	11—12	181,9	120,8	—	—
Armillaria mellea	einige Prozent	179,6	89,1	94,2	—
Marasmius scorodonis	—	—	—	—	—
Amanita muscari	6	227	177	82	4,4
Nematoloma fasciculare (= *Hypholoma fasc.*).	2,4	150,8	112,5	—	—
Inoloma alboviolaceus	—	128	123	—	—
Pholiota sqarrosa	~3,8	168,3	51,3	—	—

Tabelle 15. *Säurezahlen des Fettes von Amanita muscaria.* (Nach J. ZELLNER 1905.)

	Säurezahl
Ganz junge, frische Pilze	38,22
Ältere, frische Pilze	60,61
Pilze, 4 Wochen trocken gelagert	69,46
Pilze, 2 Monate gelagert.	125,20
Pilze, 4 Monate gelagert.	177,00
Pilze, 12 Monate gelagert	180,00

vorliegen, deren Menge dann mit steigendem Alter zunimmt. Es dürfte also schon intra vitam eine starke Fettspaltung einsetzen, die sich autolytisch nach dem Absterben fortsetzt. Daß indes auch die Sporen, also Dauerzellen, von höheren Pilzen viel freie Fettsäuren enthalten, möge die folgende Tabelle 16 zeigen, die zugleich einige Anhaltspunkte über den Fettgehalt der Sporen bringt. Alle drei durch ZELLNER untersuchten Arten von Brandsporen enthalten übrigens kräftig wirksame Lipasen.

von höheren Pilzen. (Nach J. ZELLNER und Mitarbeiter.)

Fettsäuren	Phosphatide	Unverseifbares	Literatur
i-Valeriansäure, Palmitinsäure, Stearinsäure, viel Ölsäure	—	Ergosterin, Fungisterin, „Calocerol"	N. FRÖSCHL und J. ZELLNER (1928)
—	+ (PO_4)	27,26% (Ergosterin, Harz)	J. ZELLNER (1918)
—	—	9,04% (Ergosterin)	J. ZELLNER (1915)
—	+ (PO_4, Cholin)	15,25% (Ergosterin)	J. ZELLNER (1915)
Stearinsäure, Palmitinsäure, Ölsäure	+ (PO_4, Cholin)	25,16% (Ergosterin, Fungisterin)	J. ZELLNER und E. ZIKMUNDA (1930)
—	—	17,5% (Ergosteringemisch, Cerebrin, Harz, „Polyporol")	J. ZELLNER (1913)
—	—	32,26% (Sterine)	J. ZELLNER (1915)
gesättigte und ungesättigte Fettsäuren (Ölsäure)	—	2 Sterine der Ergosteringruppe	J. ZELLNER (1908a)
nur flüssige Fettsäuren (Ölsäure, wenig Linolsäure)	—	Ergosterin, Fungisterin	N. FRÖSCHL und J. ZELLNER (1928)
Palmitinsäure, Ölsäure, etwas Linolsäure	+ (PO_4)	7,4% (Ergosterin, Cerebrin)	L. BARD und J. ZELLNER (1923)
Stearinsäure, keine „Lactarsäure" (C_{15})	+ (PO_4)	—	J. ZELLNER (1913)
viel feste Fettsäuren (Stearinsäure)	+ (PO_4, Cholin)	Ergosterin	J. ZELLNER (1915)
—		4,5% (Ergosteringemisch)	J. ZELLNER (1913)
Palmitinsäure, Stearinsäure, feste Fettsäuren $C_{>18}$, flüssige Fettsäuren		Ergosterin, Fungisterin, Cerebrin, Paraffin (Fp. 55°)	N. FRÖSCHL und J. ZELLNER (1928)
Palmitinsäure, Ölsäure, keine Linolensäure	7,42% (ber. aus P)	wenig (Ergosterin, Cerebrin)	W. HEINISCH und J. ZELLNER (1904), J. ZELLNER (1911a)
meist flüssige Fettsäuren	+ (PO_4, Cholin)	25% (Ergosterin, Cerebrin)	J. ZELLNER (1911b)
Palmitinsäure, Ölsäure	+ (PO_4)	ergosterinartige Stoffe	L. BARD und J. ZELLNER (1923)
—	—	12,9% (Ergosterine)	J. ZELLNER (1913)

Tabelle 16. *Fette von Sporen höherer Pilze.*

Sporen von	Fett % Trockengewicht	Verseifungszahl	Säurezahl	Literatur
Coprinus atramentarius . . .	0,25			H. BRACONNOT (1938)
Elaphomyces cervinus	0,23			G. ISSOGLIO (1917)
Elaphomyces hirtus	1,6			H. LUDWIG (1869b)
Ustilago maydis	1,4	166,8 (196,2)[1]	84,0 (98,8)[1]	J. ZELLNER (1910)
Tilletia levis	1,5	182,0	106,3	J. ZELLNER (1911c)
Tilletia tritici	1,5	184,8	104,6	J. ZELLNER (1911c)

[1] Auf Reinfett berechnet.

Hinsichtlich der *Art der Fettsäuren* scheinen die Fette der höheren Pilze nichts besonderes zu bieten: Palmitin-, Stearin- und Ölsäure werden am häufigsten angegeben. Die Lactarsäure ($C_{15}H_{30}O_2$), welche W. THÖRNER (1879) für *Russula integra („Agaricus integer")*, TH. BISSINGER (1883) und R. CHODAT und PH. CHUIT (1889) für *Lactarius piperatus* angegeben haben, ist zu streichen. Es handelt sich nach J. BOUGAULT und C. CHARAUX (1912) und J. ZELLNER (1920) um Stearinsäure. Dagegen scheint die von J. BOUGAULT und C. CHARAUX (1910a, b, 1911, 1912) gefundene und von J. ZELLNER (1920) bestätigte Lactarsäure (ε-Ketostearinsäure) als Milchsaftbestandteil in vielen, aber nicht allen *Lactarius*arten vorzukommen[1].

Phosphatide sind in vielen Fällen wenigstens indirekt (als Lipoid-P) nachgewiesen. Man wird annehmen dürfen, daß sie ganz allgemein vorhanden sind.

Auffallend hoch ist in vielen Fällen der unverseifbare Anteil des Rohfettes von höheren Pilzen. Daß hierbei *Sterine* eine wichtige Rolle spielen, ist sicher. R. BOEHM (1885) scheint als erster bei *Boletus luridus* und *Amanita pantherina* „cholesterinähnliche Substanzen" angegeben zu haben. Durchweg dürfte das *Ergosterin* das Hauptsterin darstellen (E. GÉRARD 1891, E. OPITZ 1891). Als Begleiter ist das *Fungisterin* sehr häufig (vgl. z. B. A. GORIS und M. MASCRÉ 1911, M. T. ELLIS 1918). *Cerevisterin* haben H. WIELAND und W. COUTELLE (1941) neben viel Ergosterin und wenig Fungisterin im Knollenblätterpilz *(Amanita phalloides)* gefunden. Das *Mycosterin*, welches T. IKEGUCHI (1914, 1919) aus *Collybia shiitake* angibt, ist nach den Konstanten ($\alpha_D^{20} = -129{,}4$, Fp. 159—160°) doch wohl nur Ergosterin.

Unter den Triterpenen ist vor allem *Squalen* ($C_{30}H_{50}$) zu nennen, welches H. WIELAND und G. COUTELLE (1941) für den Knollenblätterpilz *(Amanita phalloides)* angeben (18 mg-%). In voller Entwicklung begriffen ist zur Zeit die Erforschung der *tetra-* und *pentacyclischen Triterpene* aus höheren Pilzen.

Es seien genannt:

Polyporensäure A	$C_{30}H_{46}O_3$	*Polyporus betulinus*
Polyporensäure C	$C_{31}H_{46}O_4$	
Eburicosäure	$C_{31}H_{50}O_3$	*Fomes officinalis* *Polyporus anthracophilus* *P. eucalyptorum, P. sulphureus* *Lentinus dactyloides*
Dihydroeburicosäure	$C_{31}H_{48}O_3$	
Tumulosussäure	$C_{31}H_{50}O_4$	*Polyporus tumulosus, P. australiensis, P. betulinus, Poria cocos*
Pinicolsäure A	$C_{30}H_{46}O_3$	*Polyporus pinicola*
Trametenolsäure A	$C_{30}H_{48}O_3 \pm CH_2$	*Trametes odorata*

Im übrigen sei auf die zusammenfassenden Darstellungen bei O. JEGER (1950), M. STEINER und H. HOLTZEM (1955) und E. R. H. JONES und T. G. HALSALL (1955) verwiesen.

Die Pilzcerebrine werden weiter unten (S. 81 ff.) im Zusammenhang kurz behandelt.

[1] Nach J. ZELLNER (1920) enthält der *Milchsaft* von *Lactarius vellereus* FR. 19,5% Trockensubstanz, von der rund drei Viertel in Äther löslich sind. Aus der Ätherlösung fallen sofort Kristalle von reiner Stearinsäure. Neutralfette und Ergosterin fehlen. *Stearin*säure scheint für die Milchsäfte folgender *Lactarius*arten kennzeichnend zu sein: *azonites* Bull., *controversus* Pers., *vellereus* Fr., *piperatus* L., *subdulcis* Bull. (dunkle Varietät), *torminosus* Schoeff., *scrobiculatus* Scop., *deliciosus* L.; *Lactarin*säure (=Ketostearinsäure) für *L. uvidus* Fr., *theiogalus* Bull., *lilacinus* Lasch, *subdulcis* Bull. (blasse Varietät), *plumbeus* Bull., *rufus* Bull., *pallidus* Pers.

VII. Lichenisierte Pilze (Flechten).

Allgemein gesehen, gehören die Flechten zu den chemisch am besten studierten Gruppen des Kryptogamenreiches. Die meisten Untersuchungen beschränken sich aber so gut wie ausschließlich auf die charakteristischen „Flechtenstoffe" (W. ZOPF 1907, Y. ASAHINA und S. SHIBATA 1954). Ältere Untersuchungen des Nahrungswertes einiger Flechten vermitteln einige Daten über den Rohfettgehalt: *Cetraria islandica* 0,40%, *Cetraria nivalis* 0,99% (B. HANSTEEN 1906[1]), *Cetraria islandica* 4,30%, *Cladonia rangiferina* 2,59% Rohfett in Trockengewicht (E. SALKOWSKI 1919). Erst J. ZELLNER und Mitarbeiter bringen einige Analysen, in denen auch die *Zusammensetzung* von Flechtenfetten berücksichtigt wird:

Peltigera canina: fette und flüssige Fettsäuren; Ergosterin (J. ZELLNER 1932).

Alectoria ochroleuca: Ölsäure, Linolsäure, wenig gesättigte Fettsäuren; Sterine, Kohlenwasserstoff $C_{30}H_{62}$ (J. KLIMA 1933).

Parmelia physodes: Rohfett 1,3% vom Trockengewicht, flüssige und feste Fettsäuren; Ergosterin, ein Paraffin (J. ZELLNER 1934).

Parmelia furfuracea: flüssige und feste Fettsäuren; Ergosterin, Fungisterin (J. ZELLNER 1935).

Umbilicaria (Gyrophora) Dillenii: Ergosterin (J. ZELLNER 1935).

Ergosterin war übrigens schon von E. GÉRARD (1895) in *Lobaria pulmonaria* festgestellt worden.

Mit wenigen Worten muß an dieser Stelle der Fette gedacht werden, die in ansehnlicher Menge in endolithischen Hyphen von Kalkkrustenflechten, nicht aber bei silikatbewohnenden Arten vorkommen. Sie finden sich in kugelig aufgetriebenen „Sphäroidzellen". H. ZUKAL (1886), J. M. HULTH (1891) und E. BACHMANN (1890, 1892) haben darüber zuerst berichtet. H. ZUKAL (1886) wies bereits durch Färbbarkeit, Löslichkeit und durch die mikrochemische Verseifung nach, daß es sich um echte Fette handelt. Er betrachtet sie als Reservestoffe, die letzten Endes natürlich auf Photosyntheseprodukte der Gonidienalgen zurückgehen. Dem widersprach sehr energisch M. FÜNFSTÜCK (1895, 1896, 1899). Er weist nicht nur darauf hin, daß die Fetteinschlüsse vor allem in gonidienfreien Partien des Flechtenkörpers vorkommen, sondern glaubt auch durch Kulturversuche bewiesen zu haben, daß sie sich auch im Dunkeln bilden, ebenso in Flechten, deren Gonidienschicht operativ entfernt wurde. Er hält es „für im höchsten Grade wahrscheinlich, daß die durch die Zersetzung der kohlensauren Salze von seiten der Flechtensäuren frei werdende Kohlensäure das Ausgangsmaterial für die Ölbildung", und zwar durch die Pilzhyphen(!) darstellt. Es dürfte sich erübrigen, diese offensichtliche Fehlinterpretation ausführlich zu widerlegen.

VIII. Zur Frage der Pilzcerebrine.

Bei der Besprechung der Pilzlipoide wurden mehrfach die Cerebrine erwähnt, so bei der Hefe (S. 68), bei den Schimmelpilzen (S. 73), bei den Fruchtkörpern der höheren Pilze (S. 80). Da es sich hierbei um lipoide Substanzen handelt, die im gesamten Pflanzenreich bisher anscheinend nur bei den Pilzen gefunden wurden, dürfte im Rahmen dieses Beitrages ein kurzes Eingehen darauf berechtigt sein.

[1] Die Analysen HANSTEENS wurden mit entbittertem Material durchgeführt. Vielleicht erklärt sich damit der auffallend niedrige Fettgehalt, der bei *Cetraria islandica* gefunden wurde.

Tabelle 17. *Spaltprodukte einiger Pilzcerebrine.*

Pilz	Base	Säure	Autor
Hefe	$CH_3(CH_2)_{12}CH{=}CH \cdot CH(OH) \cdot CH(NH_2) \cdot CH_2OH$ (Sphingosin)	α-Oxy-n-octakosansäure	E. RUPPOL (1937b)
Hefe	$CH_3(CH_2)_{14}CH(OH) \cdot CH(NH_2) \cdot CH(OH) \cdot CH_2 \cdot CH_2OH$	α-Oxy-n-hexakosansäure	F. REINDEL und Mitarbeiter (1940)
Hefe	—	α-Oxy-n-hexakosansäure und (<10%) α-Oxy-n-tetrakosansäure (Cerebronsäure)	A. C. CHIBNALL, S. H. PIPER und E. F. WILLIAMS (1953)
Aspergillus citromyces 2 Cerebrine a) Fp 108—110° b) Fp 147—149°	Sphingosin (s. oben)	α-Oxy-n-tetrakosansäure	E. RUPPOL (1944)
Penicillium sp. a) „Cerebrin" b) „Oxycerebrin" c) „Desoxycerebrin"	$CH_3(CH_2)_{13}CH(OH) \cdot CH(OH) \cdot CH(NH_2) \cdot CH_2OH$	α-Oxy-n-tetrakosansäure α,β-Dioxy-n-tetrakosansäure n-Tetrakosansäure (Lignocerinsäure)	T. ODA (1952a, b, c)

Zum ersten Male erwähnen A. BAMBERGER und A. LANDSIEDI (1905) für *Calvatia maxima* PERS (= *Lycoperdon bovista* L.) neben „2 Ergosterinen" eine „zur Gruppe der Cerebroside" gehörige Substanz. J. ZELLNER (1911b) fand den gleichen oder einen ähnlichen Stoff wieder, als er den Fliegenpilz (*Amanita muscaria* L.) untersuchte, ebenso in *Hypholoma fasciculare* (J. ZELLNER 1911a), später in *Boletus satanas* (L. BARD und J. ZELLNER 1923), in *Marasmius scorodonius* (N. FRÖSCHL und J. ZELLNER 1928), in *Polyporus pinicola* (E. HARTMANN und J. ZELLNER 1928). R. ROSENTHAL (1922), ein Schüler ZELLNERs nahm sich der „Cerebrine" an *Amanita muscaria* und *Hypholoma fasciculare* näher an, er stellt an kristallisierter Substanz die Identität beider fest und bestätigte die Angaben ZELLNERs daß ein N-haltiges Lipoid vorliegt, welches aber im Gegensatz zu den Cerebrosiden des Tierkörpers keine Zuckerkomponent enthält. F. REINDEL (1930) begegnete einem Cerebrin in de Hefe, als er das Ergosterin und seine Begleitsterine studierte. Er kennzeichnete die Reinsubstanz und konnte durch Spaltung mit methanolischer HCl α-Oxyhexakosansäure und eine Base $C_{17}H_{35}NO_2$ erhalten. Später wir von F. REINDEL, A. WEICKMANN S. PICARD, K. LUBER und I TURULA (1940) $C_{20}H_{41}NO_2$ a Formel der Spaltbase angegeben Ehe auf die Konstitution einge gangen wird, sei zunächst d Liste der bisher festgestellte Cerebrinvorkommen unter de Pilzen vervollständigt.

Aspergillus sydowi, Cerebri nur nach vorheriger Autolys des Mycels isolierbar (N. B HONOS und W. H. PETERSO 1943).

Aspergillus citromyces (E. RUPPOL 1943).
Penicillium sp. (Penicillin produzierende Art) (T. ODA 1952a, b, c).
Claviceps purpurea (Mutterkorn) (E. RUPPOL 1949).

Das Verhältnis von Cerebrosiden und Cerebrinen sei zunächst durch die folgende Formel klargestellt (vgl. das Sammelreferat von W. D. CELMER und H. E. CARTER 1952).

$$CH_3\cdot(CH_2)_{12}\cdot CH{=}CH\cdot \underset{OH}{CH}\cdot \underset{NH}{CH}OR_2$$

Sphingosin

$$\boxed{H \quad HO}\,\underset{\overset{\|}{O}}{C}\cdot R_1$$

R_1 = Fettsäurerest
R_2 = H — *Cerebrin*
= $C_6H_{11}O_5$ (Galaktose) — *Cerebrosid*

Durch die Angaben der verschiedenen Autoren über die beiden amidartig verknüpften Komponenten: die Base (Sphingosin o. a.) und die Fettsäure (R_2), sei in Tabelle 17 der gegenwärtige Stand der Forschung kurz gekennzeichnet.

Literatur.

ABE, Y.: Chemical composition of molds. II. Fatty material of *Penicillium chrysogenum*. Proc. Eng. Faculty Keiogizuku Univ. **2**, 15—20 (1949). Ref. Chem. Abstr. **1953**, 4949—4950.— Constitution of fatty oil of *Penicillium chrysogenum* Q 176. J. Oil. chem. Soc. Jap. **1**, 132 bis 135 (1952). Ref. Chem. Abstr. **47**, 2511 (1953). — ANGELETTI, A., and G. TAPPI: The biosynthesis of sterols by the action of some strains of *Penicillium*. Gazz. chim. ital. **77**, 112—114 (1947). Ref. Chem. Abstr. **41**, 7439 (1947). — ASAHINA, Y., and S. SHIBATA: Chemistry of lichen substances. Japan. Soc. for the Promotion of Science, Ueno, Tokyo, Japan 1954, 240 S. — ASO, K.: Die chemische Zusammensetzung der Sporen von *Aspergillus oryzae*. Bull. Coll. Agricult. (Tokyo) **4**, 81—96 (1900). Ref. Chem. Cbl. **1900 II**, 55. — AUSTIN, W. C.: Phospholipins in yeast. J. of Biol. Chem. **59**, 111 (1924).

BACHMANN, E.: Die Beziehungen der Kalkflechten zu ihrem Substrat. Ber. dtsch. bot. Ges. **8**, 141—145 (1890). — Der Thallus der Kalkflechten. Ber. dtsch. bot. Ges. **10**, 30—37 (1892). — BAMBERGER, A., u. A. LANDSIEDL: Beiträge zur Chemie der Sclerodermeen. Mh. Chem. **26**, 1109—1118 (1905). — BARBER, H. H.: The production of fat from carbohydrates and similar media by a species of *Penicillium*. Biochemic. J. **23**, 1158—1164 (1929). — BARD, L., u. J. ZELLNER: Zur Chemie der höheren Pilze. XVII. Mitt. Über *Amanita muscaria* L., *Inoloma alboviolaceum* Pers., *Boletus satanas* Lenz. und *Hydnum versipelle*. Mh. Chem. **44**, 9—17 (1923). — BARTON, D. H. R., and T. BRUUN: A new sterol from a strain of *Aspergillus niger*. J. Chem. Soc. (Lond.) **1951**, 2728—2733. — BARTON, D. H. R., and J. D. COX: The application of the method of molecular rotation differences to steroids. Part VI. New Sterol. J. Chem. Soc. (Lond.) **1948**, 1357—1362. — BARY, A. DE: Morphologie und Physiologie der Pilze, Flechten und Myxomyceten. Leipzig: Wilhelm Engelmann 1884. — BAUGHMAN, W. F., and G. S. JAMIESON: Chemical composition of ergot oil. Oil a. Fat Industr. **5**, 85—90 (1928). Ref. Chem. Abstr. **22**, 1865 (1928). — BAYER, J.: Zit. nach REINDEL, WALTER und RAUCH, Liebigs Ann. **452**, 34 (1927). — BERNHARD, K., u. H. ALBRECHT: Die Lipoide von *Phycomyces Blakesleeanus*. Helvet. chim. Acta **31**, 977—988 (1948). — BERNHAUER, K., u. G. POSSELT: Über Schimmelpilzlipoide. II. Mitt. Die Zusammensetzung eines *Aspergillus-niger*-Fettes. Biochem. Z. **294**, 215—220 (1937). — BIRKINSHAW, J. H., R. K. CALLOW and G. F. FISCHMANN: Studies in the biochemistry of microorganisms. XXII. The isolation and characterisation of ergosterol from *Penicillium puberulum* BAINIER grown on a synthetic medium containing glucose as sole source of carbon. Biochemic. J. **25**, 1977—1980 (1931). — BIRKINSHAW, J. H., and H. RAISTRICK: The biochemistry of microorganisms. XXXVIII. The metabolic products of *Penicillium miniluteum* DIERCKX minioluteic acid. Biochemic. J. **28**, 828—836 (1934). Ref. Chem. Abstr. **28**, 7248—7249 (1934). — BISSINGER, TH.: Über Bestandteile der Pilze *Lactarius piperatus* und *Elaphomyces granulatus*. Arch. Pharmaz. **221**, 321—344 (1883). — BLINC, M., u. M. BOJEC: Fett aus *Mucor mucedo*. Arch. Mikrobiol. **12**, 41—45 (1942). — BOEHM, R.: Beiträge zur Kenntnis der Hutpilze in chemischer und toxikologischer Beziehung. I. *Boletus luridus*. II. *Amanita pantherina*. Arch. exper. Path. u. Pharmakol. **19**, 60—100 (1885). — BOHONOS, N., and W. H. Peterson: Chemistry of mold tissue. XVI. The isolation of fungus cerebrin from the mycelium of *Aspergillus Sydowi*. J. of Biol. Chem. **149**, 295—300 (1943). Ref. Chem. Abstr. **37**, 5447 (1923). — Bougault, J., et C. Charaux: Sur l'acide lactarinique, acide cétostéarique retiré de quelques champignons du genre *Lactarius*. J. Pharmacie, VII. s. **4**, 332—343 (1910a). — Sur l'acide lactarinique, acide céto-stéarique retiré de quelques champignons du genre *Lactarius*. J. Pharmacie, VII. s. **4**, 489—494 (1910b). — Sur l'acide lactarinique, acide cétostéarique retiré de quelques champignons du genre *Lactarius*. C. r. Acad. Sci. Paris **153**, 572—573 (1911). — Acide lactarinique, acide lactarique et acide stéarinique dans les champignons. J. Pharmacie, VII. s. **5**, 65—71 (1912). — Bouillon-Lagrange: Examen chimique de la Truffe *Lycoperdon Tuber* Linn. Ann. de chim. **46**, 191—222 (1801). — Analyse de deux espèces d'Agaric, le *boletus larix* et le *boletus igniarius* (Lin.). Ann. de chim. **51**, 75—96 (1801). — Braconnot, H.: Recherches analytiques sur la nature des champignons. Ann. de chim. **79**, 265—304 (1811a). — Recherches analytiques sur la nature des champignons. Ann. de chim. **80**, 272—292 (1811b). — Nouvelles recherches analytiques sur les champignons, pour servir de suite à celles qui ont été insérées dans les tom. LXXIX et LXXX des Annales de chimie. Ann. de chim. **87**, 237—270 (1813). — Examen chimique des sporules de l'*Agaricus atramentarius*. Ann. de chim. (2) **69**, 434—444 (1838).

Campbell, P. N., D. H. Simmonds and T. S. Work: The occurrence of glycerylphosphorylethanolamine in extracts of liver and yeast. Biochemic. J. **49**, XVI—XVII (1951). — Castille, A., et E. Ruppol: Ergostérol et levures. Bull. Acad. Mèd. Belg., V. s. **13**, 48—60 (1933). — Cavallito, C. J.: Ergosterol from some species of *Penicillium*. Science (Lancaster, Pa.) **100**, 333 (1944). — Celmer, W. D., and H. E. Carter: Chemistry of phosphatides and cerebrosides. Physiologic. Rev. **32**, 166—196 (1952). — Chapman, A. C.: Spinacene and some of its derivatives. J. Chem. Soc. (Lond.) **113**, 458—460 (1917). — Spinacene a new hydrocarbon from certain fish liver oils. J. Chem. Soc. (Lond.) **1947**, 56—69. — Chiapella, A. R.: Über einen wenig bekannten eßbaren Pilz. Z. Unters. Nahrungsmitt. usw. **13**, 384—389 (1907). Ref. Chem. Zbl. **1907 I**, 457—458. — Chibnall, A. C., S. H. Piper and E. F. Williams: The α-hydroxy-n-fatty acids of yeast cerebrin. Biochemic. J. **55**, 711—714 (1953). — Chodat, R., et Ph. Chuit: Untersuchungen über *Lactarius piperatus*. Arch. Sci. physiques et natur., Genève **5**, 385—403 (1889). Ref. Chem. Zbl. **2**, 144—145 (1889). — Clautriau, G.: Les réserves hydrocarbonées de Thallophytes. Miscel. biol. stat. Zool. Wimereux, Paris **1899**, 114. Zit. nach Ch. Pontillon, Rév. gén. Bot. **45**, 21—52 (1933). — Clutterbuck, P. W., H. Raistrick and M. L. Rintoul: Biochemistry of microorganisms. XVI. Production from dextrine by *Penicillium spiculisporum*, Lehmann of a new polybasic fatty acid, $C_{17}H_{28}O_6$ (the lactone of γ-hydroxy-β-δ-dicarboxypentadecoic acid.). Trans. Roy. Soc. (Lond.) B **220**, 301—330 (1931). Ref. Chem. Abstr. **26**, 2486 (1932). — Cramer, E.: Die Zusammensetzung der Sporen von *Penicillium glaucum* und ihre Beziehung zu der Widerstandsfähigkeit derselben gegen äußere Einflüsse. Arch. f. Hyg. **20**, 197—210 (1894). — Czapek, F.: Biochemie der Pflanze, 2. Aufl., Bd. 1—3. Jena: Gustav Fischer 1913.

Damm, H.: Eine neue biochemische Fettsynthese. Chemiker-Ztg **1943**, 47—49. — Daubney, C. G., and J. Smedley-MacLean: The carbohydrate and fat metabolism of yeast IV. The nature of the phospholipins. Biochemic. J. **21**, 373—385 (1927). — Diemair, W. u. J. Koch: Beitrag zur Kenntnis der Hefephosphatide. Angew. Chem. A **60**, 155—157 (1948). — Diemair, W., u. W. Poetsch: Phosphatide als Fettbegleitstoffe der Nährhefen Biochem. Z. **319**, 571—591 (1949). — Dieterle, H., H. Diester u. T. Thiemann: Beitrag zur Kenntnis des fetten Öles von *Secale cornutum* und der in diesem Öl enthaltenen Daturinsäure. Arch. Pharmaz. **265**, 171—187 (1927). — Dirr, K., u. A. Ruppert: Über den Wert der Wuchshefen für die menschliche Ernährung. VI. Die Phosphatide der Wuchshefen Biochem. Z. **319**, 163—173 (1948). — Dirr, K., u. O. v. Soden: Über den Wert der Wuchshefen für die menschliche Ernährung. IV. Über die Lipide der Wuchshefen. Biochem. Z **312**, 263—275 (1942). — Dubaquié: Recherches sur les matières grasses des végétaux inférieurs. Thèse Sciences Bordeaux 1909, 88 p. Zit. nach Pontillon, 1933.

Ellis, M. T.: Contributions to our knowledge of the plant sterols. Part I. The sterol content of wheat *(Triticum sativum)*. Biochemic. J. **12**, 154—172 (1918). — Contribution to our knowledge of the plant sterols. Part II. The occurrence of phytosterol in some of the lower plants. Biochemic J. **12**, 173—177 (1918).

Ficinus, O.: Über den Fettgehalt des Mutterkorns. Arch. Pharmaz. **203**, 219 (1873). — Fink, H., H. Haehn u. W. Hoerburger: Über die Versuche zur Fettgewinnung mittels Mikroorganismen mit besonderer Berücksichtigung der Arbeiten des Instituts für Gärungsgewerbe. Chemiker-Ztg. **61**, 689—693, 723—726, 744—747 (1937). — Fink, H., G. Haeseler u. M. Schmidt: Zur Frage der Fettgewinnung mit Hilfe von Mikroorganismen. Über das Fettbildungsvermögen verschiedener Stämme von *Oidium lactis (Oospora lactis)*. Z. Spiritusindustr. **60**, 74—77, 82—83 (1937). — Fiore, J. V.: Mechanism of enzyme action XXXII. Fat and sterol in *Fusarium lini*, *F. lycopersici*, and *F. solani* D_2 purple. Arch. of Biochem. **16**, 161—168 (1948). — Flückiger, F. A.: Pharmakognosie des Pflanzenreiches

3. Aufl. Berlin: R. Gaertner 1891. — FRITSCH, K.: Beiträge zur chemischen Kenntnis einiger Basidiomyceten. Diss. Erlangen 1889. — Arch. Pharmaz. **227**, 193—222 (1889). — FRÖSCHL, N., u. J. ZELLNER: Zur Chemie der höheren Pilze. XX. Mitt. Über *Omphalaria campanella* BATSCH., *Marasmius scorodonius* FR., *Boletus cavipes* OPAT. und *Calocera viscosa* PERS. Mh. Chem. **50**, 201—210 (1928). — FÜNFSTÜCK, M.: Die Fettabscheidungen der Kalkflechten. Fünfstücks Beitr. wiss. Bot. **1**, 157—220 (1895). — Die Fettabscheidungen der Kalkflechten (Nachtrag). Fünfstücks Beitr. wiss. Bot. **1**, 316—321 (1896). — Weitere Untersuchungen über die Fettabscheidungen der Kalkflechten. Festschr. für SCHWENDENER, S. 341—356. Berlin: Gebrüder Bornträger 1899.

GANSER, J. B.: Untersuchung der Bestandteile des Mutterkorns, ihre chemischen und physikalischen Eigenschaften sowie ihre medizinischen Wirkungen; unter Berücksichtigung der alten Arbeiten sowohl, als auch der neueren Angaben von W. T. WENZELL über Ecbolin, Ergotin und Ergotsäure. Arch. Pharmaz. **194**, 195—212 (1870). — GÉRARD, E.: Sur les matières grasses de deux champignons appartenant à la famille des Hyménomycètes. J. Pharmacie, V. s. **21**, 408—414 (1890). — Sur les matières grasses de deux champignons appartenant à la famille des Hyménomycètes. J. Pharmacie, V. s. **23**, 7—12 (1891). — Sur les cholestérines des végétales. C. r. Acad. Sci. Paris **114**, 1544—1546 (1892a). — Cholestérines des Champignons. Bull. Soc. mycol. France 8, 169 (1892b). Zit. nach CH. PONTILLON, Rév. gén. Bot. **45**, 21—52 (1933). — Sur les cholestérines des Cryptogames. C. r. Acad. Sci. Paris **121**, 723—726 (1895). — Sur une lipase végétale extraits des *Penicillium glaucum*. C. r. Acad. Sci. Paris **124**, 370—371 (1897). — Sur les cholestérines des végétaux inférieurs. C. r. Acad. Sci Paris **126**, 909—911 (1909). — GÉRARD, E., et P. DAREXY: Untersuchungen über die Fettsubstanz der Bierhefe. J. Pharmacie (6) **5**, 275—280 (1897). Ref. Chem Zbl. **1897**, 1, 870. — GOBLEY, M.: Recherches chimiques sur les champignons vénéneux, premier mémoire, lu à l'Académie impériale de médecine. Analyse du champignon comestible des couches. J. Pharmacie, III. s. **24**, 81—91 (1856). — GOODWIN, T. W.: Carotinoids. In K. PAECH u. M. V. TRACEY, Moderne Methoden der Pflanzenanalyse, Bd. 3, S. 272—311. Berlin-Göttingen-Heidelberg: Springer 1955. — GORIS, A., et M. MASCRÉ: Sur la composition chimique de quelques champignons. C. r. Acad. Sci Paris **153**, 1082—1084 (1911). — GOUPIL, R.: Recherches sur les composés phosphorés formés par l'*Amylomyces rouxii*. C. r. Acad. Sci. Paris **156**, 959—962 (1913). — GUICHARD, P.: Composition et analyse de la levure. Bull. Soc. Chim. Paris, III. s. **11**, 230—239 (1894).

HAEHN, H., u. W. KINTTOF: Beitrag über den chemischen Mechanismus der Fettbildung aus Zucker. Chem. Zelle u. Gewebe **12**, 115—156 (1925). — HANAHAN, D. J., and M. E. JAYKO: The isolation of dipalmitoleoyl-L-a-glycerylphosphorylcholine from yeast. A new route to Dipalmytoyl-L-a-lecithin. J. Amer. Chem. Soc. **74**, 5070—5072 (1952). — HANSTEEN, B.: Nordische Flechten als Nahrungsmittel. Chemiker-Ztg **30**, 638 (1906). — HART, M. C., and F. W. HEYL: The sterols of ergot. J. Amer. Chem. Soc. **52**, 2013—2015 (1930). — HARTMANN, E., u. J. ZELLNER: Zur Chemie der höheren Pilze. XIX. Mitt. Über *Polyporus pinicola* FR. Mh. Chem. **50**, 193—200 (1928). — HAXO, F. T.: Some biochemical aspects of fungal carotenoids. Fortschr. Chem. organ. Naturstoffe **12**, 169—197 (1955). — HÉE, A.: Contribution à l'étude de la respiration chez les végétaux. Thèse sciences, Paris 1930, 244 p. Zit. nach CH. PONTILLON, Rév. gén. Bot. **45**, 21—52, (1933). — HEIDUSCHKA, A., u. P. LINDNER: Über den Ergosteringehalt der Hefe. Hoppe-Seylers Z. **181**, 15—23 (1929). — HEINISCH, W., u. J. ZELLNER: Zur Chemie des Fliegenpilzes (*Amanita muscaria* L.). Mh. Chem. **25**, 537—544 (1904). — HENNEBERG, W.: Studien über das Verhalten einiger Kulturheferassen bei verschiedenen Temperaturen. Ein Beitrag zur Enzymtätigkeit, zur Lebensdauer, Haltbarkeit und zum Absterben der Hefen. Z. Spiritusindustr. **27**, 96—97, 105—106, 116—117, 126—127, 135—137, 146—147, 160—161, 173, 182—183, 194—195, 205—206, 213—216, 226, 239—260 (1904). — HERMANN, J. C.: Beiträge zur chemischen Kenntnis des Mutterkorns. Diss. München 1869. 19 S. — HILDITCH, T. P., and R. K. SHRIVASTAVA: The component fatty acids of a yeast fat. Biochim. et Biophysica Acta **2**, 80—85 (1948). — HILDITCH, T. P., I. C. SIME, Y. A. H. ZAKY and M. L. MEARA: The component acids of various vegetable fats. J. Soc. Chem. Industr. **63**, 112—114 (1944). — HINSBERG, O., u. E. ROOS: Über einige Bestandteile der Hefe. Hoppe-Seylers Z. **38**, 1—15 (1903). — Nachtrag zur der Abhandlung über einige Bestandteile der Hefe. Hoppe-Seylers Z. **42**, 189—192 (1904). — HOFMANN, J.: Über die chemischen Bestandteile einiger Pilze. Diss. Univ. Zürich 1901, 84 S. — HOLMBERG, J.: Yeast lipids. I. Components acids of *Rhodotorula gracilis* fat. Svensk. kem. Tidskr, **60**, 14—20 (1948). — HOPPE-SEYLER, F.: Über Lecithin und Nuclein in der Bierhefe. Hoppe-Seylers Z. **2**, 427—429 (1879a). — Über Lecithin in der Hefe. Hoppe-Seylers Z. **3**, 374—380 (1879b). — HULTH, J. M.: Über Reservestoffbehälter von Flechten. Bot. Zbl. **45**, 209—210, 269—270 (1891).

IKEGUCHI, T.: Über die Pilzsterine. I. Über eine sterinähnliche Substanz aus *Lykoperdon gemmatum*. Hoppe-Seylers Z. **92**, 256—260 (1914). — A new sterol. J. of Biol. Chem. **40**,

175—182 (1919). — Issoglio, G.: Chemische Untersuchungen über *Elaphomyces hirtus*. Gazz. chim. ital. **47**, 31—48 (1917).

Jeger, O.: Über die Konstitution der Triterpene. Fortschr. Chem. organ. Naturstoffe **7**, 1—86 (1950). — Jones, E. R. H., u. T. G. Halsall: Tetracyclic triterpenes. Fortschr. Chem. organ. Naturstoffe **12**, 45—130 (1955).

Kappes, H. C.: Analyse der Massenkulturen einiger Spaltpilze und der Soorhefe. Diss. Leipzig 1890. 55 S. — Kaibara, I. T.: Die Fettproduktion von Schimmelpilzen. J. Agricult. Chem. Soc. Japan **22**, 89 (1948). Ref. Fette u. Seifen **54**, 706 (1952). — Kaufmann, H. P., u. O. Schmidt: Über die Zusammensetzung des Fettes von *Oidium lactis (Oospora lactis)*. Vorratspflege u. Lebensmittelforsch. **1**, 166—169 (1938). — Kiesel, A.: Untersuchungen über Protoplasma. Über die chemischen Bestandteile des Plasmodiums von *Reticularia lycoperdon*. Hoppe-Seylers Z. **150**, 149—176 (1925). — Untersuchungen über das Protoplasma. II. Über die chemischen Bestandteile des Plasmodiums von *Lycogala epidendron* und die Veränderung derselben während der Sporendifferenzierung. Hoppe-Seylers Z. **164**, 103—111 (1927). — Die Plasmodien der Myxomyceten als Objekt der chemischen Protoplasma-Untersuchungen. Protoplasma (Berl.) **6**, 332—369 (1929). — Klima, J.: Zur Chemie der Flechten. II. *Alectoria ochroleuca* Ehrh. Mh. Chem. **62**, 209—213 (1933). — Koch, W.: Die Lecithane und ihre Bedeutung für die lebende Zelle. Hoppe-Seylers Z. **37**, 161—188 (1902). — Kohlrausch, O.: Über die Zusammensetzung einiger eßbarer Pilze mit besonderer Berücksichtigung ihres Nahrungswertes. Diss. Göttingen 1867. 35 S. — Konrad, P.: Notes critiques sur quelques champignons du Jura. Bull. Soc. mycol. France, V. s. **47**, 129 (1931). Zit. nach Ch. Pontillon, Rév. gén. Bot. **45**, 21—52 (1933). — Kordes, H.: Biologische Untersuchungen über das in Dauerzellen und Hyphen verschiedener Pilze auftretende Fett. Bot. Archiv **3**, 282—311 (1923). — Kroeker, E. H., F. M. Strong and W. H. Peterson: The chemistry of mold tissue. VII. The lipids of *Penicillium aurantio-brunneum*. J. Amer. Chem. Soc. **57**, 354—356 (1935).

Laborde, J.: Recherches physiologiques sur une moisissure nouvelle l'*Eurotiopsis Gayoni*. Ann. Inst. Pasteur **11**, 1—43 (1897). — Lefort, M. J.: Études chimiques du champignon comestible, suivies d'observations sur sa valeur nutritive. J. Pharmacie III. s. **29**, 190—203 (1856). — Lepeschkin, W. W.: Über die chemische Zusammensetzung des Protoplasmas des Plasmodiums. Ber. dtsch. bot. Ges. **41**, 179—187 (1923). — Über die chemische Zusammensetzung der lebenden Materie. Biochem. Z. **171**, 126 (1926). — Lettré, H., u. R. Tschesche: In Lettré-Inhoffen-Tschesche, Über Sterine, Gallensäuren und verwandte Naturstoffe, 2. Aufl., Bd. 1. Stuttgart: Ferdinand Enke 1954. — Lim, H.: Chemical studies on *Rhizopus japonicus*. J. Faculty Agricult. Hokkaido Univ. **36**, 165—209 (1935). — Lösecke, A. v.: Beiträge zur Kenntnis eßbarer Pilze. Arch. Pharmaz. **209**, 133—146 (1876). — Loew, O.: Über den Nachweis des Lecithins. Pflügers Arch. **19**, 342—346 (1878). — Ludwig, H.: Chemisches über das Mutterkorn *(Secale cornutum)*. Arch. Pharmaz. **164**, 193—218 (1863). — Vorkommen von Cholesterin im Mutterkorn. Arch. Pharmaz. **187**, 36 (1869a). — Über einige Bestandteile der Hirschtrüffel. Arch. Pharmaz. **189**, 24—34 (1869b). — Lukacs, L., u. J. Zellner: Zur Chemie der höheren Pilze. (XXII. Mitt.). Über *Ganoderma lucidum* Leiss., *Hydnum imbricatum* L. und *Cantharellus clavatus* Pers. Mh. Chem. **62**, 214—219 (1933).

Marcelet, H.: Présence de carbures d'hydrogène dans le produit enlevé par la desodoration dans le raffinage de l'huile d'olive. C. r. Acad. Sci. Paris **202**, 867—869 (1936). — Margewicz, K.: Eßbare Pilze. Bestimmung der Nährstoffe in ihnen. Diss. Kaisl. Med. Akad. St. Petersburg 1883. 54. S. Ref. Justs Bot. Jber. **1885 I**, 85. — Marschall: Über die Zusammensetzung des Schimmelpilz-Mycels. Arch. f. Hyg. **28**, 16—29 (1897). — Mjoen, J. A.: Zur Kenntnis des in *Secale cornutum* enthaltenen fetten Öls. Arch. Pharmaz. **234**, 278—283 (1896).

Nägeli, C. v.: Über die chemische Zusammensetzung der Hefe. J. prakt. Chem. **17**, 403—428 (1878). — Über die Fettbildung bei den niederen Pilzen. Sitzgsber. bayer. Akad. Wiss., Math.-naturwiss. Kl. **9**, 288—316 (1879). — Nägeli, C. v., u. O. Loew: Über die chemische Zusammensetzung der Hefe. Liebigs Ann. **193**, 322—348 (1878). — Über die Fettbildung bei niederen Pilzen. J. prakt. Chem. (2) **21**, 97—114 (1880). — Neville, A.: The fat of yeast. Biochem. J. **7**, 341—348 (1913). — Newman, M. S., and R. J. Anderson: The chemistry of the lipids of yeast. I. The composition of the acetone-soluble fat. J. of Biol. Chem. **102**, 219—228 (1933a). — The biochemistry of the lipids of yeast. II. The composition of the phospholipids. J. of Biol. Chem. **102**, 229—235 (1933b). — Nilsson, R., L. Enebo u. E. Brunius: Über Vitamin A-Vorkommen bei *Rhodotorula*. Svensk. kem. Tidskr. **54**, 134 bis 135 (1942).

Oda, T.: Components of penicillin-producing molds. II. Fungus cerebrin. J. Pharmaceut. Soc. Japan **72**, 136—139 (1952a). — Components of penicillin-producing molds. III. Fungus cerebrin. J. Pharmaceut. Soc. Japan **72**, 139—141 (1952b). — Components of penicillin-producing molds. IV. Fungus cerebrin. J. Pharmaceut. Soc. Japan **72**, 142—145

(1952c). Ref. Chem. Abstr. **46**, 6192 (1952). — OPITZ, E.: Über das Fett von *Amanita pantherina*. Arch. Pharmaz. **229**, 291—292 (1891). — OXFORD, A. E., and H. RAISTRICK: Studies in the biochemistry of microorganisms. XXXIII. The mycelial constituents of *Penicillium brevicompactum* DIERCKX and related species. Part I. Ergosteryl palmitate. Biochemic. J. **27**, 1176—1180 (1933). — Studies in the biochemistry of microorganism. XXXIX. The metabolic products of *Penicillium crateriforme* GLIMAEUS and ABBOTT. Biochemic. J. **28**, 1321—1324 (1934). Ref. Chem. Abstr. **29**, 132 (1935).

PAYEN: Mémoire sur le développement des végétaux. 3. mémoire. Mém. savants étrangers **9**, 1 (1846). Zit. nach CH. PONTILLON, Rév. gén. Bot. **45**, 21—52 (1933). — PECK, R. L., and C. L. HAUSER: Chemical studies of certain pathogenic fungi. I. The lipids of *Blastomyces dermatitidis*. J. Amer. Chem. Soc. **60**, 2599—2603 (1938). — Chemical studies on certain pathogenic fungi. II. The lipids of *Monilia albicans*. J. Amer. Chem. Soc. **61**, 281—284 (1939). — PENAU, H., et G. TANRET: Sur un stérol dextrogyre de la levure, le zymostérol. Bull. Soc. Chim. biol. Paris **11**, 929—936 (1929). — PERRIER, A.: Sur la formation et le rôle des matières grasses chez les champignons. C. r. Acad. Sci. Paris **140**, 1052—1054 (1905). — PETTENKOFER, D.: Chemische Versuche mit Mutterkorn. Repert. Pharmacie **3**, 66—76 (1817). Zit. nach CH. PONTILLON, Rév. gén. Bot. **45**, 21—52 (1933). — PONTILLON, CH.: Contributions à l'étude physiologique des lipides du *stérigmatocystis nigra*. Rév. gén. Bot. **44**, 417—449, 465—483, 526—560 (1932); **45**, 21—52 (1933).

RATHJE, A.: Neuere Untersuchungen der Fette von *Lycopodium, Secale cornutum, Semen Arecae* und *Semen Aleuritis cordatae*. Arch. Pharmaz. **246**, 692—709 (1908). — REICHERT, R.: Über die Lipoide der Wuchshefe (*Torula utilis*). I. Die Zusammensetzung der acetonlöslichen Lipoide. Helvet. chim. Acta **28**, 484—495 (1945). — REINDEL, F.: Über das Ergosterin der Hefe. III. Liebigs Ann. **466**, 131—147 (1928). — Über Pilzcerebrin. Liebigs Ann. **480**, 76—92 (1930). — REINDEL, F., u. A. DETZEL: Über das Ergosterin der Hefe. IV. Liebigs Ann. **475**, 78—86 (1929). — REINDEL, F., u. E. WALTER: Über das Ergosterin der Hefe. II. Liebigs Ann. **460**, 212—224 (1927). — REINDEL, F., E. WALTER u. H. RAUCH: Über das Ergosterin der Hefe. I. Liebigs Ann. **452**, 34—46 (1927). — REINDEL, F., u. A. WEICKMANN: Zur Kenntnis des Zymosterins. Liebigs Ann. **475**, 86—100 (1929). — REINDEL, F., A. WEICKMANN, S. PICARD, K. LUBER u. P. TURULA: Über Pilzcerebrin. II. Liebigs Ann. **544**, 116—137 (1940). — REINKE, J.: Über die Zusammensetzung des Protoplasma von *Aethalium septicum*. Bot. Ztg **38**, 815 (1880). — REINKE, J., u. H. RODEWALD: Studien über das Protoplasma. I. Die chemische Zusammensetzung des Protoplasmas von *Aethalium septicum*. Unters. Bot. Laborat. Göttingen **2**, 1—69 (1881a). — Über Paracholesterin aus *Aethalium septicum*. Unters. bot. Laborat. Göttingen **2**, 70—75 (1881b). — Über Paracholesterin aus *Aethalium septicum*. Liebigs Ann. **207**, 229—235 (1881c). — RÉMOND et H. LASALLE: Production de cholestérine par un champignon. C. r. Soc. Biol. Paris **43**, 426 (1925). — REWALD, B.: Über den Phosphatidgehalt von Hefe. Biochem. Z. **218**, 481—486 (1930). — RIEGEL, E.: Beitrag zur Untersuchung der Familie der Schwämme. Jb. prakt. Pharm. **7**, 222 (1843). Zit. nach CH. PONTILLON, Rév. gén. Bot. **45**, 21—52 (1933). — RIPPEL-BALDES, A., A. PIETSCHMANN-MEYER u. W. KÖHLER: *Candida (Nectaromyces) Reukaufii* als Fettbildner. Arch. Mikrobiol. **14**, 113—127 (1950). — ROSENTHAL, R.: Zur Chemie der höheren Pilze. XVI. Mitt. Über Pilzlipoide. Mh. Chem. **43**, 237—253 (1922). — RUPPOL, E.: Chemical composition of the fat from *Aspergillus citromyces*. J. Pharmacie Belg. **19**, 63—68 (1937a). Ref. Chem. Abstr. **31**, 6687 (1937). — Constitution of the cerebrine of beer yeast. Bull. Soc. Chim. biol. Paris **19**, 1165—1172 (1937b). Ref. Chem. Abstr. **32**, 189—190 (1938). — Plant cerebrins. I. Cerebrins of *Aspergillus citromyces*. Bull. Soc. Chim. biol. Paris **25**, 57—66 (1943). Ref. Chem. Abstr. **38**, 2361 (1944). — Note on some constituents of the unsaponifiable portion of ergot oil. J. Pharmacie Belg., N. S. **4**, 55—58 (1949). Ref. Chem. Abstr. **43**, 6367 (1949). — RUTHNER, O., u. J. ZELLNER: Zur Chemie der höheren Pilze. XXIII. *Geaster fimbriatus* FR. und *Polystictus velutinus* PERS. Mh. Chem. **66**, 76—80 (1935). — RUZICKA, L., R. DENSS u. O. JEGER: Zur Kenntnis der Triterpene. 96. Mitt. Beweis der Identität von Lanosterin und Kryptosterin. Helvet. chim. Acta **28**, 759—766 (1945).

SAITO, A.: The chemical composition of mold. I. Sterols isolated from the mycelium of *Penicillium*. J. Fermentat. Technology (Japan) **27**, 207—210 (1949a). Ref. Chem. Abstr. **47**, 2257 (1953). — The chemical composition of mold. II. Sterols isolated from molds. J. Fermentat. Technology (Japan) **27**, 328—330 (1949b). Ref. Chem. Abstr. **47**, 2257 (1953). — SALISBURY, L. F., and R. J. ANDERSON: The chemistry of the lipids of yeast. III. Lecithin and Cephalin. J. of Biol. Chem. **112**, 541—550 (1936). — SALKOWSKI, E.: Über den Kohlenhydratgehalt der Flechten und den Einfluß der Chloride auf die Alkoholgärung. Hoppe-Seylers Z. **104**, 105—128 (1919). — SANI, G.: Untersuchungen über das Oleasten. Ein in der Frucht des Ölbaumes vorkommender Kohlenwasserstoff. Atti Accad. naz. Lincei (6) **12**, 238—242 (1930). Ref. Chem. Zbl. **1931 I**, 1297—1298. — SAVARD, K., and G. A. GRANT: Ergosterol from the mycelium of *Penicillium notatum* (submerged culture). Science (Lancaster, Pa.) **104**, 459—460 (1946). — SCHINDELMEISER, J.: Enzyme im Mutterkorn.

Apothker-Ztg. **24**, 837—838 (1909). — SCHMIEDER, J.: Über Bestandteile des *Polyporus off.* FR. Ein Beitrag zur chemischen Kenntnis der Pilze. Diss. Erlangen 1886. 67 S. — Über die chemischen Bestandteile des *Polyporus officinal.* (*Agaricus alb.* der Officinen.) Arch. Pharmaz. **224**, 641—668 (1886). — SEDLMAYR, TH.: Beiträge zur Chemie der Hefe. Z. ges. Brauwes. **26**, 381—387 (1903). Ref. CheN. Zbl. **1903 II**, 258. — SIEBER, N.: Beiträge zur Kenntnis der chemischen Zusammensetzung der Schimmelpilze. Z. prakt. Chem., N. F. **23**, 412—421 (1881). — SIEGEL, O.: Beiträge zur Kenntnis eßbarer Pilze. Diss. Göttingen 1870, 38 S. — SINGH, J., T. K. WALKER and M. L. MEARA: The component fatty acids of the fat of *Aspergillus nidulans.* Biochemic. J. **61**, 86—88 (1955). — SMEDLEY-MACLEAN, I.: The isolation of a second sterol from yeast fat. Biochemic. J. **22**, 22 (1928). — SMEDLEY-MACLEAN, I., and E. M. THOMAS: The nature of yeast fat. Biochemic. J. **14**, 483—493 (1928). — STEINER, M., u. H. HOLTZEM: Triterpene und Triterpen-Saponine. In K. PAECH u. M. TRACEY, Moderne Methoden der Pflanzenanalyse, Bd. 3, S. 58—140. Berlin-Göttingen-Heidelberg: Springer 1955. — STOHMER, F.: Über den Nährwert der eßbaren Schwämme. Zschr. f. Nahrungsm. Untersuchg. **1**, 4—6 (1887). Ref. Bot. Cbl. **30**, 210—211 (1887). — STOLL, A., u. E. JUCKER: Phytosterine, Steroidsaponine und Herzglykoside. In K. PAECH u. M. TRACEY, Moderne Methoden der Pflanzenanalyse, Bd. 3, S. 141—271. Berlin-Göttingen-Heidelberg: Springer 1955. — STRONG, F. M., and W. H. PETERSON: The chemistry of mold tissue. IV. The lipids of *Aspergillus Sydowi.* J. Amer. Chem. Soc. **56**, 952—955 (1934). — SULLIVAN, M. X.: Some organic constituents of the culture solution and the mycelium of molds from soil. Biochem. Bull. **3**, 86—87 (1913). — Some organic constituents of the culture solution and the mycelium of molds from soil. Science (Lancaster, Pa.) **38**, 678 (1913). — SUMI, M.: Über die chemischen Bestandteile der Sporen von *Aspergillus oryzae.* Biochem. Z. **195**, 161—174 (1928). — Nachtrag zur Arbeit „Über die chemischen Bestandteile der Sporen von *Aspergillus oryzae*" von MIDZUHO SUMI. Biochem. Z. **204**, 412 (1929).

TÄUFEL, K., H. THALER u. H. SCHREYEGG: Squalen als Bestandteil des Hefefettes. Fette u. Seifen **43**, 26—29 (1936a). — Zur Kenntnis des Fettes der Hefe (*Saccharomyces* sp.). Z. Unters. Lebensmitt. **72**, 394—404 (1936b). — TAKATA, R.: Utilization of microorganism as human food. J. Soc. Chem. Industr. Japan **32**, 497—550, 544—557 (1929); **32**, Suppl., 155—158B, 169—172B (1929). Ref. Chem. Abstr. **23**, 4748 (1929). — The utilization of microorganism as human food. XI. The sterols of mycelium of *Aspergillus oryzae.* J. Soc. Chem. Industr. Japan **32**, Suppl., 268—269 (1929). Ref. Chem. Abstr. **24**, 2206 (1930). — TANRET, C.: Sur un nouveau principe immédiat de l'ergot de seigle, l'ergosterine. C. r. Acad. Sci. Paris **108**, 98—100 (1889). — Sur l'ergostérine. C. r. Acad. Sci. Paris **147**, 75—77 (1920). — TAPPI, G.: The biosynthesis of sterols by action of some strains of *Penicillium.* II. Ergosterol from a strain of *Penicillium notatum.* Gezz. chim. ital. **78**, 311—316 (1948). Ref. Chem. Abstr. **42**, 8867 (1948). — THÖRNER, W.: Über eine neue, in *Agaricus integer* vorkommende, organische Säure. Ber. dtsch. chem. Ges. **12**, 1636—1637 (1879). — TSCHESCHE, R.: Neuere Vorstellungen auf dem Gebiete der Biosynthese der Steroide und verwandter Naturstoffe. Fortschr. Chem. organ. Naturstoffe **12**, 131—168 (1955). — TSCHESCHE, R., u. F. KORTE: Zum biochemischen Syntheseweg der Steroide. Angew. Chem. **66**, 32—33 (1954). — TSCHIRCH, A.: Handbuch der Pharmakognosie, Bd. III/1, S. 139—165. Leipzig: H. Tauchnitz. — TSUJMOTO, M.: Ein stark ungesättigter Kohlenwasserstoff im Haifischleberöl. J. Industr. a. Engng. Chem. 8, 889—896 (1916). Ref. Chem. Zbl. **1918 I**, 638. — Squalen, ein stark ungesättigter Kohlenwasserstoff im Haifischleberöl. J. Industr. a. Engng. Chem. **12**, 63—72 (1920). Ref. Chem. Zbl. **1920 I**, 862. — TURPEINEN, O.: Studies on microbiological lipides. The formation and chemical composition of lipides in *Geotrichoides* sp. LANGERON et TALICE. Ann. Acad. Sci. fenn., Ser. A **46**, 110 (1936).

USDIN, V. R., and R. C. BURELL: Preliminary study of the lipides of *Rhodotorula gracilis.* Arch. of Biochem. a. Biophysics **36**, 172—177 (1952).

VANDERMEULEN, H.: Chemical composition of the oil of ergot of rye. J. Pharmacie Belg. **21**, 195—197, 213—217, 237—241 (1939). Ref. Chem. Abstr. **33**, 8358 (1939). — VAUQUELIN: Expériences sur les champignons. Ann. de chim. **85**, 5—25 (1813). — Analyse du seigle du bois de Boulogne près Paris. Ann. de Chim. (2) **3**, 337—349 (1816).

WARD, G. E., and G. S. JAMIESON: The chemical composition of the fat produced by *Penicillium javanicum* VAN BEIJMA. J. Amer. Chem. Soc. **56**, 973—975 (1934). — WEICHHERZ, J., u. R. MERLÄNDER: Untersuchungen über das Hefefett. Biochem. Z. **239**, 21—27 (1931). — WEIS, G.: Untersuchungen über das Hefefett. I. Biochem. Z. **243**, 269—273 (1931). — WIELAND, H., u. M. ASANO: Zur Kenntnis der Sterine der Hefe. Liebigs Ann. **473**, 300—313 (1929). — WIELAND, H., u. G. COUTELLE: Zur Kenntnis des Fungisterins und anderer Inhaltsstoffe in Pilzen. Liebigs Ann. **548**, 270—283 (1941). — WIELAND, H., u. W. M. STANLEY: Zur Kenntnis der Sterine der Hefe. III. Liebigs Ann. **489**, 31—42 (1931). — WIGGERS, H. A. L.: Untersuchung über das Mutterkorn, *Secale cornutum.* Liebigs Ann. **1**, 129—182 (1832). — WINCKLER, F. L.: Versuche über das Mutterkorn. Arch. Pharmaz. **23**, 148 (1827). Zit. nach CH. PONTILLON, Rév. gén. Bot. **45**, 21—52 (1933). — WINDAUS, A.,

u. W. GROSSKOPF: Über das Ergosterin der Hefe. Hoppe-Seylers Z. **124**, 8—14 (1923). — WINDAUS, A., u. R. TSCHESCHE: Über das sogenannte „Isocholesterin" des Wollfettes. Hoppe-Seylers Z. **190**, 51—61 (1930). — WOOLLEY, D. W., F. M. STRONG, W. H. PETERSON and E. A. PRILL: The chemistry of mold tissue. X. The phospholipides of *Aspergillus sydowi*. J. Amer. Chem. Soc. **57**, 2589—2591 (1935).

ZEEH, H.: Weitere Beiträge zur chemischen Kenntnis einiger Bestandteile aus *Secale cornutum*. Diss. Erlangen 1894. 29 S. — ZEGA, A.: *Agaricus campestris*. Chemiker-Ztg **24**, 285—286 (1900). — Eßbare Pilze. Chemiker-Ztg **26**, 10 (1902). — ZELLNER, J.: Zur Chemie des Fliegenpilzes (*Amanita muscaria* L.). II. Mitt. Mh. Chem. **26**, 727—747 (1905). — Zur Chemie des Fliegenpilzes (*Amanita muscaria* L.). III. Mitt. Mh. Chem. **27**, 281—293 (1906a). — Über das fettspaltende Ferment der höheren Pilze. Mh. Chem. **27**, 295—304 (1906b). — Chemie der höheren Pilze. Leipzig: Wilhelm Engelmann 1907. — Zur Chemie der höheren Pilze. I. Mitt. *Trametes suaveolens* FR. Mh. Chem. **29**, 45—54 (1908a). — Zur Chemie der höheren Pilze. II. Mitt. *Polyporus igniarius* FR. Mh. Chem. **29**, 1171—1187 (1908b). — Zur Chemie der höheren Pilze. III. Mitt. Über Pilzdiastasen. Mh. Chem. **30**, 231—246 (1909). — Zur Chemie der höheren Pilze. V. Mitt. Über den Maisbrand (*Ustilago Maydis* TULASNE). Mh. Chem. **31**, 617—634 (1910). — Zur Chemie der höheren Pilze. VII. Mitt. *Hypholoma fasciculare* HUDS. Mh. Chem. **32**, 1057—1063 (1911a). — Zur Chemie des Fliegenpilzes (*Amanita muscaria* L.). IV. Mitt. Mh. Chem. **32**, 133—142 (1911b). — Zur Chemie der höheren Pilze. VIII. Mitt. Über den Weizenbrand (*Tilletia levis* KÜHN und *tritici* WINTER). Mh. Chem. **32**, 1065—1074 (1911c). — Zur Chemie der höheren Pilze. IX. Mitt. Über die durch *Exobasidium vaccinii* WORON. auf *Rhododendron ferrugineum* erzeugten Gallen. Mh. Chem. **34**, 311—319 (1913a). — Zur Chemie der höheren Pilze. X. Mitt. Über *Armillaria mellea* VAHL., *Lactarius piperatus* L., *Pholiota squarrosa* MÜLL. und *Polyporus betulinus* FR. Mh. Chem. **34**, 321—336 (1913b). — Zur Chemie der höheren Pilze. XI. Mitt. Über *Lactarius scrobiculatus* SCOP., *Hydnum ferrugineum* FR., *Hydnum imbricatum* L. und *Polyporus applanatus* WALLR. Mh. Chem. **36**, 611—632 (1915). — Zur Chemie der höheren Pilze. XII. Mitt. Über *Lenzites sepiaria* SW., *Panus stypticus* BULL. und *Exidia auricula judae* FR. Mh. Chem. **38**, 319—330 (1917). — Zur Chemie der höheren Pilze. XIII. Mitt. Über *Scleroderma vulgare* FR. und *Polysaccum crassipes* D. C. Mh. Chem. **39**, 603—615 (1918). — Zur Chemie der höheren Pilze. XIV. Mitt. Über *Lactarius rufus* SCOPOL., *Lactarius pallidus* PERS. und *Polyporus hispidus* FR. Mh. Chem. **41**, 443—453 (1920a). — Über dem Milchsaft von *Lactarius vellereus* FR. Hoppe-Seylers Z. **111**, 293—296 (1920b). — Zur Chemie der Flechten. I. Mitt. Über *Peltigera canina*. Mh. Chem. **59**, 300—304 (1932). — Zur Chemie der Flechten. III. Mitt. *Parmelia physodes* L. Mh. Chem. **64**, 6—11 (1934). — Zur Chemie der Flechten. IV. Mitt. *Gyrophora Dillenii* (TUCK.) MÜLL. ARG. und *Parmelia furfuracea* L. Mh. Chem. **66**, 81—86 (1935). — ZELLNER, J., u. E. ZIKMUNDA: Zur Chemie der höheren Pilze. XXI. Mitt. Über *Polyporus sulfureus* L. und *Lentinus squamosus* SCHROET. Mh. Chem. **56**, 200—203 (1930). — ZOOK, H. D., T. S. OAKWOOD and F. C. WHITMORE: Isolation of ergosterol from *Penicillium notatum*. Science (Lancaster, Pa.) **99**, 427—428 (1944). — ZOPF, W.: Die Pilze. Breslau: Trewendt 1890. — Die Flechtenstoffe in chemischer, botanischer, pharmakologischer und technischer Beziehung. Jena: Gustav Fischer 1907. — ZUKAL, H.: Über das Vorkommen von Reservestoffbehältern bei Kalkflechten. Bot. Ztg **44**, 762—770 (1886).

Nachtrag zu S. 70 (während der Korrektur).

Vor kurzem berichteten J. SINGH, S. SHAH und T. K. WALKER [Biochemic. J. **62**, 222—224 (1956)] über das Fett von *Penicillium lilacinum*. Verseifungsäquivalent 286,9, Jodzahl 63,7; 0,2% freie Fettsäuren, 2,9% Unverseifbares. Das Gemisch der Fettsäuren enthielt folgende Prozentanteile: gesättigt C_{14} 0,1, C_{16} 23,3, C_{18} 9,4, C_{20} 1,4; Hexadecensäure 3,4, Ölsäure 38,6, Linolsäure 13,4, ungesättigt C_{20} 1,4.

Les lipides bactériens.

Par

J. Asselineau.

La chimie des lipides bactériens a fait l'objet d'une revue récente (ASSELINEAU et LEDERER 1953b), et plusieurs revues limitées aux lipides des Mycobactéries ont déjà été publiées (ASSELINEAU 1952, LEDERER 1952a et b, SEIBERT 1951).

Les Bactéries constituent un groupe botanique très hétérogène dont les lipides, à l'heure actuelle, n'ont encore été étudiés que d'une manière très fragmentaire: il est donc difficile de tirer des conclusions générales. Cependant, les lipides bactériens se distinguent de ceux des autres représentants du règne végétal, en particulier par l'*absence de stérols.* La recherche des stérols a presque toujours eu un résultat négatif (voir ANDERSON, SCHOENHEIMER, CROWDER et STODOLA 1935, en ce qui concerne les Mycobactéries). L'unique cas où la présence d'un stérol (non identifié) a pu être démontrée, est celui de *Azotobacter chroococcum,* bactérie fixatrice de l'azote (SIFFERD et ANDERSON 1936).

La présence dans les lipides bactériens d'*acides gras à chaîne ramifiée,* quoique fréquente, ne semble pas être générale: *Streptococcus hemolyticus* constitue un exemple où leur absence semble établie (HOFMANN et TAUSIG 1955a). Cependant, un certain nombre d'acides ramifiés ont été isolés des bactéries (voir tableau 3, page 101), tandis que l'acide sterculique constitue le seul représentant connu chez les végétaux supérieurs (en dehors des acides ω-substitués du type hydnocarpique et chaulmogrique) et que leur présence chez les animaux est généralement limitée à des sources très particulières.

De même, la richesse des lipides bactériens en *glycolipides* est fréquente (particulièrement chez les Mycobactéries), mais elle ne peut pas être considérée comme générale: *Agrobacterium tumefaciens* par exemple ne renferme pas de glycolipides en proportion notable. Cependant, le caractère structural le plus spécifique des bactéries semble être la présence de *complexes lipo-polysaccharido-protéiques,* qui possèdent souvent d'importantes propriétés biologiques (antigénicité, toxicité, ...). En ce qui concerne les glycolipides, la revue de LOVERN dans le chapitre VIII de cet ouvrage (page 376) pourra être consultée avec profit.

Deux groupes de bacilles se distinguent nettement en ce qui concerne la singularité de leurs lipides: ce sont, d'une part le groupe Bacille tuberculeux-Bacille diphtérique[1], et d'autre part les bacilles du genre *Bacillus.* Les premiers présentent une complexité remarquable de leurs lipides, tant par la richesse du nombre de constituants que par la complexité de structure des substances isolées. Les lipides des bacilles du genre *Bacillus* présentent la singularité inverse: ils ne sont constitués pratiquement que par une seule substance, le lipide β-hydroxy-butyrique.

[1] Pendant longtemps, le genre *Corynebacterium* a été classé avec le genre *Mycobacterium* dans la même famille des *Mycobacteriaceae* (WILSON et MILES 1948). Dans la dernière édition du manuel de BERGEY (1953), le genre *Corynebacterium* est inclu dans le sous-ordre des *Eubacteriinae,* ordre des *Eubacteriales.* Rappelons d'autre part qu'il existe une réaction croisée entre un haptène lipidique diphtérique et des anticorps tuberculeux (MACHEBOEUF et CASSAGNE 1935).

A. Isolement des fractions lipidiques.

Les bacilles présentent une teneur en lipides très variable selon les espèces. Au cours d'une des rares investigations systématiques effectuées dans ce domaine, Nicolle et Alilaire (1909) ont déterminé la teneur en lipides libres totaux de 13 espèces bactériennes cultivées sur milieu gélosé, et observé que cette teneur peut varier de 1,48% (bactérie du charbon) à 11,77% *(Escherichia coli)*.

Il est connu depuis longtemps que la composition du milieu de culture influe considérablement sur la teneur en lipides[1]. Barbier et Lederer (1953), étudiant les effets de l'addition de 1,4% d'acétate de sodium au milieu de culture (Sauton) de diverses mycobactéries, ont observé par exemple qu'une souche de *Mycobacterium phlei* produit (après 4 semaines de culture) 150% du poids de bacilles et 200% de lipides par rapport aux témoins; la composition qualitative des lipides reste apparemment inchangée. Aucun effet stimulateur de l'acétate de sodium sur la croissance de *Mycobacterium Pellegrini* n'a été observé (Goodwin et Jamikorn 1956).

Tableau 1. *Schéma du procédé d'extraction des lipides des Mycobactéries* (Anderson 1939; légèrement modifié: Aebi, Asselineau et Lederer 1953).

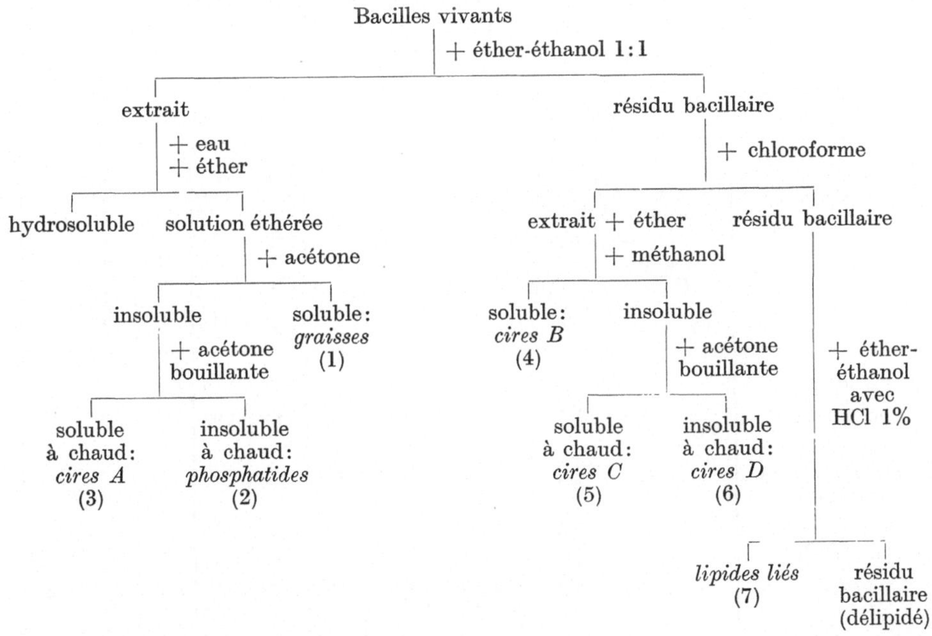

(1): Renferment des acides gras libres, des esters de tréhalose et d'acides gras, une fraction insaponifiable, des pigments et des acides aromatiques.
(2): Acides glycéro-phosphorique et inosito-phosphorique liés à des acides gras et à des polysaccharides.
(3): Constituées essentiellement de mycocéranate de phtiocérol et d'alcools apparentés.
(4): Renferment des glycérides.
(5): Mélange de mycocéranate de phtiocérol, et d'esters d'acides mycoliques et de sucres (en particulier, «cord factor») ou de glycérol.
(6): Esters d'acides mycoliques et de polysaccharides (avec 3 acides aminés dans le cas des souches humaines: alanine, acide glutamique et acide α,ε-diamino-pimélique).
(7): Esters d'acides mycoliques et de polysaccharides.

[1] Voir Steiner, p. 218—240 de cet ouvrage.

Les variations qualitatives de la composition en lipides suivant celles du milieu de culture [voir par exemple Creighton, Chang et Anderson (1944), Čmelik (1954b)] soulèvent le problème de la composition des lipides de bacilles pathogènes dans les lésions. *Bacillus anthracis* fournit un exemple particulièrement démonstratif des différences que peut présenter un bacille isolé d'un animal infecté par rapport au bacille de même espèce cultivé *in vitro* (Smith, Keppie et Stanley 1953). Rappelons dans le cas des Mycobactéries, la différence de coloration des bacilles avec le Noir Soudan B suivant leur croissance *in vitro* ou *in vivo* (Sheehan et Whitwell 1949), les résultats négatifs obtenus par Anderson *et ses collaborateurs* (1943) lors de la recherche dans des lésions pulmonaires de certains constituants lipidiques caractéristiques du bacille tuberculeux de culture, et les récentes observations de Segal et Bloch (1956) sur le métabolisme de bacilles isolés de souris infectées.

L'analyse des lipides bactériens est compliquée par les risques de lipolyses qui peuvent se produire au cours de la récolte des bacilles et de l'extraction. Ces risques sont limités le plus possible en soumettant les bacilles *vivants* à l'action des solvants organiques. Mais Hanahan (1952) a montré la possibilité d'hydrolyse enzymatique de phosphatides au sein de l'éther humide, et la présence dans les bactéries d'enzymes capables d'hydrolyser les phospholipides est bien connue (*cf.* Hayaishi et Kornberg 1954). Il est difficile d'apprécier l'importance de ces phénomènes, qui peut varier d'une souche à une autre. Nous avons

Tableau 2. *Pourcentages (par rapport au poids*

<table>
<tr><th rowspan="4"></th><th colspan="5">Mycobactéries</th><th colspan="2" rowspan="4">Corynebacterium diphtheriae</th><th colspan="2">Agrobacterium tumefaciens</th><th colspan="3">Lactobacillus</th></tr>
<tr><th colspan="3">M. tuberculosis</th><th rowspan="3">M. avium</th><th rowspan="3">M. phlei</th><th rowspan="3">milieu glycériné</th><th rowspan="3">milieu glucosé</th><th rowspan="3">acidophilus</th><th rowspan="3">arabinosus</th><th rowspan="3">casei</th></tr>
<tr><th colspan="2">hominis H-37</th><th rowspan="2">bovis (Vallée)</th></tr>
<tr><th>Rv</th><th>Ra</th></tr>
<tr><td>Graisses</td><td rowspan="3">12,7</td><td>5,6</td><td>3,3</td><td>2,2</td><td>3,0</td><td>3,97</td><td>8,5</td><td>0,92</td><td>2,14</td><td>4,75</td><td></td><td></td></tr>
<tr><td>Phosphatides</td><td>1,8</td><td>1,5</td><td rowspan="2">2,2</td><td>0,74</td><td>0,41</td><td>3,3</td><td>0,88</td><td>3,95</td><td>2,25</td><td></td><td></td></tr>
<tr><td>Graisses ou cires accompagnant les phosphatides</td><td>2,2</td><td>1,2</td><td>0,03</td><td>0,25</td><td>0,5</td><td></td><td></td><td></td><td></td><td></td></tr>
<tr><td>cires extraites par $CHCl_3$</td><td>8,4</td><td>6,5</td><td>8,3</td><td>10,8</td><td>2,9</td><td>0,27</td><td>0,4</td><td>0,02</td><td>0,01</td><td></td><td></td><td></td></tr>
<tr><td>Lipides libres totaux</td><td>21,1</td><td>16,1</td><td>14,3</td><td>15,2</td><td>7,0</td><td>4,9</td><td>12,7</td><td>1,82</td><td>6,10</td><td>7,0</td><td>0,5</td><td>0,61</td></tr>
<tr><td>Lipides liés</td><td></td><td></td><td></td><td>10,8</td><td>18,0</td><td>5,1</td><td></td><td></td><td>1,0</td><td></td><td>2,0</td><td>2,29</td></tr>
<tr><td>Références</td><td colspan="3">Asselineau (1952)</td><td>Anderson (1939)</td><td>Barbier (1954)</td><td>Chargaff (1931)</td><td>Macheboeuf et Cassagne (1935)</td><td colspan="2">Geiger et Anderson (1939)</td><td>Crowder et Anderson (1934a)</td><td>Hofmann, Lucas et Sax (1952)</td><td>Hofmann et Sax (1953)</td></tr>
</table>

eu l'occasion d'observer dans le cas de bacilles tuberculeux, des trans-estérifications (AEBI, ASSELINEAU et LEDERER 1953), et des hydrolyses de phosphatides (ASSELINEAU 1951a, p. 58) et de cires (ASSELINEAU 1952, p. 22). D'autre part, un autre phénomène peut intervenir, difficilement contrôlable: la variation des souches.

L'âge des cultures présente une grande influence quantitative sur la teneur en lipides, étudiée notamment par CASSAGNE (1939) chez le Bacille diphtérique. La variation avec l'âge affecte de façon très différente les diverses fractions lipidiques, ainsi que l'ont observé LEMOIGNE, MILHAUD et CROSON (1949) chez *Bacillus megatherium*, ASSELINEAU (1951b) chez *Mycobacterium tuberculosis*, et ČMELIK (1955b) chez *Salmonella typhosa*.

L'extraction des lipides bacillaires est généralement réalisée par la technique développée par ANDERSON dans le cas des *Mycobactéries*; le schéma de ce procédé, avec quelques modifications, est représenté dans le tableau 1. Après épuisement des cellules bactériennes par un mélange d'éther et d'alcool, puis par le chloroforme, les résidus bacillaires renferment encore des *lipides liés*, qui ne sont pas extractibles par les solvants organiques neutres. Leur isolement nécessite l'action préliminaire d'un acide ou d'un alcali. Dans le cas de *Corynebacterium diphtheriae*, l'action prolongée du chloroforme a permis d'extraire, au moins en partie, les lipides liés (GUBAREV, GABRILOVICH et SHTULBAUM 1952); ce résultat peut être dû à une lente libération d'acide chlorhydrique par le chloroforme.

de bacilles secs) en diverses fractions lipidiques.

Brucella			*Bacillus*		*Salmonella*			*Escherichia*		*Malleomyces mallei*	*Haemophilus pertussis*
abortus	*suis*	*melitensis*	*megatherium*	*subtilis*	*typhi*	*ballerup*	*paratyphi C*	*coli*	*communior*		
0,90	2,40	4,80		2,0	1,3	1,0	2,8			5,1 à 6,7	21,7
4,80	2,70	0,40	} 1,97	0,8	0,2	1,0	0,8	4 à 5,5	4,3 à 4,9	environ 0,5	2,37
0,40	0,20	0,10	19,1	3,3						0,3 à 0,6	
6,10	5,60	5,30	21,0	6,1	1,5	2,0	3,6	4,5 à 8	4,6 à 6,7	6,1 à 7,5	24
0,33											
HUSTON, HUDDLESTON et HERSHEY (1934)			LEMOIGNE, MILHAUD et CROSON (1949)	AKASI (1940)	ČMELIK (1952, 1954c)	ČMELIK (1953)	ČMELIK (1954a)	WILLIAMS, BLOOR et SANDHOLZER (1939)		UMEZU (1940)	IDA et OTANI (1950)

Le tableau 2 rassemble les principales données de la littérature sur la teneur des bacilles en diverses fractions lipidiques. Aucune comparaison ne peut être faite, étant donné la diversité des conditions de culture utilisées. Ce tableau souligne la faible teneur en lipides généralement présente chez les bactéries, sauf dans le cas des Mycobactéries et de certains *Bacillus*. Un autre fait est également mis en évidence: en dehors des Mycobactéries, les bacilles ne renferment qu'un très faible pourcentage en fractions «cireuses» extractibles par le chloroforme (le cas de *B. megatherium* qui ne renferme guère que le lipide β-hydroxybutyrique[1] insoluble dans l'éther, étant mis de côté). Or, ce sont ces fractions cireuses qui jouent un rôle important dans l'activité pathogène particulière des Mycobactéries (Rich 1951, Asselineau 1952).

B. Composition des diverses fractions lipidiques.

1. Graisses.

Les graisses renferment souvent une proportion élevée d'*acides gras libres*, comme par exemple chez *Malleomyces mallei* (Umezu 1940), chez les Mycobactéries (Fethke 1938) et à un degré plus élevé, chez *Corynebacterium diphtheriae* (Chargaff 1931, Pudles 1954). A partir des graisses neutres, aucune fraction ne semble avoir été isolée à l'état pur: la saponification fournit toujours, à côté des acides gras et des constituants hydrosolubles, une fraction insaponifiable non-stérolique, mélange de substances amorphes difficiles à purifier.

Les *glycérides*, mis en évidence par la recherche du glycérol dans l'hydrosoluble, ont été observés chez *Lactobacillus acidophilus* (Crowder et Anderson 1934), *Agrobacterium tumefaciens* (Velick et Anderson 1944), diverses espèces de *Brucella* (Pennell 1950), *Malleomyces mallei* (Umezu 1940), et diverses espèces de *Salmonella* (Čmelik 1952, 1953, 1954a).

Les bacilles du genre *Bacillus* ne renferment qu'au plus des traces de glycérides (Lemoigne, Milhaud et Croson 1949); les Mycobactéries renferment des glycérides, mais le principal constituant hydrosoluble de leurs graisses est le *tréhalose* (Anderson et Newman 1933a). Asano et Takahashi (1945) ont caractérisé la *dihydroxy-acétone* dans les graisses de *Corynebacterium diphtheriae*.

Signalons encore la présence dans les graisses de Mycobactéries, de *composés naphtoquinoniques* [phtiocol (I) (Anderson et Newman 1933b, c)] et un homologue supérieur de la vitamine K_2 (Snow 1952)], et de faibles quantités d'*acides*

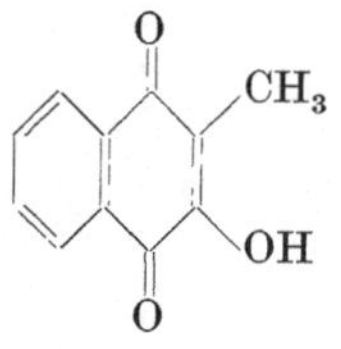

O
CH$_3$
OH
O

Phtiocol (I)

aromatiques [acide anisique (Anderson et Newman 1933b), acides salicylique et phényl-acétique (Stendal 1934), acide phtalique (Aebi, Asselineau et Lederer 1953)]. A partir d'un «bacille de la lèpre», Crowder, Stodola et Anderson (1936) ont isolé les α- et β-*léprosols*, qui sont des dérivés trisubstitués du *m*-méthoxy-phénol (Butenandt et Stodola 1939); Noll (1956) a caractérisé dans une souche bovine de *Mycobacterium tuberculosis* un composé aromatique qui pourrait présenter une structure voisine.

[1] Voir formule (VI) p. 102.

2. Phosphatides.

Un chapitre de l'ouvrage de WITTCOFF (1951) a été consacré aux phosphatides bactériens, et une revue des phosphatides des végétaux a été écrite par LOVERN (chapitre VIII, page 376).

La teneur en phosphatides, très variable selon les espèces bactériennes, présente comme valeurs extrêmes, une teneur pratiquement nulle chez divers *Bacillus* (LEMOIGNE, MILHAUD et CROSON 1949), et une teneur de plus de 5% (par rapport au poids de bacilles secs) chez *Escherichia coli* (WILLIAMS, BLOOR et SANDHOLZER 1939). Cependant, les chiffres de pourcentage trouvés dans la littérature correspondent souvent à des préparations de degré de purification très variable. Dans le cas de *Salmonella*, ČMELIK (1954c) a observé que des souches virulentes renferment nettement moins de phosphatides que des souches peu ou pas virulentes.

Les phosphatides bactériens sont souvent constitués par des combinaisons de phospholipides et de polysaccharides, ce qui conduit à des composés de poids moléculaire élevé, difficiles à purifier. ANDERSON, LOTHROP et CREIGHTON (1938) ont isolé du phosphatide de *Mycobacterium tuberculosis* (2,9% P et 0,4% N) un polysaccharide constitué de mannose et d'inositol (2:1), auquel ils ont donné le nom de *manninositose*. A partir d'une autre préparation de phosphatide de Bacille tuberculeux (type humain), SÜTÖ-NAGY et ANDERSON (1947) ont isolé, à côté d'acide glycéro-phosphorique, les acides *inosito-monophosphorique* et *inosito-glycéro-diphosphorique*. MACHEBOEUF et FAURE (1939) considèrent que les phosphatides du Bacille tuberculeux sont des acides glycéro-phosphoriques liés par estérification, d'une part à des acides gras, et d'autre part à des polyalcools non azotés. A l'aide de méthodes chromatographiques, VILKAS et LEDERER (1956) ont isolé du phosphatide de Bacille tuberculeux (type humain) un sel de magnésium d'un *phosphatidyl-inosito-di-D-mannoside*.

La présence de polysaccharides, également observée dans les phosphatides de *Mycobacterium phlei* [mannose, inositol et glucose (PANGBORN et ANDERSON 1932)] et de *Lactobacillus acidophilus* [galactose, glucose et fructose (CROWDER et ANDERSON 1934b)] n'a d'ailleurs rien d'exceptionnel: des exemples analogues sont rencontrés dans les phosphatides végétaux [phosphatides de la carotte par exemple (HANAHAN et CHAIKOFF 1947)], et récemment, HAWTHORNE et CHARGAFF (1954) ont isolé du soja une préparation de phosphatide qui renferme de l'acide inosito-phosphorique lié à de l'arabinose ou du galactose. Citons encore la mise en évidence d'*acide galactose-phosphorique* dans une fraction de phosphatide de *Salmonella typhi* (ČMELIK 1955c).

BLOCH (1936), MACHEBOEUF et FAURE (1939) et VILKAS et LEDERER (1956) ont montré la présence dans les phosphatides de Bacille tuberculeux de sels de magnésium d'*acides phosphatidiques*; ČMELIK (1955c) a caractérisé de tels acides chez *Salmonella typhi*. Ces composés, très sensibles à l'hydrolyse (voir par exemple OLLEY 1954), peuvent facilement échapper à l'observation.

Le répartition de bases azotées telles que choline ou éthanolamine fournit une bonne illustration de la diversité de composition des lipides bacillaires. La *choline* seule a été mise en évidence dans le phosphatide de *Lactobacillus acidophilus* (CROWDER et ANDERSON 1934b) et de *Neisseria gonorrhoae* (STOKINGER, ACKERMAN et CARPENTER 1944). Une substance du type choline a été caractérisée par PUDLES (1954) chez *Corynebacterium diphtheriae*. L'*éthanolamine* seule a été décelée dans le phospholipide A de *Escherichia coli* (IKAWA, KOEPFLI, MUDD et NIEMANN 1953b) et dans des fractions lipidiques phosphorées de *Brucella abortus* (STAHL 1941), de *Pseudomonas denitrificans* et de *Staphylococcus aureus*

(Few 1955), de *Salmonella typhi* (Čmelik 1955a et c) et de *Mycobacterium marianum* (Michel et Lederer 1955). La présence simultanée de choline et d'éthanolamine a été observée chez *Agrobacterium tumefaciens* (Geiger et Anderson 1939). Enfin, aucune base azotée de ce type n'a été trouvée dans la plupart des souches de Mycobactéries et de *Salmonella*; une souche de *Escherichia coli* a même fourni un phosphatide dépourvu d'azote (Williams, Bloor et Sandholzer 1939). Mais la possibilité de scission enzymatique des constituants azotés au cours des opérations de récolte et d'extraction des bacilles ne doit pas être écartée.

D'autre part, des *acides aminés* sont fréquemment rencontrés comme constituants azotés de ces phosphatides. Velick et Anderson (1944) ont signalé la présence d'un acide aminé (non identifié) dans le phosphatide de *Agrobacterium tumefaciens*. Plus récemment, Barbier et Lederer (1952) ont caractérisé l'hydroxylysine dans un échantillon de phosphatide de *Mycobacterium phlei*, mais cet acide aminé n'a pu être retrouvé dans des préparations ultérieures de phosphatides de la même souche (Barbier 1954). Gendre et Lederer (1955) ont isolé l'ornithine à partir de phosphatides de Bacille tuberculeux, et l'ont caractérisé dans les phosphatides de plusieurs souches de *Mycobacterium tuberculosis*, *M. avium* et *M. phlei*. Des acides aminés sont également présents dans les phosphatides de *Salmonella ballerup*, d'une souche virulente de *S. typhi* (Čmelik 1953, 1955a et c) et de *S. paratyphi C* (Čmelik 1954a). Le phospholipide A de complexes lipopolysaccharidiques de *Escherichia coli* (Ikawa, Koepfli, Mudd et Niemann 1953b, Westphal et Lüderitz 1954) et de *Salmonella abortus equi* (Werner, Lüderitz et Westphal 1954) renferme des acides aminés dicarboxyliques qui, d'après Westphal et Lüderitz (1954), interviennent dans la liaison du phospholipide A avec un composant protéïque.

Signalons enfin l'isolement, à partir du phospholipide A de *Escherichia coli* (Ikawa, Koepfli, Mudd et Niemann 1953b) d'une diamine à longue chaine, la *nécrosamine* (II).

$$CH_3{-}(CH_2)_{14}{-}\underset{\displaystyle NH_2}{\underset{|}{CH}}{-}\underset{\displaystyle NH_2}{\underset{|}{CH}}{-}(CH_2)_2{-}CH_3$$

Nécrosamine (II)

3. Cires vraies.

Les cires (esters d'alcools supérieurs avec des acides gras supérieurs) ne semblent guère exister que dans les cires A de Mycobactéries. On trouve en effet dans les souches humaines et bovines de *Mycobacterium tuberculosis* des esters de phtiocérol (Stodola et Anderson 1936) et alcools apparentés (Demarteau-Ginsburg, Ginsburg et Lederer 1953) [*phtiocérol* $C_{34}H_{67}(OH)_2(OCH_3)$; *phtiotriol* $C_{34}H_{67}(OH)_3$; *phtiodiolone* $C_{33}H_{66}(CO)(OH)_2$; *phtiocérolone* $C_{33}H_{66}(CO)(OH)(OCH_3)$], principalement avec l'acide ramifié mycocéranique $C_{31}H_{62}O_2$ (voir page 101) (Philpot et Wells 1952, Asselineau 1954a).

A la suite des travaux de Ställberg-Stenhagen et Stenhagen (1948, 1950), de Hall, Lewis et Polgar (1954, 1955), de Demarteau-Ginsburg et Lederer (1955) et de Ryhage, Stenhagen et von Sydow (1956), la formule (III) peut être envisagée pour le phtiocérol:

$$CH_3{-}(CH_2)_{22}{-}CHOH{-}CH_2{-}CHOH{-}(CH_2)_3{-}\underset{\displaystyle CH_3}{\underset{|}{CH}}{-}CH(OCH_3){-}CH_2{-}CH_3$$

Phtiocérol (III)

Les cires de certaines souches de *Mycobacterium avium* et de *M. phlei* contiennent des esters de (+)eicosanol-2 et de (+)octadécanol-2 avec des acides gras supérieurs (Reeves et Anderson 1937, Pangborn et Anderson 1936).

La présence d'*alcool corinnique* $C_{35}H_{70}O$ dans une souche de *Corynebacterium diphtheriae* (Gubarev et Vakulenko 1945) rend possible la présence de cires de ce type dans ces bacilles.

4. Microsides.

Sous ce terme, proposé par Gubarev (1951), sont groupés les esters d'acides gras avec des constituants osidiques, d'origine bactérienne. Certaines fractions continuent à être improprement appelées «cires» en raison du long usage de cette dénomination.

De telles substances sont très répandues dans les lipides bactériens, principalement chez les Mycobactéries. La présence d'esters de tréhalose dans les graisses de *Mycobacterium tuberculosis* a déjà été mentionnée (voir page 94). L'acide mycolique (voir page 103), constituant important des Mycobactéries, est présent principalement sous forme d'esters, soit avec un polysaccharide (renfermant chez les souches humaines: arabinose, galactose et mannose, ainsi que les trois acides aminés alanine, acide L-glutamique et acide α,ε-diaminopimélique [Asselineau, Choucroun et Lederer 1950]), soit avec le tréhalose chez *M. avium* (Reeves et Anderson 1937) et chez *M. phlei* (Pangborn et Anderson 1936).

Le «*cord factor*», facteur toxique isolé des souches de Mycobactéries qui poussent en formant des cordes (c'est à dire souches virulentes ou atténuées) consiste en dimycolyl-6,6' tréhalose (IV) (Noll, Bloch, Asselineau et Lederer 1956)[1]; le même type de toxicité est retrouvé chez des esters synthétiques d'acides mycoliques et de glucose ou glucosamine (liés à l'hydroxyle 6 du composant osidique) (Asselineau, Bloch et Lederer 1954, Asselineau et Lederer 1955b).

$$HO)-C_{60}H_{120}-CH(OH)-CH(C_{24}H_{49})-COO-CH_2 \quad \ldots \quad CH_2-OOC-CH(C_{24}H_{49})-CH(OH)-C_{60}H_{120}(OH)$$

Cord factor (IV)[2]

La présence d'esters de polysaccharides (à base de galactose et de mannose) a été mise en évidence chez *Corynebacterium diphtheriae* (Gubarev et Vakulenko 1945, Gubarev et Lubenets 1948, Gubarev, Lubenets, Kanchukh et Galaev 1951, Lubenets 1954, Kanchukh 1954), ainsi que des esters de tréhalose avec des hydroxy-acides (Gubarev, Lubenets et Galaev 1953, Alimova 1955), susceptibles de présenter une certaine analogie avec le «cord factor» des Mycobactéries.

[1] Les acides mycoliques pouvant présenter de légères différences de structure d'une souche de Mycobactérie à une autre, il en résulte l'existence de cord factors, différant par la nature de l'acide mycolique qu'ils renferment.

[2] La synthese du cord factor vient d'être réalisée par T. Gendre et E. Lederer (Bull. Soc. chim. France **1956**, 1478—1482).

Signalons enfin l'isolement à partir de *Pseudomonas*, de composés renfermant de l'acide (—)β-hydroxy-*n*-décanoïque et du L-rhamnose liés sous forme de *glycoside* (BERGSTRÖM, THEORELL et DAVIDE 1946, JARVIS et JOHNSON 1949). Le rhamnoside isolé de *Ps. aeruginosa* (JARVIS et JOHNSON 1949) possède la structure probable (V).

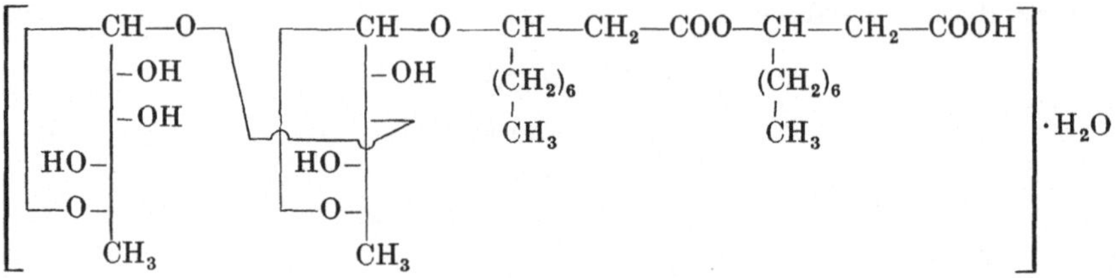

```
┌ ┌──CH—O─┐  ┌───CH—O—CH—CH2—COO—CH—CH2—COOH ┐
│ │   |-OH  \ │    |-OH  (CH2)6       (CH2)6   │
│ │   |-OH   \│    |     CH3          CH3      │ ·H2O
│ HO-|        HO-|                             │
│ └──O-|      └──O-|                           │
└     CH3          CH3                         ┘
```

Rhamnoside de *Ps. aeruginosa* (V)

L'acide (—)β-hydroxy-décanoïque a été d'autre part isolé d'un polypeptide antibiotique, la viscosine (OHNO, TAJIMA et TOKI 1953), produit par *Pseudomonas viscosa* et un acide hydroxylé probablement identique vient d'être isolé sous forme d'amide avec la L-serine *(acide serratamique)* de souches de *Serratia* (CARTWRIGHT 1955).

5. Lipo-polysaccharides de bacteries gram-negatives.

Ces complexes qui représentent 5 à 10% du poids des bacilles et possèdent d'importantes propriétés biologiques, ont fait l'objet d'une revue récente (WESTPHAL et LÜDERITZ 1954). Deux fractions lipidiques distinctes peuvent en être isolées: l'une, facilement éliminable sans perte des propriétés biologiques essentielles, constitue le *phospholipide B* (environ 10% du complexe), et l'autre, isolable seulement par dégradation poussée accompagnée de la perte de certaines propriétés biologiques, est appelée *phospholipide A* (5 à 15% du complexe). L'hydrolyse acide du phospholipide A de *Escherichia coli* (IKAWA, KOEPFLI, MUDD et NIEMANN 1953a, b) fournit, d'une part les acides laurique, myristique, palmitique, et D(—)β-hydroxy-myristique, et d'autre part la D-glucosamine, l'éthanolamine, la nécrosamine, l'acide aspartique et l'acide glycéro-phosphorique. Les fractions hydrosolubles obtenues par hydrolyse des phospholipides A d'une autre souche de *Escherichia coli* et de *Salmonella abortus equi*, étudiées par WERNER, LÜDERITZ et WESTPHAL (1954), présentent une composition semblable. Le phospholipide A, couplé à une protéine inactive (caséïne par exemple), est capable de lui conférer des propriétés fortement toxiques et pyrogènes. «Il semble sûr que la toxicité des endotoxines de bactéries Gram-négatives est liée à la présence du composant lipidique A dans le complexe initial» (WESTPHAL et LÜDERITZ 1954).

Parmi les travaux récents dans ce domaine, citons encore l'étude de l'antigène somatique de *Shigella dysenteriae* (DAVIES, MORGAN et RECORD 1955), celle d'un lipopolysaccharide isolé de *Shigella sonnei* phase II (JESAITIS et GOEBEL 1955), et l'isolement à partir de *Shigella sonnei*, d'un antigène lipopolysaccharidique de faible toxicité (JOHNSON 1955).

6. Acides gras.

Les principaux acides gras isolés de bactéries, avec la formule développée qui leur est attribuée, ont été classés dans le tableau 3.

Il semble qu'en ce qui concerne les *acides à chaîne normale*, il n'existe pas de terme supérieur à C_{26}. ASANO et TAKAHASHI (1945) font mention d'un acide

Tableau 3. *Acides gras isolés des bactéries.*

Nom de l'acide et formule	Souche bactérienne	Référence
I. *Acides normaux saturés* ($>C_{10}$): $CH_3—(CH_2)_n—COOH$.		
acide laurique, n = 10	*E. coli*	IKAWA, KOEPFLI, ... (1953a)
	L. acidophilus	CROWDER, ANDERSON (1934a)
	S. ballerup	ČMELIK (1953)
	S. typhimurium	AKASI (1939)
	Str. hemolyticus	HOFMANN, TAUSIG (1955a)
acide myristique, n = 12	*C. diphtheriae*	ASANO, TAKAHASHI (1945)
	E. coli	IKAWA, KOEPFLI, ... (1953a)
	L. acidophilus	CROWDER, ANDERSON (1934a)
	«*M. leprae*»	ANDERSON, REEVES, CROWDER (1937)
	S. typhimurium	AKASI (1939)
	Str. hemolyticus	HOFMANN, TAUSIG (1955a)
acide palmitique, n = 14	*A. tumefaciens*	GEIGER, ANDERSON (1939)
	B. dysenteriae	MORGAN, PARTRIDGE (1940)
	C. diphtheriae	CHARGAFF (1931), MACHEBOEUF, CASSAGNE (1935), ASANO, TAKAHASHI (1945), PUDLES (1954), PUSTOVALOV (1956)
	E. coli	IKAWA, KOEPFLI, ... (1953a)
	L. acidophilus	CROWDER, ANDERSON (1934a, b)
	M. tuberculosis[1]	ANDERSON (1939) ASSELINEAU (1953a)
	«*M. leprae*»	ANDERSON, REEVES, CROWDER (1937)
	S. paratyphi C	ČMELIK (1954a)
	S. typhimurium	AKASI (1939)
	Str. hemolyticus	HOFMANN (1953)
acide stéarique, n = 16	*L. acidophilus*	CROWDER, ANDERSON (1934a, b)
	L. arabinosus	HOFMANN, LUCAS, SAX (1952)
	L. casei	HOFMANN, SAX (1953)
	M. tuberculosis[1]	ANDERSON (1939)
	«*M. leprae*»	ANDERSON, REEVES, CROWDER (1937)
	S. typhi	ČMELIK (1952)
	S. ballerup	ČMELIK (1953)
	Str. hemolyticus	HOFMANN (1953)
acide docosanoïque, n = 20	*C. diphtheriae*	ASANO, TAKAHASHI (1945)
	«*M. leprae*»	ANDERSON, REEVES, CROWDER (1937)
acide tétracosanoïque, n = 22	*C. diphtheriae*	ASANO, TAKAHASHI (1945)
	L. acidophilus	CROWDER, ANDERSON (1934b)
	«*M. leprae*»	ANDERSON, REEVES, CROWDER (1937)
	M. phlei, smegmatis	BARBIER, LEDERER (1954)
	M. bruynoghi	ASSELINEAU (1954b)

[1] Sauf mention spéciale, il s'agit de tous les types de *M. tuberculosis*, ainsi que de *M. avium* et *M. phlei*.

Tableau 3. (Suite.)

Nom de l'acide et formule	Souche bactérienne	Référence
acide hexacosanoïque, n = 24	*M. tuberculosis, var. hominis*	ANDERSON (1939), ASSELINEAU (1953a)
	var. bovis	ANDERSON (1939) MACHEBOEUF, FETHKE, LEVY (1934)
II. *Acides normaux insaturés.*		
acide crotonique $CH_3—CH=CH—COOH$	*M. tuberculosis var. hominis*	GORIS, SABETAY (1946)
acide tétradécénoïque $CH_3—(CH_2)_x—CH=CH—(CH_2)_y—COOH$ $x+y=10$	*C. diphtheriae*	ASANO, TAKAHASHI (1945)
	«*M. leprae*»	ANDERSON, REEVES, CROWDER (1937)
acide hexadécène-9 oïque (acide palmitoléïque) $CH_3—(CH_2)_5—CH=CH—(CH_2)_7—COOH$	*C. diphtheriae*	CHARGAFF (1931), ASANO, TAKAHASHI (1945), PUDLES (1954) PUSTOVALOV (1956)
	«*M. leprae*»	ANDERSON, UYEI (1932)
	S. typhimurium	AKASI (1939)
	Str. hemolyticus	HOFMANN, TAUSIG (1955a)
acide hexadécène-11 oïque $CH_3—(CH_2)_3—CH=CH—(CH_2)_9—COOH$	*Str. hemolyticus*	HOFMANN, TAUSIG (1955a)
acide hexadécène-x oïque $CH_3—(CH_2)_x—CH=CH—(CH_2)_y—COOH$ $x+y=12$	*S. typhi*	ČMELIK (1952)
acide octadécène-2 oïque $CH_3—(CH_2)_{14}—CH=CH—COOH$	*M. phlei*	SNOW (1954a)
acide octadécène-9 oïque (acide oléïque) $CH_3—(CH_2)_7—CH=CH—(CH_2)_7—COOH$	*B. dysenteriae*	MORGAN, PARTRIDGE (1940)
	M. tuberculosis	GORIS (1920) ANDERSON (1939)
	S. ballerup	ČMELIK (1953)
	S. typhimurium	AKASI (1939)
	Str. hemolyticus	HOFMANN, TAUSIG (1955a)
acide octadécène-11 oïque (acide *cis*-vaccénique) $CH_3—(CH_2)_5—CH=CH—(CH_2)_9—COOH$	*A. tumefaciens*	HOFMANN, TAUSIG (1955b)
	L. arabinosus	HOFMANN, LUCAS, SAX (1952)
	L. casei	HOFMANN, SAX (1953)
	Str. hemolyticus	HOFMANN, TAUSIG (1955a)
acide octadécène-x oïque $CH_3—(CH_2)_x—CH=CH—(CH_2)_y—COOH$ $x+y=14$	*C. diphtheriae*	PUSTOVALOV (1956)
	L. acidophilus [1]	CROWDER, ANDERSON (1943a)
	S. typhi	ČMELIK (1952)
	S. paratyphi C	ČMELIK (1954a)
acide octadéca-diénoïque	*M. tuberculosis*	ANDERSON (1939)
acide eicosénoïque $CH_3—(CH_2)_x—CH=CH—(CH_2)_y—COOH$ $x+y=16$	*C. diphtheriae*	PUSTOVALOW (1956)
	«*M. leprae*»	ANDERSON, REEVES, CROWDER (1937)
acide heneicosénoïque, $x+y=17$	*C. diphtheriae*	ASANO, TAKAHASHI (1945)
acide hexacosénoïque, $x+y=22$	*M. tuberculosis, var. hominis*	WIEGHARD, ANDERSON (1938)
acide octacosénoïque, $x+y=24$	*C. diphtheriae*	ASANO, TAKAHASHI (1945)

[1] Dans ce cas particulier, il s'agit probablement d'acide *cis*-vaccénique (voir HOFMANN 1953).

Tableau 3. (Suite.)

Nom de l'acide et formule	Souche bactérienne	Référence
III. *Acides à chaîne ramifiée.*		
acide méthyl-x heptanoïque	*B. polymyxa*	HAUSMANN, CRAIG (1954)
acide (+)méthyl-6 octanoïque CH_3—CH_2—$CH(CH_3)$—$(CH_2)_4$—COOH	*B. polymyxa* *B. colistinus*	BELL, BONE, ENGLISH, ... (1948) ODA, UEDA (1954)
acide tuberculostéarique CH_3—$(CH_2)_7$—$CH(CH_3)$—$(CH_2)_8$—COOH (acide D(—)méthyl-10 stéarique)	*M. tuberculosis, var. hominis* *var. bovis*	ANDERSON (1939) ASSELINEAU (1953a) CASON, SUMRELL, ... (1953) ANDERSON (1939)
acide lactobacillique CH_3—$(CH_2)_5$—CH—(CH_2)—CH—$(CH_2)_9$—COOH (cycle CH_2 entre les deux CH)	*L. arabinosus* *L. casei*	HOFMANN, LUCAS, SAX (1952) MARCO, HOFMANN (1956) HOFMANN, SAX (1953)
acide phytomonique = acide lactobacillique	*A. tumefaciens*	VELICK (1944) HOFMANN, PANOS, TAUSIG (1954)
acide C_{27}-phtiénoïque (ou acide mycolipénique)[1] CH_3—$(CH_2)_{17}$—$CH(CH_3)$—CH_2—$CH(CH_3)$—CH=$C(CH_3)$—COOH	*M. tuberculosis, var. hominis*	ANDERSON (1939)[1] POLGAR (1954a) CASON, FREEMAN, SUMRELL (1951 AEBI, ASSELINEAU, LEDERER (1953) ASSELINEAU, ASSELINEAU, STÄLLBERG-STENHAGEN, (1956)
acide mycocéranique (ou acide mycocérosique, ou C_{31}-mycosanoïque) CH_3—$(CH_2)_n$—$CH(CH_3)$—CH_2—$CH(CH_3)$—CH_2—$CH(CH_3)$—COOH $n \sim 21$	*M. tuberculosis, var. hominis*	GINGER, ANDERSON (1945) POLGAR (1953, 1954b) MARKS, POLGAR (1955) ASSELINEAU (1953a) CASON, FONKEN (1956)
acide diphtérique, $C_{35}H_{68}O_2$	*C. diphtheriae*	CHARGAFF (1931)
acide corinnique, $C_{35}H_{68}O_2$	*C. diphtheriae*	GUBAREV, VAKULENKO (1949)
IV. *Acides hydroxylés.*		
1. à chaîne normale.		
acide D(—)-β-hydroxy-butyrique CH_3—CHOH—CH_2—COOH	*B. megatherium et autres B.*	LEMOIGNE (1946)
acide D(—)β-hydroxy-décanoïque CH_3—$(CH_2)_6$—CHOH—CH_2—COOH	*Ps. pyocyanea* *Ps. aeruginosa* *Ps. viscosa*	BERGSTRÖM, THEORELL, DAVIDE (1946) JARVIS, JOHNSON (1949) OHNO, TAJIMA, TOKI (1953)
acide D(—)β-hydroxy-myristique CH_3—$(CH_2)_{10}$—CHOH—CH_2—COOH	*E. coli*	IKAWA, KOEPFLI, ... (1953a)
acide dihydroxy-stéarique, $C_{17}H_{33}(OH)_2$ (COOH)	*L. acidophilus*	CROWDER, ANDERSON (1932)

[1] La présence d'homologues en C_{25}, C_{26} et C_{29} vient d'être observée par CASON et FONKEN (1956); les souches avirulentes de *M. tuberculosis var. hominis* sont dépourvues de tels acides (CASON, ALLEN, DEACETIS, FONKEN 1956, ASSELINEAU 1953a).

Tableau 3. (Suite.)

Nom de l'acide et formule	Souche bactérienne	Référence
lipide β-hydroxy-butyrique (VI)	*B. megatherium* et *autres B.*	LEMOIGNE (1946) KEPES, PEAUD-LENOËL (1952)
	Az. chroococcum	LEMOIGNE, GIRARD (1943)
	Mic. halodenitrificans	SMITHIES, GIBBONS, BAYLEY (1955)

$$\text{(VI)}\quad HO-\underset{\underset{CH_3}{|}}{CH}-CH_2-\underset{\underset{O}{\|}}{C}-O-\underset{\underset{CH_3}{|}}{CH}-CH_2-\underset{\underset{O}{\|}}{C}-O-\underset{\underset{CH_3}{|}}{CH}-CH_2-\underset{\underset{O}{\|}}{C}-\cdots-O-\underset{\underset{CH_3}{|}}{CH}-CH_2-COOH$$

2. à chaine ramifiée.

Nom de l'acide et formule	Souche bactérienne	Référence
acide hydroxy-3 méthyl-2 pentanoïque $CH_3-CH_2-CHOH-CH(CH_3)-COOH$	*M. phlei*	SNOW (1954b)
acide coryno-mycolique $CH_3-(CH_2)_{14}-CHOH-\underset{\underset{C_{14}H_{29}}{\vert}}{CH}-COOH$	*C. diphtheriae*	LEDERER, PUDLES (1951) LEDERER, PUDLES, BARBEZAT, TRILLAT (1952) PUSTOVALOV (1956)
acide coryno-mycolénique $CH_3-(CH_2)_5-CH{=}CH-(CH_2)_7-CHOH-\underset{\underset{C_{14}H_{29}}{\vert}}{CH}-COOH$	*C. diphtheriae*	PUDLES, LEDERER (1953, 1954a, b)
acide corynolique (corynine) $CH_3-CHOH-(CH_2)_7-\underset{\underset{CH_3}{\vert}}{CH}-CHOH-\underset{\underset{CH_3}{\vert}}{CH}-\underset{\underset{CH_3}{\vert}}{CH}-\underset{\underset{COOH}{\vert}}{CH}-(CH_2)_{17}-\underset{\underset{CH_3}{\vert}}{CH}-(CH_2)_{14}-CH_3$	*C. diphtheriae*	CHARGAFF (1931) TAKAHASHI (1948)
acide «C_{60} Brévannes» $C_{59}H_{118}(OH)(COOH) \pm 3\ CH_2$	*M. tuberculosis, var. hominis*	ASSELINEAU (1954a)
acides mycoliques $R-CHOH-\underset{\underset{C_{24}H_{49}}{\vert}}{CH}-COOH$ $R = C_{60}H_{120}(OCH_3)$ ou $C_{60}H_{120}(OH)$ ou $C_{60}H_{121}$ $\}\ \pm 5\ CH_2$	*M. tuberculosis, var. hominis*	STODOLA, LESUK, ANDERSON (1938) ASSELINEAU, LEDERER (1951, 1953b, 1955a) ASSELINEAU, GENDRE (1954)
	var. bovis	CASON, ANDERSON (1938) DEMARTEAU (1951)
$R-CHOH-\underset{\underset{C_{22}H_{45}}{\vert}}{CH}-COOH$ $R = C_{58}H_{116}(OH) \pm 5\ CH_2$	*M. phlei* *M. smegmatis*	BARBIER, LEDERER (1954)
	M. bruynoghi	ASSELINEAU (1954b)
acides mycoloniques $R-CHOH-\underset{\underset{C_{24}H_{49}}{\vert}}{CH}-COOH$ $R = C_{59}H_{119}(C{=}O) \pm 5\ CH_2$	*M. tuberculosis, var. hominis* *var. bovis*	ASSELINEAU (1953b) DEMARTEAU, LEDERER (1952) GINSBURG, LEDERER (1952)
acides mycoliques dicarboxyliques, $C_{76}H_{148}O_6$	*M. avium*	ANDERSON, CREIGHTON (1939)
acide mycolonique dicarboxylique, $C_{68}H_{132}O_6$	*M. phlei*	CLERMONTÉ, LEDERER (1956) PECK, ANDERSON (1941)

octacosénoïque, dans les lipides du bacille diphtérique; cet acide n'a pas été retrouvé par PUDLES (1954). Des mélanges complexes de termes homologues, tels que ceux que l'on rencontre dans les cires végétales, semblent rares dans les lipides bactériens: dans le cas des acides normaux saturés de *Mycobacterium tuberculosis* par exemple, on n'observe aucun terme intermédiaire entre C_{18} et C_{26} (ANDERSON 1939, ASSELINEAU 1953a)[1]; le principal acide saturé du bacille diphtérique est l'acide palmitique, et PUDLES (1954) n'a trouvé aucun intermédiaire entre cet acide et l'hydroxy-acide ramifié coryno-mycolique en C_{32}. [Cependant, PUSTOVALOV (1956) mentionne l'isolement d'acides octadécénoïque et eicosénoïque.]

Les *acides normaux insaturés* ont souvent été caractérisés par leur indice d'iode et par la nature de leur produit d'hydrogénation. HOFMANN (1953) vient de montrer que chez plusieurs espèces de *Lactobacillus*, la fraction acide monoinsaturée en C_{18} est constituée par l'acide *cis*-vaccénique (acide octadécène-11 oïque); par suite, certaines «identifications» d'acide oléïque trouvées dans la littérature sont sujettes à caution. Très peu d'acides fortement insaturés ont été caractérisés.

La coexistence chez un même bacille *(Lactobacillus arabinosus*, ou *L. casei)* d'acide *cis*-vaccénique (ou octadécène-11 oïque) et d'acide lactobacillique (ou méthylène-11,12 octadécénoïque) pourrait permettre d'envisager une relation biogénétique entre ces deux acides.

Tous les *acides hydroxylés* dont la structure est connue sont *β-hydroxylés*, sauf l'acide corynolique (TAKAHASHI 1948) isolé du bacille diphtérique; or, la structure très complexe attribuée à cet acide rendrait nécessaire certains examens pour obtenir sa confirmation. Dans le cas de deux autres acides, la position de l'hydroxyle n'a pas été déterminée: il s'agit de l'acide dihydroxy-stéarique de *Lactobacillus acidophilus* (CROWDER et ANDERSON 1932), et d'un hydroxy-acide en C_{60} isolé d'une souche humaine de *Mycobacterium tuberculosis* (ASSELINEAU 1954a) (cependant, le spectre infra-rouge de ce dernier acide est compatible avec la présence de l'hydroxyle en β). Les acides β-hydroxylés dont la configuration a été déterminée appartiennent à la série D (ainsi que les alcools secondaires (+)eicosanol-2 et (+)octadécanol-2 mentionnés p. 97) (cf. SERCK-HANSSEN et STENHAGEN 1955).

Les *acides mycoliques*, acides β-hydroxylés à haut poids moléculaire ($C_{88}H_{176}O_4$, $C_{87}H_{174}O_4$ et $C_{87}H_{174}O_3$)[2], acido-résistants, sont isolés à partir de chaque souche de Mycobactéries sous forme de mélanges souvent complexes d'acides chimiquement très voisins. Ces acides, séparables par chromatographie, diffèrent entre eux, en particulier, par la nature des fonctions oxygénées présentes en dehors du carboxyle. Les acides mycoliques en O_3 possèdent un hydroxyle en position 3; les acides en O_4 possèdent, en plus de l'hydroxyle en position 3, un hydroxyle, ou un méthoxyle, ou un groupe carbonyle. La présence d'un hydroxyle en position 3 et d'une longue ramification en position 2, entraine pour ces acides la propriété de se couper par pyrolyse avec libération, d'*acide n-hexacosanoïque* dans le cas des acides de souches humaines et bovines, et d'*acide n-tétracosanoïque* dans le cas des acides de souches aviaires et saprophytes. Des formules

[1] Cependant, CASON et FONKEN (1956) ont observé la présence, dans les lipides de bacilles tuberculeux (souches humaines virulentes), de mélanges complexes d'acides ramifiés supérieurs, et en particulier, d'acides ramifiés α,β-insaturés du type *phtiénoïque* en C_{25}, C_{26}, C_{27}, et C_{29}.

[2] Ces formules, sujettes à une incertitude de $\pm 5\,CH_2$, sont employées par convention dans le cas des acides possédant un méthoxyle (C_{88}), ou dépourvus de méthoxyle (C_{87}) (ASSELINEAU et LEDERER 1953b).

développées ont été proposées pour plusieurs acides mycoliques (pour la chimie des acides mycoliques, voir ASSELINEAU et LEDERER 1953b et 1955a).

L'examen du tableau 3 montre que deux familles bactériennes se distinguent nettement des autres d'après les propriétés de leurs *acides ramifiés*: ce sont les Mycobactéries et *Corynebacterium diphtheriae*. Les acides ramifiés isolés des autres espèces bactériennes ne dépassent pas C_{20} et ne renferment dans leur molécule que des ramifications méthyles. Au contraire, les lipides des Mycobactéries et du Bacille diphtérique sont riches en acides à haut poids moléculaire (C_{27}, C_{31}, C_{32}, C_{52}, C_{88}). Les acides coryno-mycolique et mycoliques possèdent des ramifications de 14 à 24 atomes de carbone, et semblent constituer les seuls exemples de produits aliphatiques naturels ayant de si longues chaînes latérales. Il serait intéressant à ce point de vue de connaître la composition d'autres espèces de Corynebactéries.

Bibliographie.

AEBI, A., J. ASSELINEAU et E. LEDERER: Sur les lipides de la souche humaine «Brévannes» de *Mycobacterium tuberculosis*. Bull. Soc. chim. biol. Paris **35**, 661—684 (1953). — AKASI, S.: On the fatty substance obtained from the typhoid bacillus of mice. J. of Biochem. (Tokyo) **29**, 13—20 (1939). — The chemical components of *Bacillus subtilis*. Lecture Coll. Res. Inst. Kyoto Univ. **12**, 195 (1940). — ALIMOVA, E. K.: The chromatographic study of the carbohydrates of microsides (esters of carbohydrates and fatty acids). Biokhimiya **20**, 516—521 (1955). — ANDERSON, R. J.: The chemistry of the lipoids of the tubercle bacillus and certain other microorganisms. Fortschr. Chem. organ. Naturstoffe **3**, 145—202 (1939). — ANDERSON, R. J., and M. M. CREIGHTON: The chemistry of the lipoids of tubercle bacilli. LVII. The mycolic acids of the avian tubercle bacillus wax. J. of Biol. Chem. **129**, 57—63 (1939). — ANDERSON, R. J., W. C. LOTHROP and M. M. CREIGHTON: LIII. Studies of the phosphatides of the human tubercle bacillus. J. of Biol. Chem. **125**, 299—308 (1938). — ANDERSON, R. J., and M. S. NEWMAN: XXXIII. Isolation of trehalose from the acetone soluble fat of the human tubercle bacillus. J. of Biol. Chem. **101**, 499—504 (1933a). — XXXIV. Isolation of a pigment and of anisic acid from the acetone soluble fat of the human tubercle bacillus. J. of Biol. Chem. **101**, 773—779 (1933b). — XXXV. The constitution of phthiocol, the pigment isolated from human tubercle bacillus. J. of Biol. Chem. **103**, 197—201 (1933c). — ANDERSON, R. J., R. E. REEVES, M. M. CREIGHTON and W. C. LOTHROP: LXV. An investigation of tuberculous lung tissue. Amer. Rev. Tbc. **48**, 65—75 (1943). — ANDERSON, R. J., R. E. REEVES and J. A. CROWDER: LII. The composition of the acetone soluble fat of *Bacillus leprae*. J. of Biol. Chem. **121**, 669—684 (1937). — ANDERSON, R. J., R. SCHOENHEIMER, J. A. CROWDER u. F. H. STODOLA: XL. Über das Vorkommen von Sterinen in Tuberkelbazillen. Z. physiol. Chem. **237**, 40—45 (1935). — ANDERSON, R. J., and N. UYEI: XXVII. The composition of the phosphatide fraction of the *Bacillus leprae*. J. of Biol. Chem. **97**, 617—637 (1932). — ASANO, M., and H. TAKAHASHI: Bacterial components of *Corynebacterium diphtheriae*. I. Studies of fats. J. Pharmaceut. Soc. Japan **65**, 17—19 (1945). — Chem. Abstr. **45**, 3906 (1951). — II. Phosphatides and wax. J. Pharmaceut. Soc. Japan **65**, 81 (1945). — Chem. Abstr. **45**, 4303 (1951). — ASSELINEAU, C., J. ASSELINEAU, S. STÄLLBERG-STENHAGEN u. E. STENHAGEN: Synthesis of racemic methyl phthienoate. Acta chem. scand. (Copenh.) **10**, 478—480 (1956). — ASSELINEAU, J.: Acides mycoliques et cires du Bacille tuberculeux. Thèse Doctorat, Paris 1950. Paris: Arnette 1951a. — Sur la variation de la teneur en lipides du Bacille tuberculeux en fonction de l'âge de la culture. Ann. Inst. Pasteur 81, 306—309 (1951b). — Lipides du Bacille tuberculeux: constitution chimique et activité biologique. Progr. Explor. Tbc. **5**, 1—55 (1952). — Sur les acides gras de la souche humaine H-37 Ra de *Mycobacterium tuberculosis*. C. r. Acad. Sci. Paris **237**, 1804—1806 (1953a). — Isolement de nouveaux acides mycoliques des souches humaines Test, R_1 et L-25 de *M. tuberculosis*; présence d'un acide mycolonique dans une souche humaine. Biochim. et Biophysica Acta **10**, 453—461 (1953b). — Sur quelques substances à 60 atomes de carbone isolées des lipides de souches humaines de *M. tuberculosis*. Bull. Soc. chim. France **1954**a, 108—112. — Résultats inédits, 1954b. — ASSELINEAU, J., H. BLOCH and E. LEDERER: Synthesis of cord factor—active substances. Biochim. et Biophysica Acta **15**, 136—137 (1954). — ASSELINEAU, J., N. CHOUCROUN and E. LEDERER: Sur la constitution chimique d'un lipo-polysaccharide antigénique isolé de *M. tuberculosis* var. *hominis*. Biochim. et Biophysica Acta **5**, 197—203 (1950). — ASSELINEAU, J., et T. GENDRE: Sur les acides mycoliques de trois souches humaines de

M. tuberculosis: H-37 Ra, H-37 Rv et un mutant Rv streptomycino-résistant. Bull. Soc. chim. France **1954**, 1226—1233. — ASSELINEAU, J., and E. LEDERER: Sur la constitution chimique des acides mycoliques de deux souches humaines virulentes de *M. tuberculosis*. Biochim. et Biophysica Acta **7**, 125—146 (1951). — Sur la structure chimique de l'acide α-mycolique de la souche humaine Test de *M. tuberculosis*. Bull. Soc. chim. France **1953**a, 335—339. — Chimie des lipides bactériens. Progr. Chim. Subst. organ. natur. **10**, 170—273 (1953b). — Chemical structure and biological activity of mycolic acids. Experimental tuberculosis. A Ciba Foundation Symposium, p. 14—38. London: J. & A. Churchill 1955a. — Synthèses de substances ayant des structures analogues à celle du «cord factor» (esters mycoliques de sucres, esters et amides mycoliques de sucres aminés). Bull. Soc. chim. France **1955**b, 1232—1240.

BARBIER, M.: Etude chimique et biochimique des lipides de *M. phlei* et de *M. smegmatis*. Thèse Doctorat, Paris 1954. — BARBIER, M., et E. LEDERER: Sur un acide aminé du phosphatide de *M. phlei*. Biochim. et Biophysica Acta **8**, 591—592 (1952). — Influence de l'acétate de sodium sur la croissance de Mycobactéries et sur leur teneur en lipides. VI. Congr. Internat. Microbiol., Rome 1953; résumé comm., p. 86. — Sur l'isolement et la constitution chimique des acides mycoliques de *M. phlei* et de *M. smegmatis*. Biochim. et Biophysica Acta **14**, 246—258 (1954). — BELL, P. H., J. F. BONE, J. P. ENGLISH, C. E. FELLOWS, K. S. HOWARD, M. M. ROGERS, R. G. SHEPHERD and R. WINTERBOTTOM: Chemical studies on polymyxin: comparison with "aerosporin". Ann. New York Acad. Sci. **51**, 897—907 (1948). — BENDICH, A., and E. CHARGAFF: The isolation and characterization of two antigenic fractions of *Proteus* OX-19. J. of Biol. Chem. **166**, 283—312 (1946). — BERGEY's Manual of determinative bacteriology, 6th edit. Baltimore: Williams & Wilkins Company 1953. — BERGSTRÖM, S., H. THEORELL and H. DAVIDE: Pyolipic acid, a metabolic product of *Pseudomonas pyocyanea* active against *M. tuberculosis*. Arch. of Biochem. **10**, 165—166 (1946). — BLOCH, K.: Über eine Phosphatidsäure aus humanen Tuberkelbazillen. Z. physiol. Chem. **244**, 1—13 (1936). — Über phosphorhaltige Lipoide aus humanen Tuberkelbazillen. Biochem. Z. **285**, 372—385 (1936). — BÖHM, P.: Über das Vorkommen von Acetalphosphatiden in Bakterien. I. Tuberkelbakterien. Z. physiol. Chem. **291**, 155—157 (1952). — BUTENANDT, A., u. F. H. STODOLA: Zur Kenntnis von α- und β-Leprosol. Liebigs Ann. **539**, 40—56 (1939).

CARTWRIGHT, N. J.: Serratamic acid, a derivative of L-serine produced by organisms of the *Serratia* group. Biochemic. J. **60**, 238—242 (1955). — CASON, J., C. F. ALLEN, W. DEACETIS and G. J. FONKEN: Fatty acids from the lipides of non virulent strains of the tubercle bacillus. J. of Biol. Chem. **220**, 893 (1956). — CASON J., and R. J. ANDERSON: LVI. The wax of the bovine tubercle bacillus. J. of Biol. Chem. **126**, 527—541 (1938). — CASON, J., and G. J. FONKEN: Complexity of the mixture of the higher fatty acids from the lipides of the tubercle bacillus. J. of Biol. Chem. **220**, 391—405 (1956). — CASON, J., N. K. FREEMAN and G. SUMRELL: The principal features of C_{27}-phthienoic acid. J. of Biol. Chem. **192**, 415—424 (1951). — CASON, J., and M. J. KALM: Assignment of geometric configuration in the 2-methyl-2-alkenoic acids. J. Organ. Chem. **19**, 1947—1959 (1954). — CASON, J., G. SUMRELL, C. F. ALLEN, G. A. GILLIES and S. ELBERG: Certain characteristics of the fatty acids from the lipids of the tubercle bacillus. J. of Biol. Chem. **205**, 435—447 (1953). — CASSAGNE, H.: Recherches biochimiques sur le Bacille diphtérique Etude de l'extraction par l'acétone de bacilles d'âges différents. C. r. Soc. Biol. Paris **131**, 693—695 (1939). — CHARGAFF, E.: Zur Chemie der Bakterien. I. Über die Lipoide der Diphtheriebakterien. Z. physiol. Chem. **201**, 191—198 (1931). — CLERMONTE, R., et E. LEDERER: Sur la structure chimique des acides phleimycoliques dicarboxyliques. C. r. Acad. Sci. Paris **242**, 2600—2603 (1956). — ČMELIK, S.: Über Bakterienlipoide. I. Die Lipoide von *Salmonella typhi*. Z. physiol. Chem. **290**, 146—150 (1952). — II. Die Lipoide von *Salmonella ballerup (Paracolobactrum ballerup)*. Z. physiol. Chem. **293**, 222—229 (1953).— III. Untersuchung verschiedener Lipoidfraktionen von *Salmonella paratyphi C*. Z. physiol. Chem. **296**, 67—73 (1954a). — Über den Einfluß verschiedener Nährlösungen auf die Verteilung der Lipoide von *Corynebacterium diphtheriae*. Schweiz. Z. Path. u. Bakter. **17**, 289—295 (1954b). — Lipoide und Virulenz bei Typhusbakterien. Experientia (Basel) **10**, 372—373 (1954c). — Über Bakterienlipoide. V. Zur Kenntnis der Lipoide von Typhusbakterien. Z. physiol. Chem. **299**, 227—234 (1955a). — Lipoidgehalt von Typhusbakterien aus Kultüren verschiedenen Alters. Naturwiss. **42**, 489 (1955b). — Über Phosphatide von Typhusbakterien. Z. physiol. Chem. **302**, 20—28 (1955c). — CREIGHTON, M. M., L. H. CHANG and R. J. ANDERSON: LXVII. The lipids of the human tubercle bacillus H-37 cultivated on a dextrose containing medium. J. of Biol. Chem. **154**, 569—579 (1944). — CROWDER, J. A., and R. J. ANDERSON: A contribution to the chemistry of *Lactobacillus acidophilus*. I. The occurrence of free, optically active dihydroxy-stearic acid in the fat extracted from *L. acidophilus*. J. of Biol. Chem. **97**, 393—401 (1932). — II. Composition of the neutral fat. J. of Biol. Chem. **104**, 399—406 (1934a). — III. The composition of the phosphatide fraction. J. of Biol.

Chem. **104**, 487—495 (1934b). — Crowder, J. A., F. H. Stodola and R. J. Anderson: XLV. The isolation of α- and β-leprosols. J. of Biol. Chem. **114**, 431—439 (1936).

Davies, D. A. L., W. T. J. Morgan and B. R. Record: The specific polysaccharide of the dominant "O" somatic antigen of *Shigella dysenteriae*. Biochemic. J. **60**, 290—303 (1955). — Demarteau, H.: Sur les acides mycoliques de la souche bovine Vallée de *M. tuberculosis*. C. r. Acad. Sci. Paris **232**, 2494—2496 (1951). — Demarteau, H., et E. Lederer: Sur un acide mycolonique, nouveau type d'acide mycolique isolé d'une souche bovine de *M. tuberculosis*. C. r. Acad. Sci. Paris **235**, 265—267 (1952). — Demarteau-Ginsburg, H., A. Ginsburg et E. Lederer: Sur trois nouvelles substances apparentées au phtiocérol. Biochim. et Biophysica Acta **12**, 587—588 (1953). — Sur la constitution chimique du phtiocérol. C. r. Acad. Sci. Paris **240**, 815—817 (1955).

Fethke, N.: Substances lipoïdiques du Bacille tuberculeux: leur constitution chimique et ses rapports avec l'infection bacillaire. Paris: Hermann 1938. — Few, A. V.: The interaction of polymyxin E with bacterial and other lipids. Biochim. et Biophysica Acta **16**, 137—145 (1955).

Geiger, W. B., and R. J. Anderson: The chemistry of *Phytomonas tumefaciens*. I. The lipids of *P. tumefaciens*. The composition of the phosphatide. J. of Biol. Chem. **129**, 519—529 (1939). — Gendre, T., et E. Lederer: Sur les substances azotées des phosphatides de quelques Mycobactéries. Ann. Acad. Sci. fenn., Ser. A, II. **60**, 313—320 (1955). — Ginger, L. G., and R. J. Anderson: LXXII. Fatty acids occurring in the wax prepared from tuberculin residues. Concerning mycocerosic acid. J. of Biol. Chem. **157**, 203—211 (1945). — Goodwin, T. W., and M. Jamikorn: Carotenoid production by a strongly chromogenic bacterium isolated from butter. Biochemic. J. **62**, 275—280 (1956). — Goris, A.: Composition chimique du Bacille tuberculeux. Ann. Inst. Pasteur **34**, 496—533 (1920). — Goris, A., et S. Sabetay: Sur la nature de la matière odorante des cultures de Bacille tuberculeux. C. r. Acad. Sci. Paris **223**, 933—934 (1946). — Gubarev, E. M.: Chemical composition of diphtheria bacteria and the biological action of hydroxy-diphtheric acids. Uspekhi Sovremennoi Biol. **31**, 108 (1951). — Gubarev, E. M., A. B. Gabrilovich and F. I. Shtulbaum: The combined lipids of diphtheria bacilli. Biokhimiya **17**, 303—306 (1952). — Gubarev, E. M., and E. K. Lubenets: Some structural peculiarities of diphtheria microbe. Dokl. Akad. Nauk USSR. **60**, 413—416 (1948). — Chem. Abstr. **42**, 7826 (1948). — Gubarev, E. M., E. K. Lubenets and Y. V. Galaev: Chemical composition of some fractions of lipides of diphtheria bacteria. Biokhimiya **18**, 37 (1953). — Gubarev, E. M., E. K. Lubenets, A. A. Kanchukh and Y. V. Galaev: Fractionation and composition of some lipide fractions of diphtheria bacteria. Biokhimiya **16**, 139—145 (1951). — Gubarev, E. M., and I. L. Vakulenko: Composition and properties of lipids from *C. diphtheriae*. Biokhimiya **10**, 285—295 (1945).

Hall, J. A., J. W. Lewis and N. Polgar: Constituents of the lipids of tubercle bacilli. VI. Phthiocerol. J. Chem. Soc. (Lond.) **1955**, 3971—3976. — Hall, J. A., and N. Polgar: The constitution of phthiocerol. Chem. a. Ind. **1954**, 1293. — Hanahan, D. J.: The enzymatic degradation of phosphatidylcholine in diethylether. J. of Biol. Chem. **195**, 199—206 (1952). — Hanahan, D. J., and I. L. Chaikoff: The phosphorus containing lipides of the carrot. J. of Biol. Chem. **168**, 233—239 (1947). — Hausmann, W., and L. C. Craig: Polymyxin B_1: Fractionation, molecular-weight determination, amino-acid and fatty acid composition. J. Amer. Chem. Soc. **76**, 4892—4896 (1954). — Hawthorne, J. N., and E. Chargaff: A study of inositol-containing lipides. J. of Biol. Chem. **206**, 27—37 (1954). — Hayaishi, O., and A. Kornberg: Metabolism of phospholipides by bacterial enzymes. J. of Biol. Chem. **206**, 647—663 (1954). — Hofmann, K.: Chemical nature of fatty acids of bacterial origin. Rec. Chem. Progr. **14**, 7—17 (1953). — Hofmann, K., R. A. Lucas and S. M. Sax: The chemical nature of the fatty acids of *Lactobacillus arabinosus*. J. of Biol. Chem. **195**, 473—485 (1952). — Hofmann, K., C. Panos and F. Tausig: On the identity of phytomonic and lactobacillic acids, and the microbiological activity of cyclopropane fatty acids. Federat. Proc. **13**, 231 (1954). — Hofmann, K., and S. M. Sax: The chemical nature of the fatty acids of *Lactobacillus casei*. J. of Biol. Chem. **205**, 55—63 (1953). — Hofmann, K., and F. Tausig: The chemical nature of the fatty acids of a group C Streptococcus species. J. of Biol. Chem. **213**, 415—423 (1955a). — On the identity of phytomonic and lactobacillic acids. A reinvestigation of the fatty acid spectrum of *Agrobacterium (Phytomonas) tumefaciens*. J. of Biol. Chem. **213**, 425—432 (1955b). — Huston, R. C., I. F. Huddleston and A. D. Hershey: The chemical separation of some cellular constituents of the Brucella group of microorganisms. Michigan Agricult. Exper. Stat. Techn. Bull. **137** (1934).

Ida, M., and T. Otani: Studies on the chemical components of *Haemophilus pertussis*. Acta Scholae Med. Kioto **28**, 25—27 (1950). — Ikawa, M., J. B. Koepfli, S. G. Mudd and C. Niemann: An agent from *E. coli* causing hemorrhage and regression of an experimental mouse tumor. III. The component fatty acids of the phospholipide moiety. J. Amer.

Chem. Soc. **75**, 1035—1038 (1953a). — IV. Some nitrogenous components of the phospholipide moiety. J. Amer. Chem. Soc. **75**, 3439—3442 (1953b).

JARVIS, F. G., and M. J. JOHNSON: A glyco-lipide produced by *Pseudomonas aeruginosa.* J. Amer. Chem. Soc. **71**, 4124—4126 (1949). — JESAITIS, M. A., and W. F. GOEBEL: Lysis of T_4 phage by the specific lipocarbohydrate of phase II *Shigella sonnei.* J. of Exper. Med. **102**, 733—751 (1955). — JOHNSON, R. B.: An immunogenic antigen of low toxicity derived from *Shigella sonnei.* J. of Immun. **74**, 286—291 (1955).

KANCHUKH A. A.: Lipides of chloroform extracts of diphtherical microorganisms. Ukrain. biochem. Ž. **26**, 186—202 (1954). — KEPES, A., et C. PEAUD-LENOËL: Sur la structure de la molécule des lipides β-hydroxy-butyriques. Bull. Soc. Chim. biol. Paris **34**, 563—575 (1952).

LEDERER, E.: Chemistry and biochemistry of Mycobacteria. Colloquium on the chemotherapy of tuberculosis, Trinity College, Dublin 1952a. — Chimie et biochimie des lipides des Mycobactéries. IIème Congrès Intern. Biochim. Symposium sur le métabolisme microbien. Paris 1952b. — LEDERER, E., et J. PUDLES: Sur l'isolement et la constitution chimique d'un hydroxy-acide ramifié du bacille diphtérique. Bull. Soc. Chim. biol. Paris **33**, 1003—1011 (1951). — LEDERER, E., J. PUDLES, S. BARBEZAT et J. J. TRILLAT: Sur la constitution chimique de l'acide coryno-mycolique du Bacille diphtérique. Bull. Soc. chim. France **1952**, 93—95. — LEMOIGNE, M.: Fermentation hydroxy-butyrique. Helvet. chim. Acta **29**, 1303—1306 (1946). — LEMOIGNE, M., et H. GIRARD: Réserves lipidiques β-hydroxy-butyriques. C. r. Acad. Sci. Paris **217**, 557—559 (1943). — LEMOIGNE, M., G. MILHAUD et M. CROSON: Sur le métabolisme lipidique du *Bacillus megatherium.* Bull. Soc. Chim. biol. Paris **31**, 1587—1591 (1949). — LUBENETS, E. K.: Complex lipide compounds in the cells of diphtheria bacteria. Biokhimiya **19**, 11—15 (1954).

MACHEBOEUF, M., et H. CASSAGNE: Etudes chimiques sur le Bacille diphthérique. Extraction fractionnée des lipides du bacille; séparation de la fraction haptène; présence de savon dans les corps bacillaires. C. r. Acad. Sci. Paris **200**, 1988—1990 (1935). — MACHEBOEUF, M., et M. FAURE: Sur l'existence dans les bacilles tuberculeux d'acides phosphatidiques complexes constitues par de l'acide glycéro-phosphorique lié par estérification, d'une part à des acides gras, et d'autre part à des polyalcools non azotés. C. r. Acad. Sci. Paris **209**, 700—702 (1939). — MACHEBOEUF, M., N. FETHKE et G. LEVY: Etudes chimiques sur le Bacille tuberculeux. Essais préliminaires d'extraction et de fractionnement des substances lipoïdiques de corps bacillaires tués par la chaleur. Ann. Inst. Pasteur **52**, 295—307 (1934). — MARCO, G. J., and K. HOFMANN: Structural studies on lactobacillic acid and other long chain fatty acids containing cyclopropane ring. Federat. Proc. **15**, 308 (1956). — MARKS, G. S., and N. POLGAR: Mycoceranic acid. Part II. J. Chem. Soc. (Lond.) **1955**, 3851—3857. — MICHEL, G., et E. LEDERER: Etude chromatographique du phosphatide de *M. marianum.* C. r. Acad. Sci. Paris **240**, 2454—2455 (1955). — MORGAN, W. T. J., and S. M. PARTRIGDE: Studies in immunochemistry. IV. The fractionation and nature of antigenic material isolated from *B. dysenteriae.* Biochemic. J. **34**, 169—191 (1940).

NICOLLE, M., et E. ALILAIRE: Note sur la production en grand des corps bactériens et sur leur composition chimique. Ann. Inst. Pasteur **23**, 547—557 (1909). — NOLL, H.: On the composition of wax fractions of *M. tuberculosis.* J. of Biol. Chem. (sous presse). — NOLL, H., H. BLOCH, J. ASSELINEAU and E. LEDERER: The chemical structure of the cord factor of *M. tuberculosis.* Biochim. et Biophysica Acta, **20**, 299—309 (1956).

ODA, T., and F. UEDA: Chemistry of Colistin. J. Pharmaceut. Soc. Japan **74**, 1246—1248 (1954). — OHNO, T., S. TAJIMA and K. TOKI: On the chemistry of Viscosine. J. Agricult. Chem. Soc. Jap. **27**, 665—669 (1953); **29**, 370 (1955). — OLLEY, J.: Instability of phosphatidic acids. Chem. a. Ind. **1954**, 1069—1071.

PANGBORN, M. C., and R. J. ANDERSON: XXV. The composition of the phosphatide fraction of the Timothy bacillus. J. of Biol. Chem. **94**, 645—472 (1932). — XLI. 1. The composition of the timothy bacillus wax. 2. The isolation of D-Eicosanol-2 and D-Octadecanol-2 from the unsaponifiable matter of timothy bacillus wax. J. Amer. Chem. Soc. **58**, 10—14 (1936). — PECK, R. L., and R. J. ANDERSON: LXIV. Concerning phlei-mycolic acid. J. of Biol. Chem. **140**, 89—96 (1941). — PENNELL, R. B.: The chemistry of Brucella organisms. In: Brucellosis, p. 37—49. Washington 1950. — PHILPOT, F. J., and A. Q. WELLS: Lipids of living and killed tubercle bacilli. Amer. Rev. Tbc. **66**, 28—35 (1952). — POLGAR, N.: The constitution of mycoceranic acid. Chem. a. Ind. **1953**, 353. — Constituents of the lipids of tubercle bacilli. III. Mycolipenic acid. J. Chem. Soc. (Lond.) **1954**a, 1008—1010. — IV. Mycoceranic acid. J. Chem. Soc. (Lond.) **1954**b, 1011—1012. — PUDLES, J.: Etude chimique des lipides du Bacille diphtérique. Thèse Doctorat, Paris 1954. — PUDLES, J., and E. LEDERER: Sur la constitution chimique d'un acide mycolique insaturé isolé du bacille diphtérique *(C. diphtheriae).* Biochim. et Biophysica Acta **11**, 163—164 (1953). — Sur l'isolement et la constitution chimique de l'acide coryno-mycolénique et de deux cétones, des lipides du bacille diphtérique. Bull. Soc. Chim. biol. Paris **36**, 759—777 (1954a). —

Essais synthétiques sur la constitution de l'acide coryno-mycolénique. Bull. Soc. chim. France **1954**b, 919—923. — PUSTOVALOV, V. L.: Higher fatty acids of the diphtheria bacillus. Biokhimiya **21**, 38—49 (1956).

RAFFEL, S.: Chemical factors involved in the induction of infectious allergy. Experientia (Basel) **6**, 410—419 (1950). — REEVES, R. E., and R. J. ANDERSON: XLVIII. The composition of the avian tubercle bacillus wax. J. Amer. Chem. Soc. **59**, 858—861 (1937). — RICH, A. R.: The pathogenesis of tuberculosis. Springfield, U.S.A.: Ch. C. Thomas 1951. — ROULET, F., u. M. BRENNER: Die Chemie des Tuberkelbazillus. Zbl. Tbk.forsch. **56**, 193—227 (1943). — RYHAGE, R., E. STENHAGEN and E. v. SYDOW: Information about the structure of phthiocerol from mass spectral and X-ray crystallographic data. Acta chem. scand. (Copenh.) **10**, 158 (1956).

SEGAL, W., and H. BLOCH: Biochemical differentiation of *M. tuberculosis* grown *in vivo* and *in vitro*. J. Bacter. **72**, 132—141 (1956). — SEIBERT, F. B.: Constituents of Mycobacteria. Annual Rev. Microbiol. **4**, 35—52 (1950). — SERCK-HANSSEN, K., and E. STENHAGEN: A general method for the synthesis of optically active β-hydroxy acids. Acta chem. scand. (Copenh.) **9**, 866 (1955). — SHEEHAN, H. L., and F. WHITWELL: The staining of tubercle bacilli with Sudan Black B. J. Bacter. **61**, 269—272 (1949). — SIFFERD, R. H., u. R. J. ANDERSON: Über das Vorkommen von Sterinen in Bakterien. Z. physiol. Chem. **239**, 269—272 (1936). — SMITH, H., J. KEPPIE and J. L. STANLEY: The chemical basis of the virulence of *B. anthracis*. I. Properties of bacteria grown in *vivo* and preparation of extracts. Brit. J. Exper. Path. **34**, 477—485 (1953). — SMITHIES, W. R., N. E. GIBBONS and S. T. BAYLEY: The chemical composition of the cell and cell wall of some halophilic bacteria. Canad. J. Microbiol. **1**, 605—613 (1955). — SNOW, G. A.: The occurrence of vitamin K in Mycobacteria. II. Congr. Internat. Biochimie, Paris 1952, résumé comm., p. 95. — Mycobactin. A growth factor for *M. johnei*. II. Degradation and identification of fragments. J. Chem. Soc. (Lond.) **1954**a, 2588—2596. — III. Degradation and tentative structure. J. Chem. Soc. (Lond.) **1954**b, 4080—4093. — STÄLLBERG-STENHAGEN, S., and E. STENHAGEN: Synthesis and X-ray investigation of methyl-substituted long-chain hydrocarbons related to phthiocerane. J. of Biol. Chem. **173**, 383—401 (1948). — Studies on hydrocarbons structurally related to phthiocerol. Synthesis of the levoratory enantiomorph of 4-methyl-tritriacontane. J. of Biol. Chem. **183**, 223—229 (1950). — STAHL, W. H.: Brucellosis. VII. Separation and study of the lipide fraction of *Brucella abortus*. Michigan Agricult. Exper. Stat., Techn. Bull. **177**, 29—34 (1941). — STENDAL, N.: Sur la présence d'acide salicylique et d'acide phénylacétique dans la graisse acétono-soluble du bacille tuberculeux. C. r. Acad. Sci. Paris **198**, 400—402 (1934). — STODOLA, F. H., and R. J. ANDERSON: XLVI. Phthiocerol, a new alcohol from the wax from the human tubercle bacillus. J. of Biol. Chem. **114**, 467—472 (1936). — STODOLA, F. H., A. LESUK and R. J. ANDERSON: LIV. The isolation and properties of mycolic acid. J. of Biol. Chem. **126**, 505—513 (1938). — STOKINGER, H. E., H. ACKERMAN and C. M. CARPENTER: Studies on the Gonococcus. I. Constituents of the cell. J. Bacter. **47**, 129—139 (1944). — SÜTÖ-NAGY, G. I. DE, and R. J. ANDERSON: LXXV. Further studies on the polysaccharides obtained from cell residues in the preparation of tuberculin. J. of Biol. Chem. **171**, 749—760 (1947). —LXXVI. Concerning inositol-glycerol-diphosphoric acid, a component of the phosphatide of human tubercle bacilli. J. of Biol. Chem. **171**, 761—765 (1947).

TAKAHASHI, H.: Bacterial components of *C. diphtheriae*. V. Phospholipides. 2. Structure of CHARGAFF's corynin. J. Pharmaceut. Soc. Japan **68**, 292—296 (1948). Chem. Abstr. **45**, 9482 (1951).

UMEZU, M.: The chemistry of *Bacillus mallei*. I. The lipides of *B. mallei*. Rep. Inst. Sci. Res. Manchoukuo **4**, 273—285 (1940). Chem. Abstr. **35**, 1825 (1941).

VELICK, S. F.: The chemistry of *Phytomonas tumefaciens*. III. Phytomonic acid, a new branched chain fatty acid. J. of Biol. Chem. **152**, 535—538 (1944). — IV. Concerning the structure of phytomonic acid. J. of Biol. Chem. **156**, 101—107 (1944). — VELICK, S. F., and R. J. ANDERSON: II. The composition of the acetone soluble fat. J. of Biol. Chem. **152**, 523—531 (1944). — VILKAS, E., et E. LEDERER: Isolement d'un phosphatidyl-inositodimannoside à partir d'un phosphatide de Mycobactéries. Bull. Soc. Chim. biol. Paris **38**, 111—121 (1956).

WERNER, G., O. LÜDERITZ u. O. WESTPHAL: Cité dans la référence suivante. — WESTPHAL, O., u. O. LÜDERITZ: Chemische Erforschung von Lipopolysacchariden gramnegativer Bakterien. Angew. Chem. **66**, 407—417 (1954). — WIEGHARD, C. W., and R. J. ANDERSON: LV. Studies on the wax of the human tubercle bacillus. J. of Biol. Chem. **126**, 515—526 (1938). — WILLIAMS, C. H., W. R. BLOOR and L. A. SANDHOLZER: A study of the lipids of certain enteric bacilli. J. Bacter. **37**, 301—313 (1939). — WILSON, G. S., and A. A. MILES: TOPLEY and WILSON's principles of bacteriology and immunity. London: Edward Arnold & Co. 1948.—WITTCOFF, H.: The phosphatides; phosphatides of microorganisms, p. 196—213. New York: Reinhold Publ. Corp. 1951.

Lipasen.

Von

Eugen Bamann und **Elsa Ullmann.**

I. Das Vorkommen pflanzlicher Lipasen.

1. Lipasen der Phanerogamen.

Anlaß zur Beobachtung einer Fettspaltung im Pflanzengewebe war das um die Mitte des letzten Jahrhunderts erwachte Interesse an pflanzenphysiologischen Fragen wie Samenkeimung und der damit verbundenen Erforschung der Mobilisierung der Reservestoffe: Fett, Stärke und Eiweiß. Sorgfältige Experimente von J. M. PELOUZE zeigten an einer Reihe ölhaltiger Samen — Flachs, Senf, Raps, Mohn —, daß durch bloßes Zerquetschen nach kurzer Zeit Spaltung des Fettes in Fettsäure und Glycerin eintritt. Andere Forscher berichten über ähnliche Beobachtungen bei der Samenkeimung (J. SACHS, E. PETERS, M. G. FLEURY, M. A. MUNTZ). Im Gange dieser Forschungen führte P. SCHÜTZENBERGER die Entstehung freier Säure in ölhaltigen Pflanzensamen auf *Fermenttätigkeit* zurück. Diese Vermutung SCHÜTZENBERGERS wurde von J. R. GREEN (1891) bestätigt, der am lipatisch stark wirksamen *Ricinussamen* die Enzymtätigkeit eindrucksvoll nachwies. Die um die Jahrhundertwende durchgeführten Untersuchungen an einer Reihe gekeimter und ungekeimter Samen durch W. SIGMUND, F. DUNLAP und W. SEYMOUR, H. MASTBAUM, K. BOURNOT sowie S. FOKIN berichten über *eine weite Verbreitung der Lipase im Pflanzenreich.* Vor allem der letztgenannte Forscher fand von 60 untersuchten Pflanzensamen etwa 30 lipatisch wirksam. Ferner wurde festgestellt, daß das Ausmaß der Lipaseaktivität quantitativ unterschiedlich ist. Anfangs nahm man an, daß Lipase nur in gekeimten, nicht aber in ruhenden Samen wirksam sei. Diese Anschauung wurde besonders für *Ricinussamen* sehr bald revidiert: Während J. R. GREEN (1891, 1905) noch das Vorkommen einer Lipase in ruhenden Ricinussamen bestritt, fanden W. CONNSTEIN und Mitarbeiter (1902) bereits, daß Säure das Ferment im ungekeimten Zustand zu „aktivieren" vermag; R. WILLSTÄTTER und E. WALDSCHMIDT-LEITZ (1924) stellten diese Ansicht richtig, indem sie die Abhängigkeit der lipatischen Wirksamkeit von der Wasserstoffionenkonzentration aufzeigten. Auch im Falle des *ruhenden Weizenkornes* wurde zunächst die Anwesenheit einer Lipase negiert (H. MASTBAUM). Es war einer geeigneteren Untersuchungsmethodik vorbehalten, esterspaltende Wirksamkeit aufzufinden (B. SULLIVAN und M. A. HOWE).

Im Zuge der Forschungen über die Veränderungen, die bei der Verarbeitung von Vegetabilien zu Nahrungs- und Genußmitteln und deren Lagerung auftreten, wurde vielfach Lipasetätigkeit nachgewiesen: So ist der wechselnde Säuregehalt im Öl des keimenden *Leinsamens* durch Lipasetätigkeit bedingt (E. R. THEIS und Mitarbeiter 1930). Lipasewirkung war auch bei der Prüfung verschiedener Varietäten der *Sojubohne* festzustellen: sie ist verantwortlich für den unterschiedlichen Säuregehalt der Öle (M. J. SMIRNOWA und M. N. LAWROWA). Bei der Untersuchung des Ablaufs der *Tabakfermentation* wurde Lipase ebenfalls als mitbeteiligt aufgefunden (TH. ANDREADIS). Im Zusammenhang mit dem Problem einer rationellen Aufbereitung und Lagerung von Müllereiprodukten sind eingehende Untersuchungen den Vorgängen gewidmet, die sich im ruhenden und keimenden *Getreidesamen* abspielen (D. MAESTRINI, B. SULLIVAN, B. SULLIVAN und M. A. HOWE, N. P. KOZMIN, L. PETT, A. KIEZEL und K. GORDIENKO, C. ENGEL, J. B. HUTCHINSON und Mitarbeiter, M. ROHRLICH und H. BENISCHKE, W. H. TEMPLETON und B. R. CARPENTER, B. C. W. HUMMEL und Mitarbeiter, M. ROTHE).

Systematische Untersuchungen führten erstmals E. BAMANN und E. ULLMANN auf breiter Basis mit Hilfe des stalagmometrischen Verfahrens der Tributyrinhydrolyse durch; dabei ließ sich ein eindrucksvolles Bild über das Vorkommen und die Verbreitung des Enzyms im Bereich der höheren Pflanzen, sowohl in den *Samen* als auch in den *vegetativen Organen* gewinnen. Aus den Untersuchungen dieser Forscher ergibt sich an wichtigen Folgerungen:

1. Unter rund 100 untersuchten Samen sind diejenigen von *Nigella sativa*, *Nigella arvensis*, *Chelidonium majus*, *Delphinium Ajacis*, *Helleborus foetidus*, *Nigella damascena*, *Bocconia cordata* und *Ricinus communis* reich an Lipase (1000—300 B-[e] je 1,0 g entfetteten und entwässerten Samenmaterials)[1].

2. Die Mehrzahl der lipatisch stark wirksamen Samen gehört der Familie der *Ranunculaceae* und der ihr verwandten Familie der *Papaveraceae* an, während der lipasereiche Ricinussamen der Familie der *Euphorbiaceae* zugehört.

3. Hoher Ölgehalt des Samens bedingt keineswegs starke lipatische Wirksamkeit. Parallelität zwischen Öl- und Enzymgehalt tritt nirgends zutage, auch nicht bei den Samen aus den Familien der *Ranunculaceae* und *Papaveraceae*.

4. Im Gewebe vegetativer Organe (von 60 Pflanzen) werden in keinem Falle Lipasewerte über 88 B-[e] je 1 g Trockengewicht gefunden. Mit diesem Enzymgehalt weist das Kraut von *Urtica dodarti* den Höchstwert auf. 60 B-[e] wurden bei *Trollius europaeus* gemessen, 40 B-[e] bei *Glaucium corniculatum*, *Chelidonium majus*, *Gossypium herbaceum* und *Datura stramonium*; 30—20 B-[e] sind in Blatt- und Stengelteilen von *Aquilegia vulgaris*, *Scrophularia nodosa*, *Nigella arvensis*, *Avena sativa*, *Helleborus foetidus*, *Brassica rapa* und *Campanula latifolia* enthalten.

Die genannten Pflanzen finden sich in den Familien der *Urticaceae*, *Ranunculaceae*, *Papaveraceae*, *Cruciferae*, *Malvaceae*, *Solanaceae*, *Scrophulariaceae*, *Campanulaceae* und *Gramineae*.

Parallelität zwischen dem Enzymgehalt der Samen und dem der vegetativen Organe derselben Pflanzenart besteht in den meisten Fällen nicht:

Die Samen von *Nigella sativa* und *Nigella damascena* z. B. sind außerordentlich lipasereich, die vegetativen Organe zeigen dagegen nur sehr geringe enzymatische Wirksamkeit.

Häufig trifft auch das Umgekehrte zu: Beispielsweise sind die Samen von *Urtica dodarti* und *Trollius europaeus* lipasearm (10—3 B-[e]), während die vegetativen Organe dieser Pflanzen einen erheblichen Enzymgehalt (88—60 B-[e]) aufweisen.

2. Lipasen der Kryptogamen.

Von den Lipasen der Kryptogamen sind diejenigen der *Pilze* und *Hefen* sowie einige *Bakterien*lipasen untersucht worden. Wesentliche Arbeitsrichtungen befassen sich vor allem mit der Wirkung von Mikroorganismen auf fetthaltige Nahrungsmittel wie Müllereiprodukte, Ölsamen und Nahrungsfette. In diesem Zusammenhang berichten neuere Arbeiten über *Eigenschaften* der Kryptogamenlipasen — *Substratspezifität*, *p_H-Wirkungsoptimum*, *Stabilität* —, sowie über diejenigen *Faktoren, die die Lipaseproduktion beeinflussen*.

a) Pilze und Hefen.

Eine sehr eingehende Untersuchung über die „Takaesterase", die Lipase aus *Aspergillus oryzae* (R. WILLSTÄTTER und A. KUMAGAWA), hat ergeben, daß dieses Enzym nur in geringem Maße befähigt ist, Neutralfett anzugreifen. Auch das Enzym der Hefe wirkt besonders auf niedere Ester wie Äthylbutyrat, Butylazetat sowie Tributyrin; Olivenöl wird dagegen nicht gespalten (P. J. FODOR). Ähnlich wie im Falle tierischer Leberesterase werden also besonders niedere Ester bevorzugt gespalten. Allgemein gilt aber diese Substratbevorzugung für die Pilzlipasen nicht; so hydrolysiert z. B. die Lipase von *Penicillium oxalicum* niedere Ester sowie Triglyceride (D. KIRSH). Das Enzym aus *Penicillium roqueforti* spaltet hinwiederum Triglyceride besser (H. A. MORRIS und J. J. JEZESKI).

[1] Für den Enzymmengenvergleich dienen Lipaseeinheiten, die im Hinblick auf das Substrat als Butyrase-[einheiten] = B-[e] bezeichnet werden (s. dazu: III 1b, S. 114).

Das lipolytische Enzym des fettreichen Pilzes *Fusarium lini* BOLLEY ist gegenüber Fetten und Triacetin gut wirksam (J. V. FIORE und F. F. NORD).

Die Angaben über das p_H-Wirkungsoptimum sind uneinheitlich; voraussichtlich sind Substrat, Puffer, Reaktionszeit und Temperatur beim Zustandekommen des jeweiligen optimalen p_H-Bereichs mitbestimmend. Die Enzyme der meisten Pilze und Hefen entfalten bei leicht *saurer Reaktion* optimale Wirksamkeit (D. KIRSH, G. GORBACH und H. GÜNTHER, R. THIBODEAU und H. MACY, J. J. PETER und F. E. NELSON, P. J. FODOR und A. CHARI, C. V. RAMAKRISHNAN und G. V. NEVGI 1950h, C. V. RAMAKRISHNAN und B. N. BANERJEE 1951c—f, 1952a, f, J. J. GHOSH). Takaesterase sowie die Lipasen aus *Aspergillus niger* und *Penicillium roqueforti* haben dagegen ihre größte lipolytische Aktivität im *schwach alkalischen Bereich* (R. WILLSTÄTTER und A. KUMAGAWA, P. J. FODOR und A. CHARI). Im Falle der Hefelipase beobachtet P. J. FODOR 2 p_H-Optima gegenüber Äthylbutyrat, eines im alkalischen bei $p_H = 7{,}4$ und eines im sauren Bereich bei $p_H = 2{,}8$, wobei die Hitzestabilität der lipatischen Wirksamkeit bei $p_H = 2{,}8$ größer ist als bei $p_H = 7{,}4$.

Weitere Arbeiten befassen sich mit Faktoren, die Pilzwachstum und Lipaseproduktion beeinflussen (C. V. RAMAKRISHNAN und B. N. BANERJEE 1951a, d—g, 1952a, d—g). Durch Adaption bilden *Aspergillus*- und *Penicillium*-Arten auf *ölhaltigen* Nährböden lipolytisches Enzym in höherer Konzentration als auf ölfreiem Substrat (J. J. GOODMAN und C. M. CHRISTENSEN). Schimmelpilze, auf lipasehaltigen Ölsamen wachsend, führen nach C. V. RAMAKRISHNAN und B. N. BANERJEE (1951c) zu Verminderung der lipolytischen Aktivität des betreffenden Samens. B. M. DIRKS und Mitarbeiter vergleichen einige Eigenschaften der Lipasen aus *Aspergillus candidus* und *Penicillium* mit *Weizen*lipase. Wesentliche Unterschiede zwischen den *Pilz*lipasen einerseits und der *Weizen*lipase andererseits ergeben sich auf Grund ihres Verhaltens gegenüber p-Chloromercuribenzoat und o-Jodosobenzoat. Diese Substanzen beeinflussen nur *Weizen*lipase im Sinne einer starken Hemmung. Untersuchungen von J. TABACHNIK und M. A. JOSLYN befassen sich mit den Bedingungen, unter denen die Esterbildung durch *Hefen* vor sich geht.

b) Bakterien.

Fettspaltende Enzyme wurden in einer großen Anzahl von Bakterienarten gefunden; beispielsweise in *Micrococcus, Bacillus, Escherichia, Corynebacterium, Pseudomonas, Achromobacter* (L. B. JENSEN).

Neuere Untersuchungsergebnisse liegen vor über die Lipasen aus *Mycobacterium phlei* und *Mycobacterium tuberculosis* (Y. YAMAMURA und Mitarbeiter, K. OGURA und Mitarbeiter) sowie über diejenigen aus *Pseudomonas fluorescens, Pseudomonas fragi, Pseudomonas aeruginosa* (D. J. LUBERT und Mitarbeiter, E. C. CUTCHINS und Mitarbeiter, M. L. GOLDMAN und M. M. RAYMAN, S. A. NASHIF und F. E. NELSON 1953a—c). Über die Kinetik der Bakterienlipase berichten A. LEMBKE und Mitarbeiter. An den wirksamen Fettspaltern *Micrococcus nacreaceus* (aus Walderde) und *Bacterium prodigiosum* (aus der Luft) studierten G. GORBACH, G. DEDIC und K. LORENZ Anreicherung und Wirksamkeitsbestimmung von Bakterienlipasen.

Ebenso wie die Lipasen der Pilze zeigen auch die Bakterienlipasen eine relative Substratspezifität: Es werden sowohl niedere Ester als auch Glyceridfette von Bakterienlipasen hydrolysiert. J. MEYER und Mitarbeiter vermerken, daß die Enzyme des KOCHschen Bacillus und anderer säurefester Bacillen Äthylbutyrat und Tributyrin besser als Olivenöl spalten. Ähnliches beobachteten G. GORBACH, G. DEDIC und K. LORENZ. Von früheren Autoren festgestellte Resistenz gegenüber hochmolekularen Fetten erklären M. L. GOLDMAN und M. M. RAYMAN mit ungenügender Emulgierung des Versuchsansatzes. Die genannten Autoren finden, daß nach gründlicher Emulgierung eine Reihe tierischer und pflanzlicher Fette von der Lipase aus *Pseudomonas fluorescens* hydrolysiert werden.

Innerhalb der einzelnen *Pseudomonas*-Arten ermittelten die Autoren eine verschieden starke lipolytische Wirksamkeit. S. Cohan und Mitarbeiter beobachteten, daß Suspensionen und Extrakte aus saprophytischen Mycobakterien (*M. tuberculosis, M. smegmatis, M. lacticola, M. phlei, M. butyricum*) β-Naphthylfettsäureester gut spalten, dagegen diejenigen aus *Pseudomonas aeruginosa, Micrococcus pyogenes, Escherichia coli, Corynebacterium diphteriae, Streptococcus pyogenes* und *Proteus vulgaris* diesen Substraten gegenüber geringe oder keine Wirksamkeit zeigen. G. Gorbach, G. Dedic und K. Lorenz finden, daß teilweise gereinigte Lipasepräparate aus *Micrococcus nacreaceus* und *Bacterium prodigiosum* nur 1—2% der lipatischen Wirksamkeit eines mittelmäßig aktiven Pankreaspräparates besitzen (Substrat: Tributyrin).

Fast alle untersuchten Bakterienlipasen wirken im p_H-Bereich von 7,8 oder darüber optimal (O. T. Avery und G. E. Cullen, F. A. Stevens und R. West, E. S. Tammisto, D. J. Lubert und Mitarbeiter, G. Gorbach, G. Dedic und K. Lorenz). Davon etwas abweichend finden A. S. Nashif und F. E. Nelson (1953a, c) das Wirkungsoptimum des Enzyms aus *Pseudomonas fragi* gegenüber Kokosnußöl bei $p_H = 7$—7,2 gegenüber Butter bei $p_H = 5{,}7$—6,6.

Die Stabilität der Bakterienlipasen gegenüber Hitzeeinwirkung ist unterschiedlich: O. T. Avery und G. E. Cullen inaktivierten die Lipase aus einer *Pneumococcus*-Art durch 10minutiges Erhitzen bei 70°; F. A. Stevens und R. West zerstörten die Lipase eines *Streptococcus* durch 10minutiges Erhitzen bei 55°. K. Ogura und Mitarbeiter beobachteten, daß von einer größeren Anzahl säureresistenter Bakterien nur *Mycobacterium avium* eine hitzestabile Esterase enthält. Auch das Enzym aus *Pseudomonas fluorescens* widerstand einer Hitzeeinwirkung von 100° während 5 min (N. L. Söhngen). Ebenso waren Enzympräparate aus *Pseudomonas fragi* noch gut wirksam nach 30minutigem Erhitzen bei 61,6 bzw. 71,6°. Vollständige Inaktivierung dieser Zubereitungen erfolgte erst nach 20minutigem Erhitzen bei 99° (S. A. Nashif und F. E. Nelson 1953a, c).

II. Über die Bedeutung der Lipasen im physiologischen Geschehen.

Histochemische Untersuchungen haben in jüngerer Zeit manches Problem auch des pflanzlichen Stoffwechsels zu fördern vermocht. Die Frage nach der *protoplasmatischen Verankerung der Katalysatoren*, nach ihrem *Wirksamwerden in der pflanzlichen Zelle*, nach der *intracellulären Regulierung* der Enzymwirkungen fand bei einzelnen Enzymen eine recht eingehende Bearbeitung (K. Süssenguth, A. Oparin 1934, 1953, K. Linderstrøm-Lang und H. Holter, J. v. Przyłęcki, F. Duspiva, H. E. Longenecker, L. Monné, L. A. Dounce, E. L. Newcomer, D. S. van Fleet 1952, M. Steiner und Mitarbeiter).

Wichtige Einsichten über die strukturelle Verankerung der Enzyme wurden vor allem an *tierischem* Material gewonnen, die aber noch der Bestätigung an pflanzlichem Material bedürfen. Solche Untersuchungen an pflanzlichen Organen sind sehr schwierig, da die zur Verfügung stehenden Zerkleinerungsmethoden häufig zu Artefakten führen und daher nur grobe Anhaltspunkte für die intracellulare Verteilung der Enzyme gewonnen werden konnten (K. Paech 1953)[1].

Günstigere Arbeitsbedingungen sind bei Untersuchungen an niederen Pflanzen, z. B. Pilzen, gegeben: Neuere Arbeiten sehen die Grana mit positiver Nadireaktion als Ort der primären Fettbildung in Pilzzellen von *Oospora lactis, Endomycopsis vernalis, Torulopsis utilis, Torulopsis lipofera* u. a. an (S. Heide,

[1] Über die Methoden der histochemischen Lokalisation siehe III 3.

M. Steiner und H. Heinemann). Nach der Auffassung von M. Steiner und H. Heinemann sind die fettbildenden Grana Erscheinungstypen der Mitochondrien.

E. H. Newcomer berichtet, daß auch die Plastiden in den höheren Pflanzen — z. B. Leukoplasten, in denen sich der Kohlenhydratstoffwechsel vollzieht, oder Oleoplasten, die Öl und Fett absondern — den Mitochondrien entstammen[1].

Diese Untersuchungsergebnisse decken sich mit der Ansicht von L. Monné, H. Holter sowie A. L. Dounce, daß die Moleküle aller Enzyme an gewisse morphologische Protoplasmaelemente gebunden und in bestimmter Weise angeordnet sind und gemäß den Bedürfnissen der lebenden Zelle aktiviert und inaktiviert werden.

In diesem Zusammenhang sind vielleicht Befunde von J. B. Hutchinson und Mitarbeiter sowie von E. Bamann und Mitarbeiter (1953a) aufschlußreich:

Die Samen des Hafers, die nach Entfernen des äußeren Perikarps lipolytisch unwirksam sind, entwickeln nach J. B. Hutchinson und Mitarbeiter im Verlaufe einer 7tägigen Keimung hohe lipatische Aktivität.

Bezüglich der Mobilisierung der Fette bei der Keimung des Samens hatte man bisher die Meinung, „in der Zelle werden die Glyceride dem Enzym dargeboten" (R. Willstätter und E. Waldschmidt-Leitz 1923). Da aber der Speicherungsort für das fette Öl nicht stets auch der Hauptsitz des fettspaltenden Enzyms zu sein scheint, nehmen E. Bamann und Mitarbeiter (1953a) für die von ihnen gefundenen löslichen Anteile der Lipase eine *Wanderung* derselben, z. B. vom Scutellum in das fettspeichernde Endosperm an.

Über die Mitwirkung von Lipasen bei Bildung oder Abbau von spezifischen Pflanzeninhaltsstoffen, die ihrer chemischen Natur nach Ester sind, wissen wir noch wenig. K. Paech (1950) sieht in der Tatsache, daß in Solanaceen Tropin trotz des großen Angebots an organischen Säuren immer nur streng spezifisch mit Tropasäure verestert auftritt, einen Anhaltspunkt für eine enzymatische Veresterung der beiden Atropinkomponenten. Auch die optische Aktivität des in der Pflanze primär gebildeten Hyoscyamins dürfte im Hinblick auf die oft beobachtete Konfigurationsspezifität der Esterasen auf enzymatische Veresterung hinweisen.

Ob der bei *Fol. Belladonnae* während des Trocknungsvorgangs beobachtete Rückgang der Alkaloidmenge auf einen Abbau der Alkaloide durch Fermente erfolgt, die die Esterbindung zwischen Tropasäure und Tropin hydrolysieren (G. P. Koch, H. Flück) ist fraglich geworden (E. Schäfer).

III. Bestimmungsmethoden und Enzymzubereitungen.

1. Die Bestimmung der Lipaseaktivität.

Die Messung der Lipaseaktivität erfolgt häufig durch direkte *Titration der gebildeten Fettsäuren.* Ebenso kann die Aktivitätsbestimmung durch Auswertung der durch die Enzymwirkung hervorgerufenen *Änderung der Oberflächenspannung* oder durch *indirekte Messung der freiwerdenden Fettsäure mit Hilfe der gasometrischen Methodik* vorgenommen werden. Kürzlich wurde auch ein *photometrisches Verfahren* ausgearbeitet.

a) Die titrimetrische Bestimmungsmethodik.

Dieses Verfahren sollte nur im Falle gut wirksamer Enzympräparate zur Anwendung gelangen. Zum Beispiel bedienen sich die Forschungen über die lipasereichen Samen von *Ricinus* (R. Willstätter und E. Waldschmidt-Leitz 1924, H. E. Longenecker und D. E.

[1] C. Engel und L. H. Bretschneider finden mit Hilfe der von ihnen angewendeten Methodik *keine* Beziehung zwischen der Anzahl der Mitochondrien und der Menge an Amylase, Proteinase, Dipeptidase und Esterase.

HALEY 1936, 1937a, b, C. E. HAGEN und Mitarbeiter, G. V. NEVGI und C. V. RAMAKRISHNAN 1950a, C. V. RAMAKRISHNAN, W. G. ROSE, C. RAVAZZONI, H. NIZAMUDDIN und B. S. KULKARNI, Y. MOULÉ, A. M. CALABRESE), *Chelidonium* (K. BOURNOT) dieser Methode. Auch J. V. FIORE und F. F. NORD bestimmen bei ihren Untersuchungen über die Lipaseaktivität des Pilzes *Fusarium lini* BOLLEY die gebildete Fettsäure titrimetrisch. R. WILLSTÄTTER und E. WALDSCHMIDT-LEITZ (1924) sowie H. E. LONGENECKER und D. E. HALEY arbeiteten Standardverfahren am Beispiel *Ricinus*lipase aus, die bei der Untersuchung dieses Enzyms sowie anderer gut wirksamer pflanzlicher Enzympräparate zur Anwendung gelangen. Nach R. WILLSTÄTTER und E. WALDSCHMIDT-LEITZ (1924) wird die Lipase an ihrer Wirkung auf Olivenöl im optimalen Reaktionsmilieu ($p_H = 4{,}7$) bei einer Temperatur von 20° C gemessen. Nach einer bestimmten Inkubationszeit stoppt man die Reaktion durch Zufügen eines Äthanol-Äthergemisches und bestimmt die freien Fettsäuren alkalimetrisch. Als Maß für die Ricinuslipase dient die Phytolipaseeinheit(Ph.-L.-E.), das ist diejenige Enzymmenge, welche nach einer Reaktionszeit von 20 min 2,5 g Olivenöl (Verseifungszahl 185,5) zu 7,5% spaltet.

Maß für die enzymatische Konzentration eines Präparates ist der Phytolipasewert = Anzahl der Phytolipaseeinheiten in 1 cg des Präparates.

Bei einer Reaktionstemperatur von 37° führen H. E. LONGENECKER und D. E. HALEY die Bestimmung der lipolytischen Aktivität in ähnlicher Weise durch[1].

In neuester Zeit beschreiben R. B. KOCH und Mitarbeiter eine Abwandlung dieses Untersuchungsverfahrens bei der Bestimmung von *Weizenkeimlipase* sowie J. MEYER und Mitarbeiter bei derjenigen von *Bakterien*lipasen: Die Methoden beruhen auf elektrometrischer Titration der durch das Enzym aus dem Substrat abgespaltenen Fettsäuren.

Eine weitere Abwandlung der titrimetrischen Bestimmungsmethodik stellt ein Verfahren dar, das T. P. SINGER und H. B. J. HOFSTEE für ihre reaktionskinetischen Messungen an der Lipase des *keimenden Weizens* anwendeten: Bei diesem Verfahren dienen als Substrate Ester flüchtiger Fettsäuren (z. B. Monobutyrin); die mit Hilfe einer Wasserdampfdestillation abgetrennten Fettsäuren werden durch alkalimetrische Titration bestimmt.

Bei ihren Arbeiten über die Substratspezifität isodynamer pflanzlicher Lipasen verfolgten E. BAMANN und Mitarbeiter die Spaltung des Buttersäuremethylesters im Versuchsansatz von E. BAMANN und J. N. MUKHERJEE nach der Methode von R. WILLSTÄTTER und F. MEMMEN (1924).

b) Die stalagmometrische Methode der Tributyrinbestimmung.

Wegen seiner großen Empfindlichkeit ist für die Bestimmung pflanzlicher Lipasevorkommen das stalagmometrische Verfahren der Tributyrinhydrolyse (E. BAMANN und Mitarbeiter 1942, 1953a) besonders geeignet.

Die Messungen werden mit einem OSTWALDschen geraden Stalagmometer durchgeführt, das bei 20° einen Wasserwert von 20 Tropfen zeigt.

Als Substrat dient eine wäßrige mit Tributyrin „Merck" gesättigte Lösung, die nach den von R. WILLSTÄTTER und F. MEMMEN (1923) angegebenen Richtlinien hergestellt wird. Zu 57 cm³ der Substratlösung gibt man nach E. BAMANN und E. ULLMANN 2 cm³ Pufferlösung zur Einstellung des gewünschten p_H und 1 cm³ des Enzympräparates. Bei einer Temperatur von 20° wird nunmehr die Tropfenzahl zu Beginn des Versuchs und nach gewissen Intervallen bestimmt.

Nach dem Vorbild der Untersuchung von R. WILLSTÄTTER und F. MEMMEN (1923) leiten E. BAMANN und E. ULLMANN als Maß für den Lipasegehalt Einheiten ab, die im Hinblick auf das Substrat als Butyrase-[einheiten] = B-[e] bezeichnet werden[2]. Eine B-[e] stellt diejenige Enzymmenge dar, die unter den Bedingungen der Methode in 50 min eine Abnahme von 20 Tropfen bewirkt.

Vergleichende Bestimmungen mit Lipasepräparaten verschiedener Pflanzen im p_H-Wirkungsoptimum ergaben, daß in den meisten Fällen innerhalb des Bereiches, das für Aktivitätsmessungen in Betracht kommt (0,2—4 B-[e]), Proportionalität zwischen Reaktionszeit und Enzymmenge besteht; nur gelegentlich zeigen sich kleine Abweichungen. Dieser Umstand ermöglicht über eine qualitative Erfassung des Lipasevorkommens hinaus auch eine angenäherte

[1] Einzelheiten über die Untersuchungstechnik nach R. WILLSTÄTTER und E. WALDSCHMIDT-LEITZ sowie nach H. E. LONGENECKER und D. E. HALEY siehe E. WALDSCHMIDT-LEITZ und A. SCHÄFFNER.

[2] Durch diese Schreibweise kommt zum Ausdruck, daß der quantitative Vergleich ein angenäherter ist und ihm nur scheinbare Einheiten zugrundeliegen.

mengenmäßige Bestimmung der Lipase, so daß Enzymmengenvergleiche zwischen dem Lipasegehalt der verschiedenen Organe einer Pflanze oder ein und demselben Organ verschiedener Pflanzen möglich sind.

Vergleicht man die stalagmometrische Methode mit der vielfach angewendeten Ölspaltungsmethode hinsichtlich der Eignung, so gewinnt man die Überzeugung, daß das letztgenannte Verfahren, bei dem die Verseifung eines fetten Öles, z. B. des Olivenöles, meistens alkalimetrisch verfolgt wird, für die Bestimmung geringer Lipasewirksamkeit nicht zu befriedigen vermag. In diesen Fällen bleiben die Werte für die Aciditätszunahme sehr niedrig (E. R. THEIS und Mitarbeiter 1930, S. FOKIN, K. G. FALK), auch bei langen, gelegentlich sogar mehrere Tage dauernden Versuchszeiten; unvermeidbare oder unterlaufene Ungenauigkeiten entstellen daher das Ergebnis.

Ein weiterer, nicht zu entbehrender Vorteil der stalagmometrischen Methodik besteht darin, daß das Reaktionssystem für vergleichende Messungen bei verschiedener Wasserstoffionenkonzentration geeignet ist. Dies ist dadurch gegeben, daß das Substrat — im Gegensatz zu dem p_H-abhängigen Emulsionssystem der Ölspaltungsansätze — in Wasser gelöst ist.

c) Die manometrische Methode.

Eine weitere empfindliche Bestimmungsmöglichkeit stellt das von P. RONA und A. LASNITZKI für die Bestimmung von Lipasen in tierischen Geweben und Körperflüssigkeiten entwickelte manometrische Verfahren dar, das wäßrige Emulsionen bzw. Lösungen von Tributyrin oder Mono- und Tryglyceriden niederer Fettsäuren als Substrate verwendet, wobei die durch die Enzymtätigkeit abgespaltene Fettsäure aus Bicarbonat Kohlensäure entbindet, die manometrisch gemessen wird. Dieses Untersuchungsverfahren benutzten H. W. NICOLAI zur Bestimmung von Lipase in *keimenden Kiefernsamen*, T. P. SINGER und H. B. J. HOFSTEE bei ihren Arbeiten über die *Weizenkeim*lipase sowie H. F. MARTIN und G. F. PEERS bei ihren Untersuchungen über Spezifität und Anreicherung der *Hafer*lipase.

d) Die photometrische Methode.

A. PURR arbeitete ein photometrisches Bestimmungsverfahren aus, wobei die Lipasen aus tierischen (Pankreas, Leber) und pflanzlichen (Hafer) Geweben aus Phenolphthaleindibutyrat Phenolphthalein abspalten, das photometrisch gemessen wird.

e) Eine halbquantitative colorimetrische Methode.

Eine halbquantitative colorimetrische Methode zum Nachweis von Esterasen in Bakterien- und Pilzkulturen beschreibt M. K. MUFTIC: Dem Nährboden wird Indoxylbutyrat zugegeben. Unter aeroben Bedingungen findet eine Oxydation eventuell freigesetzten Indoxyls zu Indigoblau statt, dessen Ausbreitungszone gemessen wird.

2. Herstellung von Enzympräparaten.

Während die Methodik der Fermentisolierung und -reinigung auf dem Gebiete der tierischen Lipasen besonders durch die Arbeiten von R. WILLSTÄTTER und seiner Schule sowie E. BAMANN und Mitarbeiter (s. E. WALDSCHMIDT-LEITZ und A. SCHÄFFNER) weit vorangetrieben wurde, bereitete die Herstellung gereinigter Fermentpräparate aus Pflanzen erhebliche Schwierigkeiten und wurde daher selten angestrebt. Über einige erfolgreiche Reinigungsvornahmen wird in Abschnitt V berichtet. Bei den meisten Untersuchungen diente als Enzymquelle das *unvorbehandelte* oder ein mit Hilfe von Aceton, Äther oder Petroläther *getrocknetes* und *entfettetes Pflanzenpulver* (K. BOURNOT, B. SULLIVAN und M. A. HOWE, H. E. LONGENECKER und D. E. HALEY 1935, E. SCHREIBER 1940, G. GORBACH, S. OLCOTT und T. D. FONTAINE, E. BAMANN und E. ULLMANN, C. V. RAMAKRISHNAN und Mitarbeiter, Y. MOULÉ, A. M. CALABRESE).

Da die Lipase des *Ricinussamens* in fettfreier Form sehr empfindlich gegenüber Wasser und wäßrigen Reagentien ist, arbeiteten R. WILLSTÄTTER und E. WALDSCHMIDT-LEITZ (1924) ein Herstellungsverfahren für eine *fetthaltige Sahne* aus, die durch Zentrifugieren einer durch Anteigen des geschälten Samens mit Wasser erhaltenen Samenmilch angefertigt wird. Nach dieser Methode wird die Bereitung von Enzympräparaten aus *Ricinus* auch in neueren Untersuchungen vorgenommen (G. M. ROSE, C. RAVAZZONI).

Ferner versuchte man, wäßrige Kochsalz- oder Glycerin-Extrakte im Falle der Ricinuslipase der Bestimmung der lipolytischen Aktivität zugrunde zu legen (K. G. FALK und K. SUGIURA). Die Möglichkeit der Überführung von Ricinussamenlipase in wäßrige Lösung wurde aber unter Berufung auf eine eingehende Untersuchung von R. WILLSTÄTTER und E. WALDSCHMIDT-LEITZ (1924) immer wieder in Abrede gestellt.

In jüngster Zeit zeigten allerdings in systematisch durchgeführten Untersuchungen E. BAMANN und Mitarbeiter (1953a), daß — entgegen den bisherigen, besonders an Ricinusenzym gesammelten Erfahrungen — sich auch *die Lipase höherer Pflanzen in Lösung überführen und sogar anreichern läßt:* Für das Herauslösen der löslichen Anteile an Lipase in den Samen kann man *Wasser* verwenden. Dabei erhält man im Falle mancher Samen bis zu 80 und noch mehr Prozent vom Lipasegehalt in Lösung. Wirksamer als Extraktionsmittel ist eine *n/40-Ammoniaklösung:* Die Ausbeute in den Lösungen erhöht sich unter diesen Bedingungen bei einzelnen Samen bis zu 96%. Auch mit *Glycerin-Wassergemischen* lassen sich lösbare Anteile der Samenlipase in Lösung überführen, mit wasserarmen weniger als mit wasserreichen. n/40-Essigsäure stellt kein Lösungsmittel für Samenlipase dar.

Für die Darstellung von Lipase*lösungen* empfehlen E. BAMANN und Mitarbeiter (1953a), die *entfetteten Äthertrockenpräparate* mit dem Lösungsmittel im Verhältnis 1:100 zu verreiben und die Suspension nach einer Extraktionszeit von 15 min einem 20minutigen Zentrifugieren bei einer Umdrehungszahl von 4000/min zu unterwerfen. Solche Lösungen sind klar, aber nicht blank, sie zeigen leichte Opalescenz. Dieselbe Beobachtung machen auch H. F. MARTIN und F. G. PEERS an Lösungen aus *Hafer*lipase.

Die Hauptmenge der lösbaren Lipase ist vielfach schon im 1. Auszug enthalten; im 2. Auszug ist noch eine verhältnismäßig geringe Menge an Lipase zu erwarten (3—5% der Gesamtmenge). Weitere Auszüge vermögen nur mehr geringe Anteile in Lösung überzuführen. Ein Restanteil der Lipase verbleibt stets im Rückstand.

3. Methoden der histochemischen Lokalisation.

Die ersten Beobachtungen über die Lokalisation der Lipase im Aleuron des Endosperms von *Ricinus communis* verdanken wir den aus dem Jahre 1891 vorliegenden Untersuchungen von J. R. GREEN. Bis zu den in neuerer Zeit durchgeführten Arbeiten an *Getreidesamen* von L. PETT, CH. ENGEL und L. H. BRETSCHNEIDER sowie J. B. HUTCHINSON und Mitarbeiter hat die Frage der Lokalisation der lipatischen Aktivität mangels geeigneter Untersuchungsmethoden kaum weitere Förderung erfahren.

Erst durch Heranziehung von Mikromethoden, die K. LINDERSTRØM-LANG und CH. ENGEL, K. LINDERSTRØM-LANG und H. HOLTER, D. GLICK und G. GOMORI (1945, 1949) für tierische Gewebe ausarbeiteten, konnten auch auf pflanzlichem Gebiet erfolgreiche Untersuchungen durchgeführt werden. L. PETT zerlegte das *Weizenkorn* und untersuchte nach dem Verfahren von K. LINDERSTRØM-LANG und CH. ENGEL Schale, Endosperm, Scutellum, Kotyledo und Radicula im Verlauf der Keimung. Die Bestimmung der lipolytischen Wirksamkeit erfolgte nach der Methode von D. GLICK gegenüber Methylbutyrat: In ungekeimtem Samen findet sich der höchste Lipasegehalt im Scutellum. Nach 12stündiger Keimung zeigt sich dort ein starker Abfall der Enzymwirksamkeit, während der Lipasegehalt im Endosperm und in den Kotyledonen ansteigt. Auch R. DAVID findet in gekeimtem Weizen die lipolytische Aktivität in Keimling und Endosperm höher als im ruhenden Korn. Über einen ähnlichen Befund im Falle des *Samens der Meeresstrandkiefer* berichtet S. BOUDON. Nach den Untersuchungen dieses Forschers findet im Keimling eine noch lebhaftere Spaltung als im Endosperm statt. Durch eine spezielle Methodik trennte CH. ENGEL bei *Getreidesamen* die Aleuron- von der Subaleuronschicht des Endosperms und beobachtete hohen Enzymgehalt in Aleuronschicht und Embryo.

Nach G. GOMORI wird die Lokalisation der Lipase im tierischen Gewebe in der Weise vorgenommen, daß die Schnitte in Anwesenheit von Calciumsalz in wasserlösliche Fettsäureester (Polyglykolsorbitanester der Fettsäuren) gelegt werden. Lipasewirksamkeit führt zur Bildung unlöslicher Calciumseifen, die in die entsprechenden Bleisalze überführt und durch Schwefelwasserstoff kenntlich gemacht werden. D. GLICK und E. E. FISCHER versuchten vergeblich mit Hilfe der beschriebenen Methode unter Verwendung von Tween 40 (= Polyglykolsorbitanester der Palmitinsäure) als Substrat die Lipase im *Weizenkorn* zu lokalisieren.

J. B. HUTCHINSON und Mitarbeiter, F. G. PEERS, H. F. MARTIN und F. G. PEERS finden im äußeren Perikarp des *Haferkorns* etwa 98% der Lipaseaktivität. Sie empfehlen, zur Gewinnung eines wirksamen Lipasepräparates die Haferkörner für 2 min in Wasser einzuweichen. Dadurch wird das lipasehaltige Perikarp vom Kern gelöst; es läßt sich leicht abtrennen.

Mit Hilfe einer neuartigen Technik gelang es F. S. VAN FLEET (1950, 1952) lipolytische Wirkung in der Endodermis etiolierter Blattstiele von *Raphanus sativus* und *Smilax herbaceae* nachzuweisen.

Beläßt man diese 20—40 Std im Dunkeln, so werden vorhandene neutrale Fettsäureglyceride hydrolysiert und freie Fettsäuren erscheinen im dermalen und perivasculären Gewebe. Der Übergang von Neutralfett zu freier Fettsäure wird mit Hilfe von Polychrommethylenblau sichtbar gemacht.

IV. Eigenschaften pflanzlicher Lipasen.

1. Verankerung und Lösbarkeit.

Lange bestand die Meinung, daß die Lipase der höheren Pflanzen in starker protoplasmatischer Verankerung und ausschließlich als Desmoenzym vorläge (K. BOURNOT, R. WILLSTÄTTER und E. WALDSCHMIDT-LEITZ 1924, M. SANDBERG und E. BRAND, E. WALDSCHMIDT-LEITZ und A. SCHÄFFNER, H. KRAUT und AE. WEISCHER, R. AMMON, R. AMMON und M. JAARMA, O. HOFFMANN-OSTENHOF) und sich somit in ihrem Verhalten sowohl von den tierischen Lipasen als auch von dem Enzym der Kryptogamen, das extrahierbar ist, wesentlich unterscheiden würde.

Vor kurzem haben E. BAMANN und Mitarbeiter (1953a) die Frage nach der *Löslichkeit* der Lipase höherer Pflanzen auf breiterer Grundlage untersucht; sie stellten dabei fest, daß die bisherigen Auffassungen in ihrer verallgemeinernden Form unrichtig sind: *Es gibt Samen, aus denen die Lipase mit Wasser, schwachen Alkalien oder auch Glycerin mehr oder weniger weitgehend in Lösung gebracht werden kann.* Beispiele dafür sind die Samen von *Nigella sativa, Nigella damascena, Delphinium Ajacis, Helleborus niger*. Selbst aus *Ricinussamen*, dessen Lipase am häufigsten als völlig unlöslich beschrieben ist, lassen sich mit den erwähnten Lösungsmitteln kleinere Mengen in Lösung überführen. Mit Hilfe von Salzlösungen, z. B. einer 1,5 n-NaCl-Lösung, gelang es N. TIETZ (unveröffentlichte Mitteilung) sogar, über 90% der Lipase des Ricinussamens zu extrahieren.

Diese Ergebnisse erfahren eine Ergänzung durch Beobachtungen, die in jüngerer Zeit an verschiedenen Samen gemacht worden sind. H. S. OLCOTT und T. D. FONTAINE fanden wäßrige Auszüge aus *Keimlingen von Baumwollsamen* lipatisch wirksam und T. P. SINGER und B. H. J. HOFSTEE unterwarfen Lipaselösungen aus *Weizenkeimlingen* erfolgreichen Reinigungsvornahmen, ferner gewannen J. B. HUTCHINSON und Mitarbeiter sowie H. F. MARTIN und F. G. PEERS Lipaselösungen aus dem Perikarp des *Haferkorns*.

Bei diesem neuen Stand unseres Wissens kann man auch einräumen, daß die Kochsalz- und Glycerinextrakte J. R. GREENS, die er aus *gekeimten Ricinussamen* herstellte, *wirklich* lipatisch wirksam waren, obgleich Nachuntersucher die Extrahierbarkeit des Enzyms und die Richtigkeit des Befundes stets in Abrede gestellt haben. Ebenso dürfen gewisse Beobachtungen von K. G. FALK, der meinte, Lipase aus *Ricinus*- und *Soja*samen in wäßrige Lösung übergeführt zu haben, heute glaubhafter erscheinen. Diese Angaben konnten bisher nicht als überzeugend angesehen werden, da der Bestätigung (J. LORBERBLATT und K. G. FALK) die begründete Ablehnung (R. WILLSTÄTTER und E. WALDSCHMIDT-LEITZ 1924) gegenüberstand. Auch einige weitere Lipaselöslichkeitsbefunde an anderem pflanzlichen Material (F. DUNLAP und W. SEYMOUR, M. VAN LAER, TH. ANDREADIS, E. TAKAMIJA) trugen zu wenig Beweiskraft in sich, als daß die allgemeine Auffassung von der *Unlöslichkeit der Lipase höherer Pflanzen* hätte beeinflußt werden können.

Tabelle 1. *Lösungsverhalten der Lipase verschiedener Samen.* (E. BAMANN und Mitarbeiter 1953a.)

Mit Äther entfettete Samenpräparate von	Bei Auszügen hergestellt mit			
	n/40-Ammoniaklösung	Wasser	86%igem Glycerin	20%igem Glycerin
	gehen von der Gesamtmenge der Lipase in Lösung			
	%	%	%	%
Nigella sativa . . .	90	53	13	61
Nigella damascena .	96	65	11	67
Delphinium Ajacis .	92	30	—	—
Helleborus niger . .	96	84	—	—
Chelidonium majus .	70	10	15	36
Ricinus communis .	20	18	26	—

Enzymlösungen sind mikroheterogene Systeme. Ihre Teilchengröße kann recht verschieden sein. Dies tritt bei der Samenlipase augenfällig in Erscheinung. Es ist durch Zentrifugieren unschwer möglich, Lösungsanteile voneinander zu trennen (E. BAMANN und Mitarbeiter 1953a).

Auf Unterschiede der Symplexe in Größe und Zusammensetzung schließen die genannten Autoren auch aus dem verschiedenartigen Verhalten in Lösungen mit verschiedenen Lösungsmitteln (Wasser, Glycerin) oder in aufeinanderfolgenden Auszügen. Das *p_H-Wirkungsoptimum* solcher Lipasefraktionen ist verschieden, ebenso ihr Verhalten gegenüber den *Zusätzen der ausgleichenden Aktivierung* (s. IV 4 e), die sich bereits bei der Untersuchung und Charakterisierung der einzelnen Fraktionen der Pankreaslipase als ein geeignetes Reagens erwiesen haben (E. BAMANN und P. LAEVERENZ). Auch ihre *Beständigkeit* ist ungleich. Daraus läßt sich die Schlußfolgerung ziehen, daß ähnlich wie im Falle des tierischen Enzyms die Lipase des Samens nicht nur in *einem* Zustand, sondern *in verschiedenen Zustandsformen* vorliegt. Hinsichtlich ihrer physiologischen Funktion rechnet sie zu den *Endo*enzymen. Was ihre ursprüngliche, in der Zelle vorliegende Form betrifft, so scheinen neben den *desmo*- auch *lyo*-Formen vorzuliegen.

2. p_H-Wirkungsoptimum.

Da die *Ricinus*lipase des ruhenden Samens gegenüber Triglyceriden ihre größte Aktivität bei $p_H = 4{,}5$—5 zeigt, wurde zunächst angenommen, daß auch andere Pflanzenlipasen optimale Wirkung in schwach saurem Bereich aufweisen (C. OPPENHEIMER). Es zeigte sich aber, daß selbst bei ein- und demselben Enzym große Streuungen auftreten können. Ihre Ursache ist wohl in *uneinheitlichen Versuchsbedingungen* und besonders in der Heranziehung von Substraten zu sehen, die in ihren physikalischen und chemischen Eigenschaften stark variieren (vgl. dazu die Arbeiten von J. LORBERBLATT und K. G. FALK, R. WILLSTÄTTER und E. WALDSCHMIDT-LEITZ 1924, H. E. LONGENECKER und D. E. HALEY 1935, M. SANDBERG und E. BRAND, B. SULLIVAN und M. A. HOWE, G. GORBACH, T. P. SINGER und B. H. J. HOFSTEE).

Auch *unterschiedlicher physiologischer Zustand* des benutzten Samenmaterials kann auf die Lage des p_H-Wirkungsoptimums Einfluß haben. Diese Feststellung trafen R. WILLSTÄTTER und E. WALDSCHMIDT-LEITZ sowie E. BRAND und M. SANDBERG und L. DI FAZIO und P. LADO.

Untersuchungen von E. BAMANN und E. ULLMANN an den Samen von etwa 100 und den vegetativen Organen von etwa 60 Pflanzen mit Hilfe der stalagmometrischen Methode haben eindeutig erwiesen, daß das p_H-Wirkungsoptimum der Lipase der vegetativen Organe und der meisten ungekeimten Samen gegenüber Tributyrin sich im *alkalischen Bereich* befindet. Allerdings erstreckt sich ihre Aktivität bis hinüber ins schwach saure Gebiet und ist bei $p_H = 4{,}5$ noch gut nachweisbar. Das Optimum selbst zeigt eine gewisse Breite und liegt meist bei $p_H = 8{,}0$—$10{,}5$.

Anders als die Mehrzahl der untersuchten Pflanzenlipasen verhält sich die Lipase des *reifen, ungekeimten Ricinussamens.* Ihre Wirksamkeit befindet sich ausschließlich im sauren Gebiet mit einem Optimum bei $p_H = 4{,}5$ (D. E. HALEY und J. F. LYMAN, R. WILLSTÄTTER und E. WALDSCHMIDT-LEITZ 1924, H. E. LONGENECKER und D. E. HALEY 1935, E. BAMANN und E. ULLMANN, G. V. RAMAKRISHNAN und Mitarbeiter). Dagegen weisen das Enzympräparat des *unreifen, noch nicht fetthaltigen Samens* sowie dasjenige der *vegetativen Organe* — bei geringerem Enzymgehalt — ein breites Wirkungsoptimum im alkalischen Bereich auf (E. BAMANN und E. ULLMANN).

Aus diesen Befunden läßt sich die Folgerung ziehen, daß die Abhängigkeit der Wirkung von der Wasserstoffionenkonzentration nicht auf das Vorkommen nur *zweier* diskreter Einzelenzyme, also einer „sauren" und einer „alkalischen" Lipase, sondern auf die Eigenart der Verankerung an das Protoplasma zurückzuführen ist. Man hat also ein wechselndes Nebeneinanderbestehen verschiedenartiger Symplexe, von denen jeder für sich eine etwas andere Abhängigkeit der Wirkung vom p_H zeigt, anzunehmen. Diese Auffassung wird den einzelnen experimentellen Befunden, so dem sehr häufig beobachteten kontinuierlichen Übergang der p_H-Abhängigkeit oder dem wechselnden Hervortreten bestimmter p_H-Wirkungsmaxima gerecht.

So liegt die optimale Wirkung von Lipaselösungen und entfetteten Präparaten von *Nigella sativa* zwar eindeutig im alkalischen Bereich, aber das *Verhältnis* der Wirksamkeit bei $p_H = 9{,}5$ und derjenigen bei $p_H = 4{,}5$ ist bei einem Auszug mit n/40-Ammoniaklösung durch den Quotienten 2,3/1, bei einem solchen mit 86%igem Glycerin durch den Quotienten 6,4/1 gekennzeichnet.

Bei dem *Vergleich aufeinanderfolgender Auszüge und des verbleibenden Rückstandes* ergibt sich hinsichtlich der Lipasewirksamkeit im alkalischen und sauren Gebiet das Bild der Tabelle 2.

Tabelle 2. *Vergleich der lipatischen Wirksamkeit bei $p_H = 8{,}5$ und 4,5 aufeinanderfolgender Auszüge sowie des verbleibenden Rückstandes.* (E. BAMANN und Mitarbeiter 1953a.)

Entfettetes Präparat von *Nigella sativa*	B-[e] gemessen bei		Wirksamkeits-quotient
	$p_H = 8{,}5$	$p_H = 4{,}5$	
1. Auszug mit n/40-Ammoniaklösung (1 + 50) .	2900	1200	2,4/1
2. Auszug mit n/40-Ammoniaklösung	137	75	1,8/1
3. Auszug mit n/40-Ammoniaklösung	62	35	1,7/1
4. Auszug mit n/40-Ammoniaklösung	30	29	1,04/1
Verbleibender *Rückstand*.	125	120	1,04/1

Danach zeigen die kleinsten Kolloidteilchen, wie sie z. B. in den Glycerinextrakten enthalten sind, in ihrer lipatischen Wirkung am ausgeprägtesten ein Wirkungsoptimum und dieses bei alkalischer Reaktion. Bei den gröber dispersen Teilchen des Rückstandes, in denen unlösliche, stark verankerte Lipase zur Wirkung gelangt, ist ein p_H-Optimum nicht mehr zu erkennen. Die Wirksamkeit bei $p_H = 4{,}5$ ist praktisch genau so groß wie bei $p_H = 8{,}5$. Zwischen diesen beiden Grenzfällen besteht ein kontinuierlicher Übergang (E. BAMANN und Mitarbeiter 1953a).

3. Spezifität.

a) Substratspezifität.

Im Falle der Zoolipasen unterscheiden R. WILLSTÄTTER und Mitarbeiter[1] auf Grund einer relativen Substratspezifität zwei Enzymgruppen: Der eine Typ, verkörpert durch das Enzym der *Leber,* eignet sich besonders zur Spaltung niederer Ester, während der andere Typ, vertreten durch das *Pankreasferment,* bevorzugt Glycerinester spaltet. Auch an einigen

[1] Siehe E. WALDSCHMIDT-LEITZ und A. SCHÄFFNER.

pflanzlichen Lipasen wurde das Problem der Substratspezifität mehr oder weniger eingehend behandelt. Die durchgeführten Arbeiten erstreckten sich im wesentlichen auf die Lipasen der Samen aus *Ricinus communis, Triticum sativum, Chelidonium majus, Nigella sativa, Nigella damascena, Helleborus niger* und *Avena sativa.*

Beim Vergleich der Wirksamkeit der *Ricinussamen*lipase gegenüber einer Anzahl pflanzlicher und tierischer Öle und Fette wie Ricinusöl, Olivenöl, Leinöl, Rapsöl, Mandelöl, Sojaöl, Baumwollsamenöl, Erdnußöl, Talg, Knochenfett, Fischöl und Tran ergibt sich, daß die genannten Fette gut verseift werden (W. Connstein und Mitarbeiter, D. E. Haley und J. F. Lyman, H. E. Longenecker und D. E. Haley 1935, Y. Moulé 1953a). Auch Tributyrin (R. Willstätter und E. Waldschmidt-Leitz 1924) und Triacetin (A. E. Taylor) sind gut spaltbare Substrate. Niedermolekulare, wasserlösliche Ester wie Methyl-, Äthyl- und Amylester der Essig-, Butter- und Malonsäure, aber auch der Benzoe-, Salicyl- und Mandelsäure werden nach H. E. Armstrong, H. E. Armstrong und E. Ormerod sowie A. Barton äußerst träge zerlegt. Ester des Cholesterins sowie Phenylester und Depside bleiben unangegriffen (F. E. Kelsey 1939a, L. Reichel und W. Reimuth). K. G. Falk und K. Sugiura sowie J. Lorberblatt und K. G. Falk finden dagegen, daß in einem wäßrigen Extrakt der Ricinusbohne die esterspaltende Wirksamkeit gegenüber Methyl- und Äthylbutyrat höher ist als gegenüber Triacetin. Sie leiten aus ihren Experimenten ab, daß im Ricinussamen neben einer ,,Lipase" auch eine ,,Esterase" enthalten wäre.

H. Nizamuddin und B. S. Kulkarni verfolgen den Verlauf der Hydrolyse durch Ricinuslipase bei Erdnuß-, Lein- und Safloröl, welche als Beispiele für nichttrocknende, halbtrocknende und trocknende Öle gewählt wurden. Die Forscher stellten einen selektiven Verlauf der Lipolyse fest: Die Abspaltung der *ungesättigten* Fettsäuren geht rascher vor sich als diejenige der *gesättigten.* G. Clément und J. Clément beobachteten dagegen, daß höher ungesättigte Triglyceride wie Sojabohnen- und Erdnußöl (J.Z. >100) von *Pankreas*lipase langsamer gespalten werden als solche aus tierischem und menschlichem Fettgewebe (J.Z. ~ 70).

Die Hydrolyse verschiedener Fette durch die Lipase des *Ricinussamens* im Vergleich zur *Pankreas*lipase untersuchten B. Ahmad und A. Bahl.

Die Lipase des *ungekeimten Weizens* wurde gegenüber Triacetin äußerst wirksam befunden, auch Buttersäuremethyl- und äthylester werden ähnlich gut gespalten; höhere Glyceride dagegen wie Tripalmitin, Tristearin oder das Samenfett des Weizenkorns sind weniger gut verseifbar (B. Sullivan und M. A. Howe). Nach den Arbeiten dieser Forscher nimmt die Wirkung der Weizenlipase nach der Keimung nur gegenüber den höheren Glyceriden zu. T. P. Singer und B. H. J. Hofstee folgern auf Grund ihrer kürzlich durchgeführten Untersuchungen wäßriger Auszüge der *Weizenkeim*lipase, daß unterschiedliche Resistenz der verwendeten Ester — Mono- und Triglyceride der niederen Fettsäuren sowie Tween 20 = Polyglykolsorbitanester der Laurinsäure — gegenüber Hydroxylionen auch im Falle der enzymatischen Hydrolyse zum Ausdruck käme und daß das Enzym lediglich die chemische Labilität des Substrates erhöhe. O. Gawron und Mitarbeiter untersuchten die Wirksamkeit wäßriger Auszüge der Weizenkeimlipase gegenüber substituierten Phenylacetaten.

Beobachtungen an lipasereichen Samen führten E. Bamann und Mitarbeiter (1953c) zu der Auffassung, daß die pflanzlichen Lipasen ebenso wie tierisches Enzym eine *ausgeprägte relative Spezifität* zeigen, die sich durch das Verhältnis Butyraseeinheiten (Wirksamkeit gegenüber Tributyrin als Vertreter der niederen Glyceride): Esteraseeinheiten (Wirksamkeit gegenüber Buttersäure-Methylester als Vertreter der Klasse der Ester einwertiger Alkohole) (B-[e]/E-[e]) kennzeichnen läßt. Es erübrigt sich damit die Annahme von K. G. FALK und K. SUGIURA, daß neben der „Lipase" eine „Esterase" vorkäme.

Der Quotient B-[e]/E-[e] beträgt bei Samentrockenpräparaten von *Nigella sativa* = 12070/1, von *Nigella damascena* = 3970/1, von *Helleborus niger* = 12000/1 und von *Chelidonium majus* = 1572/1. Vergleicht man die Substratspezifität der am leichtesten in Lösung gehenden Isodynamen und diejenige der festverankerten, unlöslichen Lipaseanteile, so findet man beim ausgesprochenen *lyo*-Enzym den „Lipase"-Charakter mehr ausgeprägt als beim *desmo*-Enzym. Ein besonders schönes Beispiel für diese Aussage ist die Lipase von *Nigella sativa*: Auszug: $\frac{\text{B-[e]}}{\text{E-[e]}} = \frac{18050}{1}$, Rückstand: $\frac{\text{B-[e]}}{\text{E-[e]}} = \frac{1060}{1}$. Zwischen der Spezifität beider extremer Lipaseanteile herrscht ein kontinuierlicher Übergang.

Bei den lipolytischen Enzymen aus *Magen*, *Pankreas* und *Leber* des Schweines ergeben sich als Quotienten die Werte: Magen = 2000/1, Pankreas = 1240/1, Leber = 4,5/1. Es wird sofort ersichtlich, daß die Lipase aus den Samen von *Nigella sativa*, *Nigella damascena*, *Helleborus niger* und *Chelidonium majus* mit ihren Quotienten 12070/1 bzw. 3970/1 bzw. 12000/1 bzw. 1572/1 in bezug auf ihre Spezifität für niedere Glyceride und einfache Ester der *Magen*- und *Bauchspeichel*lipase (Quotienten: 2000/1 bzw. 1240/1) viel näher stehen als der *Leber*esterase, die durch den Quotienten 4,5/1 hinsichtlich ihrer Spezifität charakterisiert ist.

Im Anschluß an die Arbeiten über die Substratspezifität seien Untersuchungen erwähnt, die sich die Klärung des *Wirkungsmechanismus* bei der Einwirkung von Lipasen auf Triglyceride zur Aufgabe stellen. Danach geht die Hydrolyse der Fette mit Hilfe von *Pankreas*lipase zunächst unter Abspaltung von α-, β-Diglyceriden und β-Monoglyceriden vor sich. Letztere werden nur sehr langsam gespalten (A. J. VIRTANEN und E. LINDEBERG, A. C. FRAZER, P. DESNUELLE, P. SAVARY und P. DESNUELLE, B. BORGSTRÖM). F. H. MATTSON und L. W. BECK ziehen daher den Schluß, daß *Pankreas*lipase bevorzugt die Esterbindung der primären Hydroxylgruppen des Glycerins spaltet.

Auch für pflanzliche Lipasen können ähnliche Befunde erhoben werden; H. F. MARTIN und G. F. PEERS finden, daß gereinigte *Hafer*lipase aus Tributyrin nur *ein Buttersäuremolekül* abzuspalten vermag und die verschiedenen Mono- und Dibutyrine vom Enzym nicht angegriffen werden. H. P. KAUFMANN und TH. LÜSCHING weisen bei der Analyse der Spaltungsprodukte nach der Hydrolyse von Glyceridfett durch *Ricinus*lipase Monoglyceride nach.

Ein Problem von besonderem Interesse ist die Frage nach der *Existenz von Esterasen mit absolut spezifischer Wirkung auf Atropin, Cocain und Tropacocain.* Atropinspaltende Wirkung ist bei *tierischen* Esterasen seit längerem bekannt (P. FLEISCHMANN, R. WILLSTÄTTER und E. BERNER, F. BERNHEIM und M. L. C. BERNHEIM, H. A. OELKERS und W. RAETZ, S. J. LEBEDINSKAJA, D. GLICK und Mitarbeiter, P. B. SAVIN und D. GLICK, R. AMMON und W. SAVELSBERG). Ob diese wirklich einer eigenen Gruppe *spezifischer Tropinesterasen* (D. GLICK) bzw. einer Atropinesterase mit absoluter Spezifität (R. AMMON und W. SAVELSBERG) zukommt, bedarf wohl noch weiterer Überprüfung.

In einer erst kürzlich im Institut von L. HÖRHAMMER durchgeführten Arbeit über die Stabilisierung von *Fol. Belladonnae* fand E. SCHÄFER auch *pflanzliche* Enzympräparate gegenüber Atropin wirksam. Die Atropinspaltung durch das Enzym aus den Samen von *Nigella damascena* wird von der Autorin der Samenlipase, nicht einer spezifischen Tropinesterase, zugeschrieben.

b) Die optische oder stereochemische Spezifität.

Zu den interessantesten Untersuchungen, die über die tierischen Lipasen durchgeführt sind, gehören jene über die stereochemische Spezifität dieser Enzyme (s. dazu P. RONA und R. AMMON, E. BAMANN und R. AMMON, R. AMMON und M. JAARMA). Sie haben nicht nur unseren Einblick in die Substratspezifität, die bei diesen Katalysatoren eine relative ist, erweitert, indem in den Kreis der Substrate nun auch solche treten, die zueinander wie Bild und Spiegelbild stehen, sondern auch versucht, für das Problem der Unterscheidung der Lipase verschiedener Herkunft — Tier-, Organart — einen Beitrag zu geben. So konnte man auch erwarten, daß die Untersuchung der optischen Spezifität am Enzymsystem der pflanzlichen Lipase mit seinen in Größe, Zusammensetzung und Eigenschaften verschiedenartigen Symplexen weitere und interessante Beobachtungen erlauben würde.

Aus Versuchen von E. BAMANN und Mitarbeiter (1954a) geht hervor, daß auch das System der pflanzlichen Lipase, beispielsweise in Form des makroheterogenen Verreibungspräparates aus enzymatisch gut wirksamen Samen wie *Nigella sativa*, *Nigella damascena*, *Ricinus communis* und *Chelidonium majus*, optische Spezifität zeigt. So ist bei *Chelidonium majus* das optische Auswählen der Lipase bei der Hydrolyse von racemischem Mandelsäuremethylester in der

Tabelle 3. *Die Hydrolyse von racemischem Mandelsäuremethylester durch Lipasepräparate aus einigen Pflanzensamen.* (E. BAMANN und Mitarbeiter 1954a.)
(Die Analysenproben von 50 ml enthalten 0,25 g racemischen Mandelsäuremethylester; 2 g Phosphatpuffer $p_H = 5{,}2$; bei *Ricinus*: 2 Tropfen Toluol.) $t = 37^0$.

Reaktionszeit Stunden	Entfettetes Samenmaterial							
	Nigella damascena IIb 2 g = 2600 B-[e]		*Nigella sativa* IIa 2 g = 4000 B-[e]		*Chelidonium majus* IIIb 2 g = 750 B-[e]		*Ricinus communis* Ic 2 g = 600 B-[e]	
	% Spaltung[1]	$[\alpha]_D^0$ der Mandelsäure	% Spaltung[1]	$[\alpha]_D^0$ der Mandelsäure	% Spaltung[1]	$[\alpha]_D^0$ der Mandelsäure	% Spaltung[1]	$[\alpha]_D^0$ der Mandelsäure
1/2					11,2	+15,4		
3					28,4	+30,7		
5	17,7	+71,4	6,6	−52,6			7,7	−101,7
20	36,2	+56,4	35,7	−36,6	50,2	+53,9	19,6	−124,2
40			55,3	−29,4	64,6	+30,4		
50	57,5	+34,1						
80							37,6	−110,9
100							44,3	−101,9

[1] Nach Abzug der durch Puffer bewirkten Spaltung.

Regel positiv orientiert, ebenso bevorzugt die Lipase von *Nigella damascena* den (+)-Ester. *Nigella sativa*- und *Ricinus communis*-Lipase spalten dagegen den (—)-Ester des Racemats rascher.

Aus Tabelle 3 ergeben sich auffallende Befunde, die von den an tierischem Enzym beobachteten Erscheinungen abweichen:

Bei einigen Lipasepräparaten (*Chelidonium majus, Ricinus communis*) nimmt nämlich die Spaltung der optischen Antipoden im Racemat einen ungewöhnlichen Verlauf: Die spezifische Drehung der isolierten Mandelsäure weist zu Beginn der Spaltung niedere Werte auf, steigt dann — bis zu einem gewissen Spaltungsgrad — an, um bei noch weiter fortgeschrittener Hydrolyse wieder abzufallen. Besonders ausgeprägt kommt diese Erscheinung bei *Chelidoniumsamen*lipase zum Ausdruck.

E. BAMANN und Mitarbeiter (1954a) ziehen zur Klärung dieses abweichenden Verhaltens Versuche mit gequollenem Samenmaterial heran. Sie beobachteten an einem 3 Std lang bei 37⁰ mit Phosphatpufferlösung $p_H = 5{,}2$ vorbehandelten *Chelidonium*samenpulver, daß im Vergleich zum Trockenpräparat die Umsetzungsgeschwindigkeit geringer und die Spaltung des (+)-Esters stärker ausgeprägt ist. Aus diesen Befunden sowie aus vergleichenden Beobachtungen an einem wäßrigen Auszug (lyo-Enzym) und dem erschöpfend ausgezogenen Rückstand (desmo-Enzym) des *Chelidonium*samens ergibt sich, daß im Verlauf der Vorquellung aus dem Enzymkomplex isodyname Fraktionen in Lösung gehen, die das Racemat unter viel stärkerer Bevorzugung des (+)-Esters spalten als es das desmo-Enzym tut. Daraus könnte man auf eine ausgeprägtere Spezifität des lyo-Enzyms gegenüber dem (+)-Ester und damit auf eine Abhängigkeit des optischen Auswählens vom Enzym*zustand* (lyo-, desmo-Fraktion) schließen.

Bestimmend für den außergewöhnlichen Verlauf der $[\alpha]_D$-Kurven bei Chelidoniumlipase ist aber eines der Spaltprodukte des Esters: Während der Einfluß des Methylalkohols auf den Reaktionsverlauf geringfügig ist, verursacht die Mandelsäure eine Hemmung der Racematspaltung und zugleich eine starke Bevorzugung des (+)-Esters. Dieser Einfluß der Mandelsäure auf Spaltungsgeschwindigkeit und optisches Auswählen äußert sich beim lyo-Enzym weit mehr als beim desmo-Anteil. *Es liegt also im strengen Sinne keine Abhängigkeit des optischen Auswählens vom Enzymzustand, sondern eine ungleich starke Auswirkung der Mandelsäure auf die Wirkungen von lyo- und desmo-Lipase vor.*

Diese Tatsache gibt auch die Erklärung für zwei weitere eigenartige Beobachtungen am Enzym aus *Chelidonium*samen ab, nämlich für eine bisher nicht beobachtete Abhängigkeit des sterischen Auswählens von der Enzym*menge* sowie für den Einfluß der *Substratkonzentration* auf die Spezifität des Enzyms.

Das gegenüber Tributyrin sehr aktive lyo-Enzym aus *Nigella sativa* und *Nigella damascena* ist gegenüber Mandelsäuremethylester wirkungslos (*Nigella sativa*) oder von so geringer Wirksamkeit (*Nigella damascena*), daß sich der Nachweis etwa ebenfalls ungleicher Racematspaltung durch lyo- bzw. desmo-Lipase nicht führen läßt. Im Falle beider Samenpräparate ist die gesamte bzw. der Hauptanteil der im Trockenpräparat aufgefundenen Wirksamkeit gegenüber Mandelsäuremethylester im erschöpfend ausgezogenen Rückstand wiederzufinden.

Bemerkenswert ist der Einfluß des Konservierungsmittels *Toluol* auf die Konfigurationsspezifität. Die Wirksamkeit der Samenlipase von *Nigella damascena, Nigella sativa* und *Chelidonium majus* ist in Anwesenheit von Toluol mehr oder weniger herabgesetzt. Nur im Falle der *Ricinussamen*lipase hat Zusatz von Toluol keinen wesentlichen Einfluß auf die Umsetzungsgeschwindigkeit. Im Falle des Samentrockenpräparates von *Nigella sativa* und *Chelidonium majus* beeinflußt das Konservierungsmittel aber darüber hinaus auch das optische Auswählen. Bei manchen Samentrockenpräparaten von *Nigella sativa* führt das Toluol zu einer Änderung der spezifischen Drehung der isolierten Mandelsäure nicht nur der Größe, sondern auch dem Sinne nach.

4. Faktoren, welche Wirksamkeit und Beständigkeit beeinflussen.

Neben der Wasserunlöslichkeit wurde den Phytolipasen — mit Ausnahme derjenigen der Hefen und Schimmelpilze — bisher eine größere Instabilität zugeschrieben (G. Gorbach, W. G. Rose). Diese Aussagen, die in der großen Empfindlichkeit ein Charakteristikum *pflanzlicher* Lipasen sehen wollen, ist in ihrer verallgemeinernden Form nicht zutreffend, da es neben sehr unbeständigen Samenlipasen (*Ricinus*lipase) solche bester Haltbarkeit gibt.

Nach E. Bamann und Mitarbeiter (1953a) bewirkt 14tägige Aufbewahrung wäßriger, neutraler Lipaselösungen bei 8° C aus *Delphinium Ajacis, Helleborus niger* und *Nigella sativa* einen Wirksamkeitsrückgang von nur 16 bzw. 10 bzw. 0%. Die sehr beständige Lipase aus *Nigella sativa* behält ihre Wirksamkeit über Tage hinaus unverändert auch bei alkalischer oder saurer Reaktion der Lösung. In Form von Trockenpräparaten ist die Beständigkeit der Samenlipasen eine noch weit bessere.

Man kennt Faktoren, die auch bei der pflanzlichen Lipase zu *Veränderungen* führen und ihre Wirkung und Beständigkeit beeinflussen. Hierher gehören der physiologische Zustand des Enzymmaterials, die Zubereitung und das Alter der Enzympräparate, der Einfluß von Licht, Temperatur und Substanzen chemischer Art.

a) Physiologischer Zustand des Enzymmaterials.

Es wurde bereits darauf hingewiesen, daß erstmals R. Willstätter und E. Waldschmidt-Leitz (1924) unterschiedliche Eigenschaften an der Lipase des *ruhenden und des keimenden Ricinussamens* feststellten: So verschiebt sich unter Aktivitätsverlust die Lage des p_H-Wirkungsoptimums, Olivenöl als Substrat, in der Weise, daß das Enzym nun auch die Fähigkeit erlangt, Fette im neutralen Bereich zu spalten. Durch ähnliche Unterschiede ist die Lipase des *ungekeimten* (Wirkungsoptimum $p_H = 5$) *und des gekeimten Samens* (Wirkungsoptimum $p_H = 7$) von *Soja hispida* gekennzeichnet (G. Gorbach). Vor allem aber wird an einer Reihe von Samen festgestellt, daß, anders als im Falle der Lipase des Ricinussamens, die lipolytische Aktivität im Verlauf der Keimung zunimmt (N. V. Bhide und D. L. Sahasrabuddhe). Dies zeigten auch Y. Jono an Samen von *Gerste*

Hanf, Soja, Erbse und *Kürbis*, F. Johnston und H. Sell an Samen von *Aleurites Fordii*, sowie C. V. Ramakrishnan (1954) an *Erdnußsamen*. Im Falle der *Baumwollsamen* entwickelt sich Lipaseaktivität erst während des Keimungsvorganges (H. S. Olcott und T. D. Fontaine). B. Sullivan und M. A. Howe berichten, daß während der Keimung die Enzymtätigkeit im Weizenkorn gegenüber den höheren Triglyceriden zunimmt, gegenüber den niederen Glyceriden dagegen unverändert bleibt. Auch verschiedener *Reifegrad* beeinflußt die Eigenschaften der Samenlipase. E. Bamann und E. Ullmann bestimmen mit Hilfe der stalagmometrischen Tributyrinmethode den Lipasegehalt eines *ausgereiften* Ricinussamens im optimalen Bereich von $p_H = 4,0$ mit 210 B-[e], denjenigen eines *unreifen* Samens im optimalen Bereich von $p_H = 8,5$ mit nur 10 B-[e] je 1,0 g entfetteten Trockenpräparats. Auch A. M. Calabrese (1954) stellt fest, daß mit Fortschreiten des Reifungsprozesses die lipolytische Aktivität in den Samen von *Ricinus communis* „Sanguinaria" stark zunimmt. Dagegen verzeichnen E. Theis und Mitarbeiter bei *Flachs-* und *Leinsamen* mit fortschreitender Reifung einen Rückgang des Lipasegehaltes.

b) Zubereitung und Alter der Enzympräparate.

Entfettung der Samen mit Hilfe von wasserfreien organischen Lösungsmitteln zur Bereitung von Enzymtrockenpräparaten setzt eine gewisse Stabilität des Enzyms gegenüber dem gewählten Fettlösungsmittel voraus. Nach E. Bamann und Mitarbeiter (1953a) sind bei *Nigella sativa, Nigella damascena, Delphinium Ajacis* und *Helleborus niger* die Inaktivierungsverluste bei Verwendung von Aceton, Äther und Petroläther gering. Bei der Gewinnung von Präparaten aus *Chelidonium majus*-Samen empfiehlt es sich dagegen, eine möglichst kurze Behandlung mit Petroläther vorzunehmen.

1,8 g *Chelidonium majus*-Samen (= 1 g nach Abzug des Ölgehalts) enthalten 1300 B-[e]. 1 g Trockenpräparat enthielt bei Gewinnung mit:

Petroläther (Siedepunkt 40—50°) (2mal je 7 min behandelt) . . .	1000 B-[e]
wasserfreiem Äther (2mal je 7 min behandelt)	800 B-[e]
Aceton/Äther (5 min mit Aceton, je 2mal 5 min mit Äther behandelt)	680 B-[e]
Äther (6 Std im Soxhlet-Apparat extrahiert)	150 B-[e]

Im Falle der *Hafer*lipase erfolgt durch Behandlung mit Petroläther, Benzin oder Äther *keine* Schädigung des Enzyms. Aktivitätsverlust bewirken Aceton, n-Butanol und Isobutanol, vollständige Inaktivierung tritt durch Methanol, Äthanol und Chloroform ein (H. F. Martin und F. G. Peers).

G. Gorbach beschreibt die Instabilität der *Soja*lipase gegenüber Wasser, wäßrigen Reagentien und organischen Lösungsmitteln.

Als besonders empfindlich gegenüber Wasser und wäßrigen Reagentien gilt die Lipase des entfetteten *Ricinussamens* (R. Willstätter und E. Waldschmidt-Leitz 1924, Y. Moulé 1953b). Nach Meinung von H. E. Longenecker und D. E. Haley (1937a) wirken sich auch Äther und besonders Aceton schädigend auf *Ricinus*lipase aus. Sie empfehlen Petroläther als besonders geeignet für die Entfettung des Samens. Dagegen sagen Y. Moulé (1953b) und A. M. Calabrese (1953) aus, daß Äther und auch Aceton in völlig wasserfreiem Zustand kaum schädigende Wirkung auf das Enzym besitzen, Anwesenheit geringer Mengen an Wasser allerdings zu schneller Inaktivierung des Enzyms führt. Dieser Befund wird durch noch unveröffentlichte Versuche von N. Tietz bestätigt.

Ein deutlicher Einfluß auf die Enzymwirksamkeit zeigt sich häufig bei der *Alterung* von Samen und Samenpräparaten. Nach E. Bamann und E. Ullmann geht im Laufe einer 1- bzw. 2jährigen Lagerung von Trockenpräparaten aus *Nigella sativa* und *Chelidonium majus* die enzymatische Wirksamkeit sehr erheblich zurück.

Eine noch unbestätigte Beobachtung verzeichnet dagegen im Falle eines *Helleborus niger*-Samens *Anstieg* der lipatischen Wirksamkeit im Verlauf einer $^1/_4$jährigen Aufbewahrung (E. BAMANN und Mitarbeiter 1953a). Diese Erscheinung läßt sich vielleicht mit einem älteren Befund von Y. NOGUCHI in Zusammenhang bringen, wonach die Wirksamkeit der Lipase in ruhendem *Reissamen* zunächst absinkt, um nach 12 Monaten plötzlich wieder anzusteigen.

c) Licht.

Im Zusammenhang mit Untersuchungen über das Lichtkeimproblem machte N. TIETZ an Samenpräparaten aus dem Lichtkeimer *Oenothera biennis* und den Dunkelkeimern *Nigella damascena* und *Nigella sativa* die bemerkenswerte Feststellung, daß *Licht* auf die enzymatische Fettspaltung dieser Samen einen Einfluß hat: Die Lipase des *Oenothera*samens wird durch Licht in erheblichem Maße gefördert, während diejenige der Dunkelkeimer eine Hemmung erfährt und bei längerer Lichteinwirkung stark geschädigt wird. Der Autor vermutet, daß die Anwesenheit photodynamisch wirkender Stoffe als Ursache der unterschiedlichen Wirkung des Lichtes auf die Fetthydrolyse von Licht- und Dunkelkeimern anzusehen ist. So bewirkt zugesetztes Eosin z. B. je nach Konzentration bei Belichtung einen fördernden oder hemmenden Einfluß auf die Lipasetätigkeit bei dem gegen Licht indifferenten Samen von *Chelidonium majus*, während das Damascenin, der im Samen von *Nigella damascena* enthaltene fluorescierende Farbstoff von Esternatur, in jedem Fall Hemmung der Fetthydrolyse hervorruft: Hierfür wird eine große Affinität des Damascenin zur Lipase und eine geringe Zerfallsgeschwindigkeit des Enzym-Damasceninkomplexes verantwortlich gemacht.

Bezüglich der Wirkung *energiereicher Lichtstrahlen auf Lipasepräparate* sei vermerkt, daß L. HÖRHAMMER und R. HÄNSEL die UV-Wirkung auf *Pankreas*lipase sowie *Nigella sativa*-Samenlipase näher untersuchten: Während die Schädigung einer Suspension der tierischen Lipase gemäß dem LAMBERT-BEERschen Gesetz verlief, wurde durch UV-Behandlung der pflanzlichen Lipasesuspension keine meßbare Inaktivierung erzielt. Dieses Resultat wird auf die Strahlenabsorption durch die Begleitstoffe zurückgeführt. Eine allgemeinere Bedeutung hat dieses Ergebnis vielleicht insofern, als für Stabilisierungen von Drogen das UV-Bestrahlungsverfahren wenig aussichtsreich erscheint (E. SCHÄFER).

d) Temperatur.

In den letzten Jahren hat die Frage nach der Hitzebeständigkeit der pflanzlichen Lipasen an Interesse gewonnen. Das Problem ist besonders im Zusammenhang mit der Lagerung von *Getreide* (Hafer) sowie auch an *Ricinussamen*lipase bearbeitet worden. Dabei ergab sich eine verhältnismäßig gute Stabilität der pflanzlichen Lipase; in den untersuchten Fällen überstand sie in Form ihrer entfetteten *Trocken*präparate längeres Erhitzen bei Temperaturen von mehr als 100° ohne oder mit nur geringer Schädigung ihrer Aktivität.

So ist die *Hafer*lipase gegen trockene Erhitzung ziemlich resistent. Erst bei etwa 120° beginnt ihre Wirksamkeit merklich abzunehmen. Dagegen ist sie gegenüber feuchter Erhitzung wesentlich empfindlicher. Hier liegen die Inaktivierungstemperaturen je nach dem Grad des Wassergehaltes bei 55—60° (H. JANECKE, J. B. HUTCHINSON und Mitarbeiter, T. MORAN). Eine Wirksamkeitssteigerung nach dem trockenen Erwärmen auf 55° beobachtet H. JANECKE bei Zuchthaferproben.

Auch *Ricinussamen*lipase ist gegenüber der trockenen Erhitzung recht stabil: 2stündiges Erhitzen bei 120° hat nur 50% Aktivitätsverlust gegenüber Tributyrin zur Folge (A. Schuegraf). Ein in Ricinusöl aufgeschwemmtes, entfettetes Enzympräparat verliert nach 30stündiger Erhitzung bei 100° nur etwa 40% seiner Aktivität (Y. Moulé 1953b). In Anwesenheit von Wasser wird allerdings das Enzym bei Temperaturen oberhalb 20° sehr schnell inaktiviert (Y. Moulé 1953b, C. Ravazzoni und Mitarbeiter).

e) Chemische Substanzen.

Auf Grund zahlreicher Beobachtungen aus dem Bereich *tierischer* Lipasen (R. Willstätter und F. Memmen 1923, R. Willstätter, E. Waldschmidt-Leitz und F. Memmen, R. Willstätter und E. Bamann, E. Bamann und P. Laeverenz) ist die mehr oder weniger starke Abhängigkeit der Wirksamkeit der verschiedenen Zustandsformen der Lipase von natürlichen Begleitsubstanzen und von Zusatzstoffen wohl bekannt. Diese Substanzen können sich je nach den Verhältnissen in einer Aktivierung oder Hemmung äußern. Sie erschweren vielfach eine befriedigende Charakterisierung des Enzyms, da sie beispielsweise die Lage des p_H-Wirkungsoptimums mitbestimmen oder auf den Reaktionsverlauf bei der Spaltung des Substrates Einfluß nehmen.

Das Wirkungsvermögen der untersuchten *pflanzlichen* Lipasen ist in wesentlich geringerem Maße als dasjenige der tierischen Lipasen, z. B. der *Pankreas*lipase, von äußeren Faktoren abhängig (E. Bamann und Mitarbeiter 1953a).

Während die lyo-Fraktionen der *Pankreas*lipase bei Zusatz von z. B. Albumin, Calciumchlorid und Natriumoleat = „ausgleichende Aktivierung" nach R. Willstätter[1] eine Wirkungssteigerung von mehreren Tausend Prozent erfahren (E. Bamann und P. Laeverenz, E. Bamann und W. Salzer), ist die Lösung der Samenlipase aus *Nigella sativa* und *Nigella damascena* nur in geringem Maße aktivierbar. Lipaselösungen aus Trockenpräparaten von *Delphinium Ajacis, Helleborus niger, Chelidonium majus* und *Ricinus communis* sprechen auf die „ausgleichende Aktivierung" nicht an. T. P. Singer und B. H. J. Hofstee finden $Ca^{\cdot\cdot}$, $Mg^{\cdot\cdot}$ und $Mn^{\cdot\cdot}$ ohne Einfluß auf die Aktivität von Lipaselösungen aus *Weizenkeimlingen*; nach H. F. Martin und G. F. Peers verursachen die genannten Ionen sowie Albumin, Natriumoleat, Natriumglykocholat sowie Grill 6 (= Laurinsäureester des Polyglykolsorbitans) im Falle von Enzymlösungen aus *Hafer* weder Aktivierung noch Hemmung der Wirksamkeit. Dagegen wirkt sich $CaCl_2$ aktivierend auf die Lipase des *keimenden Baumwollsamens* aus (H. S. Olcott und T. D. Fontaine). Starke Erhöhung der lipatischen Wirksamkeit im alkalischen Gebiet durch Zusatz von $Ca^{\cdot\cdot}$ ist auch im Falle des Milchsaftes von *Carica papaya* zu verzeichnen (M. Sandberg und E. Brand). Im Falle der *Weizenkeim*lipase wirken sich p-Chloromercuribenzoat, p-Aminophenylarsenoxyd und Jodoacetamid sowie o-Jodosobenzoat stark hemmend aus (T. P. Singer und B. H. J. Hofstee sowie B. M. Dirks und Mitarbeiter).

Nach einer Untersuchung von W. Heimann über den Einfluß von Naturfarbstoffen auf die fermentative Fettspaltung aktiviert das öllösliche Chlorophyll die Lipase aus *Nigella sativa*, während die wasserlösliche Verbindung dieses Enzym hemmt. (Vergleiche hierzu eine frühere Beobachtung von E. Bamann und M. Schmeller, wonach bei den Phthalophenonen mit der von der Wasserstoffionenkonzentration abhängigen Konstitutionsänderung eine aktivierende bzw. hemmende Wirkung auf gewisse tierische Lipasen verbunden ist.)

Einige interessante Beobachtungen boten sich E. Bamann und E. Ullmann bei vergleichenden Untersuchungen an natürlichen, d. h. nicht entfetteten und präparierten, mit Aceton/Äther behandelten, *Chelidoniumsamen:*

Im Falle des unvorbehandelten Samens, nicht aber am entfetteten Trockenpräparat, ergibt sich kein ausgeprägtes p_H-Wirkungsoptimum; die Wirkung nimmt mit steigender Alkalität des Versuchsansatzes kontinuierlich zu. Weiterhin

[1] Siehe dazu E. Waldschmidt-Leitz und A. Schäffner.

wirkt sich Zusatz der zweiwertigen Kationen Ca¨, Sr¨, Ba¨, Mn¨, Pb¨, Co¨, Cd¨ im alkalischen Bereich stark wirkungserhöhend aus.

Nach Zusatz des Ätherextraktionsrückstandes aus Chelidoniumsamen zu entfetteten Trockenpräparaten der gleichen oder anderer Pflanzensamen treten die oben geschilderten Erscheinungen auf.

Das verschiedenartige Verhalten ist auf die Anwesenheit eines Hemmungskörpers, nämlich der im Chelidoniumsamen reichlich vorhandenen *Fettsäuren* zurückzuführen, die sich nur im undissoziierten Zustand hemmend auswirken. Aus diesem Grunde entfällt mit Erhöhung der Alkalität des Versuchsansatzes der hemmende Einfluß. Die genannten Metallionen entfernen die störenden Fettsäuren durch Ausfällung.

Beachtenswert ist ferner der Einfluß des Antisepticums Toluol auf die Spaltungsgeschwindigkeit von Mandelsäuremethylester durch einige Samenlipasen sowie die Wirkung dieses Konservierungsmittels auf das optische Auswählen (s. IV 3 b, S. 123).

Die meisten, aber auch die widersprechendsten Untersuchungen über den Einfluß von Salzen und sonstigen Substanzen auf die Aktivität der pflanzlichen Lipase liegen im Falle der *Ricinussamen*lipase vor: So berichtet erstmals Y. TANAKA über den aktivierenden Einfluß von gewissen Neutralsalzen, darunter $MnSO_4$. Diese Erscheinung wird von K. G. FALK und M. HAMLIN, W. G. ROSE, C. V. RAMAKRISHNAN und G. V. NEVGI (1950c) sowie Y. MOULÉ (1953a) bestätigt; W. G. ROSE findet allerdings Mg¨ und Mn¨ unwirksam. Diese Ergebnisse stehen im Widerspruch zu den Beobachtungen von H. E. LONGENECKER und D. E. HALEY (1937b), die eine Hemmung durch sämtliche Salze und besonders durch Co¨ beobachteten. In einer neuen, noch unveröffentlichten Arbeit setzt sich N. TIETZ mit dem Problem des Einflusses einiger Salze auf Löslichkeit und Aktivität der *Ricinus*lipase auseinander. Auf Grund seiner Untersuchungsergebnisse bezeichnet dieser Autor den Einfluß der verschiedenen Salze auf die Aktivität der *Ricinussamen*lipase als eines der markantesten Charakteristika dieses Enzyms und führt die in der Literatur vorhandenen Widersprüche auf eine nach Zusatz stark aktivierender Salze, wie z. B. $CoSO_4$, äußerst rasch erfolgende Inaktivierung des Enzyms zurück. Längere Versuchszeiten (H. E. LONGENECKER und D. E. HALEY 1937b) täuschen daher eine Hemmung der Aktivität durch solche Salze vor.

N. TIETZ vermutet, daß der aktivierende Einfluß aller geprüften Salze auf die Lipase des *Ricinus*samens auf eine Bindung der Ionen an das Enzymeiweiß zurückzuführen ist, wie es im Falle der Enzyme mit Metallproteidcharakter angenommen wird, und daß die rasche Inaktivierung durch eine Oxydation am Enzym ausgelöst wird. Zur Stützung seiner Hypothese führt der Autor an, daß eine zu etwa 20—25% inaktivierte Anschlämmung mit NaCl- oder $CoSO_4$-Lösung durch Zusatz von Cystein-Hydrochlorid teilweise reaktiviert werden kann.

In diesem Zusammenhang ist der Befund von R. ITOH interessant, der eine Substanz in Ricinussamen fand, die in ihrer reduzierten Form die durch *Ricinus*lipase bedingte Hydrolyse aktiviert, während die oxydierte Form als Aktivator der Synthese wirkt und die Hydrolyse hemmt. Der Lipaseaktivator ist unverseifbar und gibt Reaktionen auf Sterol. Auch C. RAVAZZONI (1952b) isolierte aus dem unverseifbaren Anteil des *Ricinus*öls einen Lipaseaktivator.

Nach M. KAWASHIMA sollen Gallensäuren die hydrolysierende Wirkung der *Ricinus*lipase hemmen, die synthetisierende dagegen fördern.

Über Hemmung der *Ricinus*lipase durch Wuchsstoffe und Herbicide wie 2,4-Dichlorophenoxy-Essigsäure, Indol-Essigsäure, Naphthalin-Essigsäure, Naphthalinoxy-Essigsäure berichten kürzlich C. E. HAGEN und Mitarbeiter sowie C. RAVAZZONI (1951). L. DI FAZIO und P. LADO beobachteten eine leichte Hemmung des Enzyms durch 3,4-Benzopyren. Wasserlösliche Vitamine erhöhen nach C. V. RAMAKRISHNAN und B. N. BANERJEE (1951b) die Aktivität der *Ricinus*lipase.

Nach Y. TANAKA hat nur das Spaltprodukt Glycerin einen hemmenden Einfluß auf die enzymatische Fettspaltung durch *Ricinus*lipase. T. ONO gibt auch einen solchen durch höhere ungesättigte Fettsäuren an.

5. Synthetisierende Wirkung.

In einer Anzahl von Untersuchungen ist gezeigt worden, daß man auch mit Hilfe pflanzlicher Lipasen die Synthese von Estern aus den Spaltprodukten Säure und Alkohol außerhalb der Zelle durchführen kann. Außer den älteren Arbeiten von A. E. TAYLOR, Y. W. JALANDER, S. IWANOW, A. WELTER sowie H. E. ARMSTRONG und H. W. GOSNEY seien besonders diejenigen von K. BOURNOT (*Chelidoniumsamen*lipase) sowie von R. WILLSTÄTTER und E. WALDSCHMIDT-LEITZ (1924, *Ricinussamen*lipase), aus jüngerer Zeit diejenigen von A. MOREL und L. VELLUZ sowie L. VELLUZ und Mitarbeiter (*Ricinussamen*lipase), G. GORBACH (Sojabohnenlipase) sowie E. SCHREIBER (1942, verschiedene Samenlipasen) und aus jüngster Zeit diejenigen von G. V. NEVGI und C. V. RAMAKRISHNAN (1950b), C. V. RAMAKRISHNAN und G. V. NEVGI (1950a—c, e, g, 1951a—c, 1952, *Ricinussamen*lipase) hervorgehoben.

An wesentlichen Erkenntnissen dieser Arbeitsrichtung ergibt sich, daß im Falle der *Ricinus*lipase und *Soja*lipase die p_H-Abhängigkeit der lipatischen Fettsäure-Glyceridbildung mit jener der Fetthydrolyse übereinstimmt und die Keimlingslipase an Synthesewirkung der Samenlipase überlegen ist (R. WILLSTÄTTER und E. WALDSCHMIDT-LEITZ 1924, G. GORBACH, A. M. CALABRESE 1954). Weiterhin wurde festgestellt, daß im Falle der *Ricinus*- und *Chelidoniumsamen*-lipasen die Geschwindigkeit, mit der die Estersynthese verläuft, abhängig ist von Kettenlänge, Wertigkeit und Struktur sowie von der Konzentration der beiden Reaktionsteilnehmer Alkohol und Säure (K. BOURNOT, E. SCHREIBER 1942, C. V. RAMAKRISHNAN und G. V. NEVGI).

V. Anreicherung pflanzlicher Lipasen.

1. Darstellung gereinigter Präparate.

Die Lipase der höheren Pflanzen galt bis in die jüngste Zeit als unlösbar. Eine Konzentrierung erschien aus diesem Grunde wenig aussichtsreich. So wird lediglich für die am eingehendsten untersuchte Lipase des *Ricinussamens* von R. WILLSTÄTTER und E. WALDSCHMIDT-LEITZ (1924) das im Weiterverfolg von Versuchen früherer Forscher erzielte Anreicherungsverfahren als „Beispiel für die Reinigung eines unlösbaren Enzyms" beschrieben.

Nun hat sich aber gezeigt, daß die bisherige Auffassung von der Unlöslichkeit pflanzlicher Lipasen in ihrer verallgemeinernden Form unrichtig ist. Lipase*lösungen* aus *keimendem Weizen, Nigella sativa-, Chelidonium majus-, Helleborus niger-* und *Avena sativa*-Samen lassen sich mit den sehr einfachen Verfahren der Dialyse bzw. der Ammoniumsulfat- oder der Essigsäurefällung (T. P. SINGER und B. H. J. HOFSTEE, E. BAMANN und Mitarbeiter 1953b) oder durch mehrmaliges Gefrieren und Auftauen (H. F. MARTIN und F. G. PEERS) konzentrieren.

T. P. SINGER und B. H. J. HOFSTEE trennen aus der Lipaselösung (Ausgangsmaterial: mit Petroläther entfettete Weizenkeime) zunächst mit Hilfe von Essigsäure Ballaststoffe ab, fällen in der vorgereinigten Lösung die Lipase durch Sättigung mit Ammoniumsulfat und erhalten nach Dialyse des Niederschlags ein etwa 10fach konzentriertes Enzympräparat.

E. BAMANN und Mitarbeiter (1953b) gewinnen aus wäßrigen oder ammoniakalischen Lösungen der entfetteten Samen *(Nigella sativa, Chelidonium majus, Helleborus niger)* mit Hilfe der Dialyse oder Essigsäurefällung Präparate, deren Lipasekonzentration um ein Mehrfaches größer ist als die des Samens oder der entfetteten Samenpräparate. Im Falle des *Nigella sativa*-Samens wird durch Fällung mit *Ammoniumsulfat* sowie durch proteolytischen Abbau (R. WILLSTÄTTER und E. BAMANN) der Trägersubstanzen eine gegenüber dem unvorbehandelten Samen etwa 20fache Konzentrierung der Lipase erreicht.

H. F. MARTIN und F. G. PEERS beobachteten eine Inaktivierung der *Hafer*-lipase nach Fällung mit Essigsäure und eine teilweise Inaktivierung nach Dialyse sowie nach Sättigung mit Ammoniumsulfat. Sie finden dagegen den größten Teil der ursprünglich vorhandenen Lipaseaktivität in einem Niederschlag, der sich durch wiederholtes Gefrieren und Auftauen der Enzymlösung (Ausgangspräparat: Perikarp des Haferkorns) bildete. Das geschilderte Verfahren erweist sich als außerordentlich erfolgreich, denn die erreichte Konzentrierung beträgt das etwa 2000fache der ursprünglich im Haferkorn enthaltenen Lipaseaktivität (vgl. V 2, S. 129).

Fermentanreicherung durch Zerschäumung bewirken G. GORBACH, G. DEDIC und K. LORENZ im Falle der Bakterienlipasen: *Micrococcus nacreaceus* und *Bacterium prodigiosum*; sie erreichen mit dieser Methode im Falle der Lipase des erstgenannten Bakterienstammes eine 684fache, bei dem Lipasepräparat des letzteren eine 421fache Aktivität gegenüber derjenigen der Ausgangskulturen. Eine weitere Reinigung durch Ausziehen des getrockneten Spumats mit Hilfe NaCl-haltigen Veronalpuffers vom $p_H = 8$ ergibt eine weitere (1,73malige bzw. 4,56malige) Wirksamkeitssteigerung.

2. Analytische Befunde an gereinigten Präparaten.

Gereinigte, konzentrierte Enzympräparate aus *Weizenkeimen*, aus den Samen von *Nigella sativa* und *Avena sativa* wurden mit Hilfe verschiedener Methoden auf ihre Zusammensetzung untersucht. So stellten T. P. SINGER und B. H. J. HOFSTEE fest, daß *Weizenkeim*lipase gegenüber Thiolreagentien wie p-Chloromercuribenzoat weit empfindlicher ist als *Pankreas*lipase. Sie folgern daraus, daß man dieses Enzym in die große Familie der Sulfhydrilenzyme einreihen kann, deren Aktivität von der Unversehrtheit einer oder mehrerer SH-Funktionen abhängig ist.

Papierelektrophoretische Verfahren ziehen E. BAMANN und N. TIETZ *(Nigella sativa)* und H. F. MARTIN und F. G. PEERS *(Avena sativa)* sowie J. V. FIORE und F. F. NORD *(Fusarium lini* BOLLEY*)* heran. Eine durch Dialyse und Ammoniumsulfatlösung gereinigte Lipaselösung aus *Nigella sativa*-Samen teilt sich im elektrischen Feld in zwei dicht beieinanderliegende Fraktionen, wobei die eine Komponente leicht anodisch, die andere leicht kathodisch wandert. Nach Kohleadsorption der Ausgangslösung wird keine weitere Aufteilung im elektrischen Feld erreicht. Diese Befunde rechtfertigen die Annahme, daß die angereicherten Lipasepräparate Lösungen darstellen, welche in bezug auf ihren Proteincharakter recht einheitlich und von fremdartigen Begleitstoffen weitgehend frei sind.

Im Falle einer Lipaselösung aus *Hafer* finden H. F. MARTIN und F. G. PEERS mindestens 3 *Proteinkomponenten*, von denen nach dem Prozeß des Gefrierens und Auftauens zwei in Lösung gehen. Die dritte Komponente, die die Hauptproteinkomponente darstellt und die Lipase in etwa 2000mal höherer Konzentration enthält als das Hafermehl (vgl. V 1), wandert weder in Veronalpuffer bei $p_H = 8{,}5$, noch in einer Äthanol-Ammoniakmischung.

J. V. FIORE und F. F. NORD analysieren ein gereinigtes, konzentriertes Lipasepräparat des Pilzes *Fusarium lini* BOLLEY. Ihre Untersuchungen zeigen die Anwesenheit zweier Hauptkomponenten, von welchen eine aus mehreren nicht trennbaren Proteinen besteht.

VI. Anwendung pflanzlicher Lipase.

1. Therapeutische Anwendung.

Die an den pflanzlichen Lipasen gewonnenen Erkenntnisse haben dazu geführt, die Substitutionstherapie bei Stoffwechselstörungen, die auf relativen oder absoluten Fermentmangel zurückzuführen sind, nicht nur mit tierischen, sondern auch mit pflanzlichen Enzymzubereitungen vorzunehmen, insofern diese Vorteile, z. B. eine größere Stabilität, aufweisen. So verwendet man in jüngerer Zeit das eiweiß- und fettspaltende Papain aus *Carica Papaya*

(R. Willstätter und W. Grassmann) oder das vorwiegend kohlenhydrat- und eiweißspaltende, in gewissem Umfang auch lipolytisch wirksame Fermentgemisch aus dem Schimmelpilz *Aspergillus oryzae* (W. Grassmann und Mitarbeiter). Auf Grund neuerer eigener Untersuchungen glauben E. Bamann und Mitarbeiter (1954b), daß die Heranziehung von Lipasen höherer Pflanzen möglich und aussichtsreich ist.

2. Analytische Anwendung.

Auch zum analytischen Nachweis und zur Bestimmung von Fetten finden fettspaltende Enzyme Anwendung (H. Stetter). Nach F. E. Kelsey ist es mit Hilfe von *Ricinus*- und *Pankreas*lipase möglich, in Lipoiden Glycerinester *getrennt* zu bestimmen. Während die nach dem üblichen Verfahren mit Ammoniumsulfat gereinigte *Pankreas*lipase sowohl Glycerin- als auch Cholesterinester zerlegt, vermag *Ricinus*lipase nur die Glycerinester zu spalten (s. dazu auch IV 3a).

3. Technische Anwendung.

Nach Feststellung der bedeutenden fettspaltenden Wirkung des *Ricinussamens* arbeiteten W. Connstein und Mitarbeiter ein (heute nicht mehr angewendetes) technisches Verfahren der Fettspaltung mit Hilfe dieses Samens aus, das von E. Hoyer in E. Bamann und K. Myrbäck eingehend beschrieben ist.

Literatur.

Ahmad, B., and A. N. Bahl: Relative digestibility of common edible fats. II. Hydrolysis by pancreatic lipase. J. Sci. Ind. Res. (India) B **5**, 1—3 (1946). — Ammon, R.: Esterasen. In F. F. Nord u. R. Weidenhagen, S. 350—406. — Die synthetisierende Wirkung der esterspaltenden Enzyme. In E. Bamann u. K. Myrbäck, S. 1717—1722. — Ammon, R., u. M. Jaarma: Enzymes hydrolyzing fats and esters. In J. B. Sumner u. K. Myrbäck, S. 390—442. — Ammon, R., u. W. Savelsberg: Die enzymatische Spaltung von Atropin, Cocain und chemisch verwandten Estern. Hoppe-Seylers Z. **284**, 135—156 (1949). — Anderson, J. A.: Enzymes and their role in wheat technology. New York: Interscience Publishers 1946. — Andreadis, Th.: Untersuchungen über Vorgänge bei der Tabakfermentation. Biochem. Z. **211**, 378—394 (1929). — Armstrong, H. E.: Studies on enzyme action. Lipase. Proc. Roy. Soc. Lond., Ser. B **76**, 606—608 (1905). — Armstrong, H. E., and H. W. Gosney: Studies on enzyme action. Lipase IV. Proc. Roy. Soc. Lond., Ser. B **88**, 176—189 (1915). — Armstrong, H. E., and E. Ormerod: Studies on enzyme action. Lipase II. Proc. Roy. Soc. Lond., Ser. B **78**, 376—385 (1906). — Avery, O. T., and G. E. Cullen: Studies on the enzymes of *Pneumococcus*. Lipolytic enzymes: Esterase. J. of Exper. Med. **32**, 571—582 (1920).

Bamann, E., u. R. Ammon: Die stereochemische Spezifität der esterspaltenden Enzyme. In E. Bamann u. K. Myrbäck, S. 1704—1716. — Bamann, E., u. P. Laeverenz: Über pankreatische Lyo- und Desmolipasen. IV. Zur Kenntnis zellgebundener Enzyme der Gewebe und Drüsen in der von R. Willstätter und M. Rohdewald begonnenen Untersuchungsreihe. Hoppe-Seylers Z. **238**, 1—20 (1934). — Bamann, E., u. J. N. Mukherjee: Über die protoplasmatische Verankerung der Leberesterase. Hoppe-Seylers Z. **229**, 1—14 (1934). — Bamann, E., u. K. Myrbäck: Die Methoden der Fermentforschung. Leipzig: Georg Thieme 1941; New York: Academic Press 1945. — Bamann, E., u. W. Salzer: Lyo- und desmo-Enzyme. Erg. Enzymforsch. **7**, 28—49 (1938). — Bamann, E., u. N. Tietz: Über das Verhalten gereinigter Samenlipasepräparate bei der Elektrophorese. Biochem. Z. **324**, 502—503 (1953). — Bamann, E., u. E. Ullmann: Untersuchungen über die Lipase höherer Pflanzen. Biochem. Z. **312**, 9—40 (1942). — Bamann, E., E. Ullmann, A. Schuegraf u. R. Boshart: Verhalten einiger Samenlipasen bei der Spaltung razemischer Ester. I. Über asymmetrische Esterhydrolyse durch pflanzliche Enzyme. Biochem. Z. **325**, 170—181 (1954a). — Bamann, E., E. Ullmann u. N. Tietz: Über lösliche Lipase bei höheren Pflanzen. Biochem. Z. **323**, 489—504 (1953a). — Gewinnung angereicherter Lipasepräparate aus Pflanzensamen. Biochem. Z. **324**, 134—137 (1953b). — Über die Substratspezifität isodynamer pflanzlicher Lipasen. Biochem. Z. **324**, 249—254 (1953c). — Neue Gesichtspunkte für die Auswahl der lipatischen Komponente in therapeutisch verwendeten Enzympräparaten. Arzneimittel-Forsch. **4**, 35—36 (1954b). —Barton, A.: The lipolytic activity of the castor and soy bean. J. Amer. Chem. Soc. **42**, 620—632 (1920). — Bernheim, F., and M. L. C. Bernheim: The hydrolysis of homatropine and atropine by various tissues. J. of Pharmacol. **64**, 209—216 (1938). — Bersin, Th.: Kurzes Lehrbuch der Enzymologie. Leipzig: Akademische Verlagsgesellschaft 1951. — Bhide, N. V., and D. L. Sahasrabuddhe: A study of the enzymes present in germinating seeds. J. Univ. Bombay A **12**, 81—84 (1943). — BORGSTRÖM, B.: The mechanism of the hydrolysis of glycerides by pancreatic lipase. Acta chem. scand. (Copenh.) **7**, 557—558 (1953). — BOUDON, S.: Étude de la lipolyse au cours de la germination des graines de Pin maritime. Rev. gén. Bot. **60**, 284—313 (1953). — BOURNOT, K.: Über die Lipase der Chelidoniumsamen. Biochem. Z. **52**, 172—205 (1913); **65**, 140—157 (1914). — BRAND, E., and M. SANDBERG: On the unity of castor lipase. Proc. Soc. Exper. Biol. a. Med. **23**, 541—544 (1925/26).

CALABRESE, A. M.: Sull'attività della lipasi del *Ricinus communis* in particolari condizioni sperimentali. I. Metodo di determinazione dell'attività lipasica. Ric. Sci. **23**, 1963—1974 (1953). — Sull'attività della lipasi del *Ricinus communis* in particolari condizioni sperimentali. II. Variazioni di attività durante il cielo di maturazione del seme. Ric. Sci. **24**, 1016—1020 (1954). — CLÉMENT, G., et J. CLÉMENT: Sur le mécanisme chimique de l'hydrolyse enzymatique des huiles et graisses naturelles. I. Influence du degré de désaturation du substrat sur la vitesse d'hydrolyse, in vitro, par le suc pancréatique; étude du degré de désaturation des acides gras libérés. Bull. Soc. Chim. biol. Paris **36**, 1319—1328 (1954). — COHEN, S., J. B. KUSHNIK and C. V. PURDY: Observations of mycobacterial esterases with a series of synthetic substrates. J. Bacter. **66**, 266—273 (1953). — Inhibition of mycobacterial esterases by polymyxin B. Antibiotics a. Chemother. **4**, 18—24 (1954). — COLOWICK, S. P., and N. O. KAPLAN: Methods in Enzymology, Bd. I. New York: Academic Prεss 1955. — CONNSTEIN, W., E. HOYER u. H. WARTENBERG: Über fermentative Fettspaltung. Ber. dtsch. chem. Ges. **35**, 3988—4006 (1902). — CUTCHINS, E. C., R. N. DOETSCH and M. D. PELCZAC: The influence of medium composition on the production of bacterial lipase. J. Bacter. **63**, 267—272 (1952).

DAVID, R.: Effect of vernalization on the lipides of the shoots and endosperm of wheat seeds. C. r. Soc. Biol. Paris **233**, 428—430 (1951). Ref. Chem. Abstr. **46**, 2133e (1952). — DESNUELLE, P.: Hydrolyse enzymatique des triglycerides. II. Journées biochim. Franco-Suisse Genève 11.—13. Mai 1951. — DIFAZIO, L., and P. LADO: Lipase of *Ricinus communis* in the quiescent and in the germinating seed. Tumori **40**, 638—646 (1954). Ref. Chem. Abstr. **49**, 14923d (1955). — DIRKS, B. M., P. D. BOYER and W. F. GEDDES: Some properties of fungal lipases and their significance in stored grain. Cereal Chem. **32**, 356—373 (1955). — DOUNCE, A. L.: Cytochemical foundations of enzyme chemistry. In J. B. SUMNER u. K. MYRBÄCK, S. 187—266. — DUNLAP, F. L., and W. SEYMOUR: The hydrolytic enzyme lipase. J. Amer. Chem. Soc. **27**, 935—946 (1905). — DUSPIVA, F.: Die cytologischen Grundlagen der protoplasmatischen Verankerung der Enzyme. In F. F. NORD u. R. WEIDENHAGEN, S. 11—64.

ENGEL, C.: The distribution of the enzymes in resting cereals. III. The distribution of esterase in wheat, reye, and barley. Biochim. et Biophysica Acta **1**, 278—279 (1947). — ENGEL, C., and L. H. BRETSCHNEIDER: The distribution of the enzymes in resting cereals. IV. A comparative investigation of the distribution of enzymes and mitochondria in wheat grains. Biochim. et Biophysica Acta **1**, 357—363 (1947).

FALK, K. G.: Studies on enzyme action. IX. Extraction experiments with the castor bean lipase. J. Amer. Chem. Soc. **35**, 1904—1915 (1913). — Studies on enzyme action. XIII. The lipase of soy beans. J. Amer. Chem. Soc. **37**, 649—653 (1915). — FALK, K. G., and M. L. HAMLIN: Studies on enzyme action. III. The action of manganous sulfate on castor bean lipase. J. Amer. Chem. Soc. **35**, 210—219 (1913). — FALK, K. G., and K. SUGIURA: Studies on enzyme action. XII. The esterase and lipase of castor beans. J. Amer. Chem. Soc. **37**, 217—230 (1915). — FIORE, J. V., and F. F. NORD: On the mechanism of enzyme action. XLII. Isolation and some properties of the lipase from *Fusarium lini* BOLLEY. Arch. of Biochem. **26**, 382—399 (1950). — FLASCHENTRÄGER, B., u. E. LEHNARTZ: Lehr- und Handbuch der physiologischen Chemie. Berlin-Göttingen-Heidelberg: Springer 1951. — FLEET, D. S. VAN: The cell forms, and their common substance reactions, in the parenchyma-vascular boundary. Bull. Torrey Bot. Club **77**, 340—353 (1950). — A comparison of histochemical and anatomical characteristics of the hypodermis with the endodermis in vascular plants. Amer. J. Bot. **37**, 721—725 (1950). — Histochemical localization of enzymes in vascular plants. Bot. Review **18**, 354—398 (1952). — FLEISCHMANN, P.: Atropinentgiftung durch Blut. Arch. exper. Path. u. Pharmakol. **62**, 518—526 (1910). — FLEURY, M. G.: Recherches chimiques sur la germination. Ann. chim. phys. **4**, 38 (1865). — FLÜCK, H.: Der Einfluß verschiedener Trocknungsmethoden auf den Gesamtalkaloidgehalt von *Folia Stramonii* und *Folia Belladonnae*. Heil- u. Gewürz-Pflanzen **18**, 109—125 (1939). — FODOR, P. J.: Studies on the activity and inhibition of yeast esterase. Bull. Res. Council Israel **4**, 1—5 (1954). — FODOR, P. J., u. A. CHARI: The ester-hydrolysing systems of *Aspergillus niger* and of *Penicillium roqueforti*. Enzymologia (Den Haag) **13**, 258—267 (1949). — FOKIN, S.: Über Pflanzen, welche in ihren Samen ein Ferment enthalten, das Fett in Glycerin und Fettsäuren spaltet. J. russ. physik.-chem. Ges. **35**, 831—835 (1903). Ref. Chem. Zbl.

1903 II, 1451. — Zur Frage über die Zerlegung der Fette durch Enzyme. Chem. Rev. Fett- u. Harz-Ind. **11**, 91—92, 118—120, 139—142, 167—170, 193—195, 224—226, 244—247 (1904). Ref. Chem. Zbl. **1904 II**, 1617—1618. — Frazer, A. C.: Differentiation in the absorption of olive oil and oleic acid in the rat. J. of Physiol. **102**, 306—312, 329—333 (1943)

Gawron, O., C. J. Grelecki and M. Duggan: The effect of substituents on the hydrolysis of phenylacetate by wheat germ lipase. Arch. of Biochem. a. Biophysics **44**, 455—467 (1953). — Ghosh, J. J.: Studies on the dextrinase, amylase, protease and lipase activities of the rice-inhabiting fungi. Indian J. Physiol. **6**, 28—42 (1952). — Glick, D.: Beitrag zur enzymatischen Histochemie. VIII. Eine Mikromethode zur Bestimmung der Aktivitä lipolytischer Enzyme. Hoppe-Seylers Z. **233**, 252—256 (1934). — Properties of tropin esterase. J. of Biol. Chem. **134**, 617—625 (1940). — Glick, D., and E. E. Fischer: Studie in histochemistry. XVII. Localization of phosphatases in the wheat grain and in the epi cotyl and roots of the germinated grain. Arch. of Biochem. **11**, 65—79 (1946). — Glick, D. and S. Glaubach: Occurrence and distribution of atropinesterase and the specifity of tropin esterases. J. Gen. Physiol. **25**, 197—205 (1941). — Glick, D., S. Glaubach and D. H Moore: Azolesterase activities of electrophoretically separated proteins of serum. J. o Biol. Chem. **144**, 525—528 (1942). — Goldman, M. L., and M. M. Rayman: Hydrolysi of fats by bacteria of the *Pseudomonas* genus. Food Res. **17**, 326—337 (1952). — Gomori, G. The microtechnical demonstration of sites of lipase activity. Proc. Soc. Exper. Biol. a Med. **58**, 362—364 (1945). — Histochemical localization of true lipase. Proc. Soc. Expe Biol. a. Med. **72**, 697—700 (1949). — Goodman, J. J., and C. M. Christensen: Grain storag studies. XI. Lipolytic activity of fungi isolated from stored corn. Cereal Chem. **29**, 299—30 (1952). — Gorbach, G.: Zur Kenntnis der Sojalipase. Fette u. Seifen **48**, 308—312 (1941). — Gorbach, G., u. H. Günther: Über Hefelipase. Mh. Chem. **61**, 47—60 (1932). — Gorbach, G., G. Dedic u. K. Lorenz: Zur Anreicherung und Wirksamkeitsbestimmun von Bakterienlipasen. Arch. Mikrobiol. **21**, 237—247 (1955). — Grassmann, W., K. Maye u. E. Waldschmidt-Leitz: Über die therapeutische Bedeutung von Pilzenzymen und di ihrer Kombination mit den Fermenten des Verdauungstraktes. Ärztl. Forsch. **4**, 17—2 (1950). — Grassmann, W., u. H. Rubenbauer: Über Cellulase und Hemicellulase m besonderer Berücksichtigung ihrer therapeutischen Anwendung. Münch. med. Wschr. **193** 1817—1819. — Green, J. R.: On the germination of the seed of the castor oil plan *(Ricinus communis)*. Proc. Roy. Soc. Lond. **48**, 370—392 (1891). — Green, J. R., an H. Jackson: Further observations on the germination of the castor oil plant. Proc. Ro Soc. Lond. **77**, 69—85 (1905).

Hagen, C. E., C. O. Clagett and E. A. Helgeson: 2,4-Dichlorophenoxyacetic aci inhibition of castor bean lipase. Science (Lancaster, Pa.) **110**, 116—117 (1949). — Hale D. E., and J. F. Lyman: Castor bean lipase, its preparation and some of its propertie J. Amer. Chem. Soc. **43**, 2664—2670 (1921). — Heide, S.: Zur Physiologie und Cytolog der Fettbildung bei *Endomyces vernales*. Mit einem Beitrag zur Methodik der quantitative Bestimmung kleinster Fettmengen. Arch. Mikrobiol. **10**, 135—188 (1939). — Heimann, W Der Einfluß von Naturfarbstoffen auf die fermentative Fettspaltung. Fette u. Seifen **5** 887 (1955). — Hörhammer, L., u. R. Hänsel: Ultraviolettes Licht und Stabilisierun I. u. II. Arch. Pharmazie u. Ber. dtsch. pharmazeut. Ges. **284/56**, 164—171 (1951); **285/5** 338—344 (1952). — Hoffmann-Ostenhof, O.: Enzymologie, eine Darstellung für Chemike Biologen und Mediziner. Wien: Springer 1954. — Holter, H.: Localization of enzym in cytoplasma. Adv. Enzymol. **13**, 1—20 (1952). — Hoyer, E.: Über fermentative Fet spaltung. Hoppe-Seylers Z. **50**, 414—435 (1906/07). — Fermente in der Fettindustrie. I E. Bamann u. K. Myrbäck, S. 2816—2827. — Hummel, B. C. W., L. S. Cuendet, C. M Christensen and W. F. Geddes: Grain storage studies. XIII. Comparative changes respiration, viability and chemical composition of mould-free and mould-contaminate wheat upon storage. Cereal Chem. **31**, 143—150 (1954). — Hutchinson, J. B., H. F. Mart and T. Moran: Location and destruction of lipase in oats. Nature (Lond.) **167**, 758 (1951

Itoh, R.: Control of the enzymic action of lipase. Nature (Lond.) **137**, 783 (1936). — On the existence of a substance which controls the splitting and synthetic action of lipas J. of Biochem. (Tokyo) **23**, 299 (1936). — Studies on lipase. J. of Biochem. (Tokyo) **2** 167—175 (1937). — Iwanow, S.: Über Ölsynthese unter Vermittlung der pflanzlichen Lipas Ber. dtsch. bot. Ges. **29**, 595—602 (1911).

Jalander, Y. W.: Zur Kenntnis der Ricinuslipase. Biochem. Z. **36**, 435—476 (1911). — Janecke, H.: Über die Haferlipase. Stärke **32**, 29—34 (1951). — Jensen, L. B.: Micr biology of meats. Champaign, Illinois: Garrard Press 1945. — Johnston, F. A., and M. Sell: Changes in the chemical composition of tung kernels during germination. Unve öffentlicht. Zit. H. E. Longenecker in J. A. Anderson, S. 132. — Jono, Y.: Vergleichen Untersuchungen über den Fermentgehalt der ruhenden und keimenden Pflanzensame Acta Scholae med. Kioto **13**, 211—238 (1931). Zit. B. Sullivan in J. A. Anderson, S. 17

KAUFMANN, H. P., u. TH. LÜSCHING: In H. P. KAUFMANN, Zur Biologie der Fette. VI. Glyceride. Fette u. Seifen **55**, 673—681 u. zw. 678 (1953). — KAWASHIMA, M.: On the influence of the bile acid to the synthesizing and hydrolysing activity of the *ricinus*-lipase. Hiroshima J. Med. Sci. **3**, 153—157 (1954). — KELSEY, F. E.: The use of lipase in lipid analyses. I. Specifity of the castor bean lipase. J. of Biol. Chem. **130**, 187—193 (1939a).— The use of lipase in lipid analyses. II. Specifity of pancreatic lipase. J. of Biol. Chem. **130**, 195—198 (1939b). — The use of lipase in lipid analyses. III. The action of castor bean and pancreatic lipase on blood neutral fat. J. of Biol. Chem. **130**, 199—202 (1939c). — KIEZEL, A., i K. GORDIENKO: Über die Tätigkeit der Fermente in Weizenkorn von verschiedenem Wassergehalt während der Lagerung. Bull. Soc. Naturalistes Moscou **46**, 339—349 (1937). Zit. B. SULLIVAN in J. A. ANDERSON, S. 173. — KIRSH, D.: Factors influencing the activity of fungus lipase. J. of Biol. Chem. **108**, 421 430 (1935). — KOCH, G. P.: J. Amer. Pharmaceut. Assoc. 8, 390 (1919). Zit. H. FLÜCK, S. 113. — KOCH, R. B., A. R. FLESHER, T. H. BURTON and R. A. LARSEN: A rapid method for the determination of cereal lipase activity. Cereal Chem. **31**, 113—120 (1954). — KOZMIN, N. P.: Reifung der Weizenmehle und ihre biochemischen Grundlagen. Mühlenlab. **4**, 18—32 (1934a). — Kleberqualität und einige Faktoren, die diese beeinflussen. Mühlenlab. **4**, 109—116 (1934b). — KRAUT, H., u. AE. WEISCHER: Esterasen. In B. FLASCHENTRÄGER u. E. LEHNARTZ, S. 1070—1100 u. zw. 1073 u. 1078. — KRETOVICH, V. L., A. J. SOKOLOVA i E. N. USHAKOVA: The stable moisture content of grain and its effect on the lipase action. C. r. Acad. Sci. URSS. **27**, 701—704 (1940). Ref. Chem. Abstr. **35**, 8126 (1941).

LAER, M. H. VAN: The existence of emulsin and of lipase in malt extract. C. r. Soc. Biol. Paris **84**, 471—474 (1921). Ref. Chem. Abstr. **28**, 6742 (1921). — LEBEDINSKAJA, S. J.: Über die Bedingungen der Atropinwirkung auf Kaninchen. Arb. allruss. Inst. exper. Med. **1**, 89—92 (1936). Ref. Chem. Zbl. **1936 II**, 2161. — LEMBKE, A., L. BEUERMANN u. W. KAUFMANN: Kinetik der Bakterienlipase. Zbl. Bakter. **160**, 423—443 (1954). — LINDERSTRØM-LANG, K., u. C. ENGEL: Über die Verteilung der Amylase in den äußeren Schichten des Gerstenkorns. Enzymologia (Den Haag) **3**, 138—146 (1937). — LINDERSTRØM-LANG, K., u. H. HOLTER: Enzymatische Histochemie. Erg. Enzymforsch. **3**, 309—334 (1934). Ferner in F. F. NORD u. R. WEIDENHAGEN, S. 65—114; E. BAMANN u. K. MYRBÄCK, S. 1132—1162. — LONGENECKER, H. E.: Esterases. In J. A. ANDERSON, S. 127—152. — LONGENECKER, H. E., and D. E. HALEY: *Ricinus*lipase, its nature and specifity. J. Amer. Chem. Soc. **57**, 2019—2021 (1935). — Further studies on the nature of *Ricinus*-lipase and its action. J. Amer. Chem. Soc. **59**, 2156—2159 (1937a). — The influence of certain salts and of cholesterol on the action of *Ricinus*lipase. J. Amer. Chem. Soc. **59**, 2160—2161 (1937b). — LORBERBLATT, J., and K. G. FALK: Studies on enzyme action. XXXVI. Ester hydrolysing actions of the castor bean. J. Amer. Chem. Soc. **48**, 1655—1667 (1926). — LUBERT, D. J., M. L. SMITH and H. R. THORNTON: Estimation of lipase in dairy products. Canad. J. Res. **27**, 504—509 (1949).

MAESTRINI, D.: Enzymes of malt, apropos of a note of M. L. VAN LAER on "existence of a lipase in the extract of malt". C. r. Soc. Biol. Paris **84**, 616—617 (1921a). Ref. Chem. Abstr. **16**, 1593 (1921). — Experimental studies on the reaction velocity of plant enzymes. I. Influence of the concentration of enzymes on the reaction velocity of the enzymes of germinated barley. Arch. Farmacol. sper. **32**, 40—59 (1921b). Ref. Chem. Abstr. **16**, 1593 (1921). — MARTIN, H. F., and F. G. PEERS: Oat lipase. Biochemic. J. **55**, 523—529 (1953). — MASTBAUM, H.: Über ein fettspaltendes Enzym in der Colanuß. Chem. Rev. Fett- u. Harz-Ind. **14**, 5—7, 31—34, 44—46 (1907). Ref. Chem. Zbl. **1907 I**, 978. — MATTSON, F. H., and L. W. BECK: The digestion in vitro of triglycerides by pancreatic Lipase. J. of Biol. Chem. **214**, 115—125 (1955). — MEYER, J., J. MALGRAS et R. SCHAR: Recherches sur la spécifité de la tributyrine comme substrat réactif des esterases et des lipases. Extr. du Bull. Assoc. Diplômés Microbiol. Fac. Pharmacie de Nancy Nr 52, 3. Trim. 1953. — MONNÉ, L.: Functioning of the cytoplasm. Adv. Enzymol. 8, 1 u. zw. 48 (1948). — MORAN, T.: Lipase in oats. Food Manufact. **27**, 73—74 (1952). — MOREL, A., et L. VELLUZ: Contribution à l'étude de la synthèse biochimique glycerides. Sur la réversibilité de l'activité fermentative du cytoplasme de la graine de ricin. Bull. Soc. Chim. biol. Paris **10**, 478—488 (1928). — MORRIS, H. A., and J. J. JEZESKI: The action of microorganisms on fats. II. Some characteristics of the lipase system of *Penicillium roqueforti*. J. Dairy Sci. **36**, 1285—1298 (1953). — MOULÉ, Y.: De l'hydrolyse enzymatique par la lipase du ricin. Influence de la nature du substrat et de la présence de certains ions. Oléagineux 8, 561—563 (1953a). — Signification de l'action de la température sur l'hydrolyse de la triricinoléine par la lipase du ricin. Bull. Soc. Chim. biol. Paris **35**, 759—770 (1953b). — MUFTIC, M. K.: Ein neues Medium zum Auffinden der Esterasewirkung von Mikroorganismen. Enzymologia (Den Haag) **17**, 123—126 (1954). — MUNTZ, M. A.: Sur la germination des graines oléagineuses. Ann. chim. phys. **22**, 472—485 (1871).

Nashif, S. A., and F. E. Nelson: The lipase of *Pseudomonas fragi*. I. Characterizatior of the enzyme. J. Dairy Sci. **36**, 459—470 (1953a). — The lipase of *Pseudomonas fragi*. II. Factors affecting lipase production. J. Dairy Sci. **36**, 471—480 (1953b). — The lipase of *Pseudomonas fragi*. III. Enzym action in cream and butter. J. Dairy Sci. **36**, 481—488 (1953c). — Nevgi, G. V., and C. V. Ramakrishnan: Studies on lipase from oil seeds. I. General study of the *Ricinus*lipase from castor seeds. J. Indian Chem. Soc. **27**, 255—259 (1950a). — Studies on lipase from oil seeds. II. Synthesis of some aliphatic esters by *Ricinus*-lipase from castor seeds. J. Indian Chem. Soc. **27**, 260—262 (1950b). — Newcomer, E. H.: Mitochondria in plants. I. u. II. Bot. Review **6**, 85—147 (1950); **7**, 53—89 (1951). — Nicolai, H. W.: Die Bestimmung der Lipase in keimenden Kiefernsamen. Biochem. Z. **174**, 373—383 (1926). — Nizamuddin, H., and B. S. Kulkarni: Enzyme splitting of fats. J. Sci. Ind. Res. (India) B **12**, 390 (1953). — Noguchi, Y.: Activity of several kinds of enzymes of rice in storage. J. Sci. Agricult. Soc. (Japan) **245**, 115—120 (1923). Zit. E. Lehmann u. F. Aichele, Keimungsphysiologie der Gräser, u. zw. S. 243. Stuttgart: Ferdinand Enke 1931. — Nord, F. F., u. R. Weidenhagen: Handbuch der Enzymologie. Leipzig: Georg Thieme 1940.

Oelkers, H. A., u. W. Raetz: Untersuchungen über die Zerstörung von Cocain und Atropin im Tierkörper. Klin. Wschr. **1933**, 1985—1986. — Ogura, K., S. Imazu, M. Kato and Y. Yamamura: Esterase of tubercle bacteria and various strains of acid-fast bacteria. II. Kekkaku (Tuberculosis) **29**, 128—133 (1954a). — Esterase from acid-resistant bacteria. Symp. Enzyme Chem. (Tokyo) **9**, 104—106 (1954b). — Olcott, H. S., and T. D. Fontaine: Composition of cottonseeds. IV. Lipase of germinated seed. J. Amer. Chem. Soc. **63**, 825—827 (1941). — Ono, T.: Hydrolysis of fats and fat acid esters. III./IV. Hydrolysis of fats by enzymes. J. Agricult. Chem. Soc. (Japan) **15**, 1085—1096, 1161—1172 (1939). — Oparin, A. J.: Die Wirkung der Fermente in der lebenden Zelle. Erg. Enzymforsch. **3**, 57—72 (1934). — Variations de l'activité des enzymes dans la cellule végétale sous l'effet des facteurs exterieurs. Bull. Soc. Chim. biol. Paris **35**, 67—83 (1953). — Oppenheimer, C.: In C. Oppenheimer u. R. Kuhn, Die Fermente und ihre Wirkungen, S. 500. Leipzig: Georg Thieme 1925.

Paech, K.: Biochemie und Physiologie der sekundären Pflanzenstoffe, S. 224. Berlin-Göttingen-Heidelberg: Springer 1950. — Stoffwechsel organischer Verbindungen. Fortschr. Bot. **14**, 334—364 u. zw. 337 (1953). — Peers, F. G.: Oat lipase and tributyrin. Nature (Lond.) **171**, 981 (1953). — Pelouze, M. J.: Mémoire sur la saponification des huiles sous l'influence des matières qui les accompagnent dans les graines. Ann. chim. phys. **45**, 319—327 (1855a). — C. r. Soc. Biol. Paris **40**, 605 (1855b). — Peters, E.: Zur Keimungsgeschichte des Kürbissamens. Landwirtsch. Versuchsstat. **3**, 1—16 (1861). — Peters, J. J., and F. E. Nelson: Preliminary characterization of the lipase of *Mycotorula lipolytica*. J. Bacter. **55**, 592—600 (1948). — Pett, L.: Studies on the distribution of enzymes in dormant and germinating wheat seeds. II. Lipase. Biochemic. J. **29**, 1902—1904 (1935). — Przyłęcki, St. J. v.: Über die intracelluläre Regulierung der Enzymreaktionen mit besonderer Berücksichtigung der Amylasewirkung. Erg. Enzymforsch. **4**, 111—146 (1935). — Purr, A.: Photometrische Bestimmung von Esterase in tierischen und pflanzlichen Geweben und in Mikroorganismen. Biochem. Z. **322**, 205—211 (1951).

Ramakrishnan, C. V.: Studies on enzyme "lipase". Proc. Indian Acad. Sci., Sect. B **33**, 268—276 (1951). — Lipase and esterase activities in the groundnut seed during germination. Sci. a. Cult. **19**, 566—567 (1954a). — Effect of culture medium on lipase and esterase activities of *Aspergillus niger*. Chem. a. Ind. **1954** b, 250—251. — Purification of lipase obtained from *Aspergillus niger*. Naturwiss. **41**, 16—17 (1954c). — Ramakrishnan, C. V., and B. N. Banerjee: Studies on mold lipase. Research (Lond.) **4**, 434 (1951a). — Studies on lipase from oil seeds. XIV. Effect of some water soluble vitamins on the activity of *Ricinus*lipase from castor seeds. J. Indian Chem. Soc. **28**, 471—472 (1951b). — Destruction of lipase in oil seeds as the molds grown on them. J. Indian Chem. Soc. **28**, 591—594 (1951c). — Studies on mold lipase. Comparative study of lipases obtained from molds grown on *Carthamus tinctorius*. J. Univ. Bombay, Sect. A 20 Pt. **3**, 111—114 (1951d). — Studies on mold lipase. Comparative study of lipase obtained from molds grown on coconut. Experientia (Basel) **7**, 434—435 (1951e). — Studies on mold lipase. I. Comparative study of lipases obtained from molds grown on oil seeds. II. Investigations on the suitability of the cake medium to grow the lipolytic mold. Enzymologia (Den Haag) **15**, 33—39, 98—102 (1951f). — Studies on mold lipase. Effect of addition of vitamins and sterols to the cake medium on the growth and the activity of the lipolytic mold. Nature (Lond.) **168**, 917—918 (1951g). — Studies on lipase. Comparative study of different lipases obtained from molds grown on castor seed. J. Indian Chem. Soc. **29**, 397—399 (1952a). — Studies on lipase from oil seeds. XV. Effect of concentration of the accelerators on the hydrolysis of ground nut oil by castor seed lipase. J. Indian Chem. Soc. **29**, 400—402 (1952b). — Studies on lipase from oil seeds. XVII. Effect of buffer and enzyme concentration on the hydrolysis of amylbutyrate by

*Ricinus*lipase. J. Indian Chem. Soc. **29**, 405—406 (1952c). — Lipolytic molds. Comparative study of lipase obtained from molds grown on oil seeds. J. Amer.Oil Chem. Soc. **29**, 596—601 (1952d). — Studies on mold lipase. Comparative study of lipase obtained from molds grown on groundnut. Biochim. et Biophysica Acta **8**, 216—218 (1952e). — Studies on mold lipase. Comparative study of lipase obtained from molds grown on *Sesamum indicum*. Arch. of Biochem. a. Biophysics **37**, 131—135 (1952f). — Studies on mold lipase. Comparative study of lipase obtained from molds grown on mustard seed. Sci. a. Cult. **17**, 298—300 (1952g). — RAMAKRISHNAN, C. V., and G. V. NEVGI: Studies on lipase from oil seeds. I. Effect of different organic solvents on the synthesis and hydrolysis of esters by lipase. II. Synthetic activity of different lipases. J. Univ. Bombay, Sect. A **19** Pt 3 Sci. **28**, 31—37 (1950a). — Studies on lipase from oil seeds. III. Effect of hydroxyl groups in the alcohol on the synthesis of esters by lipase. J. Indian Chem. Soc. **27**, 263 (1950b). — Studies on lipase from oil seeds. IV. Synthesis of some esters by lipase. J. Indian Chem. Soc. **27**, 264 (1950c). — Studies on lipase from oil seeds. V. Synthesis of some esters by lipase. J. Indian Chem. Soc. **27**, 265—267 (1950d). — Studies on lipase from oil seeds. VI. Hydrolysis of some aliphatic esters by *Ricinus*lipase. J. Indian Chem. Soc. **27**, 331—332 (1950e). — Studies on lipase from oil seeds. VIII. Effect of salts on the synthesis and hydrolysis of some esters by lipase. J. Indian Chem. Soc. **27**, 334—336 (1950f). — Studies on lipase from oil seeds. IX. Comparative study of the lipases from different oil seeds. J. Indian Chem. Soc. **27**, 337—344 (1950g). — Studies on mold lipase. Comparative study of lipases obtained from molds grown on oil seeds. Ann. Biochem. a. Exper. Med. (India) **10**, 19—34 (1950h). — Studies on lipase from oil seeds. III. Effect of alcohol, acid and ester concentrations on the synthesis and hydrolysis of amylbutyrate by *Ricinus*lipase. J. Univ. Bombay, Sect. A **20** Pt. **3**, 109 (1951a). — Studies on lipase from oil seeds. X. Synthesis of some esters by *Ricinus*lipase. J. Indian Chem. Soc. **28**, 271—272 (1951b). — Studies on lipase from oil seeds. XI. Effect of H-ion concentration and the nature of the buffer on the synthesis of amylbutyrate by *Ricinus*lipase from castor seed. J. Indian Chem. Soc. **28**, 272—274 (1951c).— Studies on lipase from oil seeds. XII. Effect of H-ion concentration and the nature of the buffer on the hydrolysis of amylbutyrate by *Ricinus*lipase from castor seed. J. Indian Chem. Soc. **28**, 333 (1951d). — Studies on lipase from oil seeds. XIII. Hydrolysis of some esters by *Ricinus*lipase. J. Indian Chem. Soc. **28**, 335—336 (1951e). — Studies on lipase from oil seeds. XVI. Effect of buffer and enzyme concentrations on the synthesis of amylbutyrate by *Ricinus*lipase from castor seed. J. Indian Chem. Soc. **29**, 403—404 (1952). — RAVAZZONI, C.: Azione di sostanze di crescita ed erbicidi sui sistemi enzimatici dei vegetali. Farmaco (Pavia) **6**, 588—591 (1951). — Ricerche sulle lipasi i sulla disattivatione della lipasi del ricino. Ann. Chimica **42**, 590—597 (1952a). — Ricerche sulle lipasi vegetali. Ann. Fac. Agrar. Univ. Milano **1**, 185—189 (1952b). — RAVAZZONI, C., A. DANSI e G. BOLZONI: Ricerche sulle lipasi. Ann. Chimica **40**, 257—262 (1950). — RAVAZZONI, C., A. GARZIA et A. DANSI: Contribution à l'étude de la lipase du *Ricinus*. Congr. Internat. Biochim. Res. Comm. 1952, S. 249. Ref. Chem. Abstr. **50**, 5062h (1956). — REICHEL, L., u. W. REINMUTH: Über Wirkungsbedingungen und Spezifität der *Ricinus*lipase. Hoppe-Seylers Z. **244**, 78—80 (1936). — ROHRLICH, M., u. H. BENISCHKE: Bestimmung von Lipase in Müllereiprodukten. Getreide u. Mehl **8**, 61—65 (1951). — RONA, P., u. R. AMMON: Die stereochemische Spezifität der Esterasen und die synthetisierende Wirkung der esterspaltenden Fermente. Erg. Enzymforsch. **2**, 50—73 (1933). — RONA, P., u. A. LASNITZKI: Eine Methode zur Bestimmung der Lipase in Körperflüssigkeiten und im Gewebe. Biochem. Z. **152**, 504—522 (1924). — ROSE, W. G.: Plant lipases in emulsions of water/oil. J. Amer. Oil Chem. Soc. **28**, 47—51 (1951). — ROTHE, M.: Biochemische Studien an *Gramineen*-Lipase. Fette u. Seifen **57**, 887 (1955).

SACHS, J.: Über das Auftreten der Stärke bei der Keimung ölhaltiger Samen. Bot. Ztg **17**, 177—183, 185—188 (1859). — SANDBERG, M., and E. BRAND: On papainlipase. J. of Biol. Chem. **64**, 59—70 (1925). — SAVARY, P., et P. DESNUELLE: Étude chromatographique de l'action de la lipase pancréatique sur des triglycerides mixtes. C. r. Acad. Sci. Paris **240**, 2571—2573 (1955). — SAWIN, P. B., and D. GLICK: Atropinesterase, a genetically determined enzyme in the rabbit. Proc. Nat. Acad. Sci. U.S.A. **29**, 55—59 (1943). — SCHÄFER, E.: Zur Stabilisierung von *Folia Belladonnae*. Diss. München 1954. — SCHREIBER, E.: Über Fälle angeblichen Fehlens von Lipase in ölhaltigen Samen. Ber. dtsch. bot. Ges. **58**, 250—255 (1940). — Über den Einfluß der Glycerinkonzentration auf die enzymatische Fettsynthese vermittels Samenlipase. Hoppe-Seylers Z. **276**, 56—62 (1942). — SCHUEGRAF, A.: Zur Kenntnis der optischen Spezifität der Lipase höherer Pflanzen. Diss. München 1953. — SCHÜTZENBERGER, P.: On fermentation. Internat. scientific. Ser. XX 1876. — SIGMUND, W.: Über fettspaltende Enzyme im Pflanzenreiche. Mh. Chem. **11**, 272—276 (1890). — SINGER, T. P., and B. H. J. HOFSTEE: Studies on wheat germ lipase. I. Methods of estimation, purification, and general properties of the enzyme. II. Kinetics. Arch. of Biochem. **18**, 229—243, 245—259 (1948). — SMIRNOWA, M. J., i M. N. LAWROWA: Veränderlichkeit der chemischen Zusammensetzung in verschiedenen Varietäten der Sojabohnen.

Bull. appl. Bot. (russ.) **5**, 73—103 (1935). Ref. Chem. Zbl. **1936 I**, 3043. — SÖHNGEN, N. L.: Thermo-tolerante Lipase. Kon. Akad. Wetensch. Amsterd. Versl. van de Gewone vergad Wis. en Natuurk. Afdeel. **20**, 126—130 (1911). Zit. bei S. A. NASHIF u. F. E. NELSON 1953a. — STEINER, M., u. H. HEINEMANN: Grana mit positiver Nadi-Reaktion als Ort der primären Fettbildung in Pilzzellen. Naturwiss. **41**, 40—41 (1954a). — Über die Beziehungen zwischen den fettbildenden Grana und typischen Mitochondrien in den Zellen von *Oospora lactis*. Naturwiss. **41**, 90 (1954b). — STETTER, H.: Enzymatische Analyse. Weinheim: Verlag Chemie 1951. — STEVENS, F. A., and R. WEST: The peptase, lipase and invertase of hemolytic *Streptococcus*. J. of Exper. Med. **35**, 823—846 (1922). — SÜSSENGUTH, K.: Über das Wirksamwerden pflanzlicher Enzyme. Erg. Enzymforsch. **1**, 364—369 (1932). — SULLIVAN, B.: Esterase in relation to milling and baking. In J. A. ANDERSON, S. 153—174. — SULLIVAN, B., and M. A. HOWE: Lipase of wheat. J. Amer. Chem. Soc. **55**, 320—324 (1933). — SUMNER, J. B., and K. MYRBÄCK: The enzymes, chemistry and mechanism of action, Vol. I Part 1. New York: Academic Press 1950.

TABACHNIK, J., and M. A. JOSLYN: Formation of esters by yeast. I. The production of ethyl acetate by standing surface cultures of *Hansenula anomala*. II. Investigations with cellular suspensions of *Hansenula anomala*. J. Bacter. **65**, 1—9 (1953a). — Plant Physiol. **28**, 681—692 (1953b). — TAKAMIJA, E.: Gewinnung von Lipase. F. P. 802 772 vom 2. 3. 1936. Zit. Chem. Zbl. **1937 I**, 385. — TAMMISTO, E. S.: Untersuchungen über die Lipasen der Bakterien. Ann. Acad. Sci. fenn., Ser. A, V **38** (1933). Zit. bei S. A. NASHIF u. F. E. NELSON 1953a. — TANAKA, Y.: Einfluß der Reaktionsprodukte auf die Wirkung der Lipase. J. Coll. Engng. Tokyo **5**, 137—141 (1912a). Ref. Chem. Zbl. **1913 I**, 666. — Einfluß einiger Neutralsalze, stickstoffhaltiger Substanzen und von *Ricinus*samenextrakt auf Lipase. J. Coll. Engng. Tokyo **5**, 142—151 (1912b). Ref. Chem. Zbl. **1913 I**, 667. — TAUBER, H.: New olive oil emulsion for lipase and new observations concerning serum lipase. Proc. Soc. Exper. Biol. a. Med. **90**, 375—378 (1955). — TAYLOR, A. E.: Über die Wirkung von Lipase. J. of Biol. Chem. **2**, 87—104 (1906). Ref. Chem. Zbl. **1906 II**, 1344. — Über Synthesen durch Wirkung von Enzymen. Z. physik. Chem. **69**, 585—597 (1909). Ref. Chem. Zbl. **1910 I**, 552. — TEMPLETON, W. H., and B. R. CARPENTER: Lipase activity of certain cereal products. Analyst **78**, 726—727 (1953). — THEIS, E. R., J. S. LONG and G. F. BEAL: Studies in the drying oils. XIII. Changes in linseed oil, lipase, and other constituents of the flaxseed as it matures. Industr. Engin. Chem. **22**, 768—770 (1930). — THEIS, E. R., J. S. LONG and C. E. BROWN: Studies in the drying oils. XII. Changes in linseed oil, lipase, and other constituents of the flaxseed as it matures. Industr. Engin. Chem. **21**, 1244—1248 (1929). — THIBODEAU, R., and H. MACY: Growth and enzyme activity of *Penicillium roqueforti*. Minn. Agricult. Exper. Stat. Techn. Bull. **1942**, 152. Zit. bei S. A. NASHIF u. F. E. NELSON 1953a. — TIETZ, N.: Über mögliche Beziehungen des Fettstoffwechsels zum Lichtkeimproblem. I. Mitt. Biochem. Z. **324**, 517—529 (1953). — Die Bedeutung des Damascenin für die Fetthydrolyse. II. Mitt. Zum Lichtkeimproblem. Biochem. Z. **325**, 123—129 (1954). — Über den Einfluß einiger Salze auf Löslichkeit und Aktivität der *Ricinus*lipase. Unveröffentlicht.

VELLUZ, L.: Quelques données sur la synthèse des esters par la „lipase“ du Ricin. Bull. Soc. Chim. biol. Paris **16**, 909—916 (1934). — VELLUZ, L., et J. LE LOUS: Influence of different organic solvents on the synthesis of esters by *Ricinus*lipase. Bull. Soc. Chim. biol. Paris **21**, 814—819 (1939). Ref. Chem. Abstr. **33**, 7825 (1939). — VIRTANEN, A. J., u. E. LINDEBERG: Die enzymatische Hydrolyse des Tributyrins durch Pankreas. Acta chem. fenn. **9**, B 2, 1 (1936).

WALDSCHMIDT-LEITZ, E., u. A. SCHÄFFNER: Lipasen. In E. BAMANN u. K. MYRBÄCK. S. 1547—1584. — WELTER, A.: Beitrag zur Kenntnis der Reversibilität der Enzymwirkung. Z. angew. Chem. **24**, 385—387 (1911). — WILLSTÄTTER, R., u. E. BAMANN: Über Magenlipase. Vergleich in verschiedenen Reinheitsgraden mit Pankreaslipase. Hoppe-Seylers Z. **173**, 17—31 (1928). — WILLSTÄTTER, R., u. E. BERNER: Hydrolyse des Scopolamins. Ber. dtsch. chem. Ges. **56**, 1079—1082 (1923). — WILLSTÄTTER, R., u. W. GRASSMANN: Über die Aktivierung des Papains durch Blausäure. Hoppe-Seylers Z. **138**, 184—215 (1924). — WILLSTÄTTER, R., u. H. KUMAGAWA: Über Takaesterase; Vergleich mit Pankreaslipase und Leberesterase. Hoppe-Seylers Z. **146**, 151—157 (1925). — WILLSTÄTTER, R., u. F. MEMMEN: Zur stalagmometrischen Bestimmung der lipatischen Tributyrinhydrolyse. Hoppe-Seylers Z. **129**, 1—25 (1923). — Über die Wirkung der Pankreaslipase auf verschiedene Substrate. Hoppe-Seylers Z. **133**, 229—246 (1924a). — Vergleich von Leberesterase mit Pankreaslipase; über die stereochemische Spezifität der Lipasen. Hoppe-Seylers Z. **138**, 216—253 (1924b). — WILLSTÄTTER, R., u. E. WALDSCHMIDT-LEITZ: Über Pankreaslipase. Hoppe-Seylers Z. **125**, 132—198 u. zw. 145 (1923). — Über *Ricinus*lipase. Hoppe-Seylers Z. **134**, 161—223 (1924). — WILLSTÄTTER, R., E. WALDSCHMIDT-LEITZ u. F. MEMMEN: Bestimmung der pankreatischen Fettspaltung. Hoppe-Seylers Z. **125**, 93—131 (1923).

YAMAMURA, Y., K. OGURA and S. IMAZU: Esterase of tubercle bacilli and various strains of acid fast bacilli. Kekkaku (Tuberculosis) **28**, 51—54 (1953).

Auf Fettsäuren eingestellte Oxydasen, Dehydrasen und Reduktasen. Die α- und die β-Oxydation.

Von

Wilhelm Franke und **Helmut Frehse.**

Mit 29 Abbildungen.

I. Übersicht.

Im Anschluß an die klassischen Untersuchungen von IVANOW (1912) und MILLER (1912) ist eine umfangreiche Literatur über die Mobilisierung, den Um- und Abbau der Fette in keimenden Samen entstanden, über die an anderer Stelle dieses Bandes (ZELLER, S. 322ff.) eingehend berichtet werden wird. Untersuchungen über die *Enzyme* der vorbereitenden *hydrolytischen* Phase, die Lipasen, reichen bis in die letzten Jahrzehnte des vorausgehenden Jahrhunderts zurück (Näheres bei BAMANN und ULLMANN, S.109 ff.). Im Gegensatz dazu sind die ältesten Angaben über fettsäure-*oxydierende* Fermente in Pflanzen nicht älter als 20—25 Jahre.

Für den primären oxydativen Angriff auf das Fettsäuremolekül sind heute 2 Fermenttypen bekannt: *Lipoxydasen* und *Fettsäuredehydrasen*[1] (kurze Übersichten: FRANKE 1949, FREHSE und FRANKE 1956). Die für die höhere Pflanze offenbar typische Lipoxydase ist von den beiden weitaus am besten untersucht, während über pflanzliche Fettsäuredehydrasen bis in die letzten Jahre nur ein paar kurze Mitteilungen vorlagen. Das gleiche gilt für die zum Teil problematischen *Fettsäurereduktasen,* die ungesättigte Fettsäuren in gesättigte bzw. höher ungesättigte Fettsäuren in weniger ungesättigte überführen.

Den folgenden Ausführungen kann die Feststellung vorausgeschickt werden, daß der gegenwärtige Stand unserer *enzymatischen* Kenntnisse vom oxydativen Fettsäureabbau in der höheren Pflanze wenig befriedigend ist. Auf der einen Seite steht die nur begrenzt verbreitete und chemisch in eine Sackgasse führende *Lipoxydase*wirkung, auf der anderen die zweifellos wichtigere der *Fettsäuredehydrasen,* die in der Regel in Carboxylnähe angreifen, obwohl auch ein Angriff in der Mitte des Fettsäuremoleküls als Nebenweg wahrscheinlich gemacht worden ist. Merkwürdigerweise schien nach den neueren, wenn auch noch unvollständigen Befunden an *Pflanzensamen* der carboxylnahe Angriff im allgemeinen nicht nach dem für die tierische Zelle bewiesenen Schema der *β-Oxydation* über die *Acyl-Coenzym A-Verbindungen* (vgl. Zusammenfassungen von LYNEN 1953, 1954, 1955 und von GREEN 1954) zu erfolgen, sondern eher einer *α-Oxydation* der *freien Fettsäuren* bzw. ihrer Anionen zu entsprechen. Immerhin deuteten einige wenige Befunde über die Umsetzung von ω-aryloxysubstituierten Fettsäuren mittlerer Kettenlänge, die in den letzten Jahren an *höheren Pflanzen* erhoben worden waren (S. 179), ferner die gründliche enzymatische Durcharbeitung des

[1] Wir ziehen der Bezeichnung *Dehydrogenasen* die auf H. WIELAND zurückgehende Kurzform *Dehydrasen* vor, die nach deutschem Sprachgebrauch korrekt gebildet und unmißverständlich ist [vgl. *(de)hydrieren,* dagegen *(de)hydratisieren*].

Abbaus mittlerer Fettsäuren durch einige *Bakterien* (S. 185) und nicht zuletzt die lange bekannte *Methylketon*bildung aus Fettsäuren durch gewisse *Schimmelpilze* (S. 182) durchaus auf das Vorkommen einer normalen oder wenig abgewandelten β-Oxydation auch im Pflanzenreich hin. Zudem war bekannt, daß der Fettsäure*aufbau* im Samen in üblicher Weise aus Essigsäureresten erfolgte (Näheres s. STUMPF, S. 201), wie auch das „acetataktivierende" Enzym und offenbar allgemeiner „acylaktivierende" Fermente in höheren Pflanzen verbreitet vorzukommen schienen (S. 180). Ganz kürzlich ist nun in Erdnußmitochondrien ein komplexes System der β-Oxydation von Fettsäuren aufgefunden worden (STUMPF und BARBER 1956). Nachdem die lange Zeit stark zurückgebliebene enzymchemische Durchforschung des pflanzlichen Fettsäurestoffwechsels in den letzten Jahren jedenfalls erfreulich in Fluß gekommen ist, wird wohl in nächster Zeit mit einer Abgrenzung der bekannten Abbaumechanismen und damit einer Klärung mancher Widersprüche und einer Auffüllung unserer Kenntnislücken auf diesem Gebiete zu rechnen sein.

II. Lipoxydasen

nennt man Enzyme, die mehrfach ungesättigte Fettsäuren vom Typus der Linol- und Linolensäure sowie deren Ester unter Verlagerung einer Doppelbindung und Sauerstoffanlagerung in konjugierte Dienhydroperoxyde überführen.

1. Historisches und Allgemeines.

1932 beobachteten ANDRÉ und HOU (1932 [1, 2]), daß das in den Zellrückständen der in China üblichen „Sojamilch"-Bereitung enthaltene Öl sich durch verminderte Jodzahl, aber erhöhte Acetyl-, Verseifungs- und Säurezahl vom Öl der intakten Sojabohnen sowie der Sojamilch unterschied. Es ergab sich, daß es sich hier um thermolabile Wirkungen einer Lipase und eines erstmals als „*Lipoxydase*" bezeichneten Oxydationsfermentes handelte, das sich von den bekannten, durch Farbreaktionen nachweisbaren (Phenol-)Oxydasen unterschied. Einige Jahre später fand CRAIG (1936) mit manometrischer Methodik, daß wäßrige Samenextrakte aus weißen Lupinen die Autoxydation sameneigenen oder zugesetzten Öls verschiedener Herkunft (z. B. Lein- oder Ricinusöl) intensiv katalysierten; nur wenig CO_2 wurde bei der mit steigender Reaktionstemperatur rasch zum Stillstand kommenden Enzymreaktion gebildet.

1939 berichteten SUMNER und DOUNCE über eine in Sojabohnen vorkommende „*Carotinoxydase*", die für das Ausbleichen von Carotin in Gegenwart ungesättigter Fette und Öle verantwortlich gemacht wurde. (Ähnliche Untersuchungen, allerdings mit rein technischer Zielsetzung, hatten schon einige Jahre vorher HAAS und BOHN ausgeführt, die 1934 in den USA ein Patent auf die Verwendung von Sojapräparaten für die Bleichung von Weizenmehl bzw. Teig daraus erhalten hatten.) SUMNER und SUMNER (1940) ergänzten die Befunde dahingehend, daß Carotin *allein* in Enzymgegenwart nur sehr langsam ausbleicht, sehr viel (z. B. 2000mal) rascher dagegen bei Anwesenheit einer optimalen (z. B. der 5fachen) Menge ungesättigten Öls; *stark* ungesättigte Öle (wie Hanf- oder Sojaöl) waren dabei wirksamer als Olivenöl oder Butter, was TAUBER (1940) bestätigte. Mineralöl war unwirksam. Offenbar war die enzymatisch katalysierte Oxydation der *Fettsäure* zu *Peroxyd* der Primärvorgang, mit dem die Carotinoxydation gekoppelt war. Da *vorgebildetes* Fettsäureperoxyd (ohne Enzym) die Ausbleichung des Carotins kaum beschleunigte (R. J. SUMNER 1942 [2]), handelte es sich bei letzterer wahrscheinlich um eine Reaktion von Peroxyd-*Radikalen*. Die Identität von Carotinoxydase und Lipoxydase konnte jedenfalls als gesichert gelten.

Zwischen 1941 und 1945 hat SÜLLMANN in der Schweiz vorzugsweise die *enzymatische* Seite der Lipoxydasereaktion untersucht. Die Aufklärung des *Chemismus* und *Mechanismus* der Enzymwirkung ist im wesentlichen durch HOLMAN in den USA und BERGSTRÖM in Schweden ab 1945 erfolgt. Die erstmalige *Reindarstellung* der Sojalipoxydase in kristallisiertem Zustand gelang THEORELL, HOLMAN und ÅKESON (1947 [1, 2]). SUMNER und seine Schule sowie HOLMAN haben weiterhin vor allem die gekoppelte *Carotinoidoxydation* (auch in *methodischer* Hinsicht) studiert; neuere Beiträge dieser Art — unter Einbeziehung von Tocopherol und anderen „Antioxydantien" — stammen von KUNKEL (1951) und TAPPEL und Mitarbeitern (1952—1956). FRANKE und Mitarbeiter haben sich zwischen 1944 und 1954 unter anderem mit dem Vergleich von enzymatischer und nichtenzymatischer Fettsäureoxydation, von Soja- und (später entdeckter) Gramineenlipoxydase sowie mit der Herstellung von Beziehungen zwischen Lipoxydase- und Fettsäuredehydrasewirkung befaßt.

An zusammenfassenden Berichten über Lipoxydasen sind (in chronologischer Ordnung) diejenigen von SÜLLMANN (1945 [2]), JEZESKI (1947), BERGSTRÖM und HOLMAN (1948 [2]), FRANKE (1949, 1951), JANECKE (1950), HOLMAN und BERGSTRÖM (1951) und FREHSE und FRANKE (1956) zu erwähnen.

2. Vorkommen.

Bisher scheint Lipoxydase mit Sicherheit nur in *höheren Pflanzen* nachgewiesen zu sein und auch dort keineswegs generell, nicht einmal in den meist untersuchten und im allgemeinen am enzymreichsten befundenen *Samen*. Allerdings wurden bisher nur Vertreter aus insgesamt 16 Familien der *Angiospermen* untersucht, so daß über die Verbreitung der Lipoxydase im *gesamten* Bereich der höheren Pflanzen Endgültiges noch nicht gesagt werden kann.

Über die Veränderungen von Fetten und Ölen durch *Mikroorganismen* liegt zwar eine nicht unbeträchtliche ältere Literatur vor (vgl. Zusammenstellungen bei HEFTER und SCHÖNFELD 1936, BÖMER und GROSSFELD 1939), doch ist es recht schwer, dieser bestimmte Angaben über *lipoxydatische* Wirkungen zu entnehmen, da diesbezügliche systematische Untersuchungen noch ausstehen und ja auch das Enzym der höheren Pflanze erst in den vierziger Jahren genauer charakterisiert wurde. Nur einige wenige neuere Arbeiten sollen daher Erwähnung finden. So haben HOROWITZ-WLASSOWA und LIVSCHITZ (1935) bei Schimmelpilzen (*Aspergillus, Penicillium*) und zahlreichen Bakterien außer der bekannten *lipatischen* Spaltung auch eine Fett*oxydation* gefunden, bei der unter Jodzahlabnahme und Refraktionszunahme Peroxyde, Oxysäuren, Aldehyde, Ketone und wasserlösliche Säuren gebildet werden. JENSEN und GRETTIE (1933, 1937) verfolgten die Oxydation von Schweinefett und partiell hydriertem Sonnenblumenöl durch verschiedene Bakterien mit Hilfe von Peroxyd- und Aldehydreaktionen (nach SCHIFF und KREIS); sterile Kontrollen zur Ausschaltung der „chemischen" Autoxydation liefen mit. Mit ähnlichen Methoden wiesen CASTELL und GARRARD (1941) die Oxydation von Triolein durch eine Reihe gramnegativer Bakterien nach, wobei sie Beziehungen zur Indophenoloxydaseaktivität der einzelnen Arten fanden. Wenig später stellen MUNDT und FABIAN (1944) mit manometrischer Methodik die Oxydation von Maisöl durch eine Reihe von Boden- und Wasserbakterien fest. Da aber Antioxydantien (Hydrochinon, Hafermehlauszug) diese bakterielle Oxydation im Gegensatz zur Autoxydation *nicht* hemmten (vgl. S. 142 und 154f.), handelt es sich hier kaum um eine Lipoxydasewirkung (sondern eher um eine Dehydrasewirkung, zumal da ein Teil der Kulturen auch Methylenblau im THUNBERG-Test in Ölgegenwart zu entfärben vermochte).

FRANKE und SCHILLINGER (1944) fanden gleichzeitig bei Untersuchung einer Reihe von Bakterien im WARBURG-Versuch keine bevorzugte Oxydation der Natriumsalze ungesättigter C_{18}-Säuren (Öl-, Linol-, Linolensäure) gegenüber Natriumstearat. Bei Gramnegativen erreichte der Umsatz der ungesättigten Säuren bestenfalls den der Stearinsäure, bei Grampositiven war er viel kleiner und dabei abnehmend mit der Zahl der Doppelbindungen. Weitere Untersuchungen über den Einfluß von Lage, Konfiguration und Zahl der Doppelbindungen auf den Fettsäureumsatz durch *Mycobakterien* ergaben, daß „der mikrobiologische Abbau der ungesättigten Fettsäuren nicht an der Doppelbindung, sondern höchstwahrscheinlich (dehydrierend) in der Nähe der Carboxylgruppe beginnt" (FRANKE, LEE und KIBAT 1949). Bei *Esch. coli* hatten schon früher MAZZA und Mitarbeiter auf Grund von Sauerstoff- und Methylenblauversuchen mit Stearinsäure und 2-Octadecensäure auf die primäre Bildung einer α,β-ungesättigten Säure aus Stearinsäure geschlossen (MAZZA und CIMMINO 1933, 1934), was sie durch spätere spektrographische Befunde bestätigt glaubten (MAZZA 1943, vgl. dagegen LANG 1952).

Ältere Beobachtungen, namentlich bei langer Versuchsdauer gemachte, verlieren unter diesen Umständen erheblich an Gewicht. Auch die Sterilitätskontrolle in den Versuchen von JENSEN und GRETTIE (s. oben) ist nicht beweisend, da im allgemeinen schon die durch lipatische Spaltung freigesetzten Fettsäuren beschleunigte Autoxydation zeigen (FRANKE 1932, KAUFMANN und FIEDLER 1939). Aber selbst wenn letztere durch die Bakterien gefördert würde, wäre die *enzymatische*, im besonderen die *lipoxydatische* Natur der Reaktion noch zu beweisen. Der letztgenannte Einwand gilt auch für die unlängst mitgeteilte Isolierung einer „Lipoxydase" aus Geflügeltuberkelbacillen (TANAKA, TANNO und HOSHISHIMA 1955): das peroxydierende Agens ist zwar thermolabil, wird aber im Gegensatz zur Samenlipoxydase durch m/1000-HCN vollständig in seiner Wirkung gehemmt, ist demnach also höchstwahrscheinlich von Häminproteidnatur (z. B. Bakteriencytochrom). Gänzlich unklar ist auch eine Angabe von MUFTIC (1955), wonach Tuberkelbacillen unter der Einwirkung von „*Cerase*, einer aus einem hefeartigen Pilz isolierten Cyclophorase-Oxydase" positive Peroxydreaktion geben sollen.

Hefe ist lipoxydasefrei (MILLER und KUMMEROW 1948), ebenso einige von SÜLLMANN (1943 [2]) geprüfte *Basidiomyceten (Psalliota campestris, Boletus viscidus, Boletus luteus).*

Die viel diskutierte Frage (BERGSTRÖM und HOLMAN 1948, FRANKE 1949, 1951, HOLMAN und BERGSTRÖM 1951) nach dem Vorkommen „echter" Lipoxydase in *tierischen Geweben* (neben den notorisch als Katalysatoren der Fettsäureautoxydation wirksamen *Häminproteiden*, vgl. z. B. FRANKE 1951) braucht hier nur kurz gestreift zu werden. Von einer einzigen Angabe für Fettgewebsextrakte (REISER 1949) abgesehen, lassen alle neueren, mit verschiedenartiger Methodik durchgeführten Differenzierungsversuche an tierischen Gewebsextrakten (WATTS und PENG 1947, TAPPEL 1952, 1953 [1], BOYD und ADAMS 1955) keinen Raum für die Annahme einer eigenen (häminfreien) tierischen Lipoxydase neben den dort stets vorhandenen Häminkatalysatoren.

Ältere orientierende Untersuchungen zur Verbreitung der Lipoxydase in *Angiospermen* stammen von STRAIN (1941) und von SÜLLMANN (1943 [2], 1945 [1]); ersterer wies das Ferment in zahlreichen *Leguminosen*, letzterer (1945) darüber hinaus in *Solanaceen* (verschiedene *Solanum-*, *Atropa-*, *Datura-*, *Nicotiana-* und *Physalis-*Arten) und *Labiaten* (*Stachys-*, *Salvia-* und *Phlomis-*Arten) nach. (Eine von KIRSSANOWA (1938) für die *Kartoffel*knolle angegebene „Carotinoxydase" war schon vor der systematischen Untersuchung der *Solanaceen* mit Lipoxydase identifiziert worden (SÜLLMANN 1943 [2].) Neuerdings wurde Lipoxydase ziemlich verbreitet in den (recht fettarmen) Samen von *Gramineen*, insbesondere von Kulturgräsern, vorgefunden (FRANKE und FREHSE 1953). Von ausgesprochenen *Ölsamen* wurde außer Soja nur *Lein* als lipoxydasepositiv erkannt, während sich unter anderen Mohn, Raps, Weißsenf, Walnuß, Hanf, Kürbis und Sonnenblume *ungekeimt* als fermentfrei erwiesen (FRANKE und FREHSE 1954). Tabelle 1 ermöglicht einen Vergleich der (an Wasserextrakten bei p_H 7 gemessenen) *Enzymaktivität* (Q_{O_2}) und des *Enzymgehaltes* (EE/g) (Einheiten s. S. 142) verschiedener *Gramineen* und *Leguminosen* (FRANKE und FREHSE 1953).

Tabelle 1. *Lipoxydaseaktivität* (Q_{O_2}) *und -gehalt (EE/g Trockenmaterial) verschiedener Gramineen- und Leguminosensamen.*

Pflanze	Aktivität Q_{O_2}	Enzym-Einheiten EE/g	Pflanze	Aktivität Q_{O_2}	Enzym-Einheiten EE/g
Gerste	150—200	etwa 6000	Erbsen	150	etwa 20000
Roggen	150—200	etwa 10000	Bohnen (weiß) . .	150	etwa 20000
Weizen	200—250	etwa 8000	Wicke.	300	etwa 30000
Hafer.	25—50	etwa 1000	Soja (p_H 7) . . .	1000	250000—500000
Mais	80—120	etwa 1500	Soja (p_H 9) . . .	2500	800000—1300000

Unter den Leguminosen sind *Sojabohnen* das mit Abstand enzymreichste Material (s. oben); aber auch die Samen von Bohne, Pferdebohne, Erbse, Lupine, Wicke, Esparsette, Luzerne, Klee und Akazie sind enzymhaltig (STRAIN 1941, 1949, SÜLLMANN 1945 [1]). Nach dem letztzitierten Autor ist das Enzym auch in den Wurzeln (besonders deren „Knöllchen"), Stengeln und Blättern der Pflanzen in meßbarer bis beträchtlicher Menge vorhanden.

HOLMAN (1948 [1]) untersuchte sowohl den *Lipoxydase*gehalt einzelner *Teile von Sojakeimlingen*, wobei sich die *Kotyledonen* am enzymreichsten erwiesen, wie auch den *zeitlichen Gang* der Lipoxydase-(und Katalase-)aktivität im Zusammenhang mit Änderungen des Gehalts an Fettsäuren und lipoidlöslichen Farbstoffen in *keimenden* Sojabohnen.

So fand er für den Lipoxydasegehalt 18 Tage alter, grüner Sojakeimlinge folgende Werte (in Enzymeinheiten/mg Trockengewicht, vgl. S. 142—143):

Gesamtpflanze	Wurzel	Sproßspitze	Blätter	Stengel	Kotyledonen
8,1	56	65	66	79	392

Über die Änderung der enzymatischen Aktivität wie auch andere stoffliche Veränderungen bei der Keimung orientiert Abb. 1. Bemerkenswert ist der starke Abfall der *Lipoxydase*-wirksamkeit schon nach 2 Tagen, wo ein Substratschwund noch nicht feststellbar ist; gleichzeitig steigt die *Katalase*aktivität stark an, um nach dem 4. Tage ebenfalls rapide abzufallen.

Auch in der keimenden *Gerste* wurde nach dem 3. Tage ein — wenn auch mehr linearer, eventuell sich später steigernder — Aktivitätsabfall beobachtet, so daß bei *etiolierten* Keimlingen im allgemeinen nach 12 Tagen keine Aktivität mehr feststellbar war (FRANKE und FREHSE 1953); in *grünen* Keimlingen sank die Aktivität langsamer ab (vgl. Abb. 23, S. 167, FRANKE und FREHSE 1954). In den früheren Keimstadien ist die Enzym*aktivität* in Wurzeln und Blättern ungefähr gleich groß, wenngleich letztere größeren Enzym*gehalt* besitzen. Bei der Kornreifung wird das Ferment im Stadium der „Milchreife" nachweisbar; bis zur Vollreife steigen Aktivität und Menge des Ferments im Samen 10- bzw. 5fach an (FRANKE und FREHSE 1953).

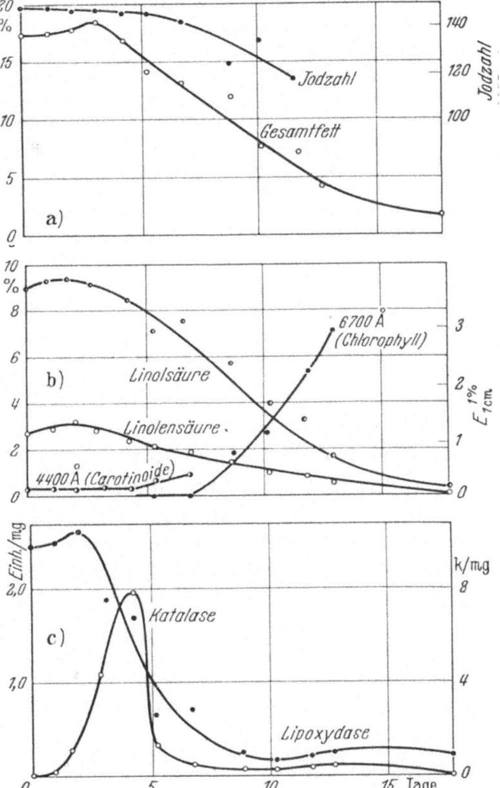

Abb. 1a—c. Stoffliche Veränderungen während der Keimung von Sojabohnen. a Gesamtfett in % der Trockensubstanz und dessen Jodzahl. b Linol- und Linolensäure in % der Trockensubstanz; Absorption des Öls ($E^{1\%}_{1\,\text{cm}}$) bei 4400 Å (Carotinoide) und 6700 Å (Chlorophyll). c Lipoxydasegehalt in Einheiten/mg fettfreie Trockensubstanz (s. S. 142f.); Katalasegehalt, ausgedrückt durch k/mg fettfreie Trockensubstanz (k = monomolekulare dekadische Reaktionskonstante). (Nach HOLMAN 1948 [1].)

Im ruhenden Samen von *Gerste* (FRANKE und FREHSE 1953) und *Weizen* (BLAIN und TODD 1955) ist die Lipoxydase im *Embryo* (und *Scutellum*) lokalisiert, das Endosperm ist nahezu enzymfrei.

BLAIN und TODD (1955) finden für Weizensorten folgende relative Enzymgehalte, auf die Gewichtseinheit bezogen:

ganzes Samenkorn	Endosperm
7,4—11	1,1—2,0
Embryo	Scutellum
62—84	57—87

Der Enzymgehalt von Weizenembryonen ist noch immer 20—30mal kleiner als der von (ganzen) Sojabohnen (vgl. auch R. J. SUMNER 1943, MILLER und KUMMEROW 1948).

Histochemische Untersuchungen (mit fettlöslichen Farbstoffen, „Nadi"-Reagens und Leukofarbstoffen, s. S. 142) über die *Verteilung von Fett und Lipoxydase* in verschiedenen Futterpflanzen sowie über den Einfluß des Nährmediums auf die Aktivität stammen von VAN FLEET (1943).

Er fand die Fermentaktivität hoch in Pflanzenteilen mit neutraler oder alkalischer Reaktion, dort, wo Wasserverlust stattfindet, unter Wundflächen und dort, wo Antioxydantien abwesend oder inaktiviert sind. Wurzeln und Stengel von Pflanzen, die in gepufferten Nährlösungen bei p_H 7,2—7,6 gewachsen waren, zeigten im Keimlingsstadium hohe Oxydaseaktivität in Fettgeweben; nach 5 Wochen waren die Vorräte an ungesättigtem Fett nur noch gering, die Fettgewebe durch niedrige Oxydaseaktivität und die Gegenwart von Fettabbauprodukten charakterisiert. In Pflanzen, die in gepufferten Medien von p_H 4,8—5,6 gewachsen waren, war der Fettvorrat größer, die Oxydaseaktivität gering (vgl. auch S. 147).

Die Spezifität von VAN FLEETS Lipoxydasetesten ist zweifelhaft (vgl. HOLMAN 1948 [1]); in fetthaltigen Geweben sprechen sie wohl primär für die Anwesenheit von Fettperoxyden, ohne diejenige von Lipoxydase zu beweisen. (Zur Kritik vgl. VAN FLEET 1952.)

Bemerkenswert war der Befund (FRANKE und FREHSE 1954), daß im ruhenden Zustande lipoxydasenegative *Ölsamen* (Raps, Hanf, Kürbis, Sonnenblume) bei der *Keimung* nach einigen Tagen Fermentgehalt zeigten (vgl. Tabelle 6, S. 167).

Für die naheliegende Annahme, daß ein natürlicher *Hemmstoff* das im ungekeimten Samen bereits vorhandene Ferment an der Entfaltung seiner Wirkung hindere, ergaben sich in einigen, doch keineswegs allen untersuchten Fällen experimentelle Hinweise: aus *Mohn*- und *Raps*samen, nicht dagegen aus den Samen von Hanf, Kürbis und Sonnenblume ließen sich wäßrige Extrakte gewinnen, die unter anderem auch die Wirkung der Lein- und Gerstenlipoxydase hemmten (FRANKE und FREHSE 1954). Diese Inhibitoren unterschieden sich durch Wasserlöslichkeit und Petrolätherunlöslichkeit von den Hemmstoffen des ruhenden *Gerstenkorns*, deren eine Komponente wohl sicher *Tocopherol* darstellt, dessen „antioxygene" Wirkung in Lipoxydasesystemen wiederholt gezeigt worden ist (Näheres S. 154f.); weiterhin wird man an *Phosphatide*, insbesondere *Kephalin* (OLCOTT und MATTILL 1936) zu denken haben. (Vgl. für *Weizen* vorliegende Befunde von PRIVETT und QUACKENBUSH 1954.) An der bekannten, besonders intensiven antioxygenen Wirkung des *Hafer*mehls scheinen hauptsächlich *Phosphatid-Proteide* beteiligt zu sein (DIEMAIR und FOX 1939, DIEMAIR, STROHECKER und REULAND 1940, TÄUFEL und ROTHE 1944).

3. Bestimmung.

Zur Bestimmung von Fermentaktivitäten bzw. -mengen ist eine Reihe verschiedener Methoden zur Anwendung gelangt. Um zuverlässige Vergleichswerte zu erhalten, ist in jedem Fall die eigenartige Kinetik der Lipoxydasewirkung (Näheres S. 146) zu berücksichtigen und im besonderen darauf zu achten, daß die Sauerstoffversorgung der Enzymansätze nicht zum geschwindigkeitsbestimmenden Faktor wird (S. 146 und 148). Die wichtigsten Methoden sind nachstehend verzeichnet. (Kurze Übersicht und Kritik bei HOLMAN 1955.)

a) Die *manometrische Methode* (nach WARBURG), bei der die Sauerstoffaufnahme unter gewissen Standardbedingungen der Temperatur, des p_H, der Substratkonzentration und des Sauerstoffpartialdruckes ermittelt wird (CRAIG 1936, SÜLLMANN 1941 [3], THEORELL, BERGSTRÖM und ÅKESON 1944). Die Enzym*aktivität* kann z. B. durch die *Atmungsgröße* $Q_{O_2} = \frac{\mu l\ O_2}{\text{mg Trockengewicht} \times \text{Stunden}}$ ausgedrückt werden, die Enzym*menge* durch *Enzymeinheiten* EE $= Q_{O_2} \times$ mg Gesamttrockengewicht (FRANKE, MÖNCH, KIBAT und HAMM 1948, FRANKE und FREHSE 1953).

b) *Peroxydbestimmungsmethoden*, bei denen die gebildeten Fettsäureperoxyde entweder mit konzentriertem wäßrigem HJ bzw. KJ-Eisessig zur Reaktion gebracht werden (SUMNER und DOUNCE 1939, FRANKE, MÖNCH, KIBAT und HAMM 1948) oder auf Ferrosalz in Gegenwart von Rhodanid einwirken (R. J. SUMNER 1943, BALLS, AXELROD und KIES 1943) oder schließlich eine Leukoverbindung (z. B. von o-Chlorphenol-indophenol oder 2,6-Dichlorphenol-indophenol) zum Farbstoff oxydieren (BALLS, AXELROD und KIES 1943; vgl. auch HARTMANN und GLAVIND 1949); im erstgenannten Fall wird das ausgeschiedene Jod titrimetrisch, in den beiden letzteren werden die gefärbten Reaktionsprodukte photometrisch bestimmt.

c) Die *colorimetrische oder photometrische Verfolgung der Carotinentfärbung* in Gegenwart von Fettsäure oder Fettsäureester, im Prinzip ebenfalls eine Peroxydbestimmungsmethode (REISER und FRAPS 1943, BALLS, AXELROD und KIES 1943, COSBY und SUMNER 1945, MITCHELL und KING 1946); gewisse Vorteile bietet die Verwendung der Carotinoid-carbonsäure *Bixin* zum gleichen Zwecke (SUMNER und SMITH 1947).

d) Die *UV-spektrophotometrische Methode*, bei der die charakteristische Absorption der konjugiert-ungesättigten Fettsäureperoxyde bei 234 mμ gemessen wird (vgl. Abb. 10 und 16, S. 151 und 157). In der von THEORELL, BERGSTRÖM und ÅKESON (1946) standardisierten Form des 2-Minuten-Versuchs ist dieses Verfahren viel angewandt worden; als *Enzymeinheit* (vgl. S. 141) wird diejenige Fermentmenge definiert, die unter bestimmten Bedingungen (0,006 m-Linolat, p_H 9,0, reiner O_2, 20°) nach 1 min eine Extinktionssteigerung ($\log I_0/I$) von 2,0 in der mit der 1,67fachen Menge Alkohol versetzten Reaktionslösung hervorruft (BERGSTRÖM und HOLMAN 1948 [2], HOLMAN 1955). Die obere Grenze genauer Bestimmungen liegt bei 2,5 Einheiten je Testansatz (1,2 ml). Exakter ist eine Variante, bei der die Extinktionszunahme in Abständen von 15 sec gemessen wird (TAPPEL, BOYER und LUNDBERG 1952, TAPPEL, LUNDBERG und BOYER 1953).

e) *Andere Methoden* sind gelegentlich verwendet worden, so z. B. die Bestimmung der *Jodzahlabnahme* (FRANKE, MÖNCH, KIBAT und HAMM 1948), die darauf basiert, daß mit den üblichen Jodzahlmethoden in *konjugiert*-ungesättigten Fettsäuren (bzw. deren Peroxyden) eine Doppelbindung zu wenig gefunden wird (KAUFMANN 1935, FRANKE 1948).

4. Reinigung und Kristallisation.

Die pflanzliche Lipoxydase ist ein relativ stabiles und leicht zu reinigendes Ferment. Die *Sojabohne* ist das gegebene Ausgangsmaterial (vgl. Tabelle 1, S. 140). Durch Wasser- oder Pufferextraktion des (mit Petroläther) entfetteten Sojamehls kann das Enzym leicht in Lösung gebracht werden, wobei man mit Wasser *mehr*, mit Puffer von p_H 4,5 *reineres* Enzym in Lösung bekommt (THEORELL, HOLMAN und ÅKESON 1947 [2]); diese Extrakte können nach den in der modernen Enzymchemie üblichen Anreicherungsverfahren weitgehend gereinigt werden. Durch amerikanische und schwedische Forscher sind ab 1943 systematische Reinigungsversuche durchgeführt worden (BALLS, AXELROD und KIES 1943, THEORELL, BERGSTRÖM und ÅKESON 1944, 1946, COSBY und SUMNER 1945), die schließlich zur Kristallisation des Ferments führten (THEORELL, HOLMAN und ÅKESON 1947 [1, 2]).

Das Verfahren der letztzitierten Autoren schließt folgende Reinigungsstufen ein: Extraktion von Sojamehl mit Acetatpuffer p_H 4,5; Fällung von Verunreinigungen mit Barium- und Bleiacetat unter Acetonzusatz; Fällung des Enzyms mit Ammonsulfat beim Sättigungsgrad 0,53; Erhitzen des gelösten Ammonsulfatniederschlags (5 min auf 63°) zur Fällung inaktiven Proteins; Ammonsulfatfraktionierung, wobei das Enzym zwischen den Sättigungsgraden 0,35 und 0,50 fällt; Alkoholfällung und erneute Ammonsulfatfraktionierung; Elektrophorese im TISELIUS-Apparat; Dialyse konzentrierter Enzymlösung gegen Ammonsulfatlösung ansteigender Konzentration, wobei Kristallisation in farblosen Blättchen oder Büscheln erfolgt.

Reinigungsversuche am *Gramineen*ferment stehen noch aus. Doch scheinen nach orientierenden Versuchen die Stabilitätsverhältnisse ähnlich zu sein wie beim Sojaferment, bis auf eine bei letzterem nicht beobachtete „Schüttelinaktivierung" (FRANKE und FREHSE 1953).

5. Eigenschaften.

(Nach THEORELL, HOLMAN und ÅKESON 1947 [2], HOLMAN 1947.)

Kristallisierte Sojalipoxydase erwies sich sowohl im Elektrophorese- wie im Sedimentations- und Diffusionsversuch als *einheitlich*. Als *Molekulargewicht* errechnet sich 102400. Der *isoelektrische Punkt* liegt bei p_H 5,4. Lipoxydase ist ein typisches *Globulin*, unlöslich in reinem Wasser, aber löslich in verdünnten Salzlösungen. Ammonsulfat fällt das Ferment annähernd bei Halbsättigung, Phosphatpuffer p_H 6,8 in der Konzentration 2,3 m. Das *Absorptionsspektrum* ist ein normales Proteinspektrum mit einer relativ hohen Absorption bei 280 mμ. Eine quantitative Aminosäurenanalyse des kristallisierten Ferments ist von HOLMAN, PANZER, SCHWEIGERT und AMES (1950) durchgeführt worden; sie hat nichts Ungewöhnliches ergeben. Über eine *prosthetische Gruppe* der Lipoxydase läßt sich einstweilen noch nichts aussagen.

Wie schon SÜLLMANN (1941 [2, 3], 1943 [2]) festgestellt hatte, verträgt das Sojaferment vielstündige Dialyse; eine Aufspaltung in Apo- und Coferment gelingt auf diese Weise also offenbar nicht. Ein zeitweise als wesentlich angesehener *Eisen*gehalt (THEORELL, BERGSTRÖM und ÅKESON 1944) hat sich später als Verunreinigung erwiesen, da erst auf mehrere Moleküle Reinferment 1 Eisenatom kommen würde (THEORELL, HOLMAN und ÅKESON 1947 [2]). Typische *Metallkomplexbildner* wie Cyanid, Azid, Pyrophosphat, Fluorid und Diäthyldithiocarbamat hemmten das *Rein*ferment selbst in hohen Konzentrationen (m/15—m/10) praktisch nicht (HOLMAN 1947). [Eine unlängst für *Roh*ferment aus Weizen angegebene partielle Enzymhemmung bei *Inkubation* mit relativ hohen HCN-Konzentrationen (m/200) ist nicht typisch für Oxydasen, sondern findet sich auch bei einigen Dehydrasen (vgl. Xanthindehydrase, DIXON und KEILIN 1936).] Nach der praktisch negativen Nitroprussidreaktion, der Aminosäurenanalyse [kein Cyst(e)in] und der geringen Empfindlichkeit gegen das SH-Gruppenreagens *p-Chloromercuribenzoat* — schwache Hemmung erst bei m/100 (HOLMAN 1947) — gehört die Lipoxydase auch nicht zur Klasse der typischen „Sulfhydrylfermente" (vgl. hierüber allgemein BARRON 1951).

Nach MICHLIN und PSCHENNOWA (1946, 1948) besteht Lipoxydase aus zwei Komponenten, deren eine als „aktive Gruppe" *peroxydierbar* sein soll; die bisher beigebrachten experimentellen Belege (z. B. Sauerstoffaufnahme ohne Substrat, Hydroperoxydreaktionen, schwache Hemmung durch Katalase) reichen aber zur Stützung dieser Auffassung nicht aus.

Einstweilen stellt die Lipoxydase einen eigenen Enzymtyp dar, der in gewisser Beziehung zwischen den (schwermetallhaltigen) *Oxydasen* und den (im allgemeinen metallfreien, zum mindesten nicht obligat metallhaltigen) *Aerodehydrasen* (zur Definition vgl. FRANKE 1940 [1, 2]) steht, von beiden sich aber dadurch unterscheidet, daß er Sauerstoff als *Peroxydgruppe* in das (isomerisierte) Substratmolekül einlagert.

Die spezifische Aktivität der Lipoxydase ist hoch: nach THEORELL, HOLMAN und ÅKESON (1947 [2]) setzt ein Fermentmolekül je Sekunde bei 20° 330 Moleküle Linolsäure um, woraus sich die beachtliche „Atmungsgröße" Q_{O_2} (S. 142) von 260000 für das Reinferment berechnen ließe.

6. Aktivatoren.

Kristallisierte Lipoxydase, im *homogenen* System auf die Alkalisalze mehrfach ungesättigter Fettsäuren (p_H etwa 9) einwirkend, zeigt ohne weitere Zusätze maximale Wirksamkeit (HOLMAN 1947). In verschiedenen Arbeiten jedoch, in denen mit nur teilweise gereinigtem Ferment und mit Substrat*emulsionen* (Seifen bei $p_H \leq 7$, Fettsäureestern) gearbeitet worden war, war eine *Aktivierung* der Fermentwirkung durch Zellpräparate verschiedener Herkunft von *Polypeptid*natur angegeben worden (BALLS, AXELROD und KIES 1943; KIES 1947 [1], THEORELL, BERGSTRÖM und ÅKESON 1944). Natur und Wirkungsweise dieser Aktivatoren sind bis heute noch nicht restlos geklärt. In der Mehrzahl der untersuchten Fälle scheint sich die Wirkung aber hauptsächlich auf das *Substrat* und nicht oder untergeordnet auf die Lipoxydase zu beziehen; eine Wirkung auf in der Lösung befindliche Begleitfermente erscheint nicht ausgeschlossen.

Offenbar handelt es sich bei den Aktivatoren um verschiedene Stoffe, die hauptsächlich aus Sojabohnen, teilweise auch aus anderem pflanzlichem Material (Gerstenmalz, Reis, Gemüsepflanzen, Hefe), ferner aus Muskel und Milch gewonnen worden waren. Ein von KIES (1947 [1]) aus Sojabohnen sogar *kristallisiert* erhaltener Aktivator schien mit einem früher beschriebenen *amorphen* Anreicherungsprodukt (BALLS, AXELROD und KIES 1943) nicht identisch zu sein.

Der SUMNERsche Arbeitskreis, der mit (meist carotinhaltigen) Fettsäureemulsionen bei p_H 6,5 und unter Zusatz von *Emulgatoren* (Gummi arabicum, Bentonit, Natrium-dodecylsulfat) arbeitete, konnte *keine* Aktivierung durch die erwähnten Präparate von BALLS bzw. THEORELL und Mitarbeitern feststellen (COSBY und SUMNER 1945, SMITH und SUMNER 1948 [2]); wohl aber fanden die Autoren eine reaktionsbeschleunigende Wirkung der *Emulgatoren*.

In diesem Zusammenhang sind auch unveröffentlichte Versuche HOLMANS (zit. nach BERGSTRÖM und HOLMAN 1948 [2], HOLMAN und BERGSTRÖM 1951) anzuführen, wonach *organische Lösungsmittel* wie Äther, Essigester und Aceton als Zusätze die Sauerstoffaufnahme in mit Ghatti-Gummi emulgierten Ferment-Fettsäure-Systemen um 25—40% steigerten. Die

Wirkung des Acetons war bei 25—35%, diejenige von Äther und Essigester bei etwa 5% Zusatz optimal und sank jäh bei Überschreitung der Löslichkeitsgrenze. Auch die Natriumsalze der *Glykocholsäure* (0,2%) und *Stearinsäure* (0,4%) erhöhten die Lipoxydasewirkung unter ähnlichen Bedingungen (um 60 bzw. 75%).

Aus dem Vergleich aller Befunde gewinnt man den Eindruck, daß die verschiedenen natürlichen Aktivatoren in der Hauptsache als oder ähnlich wie Emulgatoren in heterogenen Systemen wirken. Es erscheint durchaus möglich, daß sie *unter Zellbedingungen* — in Abwesenheit der „künstlichen" Emulgatoren des in vitro-Versuchs — eine nicht unwesentliche Rolle als Reaktionsbeschleuniger spielen. Nach HOLMAN und BERGSTRÖM (1951) mag die Wirkung die sein, „daß das oberflächenaktive Agens größere Substratdispersion erzeugt oder Oberflächenadsorption des Enzyms am emulgierten Substrat verhindert". Ob darüber hinaus eine Beeinflussung des *Reaktionsablaufs* erfolgt, wie KIES (1947 [1]) dies behauptet, bedarf weiterer Untersuchung.

In dieser Richtung könnte die Angabe sprechen, daß der kristallisierte Aktivator bei der Linolatoxydation die (schwache) Absorption bei 280 mμ (vgl. hierzu S. 151f.) weit mehr verstärkt als die (starke) bei 234 mμ. Auch wird Sauerstoffaufnahme, Peroxydbildung und sekundäre Carotinoxydation durch Aktivatorzusatz nicht im gleichen Maße beschleunigt gefunden. Eine weitere kurze Notiz (KIES 1947 [2]) besagt, daß 2 min auf 70° erhitztes Rohenzym Carotin nicht mehr in Gegenwart von Linolsäureester, wohl aber noch (nach einer gewissen Inkubationszeit) in Gegenwart von Linolat zu entfärben vermag. Ob dieser Befund im Sinne eines Mehrenzymsystems zu interpretieren oder auf eine mehr physikalische Zustandsänderung des Lipoxydasereaktionssystems zurückzuführen ist, läßt sich nach den dürftigen Angaben nicht entscheiden.

7. Spezifität.

Die eigentlichen Substrate der Lipoxydase sind die *natürlich* vorkommenden, *mehrfach* ungesättigten Fettsäuren mit *isolierten cis*-Doppelbindungen (in „methylenunterbrochener" oder „Divinylmethan"-Anordnung), im wesentlichen also *Linolsäure* (I), *Linolensäure* (II) und *Arachidonsäure* (III):

$$CH_3\cdot(CH_2)_4\cdot CH{:}CH\cdot CH_2\cdot CH{:}CH\,.\,(CH_2)_7\cdot COOH \qquad (I)$$
$$CH_3\cdot CH_2\cdot CH{:}CH\cdot CH_2\cdot CH{:}CH\cdot CH_2\cdot CH{:}CH\cdot(CH_2)_7\cdot COOH \qquad (II)$$
$$CH_3\cdot(CH_2)_4\cdot CH{:}CH\cdot CH_2\cdot CH{:}CH\cdot CH_2\cdot CH{:}CH\cdot CH_2\cdot CH{:}CH\cdot(CH_2)_3\cdot COOH \qquad (III)$$

Die für den Lipoxydaseangriff notwendige Struktureinheit ist offenbar die Gruppierung

$$\begin{array}{l} CH— \\ \| \\ CH—CH_2—CH \\ \qquad\qquad\;\; \| \\ \qquad\quad\;\; —CH \end{array}$$

in zweimaliger cis-Stellung.

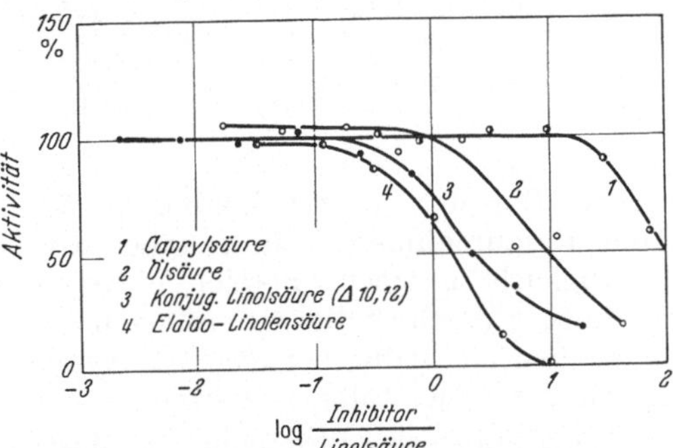

Abb. 2. Kompetitive Hemmung der mit kristallisiertem Sojaenzym bewirkten Linolatoxydation durch Fettsäurezusätze. p_H 9; 20°. (Nach HOLMAN 1947.)

Die entsprechenden *trans*-Formen (z. B. Elaidolinol- bzw. -linolensäure, HOLMAN und BURR 1945), auch *cis*-9, *trans*-12-Linolsäure (PRIVETT, NICKELL, LUNDBERG und BOYER 1955), sowie die *konjugierten* Isomeren wie Pseudolinolsäure (10,12-Octadecadiensäure), α-Eläostearinsäure (9,11,13-Octadecatriensäure) und Pseudoeläostearinsäure (10,12,14-Octadecatriensäure) (SÜLLMANN 1944, HOLMAN und BURR 1945) sind praktisch unangreifbar, ebenso *einfach* ungesättigte Fettsäuren (Öl-, Ricinol-, Erucasäure) (BALLS, AXELROD und KIES 1943, HOLMAN und BURR 1945), wenngleich in älteren Arbeiten (STRAIN 1941, R. J. SUMNER 1942 [1], SÜLLMANN 1944) für die letztere Gruppe anderslautende Befunde erhoben worden waren.

Gegenüber Fettsäuren, die auf Grund ihrer Konstitution und Konfiguration nicht angreifbar sind, zeigt Lipoxydase jedoch „*Fixierungsspezifität*" (QUASTEL und WOOLDRIDGE 1928, THUNBERG 1933), d. h. diese Säuren hemmen die Oxydation der eigentlichen Substrate, wie Abb. 2 für den Fall der Linolatoxydation zeigt.

Ähnlich wie die früher erwähnten Fettsäuren und ihre Alkalisalze werden auch deren *Ester* (HOLMAN und BURR 1945, HOLMAN 1946) und *Glyceride* (vgl. S. 138) von Lipoxydase oxydiert. Die Ester von Linol-, Linolen- und Arachidonsäure werden *enzymatisch* (nicht dagegen *chemisch*) annähernd gleich rasch angegriffen (HOLMAN und ELMER 1947).

8. Kinetik und Beeinflussung.

Die Kinetik der Lipoxydasewirkung wird kompliziert durch die starke, von den peroxydischen Reaktionsprodukten ausgehende *Fermentschädigung*, die bewirkt, daß die Reaktion im allgemeinen vor Aufnahme *eines* Moleküls Sauerstoff je Fettsäuremolekül zum Erliegen kommt (SÜLLMANN 1941 [3], BERGSTRÖM 1945 [2], FRANKE, MÖNCH, KIBAT und HAMM 1948). Daß es sich hier tatsächlich um eine Erschöpfung des *Ferments* und nicht des Substrats handelt, geht aus Abb. 3

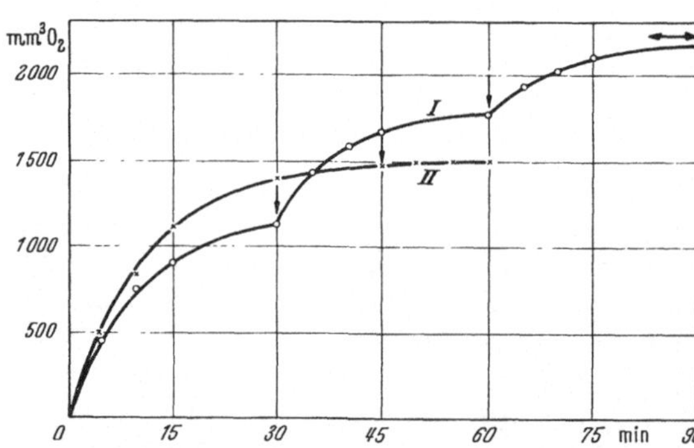

Abb. 3. Einfluß wiederholter Enzym- und Substratzugabe bei der Linolenatoxydation durch Sojaenzym. I Nach 30 und nach 60 min erneute Zugabe der Ausgangsmenge Enzymlösung. II Nach 45 min erneute Zugabe der Ausgangsmenge Substratlösung. ↔ O_2-Äquivalent des eingesetzten Substrats (2240 mm³); m/50-Linolenat; p_H 6,8; 20°. (Nach FRANKE, MÖNCH, KIBAT und HAMM 1948.)

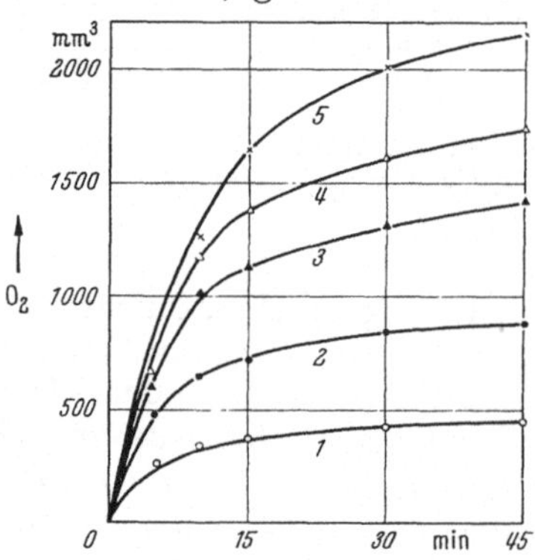

Abb. 4. Einfluß variierter Enzymkonzentration bei der Linolenatoxydation durch Sojalipoxydase. Relative Enzymmengen: *1:* 1; *2:* 2,5; *3:* 5; *4:* 10; *5:* 15. O_2-Äquivalent des eingesetzten Substrats: 2240 mm³; m/50-Linolenat; p_H 6,8; 20°. (Nach FRANKE, MÖNCH, KIBAT und HAMM 1948.)

hervor. Daß sich durch Erhöhung des Verhältnisses Enzym/Substrat die relative Sauerstoffaufnahme bis zur annähernd theoretischen steigern läßt, zeigt Abb. 4, die zugleich in einem Bereich niedriger Enzymkonzentration *Proportionalität* zur Anfangsgeschwindigkeit erkennen läßt, während bei der angewandten manometrischen Methodik bei *hohen* Enzymkonzentrationen die Diffusionsgeschwindigkeit des Sauerstoffs begrenzend wirkt (vgl. auch Abb. 7, S. 148).

Die Affinität der Lipoxydase zu ihren Substraten ist recht hoch; sowohl für das Sojaferment (HOLMAN 1947) wie für das Gerstenferment (FRANKE und FREHSE 1953) wurden Werte der MICHAELIS-*Konstanten* K_m (Dissoziationskonstante der Enzym-Substrat-Verbindung) in der Nähe von 1×10^{-3} m mit Linolat oder Linolenat festgestellt (Abb. 5).

Abweichend fanden TAPPEL, BOYER und LUNDBERG (1952) beim Sojaferment (mit der auch von HOLMAN angewandten spektrophotometrischen Methodik, S. 142) einen um zwei Größenordnungen niedrigeren Wert (2×10^{-5} m). [Ein noch tieferer, von IRVINE und ANDERSON (1953) für das Weizenenzym angegebener Wert (5×10^{-6} m) kann nicht als einwandfrei angesehen werden, da er an Fettsäure*emulsionen* bei p_H 6,5 bestimmt wurde.] Die Ursache der Diskrepanz ist noch ungeklärt; möglicherweise ist sie in der Interpretation der Substratkonzentrations-Aktivitätskurven zu suchen, die bei TAPPEL und Mitarbeitern insofern eine Anomalie aufweisen, als sie bis etwa 1×10^{-4} m-Linolat sehr steil, anschließend bis 5×10^{-3} m flacher weiter ansteigen.

Hinsichtlich des p_H-*Optimums* unterscheiden sich die Enzyme verschiedener Herkunft: für die *Gramineen*lipoxydase wird bei verschiedenen Arten (Gerste, Weizen, Mais) in annähernder Übereinstimmung ein p_H-Optimum zwischen 6,5 und 7,5 angegeben (FRANKE und FREHSE 1953, IRVINE und ANDERSON 1953 [1], FRITZ und BEEVERS 1955 [2]), für *Erbsen*ferment ebenfalls ein ausgeprägtes Optimum bei p_H 7 (ähnlich Abb. 6, rechts) gefunden (SIDDIQI und TAPPEL 1956), während *Soja*ferment bei p_H 9—10 optimal wirkt (HOLMAN 1947, FRANKE, MÖNCH, KIBAT und HAMM 1948, SMITH 1948) (Abb. 6). Letztere Angabe gilt indes nur für Fettsäure*seifen* als Substrate; für emulgierte Fettsäure*ester* und Carotinoid-Öl-Lösungen wurde auch bei Soja ein optimales p_H von 6,5 beobachtet (SUMNER und DOUNCE 1939, SUMNER und SMITH 1947, SMITH 1948).

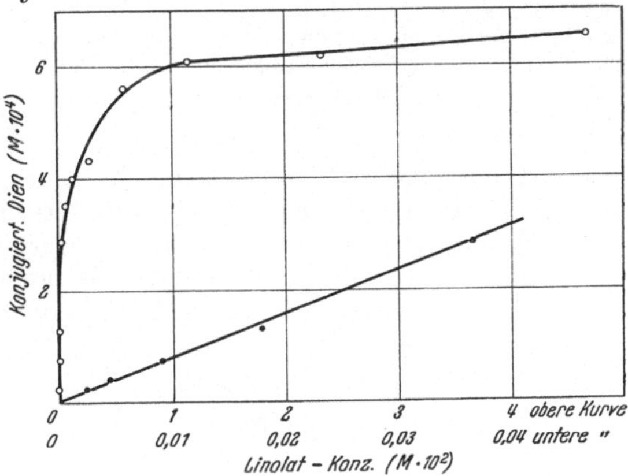

Abb. 5. Einfluß der Substratkonzentration auf die Geschwindigkeit der enzymatischen Linolatoxydation. Krist. Sojaenzym; p_H 9; spektrophotometrischer Test; untere Kurve 100fach größerer Abszissenmaßstab. (Nach HOLMAN und BERGSTRÖM 1951.)

Nach SMITH (1948) kommt das (für ein Pflanzenenzym auffallend hohe) p_H-Optimum von etwa 9,5 dadurch zustande, daß in diesem Gebiete echte Substrat*lösungen* vorliegen,

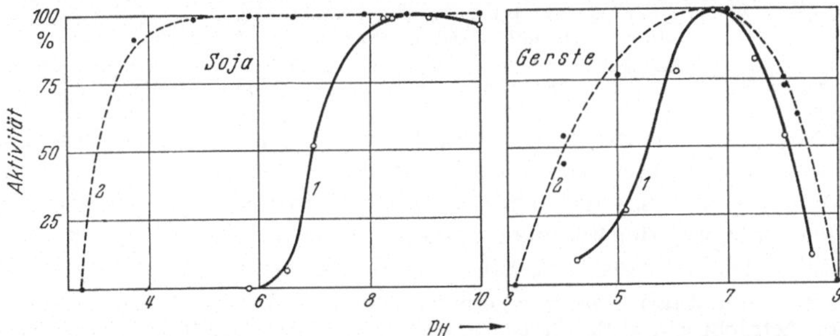

Abb. 6. p_H-Einfluß auf Aktivität (*1*) und Stabilität (*2*) von Soja- und Gerstenlipoxydase. In *1* direkte Aktivitätsbestimmung (manometrisch) beim betreffenden p_H. In *2* halbstündige Inkubation (25°) ohne Substrat beim betreffenden p_H und anschließende Aktivitätsbestimmung bei p_H 9 (Soja) bzw. 7 (Gerste). (Aktivität des nichtvorbehandelten Enzyms gleich 100 gesetzt.) (Nach FRANKE und FREHSE 1953.)

während sich bei tieferen p_H-Werten in zunehmendem Maße schwerer angreifbare *Emulsionen* bilden. Bei *wasserunlöslichen* Substraten würde diese Komplikation größtenteils wegfallen und die „wahre" p_H-Abhängigkeit des Enzyms gemessen werden.

Nach vergleichenden Untersuchungen der p_H-*Stabilität* (gestrichelte Kurven in Abb. 6) scheint das p_H-Optimum des *Gersten*enzyms von der alkalischen Seite her durch die begrenzte Enzymstabilität bedingt zu sein; für die saure Seite scheint dies nicht zu gelten. Das *Soja*enzym ist über einen weiten p_H-Bereich stabil; wenn seine Aktivität trotzdem schon im schwach sauren Gebiet stark abfällt, so kann dies durch den Lösungszustand des Substrates (SMITH 1948), eventuell auch durch den Ladungszustand des Enzyms (vgl. MYRBÄCK 1953,

Alberty 1956 [1, 2]) bedingt sein. Beim Gerstenferment erscheint das letztere wahrscheinlicher.

Fritz und Beevers (1955 [2]) unterscheiden neuerdings nach Warburg-Versuchen an Homogenaten von *Mais*keimlingen, die ohne Substratzusatz optimale Sauerstoffaufnahme bei p_H 5,0 zeigen, zwischen einem p_H-Optimum für *natives* Substrat (also hier p_H 5,0) und einem für *exogenes* Substrat (p_H 7,5 bei Linolatzusatz). Im ursprünglichen Homogenat soll eine feste Einheit zwischen Ferment und endogenem Substrat bestehen, die durch physikalische Einflüsse (wie Zentrifugieren, Frieren und Wiederauftauen) gestört wird. In dem erhöhten p_H-Optimum bei Zugabe von Substrat käme nur dessen bessere Verwertbarkeit (Löslichkeit) bei höherem p_H, nicht aber eine dem Enzym inhärente Eigenschaft zum Ausdruck.

Die Lipoxydasereaktion verläuft in Luft und in reinem Sauerstoff gleich rasch (Holman 1947, Franke, Mönch, Kibat und Hamm 1948, Franke und Frehse 1953). Die *Sauerstoffaffinität* des Ferments ist ziemlich groß, wenn auch um mehrere Größenordnungen kleiner als bei echten (metallhaltigen) Oxydasen (vgl. z. B. „Atmungsferment" = Cytochromoxydase, Haldane und Stern

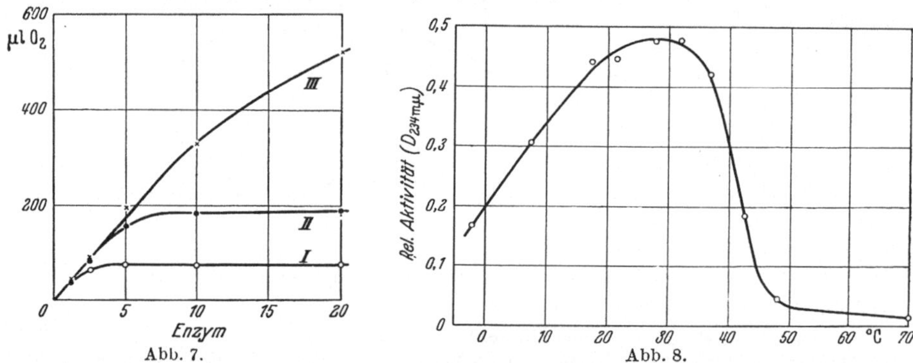

Abb. 7.　　　Abb. 8.

Abb. 7. Einfluß des Sauerstoffpartialdrucks auf die Linolatoxydation durch wechselnde Mengen Gerstenlipoxydase. O_2-Partialdruck in *I* 0,105; *II* 0,21; *III* 1,0. O_2-Aufnahmen nach 10 min; O_2-Äquivalent des eingesetzten Substrats: 1120 mm³. (Nach Franke und Frehse 1953.)

Abb. 8. Einfluß der Reaktionstemperatur auf die Anfangsgeschwindigkeit der enzymatischen Linolatoxydation. Kristallisiertes Sojaenzym; p_H 9; Spektrophotometrischer Test (S. 142). (Nach Holman 1947.)

1932). Bei Substratkonzentrationen von etwa 10^{-3} m und spektrophotometrischer Methodik finden Holman (1947) und Tappel, Boyer und Lundberg (1952) Halbsättigung des Sojaenzyms bei Sauerstoffpartialdrucken von 40 bzw. 21 mm Hg, unter Berücksichtigung der Versuchstemperaturen von 0° bzw. 20° einem K_m-Wert von 1,2 bzw. $0{,}4 \times 10^{-4}$ m-O_2 in Lösung entsprechend.

Nach Tappel und Mitarbeitern (l. c.) erhöht sich dieser Wert bei Steigerung der *Substrat*konzentration beträchtlich, z. B. 10fach bei 20fach erhöhter Linolatkonzentration. Irvine und Anderson (1953) sowie Franke und Frehse (1953) beobachteten beim Gramineenenzym mit *manometrischer* Methodik auch einen starken Einfluß der *Enzym*konzentration, worüber Abb. 7 orientiert. Ersichtlich wirkt bei größerer „potentieller" Reaktionsfähigkeit des Systems (bei gesteigerter Enzymkonzentration) die Diffusionsgeschwindigkeit des Sauerstoffs in die Flüssigkeit geschwindigkeitsbegrenzend. Dieser Umstand betont die Notwendigkeit, bei manometrischer Aktivitätsmessungen und Untersuchungen über die Beeinflußbarkeit des Ferments mit möglichst kleinen Enzymzusätzen (im sog. „Proportionalitätsbereich") zu arbeiten (vgl. S. 142, 146 und 149).

Daß das *Temperaturoptimum* der Lipoxydase zwischen 20 und 40° liegt, ist verschiedentlich für Sojaferment (Holman 1947, Franke, Mönch, Kibat und Hamm 1948) und Gramineenenzym (Franke und Frehse 1953) gezeigt worden; oberhalb 40° wird meist starker Aktivitätsabfall beobachtet (Abb. 8). Auch im Temperaturbereich optimaler *Anfangs*geschwindigkeit zeigt sich deutlich ein nachteiliger Einfluß höherer Reaktionstemperaturen auf den *Endwert* der Sauerstoffaufnahme (Abb. 9). Irvine und Anderson (1953 [1]) fanden beim Weizenferment die

Aktivierungsenergie zwischen 10 und 30° zu 6500 cal/Mol, TAPPEL, LUNDBERG und BOYER (1953) beim Sojaferment zwischen — 6 und 30° zu 4300 cal/Mol (entsprechend Temperaturkoeffizienten von 1,45 bzw. 1,28). Vom lebensmitteltechnologischen Standpunkt bedeutsam ist die Beobachtung der letztgenannten Autoren, daß unterhalb 0° die Reaktionsgeschwindigkeit im *gefrorenen* System weniger als 1% der im flüssigen Zustand bei gleicher Temperatur beobachteten beträgt.

In vergleichenden Versuchen an Soja- und an Gerstenferment wurde die „*Tötungstemperatur*" (Temperatur 50%igen Aktivitätsverlusts bei halbstündiger Einwirkung in Lösung, v. EULER 1925) zu respektiv 49° und 53° gefunden (FRANKE und FREHSE 1953). Das „ruhende" Enzym ist ersichtlich weniger temperaturempfindlich als das „arbeitende" (vgl. Abb. 8), worin höchstwahrscheinlich wieder die Peroxydschädigung (S. 146) zum Ausdruck kommt.

Angaben von IRVINE und ANDERSON (1953 [1]) für das *Weizen*enzym weichen unwahrscheinlich stark nach oben (>60°) ab; möglicherweise ist der „Proportionalitätsbereich" des Enzyms (vgl. Abb. 7) nicht genügend berücksichtigt worden.

Das *zellgebundene* Samenferment ist erheblich thermostabiler. So lieferten bei etwa 80° gedarrte helle Braumalze Extrakte, die noch 1/5—2/5 der Aktivität von Gerstenextrakten zeigten (FRANKE und FREHSE 1953).

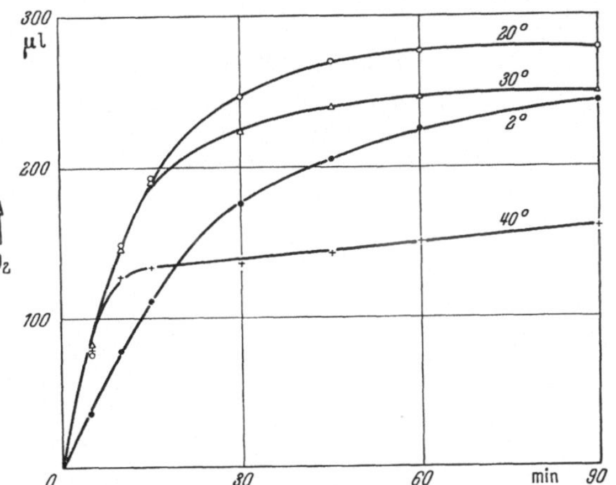

Abb. 9. Einfluß der Temperatur auf die Sauerstoffaufnahme im System Linolat-Gerstenlipoxydase. O_2-Äquivalent des Substrats: 1120 mm³; p_H 7; 30°. (Nach FRANKE und FREHSE 1953.)

Die Angaben über den *Endwert der Sauerstoffaufnahme* bei der Lipoxydasereaktion sind schwankend. Bei relativ hoher Enzym- und niedriger Substratkonzentration wird rasch und ziemlich genau *ein* Molekül O_2 je Fettsäuremolekül aufgenommen, und zwar sowohl bei Linol- wie bei Linolensäure (BERGSTRÖM 1945 [2], FRANKE, MÖNCH, KIBAT und HAMM 1948, FRANKE und FREHSE 1953, IRVINE und ANDERSON 1953 [1]) (vgl. auch Abb. 4, S. 146). Unter gewissen Bedingungen (z. B. relativ niedrige Temperatur, höherer Sauerstoffpartialdruck) und bei längeren Versuchszeiten wird aber eine sich anschließende langsamere, annähernd lineare oder sich allmählich beschleunigende, protrahierte Sauerstoffaufnahme beobachtet. So erreichten FRANKE und FREHSE (1953) in mehrstündigen Versuchen mit Gerstenferment Gesamtsauerstoffaufnahmen bis zu 1,5 Mol, IRVINE und ANDERSON (1953) mit Weizenferment bis zu 2 Mol je Mol Fettsäure. Es ist nicht entschieden, ob es sich hier um eine Enzymreaktion (noch unbekannter Natur) oder (wahrscheinlicher) um eine nichtenzymatische Folgereaktion handelt (vgl. hierzu S. 151—153).

IRVINE und ANDERSON (1953) betonen die „Zweiphasigkeit" der Lipoxydasereaktion und glauben, daß der „stationäre Zustand" der zweiten Phase gegen die meist angenommene Enzymaktivierung spreche [die aber für das Sojaferment eindeutig nachgewiesen ist (BERGSTRÖM 1945 [2], FRANKE, MÖNCH, KIBAT und HAMM 1948, vgl. Abb. 3, S. 146)]. Da „anoxydierte" Substrate stark erhöhte Autoxydabilität zeigen (FRANKE 1932), ist auch mit einer Überlagerung dieses sekundären Einflusses zu rechnen (vgl. z. B. SÜLLMANN 1941 [3]). Grenzwerte der Sauerstoffaufnahme mit *reinem* Ferment scheinen merkwürdigerweise nie bestimmt worden zu sein; vergleichende Versuche mit letzterem und mit Präparaten verschiedenen Reinheitsgrads würden wahrscheinlich aufschlußreich sein.

9. Chemismus und Mechanismus der Lipoxydasewirkung

haben im Anschluß an die 1943 von Farmer und Mitarbeitern (Farmer und Sutton 1943, Farmer, Koch und Sutton 1943) eingeleitete und später von Bolland und Koch (1945) und Bergström (1945 [1]) erfolgreich fortgeführte Entwicklung neuer Anschauungen auf dem Gebiete der Fettsäure-*Autoxydation* (vgl. Zusammenfassungen von Bergström und Holman (1948 [2]) und Franke (1950, 1951) eine recht weitgehende Aufklärung gefunden. *Grundsätzliche* Unterschiede im *Chemismus* von enzymatischer und nichtenzymatischer Oxydation ergaben sich dabei nicht; die lange Zeit angenommene Gleichartigkeit auch des *Mechanismus* (im Sinne einer Kettenreaktion) ist durch neuere Befunde zweifelhaft geworden.

a) Grundlegende Beobachtungen und Formulierungen.

Auf Grund analytischer, spektrophotometrischer und präparativer Befunde und unter Einbeziehung der erstmals wohl vom Farmerschen Arbeitskreis und etwas später von Bolland für die Fettsäure-Autoxydation schärfer formulierten *Radikalketten*-Vorstellung (Farmer, Bloomfield, Sundralingam und Sutton 1942, Farmer, Koch und Sutton 1943; Bolland 1945, 1948) wurden 1947/48 *beide* Reaktionen an Linolsäure etwa folgendermaßen formuliert (Holman 1947, Bergström und Holman 1948 [1, 2]):

Schema 1.

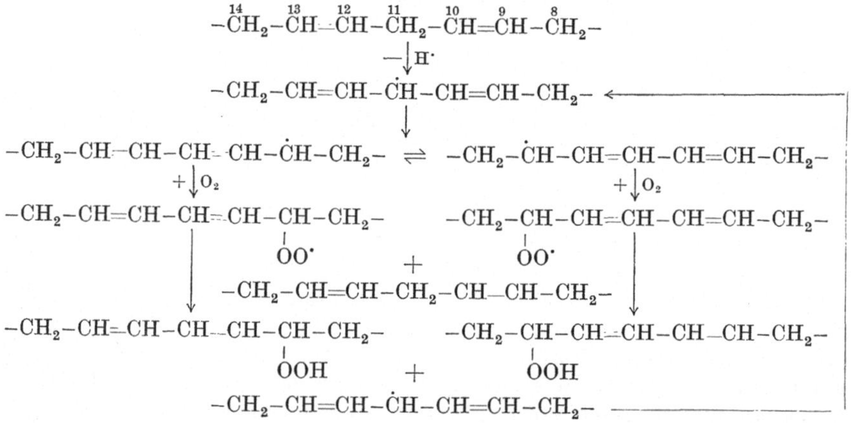

$$
\begin{array}{c}
-\overset{14}{\mathrm{C}}\mathrm{H_2}-\overset{13}{\mathrm{C}}\mathrm{H}=\overset{12}{\mathrm{C}}\mathrm{H}-\overset{11}{\mathrm{C}}\mathrm{H_2}-\overset{10}{\mathrm{C}}\mathrm{H}=\overset{9}{\mathrm{C}}\mathrm{H}-\overset{8}{\mathrm{C}}\mathrm{H_2}- \\
\downarrow -\mathrm{H^{\cdot}} \\
-\mathrm{CH_2-CH=CH-\dot{C}H-CH=CH-CH_2}- \quad \longleftarrow \\
\downarrow \\
-\mathrm{CH_2-CH=CH-CH=CH-\dot{C}H-CH_2}- \quad \rightleftharpoons \quad -\mathrm{CH_2-\dot{C}H-CH=CH-CH=CH-CH_2}- \\
\downarrow +\mathrm{O_2} \qquad\qquad\qquad\qquad \downarrow +\mathrm{O_2} \\
-\mathrm{CH_2-CH=CH-CH=CH-\underset{\displaystyle OO^{\cdot}}{\underset{|}{CH}}-CH_2}- \quad + \quad -\mathrm{CH_2-\underset{\displaystyle OO^{\cdot}}{\underset{|}{CH}}-CH=CH-CH=CH-CH_2}- \\
-\mathrm{CH_2-CH=CH-CH_2-CH=CH-CH_2}- \\
\downarrow \qquad\qquad\qquad\qquad \downarrow \\
-\mathrm{CH_2-CH=CH-CH=CH-\underset{\displaystyle OOH}{\underset{|}{CH}}-CH_2}- \quad + \quad -\mathrm{CH_2-\underset{\displaystyle OOH}{\underset{|}{CH}}-CH=CH-CH=CH-CH_2}- \\
-\mathrm{CH_2-CH=CH-\dot{C}H-CH=CH-CH_2}- \quad \text{———}
\end{array}
$$

Danach besteht die erste Phase der Autoxydation wie auch der enzymatischen Oxydation „methylenunterbrochener“ Doppelbindungssysteme in der Abspaltung eines Wasserstoffs aus der (thermisch, photochemisch oder enzymatisch) aktivierten CH_2-Gruppe unter Hinterlassung eines *Radikals*, das sich durch Resonanz in zwei mesomere *konjugiert-ungesättigte* Formen umlagert; an sie addiert sich das O_2-Molekül. Die entstandenen Peroxydradikale liefern mit neuer Fettsäure wieder das „Monodehydroradikal“.

Die Hauptargumente für diese Formulierung, insbesondere der hier hauptsächlich interessierenden *Enzym*reaktion, waren folgende:

1. Unter geeigneten Reaktionsbedingungen (nicht zu hohe Temperatur und Sauerstoffaufnahme) findet sich der *gesamte* absorbierte Sauerstoff als Fettsäure*peroxyd* vor (Bergström 1945 [2], Franke, Mönch, Kibat und Hamm 1948), ein Befund, der früher schon wiederholt für die *chemisch* katalysierte Fettsäureautoxydation erhoben worden war

(GOLDSCHMIDT und FREUDENBERG 1934, FRANKE und JERCHEL 1937, FRANKE und MÖNCH 1944, BERGSTRÖM 1945 [1]).

2. Bei der enzymatischen Oxydation von *Linol-*, *Linolen-* und *Arachidonsäure* und ihrer Ester findet starke Absorptionserhöhung im Ultraviolett bei 235 mμ, schwächere bei 270–280 mμ statt (BERGSTRÖM 1945 [2], HOLMAN und BURR 1945, HOLMAN 1946) (Abb. 10, Kurve I; zur zeitlichen Zunahme der Extinktion vgl. Abb. 16, links, S. 157). Ähnliche Verhältnisse sind bei der *Autoxydation* beobachtet worden (Literatur s. S. 150); doch hat BERGSTRÖM (1945 [1]) bei relativ niedrigen Sauerstoffaufnahmen (< 0,5 Mol je Mol Linolsäureester) und besonders schonender katalysatorfreier Oxydation (37°) die Bande bei etwa 275 mμ vermißt (Abb. 10, Kurve II) und erst nach Chromatographie des Reaktionsproduktes an Aluminiumoxyd auftreten sehen (s. S. 152f.).

Der Extinktionskoeffizient bei 235 mμ stieg proportional zu Sauerstoffaufnahme und Peroxydbildung an. Die Absorptionsbande bei 235 mμ ist typisch für ein *konjugiertes Dien* (VAN DER HULST 1935, KERNS, BELKENGREN, CLARK und MILLER 1941, MITCHELL, KRAYBILL und ZSCHEILE 1943), diejenige bei etwa 275 mμ charakteristisch für ein *konjugiertes Dienketon* (s. S. 152). Reduktion der Peroxydgruppe mit Sulfit ist ohne Einfluß auf die Ultraviolettabsorption.

3. Bei der Bestimmung des aktiven Wasserstoffs von Autoxydationsprodukten mit GRIGNARD-Reagens wurde 1 Mol CH_4 je Mol Peroxyd entwickelt (BOLLAND 1941), was für die *Hydroperoxyd-* (I) und gegen die früher angenommene *Brückenperoxyd-*Natur (II) der Oxydationsprodukte sprach:

```
—CH=CH—CH—              —CH—CH—
         |       (I)       |   |      (II)
        OOH                O———O
```

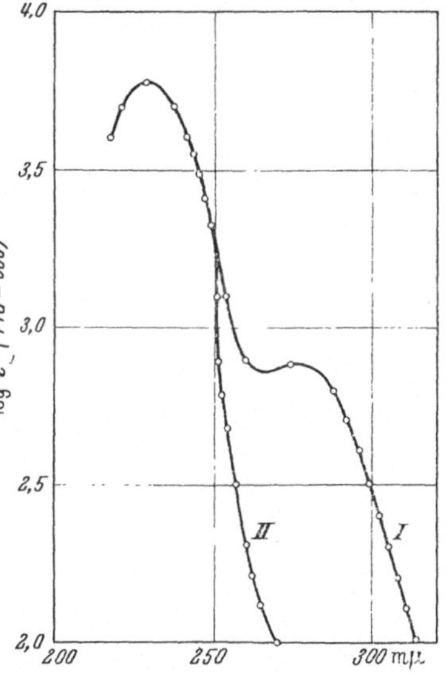

Abb. 10. Absorptionsspektren der Produkte der enzymatischen Oxydation von Linolat (*I*) und der Autoxydation von Linolsäuremethylester (*II*). O_2-Aufnahme jeweils etwa 0,5 Mol pro Mol Substrat. (Nach BERGSTRÖM 1945 [1, 2]).

In die gleiche Richtung wies der Befund, daß bei der katalytischen Hydrierung sowohl von Produkten der Autoxydation (FRANKE und MÖNCH 1944, BERGSTRÖM 1945 [1]) wie der enzymatischen Oxydation (BERGSTRÖM 1945 [2], FRANKE, MÖNCH, KIBAT und HAMM 1948) in der Hauptsache *Monooxycarbonsäuren* erhalten wurden.

Daß bei der enzymatischen Oxydation tatsächlich unter geeigneten Reaktionsbedingungen *nur* konjugierte Oxydationsprodukte entstehen, zeigten Versuche mit kristallisiertem Sojaenzym bei verschiedenen Temperaturen (HOLMAN 1947, BERGSTRÖM und HOLMAN 1948 [1,2]); durch Extrapolation wurden folgende Werte für den molaren Extinktionskoeffizienten von *Linolsäure-hydroperoxyden* bei 234 mμ gefunden (Abb. 11):

bei	0°	8°	20°	37°
	31400	27800	25600	23000

Der erste Wert entspricht der Theorie für *quantitative* Bildung eines *konjugierten Diens* (und zwar der *trans,trans*-Form, LUNDBERG 1956). Im Einklang damit hatte BERGSTRÖM (1945 [1, 2]) bei chromatographischer Auftrennung der katalytisch hydrierten Oxydationsprodukte von Linolsäure zu etwa gleichen Anteilen *9-* und *13-Oxystearinsäure*, dagegen keine 11-Oxystearinsäure erhalten.

Auffallenderweise waren bei der *Autoxydation* von Linolsäureester bei 37° auch im Bereich quantitativer Peroxydbildung aus dem aufgenommenen Sauerstoff im allgemeinen erheblich niedrigere Extinktionskoeffizienten gefunden worden, z. B. 14000 (BERGSTRÖM 1945 [1]), 19000 (HOLMAN und BURR 1946) und 22700 (BOLLAND und KOCH 1945). Doch gelang es BERGSTRÖM (1945 [1]) auch bei solchen Ansätzen nicht, nach Hydrierung 11-Oxystearinsäure zu fassen. Neueste Untersuchungen von PRIVETT und NICKELL (1956) zeigen, daß sich der Linolatautoxydation eine Weiteroxydation des Dienhydroperoxyds überlagert, wobei unter annähernder Erhaltung der Peroxydzahl *Polymerperoxyde* von niedrigerer Molarextinktion gebildet werden.

Das von BERGSTRÖM und HOLMAN aufgestellte Reaktionsschema 1 (S. 150) ist in den letzten Jahren noch hinsichtlich der mit dem Oxydationsprozeß verknüpften *sterischen Änderungen* im Fettsäuremolekül spezifiziert worden. Hauptsächlich Messungen des Infrarotspektrums ergaben, daß Autoxydation der Methylester von Linolsäure (PRIVETT, LUNDBERG, KHAN, TOLBERG und WHEELER 1953, SEPHTON und SUTTON 1956) und Linolensäure (PRIVETT, NICKELL, TOLBERG, PASCHKE und WHEELER 1954) bei 0° ein konjugiertes *cis,trans*-Dienhydroperoxyd lieferte, während bei 24° daneben auch *trans,trans*-Peroxyd — wenigstens zum Teil durch thermische Umlagerung aus ersterem — gebildet wurde. Auch die *enzymatische* Oxydation ergibt im allgemeinen die beiden genannten Peroxydformen (KHAN 1953, 1955, KHAN, LUNDBERG und HOLMAN 1954), während ein konjugiertes *cis,cis*-Hydroperoxyd bisher nicht aufgefunden wurde. Im einzelnen bestehen allerdings noch Widersprüche. So konnten die hohen, von HOLMAN (1947) beim *Reinferment* beobachteten Extinktionswerte, die für ein *trans,trans*-Dienperoxyd sprachen, von anderen Untersuchern später mit *Rohferment* nicht mehr erhalten werden (TAPPEL, BOYER und LUNDBERG 1952; PRIVETT, NICKELL und LUNDBERG, unveröffentlicht, zitiert nach LUNDBERG 1956); es wurden bestenfalls Extinktionskoeffizienten von etwa 28000, reinen *cis,trans*-Dienperoxyden entsprechend, erreicht. Bemerkenswerterweise sind die enzymatisch gebildeten Peroxyde *optisch aktiv*. LUNDBERG (1956) vertritt die Auffassung, „daß unter milden Oxydationsbedingungen und bei niedriger Enzymkonzentration Lipoxydase primär *cis,trans*-konjugierte Dienhydroperoxyde erzeugt, die sich von den Produkten der Autoxydation hauptsächlich durch ihre optische Aktivität unterscheiden, und daß *trans,trans*-konjugierte Dienhydroperoxyde wahrscheinlich im wesentlichen als Folge anschließender Sekundärreaktionen gebildet werden.“ Der hohe mit *Rein*ferment bei 0° erhaltene Extinktionswert HOLMANs (1947) ist damit allerdings nicht erklärt.

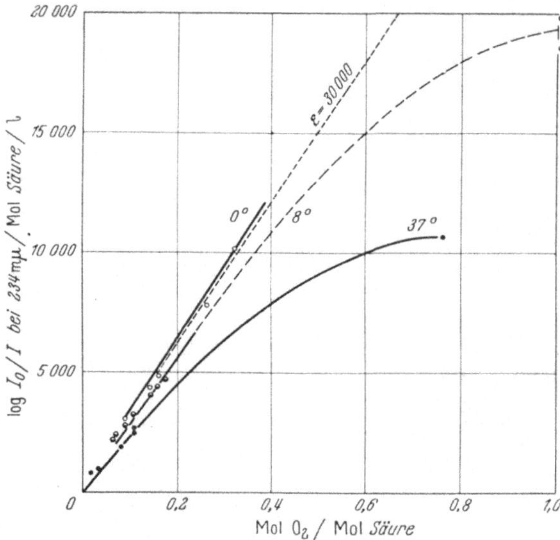

Abb. 11. Einfluß der Temperatur auf den molekularen Extinktionskoeffizienten der enzymatischen Oxydationsprodukte von Linolat bei 234 mμ. Kristallisiertes Sojaenzym; pH 9. (Nach HOLMAN 1947.)

Interessant, wenn auch nicht direkt hierher gehörend, ist der Befund, daß bei der *chlorophyll-katalysierten Photooxydation* (und nur bei dieser) auch ein *nicht*konjugiertes *trans,trans*-Peroxyd gebildet wird; es handelt sich offenbar um die bisher stets vermißte 11-Hydroperoxydoverbindung (eventuell 14-Hydroperoxydoverbindung bei Linolensäureester), die aus dem *primären* „Dehydroradikal“ des Schemas 1 (S. 150) entstanden ist (KHAN, LUNDBERG und HOLMAN 1954, KHAN 1955).

Über *Folge-* und *Nebenreaktionen* der oben skizzierten Hydroperoxydbildung ist wenig bekannt. Die bei steigender Temperatur verringerte Extinktionszunahme bei 234 mμ wird von zunehmender Extinktion bei 270—280 mμ begleitet, dem Absorptionsbereich von *Dienketonen* (Abb. 12). Danach scheint *eine* wichtige Folgereaktion in der Dehydratisierung des Primärperoxyds zu bestehen:

$$—CH{=}CH—CH{=}CH—CH(OOH)— \xrightarrow{-H_2O} —CH{=}CH—CH{=}CH—CO—.$$

Für die Entstehung von *Carbonyl*verbindungen bei der Lipoxydasewirkung hatte schon SÜLLMANN (1942) präparative Hinweise gefunden; ein exakter Nachweis steht noch aus.

Allgemein scheint bei der *enzymatischen* Oxydation die Bildung von Produkten, welche bei etwa 270 mμ absorbieren, gegenüber der Autoxydation bevorzugt zu sein (BERGSTRÖM 1945 [1, 2], HOLMAN 1946), vgl. Abb. 10. Ferner tritt bei *Linolen-* und *Arachidonsäure* die längerwellige Bande verstärkt in Erscheinung (HOLMAN und BURR 1945, HOLMAN 1946); doch ist dort fraglich, ob sie *allein* einem konjugierten Dienketon zuzuordnen ist (HOLMAN 1946).

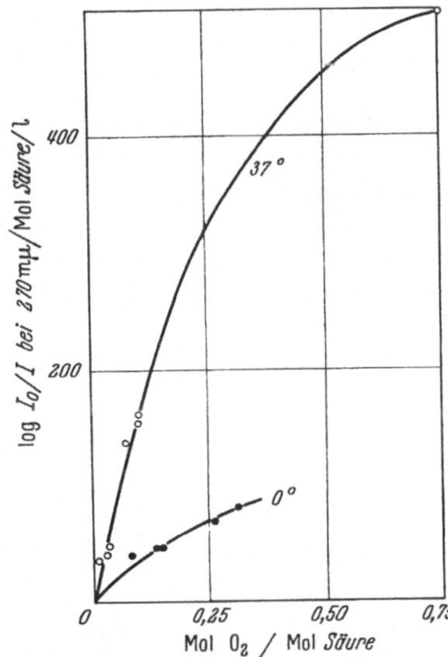

Abb. 12. Einfluß der Temperatur auf den molekularen Extinktionskoeffizienten der enzymatischen Oxydationsprodukte von Linolat bei 270 mμ. Kristallisiertes Sojaenzym; p_H 9. (Nach HOLMAN 1947.)

Es sei noch erwähnt, daß in kleinerer Menge sowohl bei Autoxydation (FRANKE und MÖNCH 1944, BERGSTRÖM 1945 [1]) wie bei der Enzymreaktion (BERGSTRÖM 1945 [2], FRANKE, MÖNCH, KIBAT und HAMM 1948) *höhere Oxydationsprodukte* gebildet werden, die bei der katalytischen Hydrierung zum Teil *α-Glykole* liefern, was auf primäre *Brückenperoxyd*bildung als Nebenreaktion zurückgeführt werden könnte. So wurde aus Hydrierungsprodukten sowohl der enzymatischen Linol- wie auch Linolensäureoxydation durch Bleitetraacetatspaltung *Capronaldehyd* ($CH_3 \cdot (CH_2)_4 \cdot CHO$) als Dinitrophenylhydrazon in einer Menge von 4% der Sauerstoffaufnahme gefaßt, was für eine bevorzugte Reaktion der 12,13-Doppelbindung sprechen würde (FRANKE, MÖNCH, KIBAT und HAMM 1948).

Möglicherweise bestehen hier Beziehungen zu einem verschiedentlich diskutierten (FARMER 1946, BOLLAND und GEE 1946), vom ursprünglichen Schema 1 (S. 150) etwas abweichenden Mechanismus der Kettenauslösung *zu Reaktionsbeginn*; danach erscheint energetisch günstiger als die direkte Ablösung eines Wasserstoffatoms aus einer α-Methylengruppe (die etwa 80 kcal je Mol erfordern würde) eine direkte Reaktion des Sauerstoffs mit einer Doppelbindung, in verkürzter Form etwa nach

Schema 2.

$$-CH_2-CH{=}CH- \xrightarrow{+O_2} \underset{\displaystyle OO^{\cdot}\ (I)}{CH_2-CH-\dot{C}H-} \xrightarrow{+\ -CH_2-CH=CH-} \underset{\displaystyle OOH\ (II)}{-CH_2-CH-\dot{C}H-} + \underset{(III)}{-\dot{C}H-CH{=}CH-}.$$

Radikal III entspricht dem Primärradikal des Schemas 1 und kann die Kette wie dort fortsetzen. Andererseits liegt die Annahme nahe, daß Radikal I sich zu einem Brückenperoxyd stabilisieren könnte. Ferner hat man nach FARMER und Mitarbeitern mit einer sekundären Reaktion von Primärperoxyden bzw. Peroxydradikalen mit Doppelbindungen und eventuell anschließender Hydrolyse zu rechnen (FARMER, BLOOMFIELD, SUNDRALINGAM und SUTTON 1942, FARMER und SUNDRALINGAM 1943, FARMER 1946):

$$\left.\begin{matrix} R \cdot OOH \\ R \cdot OO^{\cdot} \end{matrix}\right\} + —CH{=}CH— \rightarrow \left.\begin{matrix} R \cdot OH \\ R \cdot O^{\cdot} \end{matrix}\right\} + \underset{\diagdown O \diagup}{—CH—CH—} \quad [\rightarrow —CHOH—CHOH—].$$

Vielleicht die wichtigste Folgereaktion bei der Autoxydation wie bei der enzymatischen Oxydation der ungesättigten Fettsäuren ist die *Polymerisation* der Primärperoxyde, worüber allerdings chemisch noch sehr wenig bekannt ist (vgl. S. 151f., ferner v. MIKUSCH 1956). Erste Ansätze zur erfolgversprechenden Untersuchung des Übergangs der Primär- in Polymerperoxyde liegen immerhin vor (PRIVETT und NICKELL 1956). Auch die enzymatisch (mit Rohferment) gebildeten *Polymer*peroxyde zeigen optische Aktivität. Ihr Anteil an den Oxydationsprodukten wächst mit erhöhter Enzymkonzentration und Reaktionstemperatur (PRIVETT, NICKELL, LUNDBERG und BOYER 1955). Sie enthalten zusätzlichen Sauerstoff in offenbar konstantem, von den Oxydationsbedingungen unabhängigem Verhältnis, während der Gehalt an konjugiertem Dien gegenüber dem Monomeren reduziert und stärker variabel ist (vgl. S. 149 und 151 f.).

b) Zur Frage des Kettenmechanismus. Inhibitoren und Sekundäroxydationen.

Die Formulierung der Lipoxydasereaktion als *Kettenvorgang* gründet sich im wesentlichen auf die Analogie mit der (zu Versuchszwecken meist chemisch oder photochemisch katalysierten) *Autoxydation* der ungesättigten Fettsäuren und ihrer Ester (Literatur S. 150). Dieser Fall ist durchaus ungewöhnlich, denn in *allgemeinerer* Form hat sich die des öfteren versuchte Interpretation *enzymatischer* Oxydations- und Spaltungsvorgänge als Kettenreaktionen (HABER und WILLSTÄTTER 1931, MOELWYN-HUGHES 1937, WATERS 1943, 1946) nicht durchsetzen können. (Zur Kritik vgl. HALDANE 1932, FRANKE 1934, 1940 [2], 1944.)

Zwei Argumente sind vor allem zugunsten der Kettenvorstellung anzuführen: die Hemmbarkeit auch der Enzymreaktion durch typische „*Antioxydantien*" oder „*Antioxygene*" von meist *phenolischer* Natur und die zahlreichen Beispiele von *Sekundäroxydationen*.

α) Inhibitoren. Frühzeitig beobachtet und später des öfteren systematisch untersucht wurde besonders die Hemmung der Lipoxydase durch *Phenole*; entsprechend den Unterschieden in Bestimmungsmethodik, Versuchsbedingungen und Reinheitsgrad der verwendeten Enzympräparate variieren die von den einzelnen Untersuchern angegebenen Hemmungswerte stark; so kann z. B. bei der *manometrischen* Methode eine Hemmung der Fettsäureoxydation durch Oxydation des Inhibitors teilweise oder ganz verdeckt werden. (Über eine weitere Fehlerquelle dieser Methode vgl. S. 142 und 148).

Schon STRAIN (1941) entdeckte den Hemmungseffekt von *Hydrochinon, Brenzcatechin* und *Pyrogallol*, vermißte andererseits einen derartigen Effekt bei den meta-Verbindungen Resorcin und Phloroglucin. SÜLLMANN (1943 [1]) hat eine Reihe weiterer Phenole einschließlich *α-Tocopherol* in den Kreis einer systematischen Untersuchung einbezogen. An seinen (manometrischen) Versuchen fällt auf, daß eine Hemmung in manchen Fällen nur bei längerer Versuchsdauer in Erscheinung tritt; so beträgt bei der Inhibitorkonzentration m/4000 die Hemmung der Linolensäureoxydation durch Sojalipoxydase (p_H 6,3, 37°) in Prozent:

	Nach 10 min	Nach 60 min	Nach 300 min
Mit Hydrochinon	0	47	74
Mit Brenzcatechin	32	56	76
Mit Resorcin	0	5	14
Mit Phloroglucin.	0	8	32
Mit Dioxyphenylalanin . . .	6	30	60

Bei den kinetischen Besonderheiten der Lipoxydasereaktion (vgl. S. 146 und 149) erscheint es sehr fraglich, ob die nur bei langen Versuchszeiten zu beobachtenden Hemmungen sich noch auf die *Enzym*reaktion beziehen.

Folgende Hemmungsangaben (nach TAPPEL, LUNDBERG und BOYER 1953) beziehen sich auf die (spektrophotometrisch bestimmte) *Anfangs*geschwindigkeit der Lipoxydasereaktion und zum Teil lebensmittelchemisch wichtige Antioxydantien (vgl. JANECKE 1953).

Tabelle 2. *Hemmung der Linolatoxydation mit (roher) Sojalipoxydase durch phenolische Antioxydantien (p_H 9; 30°).*

Antioxydans	Konzentration × 10^{-4} m	% Hemmung bei 25°	Antioxydans	Konzentration × 10^{-4} m	% Hemmung bei 25°
—	2	0	Hydrochinon	2	6
α-Naphthol . . .	2	94	Propylgallat.	1	73
Pyrogallol . . .	2	19	Nordihydroguajaretsäure[1].	1	100
Phloroglucin . .	2	4	α-Tocopherol	2	39

Die folgende Tabelle 3 (nach den gleichen Autoren) zeigt für dasselbe Fermentmaterial die Konzentrationsabhängigkeit (vgl. auch Abb. 15 und 19, S. 157 und 160) und die (überraschend hohe) Temperaturabhängigkeit der Inhibitorwirkung.

Tabelle 3. *Einfluß der Inhibitorkonzentration und der Reaktionstemperatur auf die Anfangsgeschwindigkeit der Linolatoxydation mit Sojaenzym bei p_H 9.*

Antioxydans	Konzentration × 10^{-4} m	% Hemmung bei		Antioxydans	Konzentration × 10^{-4} m	% Hemmung bei	
		0°	30°			0°	22°
Hydrochinon .	1	—	38,0	Nordihydroguajaretsäure	0,01	—	0
	2	0	58,7		0,1	8,0	38,0
	5	0	70,0		1	66,0	100,0
	10	0	—				
	20	53,4	90—100	Propylgallat.	0,01	—	0
					0,1	0	13,0
α-Tocopherol .	1	—	27,0		1	32,0	100,0
	2	13,0	50,0				
	8	38,3	62,6				
	16	38,3	—				

Nach SIDDIQI und TAPPEL (1956 [1, 2]) beträgt die Hemmung (in %) mit

	10^{-3} m-α-Tocopherol	10^{-3} m-Pyrogallol	10^{-4} m-Nordihydroguajaretsäure
bei Erbsenlipoxydase . . .	82	81	92
bei Luzernenlipoxydase . .	73	~100	100

Für das *Rein*ferment existieren nur wenige Angaben; HOLMAN (1947) fand bei 20° mit *α-Naphthol* 29 bzw. 61% Hemmung bei m/5000 bzw. m/1000, mit *α-Tocopherol* 25% Hemmung bei m/2500.

FRANKE (1944) hat enzymatische und nichtenzymatische (Kobalt-)Katalyse der Linolenatoxydation unter vergleichbaren Bedingungen auf ihre Beeinflußbarkeit durch eine Reihe aromatischer (meist phenolischer) Inhibitoren untersucht und Hemmungen ähnlicher Größe, jedoch keine quantitative Übereinstimmung, gefunden (Abb. 13). Ähnliche Befunde konnten nur noch bei *Katalase* und *Oxaloxydase* (bzw. den entsprechenden rein chemischen Katalysatoren) erhoben werden, während eine Reihe anderer Oxydationsfermente eine *generelle* Hemmbarkeit dieser Art nicht zeigte. Die geringe Empfindlichkeit der meisten Oxydationsfermente gegenüber Antioxydantien haben TAPPEL und MARR (1954) später für *α-Tocopherol, Propylgallat* und *Nordihydroguajaretsäure* bestätigt. Dies spricht dafür, daß die Lipoxydase zu den wenigen Fermenten gehört, für deren Wirkung ein Kettenmechanismus überhaupt in Betracht gezogen werden kann.

---

[1] $(HO)_2C_6H_3—CH_2—CH(CH_3)—CH(CH_3)—CH_2—C_6H_3(OH)_2$.

In einigen Fällen ist eine *Oxydation* des Inhibitors während seiner Funktion nachgewiesen worden, z. B. bei der *Nordihydroguajaretsäure* (TAPPEL, BOYER und

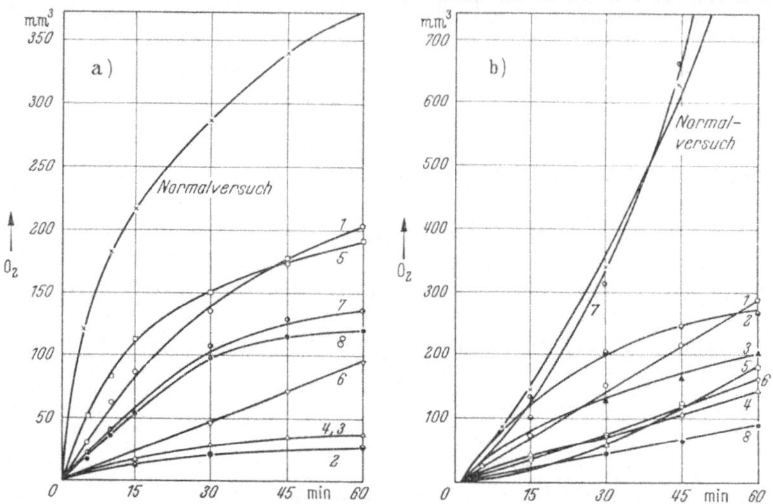

Abb. 13 a u. b. Hemmung von enzymatischer (a) und nichtenzymatischer (b) Oxydation der Linolensäure durch eine Reihe von Antioxydantien in der Konzentration m/500. a Sojalipoxydase bei p_H 6,3; b m/2500-$Co(NO_3)_2$ bei p_H 4,7 *1* Phenol; *2* Brenzcatechin; *3* Hydrochinon; *4* Chinon; *5* Resorcin; *6* Pyrogallol; *7* Phloroglucin; *8* Diphenylamin. (Nach FRANKE 1944.)

LUNDBERG 1952) (vgl. Abb. 19, S. 160) und bei *Tocopherol* (KUNKEL 1951, TAPPEL, LUNDBERG und BOYER 1953) (Abb. 14). Es bestehen hier also enge Beziehungen zu den Erscheinungen der *Sekundäroxydation* (s. S. 157f.).

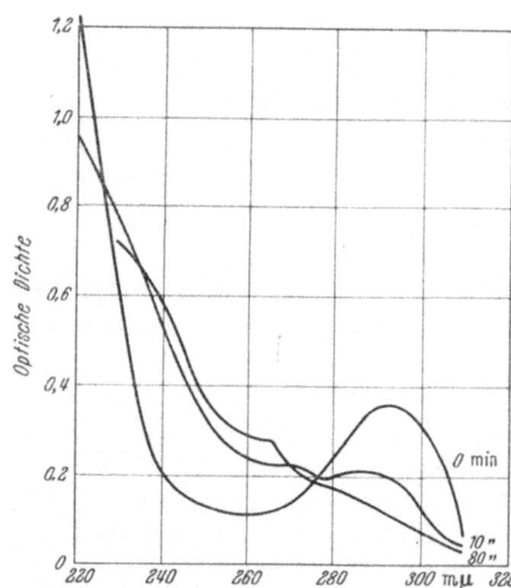

Abb. 14. Durch Oxydation von α-Tocopherol bedingte spektrale Änderungen im Linolat-Lipoxydase-System. 7×10^{-3} m-Linolat; 6×10^{-4} m Tocopherol; p_H 9; 30°. Absorptionsmaximum des α-Tocopherols: 290–295 mμ, Absorptionsmaximum des α-Tocopherylchinons: 260—270 mμ. (Nach TAPPEL, LUNDBERG und BOYER 1953.)

Nach BOLLAND und TEN HAVE (1947 [1, 2]) erfolgt der Kettenabbruch auf der Stufe des Fettsäureperoxydradikals ($R\cdot OO^\cdot$) durch *Phenole* und verwandte Verbindungen (AH_2) nach

$$R\cdot OO^\cdot + AH_2 \rightarrow R\cdot OOH + AH^\cdot.$$

Das Inhibitorradikal $AH^\cdot$ (von der Natur eines „Semichinons", vgl. MICHAELIS 1933, 1945; MICHAELIS und SCHUBERT 1938; für *Tocopherol* MICHAELIS und WOLLMAN 1949) vermag nicht als „Wasserstoffacceptor" der reaktionsfähigen CH_2-Gruppen des Fettsäuremoleküls (vgl. Schema 1, S. 150) zu fungieren, wodurch die Kette unterbrochen wird; es verschwindet a) durch Disproportionierung (zu Phenol + Chinon) oder b) Zusammenstoß mit einem Fettsäureperoxydradikal (unter Chinonbildung):

a) $$2\,AH^\cdot \rightarrow A + AH_2;$$
b) $$R\cdot OO^\cdot + AH^\cdot \rightarrow R\cdot OOH + A.$$

Auch in Abwesenheit von Inhibitoren kann schließlich Kettenabbruch erfolgen, nämlich nach

$$\left.\begin{array}{r} R^\cdot + R^\cdot \rightarrow \\ R\cdot OO^\cdot + R^\cdot \rightarrow \\ R\cdot OO^\cdot + R\cdot OO^\cdot \rightarrow \end{array}\right\}\ \text{inaktive Produkte}$$

(BOLLAND 1945, 1948); die Bildung *dimerer* Nebenprodukte erscheint so leicht erklärlich (vgl. S. 151f. und 153f.).

Abb. 15 zeigt den Einfluß eines *Tocopherol*-Zusatzes auf die Extinktion des Fettsäureoxydationsproduktes bei 232,5 mμ während der enzymatischen Oxydation von Linolsäuremethylester (KUNKEL 1951). Da die Extinktion beim höchsten Tocopherolzusatz nur mehr 8% von der des Kontrollansatzes beträgt, wird geschlossen, daß in letzterem wenigstens 90% des Umsatzes auf Rechnung von Reaktionsketten gehen.

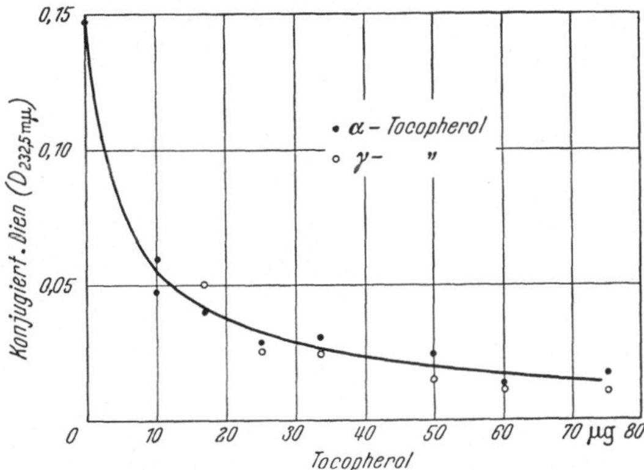

Abb. 15. Einfluß wechselnder Zusätze von α- und γ-Tocopherol auf die enzymatische Oxydation von Linolsäuremethylester nach Extinktionsmessungen bei 232,5 mμ. 5 mg Substrat in 4 ml Gesamtvolumen; p_H 6,5; reiner O_2; 20°; Reaktionsdauer 2 min. (Nach KUNKEL 1951.)

β) **Sekundäroxydationen** zahlreicher zellvertrauter Stoffe in Gegenwart ungesättigter Fettsäuren, ihrer Ester und Glyceride sind in der Literatur beschrieben worden und gehören zum Teil zu den ältesten Beobachtungen über Lipoxydasewirkungen überhaupt. In erster Linie sind hier zu erwähnen die *Carotinoide* (ältere Literatur S. 138 und 142, außerdem BERNSTEIN und THOMPSON 1947, SMITH und SUMNER 1948 [1], HOLMAN 1948 [2], KUNKEL 1951), ferner *Tocopherol* (Literatur S. 154f.), *Chlorophyll a* und *b* (STRAIN 1941, 1949), *Hämin* (KIES 1946), *Ascorbinsäure* (STRAIN 1941, 1949), *SH-Glutathion* (MAPSON 1953, MAPSON und MOUSTAFA 1955), *Adrenalin* (SÜLLMANN 1945 [2]) und *Dopa* (Dioxyphenylalanin) (STRAIN 1941, SÜLLMANN 1945 [2]). Im System Lipoxydase — Linolsäure —

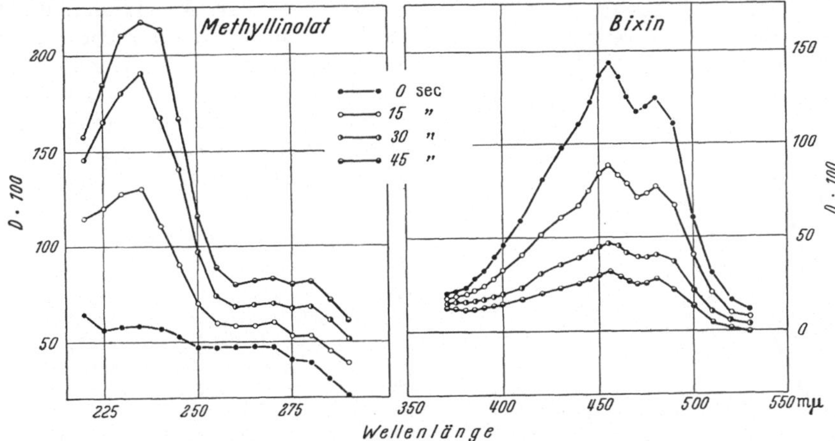

Abb. 16. Korrespondierende Änderungen der Absorptionsspektren von Linolsäure-methylester und Bixin während der gekoppelten enzymatischen Oxydation. 10 mg Methyllinolat + 0,035 mg Bixin in Alkohol/Wasser 1:3; p_H 6,5; 25°; gereinigtes Sojaenzym. (Nach SMITH und SUMNER 1948 [1].)

Polyphenol — Aminosäure sind sogar *tertiäre* Oxydationen von *Aminosäuren* (Glykokoll, Alanin, Leucin, Glutaminsäure) unter NH_3- und CO_2-Bildung beschrieben worden (MICHLIN und PSCHENNOWA 1949, HEIMANN, MATZ, GRÜNEWALD und HOLLAND 1955).

Nur die Oxydation der Carotinoide *Carotin* und *Bixin* hat *eingehendere* Bearbeitung gefunden. SMITH und SUMNER (1948 [1]) haben zuerst die spektralen Veränderungen von *Bixin* und *Linolsäuremethylester* bei der gemeinsamen enzymatischen Oxydation in quantitative Beziehung gesetzt. Während die Absorption des Fettsäureesters bei 232 und 270 mμ zunahm (S. 151), nahm diejenige des

Farbstoffs bei 455 und 480 mμ entsprechend ab (Abb. 16). Noch eingehender hat HOLMAN (1948 [2]) die Verhältnisse bei der sekundären *Carotin*oxydation mit *reinem* Sojaenzym untersucht. Abb. 17 zeigt in zwei verschiedenen Versuchsserien, daß sowohl oxydierte (entfärbte) Carotinmenge wie auch gebildetes konjugiertes Dien einerseits der Zeit, andererseits der Enzymmenge parallel gehen (vgl. Lipoxydasebestimmungsmethoden S. 142). Abb. 18, gleichfalls nach HOLMAN (1948 [2]), belegt zunächst sowohl für *Bixin* (I) wie für *Carotin* II), daß der Farbstoffschwund bei nicht zu hohen Konzentrationen proportional der Farbstoffkonzentration zunimmt. Weiterhin erkennt man aus den Kurven III und IV, daß mit steigender Carotinoidkonzentration die Menge des gebildeten konjugierten Fettsäuredienperoxyds *abnimmt*. Aus den Neigungen beider Kurvenpaare läßt sich berechnen, daß die Oxydation *eines* Bixin- bzw. Carotinmoleküls die Bildung von 26 bzw. 43 Molekülen Dienperoxyds *verhindert*. Man kann danach, in Analogie zur Antioxydanswirkung (S. 156f., im besonderen Abb. 15), auch die Oxydation der Carotinoide formal mit dem Abbruch von Fettsäureketten erklären; die obigen Molekülzahlen würden dann einen ungefähren Anhaltspunkt für die Zahl der Kettenglieder bei der normalen Fettsäureoxydation durch Lipoxydase geben.

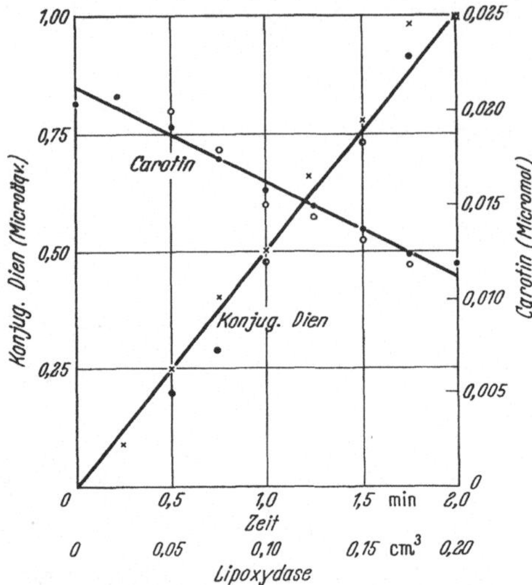

Abb. 17. Gekoppelte enzymatische Oxydation von Linolat und Carotin. Kristallisiertes Sojaenzym; p_H 9; spektrophotometrische Methodik. ● × Zeitliche Mengenänderungen an Carotin bzw. konjugiertem Linolsäurehydroperoxyd bei konstanter Enzymkonzentration. ○ ◐ Entsprechende Mengenänderungen im 2 min-Versuch bei variierter Enzymkonzentration. (Nach HOLMAN 1948 [2]).

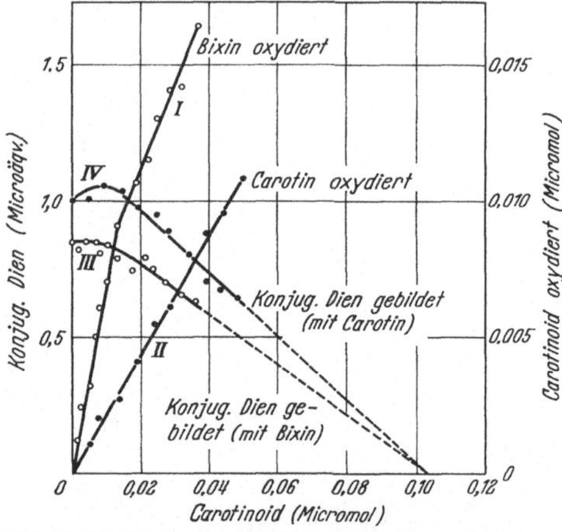

Abb. 18. Einfluß der Carotinoidkonzentration auf die enzymatische Carotinoidoxydation (*I, II*) und die Bildung konjugierten Linolsäurehydroperoxyds (*III. IV*). Kristallisiertes Sojaenzym; p_H 9; spektrophotometrische Methodik. (Nach HOLMAN 1948 [2].)

Freilich ist eine „H-Donator"-Wirkung von Carotinoidmolekülen gegenüber Fettsäureperoxydradikalen, wie sie BERGSTRÖM und HOLMAN (1948 [2]), einem früher (S. 156) gegebenen Schema entsprechend, annehmen, chemisch nicht sehr plausibel. Vielleicht bestehen eher Beziehungen zwischen *Carotinoidepoxyd*bildung (s. S. 159) und Kettenabbruch (vgl. S. 153), deren Formulierung aber noch aussteht.

KUNKEL (1951) erhielt bei der sekundären *Bixin*oxydation (gemessen durch die Absorptionsabnahme bei 457 mμ, vgl. Abb. 16, S. 157) eine logarithmische Beziehung einerseits zur *Reaktionszeit*, andererseits zur *Quadratwurzel* der Fermentkonzentration. Doch sind seine Befunde bestritten (s. S. 160).

Über die *spektralen* und *chemischen* Veränderungen von *Carotin* (HUNTER und KRAKENBERGER 1946, HOLMAN 1949), *Vitamin A* (BOLOMEY 1947 [1, 2] TROITZKI 1948, HOLMAN 1950) und *Bixin* (KUNKEL und NELSON 1949) bei der *nichtenzymatischen* Autoxydation bzw. Sekundäroxydation ist wiederholt gearbeitet worden; über die Verhältnisse bei der *enzymatischen* Oxydation liegt jedoch nur eine kurze Notiz vor.

Auf Einzelheiten kann hier nicht eingegangen werden (Literatur s. FRANKE 1951). Bemerkenswert ist, daß schon bei nur 10%iger Autoxydation des eingesetzten Linolsäureesters Carotin, Vitamin A oder Bixin praktisch vollständig zerstört sind (HOLMAN 1949, 1950, KUNKEL und NELSON 1949). Die Fassung von Zwischenprodukten des Carotinoidabbaus ist in diesen *komplexen* Oxydationssystemen kaum möglich. Doch ist durch *Autoxydationsversuche* an β-Carotin in Benzol oder Erdnußöl (HUNTER und KRAKENBERGER 1946) folgende Reaktionsfolge durch chromatographische Analyse wahrscheinlich gemacht worden:

```
H3C     CH3                              H3C     CH3
   \   /                                    \   /
     C                                        C
   /   \                                    /   \
H2C     C—CH=CH—C=        →              H2C     C—CH=CH—C=        →
 |      ||       |                        |      | \O     |
H2C     C       CH3                      H2C     C /     CH3
   \   / \                                  \   / \
    CH2   CH3                                CH2   CH3
        I                                        II

H3C     CH3                              H3C     CH3
   \   /                                    \   /
     C                                        C
   /   \                                    /   \
H2C     C=====CH   CH3        →          H2C     C—CH=CH—C=
 |      |     |     |                     |   O//  \O     |
H2C     C     CH—C=                      H2C     C /     CH3
   \   / | \O/                              \   / \
    CH2  CH3                                 CH2   CH3
        III                                      IV
```

Laufen die Vorgänge nur an *einem* β-Iononring ab, so kommt man vom *β-Carotin* (I) über *β-Carotin-monoepoxyd* (II) und *Mutatochrom* (III) zu *Semi-β-Carotinon* (IV); erfolgen sie symmetrisch an *beiden*, dann erhält man *β-Carotin-diepoxyd* (II), *Luteochrom*[1], *Aurochrom* (III) und schließlich *β-Carotinon* (IV). (Vgl. auch KARRER und JUCKER 1948.)

HOLMAN (1949) sah bei der spektrophotometrischen Untersuchung des viel aktiveren enzymfreien Systems Carotin-Linolsäureester in der Phase raschen Carotinabbaus allerdings nur vorübergehend eine Bande bei 337,5 mμ auftreten, als einzigen Hinweis auf die Bildung eines Zwischenprodukts auf dem Wege vom Carotin (Hauptabsorption bei 460 mμ) zu den Endprodukten, die — ganz ähnlich wie die Hauptprodukte der Fettsäureautoxydation (S. 151) — bei 232,5 und 275 mμ absorbieren, demnach nur noch wenige Doppelbindungen enthalten. FRIEND (1956) hat kürzlich im Ätherextrakt der Reaktionsprodukte des Systems Lipoxydase-Natriumlinolat-β-Carotin chromatographisch Epoxyde mit einem *aurochrom*ähnlichen Spektrum — möglicherweise *cis,trans*-Isomere dieses Farbstoffs — nachgewiesen. Die Hauptprodukte der enzymatischen Reaktion zeigen Spektren ähnlich denen der *β-Apocarotinale*[2], die durch partiellen Abbau von β-Carotin mit H_2O_2 und OsO_4 erhalten werden; doch weist die langsame Reduktion der CO-Gruppe und ihr Unvermögen, mit Methylamin zu reagieren, eher auf *Ketone* hin.

γ) Einwände gegen die Kettenvorstellung. Neuerdings haben sich TAPPEL, BOYER und LUNDBERG (1952) mit beachtlichen, wenn auch noch nicht voll ausreichenden Argumenten gegen den Versuch gewandt, die bei der Deutung des *Autoxydationsvorgangs* an ungesättigten Fettsäuren bewährte Vorstellung einer Kettenreaktion [vgl. neuere Zusammenfassungen von FRANKE (1950) und KERN und WILLERSINN (1955)] auch auf die *Fermentreaktion* zu übertragen.

Es handelt sich im wesentlichen um folgende Punkte:

1. Die Anfangsgeschwindigkeit der Linolatoxydation wird im allgemeinen proportional der Enzymkonzentration (vgl. Abb. 7, S. 148) gefunden (BALLS, AXELROD und KIES 1943, THEORELL, BERGSTRÖM und ÅKESON 1944, HOLMAN 1947, FRANKE, MÖNCH, KIBAT und HAMM 1948, FRANKE und FREHSE 1953, IRVINE und ANDERSON 1953 [1], TAPPEL, LUNDBERG und

[1] *Luteochrom* zeigt am einen Ring die Epoxydstruktur II, am anderen die furanoide Konfiguration III.

[2] Die bisher bekannten *β-Apocarotinale* enthalten einen β-Iononring mit längerer (3 bis 4 Isoprenreste umfassender) aldehydischer Seitenkette (vgl. KARRER und JUCKER 1948).

BOYER 1953), während für eine Kettenreaktion nach HALDANE (1932) Quadratwurzelabhängigkeit zu erwarten wäre (die zwar von KUNKEL (1951) angegeben (s. S. 158), von TAPPEL, BOYER und LUNDBERG (1952) aber auf nicht ausreichende Dispersion des von ihm als Substrat verwendeten Linolsäure-methylesters zurückgeführt wird.

2. Abweichend vom Autoxydationsvorgang zeigt die enzymatische Oxydation von Linolat keine Induktionsperiode; entgegenstehende Befunde HOLMANS (1948 [2]) an der *Bixin*oxydation in Linolatgegenwart (vgl. auch HOLMAN und BERGSTRÖM 1951) sollen auf Besonderheiten dieses *komplexen* Reaktionssystems zurückgehen.

3. Die Lipoxydasewirkung zeigt die übliche Substratkonzentrationsabhängigkeit einer Enzymreaktion (vgl. Abb. 5, S. 147), die durch Annahme einer dissoziablen Enzym-Substratverbindung (MICHAELIS und MENTEN 1913) oder Adsorptionsbindung zwischen Enzym und Substrat (HITCHCOCK 1926, WEIDENHAGEN 1932) deutbar ist, während die Autoxydationsreaktion bei hohem Sauerstoffdruck Proportionalität zur Linolatkonzentration aufweist (BOLLAND 1949).

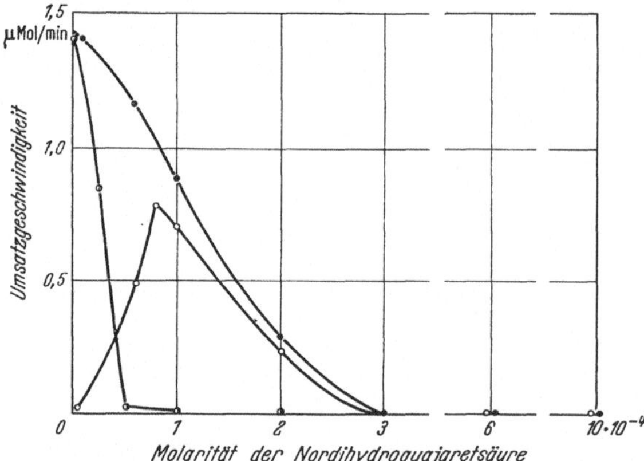

Abb. 19. Lipoxydase-katalysierte Sauerstoffaufnahme (●), Nordihydroguajaretsäureoxydation (○) und Linolatoxydation (◑) in Abhängigkeit von der Nordihydroguajaretsäurekonzentration. $1{,}8 \times 10^{-3}$ m-Linolat; p_H 7; 25°. Oxydation des *Linolats* verfolgt durch Absorptionsmessung bei 232,5 mµ, der *Nordihydroguajaretsäure* durch Absorptionsmessung bei 415 mµ (Absorptionsmaximum des (o-chinoiden) Oxydationsprodukts). (Nach TAPPEL, BOYER und LUNDBERG 1952.)

4. Die *enzymatisch* gebildeten Fettsäurehydroperoxyde haben einen höheren molaren Extinktionskoeffizienten (vgl. S. 151) als die autoxydativ entstandenen (Zusammenstellung von Literaturwerten bei HOLMAN 1946), was auf Struktur- oder Konfigurationsunterschiede und einen „richtenden" Einfluß des Ferments hinweist. (Dazu kommt neuerdings noch die *optische Aktivität* sowohl der Primär- wie der Polymerperoxyde, vgl. S. 152 und 154.)

5. Die Wirkung verschiedener Antioxydantien entspricht nach Versuchen von TAPPEL, BOYER und LUNDBERG (1952) nicht der üblichen Vorstellung eines „Kettenabbruchs". So verhinderte in Ansätzen mit α-*Tocopherol*, in denen das Ausmaß von Linolat- und Tocopherol-Oxydation vergleichend spektrophotometrisch verfolgt wurde, 1 Molekül Tocopherol die Oxydation von nur 0,3—0,4 Molekülen Linolat, während nach einem Kettenmechanismus die Oxydation von wesentlich mehr als 1 Molekül Linolat hätte verhindert werden sollen (vgl. hierzu S. 157f.). Ferner kann *Nordihydroguajaretsäure* (Formel S. 155) in Gegenwart von Lipoxydase + Linolat oxydiert werden, auch wenn im Endeffekt keine (durch Absorptionsmessung bei 232,5 mµ bestimmte) *Linolat*oxydation erfolgt ist (Abb. 19); andererseits wird in Abwesenheit von Linolat das Antioxydans nicht angegriffen. Nach TAPPEL und Mitarbeitern (1952) wirkt die Verbindung direkt auf das Enzym hemmend, während dem Linolat eine „cofermentartige" Funktion zugeschrieben wird (vgl. jedoch S. 162).

TAPPEL, BOYER und LUNDBERG (1952) schlagen folgendes Schema 3 (S. 161) für den Mechanismus der Linolatoxydation (a) und der Oxydation eines Antioxydans (AH_2) in Linolatgegenwart (b) durch Lipoxydase vor.

Nach Bildung eines O_2-Linolat-Lipoxydasekomplexes (I) entsteht — unter Übertragung eines Elektrons und eines H-Ions vom Linolat auf das Sauerstoffmolekül — auf der Enzymoberfläche ein Biradikal (II), das, einer Vorstellung von MICHAELIS (1946) entsprechend, durch das Enzym stabilisiert werden könnte. Noch unter dem steuernden Einfluß des Enzyms reagiert es weiter unter Bildung eines konjugierten Peroxyds (III), das dann vom Enzym abdissoziiert (IV). In Gegenwart eines Antioxydans (AH_2) reagiert das Biradikal zweiphasig unter Abgabe von Wasserstoff und intermediärer Bildung des Radikals $AH^{\cdot}$. Am Enzym soll Linolat regeneriert und H_2O_2 gebildet werden. Die Autoren lassen die Möglichkeit offen, daß unter geeigneten Bedingungen auch freies Linolat, ähnlich wie ein Antioxydans, mit dem Biradikal reagieren könne, wodurch als *Nebenreaktion* die Bildung *kurzer* Ketten denkbar erscheint.

Es ist schwer, bei dem bisher vorliegenden bescheidenen experimentellen Material endgültig Stellung zu den Vorstellungen der amerikanischen Autoren zu nehmen, die zweifellos Beachtung verdienen. Eine Schwierigkeit scheint darin zu bestehen, daß in Gegenwart des Antioxydans nach Schema 3 nicht die ursprüngliche, sondern eine konjugierte Linolsäure „regeneriert" wird, die beobachtete Reduktion der Extinktion bei 234 mμ also nicht erfolgen dürfte. Die Zusatzannahmen, die TAPPEL, BOYER und LUNDBERG hier machen müssen (Abdissoziation dieser „Pseudolinolsäure" vom Enzym erst bei der Reaktion mit Sauerstoff, Stabilisierung eines Linolatradikals mit freiem Elektron an C^{11} in Gegenwart des Antioxydans), wirken nicht sehr überzeugend. Man wird auch

Schema 3.

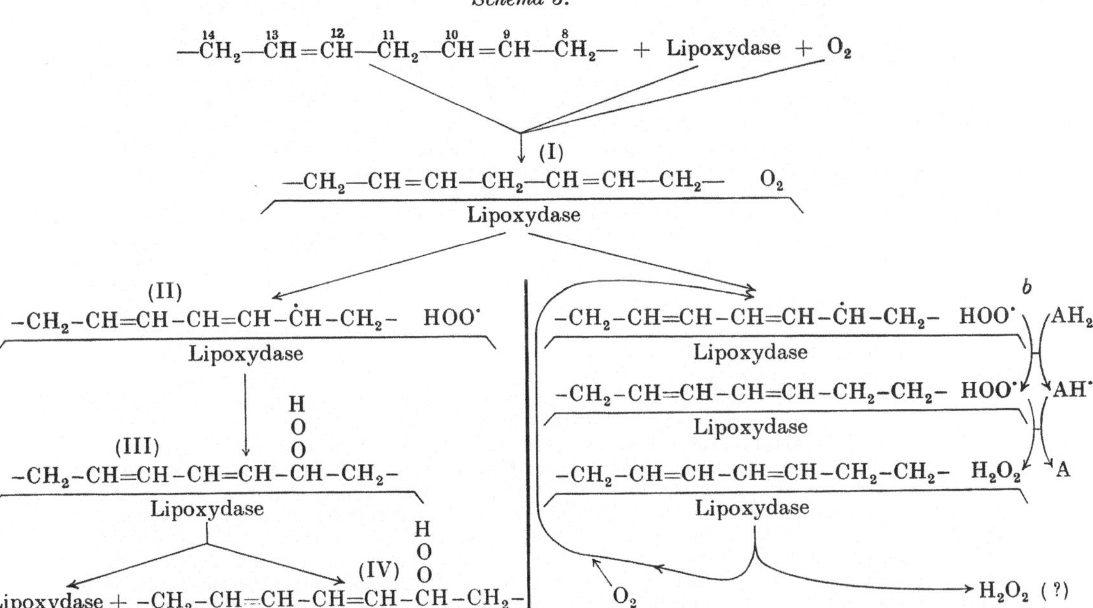

das Verhalten leicht oxydierbarer Sekundärsubstrate ohne typische Antioxygenwirkung (aber mit Wirkung auf die „Konjuen"-Bildung, vgl. S. 157f.) und ohne vorstellbare Wirkung auf das Enzym (z. B. Carotinoide) in die Untersuchung vergleichend einzubeziehen haben; eine Formulierung nach Schema 3b erscheint hier *chemisch* wenig plausibel (vgl. S. 158). Im übrigen bleibt die bemerkenswerte Tatsache bestehen, daß die gegen Inhibitoren sonst so beständige Lipoxydase (S. 144) gerade gegenüber *phenolischen* und ähnlichen *aromatischen Antioxydantien* empfindlich ist, die nach vergleichenden Untersuchungen keineswegs generelle oder auch nur häufige Inhibitoren von Oxydationsfermenten darstellen (S. 155). Die besondere Beziehung dieser Stoffe zum Oxydations*substrat* (vielleicht in der enzymgebundenen Radikalform) bleibt bestehen.

LUNDBERG (1956) formuliert die heutige widerspruchsvolle Situation unter Mitverwertung eigener, noch unveröffentlichter Versuche folgendermaßen: „Obwohl diese Beobachtungen darauf hinweisen, daß der Kettenreaktionsmechanismus der Autoxydation im Fall der Lipoxydaseoxydation nicht anwendbar ist, schließen sie doch nicht die Möglichkeit eines anderen Typs eines Kettenmechanismus aus, wie man diesen für andere Enzyme angenommen hat. Tatsächlich sind derartige Kettenmechanismen sehr wahrscheinlich. Unsere derzeitige Ansicht besagt jedoch, daß in einem solchen Mechanismus das Enzym bei der Bildung eines jeden Hydroperoxydmoleküls bzw. polymerer Produkte beteiligt ist. Die Beobachtung, daß bei niedrigeren Sauerstoff- und höheren Lipoxydase-Konzentrationen mehr Polymere gebildet werden, steht also völlig in Übereinstimmung mit dieser Ansicht."

10. Zellbedeutung der Lipoxydase.

Unsere Kenntnisse in diesem Punkte sind noch sehr mangelhaft, was unter anderem darin zum Ausdruck kommt, daß auch so erfolgreiche Bearbeiter des Ferments wie HOLMAN und BERGSTRÖM in ihrer letzten Monographie des Ferments (1951) sich mit der Feststellung begnügen: "The role of lipoxidase in the metabolism is unknown." In der Tat bestehen über die physiologische Bedeutung der Lipoxydase, die ja nur sporadisch verbreitet ist und deren Wirkung im *chemischen* Sinne eigentlich keinen Fett*abbau* darstellt, bis heute kaum mehr als einige — sich zudem zum Teil widersprechende — Vermutungen.

HOLMAN (1948) hatte auf Grund des Befundes, daß in keimender Soja die Lipoxydaseaktivität bereits zu einem Zeitpunkt stark abfällt, wo ein Substratschwund noch kaum eingesetzt hat (Abb. 1, S. 141), geschlossen, daß die Bedeutung der Lipoxydase möglicherweise nur in der *Auslösung* der Fettsäureoxydation bestehe, die dann als Kettenreaktion autokatalytisch weiterlaufe. HICKMAN (1948) hat die — im Prinzip in entgegengesetzte Richtung gehende — Vermutung ausgesprochen, daß es eine der physiologischen Funktionen des *Tocopherols* sei, den Ablauf der erwünschten Oxydation durch Lipoxydase zu gestatten und unerwünschte Neben- und Folgereaktionen zu unterdrücken. [Vgl. ähnliche Vorstellungen einer *Schutzwirkung* von Tocopherol gegen die *Häminproteid*katalyse im *tierischen* Organismus (TAPPEL 1953 [2, 3], 1954, COLLIER und McRAE 1955).] Auch die zahlreichen, in der Literatur beschriebenen *Sekundäroxydationen* zellvertrauter Stoffe, von denen früher (S.157f.) schon die Rede gewesen war, sind bisweilen als physiologisch bedeutsam angesehen worden. In diesem Zusammenhang verdient die neuerdings aufgezeigte Möglichkeit einer „coferment-artigen" Wirkung der ungesättigten Fettsäure bei der Oxydation andersartiger Substrate (S. 160) erhöhte Beachtung (HOLMAN 1953). Allerdings ist eine derartige Wirkung von Linolat bei der „cofermentgemäßen" Konzentration von 10^{-4} m beim Erbsenferment in einer Reihe von Fällen (Cytochrom c-oxydase, DPNH-oxydase, Diaphorase, Cystein- und Glutathion-oxydase) vermißt worden (SIDDIQI und TAPPEL 1956 [1]).

FRANKE und FREHSE (1954) hatten einen auffallenden Parallelismus in Verbreitung und Auftreten von Aktivitätsänderungen bei der Keimung einerseits von Lipoxydase, andererseits einer auf die *Dehydrierung* ungesättigter Fettsäuren eingestellten „*Lipodehydrase*" festgestellt (Näheres S. 165 und 167). Der Eindruck eines Zusammenwirkens beider Enzyme verstärkte sich durch den weiteren Befund, daß lipoxydatisch „anoxydierte" Fettsäuren erheblich leichter dehydriert wurden als unvorbehandelte Substrate; über die Ursachen dieser Erscheinung wird noch zu sprechen sein (S. 168). Jedenfalls könnte eine plausible, wenn auch im Mechanismus noch undurchsichtige Rolle der Lipoxydase darin bestehen, den Abbau der ungesättigten Fettsäuren in Carboxylnähe durch begrenzte Oxydation in Doppelbindungsnähe einzuleiten bzw. zu beschleunigen.

Schließlich haben unlängst SIDDIQI und TAPPEL (1956 [1]) die Vermutung geäußert, daß Lipoxydase an der Bildung des Lipoidfilms (*Cutin*), welcher die der Luft ausgesetzten Teile der Pflanze bedeckt (PRIESTLEY 1943), beteiligt sein könnte. Danach würden von den Pflanzenzellen ausgeschiedene Fettsäuren u. ä. enzymatisch zu Peroxyden oxydiert, die nach Wanderung entlang den Zellwänden in der Außenschicht einer Polymerisation unterlägen, die Cutin liefert. Diese Vorstellung muß sich wieder mit der Tatsache auseinandersetzen, daß Lipoxydase keineswegs generell in Pflanzen vorkommt (S. 139f.). Die unlängst von MATIC (1956) aus dem Cutin von *Agave americana* isolierten Fettsäuren (9,10,18-Trioxy-, 10,18-Dioxy-, 18-Oxy-octadecansäure, 18-Oxyoctadec-*cis*-9-ensäure und 10,16-Dioxyhexadecansäure) sind nach ihrer Konstitution kaum als Produkte einer Lipoxydasewirkung anzusprechen.

11. Praktische Bedeutung.

Fälle einer Verwertung der Lipoxydasewirkung im positiven Sinne sind sehr selten. Meist wird versucht, die Enzymwirkung als nachteilig zu vermeiden bzw. zu hemmen.

Ein Beispiel der erstgenannten Art ist die schon 1934 in den USA patentierte Verwendung von Sojapräparaten für die *Bleichung* von Weizenmehl bzw. daraus bereiteten Teigen (HAAS und BOHN 1934). Neuere Befunde zeigen, daß die sameneigene Lipoxydase des Weizenkorns am Bleichprozeß im Teig bei Fettgegenwart einen maßgeblichen Anteil haben kann (TODD, HAWTHORN und BLAIN 1954). Andererseits können aber auch die durch Lipoxydase oxydierten Fettsäuren die Teigbeschaffenheit erheblich verschlechtern und einen ranzigen

Geschmack hervorrufen (Miller und Kummerow 1948, Neumann-Pelshenke 1954). Auch wird zur Makkaroniherstellung zwecks *Erhaltung* der gelben Carotinoidfarbstoffe die Verwendung lipoxydasearmer Weizensorten empfohlen (Irvine und Anderson 1953 [2]). Lipoxydasebestimmungen in Hartweizen oder Hartgrießproben ermöglichen gewisse Voraussagen über die Farbe der daraus herzustellenden Makkaroni-Erzeugnisse (Irvine 1955).

Frühe Beobachtungen bezogen sich auf die starken Carotin- bzw. Vitamin A-Verluste bei der üblichen Bereitung von *Luzernen*heu und ihre weitgehende Vermeidbarkeit durch Schnelltrocknung im Heißluft- oder Abgasstrom (Hauge und Aitkenhead 1931, Hauge 1935). Erst wesentlich später konnte *Lipoxydase* als Hauptursache der Carotinverluste festgestellt werden (Mitchell und Hauge 1946 [1, 2], Mitchell und King 1946). Bernstein und Thompson (1947) beobachteten in *Bohnen*blättern beim Trocknen sowohl *photochemische* als *enzymatische* Carotinzerstörung. Erstere ist maximal bei 68—80% Wasserverlust, letztere bei 80—95%; in *vollständig* entwässertem Material ist nur mehr eine minimale Photoreaktion nachweisbar. Der Temperaturkoeffizient der Enzymreaktion liegt bei 1,65 (4—25°), derjenige der Lichtreaktion bei 1,25 (24—64°). Bemerkenswert ist der Befund, daß in teilweise getrockneten und besonders in vorgefrorenen und wiederaufgetauten Blättern etwa die Hälfte des Carotins *rasch*, der Rest viel langsamer verschwindet (was wohl auf Enzyminaktivierung zurückgehen dürfte, vgl. S. 146). Ähnliche Befunde sind neuerdings von Budnitzkaja (1955) für Futterpflanzen wie Klee und Wicke erhoben worden.

Der neuerdings erbrachte Nachweis von Lipoxydase in *Gramineen* (S. 140) läßt die Frage des enzymatisch bedingten Carotinverlusts auch für den Fall des *Gras*heus bedeutsam erscheinen.

Auf den nachteiligen Einfluß von Enzymwirkungen bei der Aufbereitung von *Ölsamen* hat Lüde schon 1943 hingewiesen. Auch die neuerlich beobachtete starke Abnahme des Fettgehalts von *Milo-(Sorghum-)Keimen* bei der Lagerung (Lindemann 1953) dürfte wenigstens zum Teil enzymatischer Natur sein. Nach neuen Untersuchungen (Rothe 1953, 1954, 1955, Thomas und Rothe 1954) ist auch beim Bitterwerden von *Hafer*produkten während der Lagerung die Lipoxydase beteiligt. Der Bitterstoff soll aus zwei Komponenten bestehen, einem fettlöslichen Autoxydationsprodukt ungesättigter Fettsäuren und einer wasserlöslichen saponinähnlichen Substanz (Mohr 1954).

Schließlich sei noch anhangsweise erwähnt, daß Fettsäureperoxyde (tierische) Oxydationsfermente wie Cytochrom-, Succino-, Cholin-, Amino-oxydase (Bernheim, Wilbur und Kenaston 1952, Ottolenghi, Bernheim und Wilbur 1955) wie auch Pankreas-Lipase (Hérisset 1954) zu hemmen vermögen. In Beziehung hierzu stehen wahrscheinlich Beobachtungen über die Oxydation von SH-Gruppen von Proteinen durch Fettsäureperoxyde (Dubouloz und Fondarai 1953, Dubouloz, Fondarai und Pracchia 1954).

III. Fettsäuredehydrasen.

1. Übersicht.

In der älteren Literatur existieren über pflanzliche Fettsäuredehydrasen nur wenige Mitteilungen. Als erster berichtete Grande (1934) aus dem Thunbergschen Institut über Samenextrakte, die Palmitin- und Stearinsäure mit Acceptorfarbstoffen (hauptsächlich 1-Naphthol-2-sulfonat-indo-2,6-dichlorphenol) zu dehydrieren vermochten.

Als positiv wurden befunden: *Acacia communis, Agrostemma coronaria* und *A. coelirosa, Amaranthus caudatus atropurpureus, Campanula speculum, Convolvulus tricolor, Evonymus europaea, Galega officinalis, Papaver somniferum* und *P. rupifragum, Trifolium pratense*; am wirksamsten waren *Agrostemma, Amaranthus* und *Papaver*.

Etwas später wies Zeller (1935) in *Kürbis*samen Dehydrierung der beiden gesättigten Fettsäuren (C_{16} und C_{18}) wie auch der Ölsäure mittels der Thunberg-Methodik (mit Methylenblau) nach und Yosii (1937) fand Stearinsäuredehydrierung bei Soja-, Mungo- und schwarzen Bohnen. Eine Dehydrierbarkeit *mehrfach* ungesättigter Fettsäuren ist entweder nicht untersucht oder [wie von Zeller (1935)] vermißt worden.

In den letzten Jahren haben einerseits Stumpf und Mitarbeiter in den USA, andererseits Franke und Frehse in Deutschland das Problem des enzymatischen Fettsäureabbaus in höheren Pflanzen auf breiterer Basis wieder aufgenommen. Während die erstgenannten Autoren nur mit Erdnußkotyledonen und C^{14}-markierter Palmitinsäure arbeiteten und sich auf die Bestimmung des oxydativ gebildeten $C^{14}O_2$ beschränkten, haben die letztgenannten Untersucher an einer größeren Zahl von Pflanzensamen das Verhalten gesättigter *und* ungesättigter (zum Teil auch brom- und hydroxylsubstituierter) höherer Fettsäuren nach der THUNBERG-Methodik mit Farbstoffacceptoren vergleichend geprüft. (Vgl. kurze Zusammenfassung von FREHSE und FRANKE 1956.) Im vorliegenden Abschnitt „Fettsäuredehydrasen“ sollen im wesentlichen nur die *Dehydrierungs*befunde der letztgenannten Autoren gebracht werden, während die Beobachtungen von STUMPF und Mitarbeitern am *komplexen* Oxydationssystem der Erdnuß*mikrosomen* — das den von FRANKE und FREHSE studierten *primären* Dehydrierungsschritt zweifellos als Teilreaktion leistet — in einem folgenden Abschnitt „α-Oxydation“ (nach dem einstweilen hervorstechendsten chemischen Charakteristikum dieses Abbaus) besprochen werden sollen. Das von STUMPF und BARBER (1956) neuerdings in Erdnuß*mitochondrien* aufgefundene, noch kompliziertere System der *β-Oxydation* (dessen Primärreaktion noch nicht untersucht ist) wird im entsprechenden späteren Abschnitt (S. 179f.) behandelt werden[1].

2. Dehydrasen gesättigter und ungesättigter höherer Fettsäuren (Stearat- und „Lipo“dehydrase) in Pflanzensamen.

a) Verbreitung und relative Aktivität.

Die Nachweisbarkeit von Dehydrierungseffekten an Fettsäuren hängt weitgehend vom Redoxpotential des verwendeten Acceptorfarbstoffs ab, das für die energetisch anspruchsvolle Fettsäuredehydrierung möglichst hoch liegen soll (FRANKE und FREHSE 1954).

Das „Gruppenpotential“ (E'_0 bei p_H 7) für die Dehydrierung einer gesättigten zur entsprechenden ungesättigten Carbonsäure liegt zwischen —0,025 und +0,025 V (KALCKAR 1941). *Methylenblau* mit einem E'_0-Wert vom +0,011 V nähert sich bereits der thermodynamischen Grenze der Acceptorwirkung; in der Tat waren die Effekte, die FRANKE und FREHSE (1954) mit diesem Farbstoff bei Samenextrakten erhielten, in den meisten Fällen sehr schwach. Besser wurden die Dehydrierungseffekte schon bei *2,6-Dichlorphenol-indophenol* (TILLMANS-Reagens, $E'_0 = +0,217$ V) und in einer mit den sonst meist verwendeten Farbstoffen nie erzielten Deutlichkeit traten sie hervor, als zu dem Indaminfarbstoff BINDSCHEDLERS *Grün* $\left[(CH_3)_2N\text{—}C_6H_4\text{—}N\!=\!C_6H_4\!=\!N(CH_3)_2\right]Cl$ von noch etwas höherem Potential übergegangen wurde. Man erkennt dies deutlich aus der folgenden Tabelle 4, in der für eine Farbstoffreihe neben den Potentialwerten die zugehörigen Entfärbungszeiten ohne und mit Substrat (t_0 bzw. t) und die daraus berechneten „Dehydrierungsintensitäten“ I verzeichnet sind.

Tabelle 4. *Dehydrierung von Linolat (m/140) durch Gerstenextrakt in Gegenwart verschiedener Acceptorfarbstoffe (m/6500) bei p_H 7,4.*

Farbstoff	E'_0 (p_H 7)	Entfärbungszeiten (min.)		I
		ohne Substrat t_0	mit Substrat t	
BINDSCHEDLERS Grün . . .	0,224	92	1,5	65,5
2,6-Dichlorphenol-indophenol	0,217	66	12	6,8
Toluylenblau	0,115	> 120	120	< 0,8
Thionin	0,063	> 120	120	< 0,8
Methylenblau	0,011	≫ 120	> 120	~0

Nach THUNBERG (1929) ist $I = 100\,(1/t - 1/t_0)$. Bisweilen findet sich im folgenden auch die „spezifische Dehydrierungsintensität“ $I_0 = I$/mg Enzymtrockengewicht angegeben. Um die *enzymatische* Natur der beobachteten Donatoreffekte sicherzustellen, wurden des öfteren auch vollständige Ansätze mit inaktiviertem (gekochtem) Enzym als Kontrollen mitgeführt: die damit erhaltenen Entfärbungszeiten t_i wurden als Bezugswerte statt t_0 genommen, wenn sie kürzer waren als letztere.

[1] Siehe Nachtrag S. 200.

In Tabelle 5 sind die Entfärbungszeiten für den besonders geeigneten Acceptorfarbstoff BINDSCHEDLERS Grün in Ansätzen ohne und mit Substrat (zum Teil auch gekochtes Enzym enthaltend) zusammengestellt, die an Phosphatpufferextrakten verschiedener Samen — die mit L bezeichneten waren schon ungekeimt *lipoxydase*positiv (vgl. S. 140) — beobachtet wurden. In Abb. 20 sind die aus den Entfärbungszeiten und Trockengewichten berechneten I_0-Werte für die C_{18}-Fettsäuren in Abhängigkeit von der Zahl der Doppelbindungen (F) aufgezeichnet.

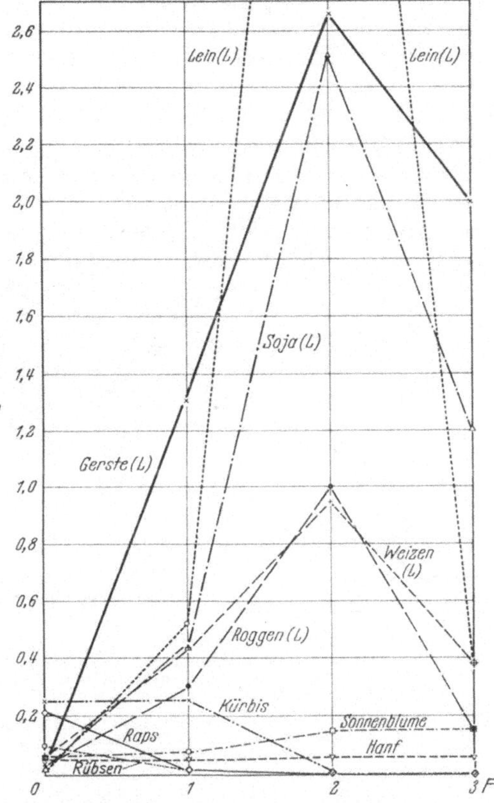

Abb. 20. Spezifische Dehydrierungsintensität (I_0) in verschiedenen Samenextrakten gegenüber den C_{18}-Fettsäuren mit *0*, *1*, *2* und *3* Doppelbindungen (F). I_0 für Linolsäure bei Lein: etwa 6. (Nach FREHSE und FRANKE 1956.)

Tabelle 5 und Abb. 20 lassen zunächst erkennen, daß das Vermögen zur Dehydrierung *gesättigter* Fettsäuren bei Pflanzensamen offenbar recht verbreitet, wenn auch meist schwach ist. Auch *Ölsäure* wird von den untersuchten Samen, Rübsen (und vielleicht Raps) ausgenommen, angegriffen. Überraschend war das Verhalten der *mehrfach ungesättigten* Fettsäuren: während sie von einigen Samen (Kürbis, Raps, Rübsen) überhaupt nicht dehydriert werden, stellen sie bei den übrigen gute bis ausgezeichnete, den gesättigten Fettsäuren oft um 1—2 Größenordnungen überlegene Wasserstoffdonatoren dar. Es ist bemerkenswert, daß sich als besonders aktiv in dieser Hinsicht die schon in ruhendem Zustande *lipoxydasepositiven* Samen erwiesen (vgl. S. 140). Linolensäure wird günstigstenfalls mit gleicher, meist aber mit geringerer Intensität angegriffen als Linolsäure. Nach den vorliegenden Befunden liegt es nahe, wenigstens *zwei* Fettsäuredehydrasen zu unterscheiden: eine vorwiegend auf höhere gesättigte Fettsäuren eingestellte

Tabelle 5. *Dehydrierung von Fettsäuren (m/140)* mit BINDSCHEDLERS *Grün (m/10000) durch Samenextrakte (p_H 7,4, 30°).*

Pflanze	Enzymtrockengewicht	t_0	Substrate						
			Palmitinsäure	Stearinsäure	Ölsäure	Linolsäure		Linolensäure	
	mg		t	t	t	t	t_i	t	t_i
Gerste (L) . . .	25	26	22	22	3	1,5	120	2	120
Roggen (L) . . .	44	51	38	38	9	2,5	—	11	—
Weizen (L) . . .	35	30	14	16	5	3	120	6	120
Soja (L)	51,5	53	24	36	4	0,75	25	1	22
Lein (L)	21	300	180	170	8	0,75	—	12	110
Hanf	23,5	60	35	40	45	35	84	35	75
Sonnenblume . .	25,5	24	15	19	17	12	35	12	39
Kürbis	22	29	11	11	11	40	—	38	—
Raps	36	19	10	8	16	33	—	19	37
Rübsen	35	27	20	14	40	35	—	30	—

Stearatdehydrase („*Stearicodehydrase*“ nach GRANDE 1934) und ein vorwiegend auf mehrfach ungesättigte Fettsäuren eingestelltes Ferment, das FRANKE und FREHSE (1954) in Anlehnung an die weitgehend substratanaloge Lipoxydase (S. 145) ad hoc als „*Lipodehydrase*“ bezeichnet haben; vielleicht überschneiden sich bei der Ölsäure die beiden Spezifitätsbereiche.

b) Lipodehydrase.

α) Eigenschaften. Einstweilen liegen nur einige Beobachtungen an Rohextrakten von Gerste (zum Teil auch Soja) vor (FRANKE und FREHSE 1954). Die Gerstenlipodehydrase erwies sich als ein *lösliches* Ferment, das auch hochtouriges Zentrifugieren (15000 Touren/min, entsprechend 20000 × g) ohne Aktivitätsminderung vertrug. Ebensowenig beeinflußte 24stündige Dialyse die Extraktaktivität.

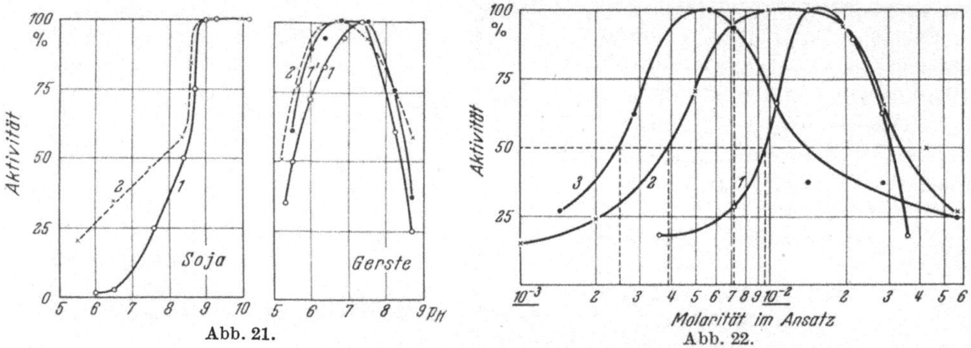

Abb. 21. Abb. 22.

Abb. 21. p_H-Einfluß auf Aktivität (*1* Linolat, *1'* Oleat) und Stabilität (*2* Linolat) von Soja- und Gerstenlipodehydrase. In *1* und *1'* direkte Aktivitätsbestimmung (mit BINDSCHEDLERS Grün) beim betreffenden p_H; in $2^1/_4$stündige Inkubation (20°) ohne Substrat beim betreffenden p_H und anschließende Aktivitätsbestimmung bei p_H 7,4. (Aktivität des nichtvorbehandelten Enzyms = 100.) (Nach FRANKE und FREHSE 1954.)

Abb. 22. Aktivitäts-pS-Kurven der Gerstenlipodehydrase für Oleat (*1*), Linolat (*2*) und Linolenat (*3*). m/10000-BINDSCHEDLERS Grün; p_H 7,4; 30°. Durchgehende gestrichelte Vertikallinie entspricht der Substratkonzentration (m/140) vorausgegangener Versuche; übrige gestrichelte Linien zur Ermittlung der MICHAELIS-Konstanten K_m. (Nach FRANKE und FREHSE 1954.)

Zusatz von Codehydrase I, Adenylsäure, Adenosintriphosphorsäure (ATP), Hypoxanthin, Coenzym A (CoA) und CoA + ATP zu dialysierten Extrakten hatte keinen eindeutig aktivitätssteigernden Einfluß. Dieser Umstand unterscheidet die pflanzliche Lipodehydrase in sehr bemerkenswerter Weise von den in der Literatur beschriebenen (meist an gesättigten Substraten untersuchten) Fettsäuredehydrasen *tierischer* Herkunft (LANG 1939, LANG und MAYER 1939 [1,2], SHAPIRO und WERTHEIMER 1943, BURTON 1948, LE BRETON und CHAMPOUGNY-CLÉMENT 1948, FANTL und LINCOLN 1949, DRYSDALE und LARDY 1953, KORNBERG und PRICER jr. 1953, MAHLER, WAKIL und BOCK 1953). Insbesondere geben diese Untersuchungen (ebenso wie die fast gleichzeitig ausgeführten Versuche des STUMPFschen Arbeitskreises am „Palmitatoxydase“-System der Erdnußmikrosomen, vgl. S. 170) keinen Hinweis dafür, daß die Dehydrierung in derartigen Samenextrakten, wie neuerdings für das wichtigste *tierische* Enzymsystem der *β-Oxydation* beschrieben (DRYSDALE und LARDY 1953, KORNBERG und PRICER jr. 1953, MAHLER, WAKIL und BOCK 1953, weitere Literatur S. 177f.), an den *Acylderivaten* des Coenzyms A erfolgt; sie sprechen vielmehr für eine Dehydrierung des freien *Fettsäureanions* (vgl. dagegen das Mitochondrienenzym S. 181).

Die Lipodehydrase der Gerste ist in mancher Beziehung stabiler als die Lipoxydase gleicher Herkunft; so zeigt sie nicht die bei letzterer beobachtete „Schüttelinaktivierung“ (S. 143) und weist eine „Tötungstemperatur“ von 57° (gegen 53° bei Lipoxydase, S. 149) auf.

Der *p_H-Einfluß* auf Aktivität und Stabilität der Lipodehydrase wurde vergleichend an Gersten- und an Sojaenzym untersucht (Abb. 21). Die schon bei

Lipoxydase je nach Herkunft beobachteten erheblichen Unterschiede (vgl. Abb. 6, S. 147) kehren hier in ähnlicher Form wieder; allerdings vermißt man die gegenüber der Aktivität stark vergrößerte Spannweite der *Stabilität*, die bei der Lipoxydase (besonders aus Soja) im sauren Gebiete in Erscheinung getreten war.

Den *Einfluß der Substratkonzentration* zeigt am Gerstenenzym für Öl-, Linol- und Linolensäure Abb. 22. Die glockenförmigen, ziemlich symmetrischen Aktivitäts-pS-Kurven weisen sehr ausgesprochene Konzentrationsoptima bei resp. m/70, m/100 und m/200 (MICHAELIS-Konstanten K_m von resp. 9,6, 3,9 und $2{,}5 \times 10^{-3}$ m entsprechend) auf, worin sowohl eine mit steigender Zahl der Substratdoppelbindungen zunehmende Affinität des Substrats zum Enzym als auch eine zunehmende Hemmung durch erhöhte Substratkonzentration, wahrscheinlich infolge Bildung eines inaktiven Enzym-Substratkomplexes höherer Ordnung, zum Ausdruck kommt (vgl. z.B. HALDANE und STERN 1932, LINEWEAVER und BURK 1934).

Tabelle 6. *Lipoxydaseaktivität (Q_{O_2}, S. 142) und Fettsäuredehydrasenaktivität (I_0 mit m/10000-BINDSCHEDLERS Grün, S. 164) in Phosphatextrakten aus ruhenden Samen und Keimlingen.*

Pflanze	Q_{O_2} (Lipoxydase)	I_0 Stearinsäure	I_0 Ölsäure	I_0 Linolsäure
Raps, Samen	0	0,22	0,02	0
Keimling (1 cm lang)	20	0,42	0,25	0,20
Kürbis, Samen	0	0,26	0,26	0
Keimling (5 cm lang)	120	—	—	0,29
Hanf, Samen	0	0,04	0,03	0,05
Keimling (2 cm lang)	85	0,25	0,38	0,85

β) **Verhalten bei der Samenkeimung. Beziehungen zur Lipoxydase.** Aus Tabelle 5 und Abb. 20 (S. 165) war bereits hervorgegangen, daß hohe Lipoxydase- und Lipodehydraseaktivität in Samen nahe vergesellschaftet vorkommen. Da nach früher (S. 142) erwähnten Versuchen Lipoxydase in manchen Pflanzen erst bei der *Keimung* der Samen auftritt, wurde geprüft, ob dies auch für die Lipodehydrase gilt (FRANKE und FREHSE 1954). Bei Raps und Kürbis war dies in der Tat der Fall; Hanf, der im ruhenden Samen geringe Mengen Lipodehydrase, jedoch keine Lipoxydase zu enthalten schien, zeigte bei derKeimung eine entsprechende *Steigerung* der Lipodehydraseaktivität (Tabelle 6).

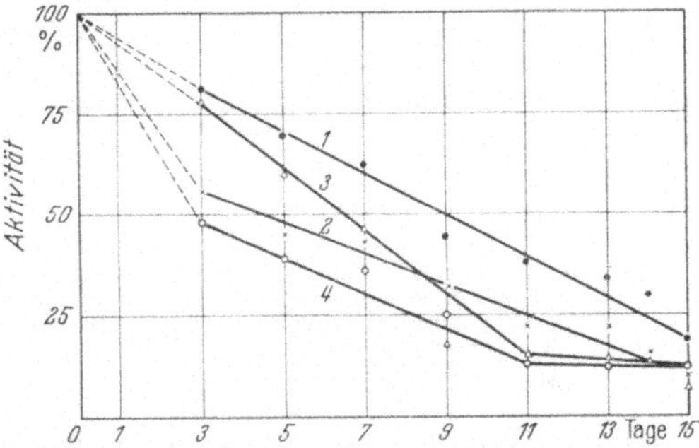

Abb. 23. Absinken der Lipoxydaseaktivität (*1*) und der Lipodehydraseaktivität (*2* Linolenat, *3* Linolat, *4* Oleat) bei der Keimung von Gerste im Lichte. In *1* m/50-Linolat, in *2–4* jeweils optimale Substratkonzentration (vgl. Abb. 22); p_H 7; 30°. (Nach FRANKE und FREHSE 1954.)

Andererseits wurde bei der im ruhenden Zustande bereits lipoxydase- und lipodehydrase-positiven *Gerste* beobachtet, daß bei längerer Keimdauer beide Enzyme mit vergleichbarer Geschwindigkeit abnahmen (Abb. 23).

Diese Befunde über eine enge Vergesellschaftung beider Fermentwirkungen legten zwei Deutungsversuche nahe: 1. Lipoxydase und Lipodehydrase sind identisch, oder 2. die Wirkung beider Fermente ist irgendwie gekoppelt. Da sich die erste Möglichkeit auf Grund verschiedener Beobachtungen über die größere Labilität der Lipoxydase z.B. gegenüber Fettsäureperoxyd, Schüttelbewegung und Temperatureinwirkung (S. 146 und 166) ausschalten ließ, blieb der Modus eines möglichen Zusammenwirkens beider Fermente zu untersuchen. FRANKE und FREHSE (1954) gaben deshalb enzymatisch anoxydiertes Linolat oder Linolenat zu üblichen THUNBERG-Ansätzen und stellten fest, daß diese „anoxydierten" Substrate im Vergleich zu „frischen" eine erheblich beschleunigte Entfärbung von Acceptorfarbstoffen — bei BINDSCHEDLERS Grün 20—30fach, bei TILLMANS-Reagens 2—3fach — bewirkten, wie sowohl für Gersten- als auch Sojaenzym nachgewiesen wurde (Tabelle 7).

Tabelle 7. *Entfärbung von Acceptorfarbstoffen durch Samenextrakte mit frischem und anoxydiertem Linolat.*
O_2-Aufnahme in a) 15%, b) 8,5%, c) 11%, d) 12% von 1 Mol O_2 je Mol Linolat.

Farbstoff	Pflanze	t_0 (min)	t (min)	
			frisches Linolat	anoxydiertes Linolat
BINDSCHEDLERS Grün (m/5000). . . .	a) Gerste	$>$ 140	12	$<$ 1
	b) Soja	$>$ 150	25	$<$ 1
Dichlorphenol-indophenol (m/20000). .	c) Gerste	100	17	5
	d) Soja	47	9	5

Abb. 24 zeigt, daß ein Autoxydationsgrad von etwa 10% der Theorie optimal wirkt und daß die aktivierende Wirkung einer „Anoxydation" des Substrats mit weiter steigender Sauerstoffaufnahme wieder nachläßt, stärker bei Linolen- als bei Linolsäure.

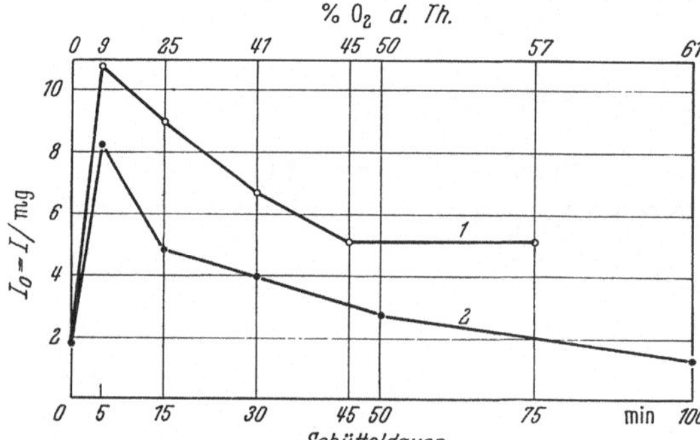

Abb. 24. Aktivierung der Dehydrierung durch enzymatische „Anoxydation" von Linolat (*1*) und Linolenat (*2*) in Abhängigkeit von der Sauerstoffaufnahme mit Gerstenenzym. Messung der Lipodehydraseaktivität mit m/50-Substrat und m/5000-BINDSCHEDLERS Grün; p_H 7,4. (Nach FRANKE und FREHSE 1954.)

Eine stichhaltige Erklärung der vorstehend beschriebenen Aktivierungseffekte kann einstweilen noch nicht gegeben werden. Grundsätzlich könnte entweder die Peroxydgruppe oder das konjugierte Doppelbindungssystem der Oxydationsprodukte (S. 150f.) maßgebend für den Aktivierungseffekt sein, was präparativ nachprüfbar erscheint. Ein orientierender „Modellversuch" mit den dreifach konjugiert-ungesättigten *Eläostearinsäuren* (α- und β-Form[1]) als Substraten ergab allerdings erheblich längere Entfärbungszeiten beim Gerstenenzym als sie bei der isomeren Linolensäure erhalten worden waren (wobei allerdings auch die viel geringere Löslichkeit der Eläostearinsäureseifen zu berücksichtigen wäre). Schließlich wäre auch an die früher (S. 152) erwähnten *sterischen* Änderungen an den Doppelbindungen während der Autoxydation bzw. enzymatischen Oxydation zu denken. Es erscheint immerhin denkbar, daß Konjugation und Konfigurationsänderung der Doppelbindungen *zusammen* zu leichter dehydrierbaren Produkten führen. Weniger wahrscheinlich ist — nach den bisherigen Erfahrungen an Oxydationsfermenten (S. 163) — eine Aktivierung der Lipodehydrase durch Fettsäureperoxyd.

Von der Möglichkeit, daß eine Aktivierung des dehydrierenden Fettsäureabbaus eine oder *die* physiologische Rolle der *Lipoxydase* darstelle, war schon S. 162 die Rede; sie wird weiter zu prüfen sein.

c) Dehydrase der höheren gesättigten Fettsäuren (Stearatdehydrase).

Entsprechend der meist geringen Aktivität des Enzyms in Pflanzensamen (S. 165) und in Anbetracht der Komplikationen, die sich durch die geringe

[1] Nach PASCHKE, TOLBERG und WHEELER (1953) *cis*-9, *trans*-11, *trans*-13- bzw. *trans*-9, *trans*-11, *trans*-13-Octadecatriensäure.

Löslichkeit der Substrate ergeben können, liegen hier einstweilen nur einige wenige Beobachtungen vor. Sie beziehen sich einerseits auf die *Spezifität* der Enzymwirkung hinsichtlich der Länge der C-Kette, andererseits auf die *Lokalisierung des Enzymangriffs* im Fettsäuremolekül.

Zur Prüfung des ersten Punktes wurden von FRANKE und FREHSE (1954) die geradzahligen gesättigten Fettsäuren von C_{18} bis herunter zu C_6 in Form der Natriumsalze (m/350) mit Gerstenextrakt und BINDSCHEDLERS Grün als Wasserstoffacceptor untersucht. Zwischen C_{18} und C_{12} bestanden nur geringe Unterschiede in der Dehydrierungsintensität, bei C_{10} war sie bereits wesentlich geringer und bei C_8 und C_6 nicht mehr mit Sicherheit feststellbar. Bemerkenswerterweise fanden auch STUMPF und Mitarbeiter mit ihrer abweichenden Methodik beim Erdnußenzym und bei Verwendung C^{14}-carboxylmarkierter Fettsäuren vergleichbare Oxydationsgeschwindigkeit zwischen C_{18} und C_{14}, während C_{12} um eine Größenordnung langsamer, C_{10} und C_8 praktisch nicht mehr angegriffen wurden, was bis auf eine Differenz von C_2 gut zu den obigen Dehydrierungsbefunden stimmt (HUMPHREYS, NEWCOMB, BOKMAN und STUMPF 1954, CASTELFRANCO, STUMPF und CONTOPOULOU 1955).

Die weitere Frage nach dem Ort des enzymatischen Angriffs im Fettsäuremolekül, d.h. die Frage, ob die Dehydrierung der gesättigten Fettsäuren immer in 2,3-Stellung (im Sinne der zunächst angenommenen *β-Oxydation*) oder eventuell auch „mittelständig" (z.B. in 9,10-Stellung bei Stearinsäure) erfolge, wurde unlängst durch Untersuchung der in 2,3- bzw. 9,10-Stellung *ungesättigten* oder *dibromsubstituierten* C_{18}-Säuren am Kürbissamenferment (das *gesättigte* Fettsäuren noch relativ am besten angreift, vgl. S. 165) geprüft (FRANKE, TASCHEN und FREHSE 1956).

Grundlage der Untersuchung waren vorausgegangene Modellversuche an der Muskel-*Succinodehydrase* gewesen, in denen sich nur Monobrombernsteinsäure (bei Substratsättigung des Enzyms etwa 3mal langsamer als Bernsteinsäure), nicht dagegen die beiden Dibrombernsteinsäuren (meso- und racemische Form) als dehydrierbar erwiesen hatten (FRANKE, TASCHEN und HOLZ 1957).

Da sich bei den beiden 9,10-Dibromstearinsäuren (meso- und racemische bzw. Oleo- und Elaidoform) *sterische* Effekte gezeigt hatten, wurden auch die entsprechenden *Dioxystearinsäuren* noch in den Kreis der Untersuchung einbezogen. Tabelle 8 (Mitte) zeigt, daß sowohl 2,3- als auch 9,10-Dibromstearinsäure (in beiden Formen) dehydrierbar sind, ebenso wie *trans*-2-Octadecensäure und die beiden 9-Octadecensäuren (Öl- bzw. Elaidinsäure).

Tabelle 8. *Dehydrierung ungesättigter und substituierter C_{18}-Fettsäuren mit Kürbissamen- und Rattenleberenzym und 2,6-Dichlorphenol-indophenol als Acceptor.*

	Kürbissamenenzym		Rattenleberenzym	
	Substratkonzentration m/200 Farbstoffkonzentration m/8000		Substratkonzentration m/300 Farbstoffkonzentration m/12000	
	t	*I*	*t*	*I*
Ohne Substrat	81,5 (t_0)	—	> 240 (t_0)	—
Stearinsäure	19	4,1	13	7,7
trans-2-Octadecensäure	20	3,8	12	8,3
Ölsäure	13,5	6,2	3	33,3
Elaidinsäure	25	2,8	14	7,2
2,3-Dibromstearinsäure	14,5	5,7	5	20,0
9,10-Dibromstearinsäure *(Oleo-)*	12,5	6,8	4	25,0
9,10-Dibromstearinsäure *(Elaido-)*	15	5,4	14	7,2
9,10-Dioxystearinsäure *(Oleo-)*	24,5	2,9	4	25,0
9,10-Dioxystearinsäure *(Elaido-)*	56,5	0,6	12	8,3

Man kann aus diesen Versuchen schließen, daß im Samenextrakt sowohl eine „carboxylnahe“ als auch eine „mittelständige“ Dehydrierung erfolgen kann. (In den ungesättigten Lipodehydrasesubstraten [vgl. S. 145] ist natürlich nur die erstere möglich.) Dieses etwas überraschende Ergebnis konnte am *Rattenleber*-enzym (dargestellt nach LANG 1939) bestätigt werden (Tabelle 8, rechts). Man hätte es danach wahrscheinlich mit zwei *verschiedenen* Stearatdehydrasen zu tun. Auffallend will an diesen Versuchen nur erscheinen, daß sich die Blockierung *eines* Abbauwegs (durch Substituenten oder eine Doppelbindung) *quantitativ* so wenig auswirkt.

Auf kleinere quantitative Unterschiede in den Angaben der Tabelle ist schon deswegen kein größerer Wert zu legen, weil die Natriumseifen der Substrate beim Versuchs-p_H (7,5) nur zum Teil und wohl in unterschiedlichem Ausmaß gelöst vorlagen (fast vollständig nur die der Ölsäure). Immerhin ist die durchgehende Überlegenheit der Öl- gegenüber der Elaidinsäure und der racemischen gegenüber den meso-9,10-Disubstitutionsprodukten sehr bemerkenswert.

IV. Die α-Oxydation.

Die Beziehungen dieses Abschnitts zum vorausgehenden über „Fettsäuredehydrasen“ sind, wie früher (S. 164) schon angedeutet, sehr eng. Die α-Oxydation stellt eine *Reaktionsfolge* dar, an deren Anfang sehr wahrscheinlich die Dehydrierung des Fettsäuremoleküls in α,β-Stellung steht. In erster Linie sollen hier die Beobachtungen des STUMPFschen Arbeitskreises zur Palmitatoxydation durch Enzyme von Erdnußmikrosomen mitgeteilt werden. Als Ergänzung folgen einige neuere Angaben zur α-Oxydation ω-aryloxysubstituierter *Fettsäurenitrile* (mit „herbicider“ Wirkung) durch höhere Pflanzen und zur α-Oxydation von *Propionsäure* durch gewisse Mikroorganismen.

1. Das „Palmitatoxydase“-System der Erdnußmikrosomen.

1952 berichteten NEWCOMB und STUMPF über Extrakte aus Erdnußkotyledonen, die aus C^{14}-markierten Palmitinsäuren (bei der niedrigen Substratkonzentration von m/30000) oxydativ $C^{14}O_2$ zu bilden vermochten; ein bei 100000 × g abzentrifugierter Anteil „submikroskopischer Partikel“ oxydierte sowohl 1- als auch 3-C^{14}-Palmitat, während der Überstand praktisch nur noch 1-C^{14}-Palmitat angriff. Später konnten HUMPHREYS und STUMPF (1955) aus der erwähnten Mikrosomen-Fraktion ein durch Natriumcholateinwirkung löslich werdendes und mit Ammonsulfat bei 80% Sättigung fällbares Enzym gewinnen, das als solches nur aus 1-C^{14}-Palmitat oxydativ $C^{14}O_2$ freisetzte; bei Zusatz von DPN (Codehydrase I) entstanden auch aus 2- und 3-C^{14}-Palmitat abgestuft kleinere Mengen $C^{14}O_2$.

Die Angaben der Autoren für *1-C¹⁴-Palmitinsäure* sind nicht einheitlich; nach HUMPHREYS und STUMPF (1955) erhöht DPN an Ammonsulfatfällungen auch bei diesem Substrat die $C^{14}O_2$-Bildung erheblich, nach anderen Angaben (HUMPHREYS, NEWCOMB, BOKMAN und STUMPF 1954, CASTELFRANCO, STUMPF und CONTOPOULOU 1955) ist es entbehrlich.

Von der *Substratspezifität* des STUMPFschen Enzymsystems hinsichtlich der Länge der C-Kette, die sich im wesentlichen mit derjenigen der Stearatdehydrase (FRANKE und FREHSE 1954) deckt, war schon die Rede (S. 169). Bemerkenswerterweise reagiert auch „Palmitatoxydase“ ähnlich wie Lipodehydrase (S. 166) und abweichend von den Fermenten der β-Oxydation (S. 177f., 181 und 185f.) primär mit *freiem* Palmitat und nicht mit Palmityl-Coenzym A (HUMPHREYS, NEWCOMB, BOKMAN und STUMPF 1954). Die Wirkungslosigkeit von m/100-Malonat als Inhibitor und das Fehlen einer stimulierenden Wirkung von Gliedern des KREBSschen *Citronensäurecyclus* (vgl. MILLERD, Bd. XII dieses Handbuchs, ferner

Zusammenfassungen von BREUSCH 1948, 1950, MARTIUS und LYNEN 1950, OCHOA 1954) auf die Fettsäureoxydation spricht auch gegen die Beteiligung des letzteren an diesem einfachen pflanzlichen Oxydationssystem (NEWCOMB und STUMPF 1952, HUMPHREYS, NEWCOMB, BOKMAN und STUMPF 1954).

Den Unterschied zwischen pflanzlicher *Mikrosomen*- und tierischer *Mitochondrien*-„Oxydase“ belegen in instruktiver Form Vergleichsversuche von

Tabelle 9. *Vergleich der Palmitatoxydation durch verschiedene pflanzliche und tierische Fermentsysteme.*

Substrat	% C^{14} als $C^{14}O_2$ erscheinend					
	A. Erdnußmikrosomen		B. Rattenlebermitochondrien			C. Erdnußmitochondrien
	I	II	I	II	III	I
1-C^{14}-Palmitat	45	42	49	17	10	45
2-C^{14}-Palmitat	15	40	4,5	2,7	1,3	27
3-C^{14}-Palmitat	6	12	30	11	14	48
15-C^{14}-Palmitat	3	0	29	10	4	35

HUMPHREYS, NEWCOMB, BOKMAN und STUMPF (1954) mit verschieden markierter Palmitinsäure (Tabelle 9). Späteren Ausführungen (S. 181) vorausgreifend, sind in die letzte Spalte unter C. bereits neueste Befunde an Erdnuß*mitochondrien* (STUMPF und BARBER 1956) mit aufgenommen worden.

Während in den Enzymansätzen aus Erdnußmikrosomen die $C^{14}O_2$-Ausbeute mit steigender Entfernung des markierten C-Atoms vom polaren Molekülende

Tabelle 10. *Aktivierung der „Palmitatoxydase“ durch α-Oxysäuren.*

Aktivator	Gesamtkonzentration in mMol	% $C^{14}O_2$
—	—	0
Glykolsäure	16,5	51
Glykolsäure	3,3	50
Glykolsäure	0,33	32
L-Milchsäure	16,5	52
L-Milchsäure	3,3	9
D-Milchsäure	16,5	0
DL-Milchsäure	16,5	50
DL-α-Oxybuttersäure . .	16,5	40
DL-α-Oxyvaleriansäure .	16,5	35

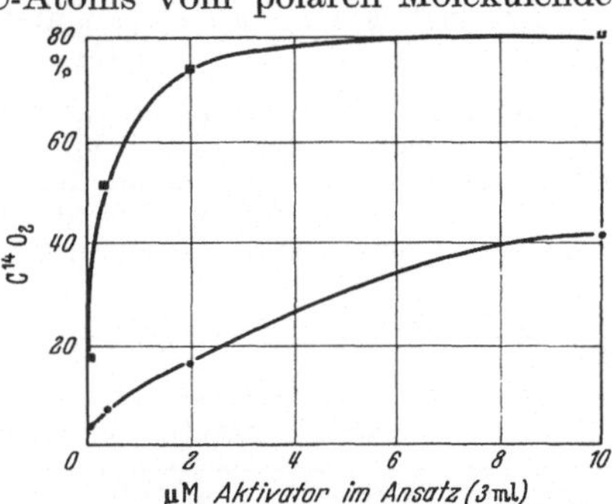

Abb. 25. Einfluß der Aktivatorkonzentration auf die Aktivierung löslicher „Palmitatoxydase“ durch Glykolat (■) und L-Lactat (●). $3,3 \times 10^{-4}$ m-Palmitat-1-C^{14} (0,1 μMol im Ansatz); p_H 7,2; 30°; Reaktionszeit 60 min (aerob). (Nach CASTELFRANCO, STUMPF und CONTOPOULOU 1955.)

kontinuierlich abnimmt, liefern die Ansätze mit tierischen Mitochondrien hohe $C^{14}O_2$-Ausbeuten aus allen ungeradzahlig markierten Palmitinsäuren, sehr geringe aus geradzahlig (2)-markierter Palmitinsäure. Erdnußmitochondrien verhielten sich ähnlich wie tierische; doch ist auch die $C^{14}O_2$-Ausbeute aus 2-C^{14}-Palmitat relativ hoch.

Von Interesse ist eine unlängst festgestellte Aktivierung der vom Mikrosomenenzym bewirkten 1-C^{14}-Palmitatoxydation durch aliphatische *α-Oxymonocarbonsäuren*, vor allem Glykolsäure und L-Milchsäure (CASTELFRANCO, STUMPF und CONTOPOULOU 1955) (Tabelle 10).

Abb. 25 zeigt die Überlegenheit der *Glykolsäure* gegenüber *Milchsäure* noch einmal besonders.

Unwirksam waren in der höchsten oben verwendeten Konzentration unter anderem Glyoxylsäure, Brenztraubensäure, α-Ketobuttersäure, β-Oxybuttersäure, Acrylsäure, Crotonsäure, Lävulinsäure, Tartronsäure, L-Äpfelsäure, Citronensäure, Fumarsäure, Glykokoll, Alanin und Serin.

Der Nachweis einer Oxydation C^{14}-markierter Glykol- oder L-Milchsäure allein oder als Zusatz zu Palmitinsäure konnte papierchromatographisch oder durch $C^{14}O_2$-Bestimmung nicht erbracht werden. Eine (in grünen Pflanzen sehr verbreitete) α-Oxysäurenoxydase (CLAGETT, TOLBERT und BURRIS 1949, NOLL jr. und BURRIS 1954; FRANKE, SCHULZ und DE BOER 1956) scheint also im untersuchten Pflanzenmaterial zu fehlen, was die Aktivatorwirkung einstweilen vollends rätselhaft macht.

Auch der *Chemismus* der Palmitatoxydation durch pflanzliches Mikrosomenenzym ist noch weitgehend ungeklärt. Die bisherigen Befunde (vgl. vor allem Tabelle 9) sprechen weit eher für eine α- als für eine β-Oxydation, wobei nur die α-*Ketopalmitinsäure* als Zwischenprodukt einigermaßen wahrscheinlich gemacht erscheint (CASTELFRANCO, STUMPF und CONTOPOULOU 1955).

Hydroxylamin, Arsenit und *Imidazol* hemmen die oxydative CO_2-Bildung aus carboxylmarkierter Palmitinsäure und zwar beträgt die Hemmung in Prozent von

	NH_2OH	H_3AsO_3	Imidazol
bei m/3300	31	69	96
bei m/600	81	85	100.

Es gelang jedoch nicht, in partiell gehemmten Ansätzen Anzeichen für das Auftreten von Zwischenprodukten zu erhalten.

Auch *Zusätze von Oxy- und Ketopalmitinsäuren* zu 1-C^{14}-Palmitinsäure hemmten die CO_2-Bildung; bei substratäquivalenter Konzentration (m/30000) betrugen die Hemmungen mit

α-Oxypalmitinsäure	α-Ketopalmitinsäure	β-Oxypalmitinsäure	β-Ketopalmitinsäure
90%	100%	50%	35%.

Die Hemmung durch die α-Derivate ist viel stärker als diejenige durch die β-Derivate; die Autoren ziehen daraus den Schluß, daß hier *kompetitive* Hemmung eines wesentlichen Reaktionsschritts der Abbaufolge durch die α-substituierten Palmitinsäuren vorliege.

Als Substrat eingesetzt lieferte 1-C^{14}-α-*Oxy*stearinsäure ohne Glykolsäurezusatz (wie auch 1-C^{14}-Palmitinsäure unter den gleichen Bedingungen) nur Spuren von $C^{14}O_2$, 1-C^{14}-α-*Keto*stearinsäure dagegen sehr viel mehr, wie Tabelle 11 zeigt. Glykolsäurezusatz aktiviert die CO_2-Bildung aus der α-Oxysäure gar nicht, aus der α-Ketosäure nur wenig. (Andererseits wird die Decarboxylierung der Ketosäure durch Arsenit und Imidazol auch nicht gehemmt.) Nicht ohne weiteres verständlich und von den Autoren nur durch Zusatzannahmen (unterschiedliches Verhalten von *freiem* und *enzymgebundenem* Substrat) deutbar ist die Reaktionsträgheit der als Durchgangsstufe einer α-Oxydation doch anzunehmenden α-Oxystearinsäure auch im Glykolatansatz.

Tabelle 11. *Vergleich der Oxydation carboxylmarkierter Präparate von Palmitin-, α-Oxypalmitin- und α-Ketopalmitinsäure durch Mikrosomenenzym ohne und mit Glykolsäurezusatz.* Konzentrationsangaben in m Mol.

Palmitinsäure	α-Oxypalmitinsäure	α-Ketopalmitinsäure	Glykolsäure	% $C^{14}O_2$
0,033	—	—	—	1
—	0,033	—	—	1
—	—	0,033	—	31
0,033	—	—	6,6	17
—	0,033	—	6,6	1
—	—	0,033	6,6	46

In diesem Zusammenhang verdienen neuere *Modellversuche* von GEYER, MARSHALL, RYAN und WESTHAVER (1955) Beachtung, weil sie dartun, daß die in vitro als so beständig geltenden höheren gesättigten Fettsäuren schon unter recht milden Bedingungen oxydativ decarboyxliert werden können.

Durch einstündige Inkubation von 1-C^{14}-Palmitinsäure oder -Stearinsäure mit einem 85—670fachen Überschuß an *Ascorbinsäure* bei p_H 7,1 und 38° wurden unter reinem Sauerstoff erhebliche $C^{14}O_2$-Mengen erhalten, maximal 21—23,5% der Theorie. Aus Laurin- und Caprylsäure entstehen unter den gleichen Bedingungen nur Spuren von $C^{14}O_2$. In Luft wird etwa 4mal weniger $C^{14}O_2$ aus höheren Fettsäuren gebildet als in Sauerstoff, unter Stickstoff überhaupt keines. Dehydroascorbinsäure ist wirkungslos. Vermutlich handelt es sich hier um eine Sekundäroxydation der Fettsäure durch bei der Autoxydation der Ascorbinsäure entstehende Peroxydradikale (HOLTZ und TRIEM 1937, HAND und GREISEN 1942, PETERSON und WALTON 1943). Wenngleich durch diese Befunde eine aktive Teilnahme von Ascorbinsäure am pflanzlichen Fettsäureabbau keineswegs erwiesen ist, glauben die Autoren doch darauf hinweisen zu müssen, daß gerade in *keimenden* Samen hohe Ascorbinsäuregehalte festgestellt worden sind (RAY 1934, LAWRENCE 1950).

2. Weitere Hinweise auf α-Oxydationen.

a) Höhere Pflanzen.

Wie später (S. 179) noch näher ausgeführt werden wird, liefern ω-aryloxysubstituierte mittlere Fettsäuren bei der Einwirkung von Pflanzengeweben entweder ein Phenol oder eine Phenoxyessigsäure, je nachdem ob die C-Zahl der Fettsäurekette ungerade oder gerade ist (FAWCETT, INGRAM und WAIN 1954,

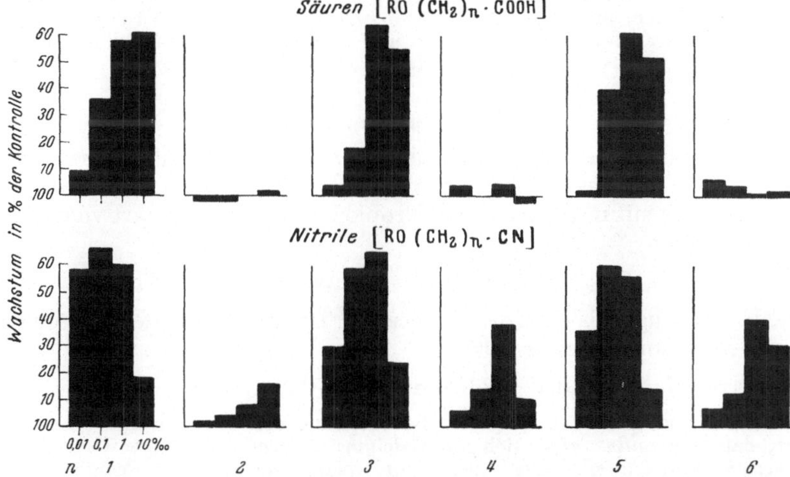

Abb. 26. Wachstumsregulierende Wirkung der ω-(2,4-Dichlorphenoxy-)fettsäuren und der entsprechenden Nitrile im Weizen-Zylindertest. (Nach FAWCETT, SEELEY, TAYLOR, WAIN und WIGHTMAN 1955.)

WAIN und WIGHTMAN 1954, 1956), in Analogie zu den klassischen Versuchen KNOOPS (1904) über den Abbau phenylsubstituierter Fettsäuren im Tierkörper (S. 175). Da den *Aryloxyessigsäuren* eine ausgesprochen wachstumsregulierende Wirkung zukommt, zeigte sich bei der Applikation höherer Homologer eine mit der C-Zahl alternierende Wirkung, wie SYNERHOLM und ZIMMERMAN (1947) schon bei der Tomate auf Grund der *epinastischen* Reaktionen (zur Methodik vgl. z.B. SÖDING 1952) nachgewiesen und auf eine *β-Oxydation* der höheren Homologen zurückgeführt hatten. Abweichende Resultate wurden nun unlängst erhalten, als mit den ω-(2,4-Dichlorphenoxy)fettsäure*nitrilen* gearbeitet wurde (FAWCETT, SEELEY, TAYLOR, WAIN und WIGHTMAN 1955). Abb. 26 zeigt eine Gegenüberstellung des Verhaltens der Fettsäuren und der entsprechenden Nitrile im Weizenzylindertest (vgl. SÖDING 1952), während Tabelle 12 qualitative Angaben über die Resultate im Weizen-Zylinder- und im Erbsen-Krümmungstest bringt.

Nach Tabelle 12 scheint Erbsengewebe die Umwandlung —CN → —COOH bei höheren Nitrilen ($n \geq 2$) nicht durchführen zu können. Dagegen kann Weizengewebe offenbar sowohl diese Hydrolyse als auch eine Verkürzung der C-Kette der Nitrile um ein C-Atom im Sinne einer α-Oxydation wie auch β-Oxydation der Säuren durchführen.

Tabelle 12. *Wachstumsregulierende Wirkung von ω-(2,4-Dichlorphenoxy-)fettsäuren und der entsprechenden Nitrile* (Cl⟨ ⟩—O—$(CH_2)_n$—COOH *bzw.* Cl⟨ ⟩—O—$(CH_2)_n$—CN, Cl) *in zwei verschiedenen Testen.*

Zahl der CH_2-Gruppen n	Weizen-Zylindertest		Erbsen-Krümmungstest		Zahl der CH_2-Gruppen n	Weizen-Zylindertest		Erbsen-Krümmungstest	
	Säuren	Nitrile	Säuren	Nitrile		Säuren	Nitrile	Säuren	Nitrile
1	aktiv$^+$	aktiv$^+$	aktiv$^+$	aktiv$^+$	4	inaktiv	aktiv$^+$	inaktiv	inaktiv
2	inaktiv	aktiv$^+$	inaktiv	inaktiv	5	aktiv$^+$	aktiv$^+$	aktiv$^+$	inaktiv
3	aktiv$^+$	aktiv$^+$	aktiv$^+$	inaktiv	6	inaktiv	aktiv$^+$	inaktiv	inaktiv

$^+$ bedeutet, daß nach Inkubation der Substrate mit Weizen- oder Erbsengewebe *2,4-Dichlorphenoxyessigsäure* als Reaktionsprodukt durch Papierchromatographie oder im biologischen Test nachgewiesen werden konnte.

Hierher gehört die weitere Beobachtung (Fawcett, Seeley, Taylor, Wain und Wightman 1955), daß Weizen, nicht aber die Erbse *3-Indolyl-acetonitril* zu 3-Indolyl-essigsäure zu hydrolysieren vermag, daß aber Gewebe beider Pflanzen aus 3-Indolyl-acetonitril nach dem Schema —CH_2CN → —COOH *3-Indolyl-carbonsäure bilden kann.*

b) Mikroorganismen.

Der Methanbildner *Methanobacterium propionicum* vollführt eine Vergärung von *Propionsäure*, die nach Versuchen einerseits mit carboxylmarkierter Propionsäure, andererseits mit nichtmarkierter Propionsäure in Gegenwart von $C^{14}O_2$ nach der Primärreaktion

$$CH_3 \cdot CH_2 \cdot COOH + 2\,H_2O = CH_3 \cdot COOH + CO_2 + 6\,H$$

abläuft; da CO_2 zugleich einziger H-Acceptor ist, resultiert die Bruttogleichung (T. C. Stadtman und Barker 1951):

$$4\,CH_3 \cdot CH_2 \cdot COOH + 2\,H_2O = 4\,CH_3 \cdot COOH + CO_2 + 3\,CH_4.$$

Mit Vorbehalt sei hier noch die ältere Beobachtung von Walker und Coppock (1928) angeführt, daß *Aspergillus niger* sich auf Calcium*propionat* als einziger C-Quelle zu entwickeln vermag und daß sich dabei zuerst (und relativ reichlicher) *Milchsäure* und später *Brenztraubensäure* in der Nährlösung nachweisen läßt. Daß die beiden letztgenannten Stoffe tatsächlich aus der unveränderten C_3-Kette der Propionsäure entstanden sind, ließe sich einwandfrei wohl nur durch Versuche mit isotopenmarkierter Propionsäure nachweisen[1].

V. Die β-Oxydation.

Schon zu Beginn dieses Jahrhunderts hat Knoop (1904) in einer vielzitierten, richtungweisenden Untersuchung die β-Oxydation als das grundlegende Prinzip des *tierischen* Fettsäureabbaus erkannt. Trotzdem hat es noch fast ein halbes Jahrhundert gedauert, bis die an der Abbaufolge beteiligten Enzyme aus der Zelle abgetrennt, charakterisiert und isoliert werden konnten. Über das Vorkommen einer β-Oxydation in der *höheren Pflanze* ist für die eigentlichen (höheren) Fettsäuren kürzlich das erste Mal in einer enzymchemischen Arbeit berichtet worden (Stumpf und Barber 1956); bis dahin belegten ihre Existenz nur einige neuere, bei Wuchs- bzw. Hemmstoffuntersuchungen gemachte Beobachtungen an ω-aryloxysubstituierten mittleren Fettsäuren. Darüber hinaus existierte eine nicht unbeträchtliche und sich immer noch vergrößernde Anzahl sporadischer Angaben über das Vorkommen einer β-Oxydation in *pflanzlichen Mikroorganismen*, wobei allerdings *enzymatische* Befunde erst in einem Teil der untersuchten Fälle vorliegen.

[1] Siehe Nachtrag S. 200.

Im folgenden soll zunächst ein kurzer Abriß unserer heutigen Kenntnisse von Chemismus und Mechanismus der β-Oxydation, wie sie ganz überwiegend durch Untersuchungen an *tierischem* Versuchsmaterial gewonnen worden sind, gegeben werden (vgl. Übersichtsreferate von KENNEDY und LEHNINGER 1952, LYNEN und OCHOA 1953, LYNEN 1953 [2], 1955, GREEN 1954, DECKER 1956); ein solcher Überblick erscheint schon als Vergleichsbasis für die in naher Zukunft zu erwartende *eingehendere* Bearbeitung der β-Oxydation in der höheren Pflanze gerechtfertigt. Anschließend sollen die einstweilen noch leicht übersehbaren, an *pflanzlichem* Material (höhere Pflanzen, Schimmelpilze, Bakterien) erhobenen Befunde gebracht werden. Da die β-Oxydation mit einer *Dehydrierung* des Fettsäuremoleküls in α,β-Stellung beginnt, ist der Anschluß an das Kapitel „Fettsäuredehydrasen" (S. 163) gegeben.

1. Chemismus und Mechanismus der β-Oxydation im tierischen Organismus.

KNOOP (1904) hatte durch Fütterungsversuche an Hunden gezeigt, daß niedere und mittlere Fettsäuren, deren vollständige Verbrennung durch Phenylsubstitution der endständigen Methylgruppe blockiert worden war, im Harn entweder als *Benzoesäure* oder als *Phenylessigsäure* (jeweils nach Kopplung an Glykokoll) ausgeschieden wurden, je nachdem, ob die aliphatische Kette der verfütterten Fettsäure eine ungerade (a) oder gerade Zahl von Kohlenstoffatomen (b) enthalten hatte:

$$\text{a)}\quad \left.\begin{array}{l} C_6H_5\text{—}CH_2 \vdots CH_2\text{—}CH_2 \vdots CH_2\text{—}COOH \\ C_6H_5\text{—}CH_2 \vdots CH_2\text{—}COOH \end{array}\right\} \rightarrow C_6H_5\text{—}COOH;$$

$$\text{b)}\quad C_6H_5\text{—}CH_2\text{—}CH_2 \vdots CH_2\text{—}COOH \rightarrow C_6H_5\text{—}CH_2\text{—}COOH$$

KNOOP schloß hieraus, daß die Fettsäuren durch β-Oxydation, d.h. durch eine Verkürzung der C-Kette um jeweils zwei C-Atome — höchstwahrscheinlich auf der Stufe der β-Ketosäure — abgebaut würden. In die gleiche Richtung wiesen Versuche von EMBDEN und Mitarbeitern, die zeigten, daß bei künstlicher Durchblutung überlebender Leber aus aliphatischen, unsubstituierten Fettsäuren mit paariger C-Zahl (C_4—C_{12}), nicht dagegen aus solchen mit unpaariger (C_5—C_{11}), reichliche Mengen *Acetessigsäure* bzw. *Aceton* entstanden (EMBDEN, SALOMON und SCHMIDT 1906, EMBDEN und MARX 1908). Über den *Chemismus* der β-Oxydation hat wohl zuerst DAKIN (1909) bestimmtere Vorstellungen geäußert, der nach Verabreichung von *Phenylpropionsäure* folgende Stoffe (zum Teil mit Glykokoll gepaart) im Harn von Hunden und Katzen nachwies: Zimtsäure (Phenylacrylsäure), Phenyl-β-oxypropionsäure, Benzoylessigsäure, Benzoesäure und Acetophenon. Die Annahme einer Abbaufolge der nachstehenden Form

$$C_6H_5\text{—}CH_2\text{—}CH_2\text{—}COOH \xrightarrow{-2\,H} C_6H_5\text{—}CH{=}CH\text{—}COOH \xrightarrow{+H_2O}$$

$$C_6H_5\text{-}CHOH\text{-}CH_2\text{-}COOH \xrightarrow{-2\,H} C_6H_5\text{-}CO\text{-}CH_2\text{-}COOH \begin{cases} \xrightarrow{-CO_2} C_6H_5\text{-}CO\text{-}CH_3 \\ \qquad\qquad\qquad \downarrow \\ \xrightarrow[+H_2O]{-CH_3COOH} C_6H_5\text{-}COOH \end{cases}$$

lag hier nahe (wenngleich DAKIN dazu neigte, die Oxysäure eher als Primär- und die ungesättigte Säure als Nebenprodukt anzusehen) und fand später durch die

Aufklärung des *Bernsteinsäure*-Abbaus in vitro im Sinne der zuerst von WIELAND (1922) klar erkannten Reaktionsfolge

$$\text{Bernsteinsäure} \xrightarrow{-2\,H} \text{Fumarsäure} \xrightarrow{+H_2O} \text{Äpfelsäure} \xrightarrow{-2\,H} \text{Oxalessigsäure} \xrightarrow{-CO_2} \text{Brenztraubensä}$$

(HAHN und HAARMANN 1928 [1, 2], 1930, HAHN, HAARMANN und FISCHBACH 1929) eine gewisse, wenn auch nur analogiemäßige *enzymatische* Fundierung.

In diesem Stadium trat für Jahrzehnte eine Stagnation unseres Wissensstands vom tierischen Fettsäureabbau ein. Im besonderen gelang es nicht, zu weiteren Aussagen über die Natur des *C_2-Bruchstücks* zu gelangen. Dies lag in erster Linie an dem Unvermögen zerkleinerter Gewebe, Fettsäuren zu oxydieren, das schon von EMBDEN für den Fall der Leber beobachtet worden war (l. c., S. 175) und das die bei der Aufklärung anderer Stoffwechselvorgänge, z. B. des Kohlenhydratabbaus, so erfolgreiche *enzymatische* Methodik hier zunächst ausschloß.

Erst als man um 1943 lernte, den Wirksamkeitsverlust beim Zerkleinern und Extrahieren von Zellen und Geweben zu vermeiden und gegen Fettsäuren aktive Gewebebreie, -homogenate und -extrakte herzustellen, wurden neue grundsätzliche Fortschritte in der Erkenntnis des Fettsäurestoffwechsels erzielt, die sich vor allem an die Namen von MUÑOZ und LELOIR (1943), LELOIR und MUÑOZ (1944), LEHNINGER (1944 [1, 2], 1945 [1, 2], 1946), BREUSCH (1943, 1944), WIELAND und ROSENTHAL (1943) knüpfen. Während LEHNINGER, auf Voruntersuchungen von MUÑOZ und LELOIR fußend, hauptsächlich mit *Lebermitochondrien* arbeitete, gelang STADTMAN und BARKER (1949 [1, 2, 3, 4, 5], 1950) erstmals die Oxydation von Capron- und Buttersäure zu Essigsäure mit Enzym*lösungen* aus anaeroben *Bakterien*. BREUSCH sowie WIELAND und ROSENTHAL beobachteten unabhängig voneinander die Bildung von *Citronensäure* aus Acetessigsäure bzw. höheren β-Ketosäuren und Oxalessigsäure. Im weiteren Verlauf dieser Untersuchungen kam man zu der Erkenntnis, daß CO_2 bei der β-Oxydation der Fettsäuren ausschließlich im Citronensäurecyclus (vgl. MILLERD, Bd. XII dieses Handbuchs) entsteht und daß das C_2-Bruchstück, das bei der β-Oxydation der *Fettsäuren* abgespalten wird, identisch ist mit dem einstweilen als „*aktivierte Essigsäure*" (LYNEN 1942, 1943) bezeichneten C_2-Körper, wie ihn die über Brenztraubensäure führende biologische *Zucker*oxydation liefert (vgl. Zusammenfassungen von MARTIUS und LYNEN 1950, KENNEDY und LEHNINGER 1952). Es hängt bei der Fettsäureoxydation von der Gegenwart oder Abwesenheit genügender Mengen von Oxalessigsäure ab, ob die „aktivierte Essigsäure" sich mit Oxalessigsäure zu *Citronensäure* oder mit einem zweiten Molekül Essigsäure zu *Acetessigsäure* kondensiert (LEHNINGER 1945 [1], 1946):

$$CH_3\cdot(CH_2)_n\cdot COOH \xrightarrow{\beta\text{-Oxydation}} CH_3\cdot COOH \text{ „aktiviert"}$$

$$\swarrow \qquad \searrow$$

$$CH_3\cdot CO\cdot CH_2\cdot COOH \qquad HOOC\cdot CH_2\cdot C(OH)(COOH)\cdot CH_2\cdot COOH$$

$$\downarrow \text{Citronensäure-Cyclus}$$

$$CO_2 + H_2O$$

Die weiteren Untersuchungen galten vor allem der Frage nach der *chemischen Natur* der „aktivierten Essigsäure" und ihrer Bildungsweise bei der β-Oxydation. Wesentlich war hier der an Leberhomogenaten und -mitochondrien erhobene Befund, daß für die Fettsäureoxydation ein unter Beteiligung von *Adenosintriphosphat* (ATP) verlaufender Aktivierungsprozeß notwendig ist (LEHNINGER

1944 [2], 1945 [2], 1946 [1], GRAFFLIN und GREEN 1948). Die Natur dieses Prozesses blieb aber zunächst unbekannt und man konnte nur eine nahe Beziehung zwischen „aktivierten Fettsäuren" und „aktivierter Essigsäure" vermuten. Erst die Entdeckung des Coenzyms A (LIPMANN 1945, NACHMANSOHN und BERMAN 1946, FELDBERG und MANN 1946, LIPTON und BARRON 1946) führte hier weiter, insbesondere die von LIPMANN (1948) entwickelte Vorstellung, daß „aktivierte" Essigsäure eine an Coenzym A gebundene Essigsäure sei. Diese durch Untersuchungen von STADTMAN (1950) und OCHOA (vgl. Zusammenfassung 1951) bestätigte Hypothese wurde schließlich durch LYNEN, REICHERT und RUEFF (1951) bewiesen, die für eine aus Hefe isolierte Substanz mit den Eigenschaften der „aktivierten Essigsäure" zeigen konnten, daß es sich um den Thioester aus Essigsäure und Coenzym A von nachstehender Konstitution handelte:

```
                                   NH—CH2—CH2—S—CO·CH3
                                   |
N=C·NH2                            CO
|   |                              |
CH  C—N                            CH2     „Aktivierte Essigsäure" =
||  ||   \CH                       |       Acetyl-Coenzym A (abgekürzt:
N—C—N/                             CH2     Acetyl-S-CoA oder Acetyl-CoA)
     |                             |
    HC———————                      NH
     |      |                      |
    HCOH    |                      CO
     |      O                      |
    HCO·PO3H2                     HCOH
     |      |                      |
    HC———————             CH3—C—CH3
     |          O      O           |
     |          ||     ||          |
    H2C—O—P—O—P—O—CH2
                |      |
                OH     OH
```

Hiermit war nicht nur das Problem der „aktivierten Essigsäure" gelöst, sondern auch eine allgemeinere Deutung der „aktivierten Fettsäuren" als *Acyl-Coenzym* A-Verbindungen nahegelegt (LYNEN 1953 [1, 2], 1954, LYNEN und OCHOA 1953).

Die *Thioester*- oder *Acylmercaptan*bindung stellt einen neuen Typ energiereicher Verbindungen dar (STERN, OCHOA und LYNEN 1952), dessen hydrolytische Spaltung mit 8300 cal/Mol ungefähr die gleiche Energie wie die Abspaltung eines Phosphatrests aus *Adenosintriphosphat* liefert (BURTON 1955). Wie weiter unten noch gezeigt werden wird, besitzt die Zelle auch die notwendigen Mechanismen, um die Energieübertragung von ATP auf die Fettsäure (unter Acyl-CoA-Bildung) reversibel durchführen zu können.

Unsere heutigen Kenntnisse vom Fettsäureabbau nach dem Prinzip der β-Oxydation wie auch vom umgekehrten Prozeß, der Fettsäuresynthese, lassen sich im nachstehenden Schema des sog. „*Fettsäurecyclus*" zur Anschauung bringen (LYNEN 1955, DECKER 1956). Bei seinem Ablauf wird im Endeffekt entweder aus einer gesättigten Fettsäure ein C_2-Rest (in Form von Acetyl-CoA) abgespalten oder umgekehrt eine vorhandene Fettsäurekette um C_2 verlängert; im Hinblick auf die unterschiedliche Länge der Fettsäurekette bei Beginn und am Ende des Cyclus wäre es wohl richtiger, von einer „*Fettsäurespirale*" zu sprechen. Der Abbau der Stearinsäure (C_{18}) beispielsweise würde 8 solcher „Umläufe" erfordern, die 9 Moleküle Acetyl-CoA liefern würden, die dann im Citronensäurecyclus vollständig zu $CO_2 + H_2O$ verbrannt werden können.

Die im Schema und nachstehend gebrauchten Fermentbezeichnungen entsprechen der neuen „rationellen" Nomenklatur, während in Klammern ältere oder Trivialnamen angegeben werden (vgl. LYNEN 1955, BEINERT, GREEN, HELE, HOFFMANN-OSTENHOF, LYNEN, OCHOA, POPJÁK und RUYSSEN 1956).

Schema 4.

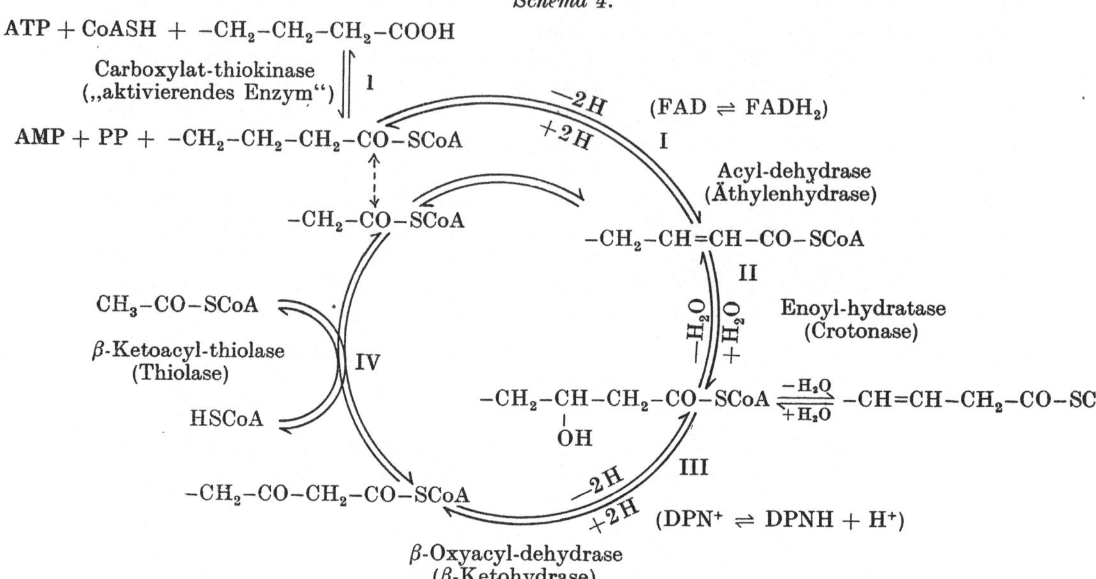

Fettsäurecyclus (nach LYNEN).

Zusammenfassend ergibt sich also folgendes Bild vom *enzymatischen Mechanismus der β-Oxydation:*

a) In der *Zubringerreaktion 1* wird die Energie für die Überführung der Fettsäure in den Thioester von Coenzym A, die eigentliche Reaktionsform der Fettsäure, aus der Spaltung von Adenosintriphosphat (ATP) bereit gestellt. Eigenartig ist die beim Studium der Acetataktivierung gemachte Beobachtung, daß Adenosin-triphosphat (ATP) hier nicht in der üblichen Weise zu Adenosin-diphosphat (ADP) und Orthophosphat, sondern zu Adenosinmonophosphat (AMP) und Pyrophosphat (PP) gespalten wird (JONES, BLACK, FLYNN und LIPMANN 1953, LIPMANN, JONES, BLACK und FLYNN 1953). Es gibt mehrere solche *Thiokinasen* („carboxylat-aktivierende Enzyme"): außer dem schon erwähnten auf Acetat eingestellten Ferment, das auch noch Propionat aktiviert, existiert ein weiteres, das Fettsäuren mit 4—11 C-Atomen umsetzt — auch ungesättigte und β-Oxysäuren reagieren — (DRYSDALE und LARDY 1953, LEHNINGER und GREVILLE 1953, MAHLER, WAKIL und BOCK 1953) und schließlich ein drittes, das höhere Fettsäuren bis C_{22} aktiviert (KORNBERG und PRICER jr. 1953)[1].

b) Im *Cyclusschritt I* wird die Acyl-Coenzym A-Verbindung unter der Einwirkung einer *Acyl-dehydrase* (Äthylenhydrase) zur entsprechenden α,β-ungesättigten Verbindung dehydriert, wobei Flavin-adenin-dinucleotid (FAD) als primärer Wasserstoffacceptor fungiert. Ein von MAHLER (1953) aus Leber isoliertes Ferment, das bevorzugt Butyryl-CoA angreift, enthält außer FAD auch noch *Kupfer* (FAD:Cu = 1:2), während zwei weitere Cu-freie Fermente bevorzugt längerkettige Acyl-CoA-Derivate umsetzen (CRANE, HAUGE und BEINERT 1955); ein relativ geringer *Eisen*gehalt (FAD:Fe $\geq$ 5:1) macht ihre Einreihung unter die „Metall-Flavoproteine" einstweilen problematisch (vgl. hierzu MAHLER und GREEN 1954; MAHLER 1956).

c) In *Cyclusphase II* wird die ungesättigte Acyl-CoA-Verbindung durch eine *Enoyl-hydratase* („Crotonase") zum β-Oxyacyl-Derivat hydratisiert. Das neuerdings rein dargestellte Ferment (STERN 1955) vermag sowohl CoA-Derivate kleiner als auch großer Kettenlänge umzusetzen. Überraschend war, daß das Reinferment nicht nur das CoA-Derivat der Crotonsäure, sondern auch der isomeren *Vinylessigsäure* hydratisiert (STERN 1955, WAKIL und MAHLER 1954). LYNEN (1955) spekuliert über die Bedeutung dieser auf eine *Isomerisierung* der Doppelbindung hinauslaufenden Erscheinung für Abbau und Biosynthese der *ungesättigten* Fettsäuren.

d) *Reaktionsschritt III* bewirkt die Dehydrierung der β-Oxyverbindung zum entsprechenden β-Ketoderivat des Coenzyms A, wobei Diphosphopyridin-nucleotid (DPN) als

[1] Siehe Nachtrag S. 200.

primärer Wasserstoffacceptor dieser *β-Oxyacyl-dehydrase* (β-Ketohydrase) wirkt. Die — hinsichtlich der C-Zahl des Substrats wieder wenig spezifische — Wirkung des isolierten Enzyms begünstigt im Gleichgewicht allerdings stark die β-Oxysäure (LYNEN, WESSELY, WIELAND und RUEFF 1952, LYNEN und WIELAND 1955).

e) In der letzten *Reaktionsphase IV* findet unter dem Einfluß des Ferments *β-Ketoacyl-thiolase* [(β-Keto-)Thiolase] eine „thioklastische" Spaltung der β-Ketoacyl-CoA-Verbindung statt, wobei einerseits Acetyl-CoA entsteht, andererseits ein um zwei C-Atome ärmerer Acylrest auf ein neu eintretendes Coenzym A-Molekül übertragen wird, womit die Voraussetzung für erneuten Ablauf des Cyclus gegeben ist. Das Gleichgewicht der Thiolasereaktion liegt ganz nach der Seite der Spaltung (STERN, COON und DEL CAMPILLO 1953 [1, 2]). Während nach Untersuchungen an rohen und teilweise gereinigten Herz- und Leberpräparaten der β-Ketoacyl-thiolase eine breite Spezifität zuzukommen schien (STERN, COON und DEL CAMPILLO 1953 [2], BEINERT, BOCK, GOLDMAN, GREEN, MAHLER, MII, STANSLY und WAKIL 1953), sprechen neuere Befunde von OCHOA (1954) am hochgereinigten Herzferment für das Vorkommen *mehrerer* Thiolasen unterschiedlicher Spezifität, von denen eine ziemlich spezifisch auf Acetyl-CoA eingestellt ist.

2. β-Oxydation in pflanzlichen Organismen.

a) Höhere Pflanzen.

Die ersten hierher gehörigen Beobachtungen wurden bei Wuchs- bzw. Hemmstoffuntersuchungen gemacht (vgl. Übersicht bei WAIN und WIGHTMAN 1956).

So fanden SYNERHOLM und ZIMMERMAN (1947) bei der Prüfung von 7 verschiedenen *ω-(2,4-Dichlorphenoxy-)fettsäuren* ($Cl—C_6H_3(Cl)—O—(CH_2)_n—COOH$), daß nur diejenigen mit ungerader Zahl von CH_2-Gruppen bei der Tomate *epinastische* Effekte (vgl. z. B. SÖDING 1952) hervorriefen. Ähnliches galt für die *ω-Naphthoxy-fettsäuren*: Essigsäure- und Buttersäure-Derivate waren wirksam, die entsprechende Propionsäure unwirksam. Die Erklärung lag nahe, daß — in Analogie zu den KNOOPschen Versuchen (S. 175) am Tier — im einen Fall durch *β-Oxydation* stets die eigentlich aktive *substituierte Essigsäure*, im anderen Falle das inaktive *Phenol* bzw. *Naphthol* entstünde.

In Abb. 26 und Tabelle 12 (S. 173f.) waren einige neuere qualitative und quantitative Ergebnisse über das Verhalten einer Reihe von ω-(2,4-Dichlorphenoxy-)fettsäuren im Weizen-Zylinder- und Erbsen-Krümmungstest (vgl. SÖDING 1952) mitgeteilt worden, die gleichfalls am einfachsten durch eine β-Oxydation — bei positiver Reaktion zum Wirkstoff *2,4-Dichlorphenoxy-essigsäure* — zu erklären waren (FAWCETT, SEELEY, TAYLOR, WAIN und WIGHTMAN 1955). Einen direkteren Beweis für die β-Oxydation erbrachten FAWCETT, INGRAM und WAIN (1954),

Tabelle 13. *Phenolgehalt* (in μg) *von Flachskeimlingen nach der Einwirkung auf Phenoxy-fettsäuren* ($C_6H_5—O—(CH_2)_n—COOH$).

Konzentration der Phenoxysäure	Zahl *n* der CH_2-Gruppen der Seitenkette									
	1	2	3	4	5	6	7	8	9	10
$1,25 \times 10^{-3}$ m	16,5	*90*	2,5	*86*	11,5	*108*	3	*57*	*20,5*	—
$2,5 \times 10^{-3}$ m[1]	7	*160*	8,8	*179*	16,2	*162,5*	4	*128*	*57,5*	*164,5*

[1] Mittel von 4 Versuchen.

die die Wurzeln von jeweils 200 Flachskeimlingen 10 Tage in *ω-phenoxyfettsäure*haltiger Lösung hielten und anschließend im Wasserdampfdestillat den *Phenol*gehalt der macerierten Keimlinge bestimmten. Das Versuchsergebnis zeigt Tabelle 13.

Nur bei gerader Zahl von CH_2-Gruppen im Substrat werden *erhebliche* Mengen *Phenol* gebildet, während bei ungerader CH_2-Zahl die entsprechende *Phenoxy-essigsäure* entsteht, wie dies in einer späteren Arbeit (FAWCETT, SEELEY, TAYLOR,

WAIN und WIGHTMAN 1955) noch besonders nachgewiesen wurde (vgl. Tabelle 12, S. 174). Folgende beiden Reaktionsfolgen sind als typisch zu betrachten:

1) C_6H_5—O—$CH_2 \cdot CH_2 \cdot COOH$ (n = 2) → [C_6H_5—O—C(=O)$\cdot CH_2 \cdot COOH$] → [C_6H_5—O—COOH] → C_6H_5—OH

2) C_6H_5—O—$CH_2 \cdot CH_2 \cdot CH_2 \cdot COOH$ (n = 3) → [C_6H_5—O—$CH_2 \cdot$C(=O)$\cdot CH_2 \cdot COOH$] → C_6H_5—O—$CH_2 \cdot COOH$

Auffallend ist in Tabelle 13 allerdings die nicht unbeträchtliche Phenolbildung aus *Phenoxycaprinsäure* (n = 9), die die Autoren durch eine — im tierischen Organismus gerade bei *mittleren* Fettsäuren beobachtete (VERKADE, ELZAS, VAN DER LEE, DE WOLF, VERKADE-SANDBERGEN und VAN DER SANDE 1933, VERKADE und VAN DER LEE 1934, BREUSCH und TULUS 1944) — *ω-Oxydation* zu erklären versuchen:

C_6H_5—O—$CH_2 \cdot (CH_2)_8 \cdot COOH$ → [C_6H_5—O—C(=O)$\cdot (CH_2)_8 \cdot COOH$] $\xrightarrow{H_2O}$ C_6H_5—OH

(wobei allerdings nicht übersehen werden sollte, daß bei den bisher beobachteten Fällen von ω-Oxydation stets eine *freie* endständige CH_3-Gruppe im Substrat vorlag.)

In einer weiteren, hier nur kurz zu erwähnenden Arbeit (WAIN und WIGHTMAN 1954) wurde die wachstumsregulierende Wirkung homologer Reihen von *ω-(4-Chlorphenoxy-)*, *ω-(2,4-Dichlorphenoxy-)*, *ω-(2,4,5-Trichlorphenoxy-)* sowie *ω-(1-Naphthyl-)Fettsäuren* vergleichend im Weizen-Zylinder-, Erbsen-Krümmungs- und Tomatenblatt-Epinastietest (ZIMMERMAN und HITCHCOCK 1942) geprüft. In den meisten Fällen, aber nicht immer, zeigte sich das bekannte Alternieren der Wirksamkeit bei steigender Kettenlänge (vgl. S. 173f.), bisweilen war aber nur das Essigsäurederivat wirksam. So zeigten z. B. 4-Chlorphenoxy- und 2,4-Dichlorphenoxy-essigsäure und ihre alternierenden Homologen Wirksamkeit in allen drei Testen, während im Falle der 2,4,5-Trichlorphenoxyreihe dieses typische Alternieren der wachstumsregulierenden Aktivität nur im Weizen-Zylindertest zu beobachten war. Wurden aber die alternierenden Homologen der letzteren Reihe mit Weizenkoleoptilgewebe vorbehandelt, dann induzierten sie Wachstumseffekte auch im Erbsen-Krümmungstest; dabei konnte der Übergang des Buttersäure- in das Essigsäurederivat papierchromatographisch nachgewiesen werden. Daraus folgt, daß die Enzymsysteme der β-Oxydation sich in den verschiedenen Pflanzengeweben hinsichtlich Spezifität und Hemmbarkeit unterscheiden.

Ein weiteres Ergebnis der Untersuchung war, daß *Alkylierung* von 1-Naphthyl-fettsäuren in β-Stellung der Seitenkette β-Oxydation und damit Wachstumswirkung verhindert.

Zur neuerdings auch bei *Asp. niger* aufgefundenen β-Oxydation von *2-Naphthoxyfettsäuren* siehe S. 184f.

Über die *enzymatische* Seite der pflanzlichen β-Oxydation war noch so gut wie nichts bekannt. Immerhin verdiente eine neuere Untersuchung von MILLERD und BONNER (1954) Beachtung, in der das verbreitete Vorkommen zunächst eines „*acetataktivierenden Enzyms*" (*Acetat-thiokinase*) in höheren Pflanzen nachgewiesen wurde, das aus Spinatblättern mehrfach angereichert werden konnte. Die in Gegenwart von Coenzym A und ATP zu Acetyl-Coenzym A verlaufende Reaktion erfolgt offenbar entsprechend der „Zubringerreaktion 1" in Schema 4 (S. 178). Das angereicherte Spinatpräparat vermochte außer Essigsäure auch eine Anzahl anderer niederer Fettsäuren (bis C_6, ausgenommen Propionsäure) zu aktivieren; so ergaben sich nach 60 min bei p_H 8 folgende relative Umsätze (in μMol/ml):

Essigsäure	Propionsäure	Buttersäure	Valeriansäure	Isovaleriansäure	β-Methylcrotonsäure	Capronsäure	Isocapronsäure	2,4-Dichlorphenoxyessigsäure
0,62	0,0	0,52	0,14	0,12	0,085	0,082	0,070	0,075

Wahrscheinlich existieren wie im tierischen Organismus (S. 178) zwei oder mehrere *Thiokinasen* verschiedenen Spezifitätsbereichs; jedoch bestand nach dieser Untersuchung noch kein Anhaltspunkt für das Vorkommen eines auf *höhere* Fettsäuren eingestellten Enzyms in Pflanzen.

Weiterhin konnte in Spinat eine *Acetacetyl-ketothiolase* nachgewiesen werden, welche die Reaktion

$$2 \text{ Acetyl-S-CoA} \rightleftharpoons \text{Acetacetyl-S-CoA} + \text{CoA—SH}$$

katalysiert, die erste synthetische Reaktionsphase des Fettsäure*aufbaus* (NEWCOMB und STUMPF 1952, 1953) wie auch der Isoprenoidsynthese und des Aufbaus der verzweigten Aminosäuren Valin und Leucin (ARREGUIN, BONNER und WOOD 1951). Der wegen der ungünstigen Gleichgewichtslage (vgl. S. 179) schwierige Nachweis der C_4-Körperbildung wird durch Zusatz von *Succinat* zum Reaktionssystem erleichtert, weil dann eine gleichfalls vorhandene *Coenzym A-Transphorase (Acetacetyl-Succinat-thiophorase)* die das Gleichgewicht störende Reaktion

$$\text{Acetacetyl-S-CoA} + \text{Succinat} \rightarrow \text{Acetacetat} + \text{Succinyl-S-CoA}$$

bewirkt.

Die noch bestehende Lücke hinsichtlich einer β-Oxydation von Fettsäuren in höheren Pflanzen scheint nun durch eine kürzlich erschienene Untersuchung von STUMPF und BARBER (1956) grundsätzlich geschlossen zu werden. Die Autoren berichten — ohne auf die früheren Arbeiten über das offenbar eine α-Oxydation bewirkende Enzymsystem der Erdnuß*mikrosomen* (S. 170f.) Bezug zu nehmen — über ein hochkomplexes Oxydationssystem in Erdnuß*mitochondrien*, das aus vorzentrifugierten Kotyledonenhomogenaten durch Abschleudern bei 10000 g abgetrennt wurde (vgl. STUMPF 1955) und eine *β-Oxydation* niederer und höherer Fettsäuren bewirkte. Der Umsatz der C^{14}-markierten Fettsäuren wurde wieder bei niedriger Substratkonzentration (meist m/22000) durch Bestimmung des oxydativ entstandenen $C^{14}O_2$ ermittelt. Aus Tabelle 9, S. 171 war bereits der Unterschied zwischen diesem und dem früher beschriebenen Mikrosomensystem deutlich geworden. Das letztere erscheint im Vergleich zum Mitochondriensystem „primitiv": zur Erzielung maximaler Aktivität müssen dem Erdnußmitochondrienpräparat ATP, CoA, DPN, TPN, $Mn^{\cdot\cdot}$, SH-Glutathion und eine Säure des Citronensäurecyclus zugefügt werden. Abweichend vom „Palmitatoxydase"-System der Mikrosomen erfolgt die Fettsäureoxydation hier über die *Acyl-Coenzym A-Verbindung* und mündet in den Citronensäurecyclus ein, weshalb *Malonat* stark hemmend wirkt (75% bei m/165). Folgende relative Umsätze (% C^{14} als $C^{14}O_2$) wurden bei den (normalen) 1-C^{14}-markierten Fettsäuren von Essigsäure (C_2) bis Stearinsäure (C_{18}) beobachtet:

C_2	C_3	C_4	C_5	C_8	C_{12}	C_{14}	C_{16}	C_{18}
36	40	49	11	33	35	59	42	22;

dagegen lagen die Zahlenwerte für einige *iso*-Säuren (C_4, C_5, C_6) zwischen 2 und 4%. Für die Abhängigkeit der Oxydationsintensität von 1-C^{14}-Buttersäure vom *Keimlingsalter* ergab sich folgende Reihe:

Tage	1	3	5	7	9	14
% C^{14}	9,8	7,7	19	45	28	1,4

Da Citronensäure durch die Enzyme des KREBS-Cyclus asymmetrich umgesetzt wird, sind wenigstens 3 Abläufe des Cyclus notwendig, ehe ein *gerad*zahliges C-Atom einer Fettsäure im Atmungs-CO_2 erscheint, während dies bei einem *ungeradzahligen* C-Atom schon nach 2 Abläufen der Fall ist (vgl. KATZ und CHAIKOFF 1955). Folgende zeitliche Abhängigkeit der CO_2-Bildung (% C^{14} als $C^{14}O_2$) wurde bei 3- und bei 2-C^{14}-Palmitat beobachtet:

Minuten	30	60	120
3-C^{14}-Palmitat	18	36	53
2-C^{14}-Palmitat	1,1	4,2	15

Diese Zahlen sind ein weiterer Beleg für die β-Oxydation, zugleich aber auch für einen relativ langsamen Durchsatz der Verbindungen durch den Citronensäurecyclus unter den gegebenen Versuchsbedingungen.

Auch die intermediäre Bildung von *Acetessigsäure* aus 1-C^{14}-Buttersäure (nachgewiesen durch aminkatalysierte Decarboxylierung) wurde von den Autoren wahrscheinlich gemacht. Der relative Anteil der Acetessigsäurebildung stieg an, wenn ihre Verwertung im Citronensäurecyclus durch Malonat blockiert wurde.

Trotz der vorläufigen Natur der Befunde von Stumpf und Barber beansprucht der Nachweis *zweier verschiedener* Abbaumechanismen für Fettsäuren im gleichen Zellmaterial erhebliches Interesse; möglicherweise dient der früher (S. 170f.) beschriebene nur dem Abbau, während der zuletzt angeführte der β-Oxydation die Voraussetzungen der Reversibilität in sich trägt.

b) Schimmelpilze.

Der Nachweis einer *Methylketonbildung* aus Fettsäuren durch Schimmelpilze der Gattungen *Penicillium* und *Aspergillus* war der erste und lange Zeit einzige Hinweis für das Vorkommen einer β-Oxydation in pflanzlichen Organismen.

Stärkle (1924) wies erstmalig nach, daß die in ranzigem *Kokosfett* aufgefundenen Methylketone (Haller und Lassieur 1910 [1, 2]) durch die Lebenstätigkeit von Schimmelpilzen aus gesättigten Fettsäuren mittlerer Kettenlänge (C_8—C_{12}) entstehen. Auf die gleichen Methylketone führte er die Ranzidität der *Butter*, für die vorher meist Buttersäure-ester verantwortlich gemacht worden waren, zurück. Auch der charakteristische Geruch gewisser *Käse*sorten (Roquefort, Gorgonzola, Stracchino, Parmesan, Stilton u. ä.), deren Reifungsprozeß durch Einimpfen von *Penicillien* in ganz bestimmte Bahnen gelenkt wird — auch *Cladosporium* spielt eine gewisse Rolle — rührt von flüchtigen Methylketonen, gebildet aus dem Milchfett, her; Methylamylketon und Methylheptylketon konnten aus Roquefortkäse isoliert werden.

Im Anschluß an Stärkles grundlegende Beobachtungen wurde in der Folgezeit der Abbau bestimmter Fette und Fettsäuren durch *Penicillium*arten (meist *Pen. glaucum*), seltener *Aspergillus*arten (z.B. *Asp. niger, Asp. fumigatus*) ziemlich eingehend studiert (Stokoe 1928, Coppock, Subramaniam und Walker 1928, Acklin 1929, Täufel, Thaler und Löweneck 1936). Die aus den einzelnen Substraten erhaltenen Methylketone sind in der folgenden Tabelle 14 zusammengestellt.

Tabelle 14. *Beim Fettsäureabbau durch Schimmelpilze nachgewiesene Methylketone.*

Fettsäure	Entsprechendes Methylketon	Fettsäure	Entsprechendes Methylketon
Buttersäure (C_4)	Aceton	Pelargonsäure (C_9)	Methylhexylketon
Isovaleriansäure (C_5)	Aceton	Caprinsäure (C_{10})	Methylheptylketon
Valeriansäure (C_5)	Methyläthylketon	Undecylsäure (C_{11})	Methyloctylketon
Capronsäure (C_6)	Methylpropylketon	Laurinsäure (C_{12})	Methylnonylketon
Önanthsäure (C_7)	Methylbutylketon	Tridecylsäure (C_{13})	Methyldecylketon
Caprylsäure (C_8)	Methylamylketon	Myristinsäure (C_{14})	Methylundecylketon

Das Optimum der Methylketonbildung liegt, wie schon erwähnt, bei den Fettsäuren zwischen C_8 und C_{12}. Eine Methylketonbildung aus niederen Fettsäuren (C_4, C_5) ist von einzelnen Untersuchern vermißt worden (Stärkle 1924, Stokoe 1928, Acklin 1929), doch zeigten spätere Untersuchungen, daß die Methylketonbildung nicht nur vom verwendeten Stamm, sondern auch von den Versuchsbedingungen (p_H, Reaktionsdauer u. ä.) stark abhängig ist (S. 183/84). *Stets* vermißt wurde Methylketonbildung bei den höheren Fettsäuren (Palmitin-, Stearin-, Ölsäure).

Der *Chemismus* der Methylketonbildung ist schon frühzeitig im Sinne einer abgewandelten β-Oxydation formuliert worden

Schema 5.

$$R{-}CH_2{-}CH_2{-}COOH \xrightarrow{-2\,H} R{-}CH{=}CH{-}COOH \xrightarrow{+\,H_2O}$$

$$R{-}CHOH{-}CH_2{-}COOH \xrightarrow{-2\,H} R{-}CO{-}CH_2{-}COOH$$

$$\downarrow$$

$$R{-}CO{-}CH_3 + CO_2,$$

ohne daß lange Zeit dafür Beweise vorlagen und ohne daß bis heute der *enzymatische* Nachweis dieser Reaktionsfolge erbracht wäre. Der Stand der Dinge ähnelt sehr demjenigen, wie er bis vor kurzem für die β-Oxydation im *tierischen* Organismus kennzeichnend war (vgl. S. 175/76).

STOKOE (1928) vertritt die Auffassung, daß auch in Aspergillaceen der *normale* Abbauweg der Fettsäuren die KNOOPsche β-Oxydation sei. Die Giftwirkung der mittleren Fettsäuren (die im Wachstumsversuch von ihm noch gesondert dargetan wurde, später eingehender von WYSS, LUDWIG und JOINER 1945) störe den normalen Umsatz der β-Ketosäure und führe zu der an sich anormalen Decarboxylierung. Resistentere Pilze wie z. B. *Oospora lactis* zeigen keine Methylketonbildung, sondern bewirken offenbar eine *Säurespaltung* der β-Ketosäuren (JENSEN 1902, STOKOE 1928). Neuere orientierende Versuche (FRANKE und HEINEN 1956, unveröffentlicht) belegen übrigens eine über die Familie des *Aspergillaceen* weit hinausreichende sporadische Verbreitung der Methylketonbildung unter den *Ascomyceten*, *Phycomyceten* und *Fungi imperfecti*.

Ab 1939 haben THALER und Mitarbeiter in einer Reihe von Arbeiten die Bedingungen der Methylketonbildung durch *Pen. glaucum*, insbesondere ihre zeitliche und p_H-Abhängigkeit, näher untersucht, wobei einfache Salz-Nährlösungen mit der Fettsäure als einziger C-Quelle verwendet wurden (THALER und GEIST 1939 [1]). Durch den Einsatz von α,β-*ungesättigten* Fettsäuren (THALER und EISENLOHR 1941) und von β-*Oxyfettsäuren* (THALER und GEIST 1939 [2]) als Substraten wurde deren Rolle als Zwischenstufen des Abbaus der gesättigten Fettsäuren zum mindesten sehr wahrscheinlich gemacht. Neuerdings ist Methylketonbildung aus den Fettsäuren mit 4—12 C-Atomen auch mit *Mycelsuspensionen* (von *Pen. roqueforti*) erhalten worden (GIROLAMI und KNIGHT 1956). Über die *Enzymatik* der Methylketonbildung ist noch sehr wenig bekannt. KARRER und HAAB (1948) wiesen in Trockenpräparaten verschiedener Penicillien ein sehr labiles Ferment nach, das mittlere β-Ketocarbonsäuren (C_6—C_9) zu *decarboxylieren* vermag. Wenig aktive, offenbar komplexe *Dehydrasen* mittlerer und höherer Fettsäuren konnten neuerdings mittels der THUNBERG-Methodik in Mycelextrakten verschiedener Aspergillen und Penicillien festgestellt werden (FRANKE und HEINEN 1956, unveröffentlicht).

Tabelle 15. *Methylketonbildung aus β-Oxymyristinsäure durch Pen. glaucum bei verschiedenem* p_H. (THALER und GEIST 1939 [2].)

p_H der Nährlösung	μg Methylketon nach Tagen								
	1	2	3	4	5	6	7	8	9
3,0	47	59	0	—	—	23	7	9	—
6,0	1	—	—	71	104	65	196	148	190
7,0	0	0	—	6	0	—	25	22	—

Abb. 27 und 28 zeigen nach Arbeiten von THALER und Mitarbeitern den Einfluß des p_H und der Versuchsdauer auf das Ausmaß der Methylketonbildung aus *Myristinsäure* bzw. *2-Tetradecensäure*, Tabelle 15 belegt das gleiche für den Fall der β-*Oxymyristinsäure*. Die Versuche sind typisch für die Methylketonbildung aus den entsprechenden Verbindungsklassen. Sie zeigen die zum Teil sehr ausgeprägten zeitlichen *Maxima* der Methylketonbildung; das nachträgliche Verschwinden der Methylketone dürfte zum Teil auf die Reduktion zu den entsprechenden sekundären Alkoholen zurückgehen, die als stete Begleiter der Methylketone seit langem bekannt sind (HALLER und LASSIEUR 1910, STOKOE 1928). Bemerkenswert ist die p_H-*Abhängigkeit* der Methylketonbildung: während das p_H-Optimum

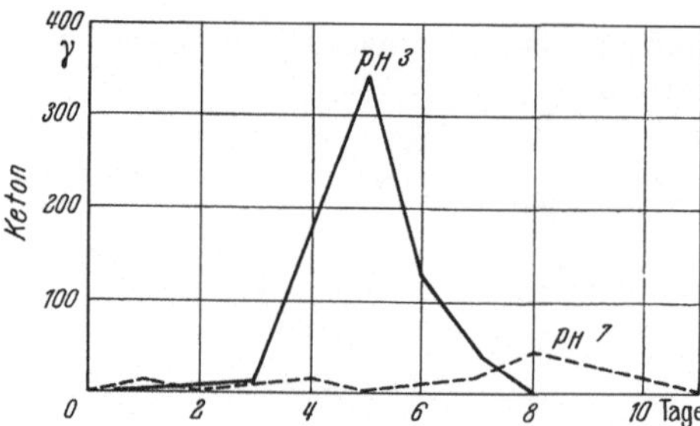

Abb. 27. Methylketonbildung aus Myristinsäure durch *Pen. glaucum* bei verschiedenem p_H (29°). Methylketonbestimmung mit der Salicylaldehydreaktion (TÄUFEL, THALER und HOHNER 1937). (Nach THALER und GEIST 1939 [1].)

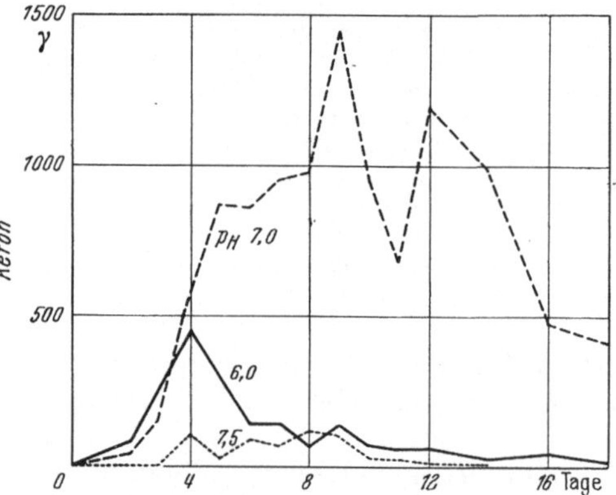

Abb. 28. Methylketonbildung aus 2-Tetradecensäure durch *Pen. glaucum* bei verschiedenem p_H (29°). (Nach THALER und EISENLOHR 1941.)

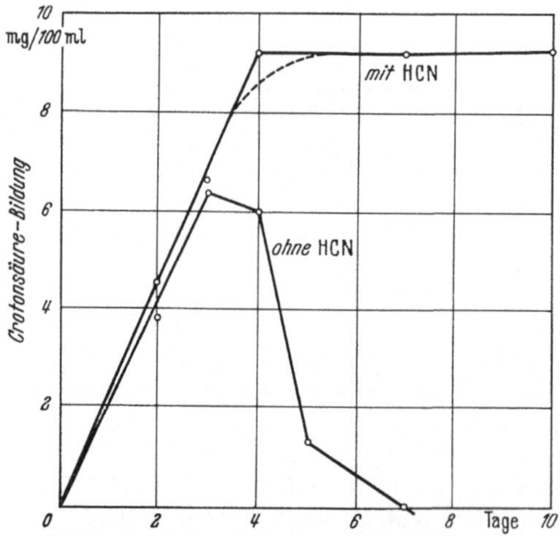

Abb. 29. Crotonsäurebildung aus Buttersäure durch wachsendes Mycel von *Asp. niger* in Abwesenheit und Gegenwart von m/65-HCN (37°). Crotonsäurebestimmung mit Hypochlorit (MUKHERJEE 1952 [3]). (Nach MUKHERJEE 1952 [2].)

für die gesättigten Fettsäuren im ziemlicl stark sauren Gebiet (p_H etwa 3) liegt, er folgt die Methylketonbildung aus den un gesättigten Fettsäuren optimal um der Neutralpunkt, aus den β-Oxysäuren im gan: schwach sauren Gebiete (p_H 6).

Von Interesse im Hinblick auf der angenommenen Reaktionsmechanis mus (S. 183) war noch die Beobachtung (THALER, SCHOTTMAYER, STÄHLIN und BECK 1949), daß wohl aus α-methyl substituierten mittleren Fettsäuren *nicht* dagegen aus *β-Methyl-β-Oxy carbonsäuren* entsprechender C-Zah Methylketone gebildet werden.

Weitere, wenn auch im einzelnen nocl nicht ganz klare Teilergebnisse zur β-Oxy dation durch Schimmelpilze bringen einige neuere Arbeiten von MUKHERJEE (1952 [1, 2]), in denen der Abbau der *Buttersäure* teils durch wachsende Kulturen, teils durch Mycelsuspensionen von *Asp. niger* hinsicht lich der gebildeten Zwischen- und Endpro dukte eingehend untersucht wurde. Sowoh Buttersäure als auch Crotonsäure und β-Oxybuttersäure liefern Acetessigsäure bzw. Aceton. Mit Buttersäure als Substrat lassen sich auch kleinere Mengen von Crotonsäure und β-Oxybuttersäure fassen. Es wird aber bezweifelt, daß der Butyratabbau *einheitlich* nach Schema 5 (S. 183) abläuft, da die Oxy dation von Crotonat als Substrat durch 0,015 m-HCN vollständig, die von Butyrat zu 46% und die von β-Oxybutyrat praktisch überhaupt nicht gehemmt wird. In Gegen wart von HCN erfolgt auch eine stärkere Anhäufung von Crotonsäure aus Buttersäure (Abb. 29). Folgendes „zweigleisige" Schema des Buttersäureabbaus durch *Aspergillus* wird vorgeschlagen (MUKHERJEE 1952 [2]):

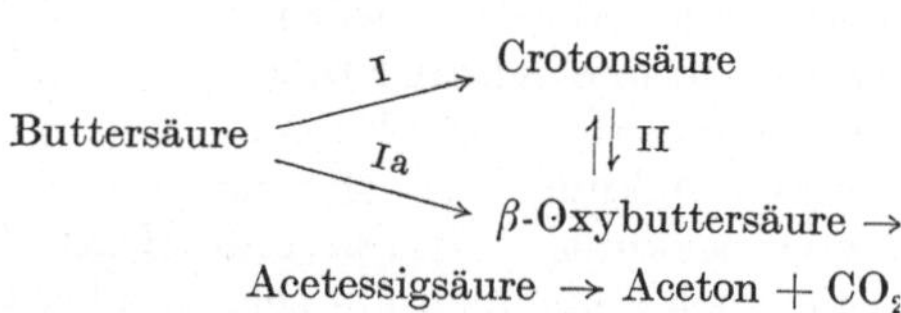

Schritt II ist der mutmaßlich HCN-emp findliche. Die Alternativreaktion Ia ist enzymchemisch einstweilen ohne Analogie. Es erhebt sich in diesem Zusammenhang aber die Frage, ob die ganze Abbaufolge des Schemas 5 überhaupt an den *freien* Fett säuren und nicht eher an den *Acylderivaten des Coenzyms A* (S. 178) abläuft; im letzteren Falle wäre Crotonsäure kein *normales* Inter mediärprodukt und der mit ihr als Substrat ausgeführte Versuch nicht stichhaltig.

Schließlich ist noch eine neue Untersuchung von BYRDE, HARRIS und WOODCOCK (1956) zu erwähnen, in der papierchromatographisch eine β-Oxydation von *ω-(2-Naphthoxy-)fettsäuren* [(Naphthyl)—$O \cdot (CH_2)_n \cdot COOH$] durch Myceldecken von *Asp. niger* nachgewiesen wurde (vgl. S. 179/80).

In diesen Versuchen erfolgte gleichzeitig eine Hydroxylierung des Naphthylrests in 6-Stellung (vgl. S. 187). Die ω-substituierten Fettsäuren, in denen n 3, 5, 7, 9 oder 11 war, lieferten als Hauptprodukt *(6-Oxy-2-naphthoxy)essigsäure*, diejenigen mit n = 2, 4, 6 oder 8 *β-(6-Oxy-2-naphthoxy)propionsäure*. Auffallend ist — beim Vergleich mit früher erwähnten Versuchen an höheren Pflanzen (S. 179/80) —, daß letztere Verbindung nicht in *2,6-Dioxynaphthalin* überging.

c) Bakterien.

Hier sind in erster Linie die schönen Untersuchungen des BARKERschen Arbeitskreises (vgl. kurze Zusammenfassung von BARKER 1951 sowie von KENNEDY und LEHNINGER 1952) an dem anaeroben *Clostridium kluyveri* zu nennen, in denen erstmals eine zellfreie β-Oxydation durchgeführt werden konnte (vgl. S. 176). In Gegenwart von Orthophosphat oxydieren Extrakte dieses Organismus *Buttersäure* hälftig zu *Essigsäure* und *Acetylphosphat* (STADTMAN und BARKER 1949 [2]):

$$CH_3 \cdot CH_2 \cdot CH_2 \cdot COOH + H_3PO_4 + O_2 = CH_3 \cdot COOH + CH_3 \cdot COO \cdot PO_3H_2 + H_2O.$$

Die Angabe, daß in Abwesenheit von Phosphat keine Oxydation stattfinde (STADTMAN und BARKER 1949 [2]), wurde später widerrufen; es wurde gezeigt, daß unter diesen Bedingungen eine Oxydation der Buttersäure zu *Acetessigsäure* erfolgen könne (KENNEDY und BARKER 1951):

$$CH_3 \cdot CH_2 \cdot CH_2 \cdot COOH + O_2 = CH_3 \cdot CO \cdot CH_2 \cdot COOH + H_2O.$$

Die Enzymausrüstung der Bakterienzelle unterscheidet sich nun in mehrfacher Beziehung von derjenigen der höheren tierischen Zelle. So besitzt die erstere das Ferment *Phosphotransacetylase*, das einen reversiblen Übergang des *energiereich* gebundenen Acetylrestes (vgl. S. 177) zwischen Phosphat (Säureanhydridbindung) und Coenzym A (Thioesterbindung) katalysiert (STADTMAN und BARKER 1950, STADTMAN, NOVELLI und LIPMANN 1951, STADTMAN 1952):

$$\text{Acetyl-P} + \text{HSCoA} \rightleftharpoons \text{Acetyl-SCoA} + \text{HP} \quad (\text{P} = \text{Phosphatrest}) \qquad \text{a}$$

Darüber hinaus wurde später in zellfreien Extrakten von *Cl. kluyveri* eine *Coenzym A-Transphorase* (vgl. S. 181) nachgewiesen (STADTMAN 1953), welche die reversible Übertragung des Coenzym A-Rests von der Acetylverbindung auf normale *Fettsäuren* mit 1—8 C-Atomen, ferner auch auf Vinylessigsäure ($CH_2 = CH \cdot CH_2 \cdot COOH$ und auf Milchsäure, bewirkt:

$$\text{Acetyl-SCoA} + R \cdot COO^- \rightleftharpoons \text{Acetat}^- + R \cdot \text{CO-SCoA} \qquad \text{b}$$

Damit war *Butyryl-Coenzym A* als eigentliche Reaktionsform der Buttersäure in den Mittelpunkt des Interesses gerückt. In der Tat kann diese Verbindung schon durch *dialysierte* Enzymlösungen leicht oxydiert werden (STADTMAN und STADTMAN 1953). Andererseits kann auch die *Butyrat*oxydation durch solche Enzymlösungen induziert werden, wenn man Acetylphosphat und das Dialysat eines gekochten Hefeextrakts zusetzt, dessen wesentliche Komponenten Coenzym A, Diphosphopyridin-nucleotid (DPN) und $Mg^{\cdot\cdot}$ zu sein scheinen (STADTMAN und BARKER 1949 [2], BARKER 1951). Unzweifelhaft laufen in diesem komplettierten System zunächst die zu Butyryl-Coenzym A führenden Reaktionen a und b (s. oben) ab. Die Einzelphasen der Butyratoxydation durch *Cl. kluyveri* sind aus dem folgenden Schema (BARKER 1951, KENNEDY und LEHNINGER 1952, STADTMAN und STADTMAN 1953) zu erkennen:

Schema 6.

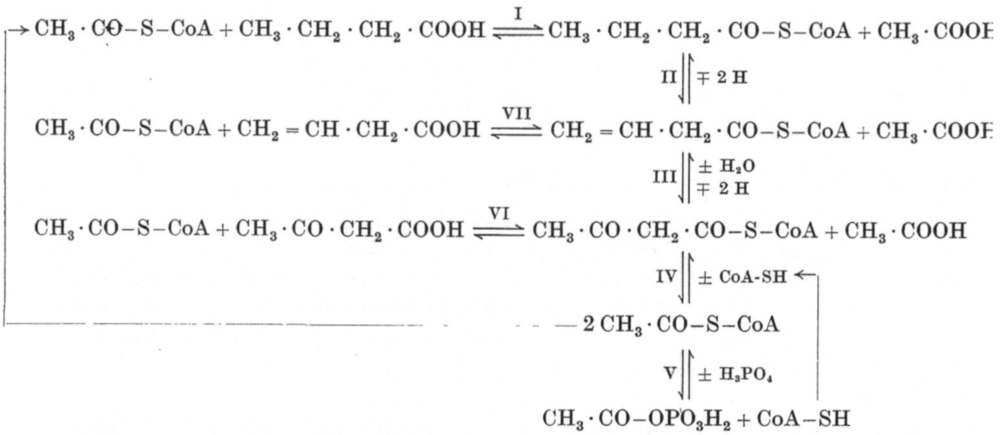

$$\rightarrow CH_3 \cdot CO-S-CoA + CH_3 \cdot CH_2 \cdot CH_2 \cdot COOH \overset{I}{\rightleftharpoons} CH_3 \cdot CH_2 \cdot CH_2 \cdot CO-S-CoA + CH_3 \cdot COOH$$

$$\text{II} \Updownarrow \mp 2\,H$$

$$CH_3 \cdot CO-S-CoA + CH_2 = CH \cdot CH_2 \cdot COOH \overset{VII}{\rightleftharpoons} CH_2 = CH \cdot CH_2 \cdot CO-S-CoA + CH_3 \cdot COOH$$

$$\text{III} \Updownarrow \pm H_2O, \mp 2\,H$$

$$CH_3 \cdot CO-S-CoA + CH_3 \cdot CO \cdot CH_2 \cdot COOH \overset{VI}{\rightleftharpoons} CH_3 \cdot CO \cdot CH_2 \cdot CO-S-CoA + CH_3 \cdot COOH$$

$$\text{IV} \Updownarrow \pm \text{CoA-SH} \leftarrow$$

$$2\,CH_3 \cdot CO-S-CoA$$

$$\text{V} \Updownarrow \pm H_3PO_4$$

$$CH_3 \cdot CO-OPO_3H_2 + CoA-SH$$

Bemerkenswert ist, daß bei diesem *bakteriellen* Butyratabbau nur *Vinylacetat* als Zwischenstufe wahrscheinlich gemacht werden konnte; die Intermediärstufen des *tierischen* Butyratumsatzes, Crotonat und β-Oxybutyrat (LYNEN, WESSELY, WIELAND und RUEFF 1952, LYNEN und OCHOA 1953), sowie eine Reihe anderer C_4-Säuren werden vom Enzymsystem des *Cl. kluyveri* nicht umgesetzt (STADTMAN und BARKER 1949 [5], KENNEDY und BARKER 1951).

Im einzelnen wäre zu Schema 6 noch folgendes zu bemerken: In *Gegenwart von Phosphat*, wenn die langsame Reaktion VI (in der Richtung nach links) durch die rascheren IV und V ausgeschaltet wird, erscheint die Butyratoxydation als ein cyclischer Prozeß, in dem Coenzym A und Acetyl-Coenzym A eine *katalytische* Funktion ausüben. Bei jedem Ablauf des Cyclus werden je ein Molekül Essigsäure und Acetylphosphat (durch Reaktion I bzw. V) gebildet. Die Regeneration der Katalysatoren Acetyl-Coenzym A und Coenzym A erfolgt durch die *β-Ketothiolase*reaktion IV (vgl. S. 179) bzw. die *Phosphotransacetylase*reaktion V; je eines der in IV gebildeten Acetyl-Coenzym A-Moleküle geht in die *Coenzym A-Transphorase*-reaktion I und die *Phosphotransacetylase*-Reaktion V ein. In *Abwesenheit von Phosphat* bestimmt die *freie* Acetessigsäure liefernde (langsame) Reaktion VI den Gesamtablauf, da Reaktion V und damit mittelbar (infolge Coenzym A-Mangels) auch IV ausfallen.

Der *phosphatfreie* Butyratabbau scheint übrigens noch nicht in allen Punkten geklärt zu sein. Mit gewissen Präparaten und unter bestimmten Bedingungen entsteht neben Acetessigsäure auch noch *Essigsäure*. STADTMAN und STADTMAN (1953) haben daher ein Alternativschema vorgeschlagen, das im Prinzip einem Umsatz der Buttersäure nach

$$2\,\text{Butyrat}^- + 3\,H_2O \xrightarrow{-8\,H} \text{Acetacetat}^- + 2\,\text{Acetat}^- + H^\cdot$$

entspricht.

Schema 7.

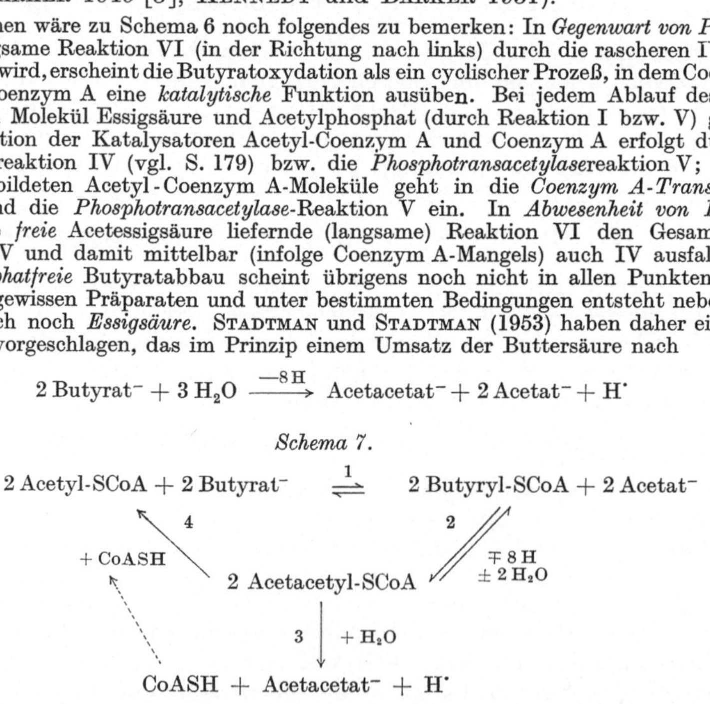

$$2\,\text{Acetyl-SCoA} + 2\,\text{Butyrat}^- \overset{1}{\rightleftharpoons} 2\,\text{Butyryl-SCoA} + 2\,\text{Acetat}^-$$

$$\text{4: } + \text{CoASH} \nwarrow \qquad \text{2: } \mp 8\,H,\ \pm 2\,H_2O \Updownarrow$$

$$2\,\text{Acetacetyl-SCoA}$$

$$\text{3} \downarrow + H_2O$$

$$\text{CoASH} + \text{Acetacetat}^- + H^\cdot$$

Beträchtliche Butyratoxydation nach diesem Schema setzt voraus, daß Reaktion 4 rascher abläuft als Reaktion 3. Nach Schema 7 würde im Gegensatz zu Schema 6 ohne

Phosphat keine Abnahme an flüchtiger Säure eintreten. Nach Versuchen von STADTMAN und STADTMAN (1953) sind möglicherweise beide Reaktionsmechanismen nach Schema 6 und 7 in variierendem Umfang am phosphatfreien Butyratabbau beteiligt, wobei 7 im allgemeinen überwiegt.

Außer Buttersäure können auch andere Fettsäuren bis zu C_8 mit gleicher oder geringerer Geschwindigkeit durch *Cl. kluyveri* oxydiert werden; für das Verhältnis der Oxydationsgeschwindigkeiten der Säuren mit 4, 5, 6, 7 und 8 C-Atomen ergab sich z. B. 1:1:0,8:0,38:0,34:0 (STADTMAN und BARKER 1949 [2]). Bei der Oxydation von *Valerian-* und *Capronsäure* mit Zelltrockenpräparaten in phosphatfreier Lösung wurden die entsprechenden *β-Ketosäuren* erhalten (LIEBERMAN und BARKER 1954). Die Sauerstoffaufnahme in den Extrakten dieses obligat Anaeroben erfolgt offenbar über *Flavin*fermente; dementsprechend bestehen Anzeichen für die intermediäre Bildung von *Hydroperoxyd* (KENNEDY und BARKER 1951). Auf die *reversible* Natur der Teilreaktionen des oxydativen Abbaus (Schema 6) sei noch besonders hingewiesen; durch reduktive Kondensation kann, von *Äthanol* (das dabei zugleich als „Acetyllieferant" fungiert) (STADTMAN und BARKER 1949 [1]) oder auch von Essigsäure bzw. Acetylphosphat + molekularem *Wasserstoff* (STADTMAN und BARKER 1949 [3]) ausgehend, Buttersäure, Capronsäure usw. *synthetisiert* werden. Auch hier ist *Vinylessigsäure* obligate Zwischenstufe (STADTMAN und BARKER 1949 [5]); daß sie nur über die Coenzym A-Verbindung, die durch Transphorasewirkung aus Acetyl- oder Butyryl-Coenzym A gebildet wird, hydriert wird, ist neuerdings noch besonders gezeigt worden (PEEL und BARKER 1956).

Weitere Hinweise auf β-Oxydationen durch Bakterien gleicher und anderer Gruppenzugehörigkeit finden sich verstreut in der neueren Literatur; doch sind in keinem Falle die Verhältnisse mit solcher Gründlichkeit untersucht worden wie bei *Cl. kluyveri* am BARKERschen Institut.

So haben französische Autoren (NISMAN, COHEN, WIESENDANGER und SZULMAJSTER 1953, SZULMAJSTER, NISMAN und COHEN 1954, SZULMAJSTER 1954) bei Untersuchungen über den Mechanismus der Buttersäuregärung folgende Enzyme in Extrakten aus *Clostridium saccharobutyricum* nachgewiesen (vgl. Schema 4, S. 178): *Butyryl-dehydrase*, *β-Oxybutyryl-dehydrase*, *Acetacetyl-thiolase* und *Phosphotransacetylase*. Auffallenderweise gelang, ähnlich wie mit *tierischem* Material (S. 178) und abweichend von *Cl. kluyveri* (s. oben), hier die Hydrierung von *Crotonat* bzw. *Crotonyl-CoA* (mit Leukosafranin); ob sie direkt erfolgt oder unter dem „isomerisierenden" Einfluß einer *Crotonase* (S. 178), ist noch ungeklärt.

Erwähnt sei weiter eine Untersuchung von WOLIN, EVANS und NIVEN jr. (1952), in der ein aerober *Butyrat*abbau durch intakte Zellen von *Streptococcus mitis* nach der Gleichung

$$CH_3 \cdot CH_2 \cdot CH_2 \cdot COOH + 2\,O_2 + 2\,H_2O = 2\,CH_3 \cdot COOH + 2\,H_2O_2$$

festgestellt wurde. Ein Nachweis von Acetylphosphat gelang — wahrscheinlich infolge Gegenwart von Phosphatase in den Zellen — nicht. Von weiteren untersuchten Substraten wurde nur noch (sehr viel langsamer) *Valeriansäure* oxydiert, nicht dagegen Isovaleriansäure und Capronsäure.

Mit den klassischen Methoden KNOOPS (1904) arbeiteten unlängst WEBLEY, DUFF und FARMER (1955) über den oxydativen Abbau *ω-phenylsubstituierter* niederer und mittlerer *Fettsäuren* durch Zellsuspensionen des Actinomyceten *Nocardia opaca*. Mit Hilfe papierchromatographischer und spektrophotometrischer Methoden wurde nachgewiesen, daß Säuren mit *ungerader* C-Zahl der Seitenkette (Phenyl-propionsäure, -valeriansäure und -heptylsäure) *Benzoesäure* lieferten, wobei *Zimtsäure* (Phenyl-acrylsäure) durch ihre charakteristische Absorption bei 278 mμ als Zwischenprodukt leicht nachweisbar war (vgl. S. 175). Bei der Prüfung von Säuren mit *gerader* C-Zahl der Seitenkette (Phenyl-essigsäure, -buttersäure, capronsäure und -caprylsäure) wurde *o-Oxyphenylessigsäure* (Benzolring mit OH)—$CH_2 \cdot COOH$ als gemeinsames Reaktionsprodukt identifiziert.

Von größerem enzymchemischem Interesse sind Untersuchungen von IVLER, WOLFE und RITTENBERG (1955) über die *Caprinsäure*-Oxydation durch zellfreie Extrakte aus *Pseudomonas fluorescens*, wobei pro Substratmolekül 1 Molekül O_2 verbraucht wurde. *Adenosintriphosphat* (ATP) und *Coenzym A* waren für den Ablauf der Reaktion notwendig, bei der ein Molekül Essigsäure gebildet wurde; merkwürdigerweise gelang der Nachweis der als zweites Reaktionsprodukt zu erwartenden *Caprylsäure* nicht, was nicht befriedigend erklärt werden konnte. Ein durch Ammonsulfat partiell gereinigter Extrakt verbrauchte nur mehr 1 Atom Sauerstoff je Substratmolekül und lieferte keine Essigsäure mehr, wohl aber bei Papierchromatographie ein noch nicht identifiziertes Produkt vom *Rf*-Wert 0,22, der den Werten für 2-Decensäure und β-Oxycaprinsäure (0,24 bzw. 0,26) nahekam. Schließlich gelang es mit Hilfe der Hydroxamsäurereaktion (LIPMANN und TUTTLE 1945, 1950, STADTMAN und BARKER 1950), in Ansätzen mit ATP und Coenzym A „aktive" Caprinsäure papierchromatographisch zu identifizieren.

Weitere Befunde dieser Art sind unlängst für *Sarcina lutea* und einen (nicht näher charakterisierten) *Vibrio* mitgeteilt worden.

In Extrakten aus *Sarc. lutea* wurde die *Thiokinase*reaktion für die Fettsäuren mit 2, 4, 5, 6, 7 und 8 C-Atomen, ferner *Dehydrasewirkung* gegenüber den Coenzym A-Verbindungen von Butter-, Capryl-, Caprin- und Laurinsäure nachgewiesen (MURRAY und DAWES 1956). Die *O_2-Aufnahme* dieser Extrakte war jedoch nur geringfügig. Im Gegensatz hierzu fand DAGLEY (1956 [1, 2]) bei *Vibrio*-Extrakten auf Zusatz von Fumarat eine fast vollständige Oxydation von *Caprylsäure*, offenbar über den Tricarbonsäurecyclus. Dialysierte Extrakte verlieren ihre Aktivität fast vollständig, sind jedoch durch Zusatz von Coenzym A, DPN und Mg" wieder reaktivierbar; allerdings vermögen sie dann nur mehr die Reaktion

$$C_7H_{15} \cdot COOH + 3\,O_2 \rightarrow 4\,CH_3 \cdot COOH$$

durchzuführen. Das Reaktionsprodukt Essigsäure wirkt hemmend; Ketosäuren konnten als Zwischenprodukte nicht entdeckt werden.

VI. Fettsäurereduktasen oder -hydrasen.

Unter diesem Begriff sollen alle Enzyme verstanden werden, welche einfach ungesättigte Fettsäuren in gesättigte oder stärker ungesättigte in weniger ungesättigte Fettsäuren überzuführen vermögen. Dabei ist grundsätzlich zu unterscheiden zwischen Fettsäuren mit „carboxylnaher" Doppelbindung, wie sie etwa im Fettsäurecyclus (S. 178) durch Dehydrierung gesättigter Fettsäuren oder Dehydratisierung von β-Oxycarbonsäuren entstehen können, und den besonders in Samen so verbreiteten Fettsäuren mit einer oder mehreren „carboxylfernen" Doppelbindungen (Ölsäure, Linolsäure, Linolensäure u. ä.).

1. Carboxylnahe Hydrierungen.

Dem Verständnis einer Hydrierung von ungesättigten Fettsäuren des erstgenannten Typs stehen keine enzymchemischen Bedenken entgegen, da zum mindesten für *tierische* Acyldehydrasen (z.B. Butyryldehydrase) eine *Reversibilität* der Wirkung (z.B. mit Leukofarbstoffen als H-Donatoren gegenüber *Crotonsäure*) exakt nachgewiesen werden konnte (SEUBERT und LYNEN 1953, LYNEN und OCHOA 1953, LYNEN 1953 [1, 2]). Auch sei an die früher (S. 187) erwähnte Reduktion von *Crotonsäure* mit Leukosafranin (SZULMAJSTER 1953) und von *Vinylessigsäure* mit molekularem H_2 (STADTMAN und BARKER 1949 [5]; PEEL und BARKER 1956) durch *Clostridien*enzyme erinnert. Leider entsprechen unserer mangelhaften Kenntnis der β-Oxydation in *höheren Pflanzen* (S. 174) noch dürftigere Kenntnisse der entsprechenden Hydrierung. Da jedoch mit Schnitten von Erdnußkotyledonen eine Synthese höherer Fettsäuren aus C^{14}-markierter Essigsäure (und wesentlich schlechter aus anderen niederen Fettsäuren) nachgewiesen werden konnte (NEWCOMB und STUMPF 1952, 1953), ist es sehr wahrscheinlich, daß diese Synthese im Sinne des Fettsäurecyclus verlief, in den ja eine ganze Anzahl von α,β-Hydrierungen eingehen. (Näheres s. STUMPF, S. 201.)

Merkwürdigerweise vermag gärende *Hefe*, deren Fähigkeit zur Durchführung einer Vielzahl „phytochemischer Reduktionen" zuerst von NEUBERG (vgl. NEUBERG und GORR 1928) und später von F. G. FISCHER (FISCHER 1939, 1940) eingehend untersucht worden ist, α,β-ungesättigte Carbonsäuren wie *Crotonsäure* und *Zimtsäure* nicht zu hydrieren. FISCHER (1940) charakterisiert die Verhältnisse folgendermaßen: „Voraussetzung für die biochemische Hydrierung einer Äthylenverbindung ist in der Mehrzahl aller bisher untersuchten Fälle ihre α,β-ständige Lage zu einem Carbonyl oder primären Hydroxyl. Olefinalkohole und -ketone, deren Äthylenbindung nicht zur charakteristischen Gruppe konjugiert gelagert ist, werden durch gärende Hefe jedenfalls nicht angegriffen, ebensowenig olefinische Säuren und sekundäre Alkohole, auch wenn sie α,β-ungesättigt sind." *α,β-ungesättigte Ketosäuren* werden unter der Einwirkung gärender Hefe offenbar hauptsächlich auf folgendem Wege umgesetzt (FISCHER und WIEDEMANN 1934):

$$R \cdot CH{=}CH \cdot CO \cdot COOH \xrightarrow{-CO_2} R \cdot CH{=}CH \cdot CHO \xrightarrow{+2H} R \cdot CH{=}CH \cdot CH_2OH \xrightarrow{+2H} R \cdot CH_2 \cdot CH_2 \cdot CH_2OH.$$

In dieser Weise reagierten z. B. *Benzyliden-* und *Sorbinyliden-brenztraubensäure* (I bzw. II): $C_6H_5 \cdot CH{=}CH \cdot CO \cdot COOH$ (I), $CH_3 \cdot CH{=}CH \cdot CH{=}CH \cdot CH{=}CH \cdot CO \cdot COOH$ (II); die beiden Doppelbindungen in 5, 6- und 7, 8-Stellung blieben im letzteren Fall unangegriffen.

2. Carboxylferne Hydrierungen.

Den Beweis eines Übergangs deuterierter ungesättigter Nahrungsfettsäuren in gesättigte haben für den *tierischen* Organismus RITTENBERG und SCHÖNHEIMER (1937) an der Maus erbracht. Ähnliche Versuche mit isotopisch markierten Substraten scheinen für *pflanzliche* Organismen zu fehlen. Doch wird die Hydrierung carboxylferner Doppelbindungen einerseits für *Pflanzensamen* (Kürbis), sogar im Enzymversuch, angegeben, andererseits für eine Anzahl von *Bakterien*, besonders solche des Pansens von Wiederkäuern. Die für Bakterien beigebrachten Belege sind zum Teil stichhaltig, während die Angaben für höhere Pflanzen problematisch sind.

Gegen die Versuche mit *Kürbiskeimlingen* (ZELLER und MASCHEK 1942), in denen teils sameneigenes Fett, teils zugesetztes Natriumoleat und -sorbinat als Substrat diente, sind sowohl methodische als auch grundsätzliche Bedenken zu erheben. Einmal scheint die angegebene Jodzahlmethode (RUPP und BRACHMANN 1926) in der verwendeten Form zur quantitativen Bestimmung von Ölsäure und besonders von Sorbinsäure ungeeignet zu sein, da die mitgeteilten Werte weit unterhalb der theoretischen liegen. [Ebenso wie *konjugierte* Doppelbindungen (vgl. S. 143) werden auch *carboxylnahe* durch die üblichen Jodzahlmethoden nicht annähernd quantitativ erfaßt, vgl. KAUFMANN 1935.] Dann aber dürften sterile Bedingungen in dem meist 7 Tage einer Autolyse bei 37° überlassenen Keimlingsbrei bei bloßer Zugabe von Chloroformwasser als Antisepticum nicht gegeben gewesen sein. Nach einigen eigenen orientierenden Versuchen (FRANKE und TASCHEN, unveröffentlicht 1952) erfolgt regelmäßig schon in den ersten Tagen intensiver Bakterienbefall der Ansätze. Da zudem für Luftausschluß in den Versuchen nicht gesorgt war, ist zwischen oxydativem Fettsäureschwund und Fettsäurehydrierung nicht zu unterscheiden. Die von den Autoren als Beweis einer „Saturase"-Wirkung angeführten Argumente (Temperatur- und p_H-Einfluß, Aktivierbarkeit und Hemmbarkeit des Doppelbindungsschwunds) gelten natürlich auch für irgendwelche Bakterienreaktionen. Die schwersten Bedenken erheben sich gegenüber den unter den Bedingungen von Enzymansätzen durchgeführten Versuchen mit Ammonsulfatfraktionen des Kürbissamenextrakts. Eine zwischen den Sättigungsgraden 0,5 und 0,85 ausfallende Fraktion soll z. B. 0,3 mMol Oleat zu rund 72% in 7 Tagen absättigen, was einem H_2-Äquivalent von 4,85 ml entsprechen würde. Es ist unverständlich, wo dieser Wasserstoff herkommen soll, da den Ansätzen kein eigener H-Donator zugesetzt wurde. Wahrscheinlich ist der Doppelbindungsschwund wieder auf Rechnung des (in den ammonsulfathaltigen Ansätzen besonders intensiven) Bakterienwachstums zu setzen. In der gleichen Richtung liegt die Beobachtung, daß mit Ammoniakwasser (n/40) hergestellte Keimlingsextrakte wesentlich bessere „Saturase"-Wirkung zeigten als Extrakte mit destilliertem Wasser, Phosphatpuffer oder Glycerin.

Dieselben Einwände gelten zum Teil auch für die mit grundsätzlich gleicher Methodik durchgeführten Versuche SCHÖNBRUNNERS (1940) an verschiedenen *Bakterien* (*Escherichia coli*, *Pseudomonas fluorescens liquefaciens*, aus faulendem Oleat isolierter Stamm, Mischkulturen aus Erd- und Schlammproben). Auch hier wurden in einfachen Inkubationsversuchen bei 24—26°

wechselnde Grade der „Absättigung" von Oleat und Sorbinat beobachtet, wobei die Versuchszeiten zwischen 4 und 34 Wochen lagen. Obwohl der Autor einleitend auf das Risiko hinweist, eine etwaige Verkürzung der Kettenlängen der Fettsäuren nicht feststellen zu können, werden die Jodzahlabnahmen im weiteren Verlauf der Untersuchung doch immer als „Absättigung" bzw. „Hydrierung" interpretiert. Dies ist bei den angewandten quasiaeroben Bedingungen, den extrem langen Versuchszeiten und dem Fehlen eines H-Donators zweifellos unzulässig.

Beweisender für eine tatsächliche *Hydrierung* erscheinen Angaben von GINSBURG-KARAGITSCHEWA und RODIONOWA (1935), die unter definiert anaeroben Bedingungen in allerdings auch wieder monatelangen Versuchen eine zum Teil starke Herabsetzung der Jodzahl bei Einwirkung von Bakterien aus *Tiefseeschlamm* auf *Delphintran*, *Sonnenblumenöl* oder *Ölsäure* beobachteten; so war nach 6—9 Monaten bei 40—45° die Jodzahl der obigen Substrate (in der angegebenen Reihenfolge) um bzw. 60, 40 und 30% gesunken. Da im allgemeinen den Ansätzen *Pepton* zugesetzt wurde, erscheint dessen Funktion als Donatorquelle bei diesen obligat Anaeroben grundsätzlich denkbar. [So vermögen z. B. *Clostridien* aus Aminosäuren Wasserstoffgas zu entwickeln oder Dismutationen zwischen Aminosäuren (STICKLAND-Reaktion, s. Übersicht von NISMAN 1954) zu bewirken.] Auf die Bedeutung dieser Befunde für die Frage der *Erdölbildung* kann hier nur kurz hingewiesen werden. (Zur bakteriellen Petroleumbildung aus Oleat vgl. LAIGRET 1947.)

In einigen neueren Arbeiten ist für durchweg *anaerobe Bakterien* auch unter üblichen und wohldefinierten Bedingungen bei relativ kurzen Versuchszeiten eine partielle Hydrierung von Linol- und Linolensäure unzweifelhaft dargetan worden. So hat ROSENFELD (1947) mit anaeroben Bakterien, die aus Meeresschlamm isoliert waren und zum Teil der Gattung *Desulfovibrio* angehörten, im System Formiat-Linolat eine partielle Hydrierung des letzteren durchführen können, wobei nach 24stündiger Inkubation im Vakuum bei p_H 7,2 der Anfangswert der Jodzahl auf $^1/_3$—$^1/_4$ zurückging. Ferner hat REISER (1951) nach 4tägiger Inkubation von Leinöl mit Panseninhalt eine Reduktion des Linolensäuregehalts von 30% auf 5% festgestellt, wobei der Linolsäuregehalt entsprechend zunahm. [Auch im Fütterungsversuch an Ziegen ist diese Hydrierung eindeutig nachgewiesen und liefert eine Erklärung dafür, daß das Depotfett der Wiederkäuer selbst nach Verfütterung hochungesättigter Fette relativ gesättigt ist (REISER und RAMAKRISHNA REDDY 1956).] Einen weiteren Beweis für eine bakterielle Hydrierung von Nahrungsfettsäuren hat man neuerdings im Vorkommen von *trans*-Formen der ungesättigten Fettsäuren im Fett der *Wiederkäuer* gesehen (HARTMAN, SHORLAND und McDONALD 1954, 1955). So wurden im Depotfett verschiedener Wiederkäuer 3,5—11,2% trans-Säuren gefunden, während der Gehalt bei Nichtwiederkäuern unterhalb 1% lag. Eine Ausnahme bei den letzteren bildeten nur die *Beuteltiere* (Känguruh, Quokka) mit trans-Fettsäuregehalten zwischen 18,1 und 21%; doch ist für den Mageninhalt dieser Tiere gezeigt worden, daß er eine pansenähnliche, wenn auch einfachere Mikrobenzusammensetzung aufweist (MOIR, SOMERS, SHARMAN und WARING 1954). Da bekannt ist, daß schon bei der rein chemischen Hydrierung mehrfach ungesättigter cis-Fettsäuren zum Teil transisomere Produkte entstehen (vgl. z. B. MARKLEY 1947), wird geschlossen, daß die trans-Fettsäuren im Fett der Wiederkäuer und der Beuteltiere durch *bakterielle Hydrierung* der höher ungesättigten Nahrungsfettsäuren im Pansen der Wiederkäuer oder im pansenähnlichen Magen der Beuteltiere entstanden sind[1].

Literatur.

ACKLIN, O.: Zur Biochemie des *Penicillium glaucum*. Ein Beitrag zum Problem der Methylketonbildung aus Triglyceriden bzw. Fettsäuren im Stoffwechsel des Schimmelpilzes. Biochem. Z. **204**, 253—274 (1929). — ALBERTY, R. A.: [1] Kinetic effects of the ionization of groups in the enzyme molecule. J. Cellul. a. Comp. Physiol. **47**, Suppl. 1, 245—281 (1956). — [2] Enzyme kinetics. Adv. Enzymol. **17**, 1—64 (1956). — ANDRÉ, E., et K. HOU: [1] Sur la présence d'une oxydase des lipides ou lipoxydase dans la graine de soja, *Glycine soja* (SIEB.). C. r. Acad. Sci. Paris **194**, 645—647 (1932). — [2] Sur les lipoxydases des graines de *Glycine soja* (SIEB.) et

[1] Siehe Nachtrag S. 200.

de *Phaseolus vulgaris* (L.). C. r. Acad. Sci. Paris **195**, 172—174 (1932). — ARREGUIN, B., J. BONNER and B. J. WOOD: Studies on the metabolism of rubber formation in the guayule. III. Experiments with isotopic carbon. Arch. of Biochem. a. Biophysics **31**, 234 —247 (1951).

BALLS, A. K., B. AXELROD and M. W. KIES: Soy bean lipoxidase. J. of Biol. Chem. **149**, 491—504 (1943). — BARKER, H. A.: Recent investigations on the formation and utilization of active acetate. Phosphorus Metabolism **1**, 204—245 (1951). — BARRON, E. S. G.: Thiol groups of biological importance. Adv. Enzymol. **11**, 201—266 (1951). — BEINERT, H., R. M. BOCK, D. S. GOLDMAN, D. E. GREEN, H. R. MAHLER, S. MII, P. G. STANSLY and S. J. WAKIL: The reconstruction of the fatty acid oxidizing system of animal tissues. J. Amer. Chem. Soc. **75**, 4111—4112 (1953). — BEINERT, H., D. E. GREEN, P. HELE, O. HOFFMANN-OSTENHOF, F. LYNEN, S. OCHOA, G. POPJÁK u. R. RUYSSEN: Nomenklatur der Enzyme des Fettsäurestoffwechsels. Biochem. Z. **328**, 75—80 (1956). — BERGSTRÖM, S.: [1] On the autoxidation of the methyl ester of linoleic acid. Ark. Kemi, Mineral. o. Geol. A **21**, Nr. 14 (1945), 18 S. — [2] On the oxidation of linoleic acid with lipoxidase. Ark. Kemi, Mineral. o. Geol. A **21**, Nr. 15 (1945), 8 S. — [3] Autoxidation of linoleic acid. Nature (Lond.) **156**, 717 (1945). — BERGSTRÖM, S., and R. T. HOLMAN: [1] Total conjugation of linoleic acid in oxidation with lipoxidase. Nature (Lond.) **161**, 55 (1948). — [2] Lipoxidase and the autoxidation of unsaturated fatty acids. Adv. Enzymol. **8**, 425—457 (1948). — BERNHEIM, F., K. M. WILBUR and C. B. KENASTON: The effect of oxidized fatty acids on the activity of certain oxidative enzymes. Arch. of Biochem. a. Biophysics **38**, 177—184 (1952). — BERNSTEIN, L., and J. F. THOMPSON: Studies on the carotene-destroying processes in drying bean leaves. Bot. Gaz. **109**, 204—219 (1947). — BLAIN, J. A., and J. P. TODD: The lipoxidase activity of wheat. J. Sci. Food Agric. **6**, 471—479 (1955). — BÖMER, A., u. J. GROSSFELD: Handbuch der Lebensmittelchemie, Bd. IV, Fette und Öle, S. 295—299. Berlin: Springer 1939. — BOLLAND, J. L.: Analytical methods in rubber chemistry. III. Determination of active hydrogen. Trans. Inst. Rubber Ind. **16**, 267—275 (1941). — Kinetic studies in the chemistry of rubber and related materials. I. The thermal oxidation of ethyl linoleate. Proc. Roy. Soc. Lond., Ser. A **186**, 218—236 (1945). — Kinetic studies in the chemistry of rubber and related materials. VI. The benzoyl peroxide-catalysed oxidation of ethyl linoleate. Trans. Faraday Soc. **44**, 669—677 (1948). — Kinetics of olefin oxidation. Quart. Rev. Chem. Soc. **3**, No 1 (1949), 21 S. — BOLLAND, J. L., and G. GEE: Kinetic studies in the chemistry of rubber and related materials. III. Thermochemistry and mechanisms of olefin oxidation. Trans. Faraday Soc. **42**, 244—252 (1946). — BOLLAND, J. L., and R. TEN HAVE: [1] Kinetic studies in the chemistry of rubber and related materials. IV. The inhibitory effect of hydroquinone on the thermal oxidation of ethyl linoleate. Trans. Faraday Soc. **43**, 201—210 (1947). — [2] Kinetic studies in the chemistry of rubber and related materials. V. The inhibitory effect of phenolic compounds on the thermal oxidation of ethyl linoleate. Discuss. Faraday Soc. **1947**, 252—260. — BOLLAND, J. L., and H. P. KOCH: The course of autoxidation reactions in polyisoprenes and allied compounds. IX. The primary thermal oxidation product of ethyl linoleate. J. Chem. Soc. (Lond.) **1945**, 445—447. — BOLOMEY, R. A.: [1] Oxidative decomposition of vitamin A. I. Stability of vitamin A towards oxidation and irradiation. J. of Biol. Chem. **169**, 323—330 (1947). — [2] Oxidative decomposition of vitamin A. II. Absorption spectrophotometry of oxidized vitamin A. J. of Biol. Chem. **169**, 331—335 (1947). — BOYD, D. H. J., and G. A. ADAMS: An assay method for lipoxidase in animal tissue. Canad. J. Biochem. a. Physiol. **33**, 191—198 (1955). — BREUSCH, F. L.: Citric acid cycle; sugar and fat breakdown in tissue metabolism. Science (Lancaster, Pa.) **97**, 490—492 (1943). — Über den Abbau der Fettsäuren in den Geweben. I. Der Abbau der β-Ketosäuren. Enzymologia (Den Haag) **11**, 169—173 (1944). — The biochemistry of fatty acid catabolism. Adv. Enzymol. 8, 343—423 (1948). — Die Verbrennung der Fettsäuren im tierischen Organismus. Angew. Chem. **62**, 66—72 (1950). — BREUSCH, F. L., u. R. TULUS: Über den Abbau der Fettsäuren in den Geweben. III. β-Oxyfettsäuren. Enzymologia (Den Haag) **11**, 352—355 (1944). — BUDNITZKAJA, JE. W.: Untersuchung der Aktivität von Futterpflanzen nach der Methode der Carotinoxydation. [Russisch.] Biochimija **20**, 614—622 (1955). — BURTON, K.: Higher fatty acid dehydrogenase of mammalian liver. Nature (Lond.) **161**, 606 (1948). — The free energy change associated with the hydrolysis of the thiol ester bond of acetyl coenzyme A. Biochemic. J. **59**, 44—46 (1955). — BYRDE, R. J. W., J. F. HARRIS and D. WOODCOCK: Fungal detoxication. The metabolism of ω (2-naphthyloxy)-n-alkylcarboxylic acids by *Aspergillus niger*. Biochemic. J. **64**, 154—160 (1956).

CASTELFRANCO, P., P. K. STUMPF and R. CONTOPOULOU: Fat metabolism in higher plants. V. A soluble palmitate oxidase requiring glycolic acid as activator. J. of Biol. Chem. **214**, 567—577 (1955). — CASTELL, C. H., and E. H. GARRARD: The action of microorganisms on fat. Canad. J. Res. C **19**, 106—110 (1941). — CLAGETT, C. O., N. E. TOLBERT and R. H. BURRIS: Oxidation of α-hydroxy acids by enzymes from plants. J. of Biol. Chem. **178**, 977 bis 987 (1949). — COLLIER, H. B., and S. C. McRAE: Antioxidants as inhibitors of linoleate oxidation catalyzed by plant lipoxidase and by hemolyzates of human erythrocytes. Canad.

J. Biochem. a. Physiol. 33, 773—779 (1955). — COPPOCK, P. D., V. SUBRAMANIAM and T. K. WALKER: The mechanism of the degradation of fatty acids by mould fungi. III. J. Chem. Soc. (Lond.) **1928**, 1422—1427. — COSBY, E. L., and J. B. SUMNER: Lipoxidase. Arch. of Biochem. **8**, 259—264 (1945). — CRAIG, F. N.: A fat oxidation system in *Lupinus albus*. J. of Biol. Chem. **114**, 727—746 (1936). — CRANE, F. L., J. G. HAUGE and H. BEINERT: Flavoproteins involved in the first oxidative step of the fatty acid cycle. Biochim. et Biophysica Acta **17**, 292—294 (1955).

DAGLEY, S.: [1] Oxidation of octanoic acid by cell-free bacterial extracts. Biochemic. J. **63**, 27P (1956). — [2] Oxidation of octanoic acid by cell-free bacterial extracts. Nature (Lond.) **177**, 1131—1132 (1956). — DAKIN, H. D.: The mode of oxidation in the animal organism of phenyl derivatives of fatty acids. IV. Further studies on the fate of phenylpropionic acid and some of its derivatives. J. of Biol. Chem. **6**, 203—219 (1909). — DECKER, K.: Der biologische Mechanismus der β-Oxydation. Fette, Seifen, Anstrichmitt. **58**, 163—167 (1956). — DIEMAIR, W., u. H. FOX: Zur Frage der Chemie der ,,natürlichen" Antioxydantien. Angew. Chem. **52**, 484—487 (1939). — DIEMAIR, W., R. STROHECKER u. K. REULAND: Beitrag zur Kenntnis und Bestimmung der ,,Antioxygene" des Hafermehles. Z. Unters. Lebensmitt. **79**, 23—42 (1940). — DIXON, M., and D. KEILIN: The action of cyanide and other respiratory inhibitors on xanthine oxidase. Proc. Roy. Soc. (Lond.) Ser. B **119**, 159—190 (1936). — DRYSDALE, G. R., and H. A. LARDY: Fatty acid oxidation by a soluble enzyme system from mitochondria. J. of Biol. Chem. **202**, 119—136 (1953). — DUBOULOZ, P., et J. FONDARAI: Sur le métabolisme des peroxydes lipidiques. II. Action des peroxydes lipidiques sur les groupements thiols protéiques. Bull. Soc. Chim. biol. Paris **35**, 819—826 (1953). — DUBOULOZ, P., J. FONDARAI et J. P. PRACCHIA: Sur le métabolisme des peroxydes lipidiques. III. Méchanisme de la destruction des peroxydes lipidiques par les tissus animaux. Bull. Soc. Chim. biol. Paris **36**, 893—900 (1954).

EMBDEN, G., u. A. MARX: Über Acetonbildung in der Leber. III. Beitr. chem. Physiol. u. Path. **11**, 318—326 (1908). — EMBDEN, G., H. SALOMON u. F. SCHMIDT: Über Acetonbildung in der Leber. II. Beitr. chem. Physiol. u. Path. 8, 129—155 (1906). — EULER, H. v.: Chemie der Enzyme, Bd. I, S. 272—274. München: J. F. Bergmann 1925.

FANTL, P., and G. J. LINCOLN: The enzymatic dehydrogenation of fatty acids. Austral. J. Exper. Biol. a. Med. Sci. **27**, 403—407 (1949). — FARMER, E. H.: Ionic and radical mechanisms in olefinic systems, with special reference to processes of double-bond displacement, vulcanisation and photo-gelling. Trans. Faraday Soc. **38**, 348—361 (1942). — Peroxidation in relation to olefinic structure. Trans. Faraday Soc. **42**, 228—236 (1946). — FARMER, E. H., G. F. BLOOMFIELD, A. SUNDRALINGAM and D. A. SUTTON: The course and mechanism of autoxidation reactions in olefinic and polyolefinic substances, including rubber. Trans. Faraday Soc. **38**, 348—356 (1942). — FARMER, E. H., H. P. KOCH and D. A. SUTTON: The course of autoxidation reactions in polyisoprenes and allied compounds. VII. Rearrangement of double bonds during autoxidation. J. Chem. Soc. (Lond.) **1943**, 541—547. — FARMER, E. H., and A. SUNDRALINGAM: The course of autoxidation reactions in polyisoprenes and allied compounds. VI. The peroxidation of rubber. J. Chem. Soc. (Lond.) **1943**, 125—133. — FARMER, E. H., and D. A. SUTTON: The course of autoxidation reactions in polyisoprenes and allied compounds. V. Observations on fish-oil acids. J. Chem. Soc. (Lond.) **1943**, 122 bis 133. — FAWCETT, C. H., J. M. A. INGRAM and R. L. WAIN: The β-oxidation of ω-phenoxyalkylcarboxylic acids in the flax plant in relation to their plant growth-regulating activity. Proc. Roy. Soc. Lond., Ser. B **142**, 60—72 (1954). — FAWCETT, C. H., R. C. SEELEY, F. TAYLOR, R. L. WAIN and F. WIGHTMAN: α-Oxidation of ω-(2:4-dichlorophenoxy)alkanenitriles and 3-indolylacetonitrile within plant tissues. Nature (Lond.) **176**, 1026—1027 (1955). — FELDBERG, W., and T. MANN: Properties and distribution of the enzyme system which synthesizes acetylcholine in nervous tissue. J. of Physiol. **104**, 411—425 (1946). — FISCHER, F. G.: Biochemische Hydrierungen. Fortschr. Chem. organ. Naturstoffe **3**, 30—80 (1939). — Neuere Methoden der präparativen organischen Chemie. VI. Die Benutzung biochemischer Oxydationen und Reduktionen für präparative Zwecke. Chemie **53**, 461—471 (1940). — FISCHER, F. G., u. O. WIEDEMANN: Biochemische Hydrierungen. I. Über die Hydrierung ungesättigter α-Ketosäuren, Aldehyde und Alkohole durch gärende Hefe. Liebigs Ann. **513**, 260—280 (1934). — FLEET, D. S. VAN: Unsaturated fat oxidase: distribution, function and histochemical identification in plant tissues. J. Amer. Chem. Soc. **65**, 740 (1943). — Histochemical localization of enzymes in vascular plants. Bot. Review **18**, 354—398 (1952). — FRANKE, W.: Zur Autoxydation der ungesättigten Fettsäuren. I. Liebigs Ann. **498**, 129—165 (1932). — Die Enzyme der Oxydation und Reduktion. In H. v. EULER, Chemie der Enzyme, Bd. II/3, S. 256—272. München: J. F. Bergmann 1934. — [1] Neuere Erkenntnisse über den Mechanismus der Atmung und Gärung (Desmolyse). Angew. Chem. **53**, 580—593 (1940). — [2] Die Enzyme der Desmolyse. In F. F. NORD-R. WEIDENHAGEN, Handbuch der Enzymologie, S. 723 bzw. 746. Leipzig: Akademische Verlagsgesellschaft 1940. — Zur Kettentheorie desmolytischer Fermentreaktionen. Zugleich zur biologischen Oxydation der Oxalsäure. III. Hoppe-Seylers Z. **281**, 162—185 (1944). — Zur biologischen Oxydation der ungesättigten Fettsäuren. Angew. Chem. **60**, 252 (1948). — Lipoxydasen. Fette u. Seifen **52**, 11—19 (1949). — Neuere Arbeiten über die autoxydativen Primärvorgänge bei der Öltrocknung. Farben, Lacke, Anstrichstoffe **4**, 301—311 (1950). — Autoxydation und enzymatische Oxydation der ungesättigten Fettsäuren. Erg. Enzymforsch. **12**, 89—172 (1951). — FRANKE, W., u. H. FREHSE: Zur Autoxydation der ungesättigten Fettsäuren. VI. Über die Lipoxydase der Gramineen, im besonderen der Gerste. Hoppe-Seylers Z. **295**, 333—349 (1953). — Über Oxydationsfermente aus höheren Pflanzen. II. Zur Kenntnis von Lipoxydase und „Lipodehydrase" und ihrer Beziehungen zueinander. Hoppe-Seylers Z. **298**, 1—26 (1954). — FRANKE, W., u. D. JERCHEL: Zur Autoxydation der ungesättigten Fettsäuren. III. Liebigs Ann. **533**, 46—71 (1937). — FRANKE, W., L. LEE u. D. KIBAT: Zum Stoffwechsel der säurefesten Bakterien. II. Über das Verhalten der ungesättigten Fettsäuren. Biochem. Z. **319**, 263—282 (1949). — FRANKE, W., u. J. MÖNCH: Zur Autoxydation der ungesättigten Fettsäuren. IV. Liebigs Ann. **556**, 200—223 (1944). — FRANKE, W., J. MÖNCH, D. KIBAT u. A. HAMM: Zur Autoxydation der ungesättigten Fettsäuren. V. Die Wirkung der Sojalipoxydase. Liebigs Ann. **559**, 221—232 (1948). — FRANKE, W., u. A. SCHILLINGER: Zum Stoffwechsel der säurefesten Bakterien. I. Orientierende aerobe Reihenversuche. Biochem. Z. **315**, 313—334 (1944). — FRANKE, W., I. SCHULZ u. W. DE BOER: Über Oxydationsfermente aus höheren Pflanzen. III. Versuche zur relativen Acceptor- und Donatorspezifität der Glykolicodehydrase. Hoppe-Seylers Z. **303**, 70—77 (1956). — FRANKE, W., G. TASCHEN u. H. FREHSE: Zur Dehydrierung der Fettsäuren. II. Orientierende Versuche über die Dehydrierbarkeit höherer gesättigter, ungesättigter sowie Brom- und Hydroxyfettsäuren durch tierische und pflanzliche Enzyme. Hoppe-Seylers Z. **303**, 58—69 (1956). — FRANKE, W., G. TASCHEN u. E. HOLZ: Zur Kenntnis der Spezifität von Dehydrasen. III. Versuche zur Spezifität von Succino- und Lacticodehydrase unter besonderer Berücksichtigung halogensubstituierter Säuren. In Vorbereitung 1957. — FREHSE, H., u. W. FRANKE: Fettoxydation durch pflanzliche Lipoxydasen und Fettsäure-Dehydrasen. Fette, Seifen, Anstrichmitt. **58**, 403—408, 593—598 (1956). — FRIEND, J.: The oxidation of β-carotene by a lipoxidase-linoleate system. Biochemic. J. **64**, 19P—20P (1956). — FRITZ, G., and H. BEEVERS: [1] Oxidation of 2,3′,6-trichloroindophenol by the lipoxidase system. Plant Physiol. **30**, 67—69 (1955). — [2] Lipoxidase and the oxygen absorption of homogenates from corn seedlings. Arch. of Biochem. a. Biophysics **55**, 436—446 (1955).

GEYER, R. P., L. MARSHALL, M. RYAN and S. WESTHAVER: Aerobic $C^{14}O_2$ formation from carboxyl-labeled stearic and palmitic acids in the presence of ascorbic acid. Arch. of Biochem. a. Biophysics **56**, 549—551 (1955). — GINSBURG-KARAGITSCHEWA, T., u. K. RODIONOWA: Beitrag zur Kenntnis der im Tiefseeschlamm stattfindenden biochemischen Prozesse. Zur Frage der Erdölbildung. Biochem. Z. **275**, 396—404 (1935). — GIROLAMI, R. L., and S. G. KNIGHT: Fatty acid oxidation by *Penicillium roqueforti*. Appl. Microbiol. **3**, 264—267 (1955). — GOLDSCHMIDT, S., u. K. FREUDENBERG: Über die Autoxydation von Linolensäure und deren Estern. Ber. dtsch. chem. Ges. **67**, 1589—1594 (1934). — GRAFFLIN, A. L., and D. E. GREEN: Studies on the cyclophorase system. II. The complete oxidation of fatty acids. J. of Biol. Chem. **176**, 95—115 (1948). — GRANDE, F.: Über das Vorkommen von Palmitico- und Stearicodehydrogenasen in einigen Pflanzensamen. Skand. Arch. Physiol. (Berl. u. Lpz.) **60**, Suppl., 189—196 (1934). — GREEN, D. E.: Fatty acid oxidation in soluble systems of animal tissues. Biol. Rev. **29**, 330—366 (1954).

HAAS, L. W., and R. BOHN: Bleaching bread dough. U. S. Pats. 1, 957, 333/7, 1934. Zit. nach Chem. Abstr. **28**, 4137 (1934). — HABER, F., u. R. WILLSTÄTTER: Unpaarigkeit und Radikalketten im Reaktionsmechanismus organischer und enzymatischer Vorgänge. Ber. dtsch. chem. Ges. **64**, 2844—2856 (1931). — HAHN, A., u. W. HAARMANN: [1] Über die Dehydrierung der Bernsteinsäure. I. Z. Biol. **87**, 107—114 (1928). — [2] Über die Dehydrierung der Äpfelsäure. I. Z. Biol. **87**, 465—471 (1928). — Über die Dehydrierung der Bernsteinsäure. II. Z. Biol. **89**, 159—166 (1930). — HAHN, A., W. HAARMANN u. E. FISCHBACH: Über die Dehydrierung der Äpfelsäure. II. Z. Biol. 88, 587—593 (1929). — HALDANE, J. B. S.: Chain reactions in enzymatic catalysis. Nature (Lond.) **130**, 61 (1932). — HALDANE, J. B. S., u. K. G. STERN: Allgemeine Chemie der Enzyme, S. 42 bzw. 105f. Dresden u. Leipzig: Theodor Steinkopff 1932. — HALLER, A., et A. LASSIEUR: [1] Étude des échappées du beurre de coco. Composition de l'essence de coco. C. r. Acad. Sci. Paris **150**, 1013—1019 (1910). — [2] Sur deux alcools actifs et une troisième cétone contenus dans l'essence de coco. C. r. Acad. Sci. Paris **151**, 697—699 (1910). — HAND, D. B., and E. C. GREISEN: Oxidation and reduction of vitamin C. J. Amer. Chem. Soc. **64**, 358—361 (1942). — HARTMAN, L., F. B. SHORLAND and I. R. C. McDONALD: Occurrence of trans-acids in animal fats. Nature (Lond.) **174**, 185—186 (1954). — The trans-unsaturated acid contents of fats of ruminants and non-ruminants. Biochemic. J. **61**, 603—607 (1955). — HARTMANN, S., and J. GLAVIND: A new sensitive method for the determination of peroxides of fats and fatty acids. Acta chem.

scand. (Copenh.) 3, 954—958 (1949). — HAUGE, S. M.: Evidence of enzymatic destruction of the vitamin A value of alfalfa during the curing process. J. of Biol. Chem. 108, 331—336 (1935). — HAUGE, S. M., and W. AITKENHEAD: The effect of artifical drying upon the vitamin A content of alfalfa. J. of Biol. Chem. 93, 657—665 (1931). — HEFTER, G., u. H. SCHÖNFELD: Chemie und Gewinnung der Fette, S. 440—448. Wien: Springer 1936. — HEIMANN, W., M. MATZ, B. GRÜNEWALD u. H. HOLLAND: Über die synergistische Wirkung von α-Alanin bei der Hemmung der Fettautoxydation durch phenolische Antioxydantien. Z. Lebensmittelunters. u. Forsch. 102, 1—6 (1955). — HÉRISSET, A.: Influence des produits d'oxydation des lipides sur l'action de la lipase pancréatique. C. r. Acad. Sci. Paris 239, 1438—1439 (1954). — HICKMAN, K.: Function of α-tocopherol in lipoxidase metabolism. Arch. of Biochem. 17, 360 (1948). — HITCHCOCK, D.: The formal identity of LANGMUIRS adsorption equation with the law of mass action. J. Amer. Chem. Soc. 48, 2870 (1926). — HOLMAN, R. T.: Spectrophotometric studies of the oxidation of fats. VII. Oxygen absorption and chromophore production in lipoxidase-oxidized fatty esters. Arch. of Biochem. 10, 519—529 (1946). — Crystalline lipoxidase. II. Lipoxidase activity. Arch. of Biochem. 15, 403—413 (1947). — [1] Lipoxidase activity and fat composition of germinating soy beans. Arch. of Biochem. 17, 459—466 (1948). — [2] Mechanism of action of lipoxidase. Trans. 3. Conference on Biol. Antioxidants. New York, J. Macy Foundation 1948. — Spectrophotometric studies of the oxidation of fats. VIII. Coupled oxidation of carotene. Arch. of Biochem. 21, 51—57 (1949). — Spectrophotometric studies of the oxidation of fats. IX. Coupled oxidation of vitamin A acetate. Arch. of Biochem. 26, 85—91 (1950). — Mode of action of lipoxidase. Trans. Amer. Assoc. Cereal Chem. 11, 135—146 (1953). — Measurement of lipoxidase activity. In D. GLICK, Methods of Biochemical Analysis, Bd. II, S. 113—119. New York: Interscience Publishers Inc. 1955. — HOLMAN, R. T., and S. BERGSTRÖM: Lipoxidase or unsaturated-fat oxidase. In J. B. SUMNER-K. MYRBÄCK, The Enzymes, Bd. II, S. 559—580. New York: Academic Press Inc. 1951. — HOLMAN, R. T., and G. O. BURR: Spectrophotometric studies of the oxidation of fats. IV. Ultraviolet absorption spectra of lipoxidase-oxidized fats. Arch. of Biochem. 7, 47—54 (1945). — HOLMAN, R. T., and O. C. ELMER: The rates of oxidation of unsaturated acids and esters. J. Amer. Oil Chem. Soc. 24, 127—129 (1947). — HOLMAN, R. T., F. PANZER, B. S. SCHWEIGERT and S. R. AMES: Crystalline lipoxidase. III. Amino acid composition. Arch. of Biochem. 26, 199—204 (1950). — HOLTZ, P., u. G. TRIEM: Über Peroxydbildung bei der Autoxydation von Ascorbinsäure und Sulfhydrylkörpern. Hoppe-Seylers Z. 248, 1—4 (1937). — HOROWITZ-WLASSOWA, L. M., u. M. J. LIVSCHITZ: Zur Frage der Wirkung der Mikroben auf Fette. Zbl. Bakter. II 92, 424—435 (1935). — HULST, S. J. N. VAN DER: Application of absorption spectra in fatty oil research. I. Rec. trav. chim. Pays-Bas 54, 639—643 (1935). — HUMPHREYS, T. E., E. H. NEWCOMB, A. H. BOKMAN and P. K. STUMPF: Fat metabolism in higher plants. II. Oxidation of palmitate by a peanut particulate system. J. of Biol. Chem. 210, 941—948 (1954). — HUMPHREYS, T. E., and P. K. STUMPF: Fat metabolism in higher plants. IV. Preparation of soluble fatty acid oxidases from peanut microsomes. J. of Biol. Chem. 213, 941—949 (1955). — HUNTER, R. F., and R. M. KRAKENBERGER: The oxidation of β-carotene in solution by oxygen. J. Chem. Soc. (Lond.) 1947, 1—4.

IRVINE, G. N.: Some effects of semolina lipoxidase activity on macaroni quality. J. Amer. Oil. Chem. Soc. 32, 558—561 (1955). — IRVINE, G. N., and J. A. ANDERSON: [1] Kinetic studies of the lipoxidase system of wheat. Cereal Chem. 30, 247—255 (1953). — [2] Note on the lipoxidase activity of various North American wheats. Cereal Chem. 30, 255—257 (1953). — The inhibition of wheat lipoxidase by cyanide. Cereal Chem. 32, 140—143 (1955). — IVANOW, S.: Über die Verwandlung des Öls in der Pflanze. Jb. wiss. Bot. 50, 375—386 (1912). — IVLER, D., J. B. WOLFE and S. C. RITTENBERG: Studies on the aerobic oxidation of fatty acids by bacteria. V. Caprate oxidation by cell-free extracts of *Pseudomonas fluorescens*. J. Bacter. 70, 99—103 (1955).

JANECKE, H.: Über das Fermentsystem der Lipoxydase. Pharmazie 5, 201—206 (1950). — Die Autoxydation von Fetten und die Anwendung von Antioxydantien. Arzneimittel-Forsch. 3, 574—586, 632—639 (1953). — JENSEN, L. B., and D. P. GRETTIE: Action of microorganisms on fats. Oil a. Soap 10, 23—32 (1933). — Action of microorganisms on fats. Food. Res. 2, 97—120 (1937). — JENSEN, O.: Studien über das Ranzigwerden der Butter. Zbl. Bakter. II 8, 107—114, 140—144, 278—281 (1902). — JEZESKI, J. J.: The enzymatic oxidation of fatty acids. Trans. Amer. Assoc. Cereal Chem. 5, 37—49 (1947). — JONES, M. E., S. BLACK, R. M. FLYNN and F. LIPMANN: Acetyl coenzyme A synthesis through pyrophosphoryl split of adenosine triphosphate. Biochim. et Biophysica Acta 12, 141—149 (1953).

KALCKAR, H. M.: The nature of energetic coupling in biological syntheses. Chem. Rev. 28, 71—178 (1941). — KARRER, P., u. F. HAAB: Über die enzymatische Decarboxylierung von β-Ketocarbonsäuren. Helvet. chim. Acta 31, 795—798 (1948). — KARRER, P., u. E. JUCKER: Carotinoide, S. 128f. Basel: Birkhäuser 1948. — KATZ, J., and I. L. CHAIKOFF:

Synthesis via the Krebs cycle in the utilization of acetate by rat liver slices. Biochim. et Biophysica Acta 18, 87—101 (1955). — KAUFMANN, H. P.: Studien auf dem Fettgebiet, S. 14f. Berlin: Verlag Chemie 1935. — KAUFMANN, H. P., u. H. FIEDLER: Untersuchungen über den natürlichen und künstlichen Oxydationsschutz der Fette. I. Zur Prüfung der Oxydationsschutzmittel. Fette u. Seifen 46, 275—277 (1939). — KENNEDY, E. P., and H. A. BARKER: Butyrate oxidation in the absence of inorganic phosphate by *Clostridium kluyveri*. J. of Biol. Chem. 191, 419—438 (1951). — KENNEDY, E. P., and A. L. LEHNINGER: The enzymatic oxidation of fatty acids. Phosphorus Metabolism 2, 253—281 (1952). — KERN, W., u. H. WILLERSINN: Die Katalyse der Autoxydation ungesättigter Verbindungen. Angew. Chem. 67, 573—581 (1955). — KERNS, D. M., R. BELKENGREN, H. CLARK and E. S. MILLER: Far-ultraviolet spectrophotometric studies on fat acids by photoelectric and spectrographic methods. J. Opt. Soc. Amer. 31, 271—279 (1940). — KHAN, N. A.: Biological oxidation. I. The infrared studies on the lipoxidase-catalyzed oxidation of linoleic acid. Arch. of Biochem. a. Biophysics 44, 247—249 (1953). — A new concept on the mechanism of autoxidation of methyl oleate, linoleate, and linolenate. Canad. J. Chem. 32, 1149—1154 (1954). — Infrared studies on initiation of the autoxidation of some fatty acid esters with and without light-sensitized chlorophyll, ultraviolet light and lipoxidase. Biochim. et Biophysica Acta 16, 159—160 (1955). — KHAN, N. A., W. O. LUNDBERG and R. T. HOLMAN: Displacement analysis of lipids. IX. Products of the oxidation of methyl linoleate. J. Amer. Chem. Soc. 76, 1779—1784 (1954). — KIES, M. W.: Effect of hemin proteins on soybean lipoxidase. Federat. Proc. 5, 141 (1946). — [1] Activation of soybean lipoxidase. J. of Biol. Chem. 170, 121—132 (1947). — [2] Complex nature of soybean lipoxidase. Federat. Proc. 6, 267 (1947). — KIRSSANOWA, W. A.: Über die Oxydation des Carotins [Russisch]. Biochimija 3, 191—201 (1938). — KNOOP, F.: Der Abbau aromatischer Fettsäuren im Tierkörper. Beitr. chem. Physiol. u. Path. 6, 150—162 (1904). — KORNBERG, A., and W. E. PRICER jr.: Enzymatic synthesis of the coenzyme A derivatives of long chain fatty acids. J. of Biol. Chem. 204, 329—343 (1953). — KUNKEL, H. O.: The coupled reactions between methyl linoleate and bixin or tocopherol during oxidation by lipoxidase. Arch. of Biochem. 30, 306—316 (1951). KUNKEL, H. O., and W. L. NELSON: The effect of bixin and carotene on the oxidation of methyl linoleate. J. of Biol. Chem. 183, 149—156 (1949).

LAIGRET, J.: Fermentation anaérobie des oléates alcalins: production de pétrole. C. r. Acad. Sci. Paris 225, 398—399 (1947). — LANG, K.: Über die tierische Fettsäuredehydrase und ihre Codehydrase. I. Hoppe-Seylers Z. 261, 240—248 (1939). — Der intermediäre Stoffwechsel, S. 262—263. Berlin-Göttingen-Heidelberg: Springer 1952. — LANG, K., u. H. MAYER: [1] Über die tierische Fettsäuredehydrase und ihre Codehydrase. II. Hoppe-Seylers Z. 261, 249—252 (1939). — [2] Über die tierische Fettsäuredehydrase und ihre Codehydrase. III. Mitt. Über die chemische Natur der „Codehydrasen". Hoppe-Seylers Z. 262, 120—122 (1939). — LAWRENCE, J. M.: Formation of reducing substances in pea seeds. Arch. of Biochem. 27, 1—5 (1950). — LE BRETON, E., et J. CHAMPOUGNY-CLÉMENT: La désaturase des acides gras supérieurs. Arch. Sci. physiol. 2, 243—255 (1948). — LEHNINGER, A. L.: [1] The relationship of the adenosine polyphosphates to fatty acid oxidation in homogenized liver preparations. J. of Biol. Chem. 154, 309—310 (1944). — [2] The relationship of the adenosine polyphosphates to fatty acid oxidation in homogenized liver preparations. J. of Biol. Chem. 157, 363—381 (1944). — [1] Fatty acid oxidation and the KREBS tricarboxylic acid cycle. J. of Biol. Chem. 161, 413—414 (1945). — [2] Activation of fatty acid oxidation. J. of Biol. Chem. 161, 437—451 (1945). — A quantitative study of the products of fatty acid oxidation in liver suspensions. J. of Biol. Chem. 164, 291—306 (1946). — LEHNINGER, A. L., and G. D. GREVILLE: Enzymic oxidation of d- and l-β-hydroxybutyrate. Biochim. et Biophysica Acta 12, 188—202 (1953). — LELOIR, L. F., and J. M. MUÑOZ: Butyrate oxidation by liver enzymes. J. of Biol. Chem. 153, 53—60 (1944). — LIEBERMAN, I., and H. A. BARKER: β-Keto-acid formation and decomposition by preparations of *Clostridium kluyveri*. J. Bacter. 68, 329—333 (1954). — LINDEMANN, E.: Untersuchungen über die Veränderung des Fettgehaltes von Milo- und Maiskeimen während der Lagerung. Stärke 5, 139—143 (1953). — LINEWEAVER, H., and D. BURK: The determination of enzyme dissociation constants. J. Amer. Chem. Soc. 56, 658—666 (1934). — LIPMANN, F.: Acetylation of sulfanilamide by liver homogenates and extracts. J. of Biol. Chem. 160, 173—190 (1945). — Biosynthesis mechanisms. Harvey Lect. 44, 99—123 (1948/49). — LIPMANN, F., M. E. JONES, S. BLACK and R. M. FLYNN: The mechanism of the ATP-CoA-reaction. J. Cellul. a. Comp. Physiol. 41, Suppl., 109—112 (1953). — LIPMANN, F., and C. L. TUTTLE: Determination of acyl phosphates. J. of Biol. Chem. 159, 21—28 (1945). — Lipase-catalyzed condensation of fatty acids with hydroxylamine. Biochim. et Biophysica Acta 4, 301—309 (1950). — LIPTON, M. A., and E. S. G. BARRON: Anaerobic synthesis of acetylcholine. J. of Biol. Chem. 166, 367—380 (1946). — LÜDE, R.: Die Gewinnung von Fetten und fetten Ölen, S. 122f. Dresden u. Leipzig: Theodor Steinkopff 1943. — LUNDBERG, W. O.: Durch Lipoxydase katalysierte Oxydation polyungesättigter Fettsäuren. Fette, Seifen, Anstrichmitt. 58, 329—331 (1956). —

LYNEN, F.: Zum biologischen Abbau der Essigsäure. I. Über die „Induktionszeit" bei ver-armter Hefe. Liebigs Ann. **552**, 270—306 (1942). — Zum biologischen Abbau der Essigsäure II. Die Wirkung von Malonsäure auf den Abbau der Essigsäure durch Hefe. Liebigs Ann **554**, 40—68 (1943). — [1] Functional group of coenzyme A and its metabolic relations especially in the fatty acid cycle. Federat. Proc. **12**, 683—691 (1953). — [2] Mécanisme de la β-oxydation des acides gras. Bull. Soc. Chim. biol. Paris **35**, 1061—1083 (1953). — Participation of coenzyme A in the oxidation of fat. Nature (Lond.) **174**, 962—965 (1954). — Der Fettsäurecyclus. Angew. Chem. **67**, 463—470 (1955). — LYNEN, F., and S. OCHOA Enzymes of fatty acid metabolism. Biochim. et Biophysica Acta **12**, 299—314 (1953). — LYNEN, F., E. REICHERT u. L. RUEFF: Zum biologischen Abbau der Essigsäure. IV. „Aktivierte Essigsäure", ihre Isolierung aus Hefe und ihre chemische Natur. Liebigs Ann. **574** 1—32 (1951). — LYNEN, F., L. WESSELY, O. WIELAND u. L. RUEFF: Zur β-Oxydation der Fettsäuren. Angew. Chem. **64**, 687 (1952). — LYNEN, F., and O. WIELAND: β-Ketoreductase In S. P. COLOWICK-N. O. KAPLAN, Methods in Enzymology, Bd. 1, S. 566—573. New York Academic Press 1955.

MAHLER, H. R.: Studies on the fatty acid oxidizing system of animal tissues. IV. The prosthetic group of butyryl coenzyme A dehydrogenase. J. of Biol. Chem. **206**, 13—2((1953). — Nature and function of metallo-flavoproteins. Adv. Enzymol. **17**, 233—29 (1956). — MAHLER, H. R., and D. E. GREEN: Metallo-flavoproteins and electron transport. Science (Lancaster, Pa.) **120**, 7—12 (1954). — MAHLER, H. R., S. J. WAKIL and R. M BOCK: Studies on fatty acid oxidation. I. Enzymatic activation of fatty acids. J. of Biol Chem. **204**, 453—468 (1953). — MAPSON, L. W.: The estimation of oxidized glutathione Biochemic. J. **55**, 714—717 (1953). — MAPSON, L. W., and E. M. MOUSTAFA: The oxidation of glutathione by a lipoxidase enzyme from pea seeds. Biochemic. J. **60**, 71—80 (1955). — MARKLEY, K. S.: Fatty Acids. Their chemistry and physical properties, S. 55f. u. 359f New York: Interscience Publishers Inc. 1947. — MARTIUS, C., and F. LYNEN: Probleme des Citronensäurecyclus. Adv. Enzymol. **10**, 167—222 (1950). — MATIC, M.: The chemistry of plant cuticles: a study of cutin from *Agave americana* L. Biochemic. J. **63**, 168—17((1956). — MAZZA, F. P.: Sulla deidrogenasi degli acidi grassi superiori. Erg. Enzymforsch **9**, 207—230 (1943). — MAZZA, F. P., e A. CIMMINO: Sull'attività deidrogenasica del *B. coli communis* sugli acidi grassi superiori. I. Rend. R. Accad. Lincei **17**, 1086—1091 (1933). — Sull'attività deidrogenasica del *B. coli communis* sugli acidi grassi superiori. II. Rend. R Accad. Lincei **20**, 113—118 (1934). — MICHAELIS, L.: Oxydations-Reduktionspotentiale 2. Aufl., S. 107—125. Berlin: Springer 1933. — Semiquinones, the intermediate steps c reversible organic oxidation-reduction. Chem. Rev. **16**, 243—286 (1935). — Fundamental of oxidation and reduction. In D. E. GREEN, Currents in Biochemical Research, S. 207—227 New York: Interscience Publishers Inc. 1946. — MICHAELIS, L., u. M. L. MENTEN: Di Kinetik der Invertinwirkung. Biochem. Z. **49**, 333—369 (1913). — MICHAELIS, L., and M. P SCHUBERT: The theory of reversible two-step oxidation involving free radicals. Chem. Rev **22**, 437—470 (1938). — MICHAELIS, L., and S. H. WOLLMAN: The semiquinone radical of toco-pherol. Science (Lancaster, Pa.) **109**, 313—314 (1949). — MICHLIN, D. M., i K. W. PSCHENNOWA: Lipoxydase [Russisch]. Biochimija **11**, 437—446 (1946). — Über den Wirkungsmechanismus eines peroxydierenden Ferments (Lipoxydase) [Russisch]. Biochimija **13**, 76—7 (1948). — Oxydation von Aminosäuren mittels des Lipoxydase-Systems [Russisch]. Biochimija **14**, 141—144 (1949). — MIKUSCH, J. D. v.: Die oxydative Trocknung von Ölen un Ölderivaten im Lichte neuer Forschungsergebnisse. Seifen, Öle, Fette, Wachse **82**, 113—11(141—145, 169—172 (1956). — MILLER, B. S., and F. A. KUMMEROW: The disposition c lipase and lipoxidase in baking and the effect of their reaction products on consumer acceptability. Cereal Chem. **25**, 391—398 (1948). — MILLER, E. C.: A physiological study of th germination of *Helianthus annuus*. II. The oily reserve. Ann. of Bot. **26**, 889—90 (1912). — MILLERD, A., and J. BONNER: Acetate activation and acetoacetate formatio in plant systems. Arch. of Biochem. a. Biophysics **49**, 343—355 (1954). — MITCHELL, H. L and S. M. HAUGE: [1] Enzymic nature of the carotene-destroying system in alfalfa. J of Biol. Chem. **163**, 7—14 (1946). — [2] Factors affecting the enzymic destruction c carotene in alfalfa. J. of Biol. Chem. **164**, 543—550 (1946). — MITCHELL, H. L., an H. H. KING: Effects of dehydration on enzymic destruction of carotene in alfalfa. J. c Biol. Chem. **166**, 477—480 (1946). — MITCHELL, J. H., H. R. KRAYBILL and F. P. ZSCHEILE Quantitative spectrum analysis of fats. Ind. Engng. Chem., Analyt. Ed. **15**, 1—3 (1943). — MOELWYN-HUGHES, E. A.: The kinetics of enzyme reactions. Erg. Enzymforsch. **6**, 23—4 (1937). — MOHR, W.: Über Haferpräparierung und Haferbitterstoffe. Angew. Chem. **6** 482 (1954). — MOIR, R. J., M. SOMERS, G. SHARMAN and H. WARING: Ruminant-like digestion in a marsupial. Nature (Lond.) **173**, 269—270 (1954). — MUFTIC, M. K.: Demonstration of fat peroxides in *Mycobact. tuberculosis* treated by cerase. Enzymologia (Den Haag **17**, 222—224 (1955). — MUKHERJEE, S.: Studies on degradation of fats by microorganism [1] II. The mechanism of formation of ketones from fatty acids. Arch. of Biochem.

Biophysics **35**, 23—33 (1952). — Studies on degradation of fats by microorganisms. [2] III. The metabolism of butyric acid by *Asp. niger*. Arch. of Biochem. a. Biophysics **35**, 34—59 (1952).— [3] A procedure for the determination of total unsaturation in the products of in vitro oxidation of fatty acids in biological systems. J. Amer. Oil Chem. Soc. **29**, 97—99 (1952). — MUNDT, J. O., and F. W. FABIAN: The bacterial oxidation of corn oil. J. Bacter. **48**, 1—12 (1944). — MUÑOZ, J. M., and L. F. LELOIR: Fatty acid oxidation by liver enzymes. J. of Biol. Chem. **147**, 355—362 (1943). — MYRBÄCK, K.: Enzymatische Katalyse, S. 18f. Berlin: W. de Gruyter & Co. 1953.

NACHMANSOHN, D., and M. BERMAN: Studies on choline acetylase. III. On the preparation of the coenzyme and its effect on the enzyme. J. of Biol. Chem. **165**, 551—563 (1946). — NEUBERG, C., u. G. GORR: Phytochemische Reduktionen. In C. OPPENHEIMER-L. PINCUSSEN, Die Methodik der Fermente, S. 1212—1221. Leipzig: Georg Thieme 1929. — NEUMANN. M. P., u. P. F. PELSHENKE: Brotgetreide und Brot, S. 305. Berlin u. Hamburg: Parey 1954. — NEWCOMB, E. H., and P. K. STUMPF: Fatty acid synthesis and oxidation in peanut cotyledons. Phosphorus Metabolism **2**, 291—300 (1952). — Fat metabolism in higher plants. I. Biogenesis of higher fatty acids by slices of peanut cotyledon in vitro. J. of Biol. Chem. **200**, 233—239 (1953). — NISMAN, B.: The STICKLAND reaction. Bacter. Rev. **18**, 16—42 (1954). — NISMAN, B., G. N. COHEN, S. WIESENDANGER et J. SZULMAJSTER: La dégradation du β-hydroxybutyrate par les extraits de *Clostridium saccharobutyricum*. C. r. Acad. Sci. Paris **237**, 1806—1809 (1953). — NOLL jr., C. R., and R. H. BURRIS: Nature and distribution of glycolic oxidase in plants. Plant Physiol. **29**, 261—265 (1954).

OCHOA, S.: Biological mechanisms of carboxylation and decarboxylation. Physiologic. Rev. **31**, 56—106 (1951). — Enzymic mechanisms in the citric acid cycle. Adv. Enzymol. **15**, 183—270 (1954). — OLCOTT, H. S., and H. A. MATTILL: Antioxidants and the autoxidation of fats. IV. Lecithin as an antioxidant. Oil a. Soap **13**, 98—100 (1936). — OTTOLENGHI, A., F. BERNHEIM and K. M. WILBUR: The inhibition of certain mitochondrial enzymes by fatty acids oxidized by ultraviolet light or ascorbic acid. Arch. of Biochem. a. Biophysics **56**, 157—164 (1955).

PASCHKE, R. F., W. TOLBERG and D. H. WHEELER: Cis, trans isomerism of the eleostearate isomers. J. Amer. Oil Chem. Soc. **30**, 97—99 (1953). — PEEL, J. L., and H. A. BARKER: The reduction of vinylacetate by *Clostridium kluyveri* and its dependence on catalytic amounts of high-energy acetate. Biochemic. J. **62**, 323—332 (1956). — PETERSON, R. W., and J. H. WALTON: The autoxidation of ascorbic acid. J. Amer. Chem. Soc. **65**, 1212—1217 (1943). — PRIESTLEY, J. H.: The cuticle in angiosperms. Bot. Review **9**, 593—616 (1943). — PRIVETT, O. S., W. O. LUNDBERG, N. A. KHAN, W. E. TOLBERG and D. H. WHEELER: Structure of hydroperoxides obtained from autoxidized methyl linoleate. J. Amer. Oil Chem. Soc. **30**, 61—66 (1953). — PRIVETT, O. S., and CH. NICKELL: Concurrent oxidation of accumulated hydroperoxydes in the autoxidation of methyl linoleate. J. Amer. Oil Chem. Soc. **33**, 156—163 (1956). — PRIVETT, O. S., CH. NICKELL, W. E. TOLBERG, R. F. PASCHKE, D. H. WHEELER and W. O. LUNDBERG: Evidence for hydroperoxide formation in the autoxidation of methyl linolenate. J. Amer. Oil Chem. Soc. **31**, 23—27 (1954). — PRIVETT, O. S., and F. W. QUACKENBUSH: Studies on the antioxygenic properties of wheat germ phosphatides. J. Amer. Oil Chem. Soc. **31**, 169—171 (1954).

QUASTEL, J. H., and W. R. WOOLDRIDGE: Some properties of the dehydrogenating enzymes of bacteria. Biochemic. J. **22** 689—702 (1928).

RAY S. N.: On the nature of the precursors of the vitamin C in the vegetable kingdom. I. Vitamin C in the growing pea seedling. Biochemic. J. **28**, 996—1002 (1934). — REISER, R.: Peroxidizing and carotene bleaching substances in bacon adipose tissue. J. Amer. Oil Chem. Soc. **26**, 116—120 (1949). — Hydrogenation of polyunsaturated fatty acids by the ruminant. Federat. Proc. **10**, 236 (1951). — REISER, R., and G. S. FRAPS: Determination of carotene oxidase in legume seeds. J. Assoc. Agric. Chem. **26**, 186—194 (1943). — REISER, R., and RAMAKRISHNA REDDY: The hydrogenation of dietary unsaturated fatty acids by the ruminant. J. Amer. Oil Chem. Soc. **33**, 155—156 (1956). — RITTENBERG, D., and R. SCHOENHEIMER: Deuterium as an indicator in the study of intermediary metabolism. VIII. Hydrogenation of fatty acids in the animal organism. J. of Biol. Chem. **117**, 485—490 (1937). — ROSENFELD, W. D.: Fatty acid transformations by anaerobic bacteria. Arch. of Biochem. **16**, 263—273 (1948). — ROTHE, M.: Über das Bitterwerden von Cerealien. I. Lipoide als Reaktionspartner. Fette u. Seifen **55**, 877—880 (1953). — Über das Bitterwerden von Cerealien. II. Zusammenhang zwischen Fettautoxydation und Bitterstoff-Bildung. Fette, Seifen, Anstrichmitt. **56**, 667—670 (1954). — Über das Bitterwerden von Cerealien. III. Zur Frage des Reaktionsmechanismus. Fette, Seifen, Anstrichmitt. **57**, 425—428 (1955). — RUPP, E., u. W. BRACHMANN: Zur Bromzahlbestimmung der Fette. Z. anal. Chem. **68**, 155—160 (1926).

SCHÖNBRUNNER, J.: Über die bakterielle Hydrierung von Ölsäure und Sorbinsäure und über ihre Beeinflussung durch Gallensäure. Biochem. Z. **304**, 26—36 (1940). — SEPHTON,

H. H., and D. A. SUTTON: The chemistry of polymerized oils. V. The autoxidation of methyl linoleate. J. Amer. Oil Chem. Soc. **33**, 263—272 (1956). — SEUBERT, W., and F. LYNEN: Enzymes of the fatty acid cycle. II. Ethylene reductase. J. Amer. Chem. Soc. **75**, 2787 (1953). — SHAPIRO, B., and E. WERTHEIMER: Fatty acid dehydrogenase in adipose tissue. Biochemic. J. **37**, 102—104 (1943). — SIDDIQI, A. M., and A. L. TAPPEL: [1] Catalysis of linoleate oxidation by pea lipoxidase. Arch. of Biochem. a. Biophysics **60**, 91—99 (1956). — [2] Alfalfa lipoxidase. Plant Physiol. **31**, 320—321 (1956). — SMITH, G. N.: Studies on lipoxidase. IV. Effect of changes in temperature and p_H on lipoxidase activity as determined by spectral changes in methyl linoleate. Arch. of Biochem. **19**, 133—143 (1948). — SMITH, G. N., and J. B. SUMNER: [1] The induced reaction between methyl linoleate and bixin during oxidation by lipoxidase. Arch. of Biochem. **17**, 75—80 (1948). — [2] On the activation of lipoxidase. Arch. of Biochem. **19**, 89—94 (1948). — SÖDING, H.: Die Wuchsstofflehre. Ergebnisse und Probleme der Wuchsstofforschung, S. 1—17. Stuttgart: Georg Thieme 1952. — STADTMAN, E. R.: Coenzyme A-dependent transacetylation and transphosphorylation. Federat. Proc. **9**, 233 (1950). — The net enzymatic synthesis of acetyl coenzyme A. J. of Biol. Chem. **196**, 535—546 (1952). — The coenzyme A transphorase system in *Clostridium kluyveri*. J. of Biol. Chem. **203**, 501—512 (1953). — STADTMAN, E. R., and H. A. BARKER: Fatty acid synthesis by enzyme preparations of *Clostridium kluyveri*. [1] I. Preparation of cell-free extracts that catalyze the conversion of ethanol and acetate to butyrate and caproate. J. of Biol. Chem. **180**, 1085—1093 (1949). — [2] II. The aerobic oxidation of ethanol and butyrate with the formation of acetyl phosphate. J. of Biol. Chem. **180**, 1095—1115 (1949). — [3] III. The activation of molecular hydrogen and the conversion of acetyl phosphate and acetate to butyrate. J. of Biol. Chem. **180**, 1117—1124 (1949). — [4] IV. The phosphoroclastic decomposition of acetoacetate to acetyl phosphate and acetate. J. of Biol. Chem. **180**, 1169—1186 (1949). — [5] V. A consideration of postulated 4-carbon intermediates in butyrate synthesis. J. of Biol. Chem. **181**, 221—235 (1949). — Fatty acid synthesis by enzyme preparations of *Clostridium kluyveri*. VI. Reactions of acyl phosphates. J. of Biol. Chem. **184**, 769—793 (1950). — STADTMAN, E. R., G. D. NOVELLI and F. LIPMANN: Coenzyme A function in acetyl transfer by the phosphotransacetylase system. J. of Biol. Chem. **191**, 365—376 (1951). — STADTMAN, E. R., and T. C. STADTMAN: Metabolism of microorganisms. Annual Rev. Microbiol. **7**, 143—178 (1953). — STADTMAN, T. C., and H. A. BARKER: Studies on the methane fermentation. VIII. Tracer experiments on fatty acid oxidation by methane bacteria. J. Bacter. **61**, 67—80 (1951). — STÄRKLE, M.: Die Methylketone im oxydativen Abbau einiger Triglyceride (bzw. Fettsäuren) durch Schimmelpilze unter Berücksichtigung der besonderen Ranzidität des Kokosfettes. Biochem. Z. **151**, 371—415 (1924). — STERN, J. R.: Crystalline crotonase from ox liver. In S. P. COLOWICK-N. O. KAPLAN, Methods in Enzymology, Bd. 1, S. 559—566. New York: Academic Press 1955. — STERN, J. R., M. J. COON and A. DEL CAMPILLO: [1] Acetoacetyl coenzyme A as intermediate in the enzymatic breakdown and synthesis of acetoacetate. J. Amer. Chem. Soc. **75**, 1517—1518 (1953). — [2] Enzymatic breakdown and synthesis of acetoacetate. Nature (Lond.) **171**, 28—30 (1953). — STERN, J. R., S. OCHOA and F. LYNEN: Enzymatic synthesis of citric acid. V. Reaction of acetyl coenzyme A. J. of Biol. Chem. **198**, 313—321 (1952). — STOKOE, W. N.: The rancidity of coconut oil produced by mould action. Biochemic. J. **22**, 80—93 (1928). — STRAIN, H. H.: Unsaturated fat oxidase: specificity, occurrence and induced oxidations. J. Amer. Chem. Soc. **63**, 3542 (1941). — Enzymic oxidation of unsaturated fats and carotenoid pigments. Acta phytochim. (Tokyo) **15**, 9—16 (1949). — STUMPF, P. K.: Fat metabolism in higher plants. III. Enzymic oxidation of glycerol. Plant Physiol. **30**, 55—58 (1955). — STUMPF, P. K., and G. A. BARBER: Fat metabolism in higher plants. VII. β-oxidation of fatty acids by peanut mitochondria. Plant Physiol. **31**, 304—308 (1956). — SÜLLMANN, H.: [1] Über den Einfluß von Lecithin und Vitamin E auf die enzymatische Oxydation des Carotins. Helvet. chim. Acta **24**, 465—472 (1941). — [2] Über die enzymatische Oxydation des Carotins. Helvet. chim. Acta **24**, 646—657 (1941). — [3] Die Aufnahme von Sauerstoff bei der enzymatischen Oxydation ungesättigter Fettsäuren. Helvet. chim. Acta **24**, 1360—1380 (1941). — Bildung von Carbonylverbindungen bei der enzymatischen Oxydation ungesättigter Fettsäuren. Helvet. chim. Acta **25**, 521—523 (1942). — [1] Inhibitoren der enzymatischen Oxydation ungesättigter Fettsäuren. Helvet. chim. Acta **26**, 1114—1124 (1943). — [2] Zur Kenntnis der Lipoxydase. Helvet. chim. Acta **26**, 2253—2263 (1943). — Über die Wirkung der Lipoxydase auf die Oxydation der Eläostearinsäure. Helvet. chim. Acta **27**, 789—793 (1944). — [1] Über die Oxydation ungesättigter Fettsäuren durch Pflanzensäfte. Experientia (Basel) **1**, 323—324 (1945). — [2] Die Lipoxydase. Fermentforsch. **17**, 610—631 (1945). — SUMNER, J. B., u. A. L. DOUNCE: Carotene oxidase. Enzymologia (Den Haag) **7**, 130—132 (1939). — SUMNER, J. B., and G. N. SMITH: The estimation of lipoxidase activity. Arch. of Biochem. **14**, 87—92 (1947). — SUMNER, J. B., and R. J. SUMNER: The coupled oxidation of carotene and fat by carotene oxidase. J. of Biol. Chem. **134**, 531—533 (1940). — SUMNER, R. J.: [1] Lipid oxidase studies.

II. The specificity of the enzyme lipoxidase. J. of Biol. Chem. **146**, 211—213 (1942). — [2] Lipid oxidase studies. III. The relation between carotene oxidation and the enzymic peroxidation of unsaturated fats. J. of Biol. Chem. **146**, 215—218 (1942). — Lipoid oxidase studies. A method for the determination of lipoxidase activity. Ind. Engng. Chem., Analyt. Edit. **15**, 14—15 (1943). — Synerholm, M. E., and P. W. Zimmerman: Preparation of a series of ω-(2,4-dichloro-phenoxy-)aliphatic acids and some related compounds with a consideration of their biochemical role as plant-growth regulators. Contrib. Boyce Thompson Inst. **14**, 369—382 (1947). — Szulmajster, J.: Sur le mécanisme de la formation des acides gras inférieurs chez les Clostridies. II. Ethylène hydrogénase. Activation du crotonate par adénosine-triphosphate et le coenzyme A. C. r. Acad. Sci. Paris **238**, 2461—2463 (1954). — Szulmajster, J., B. Nisman et G. Cohen: Sur le mécanisme de la formation des acides gras inférieurs chez les Clostridies. I. Mise en évidence de la thiolase et de la β-céto-hydrogénase. C. r. Acad. Sci. Paris **238**, 164—166 (1954).

Täufel, K., u. H. Rothe: Zur Chemie des Verderbens der Fette. XX. Zur Charakteristik des fettantioxygenen Komplexes aus Hafer. Fette u. Seifen **51**, 100—102 (1944). — Täufel, K., H. Thaler u. H. Hohner: Über den qualitativen und quantitativen Nachweis von Methylketonen. Z. Unters. Lebensmitt. **74**, 119—133 (1937). — Täufel, K., H. Thaler u. M. Löweneck: Über den Abbau von Fettsäuren durch Schimmelpilze. Fettchem. Umschau **43**, 1—4 (1936). — Tanaka, Y., H. Tanno and K. Hoshishima: Studies on the lipoxidase of the avian tubercle bacilli. (Preliminary report.) Fukushima J. Med. Sci. **2**, 1—2 (1955). — Tappel, A. L.: Linoleate oxidation catalyzed by hog muscle and adipose tissue. Food Res. **17**, 550—559 (1952). — [1] Linoleate oxidation catalysts occurring in animal tissues. Food Res. **18**, 104—108 (1953). — [2] The mechanism of the oxidation of unsaturated fatty acids catalyzed by hematin compounds. Arch. of Biochem. a. Biophysics **44**, 378—395 (1953). — [3] The inhibition of hematin-catalyzed oxidations by α-tocopherol. Arch. of Biochem. a. Biophysics **47**, 223—225 (1953). — Studies of the mechanism of vitamin E action. II. Inhibition of unsaturated fatty acid oxidation catalyzed by hematin compounds. Arch. of Biochem. a. Biophysics **50**, 473—485 (1954). — Tappel, A. L., P. D. Boyer and W. O. Lundberg: The reaction mechanism of soy bean lipoxidase. J. of Biol. Chem. **199**, 267—281 (1952). — Tappel, A. L., W. O. Lundberg and P. D. Boyer: Effect of temperature and antioxidants upon the lipoxidase-catalyzed oxidation of sodium linoleate. Arch. of Biochem. a. Biophysics **42**, 293—304 (1953). — Tappel, A. L., and A. G. Marr: Effect of α-tocopherol, propyl gallate, and nordihydroguajaretic acid on enzymic reactions. J. Agricult. Food Chem. **2**, 554—558 (1954). — Tauber, H.: Unsaturated fat oxidase. J. Amer. Chem. Soc. **62**, 2251 (1940). — Thaler, H., u. W. Eisenlohr: Zur Chemie der Ketonranzigkeit. III. Über die Bildung von Methylketonen aus α,β-ungesättigten Fettsäuren durch *Penicillium glaucum*. Biochem. Z. **308**, 88—102 (1941). — Thaler, H., u. G. Geist: [1] Zur Chemie der Ketonranzigkeit. I. Über den Abbau gesättigter Fettsäuren durch *Penicillium glaucum*. Biochem. Z. **302**, 121—136 (1939). — [2] Zur Chemie der Ketonranzigkeit. II. Über die Bildung von Methylketonen aus β-Oxyfettsäuren durch *Penicillium glaucum*. Biochem. Z. **302**, 369—383 (1939). — Thaler, H., u. I. Stählin: Zur Chemie der Ketonranzigkeit. IV. Die Identifizierung der Methylketone. Biochem. Z. **320**, 84—86 (1949). — Thaler, H., A. Schottmayer, I. Stählin u. H. Beck: Zur Chemie der Ketonranzigkeit. V. Die Bildung von Methylketonen aus α- und β-Methylfettsäuren. Biochem. Z. **320**, 87—98 (1949). — Theorell, H., S. Bergström u. Å. Åkeson: On the lipoxidase enzymes in soy bean. Ark. Kemi, Mineral. o. Geol. A **19**, Nr 6 (1944), 9 S. — Activity determination and further purification of the lipoxidase. Pharmaceut. Acta helvet. **21**, 318—324 (1946). — Theorell, H., R. T. Holman and Å. Åkeson: [1] A note on the preparation of crystalline soy bean lipoxidase. Arch. of Biochem. **14**, 250—252 (1947). — [2] Crystalline lipoxidase. Acta chem. scand. (Copenh.) **1**, 571—576 (1947). — Thomas, B., u. M. Rothe: Die Beteiligung von Fermenten an der Bildung von Fettbitterstoffen in versehrten Getreidekörnern. Getreide u. Mehl **4**, 17—19 (1954). — Thunberg, T.: Acceptormethode, Dehydrasen der Carbonsäuren, Redoxpotentiale. In C. Oppenheimer-L. Pincussen, Die Methodik der Fermente, S. 1118—1134. Leipzig: Georg Thieme 1929. — Zur Kenntnis der Spezifität der Dehydrogenasen. Biochem. Z. **258**, 48—64 (1933). — Todd, J. P., J. Hawthorn and J. A. Blain: The bleaching and improvement of bread doughs. Chem. a. Ind. **1954**, 50. — Troitzki, G. W.: Beziehung zwischen Struktur und Absorptionsspektren von Gliedern der Vitamin A-Gruppe [Russisch]. Biochimija **13**, 7—15 (1948).

Verkade, P. E., M. Elzas, J. van der Lee, H. H. de Wolf, A. Verkade-Sandbergen . D. van der Sande: Untersuchungen über den Fettstoffwechsel. I. Hoppe-Seylers Z. **15**, 225—257 (1933). — Verkade, P. E., and J. van der Lee: Researches on fat metaolism. II. Biochemic. J. **28**, 31—40 (1934).

Wain, R. L., and F. Wightman: The growth-regulating activity of certain ω-substituted lkyl carboxylic acids in relation to their β-oxidation within the plant. Proc. Roy. Soc.

Lond., Ser. B **142**, 525—536 (1954). — Chemistry and mode of action of plant growth substances. London: Butterworths Scientific 1956. — WAKIL, S. J., and H. R. MAHLER: Studies on the fatty acid oxidizing system of animal tissues. V. Unsaturated fatty acyl coenzyme A hydrase. J. of Biol. Chem. **207**, 125—132 (1954). — WALKER, T. K., and P. D. COPPOCK: Mechanism of the degradation of fatty acids by mould fungi. I. J. Chem. Soc. (Lond.) **1928**, 803—809. — WATERS, W. A.: A chemical interpretation of the mechanism of oxidation by dehydrogenase enzymes. Trans. Faraday Soc. **39**, 140—151 (1943). — Some recent developments in the chemistry of free radicals. J. Chem. Soc. (Lond.) **1946**, 409—415. — WATTS, B. M., and D. PENG: Lipoxidase activity of hog hemoglobin and muscle extract. J. of Biol. Chem. **170**, 441—453 (1947). — WEBLEY, D. M., R. B. DUFF and V. C. FARMER: β-Oxidation of fatty acids by *Nocardia opaca*. J. Gen. Microbiol. **13**, 361—369 (1955). — WEIDENHAGEN, R.: Spezifität und Wirkungsmechanismus der Carbohydrasen. Erg. Enzymforsch. **1**, 168—208 (1932). — WIELAND, H.: Über den Mechanismus der Oxydationsvorgänge. Erg. Physiol. **20**, 477—518 (1922). — WIELAND, H., u. C. ROSENTHAL: Über den Mechanismus der Oxydationsvorgänge. LIII. Weitere Versuche über den biologischen Abbau der Essigsäure. Liebigs Ann. **554**, 241—260 (1943). — WOLIN, M. J., J. B. EVANS and C. F. NIVEN jr.: The oxidation of butyric acid by *Streptococcus mitis*. J. Bacter. **64**, 531—535 (1952).

YOSII, S.: Beiträge zur Kenntnis der Fettsäureoxydation im Tierkörper. J. of Biochem. **26**, 397—424 (1937).

ZELLER, A.: Untersuchungen über die Umwandlung höherer Fettsäuren in Kohlehydrate bei der Keimung von Kürbissamen. Jb. Bot. **82**, 123—157 (1935). — ZELLER, A., u. F. MASCHEK: Über eine Saturase höherer Fettsäuren aus Kürbiskeimlingen. Biochem. Z. **312**, 354—369 (1942). — ZIMMERMAN, P. W., and A. E. HITCHCOCK: Substituted phenoxy and benzoic acid growth substances and the relation of structure to physiological activity. Contrib. Boyce Thompson Inst. **12**, 321—343 (1942).

Nachträge während der Korrektur.

Zu S. 164. Eine neuerdings beobachtete Oxydation höherer gesättigter und ungesättigter Fettsäuren durch isolierte *Chloroplasten* der Bohne ist hinsichtlich des Chemismus und Mechanismus noch unklar. In den Chloroplasten wurden 5 μg Coenzym A je Gramm Trockengewicht gefunden. [SISSAKJAN, N. M., i B. P. SMIRNOW: Synthese und Oxydation von Fettsäuren in isolierten Chloroplasten (russisch). Biochimija **21**, 273—278 (1956).]

Zu S. 174. Nach neueren Untersuchungen wird Propionsäure im *tierischen* Organismus unter Carboxylierung und Beteiligung von Coenzym A über *Methylmalonsäure* (Isobernsteinsäure) zu Bernsteinsäure umgesetzt. [KATZ, J., and L. L. CHAIKOFF: The metabolism of propionate by rat liver slices and the formation of isosuccinic acid. J. Amer. Chem. Soc. **77**, 2659—2660 (1955). — FLAVIN, M., P. J. ORTIZ and S. OCHOA: Metabolism of propionic acid in animal tissues. Nature (Lond.) **176**, 823—826 (1955).]

Zu S. 178. Die Thiokinasereaktion verläuft in zwei Phasen, z. B. im Fall der Essigsäure nach a) Acetat + ATP → Acetyl-AMP + PP; b) Acetyl-AMP + CoA → Acetyl-CoA + AMP. Im *Adenyl-acetat* (oder *Acetyl-adenylat*) bildet der Acetylrest mit dem Phosphatrest der AMP ein gemischtes Anhydrid. [BERG, P.: Acyl adenylates: an enzymatic mechanism of acetate activation. J. of Biol. Chem. **222**, 991—1013 (1956). — Acyl adenylates: the synthesis and properties of adenyl acetate. J. of Biol. Chem. **222**, 1015—1023 (1956). — TALBERT, P. T., and F. M. HUENNEKENS: Chemical synthesis and properties of butyryl adenylate. J. Amer. Chem. Soc. **78**, 4671—4675 (1956). — LEE PENG, C. H.: Butyryl adenylate and its possible function in the fatty acid activating system. Biochim. et Biophysica Acta **22**, 42—48 (1956).]

Zu S. 190. Auch in frischen Pferde*faeces* lagen 14% der ungesättigten Fettsäuren in der *trans*-Form vor, während das (wesentlich stärker ungesättigte) *Depotfett* der Pferde keine trans-Säuren enthielt. [HARTMAN, L., F. B. SHORLAND and R. J. MOIR: Occurrence of trans-unsaturated fatty acids in horse faeces. Nature (Lond.) **178**, 1057—1058 (1956).]

Biochemistry of fat formation.

By

P. K. Stumpf.

A. Synthesis of fatty acids[1].

It is a well known fact that naturally occurring fatty acids have an even number of carbon atoms. Any mechanism for fatty acid synthesis must therefore explain this fundamental observation. Isotope studies with possible precursors of fatty acids as substrates for fatty acid synthesis by intact animals or tissue slices have provided considerable evidence in support of the hypothesis that long chain fatty acids are synthesized *de novo* by an initial condensation of 2-carbon fragments (acetyl CoA) and a subsequent lengthening of the hydrocarbon chain by additional 2-carbon condensations until the complete fatty acid molecule has been constructed (CHAIKOFF and BROWN).

The actual mechanism of synthesis appears to be a reversal of the β-oxidation degradation sequence[2]. Table 1 lists the enzymes and the reactions associated with the β-oxidation cycle as developed by GREEN and LYNEN. It will be noted that all five enzyme systems which participate in this cycle are reversible. Therefore they should catalyze the synthesis of long chain fatty acids containing

Table 1. *Enzymes participating in fatty acid oxidation.* (After GREEN.)

Step	Enzyme	Reaction	$K_{eq.}$[1]
A	Activating enzyme	a C_2–C_3 acids $\underset{CoA}{\overset{ATP}{\rightleftharpoons}}$ acyl CoA + AMP + PP	1
		b C_4–C_{12} acids $\underset{CoA}{\overset{ATP}{\rightleftharpoons}}$ acyl CoA + AMP + PP	
		c C_{12}–C_{18} acids $\underset{CoA}{\overset{ATP}{\rightleftharpoons}}$ acyl CoA + AMP + PP	
B	butyryl CoA dehydrogenase	butyryl CoA $\overset{\pm 2\,H}{\rightleftharpoons}$ crotonyl CoA	10
C	unsaturated fatty acyl CoA dehydrogenase	unsaturated fatty acyl CoA $\overset{\pm H_2O}{\rightleftharpoons}$ β-hydroxy acyl CoA	1.4
D	β-hydroxyacyl CoA dehydrogenase	β-hydroxyacyl CoA $\overset{DPN^+}{\rightleftharpoons}$ β-ketoacyl CoA	6.3×10^{-11}
E	β-hydroxyacyl CoA cleavage enzyme	β-ketoacyl CoA $\overset{\pm CoA}{\rightleftharpoons}$ acyl CoA + acetyl CoA	1.4×10^4

[1] K = product/reactant.

[1] The following abbreviations are used in this contribution: AMP, ADP, ATP adenosine mono-, di-, triphosphate; CoA coenzyme A; DHAP dihydroxyacetone phosphate; DPN⁺, DPNH diphosphopyridine nucleotide, oxydized and reduced; MP peanut mitochondria; PP pyrophosphate; SPF soluble protein fraction; TCA cycle tricarboxylic acid cycle; TPN triphosphopyridine nucleotide.

[2] Discussed by FRANKE and FREHSE, this volume, p. 174—188.

an even number of carbon atoms. In establishing the conditions for the synthetic reactions to take place, the following points should be noted:

(a) Since the reverse of β-oxidation is a β-reductive condensation, a source of hydrogen atoms must be supplied by the cell. At the acyl CoA $\rightleftharpoons$ unsaturated fatty acyl CoA level (Table 1, Step B), the E_0' of the reaction is $+0.19$ V at p_H 7 and 25^0. Since either the DPN or the TPN systems which are commonly available in the cell as a source of reductive capacity apparently cannot be coupled to the reduction of unsaturated acyl CoA, a different reducing system must be present which can transfer hydrogen atoms effectively to the reducible substrate. Experimentally, the reduced form of either benzyl viologen or safronin which have very negative E_0' is coupled with acyl CoA dehydrogenase for the reduction of unsaturated acyl CoA. In animal and plant tissues, the nature of the physiological reducing system is not known. The reversal of Step D is easily accomplished by coupling the system with any of a number of DPN systems commonly found in living cells.

(b) Although the $K_{eq.}$ of Steps B, C, and D are in favor of fatty acid synthesis, Step E strongly favors the direction of degradation. Therefore, for effective synthesis of long chain fatty acids to occur, the equilibria of the entire sequence must be shifted in the direction favoring synthesis. This may be accomplished (1) by esterification of glycerol by long chain fatty acyl CoA to form neutral fats:

$$3 \text{ acyl CoA} + \text{glycerol} \rightarrow \text{triglyceride} + 3 \text{ CoA}$$

(2) by deacylation of acyl CoA to yield free fatty acids. This irreversible reaction with its loss in free energy of about 15,000 calories is of little value to the economy of the cell. Furthermore, the concentration of free fatty acids in normal plant cells is very low. (3) by the driving force of Step D where the $K_{eq.}$ is greatly in favor of synthesis, and (4) by maintaining a low free CoA concentration since a high concentration would tend to favor the direction of degradation.

(c) In plants as in animals, the major proportion of long chain fatty acids are in the even carbon series. Thus stearic acid (C_{18}), palmitic acid (C_{16}), myristic acid (C_{14}), lauric acid (C_{12}), and oleic acid (C_{18}, 1 C=C), linoleic acid (C_{18}, 2 C=C), and linolenic acid (C_{18}, 3 C=C) are the main components in neutral fats. This implies that, in fatty acid synthesis, as the chain length approaches the C_{12}—C_{18} range, elongation tends to cease. This cessation may be related to the dissociation constant of a key enzyme for its fatty acyl CoA derivative as substrate. When the CoA derivative reaches a critical chain length, the dissociation of the enzyme-substrate complex may become so great that no reaction takes place and elongation ceases.

Evidence from animal tissue. In recent years much evidence has been accumulating to support the concept of β-reductive condensation of C_2 units in animal and bacterial tissues. STADTMAN and BARKER showed that acetyl phosphate (through acetyl CoA) is rapidly condensed to butyryl CoA and then to butyric acid by a cell-free preparation from *Clostridium kluyveri*. GURIN has prepared particle-free extracts from pigeon liver mitochondria which in the presence of ATP, DPN^+, and CoA will form labeled long chain fatty acids from acetic-1-C^{14}. POPJAK has obtained similar results with mammary gland extracts. STANSLY and BEINERT, employing the enzyme systems described in Table 1 have actually demonstrated the reversal of acetyl CoA $\rightarrow$ butyryl CoA using DPNH, reduced benzyl viologen, the complete enzyme system, (from acetone powder of beef liver mitochondria) and acetyl CoA. Butyryl CoA was identified as its hydroxamic acid by the addition of hydroxylamine to the reaction mixture and the separation and identification of butyryl hydroxamate by paper chromatography. No higher acyl CoA homologs appeared in the reaction mixture, presumably because of the specificity of one of the enzymes for the C_4-CoA derivatives.

Evidence from plant tissues. At present no evidence from cell-free preparations of plant tissue is available for the synthesis of long chain fatty acids. However, in recent years the synthesis of long chain fatty acids has been studied *in vitro* by incubating slices of cotyledons from both the developing and germinating peanuts with radioactive substrates (NEWCOMB and STUMPF). Slice material was incubated aerobically with substrates for six hours, respiratory CO_2 and unused substrate was recovered, the slices pulped and saponified, the fatty acids of chain length longer than 10 carbon atoms were isolated and assayed for radioactivity. The intact cells of peanut cotyledon slices actively metabolised a variety of potential precursors of fatty acids, but incorporated into long chain fatty acids to an appreciable extent only acetate-2-C^{14} and -1-C^{14} and C^{14} evenly labeled glucose and fructose. The extent of conversion of the C^{14} of various substrates to carbon dioxide and long chain fatty acids is presented in Table 2 and Table 3. At least 7% of the added radioactivity appeared as respiratory

Table 2. *Conversion of formate-C^{14}, acetate-1-C^{14}, acetate-2-C^{14}, butyrate-1-C^{14}, valerate-3-C^{14}, caproate-1-C^{14}, and C^{14} evenly labeled glucose and fructose to carbon dioxide and long chain fatty acids by slices of maturing cotyledons.* (After NEWCOMB and STUMPF.)

Experiment No.	Substrate	Radioactivity supplied, total c.p.m.	Radioactivity in respiratory CO_2		Radioactivity recovered in fatty acids	
			Total c.p.m.	Per cent of supplied	Total c.p.m.	Per cent of supplied
1	Acetate-1-C^{14}	282,000	81,550	29.0	33,200	11.7
	Acetate-2-C^{14}	130,000	13,600	10.5	17,700	13.6
	Butyrate-1-C^{14}.	380,000	79,000	20.5	3,200	0.8
	Hexanoate-1-C^{14}.	563,000	237,400	42.0	6,300	1.1
2	Acetate-1-C^{14}	282,000	91,580	32,5	61,500	22.0
	Acetate-2-C^{14}	130,000	17,900	13.8	45,000	34,6
	Butyrate-1-C^{14}.	380,000	79,000	20.5	3,200	0.8
	Hexanoate-1-C^{14}.	563,000	267,450	48.3	24,200	4.3
3	Formate-C^{14}.	220,800	70,000	31,5	750	0.3
	Valerate-3-C^{14}	149,000	890	0.6	300	0.2
4	C^{14} evenly labeled glucose . .	706,400	112,000	15.9	43,450	6.1
	C^{14} evenly labeled fructose .	272,000	48,000	17,6	15,850	5.8

CO_2 with all substrates except valerate-3-C which yielded only 0.6%. It is clear that the compounds readily penetrated into the cells and participated in the cell metabolism. There is a greater conversion to carbon dioxide of the carboxyl carbon than of the methyl carbon of acetate. One-third of the added formate is converted to carbon dioxide. There appears to be a greater conversion to CO_2 of the carboxyl carbon of the even numbered lower fatty acids (C_2, C_4, and C_6) than of the odd numbered (C_3 and C_5).

In the extent of incorporation of labeled carbon into the higher fatty acids the compounds tested differ in a striking way. Acetate-2- and 1-C^{14} contribute to the greatest degree (22 and 34%, for example, in developing peanuts), whereas glucose and fructose furnish about 6% of their initial activity. Butyrate-1-C^{14}, caproate-1-C^{14},and valerate-1-C^{14} furnish 1–3 per cent of the initial activity while the remaining substrates contribute less than one per cent. Because of the low degree of incorporation by these substrates into the long chain fatty acids,

Table 3. *Conversion of acetate-1-C^{14}, acetate-2-C^{14}, pyruvate-1-C^{14}, propionate-1-C^{14}, valerate-1-C^{14}, and succinate-1,4-C^{14} to CO_2 and long chain fatty acids by cotyledon slices from germinating seeds.* (After NEWCOMB and STUMPF.)

Experiment No.	Substrate	Radioactivity supplied, total c.p.m.	Radioactivity in respiratory CO_2		Radioactivity recovered in fatty acids	
			Total c.p.m.	Per cent of supplied	Total c.p.m.	Per cent of supplied
1	Acetate-1-C^{14}	142,000	53,450	37.6	16,100	11.3
	Acetate-2-C^{14}	53,000	5,975	11.3	5,555	10.5
2	Acetate-1-C^{14}	89,700	50,850	56.6	6,650	7.5
	Acetate-2-C^{14}	166,800	21,720	13.2	7,225	4.3
3	Acetate-1-C^{14}	89,700	47,000	52,5	5,040	5,6
	Acetate-2-C^{14}	133,400	18,300	13.7	15,040	11.2
	Pyruvate-2-C^{14}	42,100	11,100	26,4	230	0.5
4	Propionate-1-C^{14}	277,500	18,500	6.8	610	0.2
	Valerate-1-C^{14}	194,000	19,200	10.0	4,400	2.3
	Succinate-1,4-C^{14}	35,000	22,500	64.0		0.3

it is quite probable that they do not enter as intact units but rather are first degraded to C_2-fragments which is then directed into many pathways of metabolism, one being in the direction of fatty acid synthesis. The C_1 unit, formate, is only negligibly utilized for fatty acid formation, as are pyruvate-2-C^{14} and succinate-1,4-C^{14}, despite the fact that they participate in respiration. The observation that the C_1 unit, formate, which is rapidly converted to CO_2 is quite ineffective as a fatty acid precursor suggests that CO_2 fixation is relatively unimportant in fatty acid synthesis.

Although net synthesis of fatty acids does not take place in the cotyledons of germinating seeds, the percentage incorporation of acetate into the fatty acid pool in slices from such cotyledons is similar to that which occurs in the slices of developing cotyledons. This would imply that the long chain fatty acids are in a dynamic equilibrium with their C_2 precursors and that while degradation may be the principal pathway, a not too negligible proportion of C_2 fragments are redirected to the resynthesis of long chain fatty acid.

Under anaerobic conditions, virtually no CO_2 is formed from acetate-2-C^{14}; however marked incorporation into fatty acid occurs, being 42% of the aerobic value in slices of developing cotyledons and fully equal to the aerobic slices of germinating cotyledons. 2,4-dinitrophenol inhibits fatty acid synthesis from acetate-2-C^{14} almost completely at 10^{-3} M, while the appearance of radioactivity in the CO_2 is inhibited less than 50%. At 10^{-4} M, inhibition of fatty acid synthesis is still observed, although there is no inhibition of respiratory CO_2. These results are in accord with those for animal tissues and with the hypothesis that dinitrophenol inhibits energy-requiring reactions by interfering with oxidative phosphorylation at concentrations which do not inhibit respiratory gas exchange.

In summary, these results indicate (1) the most active precursors of long chain fatty acids in cotyledons of germinating as well as developing peanuts are acetate, glucose and fructose, (2) formate as a C_1 fragment is essentially inert, (3) fatty acid synthesis is geared to energy yielding systems since 2,4-dinitrophenol greatly inhibits the synthesis mechanism, and (4) the plant is able to oxidize a rather large variety of fatty acids including the C_8, C_6, C_5, C_4, C_3 and C_2 acids. Although this type of evidence is necessarily of an indirect nature, it does in general support the concept of β-reductive condensation as suggested by the results obtained with animal and bacterial tissues.

B. Degradation and synthesis of glycerol.

Glycerol serves as the backbone of a large class of neutral fats and phospholipids. Its metabolism may be depicted as follows:

(A) ±2H: glycerol ⇌ dihydroxyacetone
(B) ATP: dihydroxyacetone → dihydroxyacetone phosphate
(D) ATP: glycerol → α-glycerophosphate
(C) ±2H: α-glycerophosphate ⇌ dihydroxyacetone phosphate
(E) −PO₄: α-glycerophosphate → glycerol
(F′) glycolysis: sugars ⇌ dihydroxyacetone phosphate
(F) glycolysis: dihydroxyacetone phosphate ⇌ pyruvic acid
(G) TCA cycle: pyruvic acid → $CO_2 + H_2O$

Although Steps A through G have been observed in bacteria and animal tissues, only Steps D, C, F, F′ and G and Step E appear to occur in plant tissues (BURTON *et al.*, DOERSCHUK, GIDEZ *et al.*, HUNTER, LINDBERG, and WIAME *et al.*). It is presumably the sequence of Steps F′, F → C → E by which glycerol is synthesized in plant tissues.

In plants the path of glycerol degradation is similar to that found in animal tissues, namely the phosphorylation of glycerol to α-glycerolphosphate (Step D), its dehydrogenation to dihydroxyacetone phosphate (Step C), which is converted to pyruvic acid (Step F). Pyruvic acid is then rapidly oxidized to CO_2 and water by the TCA system (Step G).

Evidence to support this hypothesis has been recently obtained (STUMPF). Thus, when radioactive glycerol is added to a reaction mixture containing peanut mitochondria, (obtained from germinated peanut cotyledons) soluble peanut cytoplasmic protein fraction, and a number of cofactors and cosubstrates, rapid formation of CO_2 takes place. As summarized in Table 4, the omission of any member of the complete system results in dimished or complete loss in activity. Thus, removal of either of the enzymic fractions leads to complete loss. The removal of ATP, α-ketoglutaric acid, or cocarboxylase causes a serious impairment in oxidative activity. Without Mg^{++}, and DPN there is an approximate decrease of 50% and 65% respectively. The presence of malonate and fluoride reduces oxidation to a low level. These observations can be interpreted as follows: (1) the necessary enzymes for the oxidative process are to be found in the mitochondrial and the soluble protein fractions, (2) ATP presumably serves as a source of ~P for the phosphorylation of glycerol and for the maintenance of TCA cycle activity, (3) the

Table 4. Oxidation of glycerol-1-C^{14} to $C^{14}O_2$ by extracts of peanut cotyledons. The complete reaction mixture contained 0.2 ml of peanut mitochondria (about 4 mgm protein N per ml), 0.5 ml of soluble protein fraction, 0.1 ml of glycerol-1-C^{14} containing 10^5 c.p.m. and 0.4 μmols of substrate, 0.1 ml of 0.01 M ATP, 0.1 ml of 10^{-6} M cytochrome c, 0.1 ml of 0.1 M $MgCl_2$, 0.1 ml of 0.1 M α-ketoglutaric acid, 0.1 ml of 0.1% thiamine pyrophosphate, 0.1 ml of 0.1% DPN, 0.5 ml of 0.2 M phosphate buffer at p_H 7.2, 0.2 ml of 20% KOH in center well, 0.3 ml of 10 N H_2SO_4 in side arm, and water to make a final volume of 3 ml. Incubated 2 hours, at 30°. (After STUMPF.)

Components	Total c.p.m. as $BaC^{14}O_3$
Complete	12,500
without MP	0
without SPF	0
without ATP	56
without cytochrome c	13,100
without Mg^{++}	6,500
without α-ketoglutaric acid	600
without thiamine pyrophosphate	1,420
without DPN	8,500
complete + malonate (10^{-1} M)	1,250
complete + fluoride (10^{-2} M)	60

requirement of TCA cycle acid such as α-ketoglutaric acid or any other member of the cycle implies the participation of the TCA cycle in the formation of radioactive CO_2, (4) the inhibition of the formation of radioactive CO_2 by malonate is related to the known capacity of malonate to inhibit succinic dehydrogenase, and (5) the inhibition by fluoride suggests its involvement in the inhibition of enolase, the site of fluoride inhibition in the glycolytic cycle.

Additional support are the observations of (a) an α-glycerolphosphate dehydrogenase in peanut mitochondria, (b) the accumulation of triose phosphate when α-glycerolphosphate is oxidized by mitochondrial preparation, and (c) the labeling of members of the TCA cycle when glycerol is oxidized by the complete system.

α-Glycerolphosphate dehydrogenase: The site of this activity is the mitochondrion. The soluble protein fraction is inert. α-Glycerolphosphate dehydrogenase activity can be followed by uptake of oxygen when α-glycerolphosphate is added to mitochondria, or by following the rate of reduction of redox dyes such as methylene blue, 2,6-dichlorophenolindophenol, or neotetrazolium. Fresh preparations of mitochondria are uniformly active whereas mitochondria that have been frozen, treated with acetone, lyophilized, or exposed to sonic vibration are inert. Exposure of MP to 55° for 5 minutes completely inactivates the enzyme. The p_H optimum is in the range of 6.8 to 7.3.

Table 5. Stoichiometry for oxidation of α-glycerol phosphate to triose phosphate. Reaction mixture contained 0.2 ml of MP, 0.5 ml of 0.5 M DL-α,β-glycerol phosphate, 0.5 ml of 0.1 M TRIS buffer at p_H 7.2, 0.2 ml of 20% KOH in center well and water to final volume of 3 ml. Time 2 hours, at 30°. At end of incubation time oxygen uptake was measured and the reaction stopped by adding an equal volume of 10% trichloroacetic acid to reaction mixture. Suspension was transferred to a test tube, diluted to 10 ml, filtered and aliquots analyzed for inorganic phosphate and alkaline labile phosphate. (After STUMPF.)

Experiment	Oxygen uptake μ-atoms	Triose phosphate as alkaline-labile P μmol
1	10.1	8.8
2	5.2	4.8

The stoichiometry of the reaction is established by the reaction:

$$\alpha\text{-glycerolphosphate} + {}^1/_2\, O_2 \rightarrow \text{triosephosphate} + H_2O$$

based on the data in Table 5. It is consistently observed that for one atom of oxygen taken up, one molecule of triose phosphate accumulates. It is difficult to determine whether or not the first product of oxidation is DHAP or glyceraldehyde-3-phosphate since there is evidence that a triose phosphate isomerase is present as a contaminant in the mitochondria fraction. This enzyme catalyzes the conversion of triose phosphates to essentially DHAP.

No cofactor appears to be required by the preparation. The addition of either DPN or TPN had no effect on the rate of oxidation. Cytochrome c is very rapidly reduced. Since cytochrome c need not be added for maximum dehydrogenase activity it probably occurs in mitochondria in a concentration which is not limiting. Presumably cytochrome c is the physiological redox acceptor for the dehydrogenase system. These results would indicate that plant α-glycerolphosphate dehydrogenase is similar to the particulate system found in animal tissue (TUNG).

A rather high concentration of L-α-glycerol phosphate is required to saturate the enzyme system. The approximate K_m is 0.03 M per liter.

Participation of TCA cycle: If the TCA cycle is participating in the terminal oxidation of glycerol, then each member of the TCA cycle should become labeled

Table 6. Incorporation of C^{14} into $C^{14}O_2$ and TCA cycle acids from glycerol-1-C^{14}. The reaction mixture contained 0.2 ml of peanut mitochondria, 0.5 ml of soluble protein fraction, 0.1 ml of glycerol-1-C^{14} having 10^5 c.p.m., 0.1 ml of 0.01 M ATP, 0.1 ml of 0.1 M $MgCl_2$, 0.1 ml of 0.1% thiamine pyrophosphate, 0.1 ml of 0.1% DPN, 0.1 ml of 0.1 ml citrate, 0.2 ml of 20% KOH in center well, 0.5 ml of 0.2 M phosphate buffer at p_H7.2. Time 2 hours, at 30°. (After STUMPF.)

TCA cycle acids	Total c.p.m.	TCA cycle acids	Total c.p.m.
citric acid	1,950	succinic acid	704
malic acid	3,230	fumaric acid	1,100
α-ketoglutaric acid .	380	(CO_2)	(12,000)

when glycerol-1-C^{14} is oxidized by the complete system. In Table 6 are summarized the results of an experiment in which glycerol-1-C^{14} was oxidized by the complete system for two hours in the presence of 10 μmols of citrate as the sparking system. The data clearly indicate that the terminal oxidation of glycerol proceeds through the conventional TCA cycle localized in the mitochondria since appreciable radioactivity is found in each of the members of the cycle. The variability of the radioactivity in each of the TCA cycle acids is probably related to the variability in recovery, the rates of formation and degradation in the mitochondria, and the size of the acid pools.

C. Synthesis of triglycerides.

At high concentrations of glycerol, lipase may catalyze the esterification of the alcoholic groups by free long chain fatty acids. However, under physiologic conditions, it is improbable that this mechanism plays an important role in the synthesis of triglycerides.

It is quite probable that the physiologic mechanism involves the participation of the system:

$$\begin{array}{c} \text{3 fatty acyl CoA} + \text{glycerol} \rightarrow \text{triglyceride} + \text{3 CoA} \\ \uparrow\_\_\_\_\_\_\_\_\_\_\_\_\_\_\_\_\_\_\_\_| \\ \text{ATP} + \text{long chain fatty acids} \end{array}$$

A system has been prepared from guinea pig liver which catalyzes the esterification of L-α-glycerolphosphate by 2 moles of palmityl CoA (KORNBERG and PRICER). Straight chain fatty acids with 16, 17, or 18 carbon atoms were the most effective acids. Whether triglycerides are synthesized directly from glycerol and acyl CoA by an analogous enzyme in plants or whether they first pass through a phosphatidic acid intermediary remains to be determined.

Literature.

BAALEN, J. VAN, and S. GURIN: Cofactor requirements for lipogenesis. J. of Biol. Chem. **205**, 303–308 (1953). — BURTON, R. M., and N. O. KAPLAN: Specific glycerol dehydrogenase from *Aerobacter aerogenes*. J. Amer. Chem. Soc. **75**, 1005–1006 (1953).

CHAIKOFF, I. L., and G. W. BROWN jr.: Fat metabolism and acetoacetate formation. Chapt. 7, Chemical pathways in metabolism, Vol. I. New York: Academic Press 1954.

DOERSCHUK, A. P.: Mechanism studies of glycogen and glyceride-glycerol biosynthesis. J. of Biol. Chem. **196**, 423–426 (1952).

GIDEZ, L. I., and M. L. KARNOUSKY: The metabolism of C^{14} glycerol in the intact rat. J. of Biol. Chem. **206**, 229–242 (1954). — GREEN, D. E.: Fatty acid oxidation in soluble systems of animal tissues. Biol. Rev. **29**, 330–366 (1954).

HUNTER, G. J. E.: The oxidation of glycerol by mycobacteria. Biochemic. J. **55**, 320–328 (1953).

KORNBERG, A., and W. E. PRICER jr.: Enzymatic esterification of α-glycerolphosphate by long chain fatty acids. J. of Biol. Chem. **204**, 345–357 (1953).

LINDBERG, O.: Phosphorylation of glyceraldehyde, glyceric acid, and dihydroxyacetone by kidney extracts. Biochim. et Biophysica Acta **7**, 349–353 (1951). — LYNEN, F.: Participation of coenzyme A in the oxidation of fats. Nature (Lond.) **174**, 962–965 (1954).

NEWCOMB, E. H., and P. K. STUMPF: Fat metabolism in higher plants. I. Biogenesis of higher fatty acids by slices of peanut cotyledons in vitro. J. of Biol. Chem. **200**, 233–239 1953).

POPJAK, G., and A. TIETZ: Biosynthesis of fatty acids by slices and cell-free suspensions of mammary gland. Biochemic. J. **56**, 46–54 (1954).

STADTMAN, E. R., and H. A. BARKER: Fatty acid synthesis by enzyme preparations of *Clostridium kluyveri. III.* The activation of molecular hydrogen and conversion of acetyl phosphate and acetate to butyrate. J. of Biol. Chem. **180**, 1117–1124 (1949). — STANSLY, P. G., and H. BEINERT: Synthesis of butyryl CoA by reversal of the oxidation pathway. Biochim. et Biophysica Acta **11**, 600–601 (1953). — STUMPF, P. K.: Fat metabolism in higher plants. III. Enzymic oxidation of glycerol. Plant Physiology **30**, 55–58 (1955).

TUNG, T. C., L. ANDERSON and H. A. LARDY: Studies in the particulate α-glycerol phosphate dehydrogenase of muscle. Arch. of Biochem. a. Biophysics **40**, 194–204 (1952).

WIAME, J. M., S. BOURGEOIS and R. LAMBION: Oxidative assimilation of glycerol studied with variants of *Bacillus subtilis*. Nature (Lond.) **174**, 37—38 (1954).

Die Physiologie der Fettbildung und Fettspeicherung bei niederen Pflanzen.

Von

Maximilian Steiner.

Mit 26 Abbildungen.

Der weitaus größere Teil der Untersuchungen über die Physiologie der Fettbildung niederer Pflanzen wurde mit leicht kultivierbaren Schimmelpilzen, Hefen und Bakterien, also mit *heterotrophen* Mikroorganismen durchgeführt. Erst in jüngster Zeit hat der Fettstoffwechsel der *autotrophen* Algen eine stärkere Beachtung gefunden. Die Zahl der Arbeiten, die bisher auf diesem Gebiete vorliegen, ist noch verhältnismäßig klein.

Es erschien daher zweckmäßig, den Stoff dieses Beitrages auf zwei getrennte Abschnitte zu verteilen. Der eine, größere, soll sich mit der Fettbildung durch *heterotrophe* pflanzliche Mikroorganismen beschäftigen. Alle übergeordneten Gesichtspunkte, zu denen sich die Einzelergebnisse zusammenfügen, alle Theorien und Gesetzmäßigkeiten der Physiologie der Fettbildung werden sich hier unterbringen lassen. Der Abschnitt über die *autotrophen* niederen Pflanzen wird sich dann mit kurzen Hinweisen begnügen dürfen, soweit es sich nur um die Bestätigung von Ergebnissen handelt, von denen bereits im vorangehenden Abschnitt die Rede war. Ausführlicher müssen hier nur die Fragen behandelt werden, welche die Beziehungen zwischen der Photosynthese und der Fettbildung betreffen.

I. Methodik, Bezugsgrößen und Berechnungsweisen.

Schon C. v. NAEGELI und O. LOEW (1878) haben beobachtet, daß sich nur ein Teil des Fettes aus Pilzzellen durch einfache Ätherextraktion isolieren läßt. Die Extraktion scharf getrockneter Hefe ergab 1,85% vom Trockengewicht Fett, während nach Vorbehandlung mit konzentrierter HCl 4,59% Fettsäuren, entsprechend 5,29% Fett erhalten wurden. Über ähnliche Ergebnisse, insbesondere bei Hefe, berichten zahlreiche spätere Untersucher (z. B. I. SMEDLEY-MACLEAN und D. HOFFERT 1923, P. BÉLIN 1926a, H. GEFFERS 1937, S. HEIDE 1939, O. TURPEINEN 1935, 1936, A. KLEINZELLER 1944). Die Ursachen können sowohl in der Membranbeschaffenheit (insbesondere bei Hefen) als auch in festerer Bindung oder Umhüllung der Fette vermutet werden. Reproduzierbare Werte für den „Gesamtfettgehalt" von Mikroorganismen lassen sich also nur erwarten, wenn der Extraktion ein Säureaufschluß (z. B. I. SMEDLEY-MACLEAN und D. HOFFERT 1923), ein Alkaliaufschluß (z. B. F. STOCKHAUSEN und R. ERICSEN nach H. HAEHN und W. KINTTOF 1925, P. BÉLIN 1926a) oder eine sehr gründliche mechanische Zertrümmerung der Zellen (z. B. S. HEIDE 1939, W. DIEMAIR und C. BORESCH 1950) vorausgeht. Weitere methodische Einzelheiten liegen außerhalb des Rahmens dieses Beitrages. Wegen der Isolierung und Bestimmung einzelner Fettfraktionen und wegen der Bedeutung der Fettkennzahlen sei auf das Einleitungskapitel dieses Handbuchbandes verwiesen, insbesondere S. 5—6.

Wie bei allen stoffwechselphysiologischen Arbeiten fällt auch bei Untersuchungen über Fettbildung und Fettstoffwechsel der Wahl der geeigneten Bezugsgrößen eine sehr wichtige Rolle zu; sie wird sich im Einzelfalle nach dem Versuchsziel zu richten haben.

Als unmittelbares Ergebnis der Fettbestimmung steht in der Regel der

Fettgehalt (Prozent des Trockengewichtes)

zur Verfügung. Bei bekanntem Trockengewicht der Pilzernte (je Kultur oder je Volumeinheit des Substrates) läßt sich daraus die gesamte

$$\textit{Fettmenge (mg)} = \frac{\textit{Fettgehalt (\%)} \times \textit{Trockengewicht (mg)}}{100}$$

leicht errechnen.

Die Mengenbeziehung zwischen Fettbildung und Verbrauch des Ausgangsmaterials ist durch den „*Fettkoeffizienten*" gegeben, der von A. RIPPEL (1940) in Analogie zum „*ökonomischen Koeffizienten*" eingeführt wurde. Darunter hatte W. PFEFFER (1897) die aus 100 Gewichtsteilen verbrauchter C-Quelle aufgebaute Gewichtsmenge an Trockensubstanz verstanden. In ähnlicher Weise hatten A. RIPPEL und K. NABEL (1939) die Berechnung des „*Eiweißkoeffizienten*" vorgeschlagen, wobei nach üblichem Verfahren die Eiweißmenge als „$N \times 6,25$" genommen wird. Gelegentlich wird auch in analoger Weise der „*Kohlenhydratkoeffizient*" angegeben, wobei in grober Annäherung die Kohlenhydrate als Differenz [gesamte Trockensubstanz (Fett+Eiweiß)] berechnet werden. Also

$$\textit{ökonomischer Koeffizient} = \frac{\text{Trockensubstanz (g)}}{\text{verbrauchte C-Quelle}} \times 100,$$

$$\textit{Fettkoeffizient} = \frac{\text{Fett (g)}}{\text{verbrauchte C-Quelle}} \times 100,$$

$$\textit{Eiweißkoeffizient} = \frac{(N \times 6,25)\ \text{(g)}}{\text{verbrauchte C-Quelle}} \times 100,$$

$$\textit{Kohlenhydratkoeffizient} = \frac{\text{Trockensubstanz (g)} - [\text{Fett (g)} + \text{Eiweiß (g)}]}{\text{verbrauchte C-Quelle}} \times 100.$$

Bei den Stoffwechselkoeffizienten bleiben die verschiedenen calorischen Werte der C-Quelle und der gebildeten Zellbestandteile unberücksichtigt. Da die calorischen Werte von Glucose, Fett und Eiweiß sich wie 1:2,5:1,5 verhalten, ist es selbstverständlich, daß bei gleichbleibender Ökonomie des Stoffwechsels etwa aus einer Gewichtseinheit Glucose 2,5 mal so viel Kohlenhydrate als Fette entstehen können.

Umgekehrt wird bei Fett als Nährsubstrat ein höherer ökonomischer Koeffizient zu erwarten sein als bei Glucosenährböden (z. B. O. FLIEG 1922).

A. RIPPEL (1940) hat sich eingehend mit der Ökonomie der Fettbildung befaßt (s. auch A. RIPPEL-BALDES 1952, W. KAUFMANN 1952, H. H. MARTIN 1954). Er berechnete für optimale Fettbildungsbedingungen einen Fettkoeffizienten von etwa 15 als mögliches Maximum, was mit den meisten experimentellen Ergebnissen gut übereinstimmt. Lediglich bei *Rhodotorula*arten fanden S. C. PAN, A. A. ANDREASEN und R. KOLACHOV (1949) und L. ENEBO, M. ELANDER, F. BERG, H. LUNDIN, R. NILSSON und K. MYRBÄCK (1944) Fettkoeffizienten bis 18,5.

A. KLEINZELLER (1944) berechnet den Fettertrag in ähnlicher Weise, aber auf der Basis der C-Äquivalente als

$$\textit{Konversionskoeffizient} = \frac{\text{Kohlenstoff des gebildeten Fettes (g)} - 100}{\text{verbrauchte Glucose (g)} \times 40} \times 100\,\%.$$

Auf Grund des verschiedenen calorischen Wertes der drei wichtigsten Zellfraktionen (Kohlenhydrate, Fette und Eiweiße) versuchten H. A. SPOEHR und H. W. MILNER (1949) sogar eine indirekte Fettbestimmung. Sie gehen vom *R-Wert (Reduktionswert)* aus, der für $CO_2 = O$, für $CH_4 = 100$ angesetzt wird und der sich allgemein aus der Formel

$$R = \frac{[(\%\,C \times 2,664) + (\%\,H + 7,936) - \%\,O]}{398,9} \times 100$$

berechnet. Er ist direkt proportional der Verbrennungswärme/Gramm. Werden die calorischen Näherungswerte zugrunde gelegt, so beträgt der R-Wert für Kohlenhydrate (KH) 28, ür Proteine 42, für Fette 67,5. Für die „gesamte Zellsubstanz" folgt dann,

$$\%\,\text{Protein} + \%\,\text{KH} + \%\,\text{Fett} = 100 \qquad \text{(I):}$$

wenn

$$(\%\,\text{Protein} \times 42) + (\%\,\text{KH} \times 28) + (\%\,\text{Fett} \times 67,5) = R\text{-Wert} \times 100\,\%. \qquad \text{(II)}$$

Wird der R-Wert calorimetrisch und der Proteingehalt über $N \times 6,25$ bestimmt, so läßt sich die Gl. (II) auflösen. Daß dieses Verfahren doch einige Unsicherheit in sich birgt, scheint unter anderem auch daraus hervorzugehen, daß die auf diese Weise für *Chlorella* angegebenen maximalen und minimalen Fettgehalte (4,5 und 85,6%) durch direkte Fettbestimmung (z. B. von H. G. AACH 1952) nicht reproduziert werden konnten.

Soll die Geschwindigkeit der Fettbildung oder des Fettverbrauchs oder der Substanzproduktion überhaupt bestimmt werden, so muß selbstredend der *Zeit*faktor in die Berechnung eingehen. So hat z. B. O. FLIEG (1922) den Begriff des ökonomischen Effekts eingeführt:

$$\textit{ökonomischer Effekt} = \frac{\text{ökonomischer Koeffizient}}{\text{Zeit (Tage)}}.$$

Während der ökonomische Koeffizient bei Fettkulturen von *Aspergillus niger* wesentlich höher lag als bei Glucosekultur (s. oben), verhielten sich die ökonomischen Effekte wie 1:3, d. h. die Substanzproduktion verläuft auf Fettnährböden (Triolein) wesentlich langsamer als auf Glucose. In ähnlicher Weise lassen sich natürlich die Zeitfunktionen der Fett- und Eiweißkoeffizienten ermitteln.

Speziell mit der Fettbildung beschäftigen sich einschlägig M. P. STEINBERG und Z. J. ORDAL (1954a). Sie gehen zunächst von der Voraussetzung aus, daß das Fett durch die Nichtfette der Zelle gebildet wird und berechnen dementsprechend die

$$\text{„}\textit{fat ratio}\text{“ } (FR) = \frac{\%\ \text{Fett}}{100 - \%\ \text{Fett}}.$$

Werden in einer Kurve die Werte für FR gegen die Zeit (t) aufgetragen, so entspricht die Neigung der Kurve der „fat rate“, d. i. „*fat rate*“ $= d(FR)/dt$ (vergleiche auch M. P. STEINBERG und Z. J. ORDAL 1954b).

II. Heterotrophe pflanzliche Mikroorganismen (Pilze und Bakterien).

1. Der Einfluß innerer Faktoren auf die Fettbildung.

Lipoide gehören ohne Zweifel zum obligaten Stoffbestand einer jeden lebenden Zelle. Darüber hinaus sind viele pflanzliche Organismen imstande, unter bestimmten Bedingungen größere Fettmengen zu speichern. Daß solche „zusätzlich gebildeten“ Fette von Mikroorganismen in den meisten Fällen ebenso wie etwa die Samenfette von Blütenpflanzen als energiereiche Reservestoffe angesehen werden dürfen, wird später näher zu begründen sein (s. S. 242ff.). Während bei den höheren Pflanzen in der Regel bestimmte Organe oder Gewebe der Fettspeicherung dienen, wird bei niederen Organismen das Fett einfach in den vegetativen Zellen abgelagert. E. F. TERROINE (1920) und P. BÉLIN (1926a) (s. auch E. F. TERROINE und P. BÉLIN 1927) waren wohl die ersten, die in ausgedehnten, vergleichend-physiologischen Untersuchungen bei Tieren und bei höheren und niederen Pflanzen den Unterschied zwischen dem „*élément constant*“ und dem „*élément variable*“ der Fette hervorgehoben haben. Das Fett der ersten Kategorie ist ein Konstitutionselement der Zelle, das auch unter extremsten Hungerbedingungen nicht verbraucht wird. Es wurde von BÉLIN z.B. bei *Aspergillus niger* mit etwa 1,37% vom Trockengewicht bestimmt. Was darüber liegt (bei *Apsergillus niger* in den Versuchen BÉLINS bis 12,33%) ist das „élément variable“[1].

Es muß deutlich hervorgehoben werden, daß sich unsere Kenntnisse über die Physiologie des Fettstoffwechsels und über die Zusammensetzung der Fette so gut wie ausschließlich auf dieses „élément variable“ beziehen. Über die Zusammensetzung des „élément constant“ wissen wir sehr wenig. Schon BÉLIN nahm an, daß dieses weniger aus Neutralfetten (Glyceriden), als vielmehr aus Phosphatiden, Sterolestern, Lipoproteiden zusammengesetzt sei. Bei Extraktion mit neutralen Lösungsmitteln (CCl_4, Petroläther) blieb in seinen Versuchen eine

[1] H. GYLLENBERG und A. RAITIO (1952) fanden bei einem *Penicillium* (Gruppe *Monoverticillata*) fast den gleichen Wert: 1,3—1,4% Fett vom Trockengewicht wurden auch unter extremen Hungerbedingungen nicht verbraucht.

Fettfraktion zurück, die erst nach Säureaufschluß der Analyse zugänglich wurde und die ziemlich genau an Menge dem im Hungerexperiment ermittelten „élément constant" entsprach. Nach E. F. TERROINE und P. BÉLIN (1927) hat das „élément constant" immer eine höhere Jodzahl als das „élément variable": *Aspergillus niger* Jodzahl 100 bzw. 95, *Maus* 127 bzw. 73, *Kaninchenmuskel* 160 bzw. 80, *Kaninchenleber* 140 bzw. 70, *Kaninchenniere* 134 bzw. 63, *Kaninchenlunge* 96 bzw. 50.

Die „Bereitschaft" von Mikroorganismen zur Verfettung ist von Art zu Art und selbst innerhalb der Art von Stamm zu Stamm außerordentlich verschieden. Diese Unterschiede sind ohne Zweifel eine Folge der verschiedenen genetischen Konstitution, die über eine verschiedene enzymatische Ausrüstung im Stoffwechsel der Zelle wirksam wird.

P. LINDNER und T. UNGER (1919) haben die Stämme einer sehr großen Hefesammlung (untergärige und obergärige Bierhefen, Brennerei-, Preß-, Weinhefen, wilde Hefen, Kahmhefen, *Torula*-Arten) unter für Fettbildung günstigen Bedingungen (P. LINDNER und S. CZIFER 1912) kultiviert und eine Fettbildung (mikroskopische Kontrolle) überall, aber in sehr verschiedenem Ausmaße festgestellt. L. ENEBO, M. ELANDER, F. BERG, H. LUNDIN, R. NILSSON und K. MYRBÄCK (1944) untersuchten 12 *Rhodotorula*-Arten unter gleichen Kulturbedingungen und fanden große Unterschiede im Fettgehalt (*Rh. gracilis*: 43,0%, *Rh. pallida* 11,9% Rohfett/Trockengewicht). H. GEFFERS prüfte das Fettbildungsvermögen von 39 Stämmen der Gattung *Oospora* WALLROTH. Er konnte zwei Gruppen „*M*" und „*L*" feststellen, die sich nicht nur durch den (auf Molke) erreichbaren Höchstfettgehalt, sondern auch durch andere ernährungsphysiologische und morphologische Merkmale unterschieden (Tabelle 1).

Tabelle 1. *Physiologische und morphologische Gruppenunterschiede bei Oospora* WALLROTH. (Nach H. GEFFERS 1937.)

	Gruppe M	Gruppe L
Fettbildung	gut (bis 46% Fett)	schlecht (bis 8—15% Fett)
Fettspaltung	schlecht	gut
Eiweißabbau in Milch	stark	gering
Verwertung:		
Lactose	+	—
Glycerin	—	+
Mannit	—	+
Lactat	—	+
Kolonieform	ähnlich „*Monilia*" PERS. em. SACC.	ähnlich „*Oidium lactis*" FRIES.

Zu ähnlichen Ergebnissen kamen H. FINK, G. HAESELER und M. SCHMIDT (1937), die unter 50 geprüften *Oospora*-Stämmen 10 gute Fettbildner fanden (auf Holzzucker + Bierwürze bis 23,5% Fett).

G. E. WARD, L. B. LOCKWOOD, O. E. MAY und H. T. HERRICK (1931) untersuchten unter Standardbedingungen 39 *Penicillium*- und 22 *Aspergillus*-Stämme; bei 10 Stämmen wurde >15% Rohfett, nur bei 6 >20% gefunden. Recht verschieden war auch die Fettbildung bei 9 *Fusarium*-Stämmen, welche A. KLEINZELLER und J. ŠKODA (1950) untersuchten.

Selbst innerhalb einer wohldefinierten Art treten bedeutende Rassenunterschiede auf. E. A. PRILL, P. R. WENCK und W. H. PETERSON (1935) verglichen im Parallelversuch die Leistung von 9 Stämmen des *Aspergillus Fischeri*. Der Fettgehalt schwankte von 10,5—20,5%, der Anteil freier Fettsäuren von 55,0 bis 70,3%, der unverseifbare Anteil von 5,9—13,0%, der Anteil der Sterine von 4,6—7,6%, die Jodzahl der Fettsäuren von 73—89. Dagegen fand M. RAAB (1941) bei zwei verschiedenen Herkünften von *Endomycopsis vernalis* keine Unterschiede im Fettbildungsvermögen.

Die Leistungsprüfung *verschiedener* Organismen unter *gleichen* Bedingungen vermag freilich nur einen ersten Anhaltspunkt für „gutes" oder „schlechtes" Fettbildungsvermögen zu liefern. „Gleiche" Außenbedingungen sind für verschiedene Organismen keineswegs physiologisch gleichwertig. Um etwa das maximale Fettbildungsvermögen einer Art festzustellen, bedarf es umfangreicher Reihenversuche, wie sie für eine ganze Anzahl von Organismen tatsächlich vorliegen. Darüber wird später zu berichten sein. Es ist hier lediglich darauf hinzuweisen, daß die Beurteilung eines Organismus als „guter" oder „schlechter" Fettbildner je nach den Versuchsbedingungen sehr verschieden ausfallen kann. So haben z.B. M. WOODBINE, M. E. GREGORY und T. K. WALKER (1951) die Fettbildung von 43 Schimmelpilzarten auf 5 verschiedenen Nährmedien geprüft. Von Nährmedium zu Nährmedium wechselten die Namen und die Reihenfolge der „10 besten Fettbildner". Sehr instruktiv sind die Ergebnisse von K. BERNHAUER, A. NIETHAMMER und J. RAUCH (1948), welche 13 Pilzarten durch Variation der N-Gabe (s. unten S. 227ff.) unter Bedingungen kultivierten, die für Fettbildung „günstig" bzw. „ungünstig" waren. Eine Auswahl der Resultate bringt Tabelle 2. Es zeigt sich, daß einige Arten (1, 3, 4) auf die verschiedenen Ernährungsbedingungen mit großen Differenzen im Fettgehalt reagierten, andere (2, 5) hingegen fast gar nicht. Wären die 5 Arten nur auf $^1/_1$ Harnstoff untersucht worden, so ergäbe sich für das Fettbildungsvermögen die Reihe $2 > 4 > 5 > 3 > 1$, während sie auf $^1/_{10}$ Harnstoff lautet: $3 > 4 > 2 > 1 > 5$. Ähnliche Ergebnisse hatte schon früher H. RAAF (1941) für verschiedene Hefen und Schimmelpilze mitgeteilt.

Tabelle 2. *Prozentgehalt an Rohfett und Rohprotein bei Schimmelpilzen unter günstigen und ungünstigen Fettbildungsbedingungen.* (Nach K. BERNHAUER, A. NIETHAMMER und J. RAUCH 1948.)

	Art	Kulturdauer	$^1/_1$ Harnstoff		$^1/_{10}$ Harnstoff	
			Fett	Rohprotein	Fett	Rohprotein
		Tage	%	%	%	%
1	*Mucor albo-ater*	7	4,5	23,3	14,0	12,3
2	*Mucor Ramannianus*	7	12,6	18,3	15,6	15,3
3	*Mucor circinelloides*	4	6,0	41,9	53,4	11,2
4	*Zygorrhynchus moelleri*	7	11,3	33,0	31,3	8,0
5	*Cladosporium herbarum*	7	9,1	18,3	10,3	16,3

Wenn wir die Frage stellen, *welche* Besonderheiten die physiologische Konstitution eines guten Fettbildners ausmachen, so läßt sich augenblicklich nur eine sehr fragmentarische Antwort geben. K. BERNHAUER, A. NIETHAMMER und J. RAUCH (1948) haben darauf hingewiesen, daß die *Fettpilze* sich — unter abgeänderten Ernährungsbedingungen — auch als gute *Eiweiß*bildner erweisen, während bei anderen Arten (z.B. 2 und 5 der Tabelle 2) die (freilich schlecht definierte!) „Kohlenhydratfraktion" des Mycels überwiegt („*Kohlenhydrat*-Pilze").

A. NIETHAMMER (1942a) findet beim Vergleich verschiedener *Torula*-Hefen, daß nur die *gärschwachen Arten gute Fettbildner* sind. Diese Ansicht findet man mehrfach im Schrifttum vertreten. *Endomycopsis vernalis*, einer der besten Fettbildner unter den Pilzen, produziert nach H. HAEHN und W. KINTTOF (1926) überhaupt kein Äthanol. Hierbei darf daran erinnert werden, daß auch die wenig zur Fettbildung neigende untergärige Bierhefe zu kräftiger Fettbildung veranlaßt wird, wenn man sie durch gute Belüftung und Darreichung einer assimilierbaren, aber nicht gärfähigen C-Quelle (Äthanol, Acetat) zu streng aerobem Stoffwechsel zwingt (P. LINDNER, 1919, 1921a, b, 1922, 1929, I. SMEDLEY-MACLEAN und D. HOFFERT 1923, W. HALDEN, F. BILGER und R. KUNZE 1933, W. HALDEN 1934). (Wegen der allgemeinen Unterscheidung zwischen „Gärhefe" und „Wuchshefe" vgl. S. WINDISCH 1948.)

Einen wesentlichen Fortschritt in der Analyse dieser Zusammenhänge brachten die Untersuchungen von H. H. MARTIN (1954). Er isolierte aus Blütennektar 46 Stämme von *Candida Reukaufii*, die sich nach morphologischen und physiologischen Merkmalen in zwei scharf geschiedene Typen einordnen ließen:

	Gärtyp a	Fettbildungstyp b
Bildung von Pseudomycel . . .	+	—
Kolonieform auf Würzeagar . .	bräunlich glänzend	cremefarbig glanzlos
Vergärung von Glucose	gut	schwach
In Lüftungskultur:		
ökonomischer Koeffizient . .	niedrig	hoch
prozentualer Fettgehalt . .	niedrig	hoch
prozentualer Eiweißgehalt	hoch	niedriger als bei a

Als Beispiel diene Tabelle 3, welche die Ergebnisse mit zwei solchen typenverschiedenen, aus Nektar von *Lamium album* isolierten Stämmen wiedergibt.

Tabelle 3. *Candida Reukauffii, Gärtyp und Fettbildungstyp.* (Nach H. H. MARTIN 1954.)

	Stamm 5a		Stamm 5b	
Gärvermögen ml CO_2 7 Std	12,3		2,7	
Harnstoffgabe	0,35%	0,075%	0,35%	0,075%
Fett, Prozent Trockengewicht	4,5	5,7	9,1	20,6
Eiweiß, Prozent Trockengewicht	49	12,5	35	10,7

Man wird vermuten dürfen, daß der Energiegewinn beim Gärungsstoffwechsel nicht ausreicht, um die reduktive Bildung von größeren Fettsäuremengen zu gewährleisten.

Von Stoffwechselmutanten, bei denen die Fettbildung durch einen genetischen Block unterbrochen ist, ist vor allem die acetatbedürftige *Neurospora crassa*-Mutante Y 2492-17-13-6(a)3) zu nennen (G. W. BEADLE und E. L. TATUM 1945, E. L. TATUM, R. W. BARRATT, N. FRIES und D. BONNER 1950). Sie ist nicht imstande, Glucose oder Fructose zu C_2-Körpern zu verarbeiten und bedarf zum Wachstum einer acetathaltigen Nährlösung. Mit Hilfe dieser Mutante haben R. C. OTTKE, S. SIMMONDS, E. L. TATUM, I. ZABIN und K. BLOCH (1952) die Fett- und Ergosterinsynthese aus isotopen-markiertem Acetat studiert (s. S. 257).

J. LEIN und P. S. LEIN (1950) berichten über eine acetatbedürftige *Neurospora*-Mutante, bei der die Essigsäure aber durch alle natürlich vorkommenden Fettsäuren mit Ausnahme von Palmitin- und Stearinsäure vollwertig ersetzt werden kann.

J. LEIN, T. A. PUGLISI und P. S. LEIN (1953) (s. auch J. LEIN und P. S. LEIN 1949) gelang es ferner, 5 Mutanten des gleichen Pilzes zu isolieren, bei denen die Nährlösung durch *ungesättigte* Fettsäuren vervollständigt werden muß. 4 Mutanten (S 11, S 32, S 82, S 93) sind Allele. Sie können Öl-, Linol- und Linolensäure verwenden; die fünfte Mutante (S 72) braucht Linolensäure. Gesättigte Fettsäuren oder Arachidonsäure sind in allen Fällen unbrauchbar. Die Verfasser nehmen Stoffwechselblocks nach folgendem Schema an:

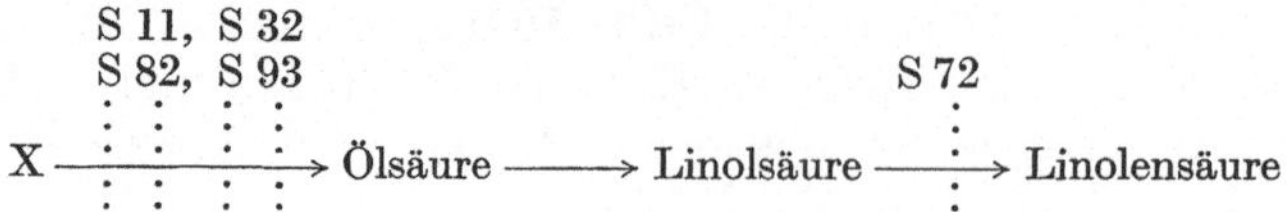

N. FRIES und H. M. YUSEF (1955) konnten durch UV-Behandlung zwei Mutanten von *Ophiostoma multiannulatum* gewinnen, die zum Unterschied von der Wildform nicht mit Glucose als einziger C-Quelle auskommen, sondern noch eine Fettsäure verlangen. Palmitin-, Stearin- und Ölsäure sind etwa gleich gut wirksam, weniger Laurin-, Linol- und Linolensäure, Säuren mit $C_{<10}$ sind unwirksam Die Stoffwechselbilanzen deuten an, daß die Mutanten sowohl Glucose wie Fettsäure für den Bau- *und* Energiestoffwechsel verwenden.

An dieser Stelle können auch die Versuche erwähnt werden, die J. HAIDASCH (1955) über Trockensubstanz- und Fettbildung an 3 genetisch verschiedenen Hefen durchführte: am Brauereistamm L und an den davon durch Polyploidierung mit Naphthylessigsäure bzw. Campher abgeleiteten Gigasstämmen 496 und 103 (s. auch R. BAUCH 1943). Wie Tabelle 4 zeigt, ist die Trockensubstanzproduktion bei den Polyploiden geringer als bei der Wildrasse, der Fettgehalt % dagegen bei Gigas 469 erhöht. Eine eingehendere physiologische Analyse steht freilich noch aus.

Tabelle 4. *Trockensubstanz- und Fettproduktion bei 2 Gigasmutanten und der Stammform (Brauhefe).* 100 ml Würze (8° Bllg.), 22°, 72 Std Lüftung, Mittelwerte aus mehreren Versuchsreihen. (Nach J. HAIDASCH 1955.)

Stamm	Trockensubstanz mg	Fett	
		mg	% vom Trockengewicht
Brauereistamm L	614,4	58,9	9,6
Gigas 469. . . .	460,1	55,3	12,1
Gigas 103. . . .	329,0	30,5	9,42

Wie zu erwarten, spielt der Gehalt der Zelle an Coenzym A (CoA) für ihr Vermögen zur Fettbildung eine ausschlaggebende Rolle (s. Beitrag W. FRANKE und H. FREHSE, dieser Handbuchband S. 174ff.). Das konnten H. P. KLEIN und F. LIPMANN (1953a) sehr schön mit Hefe zeigen, die einmal künstlich an CoA verarmt, das andere Mal (durch reichliche Pantothensäuregabe) mit CoA angereichert worden war (wegen der Methode s. H. P. SARETT und V. H. CHELDELIN 1945 und G. D. NOVELLI und F. LIPMANN 1950). Es wurde die Neubildung von Gesamtlipoid (Chloroformextrakt) und von Sterinen aus Acetat studiert. Einige Ergebnisse bringt Tabelle 5.

Tabelle 5. *Saccharomyces cerevisiae (LK2G12) ruhende Zellen in 0,1 mol Na-Acetat, 0,1 m Phosphatpuffer (p_H 7,2) 30°, 16 Std.* (Nach H. P. KLEIN und F. LIPMANN 1953a.)

Versuch Nr.	Pantothensäure-Gabe µg/ml	CoA-Gehalt (Einheiten/g Trockengewicht)	Neugebildet	
			Rohfett mg/g Trockengewicht	Sterine mg/g Trockengewicht
2	0	47	2,4	0,2
	10	280	12,7	2,5
3	0	60	1,0	0,4
	10	310	8,6	1,8
4	0	57	1,2	0,2
	10	300	9,6	1,7

Die Beziehungen zwischen CoA-Gehalt und Ausmaß der Fettsynthese sind klar erkennbar. Daß die Synthese sowohl der Fettsäure wie der Sterine tatsächlich vom dargebotenen Acetat ausgeht, zeigt ein Isotopenversuch (Tabelle 6). CoA-reiche Hefe hat also während der Versuchsdauer etwa dreimal so viel C^{14} aus dem markierten Acetat in die Fettsäure- und Sterinfraktion eingebaut als CoA-arme Hefe. Durchaus analoge Resultate wurden bei gleichartigen Versuchen mit Rattenleberpräparaten erhalten (H. P. KLEIN und F. LIPMANN 1953a). Schon früher hatten D. J. HANAHAN und S. J. AL-WAKIL (1952), gleichfalls mit markierter Essigsäure ($C^{14}H_3C^{14}OOH$ und $CH_3C^{14}OOH$), die Ergosterinsynthese CoA-armer und -angereicherter Hefen untersucht.

Eine Anreicherung bzw. Verarmung von Wirkstoffen ist auch bei der „Aktivierung“ des Fettbildungsvermögens zu vermuten, worüber A. G. C. WHITE und C. H. WERKMAN (1948) berichten. Die Zellen wurden vor der Inkubation mit

Natriumacetat und Phosphatpuffer in reinem 0,025 mol Phosphatpuffer aufgeschwemmt. Unterblieb diese Vorbehandlung, so fand im Versuch überhaupt keine Fettbildung statt, bei 18stündiger Vorbehandlung betrug die Fettzunahme ~50%, bei 24 Std ~100%, bei 48, 76, 96 Std ~25%. Trockenhefe und lyophilisierte Hefe war völlig inaktiv.

Auch in den Versuchen von I. SMEDLEY-MACLEAN und D. HOFFERT (1923) erwies sich die Intensität der Fettsynthese ruhender Zellen von Brauhefe von der Vorbehandlung abhängig. Die Fettbildung aus Acetat war um so geringer, je länger die Zellen vor dem Versuch im Eisschrank aufbewahrt worden waren.

Tabelle 6. *Saccharomyces cerevisiae (LK2G12), ruhende Zellen, 2 Std mit 2,5 μ mol $CH_3C^{14}OOK$ (Aktivität 1×10^{15} Impulse/min) inkubiert.* (Nach H. P. KLEIN und F. LIPMANN 1953a.)

	CoA-arm	CoA-reich
CoA-Gehalt (Einheiten/g Trockengewicht)	45	170
Fettsäuren (Impulse/min/mg)	67	182
Sterine (Impulse/min/mg)	12	38

H. C. GREENE und E. B. FRED (1934) stellten Unterschiede in der Fettbildung von *Aspergillus sydowi* (Stamm 1) fest, je nachdem ob Stammkulturen, von denen die Sporen gewonnen wurden, selten oder häufig überimpft wurden. Sie fanden in Parallelkulturen 13,57%, 12,18%, 10,79% Fett, wenn die Ausgangskultur alle Wochen, bzw. alle 3 bzw. 6 Monate überimpft worden war.

Nach den Beobachtungen von F. F. NORD und Mitarbeitern (insbesondere R. P. MULL und F. F. NORD 1944, S. WEISS, J. V. FIORE, F. F. NORD 1947, F. F. NORD, J. V. FIORE und S. WEISS 1948 und F. F. NORD, J. V. FIORE, G. KREITMAN und S. WEIS 1949, R. J. COLEMAN und F. F. NORD 1952) scheinen auch die von *Fusarium*-Arten gebildeten Naphthochinonpigmente beim Fettstoffwechsel eine gewisse Rolle zu spielen. Es bestehen Parallelen zwischen der Menge und Zusammensetzung des Fettes und der je nach Stamm und Kulturbedingungen verschiedenen Farbstoffproduktion. So wurde z.B. durch Zugabe von Solanion, dem Naphthochinonfarbstoff aus *Fusarium solani* (D_2 purple), zu Kulturen des farblosen *Fusarium lycopersici* dessen Fettgehalt herabgedrückt und die Jodzahl der Fette erhöht. Ein gleicher Effekt konnte freilich auch durch die Zugabe von Riboflavin oder Nicotinsäure erreicht werden.

2. Der Einfluß äußerer Faktoren auf die Fettbilduug.

a) Sauerstoffversorgung.

Schon C. v. NAEGELI und O. LOEW (1880) kamen bei ihren Versuchen mit *Penicillium* zur Erkenntnis, daß um so mehr Fett gebildet wird, je lebhafter die aerobe Atmung ist. Mit besonderem Nachdruck hat P. LINDNER (1919, 1921, 1922) auf die Notwendigkeit guter Durchlüftung für die Verfettung von Hefen und anderen Mikroorganismen hingewiesen[1], und im gleichen Sinne äußern sich viele Autoren über die Fettbildung bei Pilzen.

[1] P. LINDNER (s. a. S. 219) war von dem Gedankengang „gute Sauerstoffversorgung→Verfettung → Vermehrungsunfähigkeit" geradezu fasziniert. So glaubte er, auf dieser Grundlage das „Biosproblem" einfach lösen zu können. Nicht eine Verunreinigung des Zuckers, nicht die Menge der Hefeeinsaat, nicht die Form der Kulturgefäße sei maßgebend, sondern lediglich die Sauerstoffspannung. LIEBIG und WILDIERS hätten bei ihren Versuchen Lösungen verwandt, die nach der Dampfsterilisation schon längere Zeit gestanden und deshalb O_2-reich waren. Die Einsaatzellen verfielen sofort der Verfettung und damit der Unfähigkeit zur weiteren Teilung. PASTEUR hingegen hätte frisch sterilisierte, noch O_2-arme Lösungen verwendet, in denen die Hefezellen nicht verfetteten und deshalb vermehrungsfähig blieben!

A. RIPPEL-BALDES, K. PIETSCHMANN-MEYER und W. KÖHLER (1950) bringen Angaben über Wachstum und Fettbildung bei *Candida Reukaufii*, die in KLUYVER-Kolben bei verschieden starker Durchlüftung gezüchtet wurden (Tabelle 7). Bei zunehmender Lüftung steigt also der ökonomische Koeffizient und die absolute und auf Trockensubstanz bezogene Fettmenge. In Versuchen mit *Rhodotorula gracilis*, welche S. C. PAN, A. A. ANDREASEN und P. KOLACHOV (1949) durchführten, war es gleichgültig, ob die Durchlüftung 0,1—1 Vol. Luft/min/Vol. Lösung zuführte oder ob die Kultur geschüttelt wurde (Fettgehalt 48,2—49,6%). In „stehenden Kulturen“ wurde dagegen nur 30,2% Fett gefunden. Zu ähnlichen

Tabelle 7. *Candida Reukaufii bei verschiedener Durchlüftung.*
(Nach A. RIPPEL-BALDES, K. PIETSCHMANN-MEYER und W. KÖHLER 1950.)

	Lüftung			
	„sehr schwach“	„schwach“	„stärker“	„stark“
Trockensubstanz g/l	2,67	3,98	6,16	7,06
ökonomischer Koeffizient	2,76	4,06	5,58	6,80
Fett g/l	0,34	0,51	0,95	1,18
Fett, Prozent Trockengewicht	12,84	15,25	15,42	16,69
Fettkoeffizient	2,76	4,06	5,58	6,80

Ergebnissen kamen W. DIEMAIR und C. BORESCH bei einer nicht näher bestimmten Mucorinee (1950). Dagegen fanden E. A. PRILL, P. R. WENCK und W. H. PETERSON (1935) in Kolbenkulturen von *Aspergillus Fischeri* keinen Einfluß von Belüftung auf die Menge und Zusammensetzung des Fettes. Vermutlich war in ihren Versuchen auch ohne besondere Maßnahmen bereits eine optimale O_2-Versorgung gesichert.

Eine sehr starke Anreicherung der Lipoide und vor allem der Sterine in Hefe konnten HALDEN und Mitarbeiter (W. HALDEN, F. BILGER, R. KUNZE 1933, W. HALDEN 1934, M. SOBOTKA, W. HALDEN und F. BILGER 1935, F. BILGER, W. HALDEN, E. MAYER-PITSCH und M. PESTEMER 1937) in Hefe erreichen, wenn sie reichliche Luft und verdünnten Alkoholdampf zuführten und zugleich den Wassergehalt verminderten (s. auch F. TRAININA 1937).

Einen sehr deutlichen Einfluß der Belüftung beobachteten auch I. SMEDLEY-MACLEAN und D. HOFFERT (1923) bei ruhender Bierhefe. Wird eine Hefeaufschwemmung in Wasser durchlüftet, so nimmt der Kohlenhydratgehalt der Zellen in 45 Std um etwa $^2/_3$, der Proteingehalt um etwa $^1/_7$ ab, während das Fett eine Zunahme un 50—100% erfährt. Ohne Durchlüftung ist die Abnahme der Kohlenhydrate und Proteine geringer; eine Fettbildung findet überhaupt nicht statt. Auch die Fettsynthese aus zugesetzten Substraten (Äthanol, Na-Acetat, Zucker usw.) war bei Durchlüftung stark gefördert (Weiteres über diese Versuche S. 222).

Über das Minimum des O_2-Partialdruckes für die Fettbildung gibt es noch keine Angaben. A. KLEINZELLER (1944) hat diese untere Grenze bei seinen Versuchen mit *Torulopsis lipofera* jedenfalls nicht erreicht; er fand mit Luft, Luft + 5% CO_2, 50% O_2, 50% O_2 + 5% CO_2 keine Unterschiede im Glucoseverbrauch und in der Fettbildung ruhender *Torulopsis*-Zellen.

Für manche Pilze wird Oberflächenkultur als Erfordernis für gutes Wachstum und gute Fettbildung angegeben: so für *Endomycopsis vernalis* (H. FINK, H. HAEN und W. HOERBURGER 1937) und für *Oospora lactis* (H. GEFFERS 1937). Dagegen konnten auch mit diesen Pilzen K. BERNHAUER und J. RAUCH (1948) und B. DREWS und H. SPECHT (1950) gute Fettbildung in belüfteten Submerskulturen erzielen. Umgekehrt ist auch bei Schimmelpilzen das Schütteln der

Kultur nicht immer für die Fettbildung günstig (z. B. M. WOODBINE, M. E. GREGORY und T. K. WALKER 1951). Ebenso fand O. TURPEINEN (1936) bei *Geotrichoides sp.* Wachstum und Fettproduktion sehr stark (um rund 80%!) vermindert, wenn die Kolben während der 8tägigen Kulturdauer öfters geschüttelt wurden, während der Pilz auf eine bessere O_2-Versorgung durch Vergrößerung des Verhältnisses Oberfläche/Volumen sehr positiv reagierte.

b) C-Ernährung.

Über die Eignung verschiedener C-Quellen als Ausgangsmaterial der Fettbildung durch Mikroorganismen liegt eine große Anzahl von Untersuchungen vor. Ihr Wert liegt darin, daß sie erste Anhaltspunkte für den Weg der Fettsynthese durch den Organismus zu geben vermögen.

Zumeist wurden derartige Versuche so durchgeführt, daß im synthetischen Nährmedium die zu prüfende organische Verbindung als einzige C-Quelle gegeben wurde. Tritt Wachstum des Organismus auf, so ist damit nur gesagt, daß eine assimilierbare Kohlenstoffquelle vorliegt, aus der also im Verwendungsstoffwechsel alle Zellbestandteile aufgebaut werden können. Vorsichtige Schlüsse auf eine größere oder geringere Eignung *speziell* für die Fettbildung lassen sich erst durch die Vergleiche der auf Trockengewicht bezogenen prozentualen Fettmengen ziehen. Der andere Fall, daß eine Substanz zwar als Ausgangsmaterial für die *Fettbildung* dienen kann, für das *Wachstum* des Organismus als alleinige C-Quelle aber nicht ausreicht, läßt sich nur durch Versuche mit *ruhenden Zellen* durchführen, bei denen Wachstum durch völligen Entzug aller übrigen Nährelemente (insbesondere N) ausgeschlossen wird. Hier läßt sich dann durch genaue Bilanzierung oder durch Isotopenmarkierung eine präzise Aussage machen.

α) Versuche mit wachsenden Kulturen.

Bei *Penicillium javanicum* haben L. B. LOCKWOOD, G. E. WARD und O. E. MAY, H. T. HERRICK und H. T. O'NEILL (1934), bei *Torulopsis lipofera* L. E. DEN DOOREN DE JONG (1926a) eine große Zahl von C-Quellen auf ihre Eignung geprüft, ohne indes über ihren Einfluß auf die Fettproduktion durch den Pilz zu berichten. Die für *Endomycopsis vernalis* brauchbaren C-Quellen haben P. LINDNER (1922b), S. HEIDE (1939) und H. RAAF (1941) eingehend studiert. Einige Ergebnisse bringt Tabelle 8. Es zeigt sich, daß der prozentuale Fettgehalt des gebildeten Mycels nur geringfügig schwankt, gleichgültig, ob die gebotene Verbindung eine gute (die meisten Zucker) oder eine mäßig gute (Mannit) C-Quelle darstellt. Ameisen-, Essig-, Propion-, n-Butter-, Malon-, Äpfel-, Wein-, Palmitin- und Stearinsäure erweisen sich in Versuchen von H. RAAF als völlig unbrauchbare C-Quellen, Oxal-, Brenztrauben-, Milch-, Bernstein- und Ölsäure ergaben gelegentlich ein ganz geringfügiges Wachstum. Wurde der in Zuckerkultur verfettete Pilz auf eine neue zuckerhaltige Nährlösung gesetzt, so vermag er weiteres Fett zu bilden. Wurde dagegen in der zweiten

Tabelle 8. *Endomycopsis vernalis. Trockensubstanz und Fettbildung auf verschiedenen C-Quellen.* WÖLTJE-*Lösung mit 0,125% Asparagin, 7,5 Tage.* (Nach H. RAAF 1941.)

C-Quelle	%	Trockensubstanz mg/15 ml	Fett	
			% vom Trockengewicht	mg/15 ml
Glycerin	7,67	Spur	—	—
D-Mannit	7,59	85	42,5	36,1
Fructose	7,5	165	45,3	74,8
Glucose	7,5	185	46,3	85,8
Saccharose	7,5	196	47,0	92,2
Maltose	7,88	197	46,3	91,4
Galactose	7,5	150	47,5	71,25
Lactose	7,88	Spur	—	—
Mannose	7,5	175	47,5	83,2
Raffinose	8,26	140	42,0	58,8

Nährlösung der Zucker durch eine der oben genannten Säuren ersetzt, so setzt ein gleichgroßer Fettabbau ein wie in den Kontrollen mit destilliertem Wasser. Die Säuren können also selbst von erwachsenen Zellen *nicht* für eine Fettsynthese verwendet werden.

Auch O. TURPEINEN (1936) hatte bei *Geotrichoides sp.* gefunden, daß die Fettbildung unabhängig von der Art der C-Quelle bleibt, falls diese überhaupt verwendbar ist: Glucose, Fructose, Mannose, Galaktose, Saccharose, Maltose, Raffinose, Stärke, Erythrit, Sorbit, Mannit, Succinat, Fumarat, Citrat; als nicht verwendbar erwiesen sich: Lactose, Inulin, Arabinose, Xylose, Rhamnose, Glycerinaldehyd, Dioxyaceton, Äthanol, Glycerin, Dulcit, Acetat, Lactat, Tartrat.

M. STEPHENSON und M. D. WHETHAM (1923) stellten bei *Mycobacterium phlei* fest, daß Wachstum und Fettbildung durch Glucose und Lactat, nicht aber durch Acetat, ermöglicht wird. Acetat wird aber, gemeinsam mit Lactat geboten, sowohl für Wachstum wie für Fettbildung verwendet.

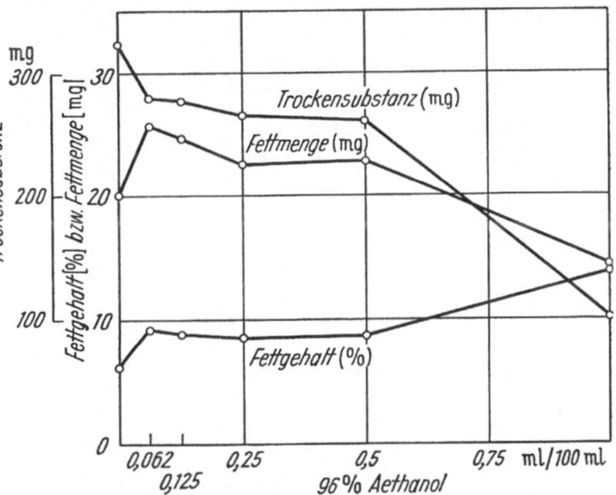

Abb. 1. *Endomycopsis vernalis.* Trockensubstanz- und Fettbildung bei verschiedener Äthanolgabe. (Nach S. HEIDE 1939.)

Von zahlreichen Autoren wurde *Äthanol* als brauchbare C-Quelle für Wachstum und vor allem für Fettbildung bei Hefen und Mycelpilzen angegeben: F. STOCKHAUSEN (1911), P. LINDNER und ST. CZIFER (1912), P. LINDNER (1919, 1921, 1922, 1929[1]), H. SCHANDERL (1936, *Racodium cellare*), W. HALDEN, F. BILGER und KUNZE (1933), W. HALDEN (1934), M. SOBOTKA, W. HALDEN und F. BILGER (1935), F. BILGER, W. HALDEN, E. MAYER-PITTSCH und M. PESTEMER (1937) (speziell Sterinanreicherung in Hefen durch Äthanol), H. FINK, J. KREBS und R. LECHNER (1937), H. FINK und F. JUST (1939) (*Torulopsis utilis*[2]). Auch S. HEIDE (1939) hat in mehreren Versuchsreihen eine stärkere prozentuale Verfettung von *Endomycopsis* bei Äthanolzugabe zum Kulturmedium gefunden. Das Ergebnis eines seiner Versuche ist in Abb. 1 dargestellt. Mit steigender Äthanolgabe nimmt zwar der prozentuale Fettgehalt zu, der Ertrag an Trockensubstanz aber ab. Im Sinne unserer späteren Überlegungen (S. 239) kann daraus noch nicht sicher auf eine Verwertung des Äthanols für die Fettsynthese geschlossen werden. Es könnte sich ebenso um eine Hemmung der Bildung von Eiweiß und anderen Zellbausteinen handeln, so daß zunehmende Mengen von Zuckerabbauprodukten für die Fettsynthese disponibel wurden.

Beweiskräftiger sind die Versuche von H. RAAF (1941) über die Fettbildung bei *Endomycopsis* aus Acetaldehyd, wofür Abb. 2 ein Beispiel gibt.

Der Pilz wurde auf WÖLTJE-Lösung mit 1% Asparagin gezogen. Der Acetaldehyd wurde je zur Hälfte 2 Std vor der Beimpfung und 3 Tage nachher gegeben. Ernte nach 6,5 Tagen.

Wie der obere Teil der Abb. 2 zeigt, bewirken steigende Aldehydmengen eine Abnahme der Trockensubstanz und des im Pilz enthaltenen Gesamt-N. Die Fettausbeute je Kultur hat ein Optimum bei etwa 0,25% Acetaldehyd. Der

---

[1] Siehe auch Fußnote 1 auf S. 216.

[2] Wegen abnehmender Äthanoltoleranz von Hefezellen bei zunehmender Verfettung vgl. W. D. GRAY (1948).

prozentuale Fettgehalt nimmt aber ständig zu (unterer Teil der Abbildung), während der prozentuale N-Gehalt der Zellen unverändert bleibt!

Die 3-C-Verbindungen des Zuckerabbaues wurden mehrfach auf ihre Eignung für Wachstum und Fettbildung geprüft. *Milchsäure* kann nach H. FINK, J. KREBS und R. LECHNER (1937) bei *Torulopsis utilis* ebenso wie *Äthanol* oder *Essigsäure* als alleinige C-Quelle für den Aufbau aller normalen Zellinhaltstoffe einschließlich der Lipoide dienen. H. FINK, H. HAESELER und M. SCHMIDT (1937) beobachteten einen deutlichen Verbrauch der *Milchsäure*, wenn fettbildende *Oospora lactis* auf Molke gezüchtet wurde. Auch H. RAAF (1941) fand eine deutliche Zunahme des Fettes, nicht aber der Pilztrockensubstanz wenn *Endomycopsis* nach Vorkultur auf Zucker auf Lactatlösung übergesetzt wurde. Die Auswertung und Bilanzierung war die gleiche wie in analogen Versuchen mit *Brenztraubensäure*, von denen Tabelle 9 ein Beispiel bringt. Während die nachgegebene Saccharosemenge in D gerade hinreicht, um die

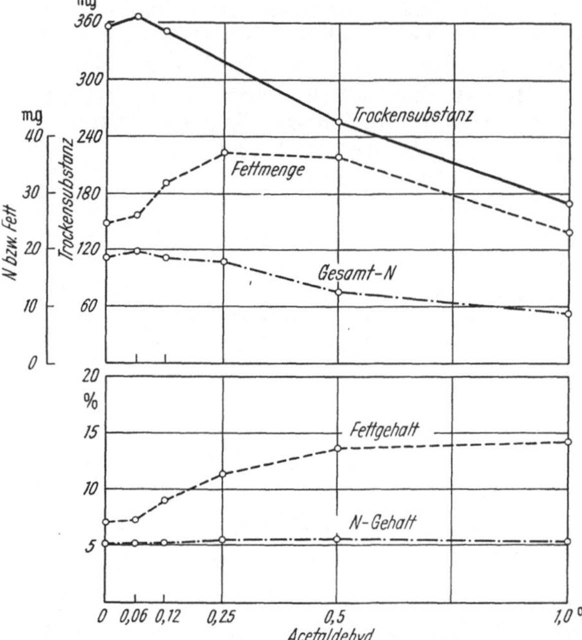

Abb. 2. *Endomycopsis vernalis*. Trockensubstanz- und Fettbildung bei verschiedener Acetaldehydgabe. (Nach H. RAAF 1941.)

vorhandenen Fettvorräte des Pilzes zu schonen, ergibt der Brenztraubensäurezusatz in C eine „echte" Neubildung von Fett, wie sich aus dem Vergleich mit

Tabelle 9. *Versuchsanordnung: Endomycopsis vernalis auf* WÖLTJE-*Lösung mit 1% Asparagin 8 Tage Vorkultur.* (Nach H. RAAF 1941.)

Am 8. Tage: 2 Schalen geerntet (A).
zu 2 Schalen 5 ml destilliertes Wasser (B).
zu 2 Schalen 5 ml 2% Brenztraubensäure in Phosphatpuffer (C).
zu 2 Schalen 5 ml 2% Saccharose (D).
(B) (C) (D) Ernte am 11. Tag, Mittelwerte jeweils von 2 Parallelkulturen.

	Trockengewicht	Fett		N		Restzucker
	mg	mg	% vom Trockengewicht	mg	% vom Trockengewicht	mg
A Kontrolle	364	23,6	6,55	19,8	5,45	28,9
B Destilliertes Wasser .	344,5	20,0	5,80	18,8	5,46	—
C Brenztraubensäure .	368	32,6	8,85	19,7	5,35	—
D Saccharose	392	24,0	6,1	20,0	5,12	—

den Kontrollen A und B deutlich zeigt. Milchsäure und Brenztraubensäure sind also „ungenügende" C-Quellen für *Endomycopsis*, die dieser aber gleichwohl zur *Fettsynthese* verarbeiten kann.

Bereits H. HAEHN und W. KINTTOF (1925) hatten durch mikroskopische Prüfung festgestellt, daß *Endomycopsis*-Zellen von Pepton-Brenztraubensäure nährböden stark verfettet waren, nicht hingegen die Kontrollen von reiner Peptonlösung.

G. GORBACH und A. SABLATNÖG (1934a, b) prüften Wachstum und Fettbildung von *Serratia marcescens (Bact. prodigiosum)* auf sehr verschiedenen C-Quellen. Die fördernden Wirkungen auf Wachstum und Fettbildung fallen nicht zusammen. Allgemeine Gesichtspunkte lassen sich nicht ableiten.

Beim Vergleich verschiedener Zuckerarten hat vielfach das Verhalten der Xylose besonders interessiert, da diese in Sulfitablaugen enthalten ist, die als Ausgangsmaterial für die biologische Fetterzeugung im technischen Maßstabe in Betracht kommen. Nach J. V. FIORE (1948) wächst *Fusarium lini* auf Xylose etwas schlechter als auf Glucose und Fructose. Der Prozentgehalt an Rohfett und Sterinen ist aber überall gleich. Für *Fusarium lycopersici* ergeben sich überhaupt keine nennenswerte Unterschiede (vgl. auch R. P. MULL und F. F. NORD 1944b). *Rhodotorula gracilis* verarbeitet Glucose und Xylose im Verhältnis 100:29. Das Wachstum auf Xylose ist schwächer und unökonomischer als auf Glucose. Diese Unterschiede können aber durch Adaptation fast völlig ausgeglichen werden (N. NIELSEN und N. G. NILLSON 1949, A. KLEINZELLER J. MÁLEK, R. PRAUS und J. ŠKODA 1952, A. KLEINZELLER und J. MÁLEK 1952).

Tabelle 10. *Rhodotorula gracilis, Fettbildung auf verschiedenen Substraten.* (Nach M. P. STEINBERG und Z. J. ORDAL 1954.)

Substrat	Trockengewicht mg/10 ml		„fat ratio" nach 38 Std
	nach 23,5 Std	nach 38 Std	
Glucose . . .	93,5	109,8	1,09
Fructose . . .	90,3	108,9	1,03
Invertzucker .	91,6	108,6	1,06
Äthanol[1] . . .	66,7	80,6	0,76
Natriumacetat[1]	66,3	80,5	0,58

[1] Äquimolekular mit den Zuckern.

M. P. STEINBERG und Z. J. ORDAL (1954) haben kürzlich gleichfalls die Verwertung verschiedener Zucker und anderer Substrate für die Fettbildung von *Rhodotorula gracilis* geprüft. Ihre Ergebnisse, ausgedrückt als „fat ratio" (s. S. 211), sind in Tabelle 10 wiedergegeben. Die einzelnen Zucker erweisen sich mithin als ziemlich gleichwertig; auch Äthanol und Acetat sind zur Fettbildung geeignet, aber weniger wirksam.

Im Zusammenhang mit dem bereits früher (S. 213) erwähnten Gegensatz zwischen Gärungsstoffwechsel und Fettbildung dürften Versuche von L. W. und

Tabelle 11. *Fettgehalt und fettfreie Trockensubstanz, Bakterien auf verschiedenen Nährböden.* (Nach L. W. und W. P. LARSON 1922.)

Nährböden: A 1,2% Liebigs Fleischextrakt, 2,9% Pepton, 1,0% NaCl.
„ B wie A, dazu 0,5% Glucose.
„ C wie A, dazu 3% Glycerin.

	C-Quelle	Vergärbarkeit	Fettfreie Trockensubstanz (g)	Fett (% Trockengewicht)
Escherichia coli	A —	—	4,38	9,59
	B Glucose	+	7,42	9,09
	C Glycerin	—	10,10	*19,9*
Staphylococcus albus	A —	—	3,71	7,97
	B Glucose	+	1,97	7,6
	C Glycerin	—	0,51	*39,9*
Bacillus megaterium	A —	—	1,53	7,15
	B Glucose	—	0,56	*18,1*
	C Glycerin	—	1,65	*33,8*
Bacillus mucosus	A —	—	3,26	13,9
	B Glucose	+	7,76	8,82
	C Glycerin	+	8,84	15,42

W. P. LARSON (1922) Erwähnung verdienen. Das Trockengewicht und der Rohfettgehalt (Äther + Aceton — Extrakt) einiger Bakterienarten wurden auf Fleischextrakt-Peptonbouillon ohne Zusatz und mit Zusatz von Glucose bzw. Glycerin untersucht. Es ergab sich, daß überall der Fettgehalt bei Zusatz *nicht vergärbarer* C-Quellen *höher* ist als mit *vergärbaren* Nährstoffen. Das zeigt Tabelle 11.

Die Beziehungen zwischen *Menge* der C-Quelle und Fettbildung werden zweckmäßigerweise im nächsten Abschnitt (S. 227ff.) im Zusammenhang mit dem Einfluß der N-Ernährung besprochen.

Nur selten wurde die Frage geprüft, ob die *Art* der *C-Quelle* die daraus synthetisierten *Pilzfette* beeinflußt. [Auf die qualitativen Beziehungen zwischen *Nahrungsfett* und *Zellfett* wird weiter unten (S. 225ff.) eingegangen.]

H. H. BARBER (1929) hat den Einfluß verschiedener Zucker (Hexosen, Pentosen) auf die Zusammensetzung des Fettes von Penicillium sp. untersucht. Es ergaben sich keine Unterschiede. Näheres hierüber S. 255f.

F. F. NORD und Mitarbeiter haben neuerdings mehrfach bei *Fusarium lini* derartige Versuche mit Glucose und Acetat durchgeführt. Wie die nachfolgende Zusammenstellung zeigt, sind die Ergebnisse uneinheitlich:

	Medium	Jodzahl des Fusarium-Fettes
R. J. COLEMAN, M. CEFOLA und F. F. NORD (1952)	3,6 g Acetat + 3,6 g Glucose 0,4 g Acetat + 6,8 g Glucose	120,0 *87,1*
R. J. COLEMAN und F. F. NORD (1952)	2,5% Acetat 2,5% Glucose	*137,2* *148,4*
R. J. COLEMAN und F. F. NORD (1953)	1,2% Acetat + 1,0% Glucose 1,87% Glucose	122,6 88,9

Auf Glucosesubstrat ist also die Jodzahl gegenüber Acetatkulturen bald erhöht, bald erniedrigt.

Unterschiede der gesuchten Art sind überhaupt nur wahrscheinlich, wenn ein verschiedenes Ausgangsprodukt für die Synthese der gesättigten und der ungesättigten Fettsäuren angenommen wird.

β) Versuche mit ruhenden Zellen.

Wie bereits früher ausgeführt, kann eine Entscheidung über die Brauchbarkeit einer von außen dargebotenen C-Quelle für die Fettsynthese am ehesten durch Versuche mit *ruhenden Kulturen* gefällt werden.

Solche Untersuchungen wurden zuerst von I. SMEDLEY-MACLEAN und D. HOFFERT (1923) durchgeführt und von L. D. MACLEOD und I. SMEDLEY-MACLEAN (1938) fortgesetzt. Als Versuchsmaterial diente Bierhefe (*Saccharomyces cerevisiae*), die in Wasser mit Zusatz verschiedener Substrate mit O_2 durchlüftet wurde. Durch Analysen wurde die Veränderung des Kohlenhydrat-, Fett- und Proteingehaltes während des Versuches ermittelt. Eine Auswahl der Versuchsergebnisse ist in Tabelle 12 dargestellt. Mit allen in der Tabelle genannten Substraten ergibt sich in den *Durchlüftungs*reihen eine deutliche Fettzunahme, die nicht auf Umwandlung aus anderen Zellbestandteilen erklärt werden kann, wie der Vergleich mit der Wasserkontrolle lehrt. *Ohne Durchlüftung* führen nur die Zucker zu einer gewissen Fettbildung, nicht aber Äthanol, Essigsäure, Milchsäure und Brenztraubensäure. Die in den zwei Zahlenreihen für Fructose angedeutete

Tabelle 12. *Fettbildung ruhender Hefezellen aus verschiedenen Substraten.* (Nach I. SMEDLEY-MACLEAN und D. HOFFERT 1923.)
Je 12,5 g Brauhefe *(Saccharomyces cerevisiae)* in 1 Liter Wasser ohne und mit Substrat, ohne und mit O_2-Durchlüftung, 25°, 45 Std. Die Zahlen (errechnet nach den gemittelten Analysenzahlen der Originalarbeit) geben die prozentuale Zu- und Abnahme der betreffenden Fraktion gegenüber der Ausgangsmenge zu Versuchsbeginn an.

Substrat	Durchlüftet			Nicht durchlüftet		
	Kohlenhydrate %	Fett %	Protein %	Kohlenhydrate %	Fett %	Protein %
Wasser	− 62	+ 59	− 14	− 39	+ 3	− 15
Äthanol 0,5%	− 40	+ 128	− 7	− 47	+ 8	− 10
Na-Acetat 0,6% . . .	+ 122	+ 220	− 13	+ 93	− 2	− 10
Na-Lactat 0,9% . . .	− 62	+ 104	− 6	− 73	− 3	− 17
Na-Pyruvat 0,88% . .	− 57	+ 145	− 1	− 39	+ 18	± 0
Glucose 1%	+ 3	+ 220	+ 3	− 8	+ 35	− 2
Fructose 1%	+ 3	+ 205	− 5	+ 4	+ 144	− 4
Fructose 5%	+ 13	+ 470	+ 16	+ 24	+ 164	+ 13
Saccharose 1% . . .	− 52	+ 196	+ 1	+ 17	+ 58	− 6
Maltose 1%	+ 19	+ 303	− 7	+ 63	+ 164	− 9

Proportionalität zwischen Substratkonzentration und neugebildeter Fettmenge ist bei allen Zuckern vorhanden.

Die *gleiche Fettzunahme* wie in der *Wasserkontrolle* ergab sich bei folgenden Substraten: Na-Format (0,1 mol), Na-Propionat (0,1 mol), Na-Butyrat (0,1 mol), iso-Propanol (0,1 mol), Aceton (0,1 mol), Glycol (0,5%), Glycerin (0,5%, 5%); *weniger Fett als in der Wasserkontrolle* fand sich bei Methanol (0,1 mol), n-Propanol (0,1 mol), n-Butanol (0,1 mol), iso-Pentanol (0,1 mol), Acetaldehyd (0,1%, 0,6%). L. D. MACLEOD und I. SMEDLEY-MACLEAN (1938) fügten dann der Reihe *indifferenter* Substrate noch folgende hinzu: Acetoin, 2:3-Butylenglycol, Methyläthylketon, Na-Citrat, Na-Succinat, Na-Gluconat; den *hemmenden* Substraten: Diacetyl, Methylvinylketon, Na-Sorbinat. Auf die Wirkung der Zugabe von Phosphat und von verschiedenen Kationen, die bei diesen Versuchen beobachtet wurde, kommen wir später (S. 234) zurück. Aus einigen Analysen von L. D. MACLEOD und I. SMEDLEY-MACLEAN geht hervor, daß an der experimentell bewirkten Lipoidzunahme die Fraktionen der Fettsäuren und des Unverseifbaren etwa gleichen Anteil haben. In Ergänzung zu den früheren Versuchen teilten I. SMEDLEY-MACLEAN und D. HOFFERT (1926) noch mit, daß ein Teil des Substrates zu CO_2 veratmet wird und daß Sulfitzugabe zu Äthanol- oder Acetatansätzen die Bildung von Fett (und von Kohlenhydraten) verhindert.

Die soeben referierten Ergebnisse konnten von H. WIELAND und F. WILLE (1935) in allen wesentlichen Punkten bestätigt werden. In verarmender — in Wassersuspension mit O_2 durchlüfteter — Hefe nehmen die Kohlenhydrate ab, die Lipoide zu. Durchlüftung mit Äthanol führt zu starker Lipoidvermehrung, Acetat ist weniger wirksam, noch weniger Lactat. Für die Äthanolversuche wird berechnet, daß im Durchschnitt von 14 Molekeln Äthanol 11 verbrannt, 2 zur Lipoidsynthese, 1 zum Kohlenhydrataufbau verwendet werden.

Ähnliche Versuche hat A. KLEINZELLER (1944) mit ruhender *Torulopsis lipofera* durchgeführt. Er fand nur mit Zuckern (Glucose, Fructose, Galactose, Maltose, Saccharose) und mit Acetaldehyd eine signifikante Fettbildung, nicht aber mit Äthanol, Acetat, Pyruvat, Lactat, Dioxyaceton, Butyrat, Ketoglutarat, Gluconat und gemischten Aminosäuren (trypsinverdautem Casein).

Die Fettbildung aus C^{13}-markiertem Acetat haben an Hefe A. G. C. WHITE und C. H. WERKMAN (1947) und S. WEINHOUSE und R. S. MILLINGTON (1947) mit positivem Erfolg geprüft. In den Versuchen der Letztgenannten wurden innerhalb von 7 Std etwa $^1/_4$—$^1/_3$ der Gesamtlipoide neu synthetisiert (Tabelle 13). Die C^{13}-Einlagerung war in den Fettsäuren stärker als im Unverseifbaren, in den gesättigten Fettsäuren wieder stärker als in den ungesättigten.

Tabelle 13. *Einbau von C^{13} in die Lipoide der Hefe aus C^{13}-Acetat.* (Nach S. WEINHOUSE und R. S. MILLINGTON 1947.)

	Menge mg	C^{13}-Überschuß
Gesamtlipoide .	184,4	0,99
Unverseifbares .	28,5	0,49
Fettsäuren . .	122,9	1,12
gesättigt . .	—	1,44
ungesättigt .	—	0,97

Weiters sind hier die Isotopenversuche von H. P. KLEIN und F. LIPMANN (1953a) zu erwähnen, in denen gleichfalls die Verwendung von Essigsäure für die Lipoidsynthese bei Hefe eindeutig bestätigt wurde. Darüber wurde bereits in anderem Zusammenhang (S. 215) berichtet.

Mit ruhendem Material von *Endomycopsis vernalis* prüften W. HAEHN und W. KINTTOF (1925) eine Reihe von C_2-, C_3-, C_4-Verbindungen, die sie als mögliche Intermediärprodukte bei der Umwandlung von Kohlenhydraten in Fette in Betracht zogen, mit positivem Ergebnis (Tabelle 14). Die fettarm vorgezogenen Pilzdecken wurden mit dem Substrat in Phosphatpufferlösung (p_H 6,8) unterschichtet. Nach 3—6 Tagen wurde im Versuchsansatz der Fettsäurengehalt durch Kaliaufschluß nach STOCKHAUSEN bestimmt (Berechnung als Ölsäure). Da keine Absolutmengen angegeben werden, könnte eine Zunahme des prozentualen Fettgehaltes auch durch Abnahme der Nicht-Fette der Zelle zustande gekommen sein. Das ist indes bei den großen Differenzen zwischen Versuch und Kontrolle ziemlich unwahrscheinlich.

Tabelle 14. *Endomycopsis vernalis. Fettbildung aus verschiedenen Substraten.* (Nach H. HAEHN und W. KINTTOF 1925.)

Substrat	Fettsäuren % Trockengewicht	
	Versuch	Kontrolle
Brenztraubensäure 1%	11,18	3,33
Milchsäure 2%	10,00	2,75
Acetaldehyd 1% (täglich erneuert) . . .	8,28	2,84
Aldol	29,28	6,78
Äthanol 4%	28,07	~3,00 (geschätzt)
Glycerin 4%	16,81	~3,00 (geschätzt)

γ) Zur Frage „Fett aus Eiweiß".

In den ältesten Arbeiten über die Physiologie der Fettbildung bei Pilzen spielt die Frage eine große Rolle, ob die Fette aus Kohlenhydraten oder aus Eiweißstoffen gebildet werden (z. B. C. v. NAEGELI 1879, C. v. NAEGELI und O. LOEW 1880, E. DUCLEAUX 1889, 1899, auch noch bei W. HENNEBERG 1926).

Zur Annahme einer Fettentstehung aus dem Eiweiß konnte vor allem der mikroskopische Befund verleiten, daß besonders in alternden Pilzzellen sich Fetttropfen „aus dem Protoplasma" bilden. Es ist kaum zweifelhaft, daß es sich dabei in der Regel nicht um Neubildung oder auch nur um eine Vermehrung der Fette, sondern nur um deren Sichtbarwerden durch Änderung des Dispersionszustandes handelte.

C. v. NAEGELI und O. LOEW (1880) fanden wechselnde Fettmengen in 2 Monate alten *Penicillium*kulturen, die auf sehr verschiedenen Nährböden (Pepton, Albumin, Rohrzucker, Tartrat, allein und in Kombination) gewachsen waren. Sie stellten zwar fest, daß die Fettbildung besonders kräftig bei Kohlenhydraternährung verläuft. Im Sinne einer Fettentstehung aus Plasmaproteinen ließ sich aber allenfalls die Zunahme des Fettgehaltes von 18,5 auf 54,5% des Trockengewichtes deuten, die sie bei der „Involution" alternder *Penicillium*hyphen zu beobachten glaubten. Es ist indes leicht nachzuweisen, daß hier eine Fehldeutung vorliegt.

Ein aliquoter Teil der Mycelien wurde 4 Wochen lang in 1%iger H_3PO_4 aufbewahrt und danach der hohe Fettgehalt festgestellt. Die Analysendaten lassen aber errechnen, daß während des Versuches das Mycel 83% seiner Trockensubstanz verloren hatte. Es war also unter extremen Autolysebedingungen vom Zellinhalt praktisch nur das Fett liegengeblieben, dessen *absolute Menge* sich aber gleichfalls um rund 54% vermindert hatte.

Man kann die Frage nach einer Fettbildung aus Eiweiß heute sinnvoll wohl nur in der Form stellen: *ob und in welchem Umfange die C-Skelete von Aminosäuren als Bausteine für die Fettsynthese verwendet werden können.* Mit der Tatsache, daß manchen Schimmelpilzen Aminosäuren (und besonders Asparagin) als einzige C- *und* N-Quelle für den Aufbau der gesamten Zellsubstanz, also *auch* von Fetten, genügen, ist die Frage wenigstens grundsätzlich positiv beantwortet (F. CZAPEK 1920).

H. RAAF (1941) hat sich bei *Endomycopsis vernalis* mit dieser Frage beschäftigt. Wurde Glykokoll, Alanin, Asparaginsäure oder Asparagin als einzige C- und N-Quelle geboten, so ging der Pilz überhaupt nicht an. Es ergaben sich lediglich Anhaltspunkte dafür, daß sich die Länge der C-Ketten, der Aminosäuren, die als *N-Quelle* gut verwertet wurden, modifizierend auf das N:C-Verhältnis im Kulturmedium und damit auf den Grad der Verfettung (s. unten S. 227ff.) auswirkt. Auch bei H. KATING (1955a) findet sich ein Hinweis in der gleichen Richtung: der Fettgehalt in einer auf reine 2%ige γ-Aminobuttersäure übertragenen Pilzdecke nahm in 20 Std von 71,8 auf 81,5 mg zu, ohne daß eine andere exogene C-Quelle zur Verfügung stand als die genannte Aminosäure.

Überraschende Ergebnisse zu unserer Frage haben nun in jüngster Zeit zwei mit Isotopen durchgeführte Arbeiten gebracht. C. H. WANG, V. H. CHELDELIN, T. E. KING und B. E. CHRISTENSEN (1951) untersuchten mit am Methyl durch C^{14} markiertem Methionin die Aufhebung der Wachstumshemmung durch 2-Chloro-4-aminobenzoesäure. Als sie die Lipoidfraktion der mit dem isotopen Methionin behandelten Bakterien untersuchten, fanden sie darin einen bedeutenden Teil der Radioaktivität und zwar sowohl im Unverseifbaren wie in den Fettsäuren, insbesondere in den C_{16}-Säuren. S. WEINHOUSE, R. S. MILLINGTON und M. STRASSMANN (1951) inkubierten Hefe mit Glykokoll, welches am *Methyl* mit C^{14} markiert war. In der Hefe fanden sich dann reichlich Fettsäuren, welche *über die ganze Kettenlänge* verteilt C^{14} enthielten. Auch im Unverseifbaren und im Methyl des Cholins wurde der markierte Kohlenstoff gefunden. Dieser Effekt blieb aber aus, wenn *carboxylmarkiertes* Glykokoll verwendet wurde. Unter der Voraussetzung einer Synthese der Fettsäuren aus C_2-Bausteinen (siehe diesen Handbuchband, S. 177f.) schlagen die Verfasser folgende Formulierung für den Weg von methylmarkiertem Glykokoll zu einer an *beiden* C-Atomen markierten Essigsäure vor:

$$C^{14} \cdot NH_2 \cdot COOH \rightarrow HC^{14}OOH$$
$$C^{14}H_2NH_2 \cdot COOH + HC^{14} \cdot OOH \rightarrow HOC^{14}H_2 \cdot C^{14}HNH_2 \cdot COOH$$
$$HOC^{14}H_2 \cdot C^{14}HNH_2 \cdot COOH \rightarrow C^{14}H_3 \cdot C^{14}O \cdot COOH$$
$$C^{14}H_3C^{14}O \cdot COOH \rightarrow C^{14}H_3 \cdot C^{14}OOH + CO_2$$
$$C^{14}H_3 \cdot C^{14}OOH \rightarrow \text{Fettsäuren}$$

δ) Zur Frage „Fett aus Fett".

Es ist eine wohlbekannte Tatsache, daß der tierische Organismus Nahrungsfette durch die Lipasen der Verdauungsdrüsen spaltet, die entstehenden Fettsäuren als solche in fein emulgierter Form resorbiert und alsbald wieder zur

Glyceridsynthese verwendet. Damit hängt es zusammen, daß das Depotfett der Tiere in seiner Beschaffenheit recht stark vom Nahrungsfett abhängig ist.

Für zahlreiche Mikroorganismen (Bakterien und Schimmelpilze) steht fest, daß sie mit Fettsäuren oder Neutralfetten als einziger C-Quelle auskommen können. Neutralfette werden in der Regel schon im Nährmedium durch Lipasen gespalten. Der weitere oxydative Angriff auf die Fettsäuren geschieht in unmittelbarem Kontakt mit der Zelle oder in dieser selbst (Näheres hierüber s. S. 258ff. dieses Beitrages).

Die wichtige Frage, ob Fette oder Fettsäuren als Ganzes in die Zelle aufgenommen werden können, und ob sie ohne vorherigen tiefgreifenden Abbau unmittelbar zur Synthese zelleigener Fette verwandt werden können, ist merkwürdig wenig untersucht und noch kaum definitiv beantwortet.

Tabelle 15. *Jodzahl der Fettsäuren von Aspergillus niger bei Kultur auf Fetten verschiedener Jodzahl und anderen C-Quellen.* (Nach P. BÉLIN 1926b.)

C-Quelle	Jodzahl der Fettsäuren der Nahrungsfette	Jodzahl der Fettsäuren des Pilzmycels	Zahl der Versuche
Glucose . . .	—	99,6	7
Pepton . . .	—	96,3	4
Leinöl . . .	183	144,8	5
Lebertran . .	159	92,5	4
Butter . . .	33—46	62,5	6
Margarine . .	27	52,8	5
Bienenwachs	12	90	1

R. H. SCHMIDT (1891) hat sich wohl als erster im Laboratorium von W. PFEFFER mit den angedeuteten Problemen befaßt. Als Versuchsobjekte dienten ihm sowohl höhere wie niedere Pflanzen. Er konnte Fettspaltung und Fettverbrauch durch *Aspergillus niger* eindeutig feststellen. Ein Eindringen der Fette oder Fettsäuren in die Pilzhyphen (z.B. nach „Markierung" des Nahrungsfettes mit Alkanna, Chlorophyll oder Anilinfarbstoffen) war nicht direkt zu beobachten. Auf Grund von Versuchen mit höheren Pflanzen zieht SCHMIDT aber den Schluß: „durch Cellulosehäute lebender Pflanzen dringen Fette mit Leichtigkeit in die Zelle ein". Günstig soll hierbei wirken, wenn dem Neutralfett eine gewisse Menge freier Fettsäuren beigemengt ist. Ebenso wie O. FLIEG (1922) beobachtete bereits SCHMIDT, daß die in die Fetttropfen des Mediums hineinwachsenden Pilzhyphen in ihrem Inneren reichlich Lipoidtropfen enthalten, die FLIEG auf Grund von Färbungsversuchen als Gemisch von Glyceriden und freien Fettsäuren ansprach. Auf die Frage, ob dieses mit dem Fett bzw. den Fettsäuren des Nährmediums identisch ist, wird von beiden Autoren keine schlüssige Auskunft gegeben.

Die einzige mir bekannte Arbeit, die in dieser Frage weiterführt, stammt von B. BÉLIN (1926b). Er kultivierte *Aspergillus niger* auf verschiedenen Fetten (als alleiniger C-Quelle), bei denen die Jodzahl der Fettsäuren ermittelt worden war. Die Lipoide der Pilzernte wurden in üblicher Weise gewonnen, verseift und an den isolierten Fettsäuren gleichfalls die Jodzahl bestimmt. Als Kontrolle dienten Mycelien, die auf Glucosenährböden (5—35% Glucose) und auf Peptonlösungen (1—3% Pepton mit und ohne Zucker) gewachsen waren, deren Fette also vom Pilz auf jeden Fall durch „Totalsynthese" gebildet worden waren. Tabelle 15 faßt die Ergebnisse zusammen. Sie sind recht klar. Beurteilt vom Sättigungsgrad (der Jodzahl) ist das Pilzfett mit dem Nahrungsfett zwar nicht identisch, von diesem aber doch weitgehend beeinflußt. Es ist stärker ungesättigt als das aus Zucker gebildete Fett, wenn ein Fett mit hoher Jodzahl (Leinöl) als C-Quelle dient, es ist umgekehrt stärker gesättigt, wenn ein Fett mit niedriger Jodzahl (Butter, Margarine) als Nahrung gegeben wird. Die Werte für Lebertran und für Bienenwachs passen nicht ganz in die Reihe. Es kann also gefolgert werden, daß zumindest ein Teil der Fettsäuren der Nahrung unverändert in den Organismus aufgenommen und dem Fettbestand der Zelle einverleibt wird. Dieser Schluß fällt und steht freilich mit der Annahme, daß es den Untersuchern tatsächlich gelungen ist, das Pilzmycel vor der Analyse durch gründliches Abspülen mit Fettlösungsmitteln vollständig vom anhaftenden Nahrungsfett zu reinigen.

Es braucht kaum betont zu werden, daß hier noch sehr wichtige Probleme ungelöst oder halbgelöst vorliegen. Der Isotopenmethode scheint sich hier ein besonders aussichtsreiches Anwendungsfeld zu bieten.

c) N-Ernährung; N:C-Verhältnis.

Es ist wenig überraschend, daß der prozentuale Fettgehalt von Mikroorganismenzellen und die je Kultur (bzw. je Volumeinheit der Nährlösung) gebildete Fettmenge mit der Menge der dargebotenen C-Quelle ansteigen. Schon die Arbeit von C. v. NAEGELI (1879) über die Fettbildung von *Penicillium glaucum* bringt dafür einige Belege. So wurde in 6 Wochen alten Kulturen auf 0,1 bzw. 0,5% Rohrzucker 15,84% Fettsäuren, auf 15% Rohrzucker 23,13% Fettsäuren im Trockenmycel gefunden.

R. E. LYONS (1897) kultivierte 3 Bakterienarten auf Fleischextraktagar mit Zusatz von 1, 5 und 10% Traubenzucker. In allen Fällen war die Menge der auf Trockengewicht bezogenen ätherlöslichen Substanz auf 5% Zucker höher als auf 1%, bei 10% allerdings wieder erniedrigt. Der Gehalt an Rohprotein sank mit steigendem Fettgehalt und umgekehrt. Ähnliches stellten auch E. F. TERROINE und J. E. LOBSTEIN (1923) bei verschiedenen Rassen von *Mycobacterium tuberculosis* fest. Auch G. GORBACH und A. SABLATNÖG (1934) fanden übrigens bei *Serratia marcescens* sehr ausgeprägte Optimumkurven für die Beziehung zwischen Fettgehalt und der Konzentration der C-Quelle im Medium (Glucose, Mannit, Brenztraubensäure, 0,05—0,5 mol). Mehrere schöne Reihenversuche mit variierter Glucosegabe bei *Aspergillus niger* sind in den Arbeiten von P. BÉLIN (1926a, b) und von E. F. TERROINE und R. BONNET (1927) (s. auch P. M. BOHN 1931) enthalten. Die Zunahme des Fettgehaltes (Fettsäure %/Trokkengewicht 4,0→9,4) bei steigender Glucosegabe (5→50%) wird von den französischen Forschern übrigens nicht so sehr als die Folge der erhöhten Zuckermenge bzw. -konzentration gedeutet, als vielmehr als Ausdruck des verschobenen Verhältnisses des Zuckers zu den übrigen Nährstoffen, in einem „milieu deséquilibré“. Über ausführliche Versuchsreihen ähnlicher Art berichten die Arbeiten von L. B. LOCKWOOD, G. E. WARD, O. E. MAY, H. T. HERRICK und H. T. O'NEILL (1934), G. E. WARD, L. B. LOCKWOOD, O. E. MAY und H. T. HERRICK (1935) (*Penicillium javanicum*); P. R. WENCK, W. H. PETERSON und E. B. FRED (1935), E. A. PRILL, P. R. WENCK und W. H. PETERSON (1935) (*Aspergillus fischeri*). Ausschließlich oder vorwiegend die Sterine werden von L. M. PRUESS, W. H. PETERSON, H. STEENBOCK und E. B. FRED (1931) und L. M. PRUESS, H. J. GORCICA, H. C. GREENE und W. H. PETERSON (1932) berücksichtigt. Als Beispiel bringt Tabelle 16 einen Auszug der Resultate von E. A. PRILL, P. R. WENCK und W. H. PETERSON (1935). *Aspergillus fischeri* wurde mit variierter Glucosegabe auf einer im übrigen unveränderten Nährlösung gezogen, die folgende Zusammensetzung hatte: 1,0 g NH_4NO_3, 0,68 g KNO_3, 0,5 g $MgSO_4 + 7\,H_2O$, 0,016 g $FeCl_3 + 6\,H_2O$, 0,005 g $ZnSO_4 + 7\,H_2O$, 100 ml destilliertes Wasser. Nach Verbrauch der Hauptmenge des eingesetzten Zuckers wurde das Pilzmycel der Analyse unterworfen, bei der nicht nur die Menge, sondern auch die Zusammensetzung des Fettes berücksichtigt wurde (Tabelle 16).

Aus der Tabelle 16 lassen sich folgende wichtige Ergebnisse ablesen: Mit steigender Zuckergabe nimmt der prozentuale Fettgehalt des Mycels stark zu,

Tabelle 16. *Menge und Zusammensetzung des Fettes von Aspergillus fischeri bei verschiedener Glucosegabe.* (Nach E. A. PRILL, P. R. WENCK und W. H. PETERSON 1935.)

Glucose-gabe g/100 ml	Kultur-dauer Tage	Glucose verbraucht g/100 ml	Öko-nomischer Koef-fizient	Rohfett % Trocken-gewicht	% des Rohfettes				Jodzahl der Fett-säuren
					Fett-säuren	Unver-seifbares	Sterine	Lipoid P	
1	4	0,84	38	10,4	51,0	16,2	5,4	2,30	113
5	7	4,65	34	10,8	67,5	10,9	4,3	0,76	94
10	8	9,5	33	13,1	75,0	9,0	6,4	0,66	80
20	12	19,2	29	18,0	85,0	6,5	4,5	0,40	78
40	16	39,4	26	28,1	84,5	3,8	3,5	0,23	76
55	22	49,5	26	33,3	85,7	3,8	2,7	0,13	75
70	28	68,8	23	36,0	85,8	3,2	2,8	0,15	73

ebenso der Anteil der Fettsäuren am Gesamtfett und der Sättigungsgrad der Fettsäuren. Der Prozentgehalt der Fette an Unverseifbarem (und Sterinen) und an Phosphatiden nimmt in gleicher Richtung ab. Es bestätigt sich also die Annahme von P. BÉLIN (1926a), daß Fettsäuren und deren Glyceride das variable Element der Fette darstellen. Das Absinken des ökonomischen Koeffizienten ist schon zum Teil durch die Mengenzunahme der calorisch hochwertigen Fette zu erklären. Die Berechnung des „calorischen Koeffizienten" würde das Gesamtbild vermutlich nur geringfügig verändern.

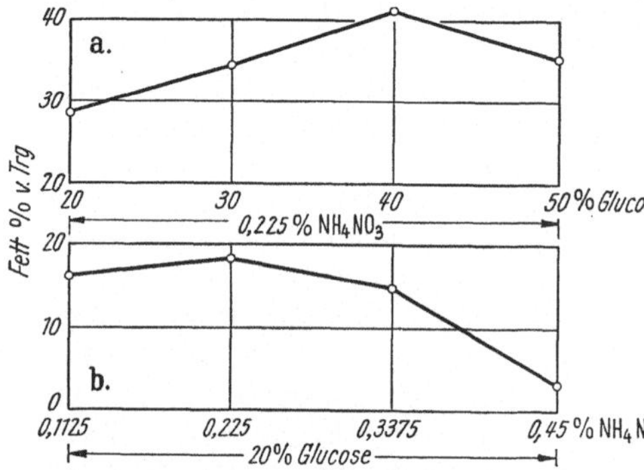

Abb. 3a u. b. Fettgehalt des Myceliums von *Penicillium javanicum* bei a variierter Glucosegabe, b variierter NH_4NO_3-Gabe. (Nach L. B. LOCKWOOD, G. E. WARD, O. E. MAY, H. T. HERRICK und H. T. O'NEILL 1934.)

Einen Schritt weiter führen die (zeitlich etwas früher liegenden) Untersuchungen von L. B. LOCKWOOD, G. E. WARD, O. E. MAY, T. H. HERRICK und H. T. O'NEILL (1934). Sie untersuchten die Fettbildung durch *Penicillium javanicum* in zwei kleinen Versuchsreihen; das eine Mal bei konstantem N- und variiertem Glucosegehalt des Mediums, das andere Mal bei konstantem Glucose- und variiertem N-Gehalt. Die Ergebnisse sind in Abb. 3a und b dargestellt. Die Kurven sind im großen und ganzen gegenläufig: Die Fettbildung wird durch zunehmende Zuckergabe bei gleicher N-Gabe gesteigert. Das ist eine Bestätigung der in Tabelle 16 enthaltenen Resultate. Sie wird aber auch durch steigende N-Gabe (bei gleichbleibender Zuckergabe) vermindert. Zum gleichen Ergebnis kam auch O. TURPEINEN (1936) bei *Geotrichoides sp.* (einer *Mycotoruloidee* aus Birkensaft).

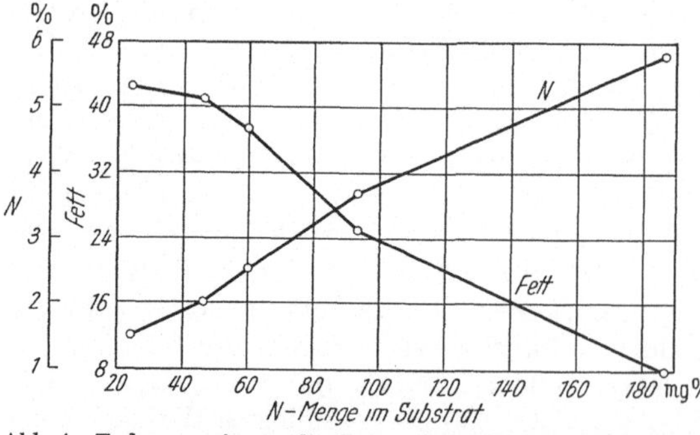

Abb. 4. *Endomycopsis vernalis.* Prozentualer Fett- und N-Gehalt der Zellen bei variierter N-Gabe (Asparagin) im Substrat. (Nach S. HEIDE 1939.)

Die nun bereits vorliegenden Elemente: steigende Menge der C-Quelle — Zunahme des Fettgehaltes (und Abnahme des Proteingehaltes der Zellen); steigende

Menge der N-Quelle — Abnahme des Fettgehaltes der Zellen, konnten durch S. HEIDE (1939) (s. auch M. STEINER 1938) zu einem einheitlichen Bild zusammengefaßt werden, als er bei *Endomycopsis vernalis* den *Fett-* und *Protein*gehalt untersuchte, wenn dem Pilz verschiedene N-Gaben bei gleichbleibender Zuckermenge geboten wurden. Wie Abb. 4 zeigt, ergeben sich gegenläufige Kurven für

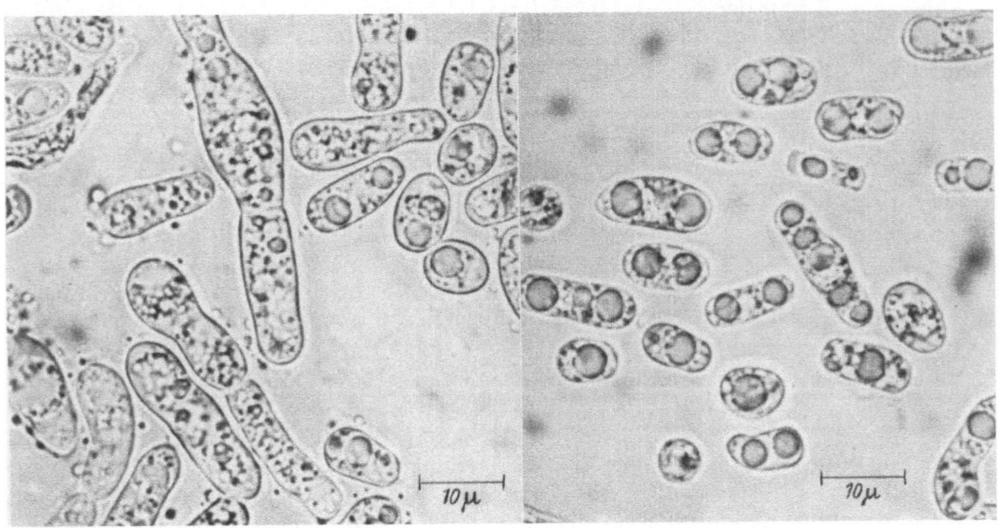

Abb. 5a u. b. *Endomycopsis vernalis.* Zellen aus dreitägigen Kulturen auf WÖLTJE-Medium. a Mit 1% Asparagin (fettarm), b mit $^1/_4$% Asparagin (fettreich). (Original.)

den N- und Fettgehalt der Zellen. Das besagt, daß der Pilz die *gebotene C-Quelle bei genügender N-Zufuhr vorzugsweise zu Eiweiß verarbeitet, bei N-Mangel dagegen vorzugsweise zu Fett* (vgl. auch Abb. 5). Für das Ausmaß der Verfettung der Zellen ist also vor allem das N:C-Verhältnis des Nährmediums maßgebend. Das zeigt die nebenstehende Tabelle 17.

Tabelle 17. *Fettgehalt von Endomycopsis vernalis in Abhängigkeit vom N:C-Verhältnis im Nährmedium.* (Nach S. HEIDE 1939.)

	Saccharose		Asparagin 0,25% (relativ 2) Fett, % Trockengewicht	Asparagin 0,125 % (relativ 1) Fett, % Trockengewicht
	%	relativ		
1	3,75	1	22,0	*41,1*
2	7,5	2	*41,1*	*44,7*
3	15,0	4	*41,7*	44,7
4	30,0	8	43,9	48,5

Schon P. LINDNER (z.B. 1919, 1922) scheint von der reziproken Abhängigkeit der Fett- und Eiweißbildung (speziell bei *Endomycopsis*) eine ziemlich klare Vorstellung gehabt zu haben. Er hatte allerdings keine quantitativen Analysen, sondern nur die mikroskopische Schätzung des Fettgehaltes zur Hand. Außerdem stiftete die Vorstellung Verwirrung, daß der Pilz grundsätzlich zuerst in einer „Eiweißphase" und dann in einer „Fettphase" wächst. Davon ist soviel richtig, daß der Pilz den gebotenen N bevorzugt aufnimmt, so daß sich das N:C-Verhältnis des Mediums im zeitlichen Verlauf der Kultur immer stärker zugunsten der Fettbildung verschiebt.

Für die *Sterinproduktion* durch *Aspergillus fischeri* hatten schon 1935 P. R. WENCK und E. B. FRED die entscheidende Bedeutung des N:C-Verhältnisses im Sinne der von HEIDE formulierten Regel festgestellt[1].

[1] K. BERNHAUER und G. PATZELT (1935) hingegen fanden den Steringehalt von *Aspergillus niger* durch variierte Zuckergabe nur wenig beeinflußt. Im übrigen kann auf die ver-

Die Untersuchungen HEIDEs wurden von H. RAAF (1941) fortgesetzt und ausgebaut. Beide Untersucher stellten fest, daß ein hoher Prozentgehalt der Zellen an Fett nicht gleichbedeutend ist mit einer maximalen Gesamtausbeute an Fett. Sehr niedrige N-Gaben drücken nämlich die Trockensubstanzproduktion stärker herab als sie den prozentualen Fettgehalt erhöhen. Die Fettausbeute in Abhängigkeit von der N-Gabe folgt dann einer Optimumkurve (Abb. 6). Sowohl HEIDE wie RAAF versuchten deshalb, den N-Gehalt der Kulturen auf einem *konstanten* niedrigen Niveau zu halten. HEIDE hatte mit Fließkulturen nicht den gewünschten Erfolg, wohl aber RAAF (s. auch M. STEINER 1941), indem er Permutit als Kationenaustauscher der Nährlösung zusetzte. Wird ein Ammonsalz als N-Quelle gegeben, so wird das NH_4^+ an den Permutit gebunden und nur in kleinen, aber konstanten Mengen für die Aufnahme durch den Pilz freigegeben. Wie Tabelle 18 zeigt, kann auf diese Weise eine Pilzernte erhalten werden, die an Gesamtmenge der Kontrolle ohne Permutit kaum nachsteht, aber viel fettreicher (und N-ärmer) ist als diese. Der gleiche Permutiteffekt wurde übrigens auch mit Asparagin als N-Quelle erhalten. Das konnte erst kürzlich von M. STEINER und H. KATING (1953) und H. KATING (1955b) damit erklärt werden, daß das Asparagin bereits außerhalb der Zelle desamidiert wird und so freies NH_4^+ für die Bindung an den Ionenaustauscher zur Verfügung steht.

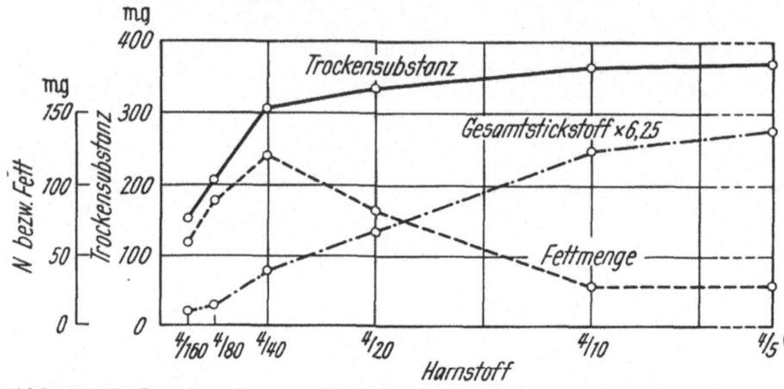

Abb. 6. *Endomycopsis vernalis*. Trockensubstanz-, Fett- und Eiweißbildung bei verschiedener Stickstoff-(Harnstoff-)gabe im Nährmedium. (Nach H. RAAF 1941.)

Tabelle 18. *Permutitkultur von Endomycopsis vernalis. N-Gabe: 184 mg Ammontartrat.* (Nach M. STEINER 1941.)

Unterlage	Trockengewicht mg	Rohfett		Gesamt-N %
		mg	%	
Quarzsand	1113	134	12,0	5,36
Permutit .	1080	290	26,9	3,75

Die von HEIDE gefundene Regel, die einen wichtigen Einblick in die Rolle der Fettbildung im Verwendungsstoffwechsel der Mikroorganismen gibt, ist seitdem bei einer Reihe von anderen fettbildenden Pilzen bestätigt worden[1].

Rhodotorula glutinis: R. NILLSON, L. ENEBO, H. LUNDIN und K. MYRBÄCK (1943).

Rhodotorula gracilis: L. ENEBO, M. ELANDER, F. BERG, H. LUNDIN, R. NILLSON und. K. MYRBÄCK (1944), L. ENEBO, L. G. ANDERSON, H. LUNDIN (1946), S. C. PAN, A. A. ANDREASEN und P. KOLACHOV (1949).

Hefe aus Erde: L. STARKEY (1946).

Torulopsis lipofera: A. KLEINZELLER (1944).

Candida Reukaufii: A. RIPPEL-BALDES, K. PIETSCHMANN-MEYER und W. KÖHLER (1950).

Saccharomyces cerevisiae (Preßhefe): F. REINARTZ und H. LAFOS (1950).

Oospora lactis, Zygorrhynchus Moelleri, Pichia farinosa, Willia anomala, Torulopsis utilis, Aspergillus niger, Penicillium glaucum: H. RAAF (1941).

Unbestimmter Schimmelpilz: W. DIEMAIR und C. BORESCH (1950).

Mucor albo-ater, M. plumbeus, M. Ramannianus, M. circinelloides, Fusarium bulbigenum. Zygorrhynchus Moelleri, Cladosporium herbarum, Trichoderma Koningi, T. lignorum, Penicillium P5 und P13, *Endomycopsis vernalis:* K. BERNHAUER, A. NIETHAMMER und J. RAUCH (1948) (s. auch Tabelle 2, S. 213).

schiedenen Untersuchungen, die sich lediglich mit den Änderungen des *Sterin*gehaltes beschäftigen, im Rahmen dieses Beitrages nicht näher eingegangen werden (s. Bd. X dieses Handbuches).

[1] Wegen der Beziehungen zwischen N-Ernährung und Fettbildung bei höheren Pflanzen sei auf den Beitrag von A. ZELLER (S. 304 dieses Bandes) verwiesen.

Mucor circinelloides: K. BERNHAUER und J. RAUCH (1950a, b) (dazu auch F. THOMAS 1950).

Aspergillus nidulans, Penicillium spinulosum, P. javanicum: M. E. GREGORY und M. WOODBINE (1953).

Es würde viel zu weit führen, auf die mannigfachen Variationen der Versuchsbedingungen einzugehen, die ein grundsätzlich gleiches Ergebnis brachten. Es sei nur erwähnt, daß die verschiedensten anorganischen und organischen N-Quellen verwendet wurden. H. RAAF fand dabei, daß verschiedene Aminosäuren in N-äquivalenter Menge die Fettbildung von *Endomycopsis* in etwas verschiedenem Ausmaße beeinflussen. Er zog zur Erklärung die verschiedene Länge der C-Kette und die daraus resultierende Verschiebung des N:C-Verhältnisses heran. H. KATING (1955a) findet deutliche Hinweise, daß die schwere oder leichte Assimilierbarkeit eine Rolle spielt. Auf Leucinlösung verfettet der Pilz stärker als z.B. auf γ-Aminobuttersäure gleichen N-Gehaltes. Da Leucin langsam und schwer verarbeitet wird, stellt sich bei dieser N-Quelle also ein gewisser „physiologischer N-Mangel“ ein.

Trotz der abweichenden Deutung, die die Verfasser selbst ihren Versuchsresultaten geben, scheint uns aus Versuchen von O. N. MASSENGALE, C. E. BILLS und P. S. PRICKETT (1931) mit Bierhefe (Typ Frohberg) folgendes hervorzugehen: In einer Lösung mit 0,35% Harnstoff bleiben Trockengewicht, Fett-, Sterin- und N-Gehalt gegenüber der Kontrolle unverändert, wenn eine nicht assimilierbare C-Quelle (Mannit, Xylose, Lactose) gegeben wird. Bei Vorhandensein einer brauchbaren C-Quelle (Glucose oder Saccharose) wächst die Hefe während der 24stündigen Versuchsdauer eiweißreich weiter: Zunahme der Trockensubstanz und des N (%), Abnahme der Fette und der Sterine (%).

Interessante Ergebnisse, welche die im Anschluß an HEIDE formulierte Regel modifizieren, hat vor kurzem H. H. MARTIN (1954) mitgeteilt. Er untersuchte bei gärkräftigen und gärschwachen Stämmen von *Candida Reukaufii* (s. auch S. 214) den Einfluß der N-Ernährung auf Fett und Eiweißbildung. Für die Resultate sind die in Tabelle 19 zusammengefaßten Zahlen repräsentativ.

Tabelle 19. *Fett- und Eiweißbildung und Stoffwechselkoeffizienten bei 2 Stämmen von Candida Reukaufii in Abhängigkeit von der N-Ernährung.* (Nach H. H. MARTIN 1954.)

Harnstoffgabe	0,5 g/100 ml	0,35 g/100 ml	0,075 g/100 ml	0,0375 g/100 ml
Schwach gärender Stamm 26b				
Ökonomischer Koeffizient	33,6	36,4	40,5	—
Fett, Prozent vom Trockengewicht	13,3	20,4	25,3	—
Fettkoeffizient	4,5	7,3	10,1	—
Eiweiß, Prozent vom Trockengewicht	25,8	26,3	8,6	—
Eiweißkoeffizient	8,7	9,6	3,5	—
Stark gärender Stamm 5a				
Ökonomischer Koeffizient	—	26,0	27,6	23,6
Fett, Prozent vom Trockengewicht	—	4,5	5,7	8,8
Fettkoeffizient	—	1,2	2,1	2,1
Eiweiß, Prozent vom Trockengewicht	—	49,0	12,5	9,3
Eiweißkoeffizient	—	12,7	3,5	2,2

Bei Stamm 5a liegen die Analysenzahlen und die Koeffizienten durchaus in der erwarteten Reihenfolge. Die Unterschiede im Fettgehalt sind bei diesem wenig zur Fettbildung neigenden Stamm freilich gering. Bei 26b liegen dagegen nur die Fettgehalte in der erwarteten Reihenfolge. Dagegen ist bemerkenswert, daß die Eiweißbildung bei Erhöhung der Harnstoffgabe von 0,35 auf 0,5% *keinen* Anstieg mehr zeigt.

Die wenigen Arbeiten über die Beziehungen zwischen N-Ernährung und Fettgehalt der *Bakterien* lassen sich kaum einheitlich zusammenfassen, was vielleicht mit der sehr verschiedenartigen Beschaffenheit der „Bakterienfette" zusammenhängt. G. GORBACH und A. SABLATNÖG (1934a, b) fanden z. B. eine stetige *Zunahme* des Ätherextraktes von *Serratia marcescens* von 2,83 auf 4,87%, wenn der Peptongehalt des Mediums stufenweise von 0,1 auf 1,0% erhöht wurde. Andererseits ist bemerkenswert, daß in den Analysenzahlen, welche E. CRAMER (1892) für 4 Bakterien angibt, bereits die „HEIDE-*Regel*" in nuce enthalten ist. Erhöhung der Peptongabe im Medium von 1 auf 5% führte zu einer Vermehrung der N-haltigen Substanz und einer Verminderung des Ätherextraktes, Zugabe von 5% Glucose zu 1%iger Peptonlösung hatte den entgegengesetzten Effekt.

d) P-Ernährung.

Daß auch eine Verringerung der *Phosphatgabe* die Verfettung von Pilzzellen fördert, wurde wohl zum ersten Male von A. L. BICHKOVSKAJA (1939) bei *Endomycopsis vernalis* gezeigt. (Melasse + 0,1% $NH_4H_2PO_4$: 37,0% Fett, 0,2% Phosphat 30,0% Fett, 0,3% Phosphat 17,7% Fett). Dieses Ergebnis wurde für *Rhodotorula glutinis* von R. NILLSON, L. ENEBO, H. LUNDIN und K. MYRBÄCK (1943), für *Rh. gracilis* von L. ENEBO, L. G. ANDERSON und H. LUNDIN (1946) und S. C. PAN, A. A. ANDREASEN und P. KOLACHOV (1949) bestätigt. Weniger klar waren die Resultate von L. B. LOCKWOOD, G. E. WARD, O. F. MAY, H. T. HERRICK und H. T. O'NEILL (1934) bei *Penicillium javanicum*. Wurde die Phosphatgabe (KH_2PO_4) zwischen 0,0375 und 38,4 g/l variiert (11 Stufen), so zeigten sich im Fettgehalt unregelmäßige Schwankungen (Maxima bei 38,4‰ Phosphat 7,2% Fett, bei 1,2‰ 7,1% und bei 0,15‰ 12,1%). Auch J. M. GARRIDO, M. WOODBINE und TH. K. WALKER (1953) fanden bei *Aspergillus nidulans*, *Penicillium javanicum* und *P. spinulosum* keine klare Abhängigkeit zwischen der Fettbildung und der zwischen 0,145 und 145,0 mg/100 ml abgestuft variierten Phosphatgabe (s. auch J. M. GARRIDO und TH. K. WALKER 1953).

Tabelle 20. *Fett- und Eiweißgehalt von Hefe nach Züchtung bei verschiedener P-Versorgung.* (Nach K. L. SCHULZE 1950.)

Relative Phosphatgabe	Trockengewicht g/l	Rohfett % vom Trockengewicht	Rohprotein, % vom Trockengewicht
100	12,6	4,2	52,52
56	12,3	6,0	45,7
2,3	10,25	24,3	29,25
1,0	13,0	22,8	31,55

O. TURPEINEN (1936) berichtet über den Einfluß einer Zugabe abgestufter Mengen von Phosphatpuffer (p_H 6,04) zur kompletten (also phosphathaltigen — 0,1% KH_2PO_4') Nährlösung auf Wachstum und Fettbildung von *Geotrichoides sp.* Bei 8tägiger Kulturdauer stieg die Trockengewichtsernte bei Zugabe von m/4 Phosphatpuffer gegenüber der Kontrolle um 27% an. Der prozentuale Fettgehalt war bei der Kontrolle, bei m/32 und bei m/16 Phosphatpuffer etwa gleich; Zugabe von m/8 und m/4 Phosphatpuffer führte zu einer Erhöhung des prozentualen Fettgehaltes um etwa $^1/_7$.

Eingehend hat sich K. L. SCHULZE (1950) mit den Zusammenhängen zwischen P-Ernährung und Fettbildung bei Hefen beschäftigt[1]. Er ging von Verfahrensfragen aus und erkannte, daß im halbtechnischen Ansatz eine starke Verfettung durch P-Mangel in der Maische noch besser als durch N-Mangel erreichbar ist (bis 30,5% Rohfett bei Stamm FS 4). Für die von SCHULZE gefundene Beziehung zwischen der P-Gabe und Fett- und Proteingehalt der geernteten Hefe gibt Tabelle 20 ein Beispiel.

[1] M. MAAS-FÖRSTER (1955) ist in einem, freilich verständlichen Irrtum, wenn sie die Phosphatversuche SCHULZES auf *Trichosporum cutaneum* bezieht, mit dem sich aber nur der erste Teil der zitierten Arbeit befaßt. Der Organismus, bei dem SCHULZE seine diesbezüglichen Versuche durchführte, ist in seiner Arbeit als „Fetthefe FS 4" bezeichnet.

Der Effekt von Phosphatmangel ist also, aufs erste gesehen, der gleiche wie bei N-Mangel. SCHULZE versucht bereits eine theoretische Erklärung für diese Befunde zu geben: Phosphatmangel hemmt erstens die Kohlenhydratassimilation unter Verlangsamung des ganzen Stoffwechsels und er führt zweitens zu einer spezifischen Hemmung der Eiweißsynthese. Diese „soll sekundär die Steigerung der Fettbildung zur Folge haben, da die von der Kohlenhydratassimilation herkommenden Bausteine mehr in den Weg der Fettsynthese gelenkt werden“.

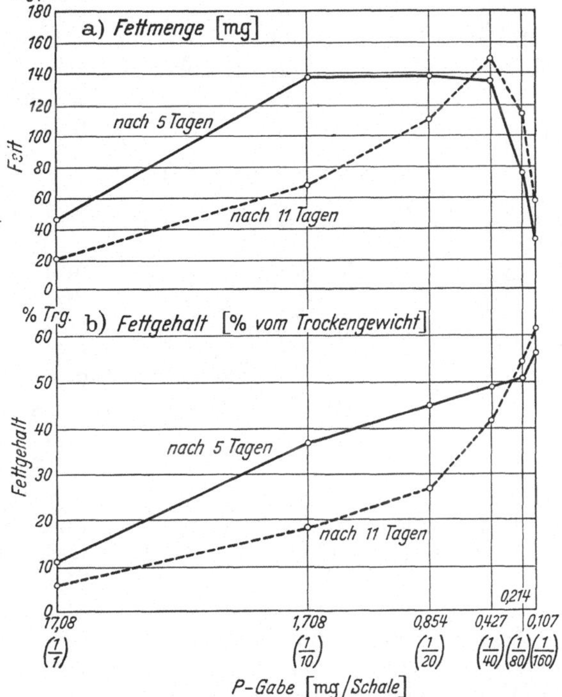

Abb. 7a u. b. *Endomycopsis vernalis*. Fettbildung bei verschiedener P-Gabe. a Fettmenge (mg), b Fettgehalt (%). (Nach M. MAAS-FÖRSTER 1955.)

Das Problem wurde neuerdings von M. MAAS-FÖRSTER (1955) wieder aufgegriffen. Sie arbeitete mit *Endomycopsis vernalis*, methodisch und in der Fragestellung im unmittelbaren Anschluß an die Arbeiten von M. STEINER (1938), S. HEIDE (1939) und H. RAAF (1941). Auch MAAS-FÖRSTER fand eine klare Beziehung zwischen Phosphatgabe und Fettbildung (Abb. 7a und b). Der prozentuale Fettgehalt der Zellen steigt von etwa 11% auf etwa 55% vom Trockengewicht an (Werte vom 5. Tag der Kultur), wenn die Phosphatdosis (KH_2PO_4) von 17,08 mg auf 0,107 mg P je Kulturschale (relativ von $^1/_1$ P auf $^1/_{160}$ P) vermindert wird (Abb. 7b). Wie bei den Versuchen mit abgestufter N-Gabe (s. oben S. 228) ergibt sich auch hier ein Optimum der gesamten Fettproduktion bei mittlerer P-Gabe (Abb. 7a). Die von SCHULZE vorgeschlagene Erklärung konnte von der Verfasserin durchaus bestätigt werden: Phosphatmangel bewirkt primär eine Hemmung der N-Assimilation. Unter den verschiedenen Ergebnissen, die zum gleichen Schlusse führten, seien nur drei erwähnt:

Tabelle 21. *Zucker- und N-Verbrauch durch Endomycopsis vernalis aus Nährlösungen mit verschiedenem P-Gehalt.* ($^1/_1$ P = 0,5% KH_2PO_4 = 17,0 mg P/Schale). In allen Kulturen 7,5% Saccharose und 1% Asparagin. (Nach M. MAAS-FÖRSTER 1955.)

Kulturdauer	Zuckerverbrauch (% der Ausgangsmenge)			N-Verbrauch (% der Ausgangsmenge)		
Tage	$^1/_1$ P	$^1/_5$ P	$^1/_{20}$ P	$^1/_1$ P	$^1/_5$ P	$^1/_{20}$ P
2,5	25	26	—	42,5	42,6	—
3,0	53	49	—	80,0	63,8	32,7
3,5	78	77	—	100,0	75,6	—
4,0	88	97	90	—	84,3	41,5
4,5	100	100	—	—	87,0	—
5 0	—	—	100	—	—	42,6
5,5	—	—	—	—	—	—
6,0	—	—	—	—	—	43,6
7,0	—	—	—	—	—	45,0

1. Verfolgt man bei verschiedener Phosphatgabe den Verbrauch des Zuckers und des Stickstoffs (Asparagin) aus der Nährlösung (Tabelle 21), so ergibt sich,

daß $^1/_5$ oder sogar $^1/_{20}$ P die *Zuckerassimilation* gegenüber $^1/_1$ P nur wenig oder gar *nicht* hemmt, während schon bei $^1/_5$ P die *N-Aufnahme* aus der Nährlösung *viel langsamer* erfolgt als bei voller P-Gabe ($^1/_1$ P). Bei $^1/_{20}$ P sind selbst am 7. Versuchstag erst 45% des zur Verfügung stehenden N verbraucht.

2. Die auf die Fettbildung steigernd wirkende P-Mangeldosis ist von der N-Gabe abhängig. Während bei voller N-Gabe ($^1/_1$ N), wie oben (Abb. 7b) gezeigt wurde, bereits $^1/_{10}$ P eine deutliche Erhöhung des Fettgehaltes bewirkt, ist bei geringer N-Gabe ($^1/_4$ N) erst bei Verminderung der P-Gabe auf $^1/_{40}$ ein ähnlicher Effekt zu verzeichnen.

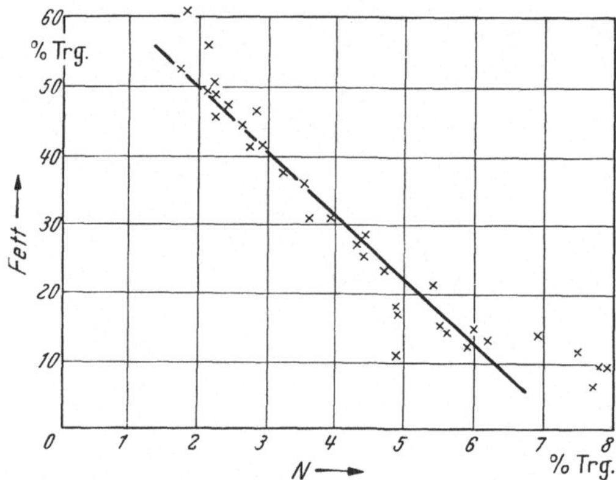

Abb. 8. *Endomycopsis vernalis.* Die Beziehung zwischen Fett- und N-Gehalt der Zellen von Kulturen mit gleicher N-, aber verschiedener P-Gabe (1% Asparagin, 0,5—0,0031% KH_2PO_4). (Nach M. MAAS-FÖRSTER 1955.)

3. In Abb. 8 schließlich ist der Fett- und N-Gehalt der Zellen aufgetragen, die sämtlich bei der *gleichen N-Gabe* (1% Asparagin), aber bei sehr *verschiedener P-Gabe* herangezogen wurden. Zwischen Fett- und N-Gehalt der Zellen ergibt sich ein fast lineares Verhältnis umgekehrter Proportionalität.

Es kann also kaum gezweifelt werden, daß P-Mangel auf dem Wege über eine Hemmung der N-Assimilation wirksam wird, daß P-Mangelkulturen also in gewisser Hinsicht unter einem physiologischen N-Mangel stehen.

Mit dieser Feststellung stehen nur scheinbar die Ergebnisse bei Bierhefe im Widerspruch, von denen I. SMEDLEY-MACLEAN und D. HOFFERT (1923, 1924) berichten. Bei Versuchen über die Fettsynthese aus verschiedenen Substraten, die bereits früher erwähnt wurden (S. 222f.), konnte auch ein sehr deutlich fördernder Effekt einer Phosphatzugabe festgestellt werden. Als Beispiel möge Tabelle 22 dienen. Die zweifellos zutreffende Erklärung der Verfasserinnen greift auf die Notwendigkeit von PO_4 für die einleitenden Schritte des Zuckerabbaus zurück. Hexosediphosphat war übrigens als Zusatz unwirksam weil es offenbar nicht in die Zelle aufgenommen wird. Durchaus gleichartige Ergebnisse erhielt A. KLEINZELLER (1944) mit *Torulopsis lipofera.* Bei Zugabe von 0,0; 0,01; 0,05; 0,1 mol KH_2PO_4 betrug die Fettbildung aus 10%iger Glucoselösung 2,10; 3,87; 3,79; 3,99 mg, der Konversionskoeffizient 9,8; 19,5; 19,8; 19,5 Durch 0,5 mmol Jodacetat wurde die Fettbildung aus Glucose bei Phosphatgegenwart vollständig sistiert.

Tabelle 22. *Einfluß der Zugabe von 4% Phosphatpuffer (p_H 7,8) auf die Fettsynthese von ruhender Hefe aus Saccharose mit und ohne Durchlüftung.* (Nach I. SMEDLEY-MACLEAN und D. HOFFERT 1923.)

Substrat	Ohne Phosphat		Mit Phosphat	
	mit O_2	ohne O_2	mit O_2	ohne O_2
	Fettzunahme		Fettzunahme	
	%	%	%	%
Saccharose, 1%	+176	+ 62	+392	+ 89
Saccharose, 5%	+212	+129	+340	+147

Die scheinbare Diskrepanz zwischen den zuerst und zuletzt beschriebenen Ergebnissen klärt sich sofort auf, wenn man berücksichtigt, daß sowohl SMEDLEY-MACLEAN und HOFFERT als auch KLEINZELLER mit *ruhenden* Zellen arbeiteten, deren Substrat keinen assimilierbaren Stickstoff enthielt. Eine Phosphatwirkung auf die N-Assimilation *konnte* also gar nicht erwartet werden. Die (selbstverständliche) Tatsache, daß Phosphat für den Zuckerabbau und somit für die Umwandlung von Kohlenhydrat in Fett notwendig ist, steht in keinem Widerspruch dazu, daß im Verwendungsstoffwechsel *wachsender Pilze* ein Phosphatmangel *zuerst* und *am stärksten* in einer Hemmung der zur Proteinsynthese führenden Vorgänge sichtbar wird.

e) S-Ernährung.

Hierfür liegen bisher hauptsächlich Angaben für *Endomycopsis vernalis* (S. HEIDE 1939) vor. Es wurde die Fettbildung bei abgestuften Na_2SO_4-Gaben zwischen 0,003 und 13,1% untersucht. Der prozentuale Fettgehalt war zwischen 0,6 und 0,03% Na_2SO_4 etwa unverändert ($\sim$22%), von da an zu 0,003% Na_2SO_4 stieg er auf etwa 35% an, wobei sich gleichzeitig die Ernte an Trockensubstanz stark verminderte. SO_4 im Minimum hat also etwa die gleiche Wirkung wie eine stark verringerte N-Gabe. Ein zweiter Anstieg der Kurve des Fettgehaltes zwischen 0,6 und 13% Na_2SO_2 dürfte auf osmotische Salzwirkungen zurückgehen. Er kann hier unberücksichtigt bleiben.

J. M. GARRIDO, M. WOODBINE und TH. K. WALKER (1953) haben bei *Aspergillus nidulans*, *Penicillium javanicum* und *P. spinulosum* die $MgSO_4 + 7\ H_2O$-Gabe zwischen 2,5 und 500 mg/100 ml variiert. Zunehmender $MgSO_4$-Gehalt der Medien führt im allgemeinen zu steigenden absoluten und prozentualen Fettgehalten. Es ist wahrscheinlich, daß sich hier die Wirkungen von Mg und von SO_4 überlagerten (s. auch J. M. GARRIDO und TH. K. WALKER 1953a, b).

f) K-Ernährung.

Der Einfluß der K-Ernährung auf die Fettbildung wurde zuerst von C. PONTILLON (1932) bei *Aspergillus niger* geprüft. Die Ergebnisse werden nur kurz referiert: Bei K-Mangel ist die Entwicklung sehr langsam, die Fettbildung ist proportional fast unverändert, die Jodzahl der Fette ist erhöht, der Gehalt an Lipoid-P und Sterinen immer sehr niedrig. Keinen klar deutbaren Einfluß hatte

Tabelle 23. *Wachstum und Zusammensetzung der Zellsubstanz bei abgestufter K-Gabe.* $^1/_1$ K: 0,5% KH_2PO_4. Bei Verminderung des K-Gehaltes auf $^1/_{10}$ usw. wurde das Phosphat durch NaH_2PO_4 zur Norm ergänzt. 4 Tage. (Nach M. MAAS-FÖRSTER 1955.)

K-Gabe	$^1/_1$ K	$^1/_{10}$ K	$^1/_{20}$ K	$^1/_{40}$ K	$^1/_{80}$ K
Zuckerverbrauch, Prozent	100	55,0	51,0	45,2	39,5
Trockensubstanz, mg	485,1	211,2	168,3	116,6	88,5
ökonomischer Koeffizient	47,8	34,2	29,4	20,3	19,9
Fett, Prozent vom Trockengewicht	18,6	9,2	8,8	9,5	9,3
Fett, mg/Kultur	90,5	19,4	14,8	11,1	8,2
N, Prozent vom Trockengewicht	5,2	6,7	6,2	6,4	6,3

eine zwischen 0,97 und 195,4 mg/100 abgestufte K-Gabe auf die Fettproduktion von *Aspergillus nidulans*, *Penicillium javanicum* und *P. spinulosum* in Versuchen von J. M. GARRIDO, M. WOODBINE und TH. K. WALKER (1953) (s. auch J. M. GARRIDO und TH. K. WALKER 1953a, b).

Sehr ausführlich hat sich zum gleichen Gegenstand M. MAAS-FÖRSTER (1955) auf Grund von Versuchen mit *Endomycopsis vernalis* geäußert: Die wichtigsten Ergebnisse einer Versuchsreihe sind in Tabelle 23 dargestellt. Sie lassen sich folgendermaßen zusammenfassen: Eine Verminderung der K-Gabe führt zu einer starken Verlangsamung des Zuckerabbaus und zu einer Verringerung der Trockensubstanzproduktion. Die letztere nimmt sogar stärker ab als der Zuckerverbrauch; dies findet in einer stetigen Abnahme des ökonomischen Koeffizienten seinen Ausdruck. Die Zusammensetzung des Pilzes — Prozentgehalt an Fett und N — ist aber zwischen $^1/_{10}$ und $^1/_{80}$ K praktisch völlig unverändert. Auch die scheinbare Ausnahme bei voller Kaliumgabe ($^1/_1$ K: erhöhter Fett- und verminderter N-Gehalt) läßt sich leicht erklären. Hier ist wegen des kräftigen Wachstums bereits nach 3,5 Tagen der gesamte Stickstoff der Nährlösung verbraucht, so daß sich am 4. Tage bereits die bekannten Symptome von N-Mangel im Verwendungsstoffwechsel bemerkbar machen.

Auf die zahlreichen Arbeiten, die sich mit der Bedeutung des Kaliums für den Kohlenhydratstoffwechsel beschäftigen, kann hier nicht eingegangen werden. Es seien nur als Beispiel die Untersuchungen von A. RIPPEL und G. BEHR (1934, 1936) angeführt, die bei *Aspergillus niger* verschlechterte Ökonomie des Stoffwechsels in Folge von Kaliummangel feststellten. Das trifft auch bei *Endomycopsis* zu. Der prozentuale Fett- und Eiweißgehalt ist, anders als bei N-, P- und S-Mangel, bei K-Mangel *unverändert.*

g) Mg-Ernährung.

L. D. MACLEOD und I. SMEDLEY-MACLEAN (1938) geben an, daß die Zugabe von Mg^{++} und Ca^{++} die Fettbildung ruhender Hefe aus Glucose und aus Acetat vermindert. A. KLEINZELLER (1944) bemerkte keinen Einfluß von Mg^{++}-Zusatz auf die Fettsynthese ruhender *Torulopsis lipofera*-Zellen. L. B. LOCKWOOD, G. E. WARD, O. E. MAY, H. T. HERRICK und H. T. O'NEILL (1934) untersuchten den Fettgehalt von *Penicillium javanicum* auf Nährlösungen, deren Gehalt an $MgSO_4$ in 9 Stufen zwischen 0,125 und 16‰ variiert war. Die prozentualen Fettgehaltsschwankungen sind ziemlich regellos. Hohe $MgSO_4$-Gaben (8—16‰) führten zu einer starken Fettverminderung.

Mit *Endomycopsis vernalis* haben neuerdings M. STEINER und U. JENDRALSKI (1956) experimentiert. Ein Beispiel für eine Anzahl gleichartig verlaufener Versuchsreihen gibt Tabelle 24.

Tabelle 24. *Trockensubstanz und Fettbildung von Endomycopsis vernalis bei variierter Mg-Gabe.* 0,25% Asparagin. $^1/_1$ Mg = 0,25% $MgSO_4 + 7\,H_2O$. Der Sulfatgehalt ist in allen Kulturen durch äquivalente Na_2SO_4-Mengen zur Norm ergänzt. (Nach M. STEINER und U. JENDRALSKI 1956.)

Nährmedium mit	$^1/_1$ Mg	$^1/_{10}$ Mg	$^1/_{20}$ Mg	$^1/_{40}$ Mg	$^1/_{80}$ Mg
Trockensubstanz, mg	265,1	246,9	236,9	166,5	143,9
Fett, mg	105,2	88,7	86,6	50,5	41,1
Fett, Prozent vom Trockengewicht	39,5	36,0	37,7	30,4	29,4

Es zeigt sich, daß sich ausgesprochener Mg-Mangel erst bei recht niedriger Gabe ($^1/_{40}$ Mg) bemerkbar macht (vgl. Trockensubstanz!). Die prozentualen Fettgehalte zeigen aber bei abnehmender Mg-Gabe eine deutlich fallende Tendenz; die Unterschiede zwischen $^1/_1$ Mg und $^1/_{80}$ Mg betragen immerhin rund 30%. Abnehmende Mg-Gaben scheinen sich also in einer *Verringerung* des Fettgehaltes der Zellen auszuwirken. Der Einfluß des *Mg-Mangels* auf die Fettbildung wäre demnach genau *umgekehrt* wie bei *N-, P- und S-Mangel.* Man wird hier vielleicht daran denken dürfen, daß Mg^{++} als notwendiger Kofaktor im Fettsäurecyclus erwiesen ist.

h) Spurenelemente.

Die wenigen Angaben, die sich auf den Einfluß von Spurenelementen auf die Fettbildung durch Pilze beziehen, vermitteln kein einheitliches Bild. L. B. LOCKWOOD und Mitarbeiter (1934) haben bei *Penicillium javanicum* insgesamt 52 Elemente in orientierenden Versuchen geprüft: Mo, Wo, Cr, Fe und Ca zeigten einen leicht fördernden Einfluß auf Zuckerverbrauch, Trockensubstanz- und Fettproduktion. Die meisten übrigen Elemente waren ausgesprochen giftig. Eine Zn-Gabe von 0,1% hat nach N. PORGES (1932) bei *Aspergillus niger* eine Erhöhung des Fettgehaltes von 2,50 auf 4,80% zur Folge. Nach G. SCHULZ (1938) kann sich dieser Effekt bei anderer Grundnährlösung ins Gegenteil verkehren. J. S. MCHARGUE und R. K. CALFEE (1931) finden bei *Aspergillus flavus* eine Steigerung des Fettgehaltes bei Zusatz von Cu, Mn und Zn zum Medium.

i) Organische Wirkstoffe.

Über das Wirkstoffbedürfnis von *Candida Reukaufii* liegen Angaben von A. RIPPEL-BALDES, K. PIETSCHMANN-MEYER und W. KÖHLER (1950) und von H. H. MARTIN (1954) vor. Nach den ersteren sind Wuchstsoffe für gutes Wachstum und für gute Verfettung nötig. Durch Zugabe von Molke als Wirkstoffquelle wurde die Trockensubstanzproduktion sehr wesentlich verbessert, aber auch der prozentuale Fettgehalt der Zelle erhöht (24,71% Fett bei 10% Molkenzusatz gegenüber 17,20% bei der Kontrolle). Nach MARTIN ermöglicht erst eine Zugabe von Biotin und/oder Aneurin ein nennenswertes Wachstum des Pilzes. Der höchste Verfettungsgrad 13,8% wurde mit 50 μg Aneurin + 1,25 μg Biotin/100 ml erreicht. Mit 1,25 μg Biotin allein wurden 11,1, mit 50 μg Aneurin 8,6% Fett erhalten. Nach M. P. STEINBERG und Z. J. ORDAL (1954) führt Zusatz von Hefeextrakt zum Substrat (Maiszucker, wohl nicht ganz wirkstofffrei!) zu einem kräftigeren Wachstum, aber zu keiner Veränderung der Fettbildungsrate von *Rhodotorula gracilis*. Auf die Fettbildung aus Glucose durch ruhende *Torulopsis lipofera*-Zellen hatte in den Versuchen A. KLEINZELLERs (1944) ein Zusatz von Aneurin oder Cholin *keinen* Einfluß.

Wenn O. TURPEINEN bei Zugabe von 5—40(!)% 6%igen Hefeextraktes zum Nährboden eine deutliche Verminderung der Verfettung von *Geotrichoides* (Kontrolle 19,3%, 40% Hefeextrakt, 13,3% Fett) feststellte, dürfte es sich mehr um eine Wirkung der verstärkten N-Gabe als um einen Wirkstoffeffekt gehandelt haben.

Der Einfluß von synthetischen Wuchsstoffen [α-Naphthylessigsäure (NE), 2,4-Dichlorphenoxyessigsäure (2-4-D) und 2,5-Dichlorphenoxyessigsäure (2-5-D)] auf Wachstum und Fettgehalt von *Torulopsis utilis* und *Rhodotorula gracilis* wurde von I. HAIDASCH (1955) untersucht. Eine Wuchsstoffwirkung auf die Fettbildung war nur bei *Rhodotorula gracilis* festzustellen, was durch die nachfolgende Tabelle 25 belegt wird. Da bei längerer Kulturdauer (mit 3facher N-

Tabelle 25. *Rhodotorula gracilis. Trockensubstanz- und Fettbildung ohne und mit Zusatz synthetischer Wuchsstoffe.* Synthetische Nährlösung (N:C = 1:100), 100 ml, Schüttelkultur 72 Std., 27°. (Nach I. HAIDASCH 1955.)

Wuchsstoff	Trockengewicht mg	Ökonomischer Koeffizient	Fett, % vom Trockengewicht	Fett-koeffizient
—	149,7	31,2	31,1	9,1
NE, 10^{-5} g/l	152,8	31,2	35,1	11,0
2-5-D, 10^{-7} g/l	163,6	31,7	36,6	11,6
2-4-D, 10^{-8} g/l	163,1	32,8	38,7	12,7

Gabe!) der Effekt einer Wuchsstoffgabe zu Kulturbeginn ausblieb, durch nachträgliche wiederholte Zufuhr von Wuchsstoff aber wieder eintritt, nimmt Verfasser zur Erklärung eine anfängliche „Störung des Stoffwechsels“ durch den Wuchsstoff an, die sich allmählich wieder normalisiert.

j) Wasserstoffionenkonzentration; Hydratur.

Über den Einfluß des p_H auf die Fettbildung bei Mikroorganismen bringt das Schrifttum nur relativ wenige Angaben. Als Beispiel seien zunächst einige Zahlen von C. PONTILLON (1932, s. auch 1930a, b) gebracht, der *Aspergillus niger* auf physiologisch neutraler (NH_4NO_3), saurer (NH_4Cl) und alkalischer (KNO_3) Stickstoffquelle kultivierte. Tabelle 26 bringt auszugsweise die Ergebnisse für eine Kulturdauer von 72 und 96 Std. was etwa dem Höhepunkt der Pilzentwicklung entspricht.

Tabelle 26. *Trockengewicht und Zusammensetzung des Mycels von Aspergillus niger bei Kultur in Medien mit physiologisch neutraler, saurer und alkalischer N-Quelle.* Daten für 72 und 96 Std. (Nach C. PONTILLON 1932.)

	Zeit	p_H	Trockengewicht	Gesamtfettsäuren		Lipoid P	Sterine
	Std		g	mg	Jodzahl	mg	mg
A NH_4NO_3 (physiologisch neutral)	72	1,6	3,23	268	33,4	3,5	4,1
	96	2,4	2,15	100	28,5	2,3	12,4
B NH_4Cl (physiologisch sauer)	72	1,2	3,24	166	67,7	2,6	8,4
	96	0,9	3,24	151	85,8	1,8	6,2
C KNO_3 (physiologisch alkalisch)	72	5,0	2,52	267	80,2	2,9	4,5
	96	5,0	2,56	146	88,2	2,7	9,0

In bezug auf Trockensubstanzbildung schneidet die physiologisch alkalische KNO_3-Kultur am schlechtesten ab, hinsichtlich der Fettbildung dagegen die physiologisch saure NH_4Cl-Kultur. Bemerkenswert ist auch die starke Variation der Jodzahl, die bei $(NH_4)NO_3$ wesentlich niedriger liegt als in physiologisch sauren oder alkalischen Medien. E. A. PRILL, P. R. WENCK und W. H. PETERSON (1935) kultivierten *Aspergillus fischeri* bei abgestuftem p_H zwischen 2 und 8. Der ökonomische Koeffizient nahm stetig nach der alkalischen Seite hin von 20 auf 27 zu, viel stärker der Prozentgehalt an Fett und der Anteil der Fettsäuren am Rohfett: p_H 3–17,0 bzw. 66,5%, p_H 6—23,9 bzw. 75,5%, p_H 8—37,0 bzw. 76,5%. Der Steringehalt fiel zwischen p_H 2 und p_H 8 von 10,3 auf 1,3% des Gesamtfettes ab. Die Jodzahlen zeigten keine starke Schwankung. Das End-p_H lag allerdings in allen Kulturen unabhängig vom Ausgangswert bei 7,5. Zugabe von Puffersubstanzen ($CaCO_3$, K-Acetat) zu physiologisch sauren N-Quellen erhöhte in allen Fällen Trockensubstanzmenge und Fettgehalt, erniedrigte aber die Jodzahl der Fettsäuren. Ähnlich, wenn auch im einzelnen etwas abweichend, verhielt sich die Trockensubstanzbildung und der Fettgehalt von *Penicillium javanicum* (L. B. LOCKWOOD, G. E. WARD, O. E. MAY, H. T. HERRICK und H. T. O'NEILL 1934) in Abhängigkeit vom p_H: flache Maxima zwischen p_H 3 und p_H 7. Für *Endomycopsis vernalis* gibt S. HEIDE (1939) ein p_H-Optimum zwischen p_H 3,5 und p_H 6,5 an; unterhalb und oberhalb dieser Grenzwerte wächst der Pilz schlecht; er bildet pathologische Zellformen mit geringem Fettgehalt aus (mikroskopischer Befund). O. TURPEINEN hatte 1936 für *Geotrichoides sp.* als Optimum für das Wachstum p_H 5,7, für die Fettbildung p_H 4,6 gefunden. Neuerdings haben M. P. STEINBERG und Z. J. ORDAL (1954) die Fettbildungsrate von *Rhodotorula gracilis*

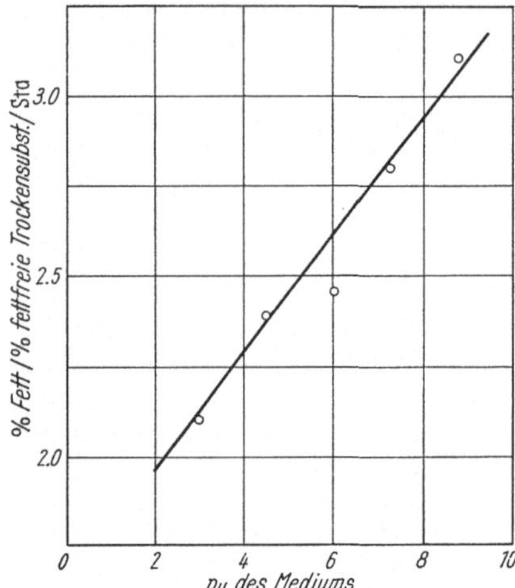

Abb. 9. *Rhodotorula gracilis.* Fettbildungsrate (% Fett/% fettfreie Trockensubstanz/Std) in Abhängigkeit vom p_H des Mediums. (Nach M. P. STEINBERG und Z. J. ORDAL 1954.)

in ihrer p_H-Abhängigkeit geprüft. Wie Abb. 9 zeigt, finden sie einen fast gradlinigen Anstieg zwischen p_H 2 und p_H 8.

Einen stetigen Anstieg der Fettbildung aus belüfteten Acetatlösungen durch Hefe zwischen p_H 5,5 und p_H 8,0 stellten auch A. G. WHITE und C. H. WERKMAN (1948) fest.

Übergeordnete Gesichtspunkte für die Physiologie der Fettbildung scheinen sich aus den obigen Befunden noch nicht ableiten zu lassen.

Daß die Fettbildung von Mikroorganismen auch von der *Hydratur* (dem Wassersättigungszustand) abhängt, ist in den Ergebnissen einiger Untersucher angedeutet. W. HALDEN, F. BILGER und R. KUNZE (1933) und M. SOBOTKA, W. HALDEN und F. BILGER (1935) konnten durch Herabsetzung des Wassergehaltes von 90% auf 75—80% bei gleichzeitiger Alkoholgabe und guter Durchlüftung eine beträchtliche Lipoid- und (insbesondere) Sterinanreicherung in Hefe erzielen. S. HEIDE (1939) konnte eine starke Erhöhung des prozentualen Fettgehaltes (bis 57%) bei gleichzeitiger Verminderung der Trockensubstanz erreichen, wenn er den osmotischen Wert des Mediums durch Zugabe von Neutralsalzen ($NaCl$, $MgSO_4$, $CaCl_2$) erhöhte. Es ist nicht ausgeschlossen, daß bei der fördernden Wirkung sehr hoher Zuckergaben auf die Fettbildung (s. z. B. Tabelle 16) der osmotische Faktor eine Rolle spielt.

k) Zusammenfassung der Ergebnisse über die Abhängigkeit der Fettbildung von der Ernährung.

In dem Schema der Abb. 10 wird der Versuch gemacht, die in den vorangehenden Abschnitten dargestellten Ergebnisse übersichtlich zusammenzufassen und einheitlich zu deuten, wobei die bei *Endomycopsis vernalis* erzielten Befunde besondere Berücksichtigung finden.

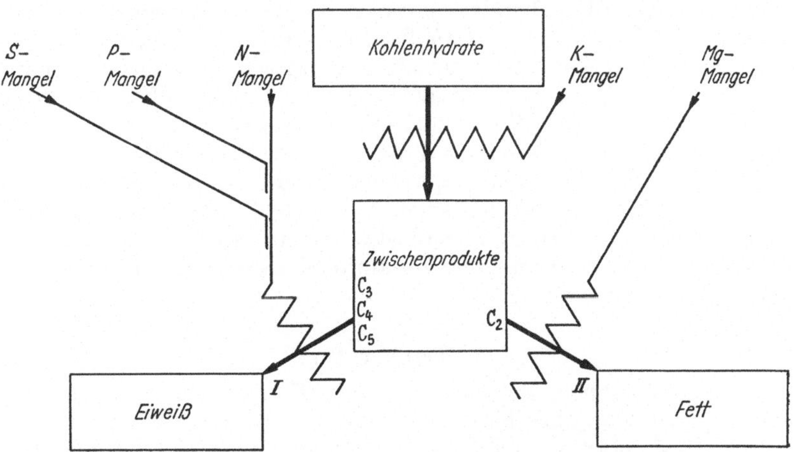

Abb. 10. Schema für die Abhängigkeit der Fettbildung von der Ernährung.

Die festgestellte Abhängigkeit der Fett- und Eiweißbildung vom Kohlenhydratgehalt, vom Stickstoffgehalt und vom C:N-Verhältnis im Substrat legen den Gedanken nahe, daß die Fett- und Eiweißsynthesen von gemeinsamen Zwischenstufen des Kohlenhydratabbaus ausgehen. Bei genügendem Angebot an assimilierbarem N wird der Weg zur Eiweißsynthese (I) bevorzugt beschritten: der Organismus wächst proteinreich, aber fettarm. Bei *N-Mangel* wird die Eiweißsynthese gedrosselt oder schließlich ganz sistiert: das Ausgangsmaterial wird vor allem für die Fettsynthese (II) verfügbar: es entstehen eiweißarme, fett·reiche Zellen.

P-Mangel hemmt die N-Assimilation (K. L. SCHULZE 1950, M. MAAS-FÖRSTEF 1955). Er hat *indirekt* die gleichen Folgen wie N-Mangel. Bei gehemmter Ei·weißsynthese kommt die gemeinsame Vorstufe in erster Linie der Fettsynthese zugute (II).

Über *S-Mangel* liegen nur wenig experimentelle Angaben vor. Man kann vermu·ten, daß auch der Mangel an assimilierbarem Schwefel durch Hemmung der Eiweiß·synthese und damit indirekt in einer Förderung der Fettbildung wirksam wird

K-Mangel hat auf den Protein- und Fettanteil der Zelle keinen Einfluß. Er vermindert nur die Trockensubstanzbildung und die Ökonomie der Kohlen·hydratveratmung. Sein Angriffspunkt ist also offensichtlich an einer Stelle zu suchen, die vor der Verzweigung der Synthesenwege liegt, die zum Eiweiß bzw Fett führen.

Mg-Mangel führt zu einer verminderten Verfettung der Zellen. Er scheint also spezifisch die eigentliche Fettsynthese (II) zu hemmen.

Über die gemeinsame Vorstufe der Protein- und Fettsynthese braucht hier nicht viel gesagt zu werden. Es handelt sich zweifellos um Verbindungen, die durch den Tricarbonsäurecyclus miteinander verbunden sind.

Speziell für *Endomycopsis vernalis* wurde durch H. KATING (1955a) (s. auch M. STEINER und H. KATING 1953) gezeigt, daß die ersten Produkte der N-Assi milation aus anorganischen und organischen N-Quellen die „Grundaminosäuren" Alanin, Asparaginsäure und Glutaminsäure sind. Für ihre Synthese durch Aminierungen und Umaminierungen müssen also die entsprechenden Ketosäuren (Brenztraubensäure, Oxalessigsäure, α-Ketoglutarsäure) als Aminoacceptoren dienen.

Der Syntheseweg (II) zum Fett wäre durch die Fettsäurenspirale nach LYNEN (s. S. 178 dieses Handbuchbandes, sowie F. LYNEN 1955) gekennzeichnet. Das Ausgangsmaterial ist die durch CoA-Bindung aktivierte Essigsäure.

Es sei schon an dieser Stelle darauf hingewiesen, daß das hier entworfene Bild von der Stellung der Fettbildung im Verwendungsstoffwechsel weitgehend dem entspricht, was J. W. FOSTER (1949) in seiner Monographie "Chemical activities of fungi" (insbesondere S. 122 und S. 164—169) als "shunt-meta bolism", also als „Stoffwechsel auf Nebengeleisen" anspricht. Da aber im Zu sammenhang damit der Autor von einem "deranged" und "pathological" "meta bolism" als Folge „abnormer Milieubedingungen" spricht, scheint es richtig an einer späteren Stelle in anderem Zusammenhange etwas eingehender auf die FOSTERsche Konzeption zurückzukommen.

l) Temperatur.

Die wenigen Angaben, die sich mit dem Einfluß der *Temperatur* auf die *Quantität der Fettbildung* beschäftigen, können sehr kurz wiedergegeben werden E. A. PRILL, P. R. WENCK und W. H. PETERSON (1935) fanden bei *Aspergillus fischeri*, daß von 20—37° der ökonomische Koeffizient stetig abnimmt (von 3: auf 27), daß der prozentuale Fettgehalt des Mycels aber bei 30° ein Maximum zeigt (20°: 25,7%, 20°: 24,4%, 37°: 20,0%). Schon früher hatte H. ZIKES (1919 bei verschiedenen Hefen festgestellt, daß die (mikroskopisch erkennbare) Ver fettung der Zellen bei 20° und 30° besser und rascher vonstatten geht als be 10° und 37°. Nach A. KLEINZELLER (1944) verläuft die Umwandlung von Glucos in Fett durch ruhende Hefe am kräftigsten bei 25°.

O. TURPEINEN fand bei *Geotrichoides sp.* für Wachstum und Fettbildung die gleiche Kardinalpunkte der Temperatur: Minimum $<+4°$, Optimum 16—20°, Maximum 26—27°

Von größerem Interesse ist die durch den *Temperaturfaktor* bedingte *Änderung* der *Qualität* der Pilzfette. Bekanntlich hat zuerst S. IVANOW (z.B. 1929) auf Grund umfangreicher Untersuchungen darauf hingewiesen, daß die Zusammensetzung der Samenfette von Blütenpflanzen durch das Klima, insbesondere die Temperatur des Wuchsortes gesetzmäßig beeinflußt wird. Bei tiefer Temperatur werden im allgemeinen stärker ungesättigte Fette gebildet als bei hohen Temperaturen. Das zeigt sich sowohl, wenn man die Fette einer bestimmten Art (z.B. von Lein) von verschiedenen Anbaugebieten gegeneinanderstellt, als auch beim Vergleich der Öle von verschiedenen Arten der warmen und der kalten Klimaten. Ähnliche Untersuchungen wurden in neuester Zeit auch von T. P. HILDITCH (z. B. 1951) und seinen Mitarbeitern mit grundsätzlich gleichem Ergebnis durchgeführt [vgl. hierzu die Beiträge von A. ZELLER (S. 284ff.) und insbesondere von M. L. MEARA (S. 29) in diesem Handbuchband].

Tabelle 28. *Menge und Kennzahlen der Fettsäuren von Aspergillus niger und Rhizopus nigricans bei Kultur unter verschiedenen Temperaturbedingungen.* (Die Zahlen sind zum Teil Mittelwerte von Parallelversuchen). (Nach L. K. PEARSON und H. ST. RAPER 1927.)

Temperatur °C	Kulturdauer Tage	Gesamte Fettsäuren mg	Mittleres Molekulargewicht	Jodzahl	Mittelwert der Jodzahl
		Aspergillus niger			
18°	17	265	302	150,2	
	56	85	327	148,1	149
25°	10	276	287	124	
	14	302	304	132,5	129
	17	254	290	131,3	
	35	252	287	127	
35°	6	468	293	92,1	
	7	527	285	92,3	95
	9	633	290	99,9	
		Rhizopus nigricans			
12°	30	1687	285	88,0	88
25°	10	1082	288	79	
	13	1168	287	77,3	78

Tabelle 27. *Jodzahl der Fettsäuren von Aspergillus niger und von Mycobacterium phlei bei Kultur unter verschiedenen Temperaturbedingungen.* (Nach E. F. TERROINE, R. BONNET, G. KOPP und J. VÉCHOT 1927.)

Jodzahl	*Aspergillus niger*		*Mycobacterium phlei*	
	17° Jodzahl	35° Jodzahl	14° Jodzahl	35° Jodzahl
	114	83	57	35
	112	87	59	35
	116	86		35
Mittel	*114*	*85*	*58*	*35*

Die „IVANOW-*Regel*", daß bei steigender Temperatur die Jodzahl der gebildeten Fette abnimmt, wurde auch bei Mikroorganismen mehrfach *bestätigt*. E. F. TERROINE, R. BONNET, G. KOPP und J. VÉCHOT (1927) untersuchten die Jodzahl der Fettsäuren, die von *Aspergillus niger* und *Mycobacterium phlei* bei 17° und bei 35° gebildet wurden. Tabelle 27 zeigt die Ergebnisse. Mit steigender Temperatur sinkt die Jodzahl der Fette. Die gleiche Regel belegten L. K. PEARSON und H. ST. RAPER (1927) durch eine etwas umfangreichere Zahlenreihe für *Aspergillus niger* und *Rhizopus nigricans* (s. Tabelle 28). Auch hier ist die Abnahme der Jodzahl mit steigender Kulturtemperatur deutlich ausgeprägt. Nur eben angedeutet ist eine Abnahme des mittleren Molekulargewichtes mit zunehmender Temperatur. Die Jodzahlen der Fette von *Aspergillus fischeri*, welche E. A. PRILL, P. R. WENCK und W. H. PETERSON (1935) untersuchten, ließen allerdings eine Temperaturabhängigkeit nicht erkennen: bei 20°, 30° und 37° wurden die Jodzahlen mit 93, 82, 88 ermittelt. Ebensowenig ergab sich eine Beziehung zwischen Temperatur und Jodzahl in den Versuchen von M. E. GREGORY und M. WOODBINE (1953) mit *Aspergillus nidulans*, *Penicillium spinulosum* und *P. javanicum*. Die Kulturen wurden allerdings anscheinend nur bei 25 und 30° untersucht. Hingegen geben die Verfasser an, daß die Jodzahl der Pilzfette mit *steigender*

Inkubationsdauer zunimmt. Y. IMAI (1950) fand wiederum die „IWANOW-Regel" bei *Penicillium chrysogenum* bestätigt. Das Mengenverhältnis flüssige/feste Fettsäuren der Nichtphosphatidfraktion war in Kulturen bei 22° 59,8/23,3, bei 16° 72,8/11,1. Und auch A. BASS und J. HOSPODKA (1952) geben für *Rhodotorula gracilis* an, daß mit zunehmender Temperatur höher gesättigte Fettsäuren von niedrigerem Molekulargewicht gebildet werden.

Ob die Temperatur den Sättigungsgrad der Fette direkt beeinflußt oder ob vielleicht die (unter anderem *temperaturabhängige*) *Geschwindigkeit* der Fettsynthese eine Rolle spielt, dürfte noch kaum zu entscheiden sein. Die letzte Erklärungsmöglichkeit wird von A. KLEINZELLER (1948) neuerdings in Betracht gezogen[1]. Im Zusammenhang damit darf daran erinnert werden, daß C. PONTILLON (1930) bei *Aspergillus niger* in den rascher verfettenden Kulturen auf NH_4NO_3 Fettsäuren niedrigerer Jodzahlen (28,5—33,4) als auf den langsamer verfettenden NH_4Cl- und KNO_3-Kulturen (54,7—87,4 bzw. 80,2—96,7) fand.

3. Die physiologische Bedeutung der Mikroorganismenfette.

Gleich wie die Depotfette der Tiere sind die Samenfette der höheren Pflanzen Reservestoffe. Schon J. SACHS (1859) hat gezeigt, daß diese bei der Keimung wieder in den Stoffwechsel einbezogen werden. Während bei den Fetttröpfchen, die in vielen Dauerorganen von Pilzen (Sporen, Zygoten usw.) mikroskopisch nachweisbar sind, die Reservestoffunktion kaum je in Frage gestellt wurde, gibt es nicht wenige Angaben in der Literatur, die diese Funktion für die in vegetativen Pilzhyphen oder in den Zellen von Hefen und dergleichen abgelagerten Fette in Zweifel ziehen oder sogar energisch in Abrede stellen. Es scheint, daß M. A. GUILLIERMOND (1900) bei seinen Untersuchungen über *Oospora lactis* die verschiedenen Bedenken gegen eine echte Reservestoffnatur der Pilzfette, die vor und nach ihm geäußert wurden, in folgenden zwei Sätzen zusammengefaßt hat: „Ces globules d'huile se forment donc surtout dans les milieux impropres à la nutrition et peuvent être considerées dans une certaine mesure comme un critérium de la mauvaise alimentation" (a.a.O. S. 474). „Ici il semblerait plutôt qu'il s'agisse d'une dégenérance du protoplasma" (a.a.O. S. 475). Damit ist zur physiologischen Dignität der Pilzfette direkt zweierlei gesagt. *Erstens*, daß sie das Resultat *abnormaler Ernährungsbedingungen* darstellen; *zweitens*, daß sie als Symptome einer *degenerativen Veränderung* der Zellen zu betrachten sind. Eingeschlossen ist damit natürlich die Vorstellung, daß sie keine normalen *Reservestoffe* darstellen. Es scheint zweckmäßig, die Frage nach der Funktion der Pilzfette im folgenden unter diesen drei Gesichtspunkten zu behandeln.

a) Sind die Mikroorganismenfette Reservestoffe?

Die Reservestoffnatur der Pilzfette müßte sich darin äußern, daß sie unter geeigneten Bedingungen, z.B. beim Fehlen einer äußeren Energiequelle, durch Einbezug in den Stoffwechsel an Menge abnehmen. Zu diesem Ergebnis kommen auch tatsächlich, mit wenigen Ausnahmen, die Untersucher, welche quantitative Fett*bestimmungen* durchgeführt haben. Die gegenteilige Meinung, daß Fettdepots von Pilzzellen auch bei C-Hunger nicht mehr verbraucht werden, wird meist auf Grund mikroskopischer Schätzungen des Fettgehaltes vertreten. Schon an dieser Stelle sei bemerkt, daß solche Schätzungen höchst unsicher sind. Das heben z.B. A. RIPPEL-BALDES, K. PIETSCHMANN-MEYER und W. KÖHLER (1950) ausdrücklich hervor; dem wird jeder mit der Materie Vertraute vorbehaltloszustimmen.

[1] Zur Abhängigkeit der Zusammensetzung der Fettsäuren von der Kulturdauer bei *Rhodotorula* vgl. A. KLEINZELLER und A. BASS (1951), bei *Aspergillus nidulans* vgl. J. SINGH und T. K. WALKER (1956, s. Nachtrag S. 279).

Der erste, der bei einem Schimmelpilz *(Eurotiopsis Gayoni)* in lang andauernden Kulturen periodisch das Trockengewicht und den Fettgehalt bestimmt hat, war A. PERRIER (1905). Seine Ergebnisse sind in Abb. 11 zur Darstellung gebracht. Es ergibt sich eindeutig, daß der Fettgehalt der Kulturen ein Maximum erreicht und *dann wieder abnimmt*. In den Äthanolkulturen, in denen das Substrat bereits am 5. Tage erschöpft ist, setzt der Verbrauch früher ein als auf

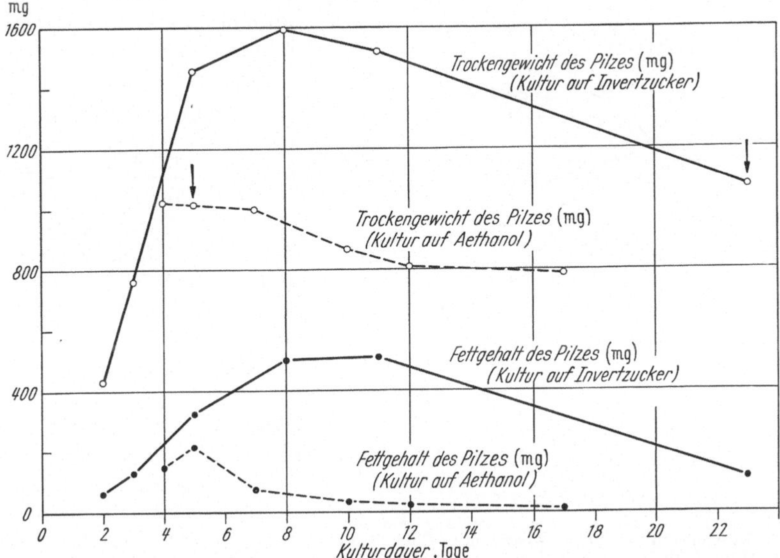

Abb. 11. Zeitkurven für Trockengewicht und Fettgehalt von *Eurotiopsis Gayoni* bei Kultur auf Invertzucker und auf Äthanol. Erschöpfung der C-Quelle des Substrates. (Nach A. PERRIER 1905.)

Invertzuckermedium; hier aber immerhin schon *vor* dem restlosen Verbrauch der äußeren C-Quelle.

Diese Ergebnisse sind auch deshalb besonders erwähnenswert, weil kurz vorher H. LABORDE (1897) mit dem gleichen Organismus experimentiert hatte. Unter dem Stichwort „dégénerance cellulosique et grasse de la plante“ erwähnt er einen Anstieg des Fettgehaltes von der normalen Höhe (4—8%) auf 29,8% vom Trockengewicht, wenn der Pilz 17 Monate lang (!) auf Hefewasser + 2‰ H_2SO_4 + 4% Äthanol (mehrfach erneuert) gehalten wurde. Er läßt es offen, ob die Fettanreicherung als „une accumulation d'un produit normal“ oder „un produit pathologique“ zu deuten ist. Er verweist dabei ausdrücklich auf die Untersuchungen von E. DUCLEAUX (1889) mit Hefen. Dieser hatte in 16 bis 17 Jahre alten (!) Kulturen von Hefe — die Zellen werden als „zum Teil noch lebend“ angegeben — erhöhte Fettgehalte (10—52%, gegenüber 5% normal) und verminderte N-Gehalte (2,7—5% gegenüber 2,5—5% normal) gefunden. Man wird nicht fehlgehen, wenn man diese Ergebnisse von LABORDE und von DUCLEAUX in gleicher Weise deutet, wie es schon bei der „Involution“ von *Penicillium* in den Versuchen von C. v. NAEGELI und O. LOEW (1880) geschah: daß es sich um das Liegenbleiben der Fette in Zellen handelt, die durch eine weit fortgeschrittene Autolyse einen Großteil ihrer Zellinhaltsstoffe verloren hatten (s. S. 225).

R. GOUPIL (1913, 1914) untersuchte bei *Mucor (Amylomyces) Rouxii* auch die Veränderung der *Zusammensetzung* der Fette während ihres Verbrauches durch die Zelle. Wenn der Prozeß der Fettbildung abgeschlossen ist und der Fettverbrauch einsetzt, fällt der Gehalt an Phosphatiden rapide fast bis auf

Null ab. Besonders rasch setzt der Fettverbrauch ein, wenn dem Pilz die ursprüngliche (wohl noch kohlenhydrathaltige) Nährlösung entzogen und stark belüftet wird. Ein solcher Versuch ist in Abb. 12 wiedergegeben. Wie der Gang der Kurven für die freien Fettsäuren zeigt, geht der Fettverbrauch durch die Zelle mit einer starken Spaltung der Glyceride Hand in Hand. Die Kurve der Verseifungszahlen läßt auf eine bevorzugte Verarbeitung der niedermolekularen Fettsäuren schließen.

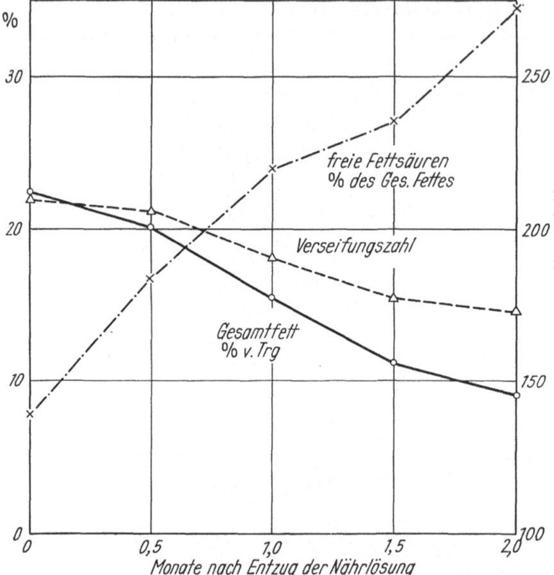

Abb. 12. Fettgehalt, Gehalt an freien Fettsäuren und Verseifungszahl in Kulturen von *Mucor Rouxii* nach Entzug des Substrates. (Freie Fettsäuren als „Ölsäure" berechnet.) (Nach R. GOUPIL 1914.)

Weitere eindeutige zahlenmäßige Beweise für den *Verbrauch* des in Pilzzellen gespeicherten Fettes lieferten L. K. PEARSON und H. ST. RAPER (1927) (*Aspergillus niger, Rhizopus nigricans*) (vgl. oben Tabelle 28); CH. PONTILLON (1930a, b, 1932) (*Aspergillus niger*), der auch die raschere Abnahme des Lipoid-P (s. auch R. GOUPIL 1913) bestätigte, während die Sterine in der Phase des Fettverbrauches nur wenig abnehmen oder sogar zunehmen. Weiter seien zitiert L. M. PRUESS, W. H. PETERSON, H. STEENBOCK und E. B. FRED (1931) (Versuche mit *Aspergillus oryzae, A. niger* und *Penicillium expansum*); E. A. PRILL, P. R. WENCK und W. H. PETERSON (1935) (*Aspergillus fischeri*); O. TURPEINEN (1936) (*Geotrichoides sp.*); H. FINK, H. HAESELER und G. SCHMIDT (1937) (*Oospora lactis*); S. HEIDE (1939), H. RAAF (1941), M. MAAS-FÖRSTER (1955) (*Endomycopsis vernalis*).

Zunächst seien die Versuchsresultate von E. A. PRILL und Mitarbeitern (1935) bei *A. fischeri* wiedergegeben, in welchen nicht nur der Verbrauch des Gesamtfettes, sondern auch einzelner Fettfraktionen berücksichtigt ist (Abb. 13). In Übereinstimmung mit den früher zitierten Forschern (CH. PONTILLON, R. GOUPIL) wird festgestellt, daß die Phosphatide im Verbrauch vorauseilen[1], während die Sterine und überhaupt die unverseifbaren Fettbestandteile zunächst träge reagieren und erst nach dem 50. Tag eine deutliche Abnahme zeigen[2].

In Abb. 14 wird das Ergebnis einer Versuchsreihe von H. FINK und Mitarbeitern wiedergegeben. Es zeigt sich, daß vom 7. zum 16. Tag des Versuches die „fettfreie Trockensubstanz" von *Oospora lactis* unverändert bleibt, daß also praktisch der gesamte Betriebsstoffwechsel auf Kosten der Fettreserven läuft. Das wird durch die Zeitkurven für Fettgehalt und Trockensubstanz durchaus bestätigt. Durchaus gleichartige Resultate für *Endomycopsis vernalis* bringen einige Kurven in der Arbeit von H. RAAF (1941).

---

[1] Auch bei *Mycobacterium tuberculosis* (BCG) fand J. ASSELINEAU (1951) eine Abnahme der Phosphatide um 50% (0,8 → 0,4% vom Trockengewicht) vom 4. zum 42. Kulturtag.

[2] Über die Mengenveränderungen einzelner Fettsäuren des Mycelfettes von *Aspergillus nidulans* im Verlaufe der Kultur haben jüngst J. SINGH und T. K. WALKER (1956) berichtet (s. Nachtrag S. 279).

Schließlich seien in Abb. 15 einige Ergebnisse von M. MAAS-FÖRSTER (1955) bei *Endomycopsis vernalis* mitgeteilt. Das Versuchsziel erforderte eine Analyse des Substrates und der Pilzernte in kurzen Zeitabständen (von 12 bzw. 24 Std). Es ergab sich, daß die Fettzunahme nur so lange andauert, als noch Kohlenhydrat im Medium vorhanden ist. Sofort nach seiner Erschöpfung setzt der Fettabbau ein, gleichgültig wie hoch der absolute und prozentuale Fettgehalt des Pilzes ist. Der maximale Fettgehalt war in dieser Versuchsreihe bei $^{1}/_{1}$ P: etwa 16%, bei $^{1}/_{5}$ P etwa 28%, bei $^{1}/_{20}$ P etwa 47%. Umgekehrt konnte S. HEIDE (1939) zeigen, daß der Fettverbrauch aufgehalten werden kann, wenn dem Pilz nach Erschöpfung des Nährmediums neuer Zucker gegeben wird. Er pflanzte am 11. Kulturtag *Endomycopsis vernalis* (mit 43% Fett vom Trockengewicht) (A) auf 7,5% Saccharoselösung, (B) auf destilliertes Wasser um. Kontrollen (C) blieben auf dem alten Substrat. Nach 9 Tagen wurden folgende Fettgehalte gefunden: A: 39,4%, B: 26,7%, C: 30,1%. Die Absolutmengen an Fett betrugen je Kulturschale: 11. Tag: 83,9 mg, 20. Tag: (A) 82,7 mg, (B) 43,3 mg, (C) 42,1 mg. Das heißt bei Kohlenhydratmangel (C), (B) hatte der Pilz in 9 Tagen rund 50% seiner Fettvorräte verbraucht, bei neuer Kohlenhydratgabe (A) so gut wie gar nichts.

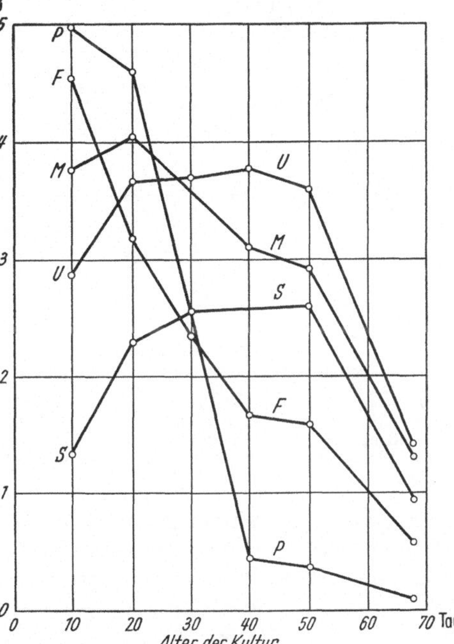

Abb. 13. Hungerwirkung auf einige Mycelbestandteile von *Aspergillus fischeri* (in g/100 ml Medium). *M* Mycelgewicht minus Fett, *F* Fett × 4, *U* Unverseifbares × 50, *S* Sterine × 50, *P* Lipoid-*P* × 1000. (Nach E. A. PRILL, P. R. WENCK, W. H. PETERSON 1935)

Auch H. GYLLENBERG und A. RAITIO (1952) berichten für ein *Penicillium sp.*, daß der maximale Fettgehalt von 9—11% vom Trockengewicht im Augenblick des völligen Glucoseverbrauchs erreicht wird und daß hiernach das Fett wieder bis auf einen Rest von 1,3—1,4% vom Trockengewicht verbraucht wird.

Die bisher dargestellten Ergebnisse lassen sich sehr eindeutig dahin zusammenfassen, daß das *in Pilzzellen gespeicherte Fett als Reservestoff angesprochen werden muß*. Hier dürfte es nun notwendig sein, in Kürze auf zwei Arbeiten einzugehen, die zu einem ausgesprochen entgegengesetzten Resultat kommen.

H. KORDES (1923) untersuchte 20 Pilzarten aus den sehr verschiedenen Systemgruppen: *Mucorineen, Plectascales, Saccharomycetales, Pyrenomycetales, Discomycetales* und *Fungi imperfecti*. Der Fettgehalt der Zellen wurde *nur* mikroskopisch geprüft. Nach den Angaben des Verfassers werden Fetteinschlüsse von Dauerzellen (Sporen, Chlamydosporen, Conidien, Gemmen) bei der Keimung *verbraucht*. Die Fetttropfen, die sich in alternden Hyphen vorfinden, bleiben hingegen unverändert liegen, selbst wenn Mycelflocken auf rein anorganische KNOP-Lösung übertragen werden. Das Mycelfett zeigt in den letzteren Fällen also das physiologische Verhalten eines *Exkretes*. Bei diesen Ergebnissen muß zunächst auffallen, daß sie sich zum Teil auch auf Arten bzw. Gattungen beziehen (*Rhizopus nigricans, Aspergillus*), bei denen andere Forscher mit quantitativen Analysen eindeutig einen Verbrauch des Mycelfettes nachweisen konnten. Besonders auffallend ist die Angabe von KORDES, daß das *gleiche* Fett, das in

Hyphen als *Exkret* fungiert, sich wie ein echter *Reservestoff* verhält (d.h. verbraucht wird), wenn es in eine *Chlamydospore* eingeschlossen wird (*Mucor hiemalis, M. Jansenii*). Selbst wenn man bereits von den oben betonten Schwierigkeiten einer einigermaßen zuverlässigen quantitativen, mikroskopischen Schätzung des Fettgehaltes absieht, sind schließlich sehr ernste Bedenken am Platze, ob die alternden „fettreichen" Hyphen in den Versuchen von KORDES, trotz der Angabe, daß sie noch „plasmolysierbar" waren, nicht eine stark reduzierte Vitalität aufweisen oder überhaupt zum größeren Teile tot waren.

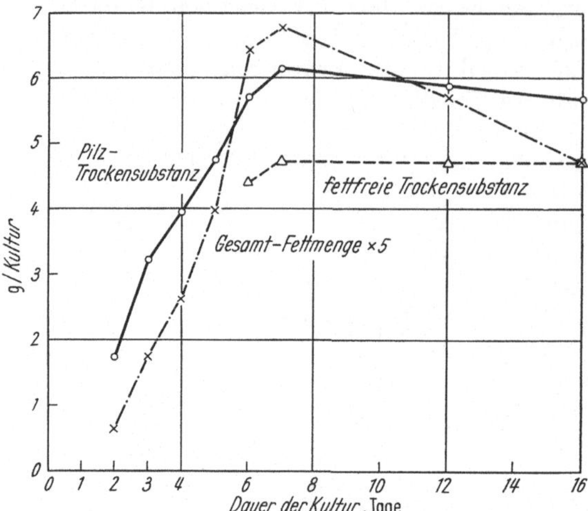

Abb. 14. Zeitkurven für Trockensubstanz und Gehalt des Mycels an Fett und fettfreier Trockensubstanz bei *Oospora lactis* (Stamm A). Kultur auf 8 l Molke, 2 g $(NH_4)_2SO_4$, 1 g KH_2PO_4, 0,5 g $MgSO_4 + 7\,H_2O$, 27°!. Am 7. Tag war alle Milchsäure und 87% des ursprünglich vorhandenen Milchzuckers verbraucht. (Nach H. FINK, H. HAESELER und M. SCHMIDT 1937.)

Die letzte Frage muß ganz besonders für die Untersuchungsobjekte erhoben werden, über die H. GEFFERS (1937) berichtete. Er untersuchte die Fettbildung von *Oospora* und kam zu dem Schluß, daß „nur ganz wenig verfettete Zellen des Pilzes ein Auskeimen und anscheinend eine Auflösung des Fettes" zeigten und daß überhaupt das Fett nicht, wie es bei einem Reservestoff der Fall sein müßte, wieder abgebaut wird. GEFFERS arbeitete mit Molkekulturen, deren Nährstoffe durch *Oospora* offenbar nur sehr langsam verbraucht werden. Trockengewicht und Fettgehalt der Ernten zeigten noch bis zum 140. (!) Tage eine Zunahme. Erst danach lassen die Tabellen eine geringfügige Abnahme des Fettes erkennen. Diese wird indes — wohl mit Recht — vom Verfasser nicht als Ausdruck eines Fettabbaues anerkannt. „Untersuchte man das Mycel (scil. solcher alter Kulturen), so ergab sich, daß es zu einer teilweisen Auflösung der Zellwände der jetzt eiweißleeren Zellen gekommen war. Hierdurch waren viel Fetttröpfchen frei geworden und gingen beim Filtrieren mit durch das Filter, sie waren also für die nachfolgende Fettbestimmung verloren. Es konnte aus diesem Zurückgehen der Fettmenge kein Schluß auf den Abbau des Fettes durch die Zellen gezogen werden." Es scheint uns, daß der Verfasser damit eine vorzügliche Beschreibung *toter Oospora*zellen gegeben hat, daß also sehr wahrscheinlich die Hauptmenge der Zellen in seinen 4 Monate alten Kulturen *tot* war. Daß diese Zellen zum Fettabbau und zur Vermehrung unfähig waren, braucht nicht wunder zu nehmen. Es ist auch kein Gegenbeweis, wenn in solchen Kulturen neue Mycelbildung und Fettsynthese einsetzte, wenn neues Substrat (Molke) zugefügt wurde; dazu genügt es, daß eine einzige lebende Zelle in der Kultur vorhanden war.

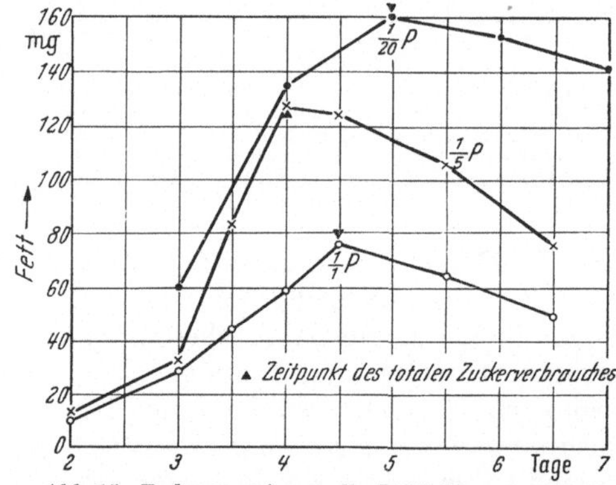

Abb. 15. *Endomycopsis vernalis*. Fettbildung und Fettverbrauch in Kulturen mit verschiedenem Phosphatgehalt des Substrates. (Nach M. MAAS-FÖRSTER 1955.)

b) Sind verfettete Pilzzellen vermehrungsunfähig?

Der Begriff der „Degeneration", den M. A. GUILLIERMOND (1900) (s. oben S. 242) auf verfettete Zellen von *Oospora lactis* anwendet, ist wenig präzis definiert. Man wird geneigt sein, anzunehmen, daß in einer „degenerierten" Zelle wichtige Lebensfunktionen gestört sind. Bei wörtlicher Interpretation wird man vor allem eine gestörte Reproduktionsfähigkeit als Ausdruck der „Degeneration" sehen. In der Tat finden sich in der Literatur zahlreiche Angaben, die stark verfetteten Pilzzellen jede Vermehrungsfähigkeit absprechen. Andere Untersucher kamen allerdings oft bei den gleichen Versuchsobjekten zum entgegengesetzten Ergebnis. Besonders P. LINDNER war ein entschiedener Anwalt der Theorie, daß verfettete Mikroorganismen vermehrungsunfähig seien (1919, 1921 a, b, 1922 a, b, P. LINDNER und T. UNGER 1919). Seine Ansichten (bezüglich der Bierhefe) werden auch von W. HENNEBERG (1926) und (für *Endomycopsis*) von H. FINK, H. HAEHN und W. HOERBURGER (1937) übernommen. W. HALDEN, F. BILGER und R. KUNZE (1933) wollen hingegen die Lipoidanreicherung der Hefezelle keineswegs als eine Degenerationserscheinung betrachtet wissen, S. HEIDE (1939) findet stark verfettete Zellen von *Endomycopsis vernalis* durchaus zur Vermehrung fähig. Fettzellen von *Anthomyces Reukaufii*, nach P. LINDNER (1922 b) ein „klassisches Beispiel von Wachstumshemmung durch Verfettung" („innerhalb 24 Std Verfettung ... vollendet", „Abimpfung auf frische Lösung ... ergebnislos") sind nach A. RIPPEL-BALDES, K. PIETSCHMANN-MEYER und W. KÖHLER (1950) „noch keimfähig". Nach H. GEFFERS (1937) (s. oben S. 246) zeigen bei *Oospora* nur „wenig verfettete Zellen ein Auskeimen", dagegen berichten H. FINK, H. HAESELER und M. SCHMIDT (1937), „die alten, stark verfetteten Oidien" (von *Oospora lactis*) „wachsen in LINDNERschen Tröpfchenkulturen schon nach kurzer Zeit aus".

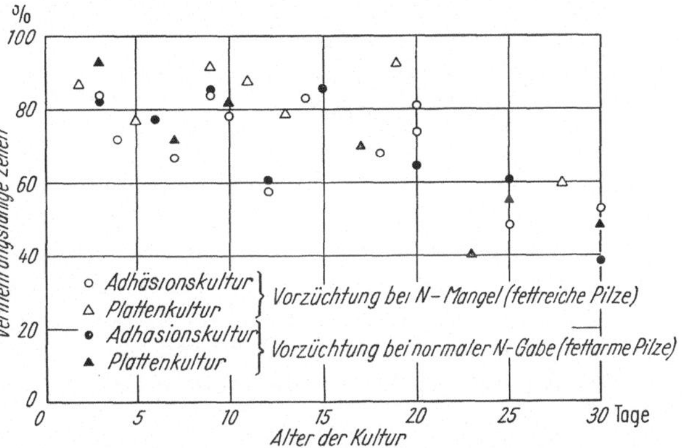

Abb. 16. *Endomycopsis vernalis*. Vermehrungsfähigkeit der Zellen in Abhängigkeit vom Alter. (Nach H. HEINEMANN 1956.)

Wie man sieht, steht oft für ein und dieselbe Art Aussage gegen Aussage. Eine Klärung schien nur durch präzise, an einem größeren Zellenmaterial gewonnene quantitative Angaben über die Beziehungen zwischen Fettspeicherung und Vermehrungsvermögen möglich. Umfangreiche und gründliche Untersuchungen in dieser Richtung wurden kürzlich von H. HEINEMANN (1956) durchgeführt. Zur Ermittlung des Prozentsatzes vermehrungsfähiger Zellen dienten zwei Methoden:

1. Das Plattenverfahren. Von dem zu untersuchenden Zellmaterial wird eine möglichst homogene Suspension in Leitungswasser hergestellt. Die absolute Zellenzahl je Volumeinheit (A) wird durch Zählung in einer THOMA-Kammer bestimmt. Aliquote Teile der gleichen Suspension werden mit Malzgelatine vermischt und in PETRI-Schalen ausgegossen. Die Zahl der Kolonien, die nach entsprechender Inkubationszeit gefunden wird, ist gleich der Zahl der keimfähigen Zellen (B), deren Prozentsatz sich nach $B/A \cdot 100$ errechnet.
2. Durch Einzellkulturen, d. i. Adhäsionskulturen auf Deckgläschen auf Hohlschliffobjektträgern.

Die nach beiden Methoden bei *Endomycopsis vernalis*, *Candida Reukaufii*, *Oospora lactis* und *Torulopsis lipofera* erhaltenen Resultate stimmen gut überein. Ein Beispiel hierfür (bei *Endomycopsis*) gibt Abb. 16. Sowohl in Platten- wie in

Adhäsionskultur zeigt sich, daß der Prozentsatz der vermehrungsfähigen Zellen mit *steigendem Alter* der Kultur *abnimmt:* im vorliegenden Falle von etwa 90% am 2.—3. Kulturtag auf etwa 50% nach 30 Tagen. Ein signifikanter Unterschied zwischen fettreichen und fettarmen Zellen ließ sich nicht feststellen. Das letztere wird durch eine weitere Versuchsreihe bestätigt (Abb. 17), in welcher etwa gleichalte Zellen *verschiedenen Fettgehaltes* auf ihr Vermehrungsvermögen geprüft wurden. Wesentliche Unterschiede sind nicht vorhanden; *von einer Vermehrungsunfähigkeit verfetteter Zellen kann keine Rede sein.* Durchaus gleichartige Ergebnisse wurden auch mit den anderen oben erwähnten Fettpilzen erhalten.

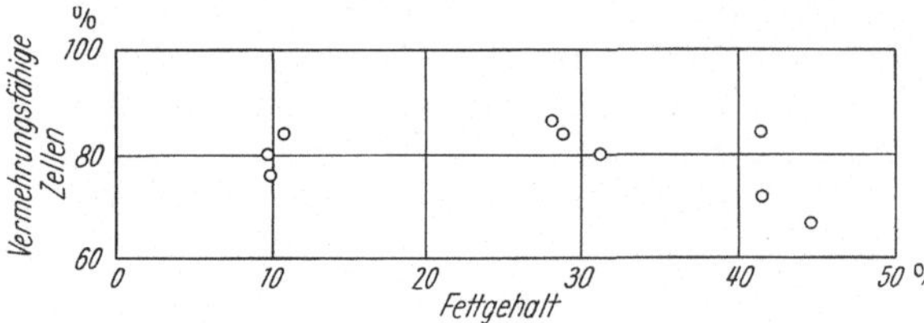

Abb. 17. *Endomycopsis vernalis.* Vermehrungsfähigkeit der Zellen in Abhängigkeit vom Verfettungsgrad. (Nach H. HEINEMANN 1956.)

Bei der direkten Beobachtung der Zellen in Adhäsionskultur ließen sich übrigens auch sehr deutliche Hinweise für die Reservestoffnatur der Hefefette finden. Wurden die Tröpfchenkulturen in kohlenhydratfreier Nährlösung ausgeführt, so konnte die Verkleinerung der Fetttropfen durch Mikrometermessung direkt verfolgt werden. So maß in einer Zelle von *Oospora lactis* der Fetttropfen bei Kulturbeginn 10,9 μ, nach 3 Tagen 7,6 μ, nach 7 Tagen 4,35 μ. Die Zahl der *Tochterzellen,* die von einer einzelnen Zelle auf kohlenhydratfreiem Medium gebildet wird, steht in deutlicher *positiver* Beziehung zum Fettgehalt der Mutterzelle. Als Beispiel hierfür diene die folgende kleine Tabelle.

Tabelle 29. *Zahl der Tochterzellen einzelner Zellen von Endomycopsis vernalis bei Tröpfchenkultur in kohlenhydratfreiem Medium.* (Nach H. HEINEMANN 1956.)

Vorkultur	Aussprossende Zellen %	Durchschnittliche Zahl der Tochterzellen nach 3 Tagen
0,25% Asparagin (etwa 40% Fett)	94	26
1,00% Asparagin (etwa 10% Fett)	59	4

Eine besondere Erwähnung erfordern auf Grund der Untersuchungen von H. HEINEMANN die Verhältnisse bei der *untergärigen Bierhefe.* Es gelang der Verfasserin ohne Schwierigkeit an Hand der von LINDNER und von HALDEN und Mitarbeitern gegebenen Vorschriften durch langdauernde Kultur bei guter Lüftung und ständiger Zufuhr von Äthanoldämpfen eine starke Verfettung der Hefezellen herbeizuführen. Werden ältere (etwa 2—3 Monate alte) Kulturen dieser Art mikroskopisch untersucht, so findet man zwei Typen von Zellen:

(A) Zellen, welche zahlreiche Fetttröpfchen enthalten, die nicht besonders deutlich sind, die aber durch Behandlung mit OsO_4 oder Fettfarbstoffen gut in Erscheinung treten (Abb. 18b).

(B) Zellen, die einen einzigen großen Fetttropfen enthalten. Dieser Zelltypus *scheint* deutlich fettreicher zu sein als (A) (Abb. 18a).

Mit zunehmendem Alter der Kultur nimmt der Mengenanteil der Zellen des Typus (B) immer mehr zu. Im Vermehrungsversuch erwiesen sich die Zellen des Typus (A) zu einem hohen Prozentsatz teilungsfähig; dagegen wurde keine einzige Zelle des Typus (B) gefunden, welche Tochterzellen bildete (Tabelle 30).

Tabelle 30. *Vermehrung von Hefezellen nach 11 Wochen dauernder Kultur in Äthanoldampf.* (Nach H. HEINEMANN.)

	Vermehrung	Keine Vermehrung
I. Zellen mit vielen kleinen Fetttröpfchen (Typ A) .	33 (77%)	10 (23%)
II. Zellen mit einem großen Fetttropfen (Typ B) . . .	0 (0%)	36 (100%)

Hier scheint also eine Bestätigung der Theorie vorzuliegen, daß stark verfettete Hefezellen vermehrungs*unfähig* sind.

Es fiel aber auf, daß die Zellen vom Typ (B) häufig etwas deformiert und durchschnittlich kleiner erschienen als die Zellen vom Typ (A). Letzteres ließ sich durch Ausmessen einer großen Anzahl von Zellen auch einwandfrei belegen. Es entstand die Vermutung, daß die Zellen des Typs (B) tot sind. Der Versuch, dies durch Fluorochromierung mit Acridinorange nachzuweisen, führte zu keinem Erfolg, da diese Methode offenbar bei stark verfetteten Zellen versagt. Dagegen gelingt es leicht, Zellen des Typs (A) in den Typ (B) überzuführen, wenn man sie durch Austrocknen, durch Behandlung mit verdünnter HCl (Abb. 18c) oder mit starkem Äthanol *abtötet*. Es kann also mit einem hohen Grad von Wahrscheinlichkeit angenommen werden, daß die Zellen des Typs (B) *tot* sind, daß also eine Vermehrungsfähigkeit bei ihnen nicht zu erwarten ist.

Im Gegensatz zu anderen Fettpilzen (*Oospora*, *Endomycopsis*) werden in den Zellen der Bierhefe offenbar die einzeln entstehenden Fetttropfen durch eine besondere Struktur des Protoplasten getrennt gehalten (vgl. auch H. WILL 1896). Erst wenn beim Absterben diese Plasmastruktur zusammenbricht, kommt es zur Vereinigung des Fettes in einem großen Tropfen. Diese Erscheinung ist offensichtlich bereits W. HENNEBERG (1926)

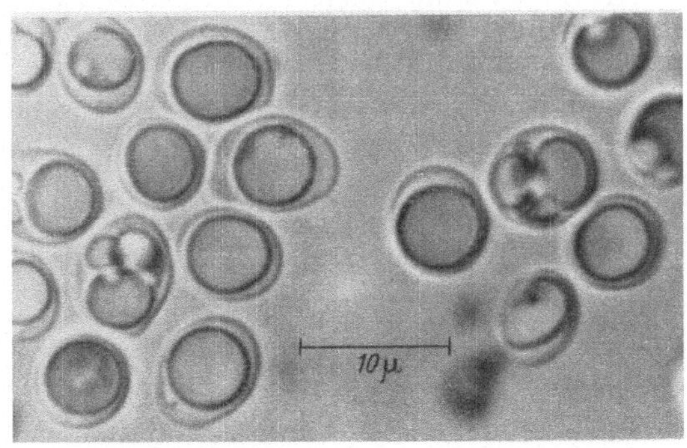

a

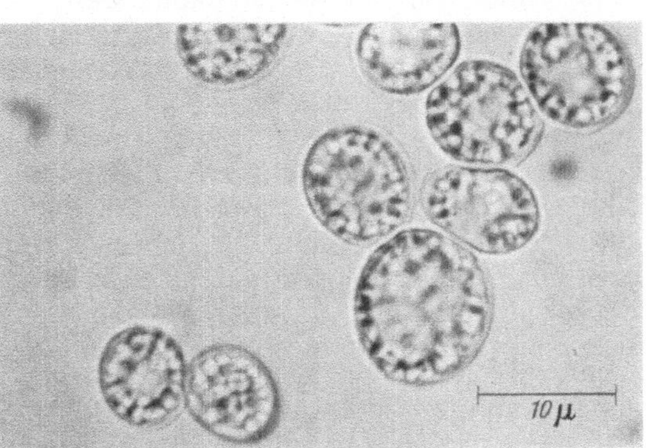

b

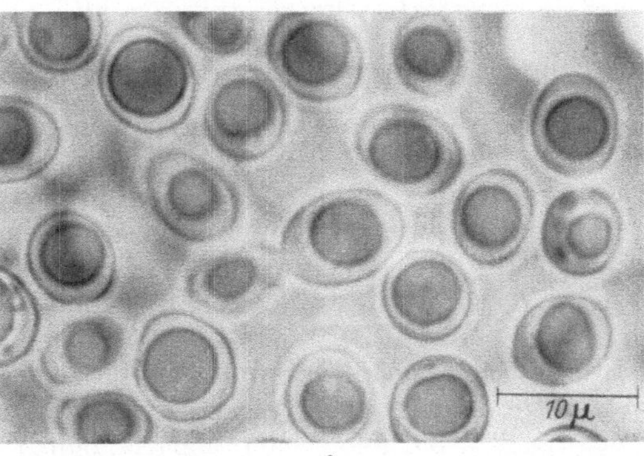

c

Abb. 18a—c. *Saccharomyces cerevisiae*, untergärige Bierhefe. Ältere Zellen aus Kulturen in Äthanoldampf bei guter O_2-Versorgung. a Vermehrungsunfähige Zellen mit einem großen Fetttropfen. („Zelltyp B"). b Vermehrungsfähige Zellen mit zahlreichen kleinen Fetttropfen. („Zelltyp A"). c Dasselbe Zellmaterial wie b nach Behandlung mit verdünnter HCl. (Nach H. HEINEMANN 1956.)

bekannt gewesen. Trotzdem haben merkwürdigerweise weder er noch P. LINDNER (1922a, b) die naheliegende Konsequenz zur Frage „Vermehrungsunfähigkeit stark verfetteter Hefezellen" gezogen. Die soeben besprochenen Tatsachen geben einen sehr eindrucksvollen Hinweis dafür, eine wie große Vorsicht bei der mikroskopischen Mengenschätzung des Fettes in Pilzzellen notwendig ist. Ein Vergleich von Abb. 18b und c zeigt, welche „optischen Täuschungen" möglich sind.

Kurz erwähnt seien in diesem Zusammenhang die Untersuchungen von C. C. LINDEGREN (1945), wonach reservestoff-(fett- oder kohlenhydrat-)reiche Hefezellen erst mit deutlicher Verzögerung („lag period") Atmung, Gärung und Sprossungsaktivität wieder aufnehmen und nur dann, wenn ein komplettes, durch alle Wirkstoffe ergänztes Medium geboten wird. Es wäre wichtig zu untersuchen, ob hier „Reservestoffreichtum" oder „Wirkstoffverarmung" die entscheidende Rolle spielt. Im letzteren, durchaus möglichen Fall ergäben sich interessante Anknüpfungspunkte an gewisse Anschauungen von P. LINDNER, die bereits früher (S. 216) kurz gestreift wurden.

c) Ist die Fettbildung eine Folge „abnormaler" Ernährungsbedingungen?

Eine dritte Gruppe von Einwänden gegen die Reservestoffnatur der Pilzfette stützt sich auf die Meinung, daß Verfettung die Folge abnormaler Ernährungsbedingungen sei (M. A. GUILLIERMOND 1900: „milieu impropres à la nutrition", P. BÉLIN 1926b, CH. PONTILLON 1930a, b, 1933: „milieu deséquilibré"). Hierzu ist das Wesentliche bereits in den Abschnitten 2a—k dieses Beitrages, insbesondere S. 239f. (vgl. auch Abb. 10) gesagt worden. Nach der dort vorgetragenen, experimentell begründeten Auffassung gehen Eiweiß- und Fettbildung von gemeinsamen Intermediärprodukten des Kohlenhydratstoffwechsels aus. Beide Prozesse befinden sich in einem Gleichgewicht, welches sehr stark nach der Seite der Eiweißbildung verschoben ist, wenn die hierfür notwendigen Nährelemente (N, P, S) in genügender Menge zur Verfügung stehen. Wird die Menge eines dieser Nährstoffe (vor allem N und/oder P) reduziert, so ist die Eiweißbildung gehemmt, das freiwerdende Material kommt in steigendem Maße der Fettbildung zugute. Eine reduzierte Kaliumgabe verschiebt das Gleichgewicht zwischen Fett- und Eiweißbildung auch dann nicht, wenn im gesamten Stoffwechsel sich bereits deutliche Symptome eines Kalium*mangels* bemerkbar machen. Verstärkte Fettbildung ist also *nicht* die Folge von *Mangelernährung schlechthin,* sondern die Folge einer durch spezifische Ursachen *gehemmten Eiweißsynthese.*

An dieser Stelle muß noch kurz zu einigen Gedankengängen von J. W. FOSTER (1949) Stellung genommen werden, die schon früher (S. 240) ganz flüchtig erwähnt wurden. FOSTER (a.a.O., S. 169) nimmt an, daß Pilze unter „normalen" Ernährungsbedingungen nur Baumaterial für die Zelle synthetisieren. Erst unter „abnormen Umweltsbedingungen", z.B. bei stark erhöhter Kohlenhydratgabe, häuft der Organismus Stoffwechselprodukte an (organische Säuren, Fette), die auf Nebengeleisen des Stoffwechsels (im „shunt metabolism") gebildet werden. Obwohl diese Produkte unter entsprechenden Bedingungen wieder im Stoffwechsel verarbeitet werden können, sieht sie FOSTER als Resultat eines abwegigen („deranged"), pathologischen („pathological") Stoffwechsels an. Auch das Fett sei ein solches Stoffwechselprodukt. FOSTER (a.a.O., S. 122) faßt ausdrücklich die Pilzfette als „insoluble shunt products", als Analoga zu den in anderen Fällen als Resultat des „shunt metabolism" entstehenden organischen Säuren auf. Weil diese aber ins Medium abgeschieden werden, hat er offensichtlich Bedenken, die Fette als Produkt eines normalen Stoffwechsels anzuerkennen.

Es scheint, daß hierzu zweierlei bemerkt werden muß.

1. Die Schimmelpilze und Hefen, die zu starker Fetteinlagerung befähigt sind, kommen in der Natur auf den verschiedensten Nährböden vor. Zu ihren „natürlichen" Substraten gehören zweifellos auch solche, die z.B. eine sehr hohe Zuckerkonzentration aufweisen (Nektarhefe!). Es scheint durchaus erlaubt, den „shunt metabolism" im Rahmen der weiten Anpassungsfähigkeit dieser Organismen an die — qualitativ und quantitativ — verschiedensten Substrate aufzufassen. Er setzt sie in den Stand, eine Vielzahl von Substraten recht „zweckmäßig" auszunutzen, entweder zur Produktion von Zellsubstanz oder aber von Stoffen, die später als „Reservematerial" für den Erhaltungsstoffwechsel verwendet werden können. Daß diese Plastizität des Stoffwechsels besondere Möglichkeiten für das physiologisch-chemische Experiment bietet, bedarf kaum einer Erwähnung.

2. Bei vielen Pilzen ist es offenbar nicht möglich, eine ganz scharfe Grenze zwischen Stoffwechselprozessen *innerhalb* und *außerhalb* der Zelle zu ziehen. Die ersten Schritte der Nährstoffassimilation werden vielfach durch Exoenzyme (Amylase, Lipasen, Proteasen) im Substrat selbst eingeleitet. Erst kürzlich konnte H. Kating (1955b, s. auch M. Steiner und H. Kating 1953) zeigen, daß bei *Endomycopsis* die einleitenden Phasen der N-Assimilation (z.B. Umaminierungen) in einer *äußeren* Stoffwechselregion ablaufen können. Um so weniger scheint ein zwingender Grund vorzuliegen, Stoffwechselprodukte wie die Fette deswegen als „abnormal" oder „pathologisch" zu betrachten, weil analoge Metabolite (wie die organischen Säuren) ins Substrat ausgeschieden werden.

Es scheint also, daß auch der von Foster sehr glücklich konzipierte Begriff des „shunt metabolism" keineswegs notwendig zur Konsequenz führen muß, die in Pilzzellen angehäuften Fette nicht als *Reservestoffe* gelten zu lassen.

4. Der Ort der Fettbildung in Pilzzellen.

Die Fettbildung in der tierischen Zelle spielt sich in den Mitochondrien ab. Die älteren histologischen und histochemischen Ergebnisse haben ihre endgültige Bestätigung durch die Feststellung erfahren, daß die Enzyme des Fettstoffwechsels in entsprechenden Partikelfraktionen enthalten sind. Die Lebermitochondrien des Schafes enthalten *alle* Enzyme des Fettsäurecyclus (H. R. Mahler 1933, s. auch F. Lynen 1955). Wie die Ausführungen von K. Stumpf (S. 201ff. dieses Handbuchbandes) zeigen, scheint auch bei höheren Pflanzenzellen der Fettstoffwechsel in den Mitochondrien lokalisiert zu sein.

Es gibt mancherlei lang bekannte Hinweise, daß auch die Fetttröpfchen in Pilzzellen nicht an „beliebiger Stelle" des Protoplasmas abgelagert werden. Die Fetttröpfchen in Pilzsporen sind oft nach Zahl und Lokalisation so charakteristisch, daß sie geradezu als taxonomisches Merkmal dienen können (z.B. E. Boudier 1914). Auch bei vegetativen Pilzzellen ist nicht selten die Zahl der bei Beginn der Fettbildung auftretenden Tröpfchen innerhalb gewisser Grenzen sippenspezifisch. Bei der Bierhefe bilden sich viele Fetttropfen, die zwar heranwachsen, die aber dauernd getrennt bleiben, so lange die Zelle nicht abstirbt (s. S. 248f.). Bei *Endomycopsis vernalis* oder *Oospora lactis* beginnt in gleicher Weise das Fett an vielen Stellen zu gleicher Zeit in Tropfenform sichtbar zu werden. Später vereinen sich die einzelnen kleinen Fettkügelchen zu einem oder wenigen großen Tropfen. Bei *Torulopsis lipofera* und *Rhodotorula*arten werden zunächst nur 2—4 Fetttröpfchen sichtbar, von denen sich in der Regel eines rasch und stark zu einer Fettkugel vergrößert. Im Endstadium der Verfettung sind neben dieser noch bis höchstens 10 kleinere Fetttröpfchen zu finden.

Mit der Frage, ob die Chondriosomen als die Orte der Fettbildung in den Pilzzellen anzusehen sind, hat sich zuerst A. GUILLIERMOND eingehend beschäftigt. In seinen ersten Arbeiten über diesen Gegenstand (1913a, b) neigt er zunächst zu einer positiven Antwort. Später aber (A. GUILLIERMOND, MANGENOT und L. PLANTEFOL 1933, A. GUILLIERMOND 1941) kommt er zur entgegengesetzten

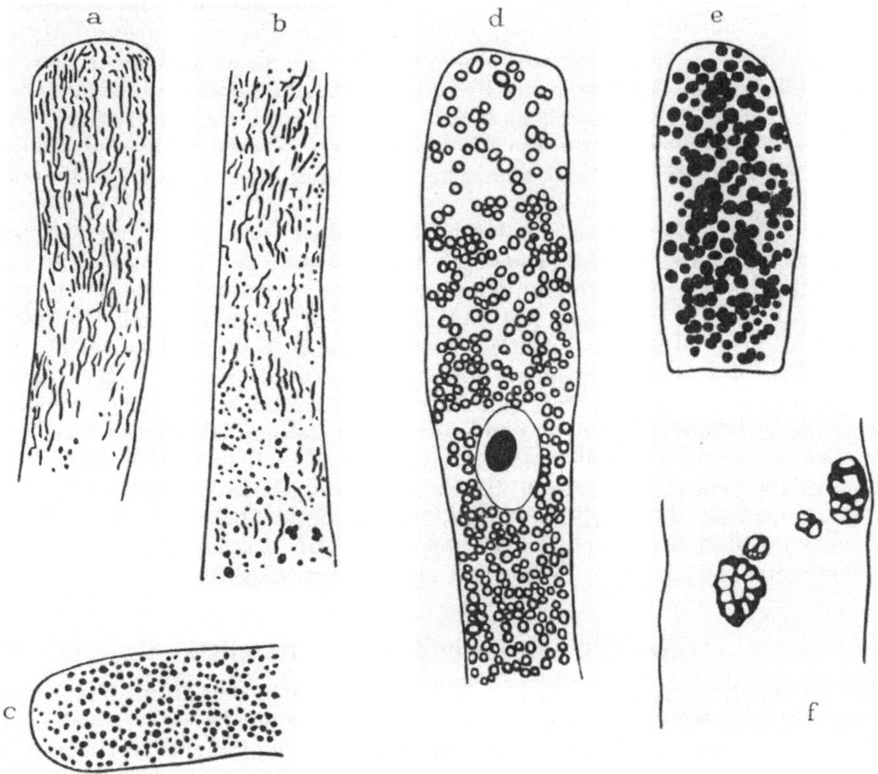

Abb. 19a—f. Aufeinanderfolgende Stadien der Fettbildung in Chondriosomen bei *Basidiobolus ranarum*. a—c und f: fixierte und mit Eisenhämolyxylin gefärbte Präparate, e: Fettfärbung. (Umgezeichnet nach H. TARWIDOWA 1938.)

Ansicht, daß das Fett in den Pilzzellen ohne Beziehung zum Chondriom entsteht. Das Problem wurde von H. TARWIDOWA (1938) neuerdings aufgegriffen. Ihre Untersuchungen an *Basidiobolus ranarum* haben zu sehr klaren Ergebnissen geführt, die in Abb. 19 veranschaulicht werden. Wurden Pilzhyphen verschiedenen Alters untersucht, so fanden sich in den jüngsten Zellpartien zahlreiche fädige Chondriokonten und wenige winzige granuläre Gebilde, die zunächst noch keine Fettreaktion gaben (Abb. 19c). Sehr bald kehrt sich das Mengenverhältnis zwischen den beiden Partikelarten um, die fädigen Formen nehmen ab, die granulären nehmen (durch Zerfall jener ?) zu (Abb. 19b). Die Granula schwellen an, sie werden stärker lichtbrechend (Abb. 19c) und färben sich als Ganzes mit Fettfarbstoffen (Abb. 19e). Durch geeignete Tinktionen läßt sich aber zeigen, daß das Fett von einer plasmatischen Hülle umgeben ist (Abb. 19d). Im Endstadium der Fettbildung finden sich in den Hyphen größere Fettklumpen, die aber bei Färbung noch ein plasmatisches Netzwerk im Inneren erkennen lassen, welches eine Entstehung dieser Gebilde durch Vereinigung mehrerer umhäuteter Fetttropfen vermuten läßt (Abb. 19f). Bis zum eigentlichen Auftreten des Fettes zeigen die granulären Gebilde die bekannten Chondriosomenfärbungen mit Janusgrün und mit reduziertem Methylenblau usw. Die Verfasserin schließt mit guten Gründen, daß bei *Basidiobolus* das Fett in den Chondriosomen gebildet wird.

Fast zu den gleichen Ergebnissen kam kürzlich H. HEINEMANN (1956) (s. auch M. STEINER und H. HEINEMANN 1954a, b), als sie typische Fettpilze, insbesondere *Oospora lactis*, untersuchte. Die Fettbildung geht von stark lichtbrechenden granulären Gebilden aus, welche positive Nadireaktion geben. Sie nehmen an Größe zu und verschmelzen schließlich zu den großen Fetttropfen, die für stark verfettete Zellen kennzeichnend sind (Abb. 20a—d). Wird bei Fetttropfen mittlerer Größe die Nadireaktion durchgeführt, so sieht man, daß sie sich auf ein etwa 1 μ großes kugeliges Gebilde beschränkt, welches dem Fetttröpfchen seitlich anliegt (Abb. 20e). Die mehrfach genannten granulären Gebilde sind vermutlich identisch mit den „Grana mit Mitochondrienfunktion" der Hefezellen, über welche H. MARQUARDT, E. BAUTZ und Mitarbeiter in mehreren Arbeiten berichtet haben (E. BAUTZ und H. MARQUARDT 1953a, b, 1954, H. MARQUARDT und E. BAUTZ 1954, E. BAUTZ 1954a, b, 1955, E. BAUTZ und U. HAGEN 1954). Diesen fettbildenden granulären Mitochondrien gehen übrigens auch bei *Oospora* fädige Chondriokonten vorauf (Abb. 20f). Diese scheinen für die eiweißbildende Phase der Zelle ebenso kennzeichnend zu sein wie jene für die fettbildende. Das läßt sich besonders schön zeigen, wenn man ihre zeitliche

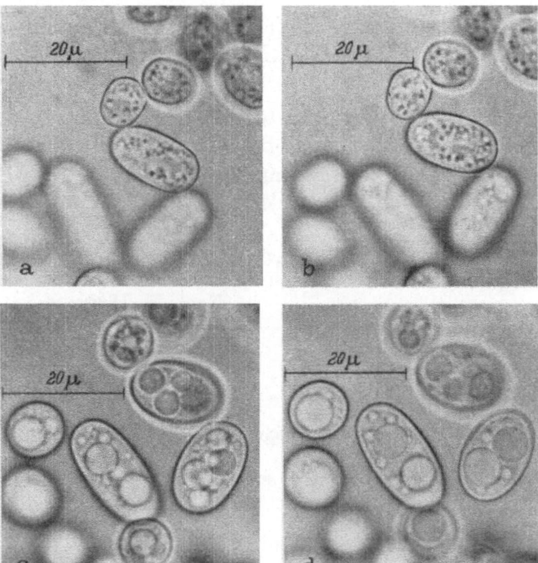

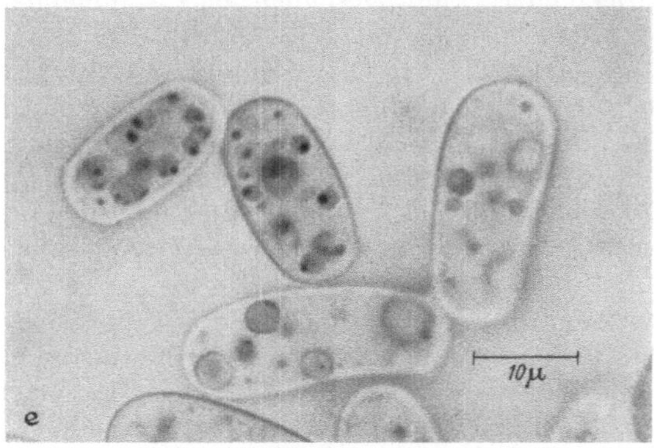

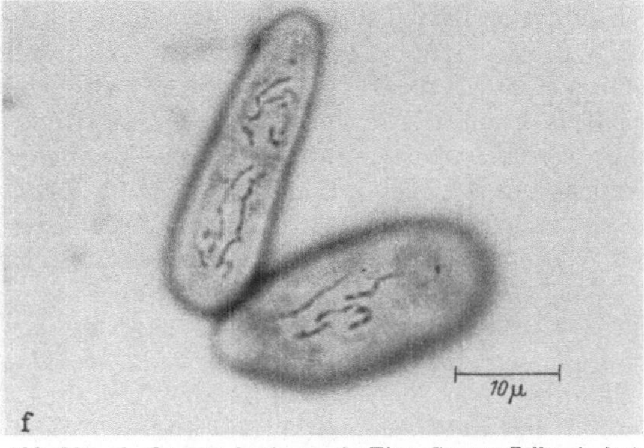

Abb. 20a—f. *Oospora lactis*. a—d: Eine Gruppe Zellen in fortschreitenden Stadien der Fettbildung. Hängetröpfchenkultur auf Malzzuckerlösung. Alter der Kultur: a 36 Std, b 45 Std, c 111 Std, d 137 Std. Ungefärbte Lebendpräparate. e Vitalfärbung mit Nadi. An den Fetttropfen sitzen blaugefärbte Granula. f Eiweißbildende Zellen mit fädigen Chondriosomen. Vitalfärbung mit Janusgrün B. (Nach H. HEINEMANN 1956.)

Aufeinanderfolge in Zellen verfolgt, bei denen durch entsprechende Variation des N:C-Verhältnisses die „Neigung“ zur Eiweiß- bzw. Fettbildung überwiegt (Tabelle 31).

Man sieht deutlich, daß bei Kultur A, in der das N:C-Verhältnis zugunsten der Fettbildung verschoben ist, sehr früh die granulären Partikel in Erscheinung treten, während im umgekehrten Falle C, wo es während der Versuchsdauer überhaupt nicht zur richtigen Verfettung kommt, lange Zeit die filären Formen den Zellaspekt bestimmen. Auch in der Arbeit von H. HEINEMANN finden sich übrigens sehr gute Indizien dafür, daß die fädigen und granulären Mitochondrienformen entwicklungsgeschichtlich zusammenhängen.

Tabelle 31. *Zeitliche Aufeinanderfolge von fädigen Chondriosomen, Chondriosomengranula und typischen Fetttröpfchen in Zellen von Oospora lactis bei verschiedener Ernährung.* (Nach H. HEINEMANN 1956.)

<table>
<tr><th rowspan="2"></th><th colspan="9">Kulturdauer Stunden</th></tr>
<tr><th>10</th><th>15</th><th>20</th><th>25</th><th>30</th><th>35</th><th>40</th><th>45</th><th>50</th></tr>
<tr><td>A 2% Malzlösung + 5% Saccharose . .</td><td colspan="9">Fäden →← Granula →← Fetttropfen ——</td></tr>
<tr><td>B 2% Malzlösung . . . .</td><td colspan="9">— Fäden →← Granula →← Fetttropfen</td></tr>
<tr><td>C 2% Malzlösung + 2% Asparagin . .</td><td colspan="9">—— Fäden —— →← Granula ——</td></tr>
</table>

Das gleichlautende Ergebnis der an verschiedenen Pilzen durchgeführten Untersuchungen von TARWIDOWA und von HEINEMANN spricht also wenigstens bei den speziellen Objekten für die Entstehung der Fetttropfen in Chondriosomen. Die grundsätzliche Bedeutung der Frage läßt Untersuchungen bei andersartigen Objekten, z. B. bei mycelbildenden Schimmelpilzen, sehr erwünscht erscheinen[1]. Offen bleibt zunächst, ob die Mitochondriengranula tatsächlich den Ort der Fett*synthese* oder der ersten Fett*ausscheidung* aus dem Protoplasma darstellen. Es spricht vieles für die erste Deutung.

Es muß weiters darauf hingewiesen werden, daß die Fette, von denen soeben die Rede war, ohne Zweifel Reservefette sind, die also dem „variablen Element“ zuzurechnen sind. Über Entstehungsart, Verteilung und chemische Bindung des „konstanten Elements“ wissen wir sehr wenig. H. WALTER (1920, 1921) konnte zeigen, daß das Plasmaeiweiß von *Fuligo varians* und von *Boletus granulatus* erst dann der Trypsinverdauung zugänglich ist, wenn eine Extraktion der Lipoide mit Äthanol, Äther oder Chloroform vorausgeht, wie es W. BIEDERMANN (z.B. 1920) bereits für das Plasma höherer Pflanzenzellen gezeigt hat. Das Protoplasma der Hefe kann allerdings nach WALTER direkt durch Pepsin und Trypsin verdaut werden, wobei eine scheinbare Vermehrung der Fetttropfen in der Zelle festzustellen ist (Lipophanerose). Beide Beobachtungen sprechen in ihrer Art für das Vorliegen lipoproteidartiger Symplexe. W. W. LEPESCHKIN (z.B. 1928) hält Verbindungen zwischen Eiweißen und Lipoiden („Vitaide“) geradezu für das Charakteristikum der lebenden Substanz, deren exothermer Zerfall den Eintritt des Zelltodes kennzeichnet. M. A. NYMANN und E. CHARGAFF (1949) konnten aus homogenisierten Hefezellen mit 0,4—0,7% eine Lipoproteinfraktion isolieren, welche zu etwa 30% aus Lipoiden besteht, die acetonlösliche und -unlösliche Anteile neben Unverseifbarem (unter anderem Ergosterin) enthalten. Die Analyse der P-Fraktionen (Lipoid-P, Ribo- und Desoxyribonucleinsäure-P,

[1] Auch in den Hyphen von *Mucor mucedo* konnte jüngst festgestellt werden, daß den in Neubildung begriffenen Fetttröpfchen plasmatische Granula mit positiver Nadireaktion seitlich anliegen (M. STEINER und H. HEINEMANN 1956, unveröffentlicht).

Phosphoprotein-P) deutet die sehr komplexe Zusammensetzung dieses Hefelipoproteins an.

Kurz erwähnt sei im vorliegenden Zusammenhang, daß R. C. THOMAS (1928, 1930) für das Vorhandensein von Kohlenhydrat-(Cellulose- oder Kallose-)Fettsäurenkomplexen in den Hyphenmembranen von *Sclerotinia*- und *Fusarium*-arten eintritt.

5. Der chemische Weg der Fettbildung.

Im Beitrag von W. FRANKE und H. FREHSE (S. 137ff., insbesondere S. 174ff. dieses Handbuchbandes) wird im Zusammenhang mit der β-Oxydation der Fettsäuren die heute gültige Vorstellung über die Synthese der Fettsäuren ausführlich dargelegt und an Hand der einschlägigen Literatur abgeleitet und begründet. Eine Wiederholung des dort Vorgetragenen erübrigt sich. Hingegen erscheint es nützlich, in einem kurzen historischen Rückblick die wichtigsten Vorstellungen über die biologischen Fettsynthesen zu erwähnen.

I. Die Theorie von E. FISCHER.

Am 2. August 1894 hielt E. FISCHER (s. E. FISCHER 1909) eine Rede zur Feier des Stiftungstages der Berliner militärärztlichen Bildungsanstalten, in der er über seine klassischen Forschungen auf dem Gebiete der Kohlenhydrate berichtete und in einem Exkurs auch auf den Weg der, damals als Tatsache bereits bekannten, Umwandlung von Kohlenhydraten in Fette einging.

Er entwickelte die Theorie, daß nach dem Schema

$$3 \times C_6 \rightarrow C_{18}$$

drei Zuckermoleküle durch Aldehydgruppen verkoppelt werden, „wie es dem Formaldehyd bei der Zuckersynthese ergeht“. Durch „Verschiebung und Wegnahme von O“ kommt es zur C_{18}-Kette der Stearin- und Ölsäure. Für die Entstehung der C_{16}-Säuren (Palmitinsäure) wird der Weg über die Vereinigung von zwei Pentosen und einer Hexose vorgeschlagen:

$$C_6 + 2 \times C_5 \rightarrow C_{16}.$$

FISCHER erkannte klar, daß hier eine Möglichkeit für die experimentelle Überprüfung seiner Hypothese vorlag. „Gerade an diesem Punkt kann aber das physiologische Experiment ansetzen. Wenn das Zuckermolekül bei der Fettbildung nicht völlig zertrümmert wird, was ich für unwahrscheinlich halte, so darf man erwarten, daß die Zucker mit verschiedenem Kohlenstoffgehalt auch anders zusammengesetzte Fette liefern werden. Mag der Versuch auslaufen wie er will, einen Rückschluß auf die natürliche Fettbildung wird er jedenfalls gestatten.“

Dieser Versuch wurde mit Mikroorganismen unseres Wissens nur einmal unternommen. H. H. BARBER (1929) kultivierte ein *Penicillium sp.* (von Maisstärkekleister isoliert) unter sonst gleichen Bedingungen auf folgenden, etwa gleich gut verwertbaren C-Quellen: Glucose, Saccharose, Xylose, Glycerin. Als Pentose wurde Xylose gewählt, weil sie am α-, β- und γ-C-Atom die gleiche Konfiguration besitzt wie Glucose. In der Pilzernte wurden das Neutralfett und die freien Fettsäuren bestimmt und die Fettsäuren beider Fraktionen durch Schmelzpunkt, mittleres Molekulargewicht und Jodzahl gekennzeichnet. Eine Zusammenfassung der Ergebnisse bringt Tabelle 32. Das Ergebnis war also negativ im Sinne der FISCHERschen Hypothese. Das Fett des Pilzes ist gleich zusammengesetzt,

Tabelle 32. *Eigenschaften der Fettsäuren von Penicillium sp. bei Kultur auf verschiedenen C-Quellen.* (Nach H. H. BARBER 1929.)

C-Quelle	Fp.	Mittleres Molekulargewicht	Gesättigte Fettsäuren (% der gesamten Fettsäuren)
Glucose, gesamte Säuren	55°	264	27
Saccharose, gesamte Säuren . . .	54°	271	29
Xylose, gesamte Säuren	54°	293	21
Glycerin, gesamte Säuren	55°	275	16
Glycerin, freie Säuren	55°	281	22

welche C-Quelle immer benützt wird; „es enthält Palmitin-, Stearin-, Öl-, α- und β-Linolsäure, frei und als Glyceride und Steride".

Die Annahme einer *direkten* Synthese von Fetten aus Zuckermolekülen lebte nur noch zweimal kurz auf, ohne daß die Richtung der experimentellen Forschung davon merklich beeinflußt wurde.

I. SMEDLEY-MACLEAN und D. HOFFERT (1926) glaubten, einige Befunde bei ihren Versuchen über Fettsynthese durch belüftete, ruhende Hefezellen aus Äthanol, Acetat, Pyruvat usw. (s. S. 222f.) so deuten zu sollen, daß der Weg zur Fettsynthese aus C_2-Körpern über eine Resynthese von Zucker läuft, nach dem Schema

Acetaldehyd → Hexose

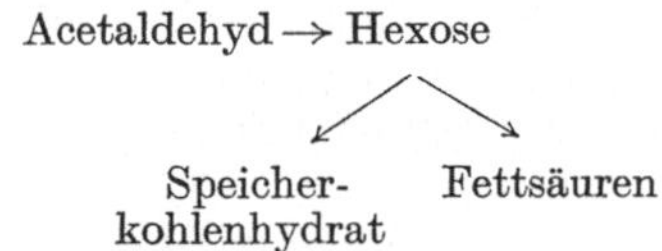

Speicher-kohlenhydrat   Fettsäuren

E. F. ARMSTRONG und J. ALLAN (1924) diskutierten — rein theoretisch — in einem Sammelreferat die Möglichkeit, daß bei der Fettbildung die „C_{18}-Einheit der Stärke" eine Rolle spielt, die über "rearrangement, oxidation, reduction, rupture of oxygen bridge, joining carbon to carbon" direkt zu Stearinsäure führen soll. Man wird sich erinnern, daß damals angesichts der SCHARDINGER-Dextrine der Aufbau der Stärke aus $3 \times C_6$- und $4 \times C_6$-Einheiten lebhaft diskutiert wurde.

II. Die Theorie von A. MAGNUS-LEVY.

A. MAGNUS-LEVY (1902) entwickelte, gleichfalls in einem Vortrag, zum ersten Male die Auffassung („zunächst nur Hypothese"), daß die Fettsäuren durch Aneinanderreihen von C_2-Körpern gebildet werden. Er stützte sich dabei auf die Tatsache, daß die natürlichen Fettsäuren fast ausnahmslos eine paarige Zahl von C-Atomen besitzen, andererseits aber auf eigene experimentelle Befunde, wonach bei der Autolyse der Leber aus Milchsäure beträchtliche Mengen Buttersäure gebildet werden, wofür er folgenden Weg vorschlug:

$$C_3 \rightarrow C_2 + C_1; \quad 2 \times C_2 \rightarrow C_4.$$

Dabei konnte er sich auf ältere Vorstellungen von M. NENCKI (1878) und von F. HOPPE-SEYLER (1878, 1879) berufen, die gleichfalls eine Entstehung der Buttersäure bei der Gärung und der Kalischmelze von Milchsäure durch Vereinigung von 2 C_2-Körpern postulierten.

Auch H. EULER (1909) nahm die Entstehung der Fettsäuren aus C_2-Bruchstücken des Zuckermoleküls an. Er dachte an wiederholte Aldolkondensation des Acetaldehyds über Crotonaldehyd (C_4) zu Sorbinaldehyd (C_6), von dem 3 Moleküle zum C_{18}-Skelet der Stearinsäure zusammentreten könnten.

H. HAEHN und W. KINTTOF (1925) knüpfen in ihren Vorstellungen über den Weg der Fettbildung an das NEUBERGsche Gärungsschema an. Sie lassen die Fettsäurebildung vom Acetaldehyd und dem Methylglyoxal ausgehen:

$CH_3CHO + CH_3CHO$ Acetaldehyd	$= CH_3CHOH \cdot CH_2CHO$ Aldol
$CH_3CHOH \cdot CH_2CHO - H_2O$	$= CH_3CH = CHCHO$ Crotonaldehyd
$CH_3CH = CH \cdot CHO$ + $CH_3CO \cdot CHO$ (H_2 / O)	$CH_3CH_2CH_2CHO$ = Butyraldehyd $CH_3CO \cdot COOH$ Brenztraubensäure ↙ ↘
Methylglyoxal	$= CH_3CHO$ CO_2
$CH_3CH_2CH_2CHO + CH_3CHO$	$= CH_3CH_2CH_2CHOHCH_2CHO$ β-oxy-Capronaldehyd
$CH_3CH_2CHOHCH_2CHO - H_2O$	$= CH_3CH_2CH_2CH = CH \cdot CHO$ α,β-Hexylenaldehyd
$CH_3CH_2CH_2CH = CH \cdot CHO$ + CH_3COCHO (H_2 / O)	$CH_3CH_2CH_2CH_2CH_2CHO$ = Capronaldehyd CH_3COCHO

und weiter: $C_6 + C_2 \ldots C_8$
$C_8 + C_2 \ldots C_{10}$
bis C_{18} (Ölsäure, Stearinsäure)

Zur Prüfung dieses Schemas führten die Verfasser bei *Endomycopsis vernalis* Fütterungsversuche mit einigen der vermuteten Zwischenprodukte durch. Sie fanden, daß Acetaldehyd, Äthanol, Aldol, Brenztraubensäure und Milchsäure zur Fettsynthese verwandt werden (Weiteres hierüber S. 224).

L. REICHEL und O. SCHMID (1939) haben anknüpfend an die vorige Arbeit beim gleichen Organismus die Verarbeitung höherer Aldehyde geprüft. Octyl- und Decylaldehyd wurden nur zu den entsprechenden Säuren oxydiert. Dagegen wurde aus Hexadienal eine Säure etwa vom Äquivalent der Ölsäure (C_{18}), aus Octotrienal (und aus Crotonaldehyd und Acetaldol) eine Säure von $\sim C_{16}$ gefunden. Die Ausbeuten der Säuren sind klein, 2—20 mg je Kultur, Bilanzen werden nicht angegeben. Die Versuche bedürften zweifellos der Bestätigung. Sie lassen sich schwer in unsere heutigen Vorstellungen über die Fettsynthese einordnen.

Einen wesentlichen Schritt nach vorwärts bedeuteten Isotopenversuche zur Fettsynthese aus C_2-Körpern (Versuche mit Deuterium-indizierter Essigsäure bzw. H_2O: R. SONDERHOFF und H. THOMAS 1937, R. SCHOENHEIMER und D. RITTENBERG 1937, K. BERNHARD und H. ALBRECHT 1948, R. C. OTTKE, S. SIMMONDS und E. L. TATUM 1950; mit C^{13} bzw. C^{14}-Markierung: K. BLOCH und D. RITTENBERG 1944, D. RITTENBERG und K. BLOCH 1944, R. C. OTTKE, E. L. TATUM, I. ZABIN und K. BLOCH 1952).

Wichtige Beiträge zur Aufklärung des Weges des Fettabbaues und -aufbaues brachte das Studium des von H. A. BARKER und A. M. TAHA (1942) entdeckten *Clostridium Kluyveri*, welches zur Synthese von C_4—C_8-Fettsäuren aus Äthanol und Essigsäure befähigt ist, Kohlenhydrate aber überhaupt nicht zu verarbeiten vermag. E. R. STADTMAN und A. H. BARKER (1949a—e, E. R. und T. E. STADTMAN und A. H. BARKER (1949, 1950) gelang es, mit einem Enzympräparat aus diesem Bacterium eine zellfreie reversible Oxydation von Fettsäuren durchzuführen. Damit kann diese Übersicht schließen, da der eben gegebene Hinweis den unmittelbaren Anschluß an die Darstellungen von W. FRANKE und H. FREHSE (S. 174ff. dieses Bandes) und von P. K. STUMPF (S. 201ff.) herstellt.

6. Fette als Nährsubstrat für Mikroorganismen.

In anderen Beiträgen dieses Handbuchbandes werden die Lipasen (E. BAMANN und E. ULLMANN, S. 109) und die Enzyme des Fettsäurenabbaues (W. FRANKE und H. FREHSE, S. 137) ausführlich behandelt. Die nachfolgende Darstellung darf sich deshalb auf einige ergänzende Hinweise ernährungsphysiologischen Inhaltes beschränken.

In seinen drei Arbeiten „Recheres sur la vie dans l'huile" hat PH. VAN TIEGHEM (1880, 1881a, b) wohl die ersten Angaben über pflanzliche Mikroorganismen gemacht, die Fette als Nährsubstrat ausnützen können. Er beobachtete eine mit Bierhefe nicht identische Hefe (*„Saccharomyces olei"*), die sich in Olivenöl submers kräftig entwickelt. Er konnte ferner durch Einbringen gequollener anorganischer und organischer Körper (Holz, Samen, Papier, Erde) die Entwicklung der im Öl vorhandenen Sporen verschiedener Schimmelpilze anregen. Die Mycelentwicklung erfolgte submers bis zur Sporenbildung. Das Wasser des Pilzkörpers ist Verbrennungswasser des Fettes. Auch andere pflanzliche und tierische Öle (Mandel-, Erdnuß-, Hanf-, Mohnöl, Lebertran) geben das gleiche Resultat, nicht aber durch H_2SO_4 gereinigtes Rapsöl, gekochtes Leinöl oder ausgeschmolzener Hammeltalg. Letztere enthalten nicht mehr die für das Pilzwachstum notwendigen N-Quellen. An der Ausscheidung von Fettsäurekriställchen wurde erkannt, daß die Neutralfette durch Schimmelpilze (*Penicillium* und andere) verseift werden.

Eine Reihe von Arbeiten der Jahrhundertwende und späterer Jahre ging zumeist von *praktischen Gesichtspunkten* aus. Sie untersuchten die Qualitätsminderung, denen fetthaltige Naturprodukte beim Befall durch Mikroorganismen unterliegen. So stellten z. B. H. RITTHAUSEN und BAUMANN (1896) fest, daß in gepulverten Preßkuchen von Rübsen (mit einem Wassergehalt von 12,4%) der Fettgehalt von 9,5% auf 1,9% abfiel, wenn sie 2 Jahre lang in verschlossenen Flaschen, stark von Schimmelpilzen befallen, aufbewahrt wurden. O. KLEIN (1900) zeigte, daß Olivenpreßlinge (Bagassen) sehr leicht verpilzen und daß dabei, je nach Wassergehalt, eine starke Fettspaltung und -verminderung eintritt. In festgestampftem Preßkuchen mit 30% Wasser sank in 4 Monaten der Fettgehalt von 17,5 auf 9,1% des Trockengewichtes, der Anteil freier Fettsäuren stieg gleichzeitig von 1,2 auf 73% an.

Daß überhaupt die Massenentwicklung und die Stoffwechselaktivität der fettverzehrenden Mikroorganismen durch einen entsprechenden Wassergehalt sehr gefördert wird, haben vor allem die Untersuchungen von J. KÖNIG, A. SPIECKERMANN und W. BREMER (1901) gezeigt. Das Optimum der Fettspaltung in Baumwollsaatmehl lag bei einem Wassergehalt von 25%. Es konnte weiters festgestellt werden, daß in der natürlichen Mikroflora bei geringem H_2O-Gehalt (14—20%) vor allem Schimmelpilze (*Eurotium*-Arten), bei mittlerem H_2O-Gehalt (25—30%) *Oidium*-Formen, bei 33% H_2O und darüber dagegen *Bacillus*-Arten überwiegen. Die Schimmelpilze bevorzugten auf sterilem Baumwollsaatmehl deutlich die Fette, die Bakterien mehr die N-haltige Substanz. 13 Schimmelpilze vermochten als Reinkultur mit Baumwollsaatöl als einziger C-Quelle (neben Pepton) zu wachsen, nicht aber *Monilia sp. div.*, *Saccharomyceten* und *Mycoderma*arten. Von 2,5 g Öl [$(NH_4)_2HPO_4$ als N-Quelle] wurden nach 17tägiger Kultur bei *Penicillium glaucum* 0,3 g, bei *Aspergillus flavus* 0,65 g als Rest wiedergefunden. Über ähnliche Versuchsergebnisse berichten auch E. HASELHOFF und F. MACH (1906), die den Fettschwund in verschimmeltem Reismehl bestimmten. Aus dem Ausgangsmaterial wurden *Penicillium glaucum* und *Aspergillus oryzae* isoliert und als kräftige Fettverbraucher erkannt. Aus sterilisierten natürlichen pflanzlichen Futtermitteln (Kleie, Futtermehle, Preßkuchenmehle, Biertreber) verschwand nach Beimpfung durch *Aspergillus oryzae* in einem Monat zwischen 75—95% der ätherlöslichen Substanz. Auch S. A. WAKSMAN (1931) untersuchte den Abbau der Fette und anderer Bestandteile pflanzlicher Materialien (Stroh, Laub u. dgl.) durch Rein- und Mischkulturen von Mikroorganismen unter verschiedenen Bedingungen. Actinomyceten wurden unter anderem als wirksame Fettzerstörer festgestellt. Auch L. B. JENSEN (1937) hebt hervor, daß *reine*, wasserfreie Fette keinen geeigneten Nährboden für Mikroorganismen darstellen, daß aber (L. B. JENSEN und D. P. GRETTER 1937) schon 0,3% Feuchtigkeit in Tierfetten eine gewisse Bakterienentwicklung möglich macht. Daß Fette und Fettsäuren, die im Boden verteilt werden, mikrobiell rasch abgebaut werden, hat zuerst M. RUBNER (1900) gezeigt. Unter Umständen kann Mikroorganismenbefall freilich sogar zu einer Zunahme des Fettgehaltes führen. Das zeigte z. B. A. HEBEBRAND (1893) bei Brot, das durch *Penicillium glaucum* und *Mucor mucedo* befallen war. Hier hatten offenbar die Pilze aus den Kohlenhydraten des Substrates Fett gebildet.

Besonders intensiv wurden schon früh die durch Mikroorganismen verursachten *Veränderungen der Butter* studiert (F. LAFAR 1891, J. HANUŠ, 1900, J. HANUŠ und A. STOCKÝ 1900, R. REINMANN 1900, O. LAXA 1902, O. JENSEN 1902). Wegen ihres Gehaltes an Proteinen, an Wasser und Nährsalzen ist die Butter ein ausgezeichneter Nährboden für Mikroorganismen, Schimmelpilze (*Mucor sp. sp.*, *Verticillium glaucum*, *Penicillium glaucum*, *Aspergillus niger*, *Oospora lactis*) und Bakterien (insbesondere *Pseudomonas fluorescens*, *Escherichia coli*, *Serratia marcescens*, *Achromobacter*). 1 g gewöhnlicher Marktbutter enthält nach LAFAR 2—47 Millionen Keime. *Pseudomonas fluorescens*, regelmäßig in der Butter vorhanden, spaltet das Butterfett rasch; das gleiche gilt für *Oospora lactis*, *Monilia* und gewisse Schimmelpilze. Gerade bei der Butter, deren Glyceride viel niedermolekulare Fettsäuren (Buttersäure und andere) enthalten, führt, im Gegensatz zu den meisten anderen pflanzlichen und tierischen Fetten, eine partielle Verseifung bereits zu erheblicher Qualitätsminderung. Daneben spielen auch nichtmikrobielle Vorgänge (Photooxydation) beim Zustandekommen der Ranzigkeit eine Rolle, wie Versuche mit völlig keimfreier Butter bewiesen haben.

Die schon in den erwähnten Arbeiten begonnene *physiologische Analyse* wurde dann von zahlreichen Forschern fortgesetzt und ausgebaut, wobei man fast immer mit Reinkulturen und mit synthetischen Substraten wohldefinierter Zusammensetzung experimentierte. Es seien hier, ohne Anspruch auf Vollständigkeit, erwähnt: R. H. SCHMIDT (1891), K. SCHREIBER (1902), O. RAHN (1906), A. ROUSSY (1909, 1911), O. FÜRTH und C. SCHWARZ (1909), A. SPIECKERMANN (1912, 1914), O. FLIEG (1922), H. KORDES (1923), E. F. TERROINE, R. BONNET und P. DUQUÉNOIS (1927), E. F. TERROINE, R. BONNET, G. KOPP und J. VÉCHOT (1927), F. E. HAAG (1928), K. KÜSSNER (1928), L. M. HOROWITZ-WLASSOWA und M. J. LIVSCHITZ (1935), L. B. JENSEN (1937), H. J. PEPPLER (1941), C. H. CASTELL und Mitarbeiter (1940, 1941), J. O. MUNDT und F. W. FABIAN (1944), A. M. STERN, Z. J. ORDAL, H. O. HALVORSON (1954).

Im folgenden wird versucht, einige Ergebnisse kurz wiederzugeben.

1. Für den raschen und wirksamen Zugriff der Mikroorganismen ist es wesentlich, daß die *Fette in feiner Verteilung* vorliegen. In Flüssigkeitskulturen darf das Fett nicht die Oberfläche des wäßrigen Nährmediums in geschlossener Schicht überziehen. Es müssen zumindest einzelne, getrennte Fettinseln vorhanden sein. Gut bewährt hat sich der Zusatz eines Emulgators (z.B. Gummi arabicum, K. SCHREIBER 1902). Gelnährböden werden am besten so hergestellt, daß die durch Erwärmen verflüssigte Agarlösung mit dem Fett geschüttelt wird, so daß

dieses nach dem Erstarren in feinen Tröpfchen vorliegt. A. SPIECKERMANN (1912, 1914) hat für seine Kulturen mit gutem Erfolg Kieselgur verwendet, die er mit den Fetten oder Fettsäuren vermischte und mit einer Lösung der übrigen Nährstoffe versetzte.

Nach A. ROUSSY (1909, 1911) zeigt das Wachstum von Schimmelpilzen *(Rhizopus, Phycomyces, Aspergillus niger)* in Abhängigkeit vom Fettgehalt des Agars ausgesprochene Optimumkurven: Minimum bei etwa 2%, Optimum bei 6—10%, Maximum bei etwa 40%. Der Relativanteil der Lipo- und Hydrophase dürfte hierbei wohl die ausschlaggebende Rolle spielen.

2. Es gibt zahlreiche Bakterien und Pilze, die auf Nährböden mit Fetten oder Fettsäuren als *einziger C-Quelle* wachsen können.

Unter den Schimmelpilzen seien z.B. Arten von *Mucor (M. mucedo, M. racemosus), Phycomyces nitens, Rhizopus nigricans, Aspergillus niger, Penicillium (P. „glaucum", P. luteum, P. javanicum), Oospora* genannt; unter den Bakterien Arten von *Pseudomonas (Ps. fluorescens, Ps. aeruginosa), Serratia marcescens, Alcaligines viscosus* und *A. lipolyticus*, Arten von *Achromobacter, Sarcina, Micrococcus, Staphylococcus* (z.B. *S. aurens*), einige *Bacillen* der *Subtilis*-Gruppe.

Neben vielen aeroben gibt es auch *anaerobe* fettspaltende Bakterien, z.B. Desulfurikanten (vgl. W. D. ROSENFELD 1946, hier auch weitere Literatur).

Nur wenige Hefen scheinen auf Fettnährböden zu gedeihen.

Alle fettspaltenden und fettverzehrenden Mikroorganismen vermögen anscheinend auch auf den „allgemeinen" Nährböden zu wachsen (Pepton, Zucker usw.)[1]. Der Vorschlag von E. DE KRUYFF (1906), die Fettspaltung und -oxydation als gattungsbegründendes Merkmal („Lipobacter") zu betrachten, hat mit Recht keinen Anklang gefunden.

3. Der erste Angriff der Mikroorganismen auf Fette des Substrates besteht fast immer in der *Lipolyse*, d.h. in der Spaltung der Neutralfette durch Lipasen.

Die Fett*spaltung* kann quantitativ durch Ermittlung der Säurezahl nachgewiesen werden. Auf Agarplatten läßt sie sich durch die Nilblausulfatprobe (Neutralfette rosa, Fettsäuren blau) oder durch die $CuSO_4$-Reaktion (Bildung blauer Kupferseifen mit freien Fettsäuren) feststellen. Bei mikroskopischer Betrachtung wurden als Produkte der Lipolyse in der Regel Kristalle von Fettsäuren oder Seifen festgestellt (R. H. SCHMIDT 1891, O. FLIEG 1922, A. SPIECKERMANN 1912, 1914, insbesondere C. H. CASTELL und E. H. GARRARD 1941).

Die ersten Angaben über Pilzlipasen (*Penicillium glaucum* und *Aspergillus niger*) stammen von L. CAMUS (1897a, b).

Fettverarbeitende Bakterien, bei welchen keine Lipasebildung nachweisbar ist, scheinen relativ selten zu sein. N. L. SÖHNGEN (1913) gibt dieses Verhalten für einige *Mycobacterien*arten *(Mycobacterium phlei, M. lacticola, M. rubrum* und andere) an, L. M. HOROWITZ-WLASSOWA und M. J. LIVSCHITZ (1935) für *Aerobacter aerogenes* und andere. C. H. CASTELL und E. H. GARRARD (1941) prüften 40 Bakterienarten auf ihr Vermögen zur Fettspaltung (Triolein) und zur Fettsäureoxydation. Meist gehen beide Vorgänge parallel. Einige Arten gaben aber die Reaktionen für Fettsäureoxydation (KREIS-Test, SCHIFF-Test, Oxydasereaktion), ohne nachweisbar lipolytisch zu sein (z.B. *Alcaligenes faecalis, Achromobacter putrefaciens, Erwinia carotovora, Rhizobium leguminosorum, Salmonella pullorum, Staphylococcus capsulatus, Escherichia coli*). Lipolyse war vor allem in den Gattungen *Pseudomonas, Phytomonas, Alcaligenes, Achromobacter* und bei *Staphylococcus aureus* sehr deutlich nachweisbar.

---

[1] *Pityrosporum ovale* scheint einer der wenigen, *obligat* lipophilen Pilze zu sein (R. W. BENHAM 1939, 1940). (Über Mikroorganismen, deren Nährsubstrat der *Ergänzung* durch „essentielle Fettsäuren" bedarf, vgl. S. 263f.)

4. Das bei der *Fettspaltung auftretende Glycerin* wird meist sehr rasch verbraucht (R. H. SCHMIDT 1891, A. SPIECKERMANN 1912, A. ROUSSY 1911), obgleich es, z. B. für viele Schimmelpilze (A. ROUSSY 1911, H. KORDES 1923), keine besonders gute C-Quelle darstellt. Reine Fettsäuren werden bei Glycerinzugabe oft besser verarbeitet als ohne eine solche (O. FLIEG 1922, H. KORDES 1923).

5. Die *relative Eignung*, d.h. der raschere oder langsamere Verbrauch der verschiedenen *Fettsäuren* sollte vor allem aus der Änderung der Kennzahlen natürlicher Fette oder Fettsäurengemische unter der Einwirkung wachsender Mikroben feststellbar sein. Ganz allgemein wird die selbstverständliche Erhöhung der Säurezahl festgestellt, wenn Neutralfette als Substrat verwendet werden (Lipolyse). Im übrigen gehen die Angaben der Literatur sehr weit auseinander.

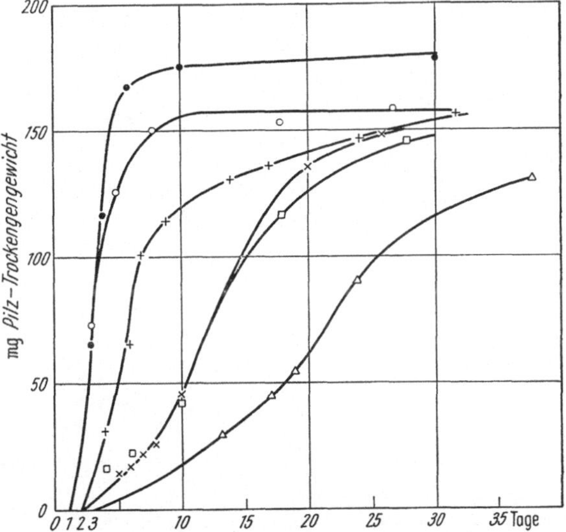

Abb. 21. Trockensubstanzbildung von *Aspergillus niger* auf Fettsäuregemischen verschiedenen Sättigungsgrades. ● Fettsäuren des Leinöls (Jodzahl 180), + des Lebertrans (Jodzahl 180), □ des Erdnußöls (Jodzahl 96), △ des Butterfettes (Jodzahl 33), ○ Ricinolsäure (Jodzahl 85), × Ölsäure (Jodzahl 90). (Nach E. F. TERROINE, R. BONNET, G. KOPP und J. VÉCHOT 1927.)

O. RAHN (1906 b) berichtet, daß Bakterien anscheinend die Fettsäuren ohne Auswahl verzehren, Schimmelpilze dagegen niedermolekulare Säuren bevorzugen, was sich in einer Verminderung der Verseifungs- bzw. Neutralisationszahl und der REICHERT-MEISSL-Zahl und eventuell in einer Verminderung der Jodzahl äußert. J. S. HEPBURN (1909, 1910) faßt die bis dahin vorliegenden Ergebnisse überhaupt im letzteren Sinne zusammen, P. BÉLIN (1926) fand bei Kultur von *Aspergillus niger* auf verschiedenen Fetten Verseifung, aber keine Auswahl unter den Fettsäuren, also keine Veränderungen der für diese kennzeichnenden Konstanten.

Auch L. M. HOROWITZ-WLASSOWA und M. J. LIVSCHITZ (1935) konnten nach einmonatigem Wachstum von *Pseudomonas aeruginosa* auf Sojaöl nur eine starke Zunahme der Säurezahl, sonst aber keine nennenswerten Veränderungen (Jodzahl, Verseifungszahl, Acetylzahl) feststellen.

Sehr eingehend hat sich mit der selektiven Aufnahme aus Fettsäurengemischen A. SPIECKERMANN (1912, 1914) in Versuchen mit *Penicillium glaucum* befaßt. Er stellte zunächst fest, daß bei der Verarbeitung im allgemeinen die Reihenfolge gilt: Na-, K-Salze $<$ NH_4-, Ca-Salze $<$ freie Fettsäuren, Glyceride. Löslichkeitsunterschiede und p_H-Verschiebungen dürften dabei eine Rolle spielen. Es wurden gesättigte und ungesättigte Säuren (Doppel- und Dreifachbindung), Keto- und Diketo-, Oxy- und Dioxysäuren untersucht. Im allgemeinen nimmt die Assimilierbarkeit mit steigender Zahl der C-Atome ab. Ungesättigte Säuren werden rascher verbraucht als gesättigte mit gleicher C-Zahl. So ergibt sich z. B. die Reihe Laurin-, Myristinsäure $>$ Öl-, Elaidinsäure $>$ Stearin-, Arachinsäure, die durch Versuche auf Nährböden mit Gemischen von zwei Säuren sehr schön belegt werden konnte.

E. F. TERROINE, R. BONNET, G. KOPP und J. VÉCHOT (1927) prüften die Trockensubstanz von *Aspergillus niger* auf Nährböden, die natürliche Fettsäurengemische mit bekannter Jodzahl als einzige C-Quelle enthielten. Wie Abb. 20 zeigt, verläuft die Zeit-Wachstumskurve um so steiler, je stärker ungesättigt die Fettsäuren des Substrates sind[1].

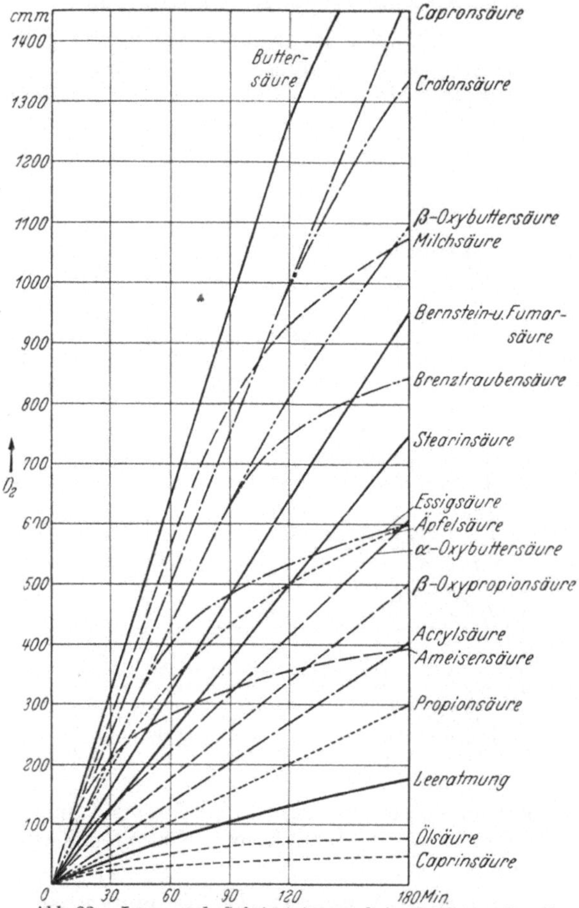

Wie wenig zulässig es ist, aus Ergebnissen bei einer Organismenart zu verallgemeinern, zeigen z.B. die umfangreichen Versuche von L. E. DEN DOOREN DE JONG (1926), der 258 verschiedene organische Substanzen, darunter 24 Fettsäuren (C_1—C_{18}) auf ihre Eignung als Nährstoff für ein Dutzend Bakterienarten prüfte. Auf Palmitin- und Stearinsäure wuchs kein einziger Stamm; *Serratia marcescens* wuchs auf gesättigten C_2- und C_6—C_{10}-Säuren, *Mycobacterium phlei* auf gesättigten C_2—C_{10}-Säuren, auf Öl- und (schwächer) auf Elaidinsäure, *Sarcina lutea* auf C_2—C_5-, Öl- und Elaidinsäure. Von 15 verschiedenen Stämmen der *Pseudomonas putrida* wuchsen alle auf C_3—C_5-, nur 4 auf Ameisensäure (C_1), nur einer auf Laurinsäure (C_{12}) usw.

Abb. 22. Leer- und Substratatmung bei *Alcaligenes faecalis* (= *B. alcaligenes*). (Nach W. FRANKE und W. PERIS 1938.)

E. F. TERROINE fand nur die natürlichen cis-Formen der Öl- und Erucasäure durch Bakterien verwertbar, nicht die trans-Formen (Elaidin- und Brassidinsäure). Für *Penicillium glaucum* erwies sich in den Versuchen A. SPIECKERMANNs (1914) Öl- und Elaidinsäure etwa gleichwertig, Brassidinsäure aber der Erucasäure unterlegen.

Die Sauerstoffaufnahme ruhender Bakterienzellen auf Fettsäurensubstraten haben im WARBURG-Apparat W. FRANKE und W. PERIS (1938), W. FRANKE und A. SCHILLINGER (1944) und W. FRANKE, L. LEE und I. KIBAT (1948) sehr eingehend studiert.

Es ergaben sich Einflüsse der Kettenlänge und des Sättigungsgrades (Abb. 22). Diese Ergebnisse stehen also zum Teil mit den Resultaten bei wachsenden Bakterien in Widerspruch. Besonders erwähnenswert ist die von den genannten Forschern vielfach festgestellte „Fettsäurelücke“; auf Fettsäuren mittlerer Kettenlänge (C_{10}—C_{12}) ist die Atmung nicht nur geringer als auf Säuren mit kleinerer

[1] F. E. HAAG (1926) glaubte aus seinen Versuchen mit paraffinzersetzenden Bakterien den Schluß ziehen zu sollen, daß Äthylenbindungen überhaupt eine notwendige Voraussetzung für die Verwertbarkeit eines Kohlenwasserstoffes oder einer Fettsäure sind.

oder größerer C-Zahl, sondern vielfach sogar unter den Wert der Leeratmung herabgedrückt (Abb. 23). Die Caprinsäurehemmung bei *Sarcina lutea* erwies sich im Auswaschungsversuch sogar als irreversibel. Wegen weiteren Einzelheiten und deren möglicher Deutung muß auf die Originalarbeiten verwiesen werden. (Hierzu auch H. J. PEPPLER 1941 und J. O. MUNDT und F. W. FABIAN 1944.)

6. Daß *Zwischenprodukte* des Fettsäureabbaus im Nährmedium von Pilz- und Bakterienkulturen auf Fetten und Fettsäuren in nennenswerten Mengen zumeist nicht gefunden werden, geht fast aus allen einschlägigen Angaben hervor. Der positive Ausfall der sehr empfindlichen KREIS- und SCHIFF-Teste auf „anoxydierte Fettsäuren", der sich bei Kultur fettoxydierender Bakterien alsbald einstellt (z. B. C. H. CASTELL 1940, 1941, C. H. CASTELL und E. H. GARRARD 1940 a, b, 1941 a, b), ist dazu kein Widerspruch.

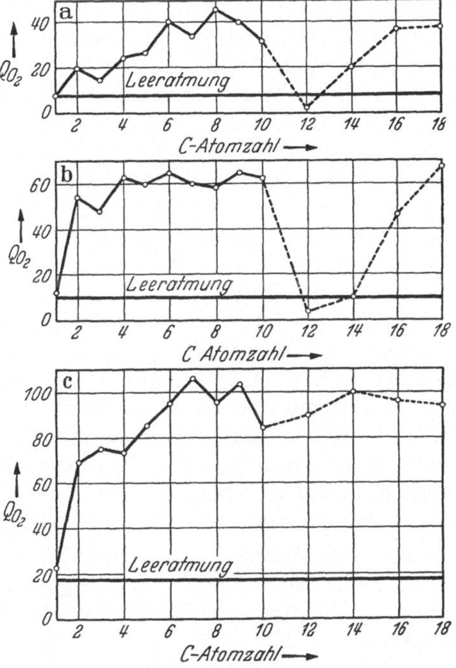

Abb. 23a u. b. Abhängigkeit der Atmungsgröße (Q_{O_2}) von der C-Atomzahl gesättigter Fettsäuren. a *Sarcina lutea*, b *Pseudomonas turcosa (B. turcosum)*, c *Pseudomonas aeruginosa (B. pyocyaneum)*. (Nach W. FRANKE und W. PERIS 1938.)

Was die Absättigung ungesättigter Säuren im Kulturmedium betrifft, so sind die Angaben uneinheitlich. Nach A. SPIECKERMANN (1914) sinkt die Jodzahl reiner Ölsäure in *Penicillium*kulturen von 88—89 auf 58—84, J. SCHÖNBRUNNER (1940) findet eine recht kräftige Oleatabsättigung in Bakterienkulturen (z. B. 32% durch *Pseudomonas fluorescens*, 15 Wochen). H. REYNOLDS und E. W. HOPKINS (1938) finden dagegen bei Bakterien- und Pilzkulturen die unverbrauchte Ölsäure mit unveränderter Jod- und Verseifungszahl wieder; nur *Clostridium acetobutylicum* reduzierte Ölsäure in geringem Ausmaße.

Die *Methylketonbildung* stellt den einzigen bisher genau studierten Fall einer wirklichen Anhäufung von „Intermediärprodukten" des Fettsäureabbaues in Pilzkulturen dar. Hierüber wird im Beitrag von W. FRANKE und H. FREHSE (S. 182 ff. dieses Bandes) ausführlicher berichtet, so daß dieser kurze Hinweis genügen dürfte.

Nach den modernen Vorstellungen „brennt nach einer Zündung die Fettsäurekette normalerweise bis zum Ende ab" (C. MARTIUS 1953), so daß eigentliche Zwischenprodukte gar nicht zu erwarten sind.

7. Zur Frage der *essentiellen Fettsäuren* bei Mikroorganismen sei vor allem auf die Untersuchungen von M. R. POLLOCK, G. H. HOWARD und B. W. BOUGHTON (1949 a, b) und von B. W. BOUGHTON und M. R. POLLOCK (1952) hingewiesen. Ein Bakterium aus der Diphtheriegruppe „Corynebacterium Q" benötigt zum Wachstum Ölsäure. Zwischen 0,5 und 10 $\mu\gamma$ sind Ölsäuregabe und Zellenzahl annähernd proportional.

Ölsäure ließ sich fast gleichwertig durch Elaidin-, Petroselin-, Linol-, Linolen- und cis-Palmitoleinsäure ersetzen. Auch cis- und trans-Vaccensäure, cis- und trans-Heptadec-9-ensäure, 9- und 12-Methylölsäure sind wirksam. Alle gesättigten Fettsäuren, Brassidinsäure, α-β-Ölsäure, cis- und trans-Myristolsäure, cis-Undec-9-ensäure, cis-Undec-10-ensäure, cis und trans-Octadec-4-ensäure, cis- und trans-Gadoleinsäure, 7-11-Dimethylölsäure, 5,7,13,17-Tetramethylölsäure erwiesen sich als inaktiv.

Recht verwickelt liegen die Verhältnisse bezüglich der bald fördernden, bald hemmenden Wirkung ungesättigter Fettsäuren bei der Gattung *Lactobacillus* und anderen Bakterien. B. L. HUTCHINGS und E. BOGGIANO (1947) fanden bei *Lactobacillus acidophilus* und *L. Delbrueckii* eine Förderung, bei *L. plantarum* und *L. Leichmannii* eine Hemmung durch Oleat. A. R. WHITEHILL, J. J. OLESON und Y. SUBBA ROW (1947) isolierten aus dem Rattenblinddarm einen Lactobacillus, dessen Ölsäure-(Na-Oleat-)bedarf nicht durch Acetat, Biotin, Riboflavin, Pantothensäure usw. befriedigt werden konnte. Auch nach V. R. WILLIAMS, H. P. BROQUIST und E. E. SNELL (1949) (s. auch V. R. WILLIAMS und E. A. FIEGER 1949) ist für *Lactobacillus bulgaricus, L. acidophilus, L. helveticus* und *Streptococcus lactis* Ölsäure notwendig. Sie kann aber mit noch besserem Erfolg durch Laurinsäure- und Ölsäuretween, durch Lecithin und sogar durch Saponin ersetzt werden. Diese Ergänzung des Nährbodens ist auch nur bei Abwesenheit von Biotin notwendig. Das letztere bestätigen G. M. SHULL, R. W. THOMA und W. H. PETERSEN (1949) für *Clostridium welchii*. Die bei Biotinmangel nötige Ölsäure kann durch Vaccen-, Linol- und Ricinolsäure oder durch Netzmittel (Tween 80) ersetzt werden. Nach E. KODICEK und A. N. WORDEN (1945) ist eine hemmende Wirkung von Öl-, Linol- und Linolensäure bei *L. helveticus* durch Lecithin, Cholesterin, Calciferol, aber auch durch $CaCl_2$ fast völlig aufzuheben. R. H. DEIBEL und C. F. NIVEN jr. (1955) finden, daß die für hämolytische *Streptococcus*-Stämme und *Lactobac. Leichmannii* notwendige Ölsäure durch einen nichtlipoiden Bestandteil des Hefeextraktes und durch CO_2(!) ersetzbar ist. Es ist wohl sicher, daß in allen diesen Fällen keine eigentliche Ernährungswirkung vorliegt, sondern eher, z.B. mit V. R. WILLIAMS und E. A. FIEGER (1949), an eine Permeabilitätsregulation gedacht werden muß.

Der Pilz *Pityrosporum ovale* braucht nach R. W. BENHAM (1939, 1940) Ölsäure (oder Butter oder Lanolin) neben Glucose; auch *Polyporus*- und *Fomes*-arten werden nach H. M. YUSEF (1953) durch Ölsäure im Wachstum stark gefördert.

8. Daß auch *Paraffinkohlenwasserstoffe* als alleinige C-Quelle für Mikroorganismen (Schimmelpilze und Bakterien) dienen können, hat zuerst O. RAHN (1906) bewiesen. N. L. SÖHNGEN (1913) hat gezeigt, daß die zum Paraffinabbau befähigten Bakterien (*Pseudomonas-, Micrococcus-, Mycobacterium*arten) — Schimmelpilze (*Aspergillus-, Penicillium*arten) sind weniger wirksam — sämtlich auch Fette und Fettsäuren angreifen. In der Folge haben dann die Untersuchungen von W. O. TAUSSON (1925, 1927, 1928a, b, 1929) wichtige Beiträge geliefert: Auch aromatische Kohlenwasserstoffe, Benzol, Toluol, Xylol, Naphthalin, Phenanthren, können durch Bakterien abgebaut werden. Mikroorganismen, welche Kohlenwasserstoff verarbeiten, sind weit verbreitet; besonders zahlreich finden sie sich im Boden und Wasser der Erdölgebiete und im Erdöl selber. Wegen der Probleme und Ergebnisse der „Erdölmikrobiologie" sei auf die zusammenfassenden Berichte von A. MÜLLER und W. SCHWARTZ (1948) und D. KNÖSEL und W. SCHWARTZ (1954) und von E. BEERSTECHER (1954) verwiesen.

Im Zusammenhang mit dem Hauptthema unseres Beitrags sei vermerkt, daß F. JUST und W. SCHNABEL (1948) über erfolgreiche Versuche berichteten, feste und flüssige Paraffine als Ausgangsmaterial für die mikrobiologische Fettsynthese zu verwenden. Sie konnten in 4—5 Liter Kultur auf je 1 g verbrauchten Paraffins 1 g Bakterientrockensubstanz mit 46% Rohfett erhalten.

III. Autotrophe Mikroorganismen (Algen).

1. Allgemeine Vorbemerkungen.

Daß nicht nur Dauerzellen (Zygoten, Dauersporen), sondern auch vegetative Zellen von Algen zur Bildung mikroskopisch sichtbarer Fetttröpfchen befähigt sind, ist seit langem bekannt (z.B. F. SCHMITZ 1883, F. STEINECKE 1923, O. KOPETZKI-RECHTPERG 1935). Besondere Beachtung haben Algenfette gefunden, die durch ihre Beimengung lipophiler Carotinoidpigmente auffallen, z.B. bei *Haematococcus* oder *Trentepohlia* (L. GEITLER 1923); oder aber die Fette von Algensippen, die keine Stärke, sondern eben Fette als normales „Assimilationsprodukt" ablagern, wie die Diatomeen (z.B. W. BENECKE 1900, C. MERESCHKOWSKY 1903, O. RICHTER 1909, K. HÖFLER 1940) oder *Vaucheria* (A. MEYER 1918). Alle Schlußfolgerungen, die über die Physiologie der Fettbildung oder über die physiologische Rolle der Fette auf Grund mikroskopischer Befunde vorgebracht wurden, sind mit großer Skepsis zu betrachten; die mikroskopische Untersuchung allein vermag nur in seltenen Fällen zu entscheiden, ob es sich wirklich um Fettbildung oder aber um „Lipophanerose (E. KÜSTER 1935), d.h. ein Sichtbarwerden durch Veränderung des Dispersitätsgrades präexistenter Zellipoide (z.B. in den Plastiden) handelt. Daß reversible Lipophanerose gerade bei Algen vorkommt, hat z.B. K. EIBL (1940) bei *Spirogyra* gezeigt.

Die Zahl der Untersuchungen über die Fettbildung der Algen, die auf der tragfähigen Grundlage quantitativer Fettbestimmungen stehen, ist recht gering. Es wird versucht, ihre Ergebnisse im folgenden in drei Kapiteln zusammenzufassen. Die ersten beiden behandeln Fragen, über die schon bei den heterotrophen Mikroorganismen ausführlich gesprochen wurde: die *Abhängigkeit der Fettbildung* bei Algen von ‚inneren" und „äußeren" Faktoren und die *Reservestoffnatur* der Algenfette. Der dritte Abschnitt geht von der Tatsache aus, daß alle Algenfette letzten Endes — direkt oder indirekt — auf die CO_2-Assimilation der Algenzelle selbst zurückgehen und nicht — wie die Fette der Pilze und Bakterien — aus exogenen organischen C-Quellen entstehen. Es wird also von den Algenfetten als „Produkten der Photosynthese" die Rede sein.

2. Innere und äußere Bedingungen für die Fettbildung bei Algen.

D. v. DENFFER (1948a) hat Zellvermehrung und Fettbildung der Diatomee *Nitzschia palea* in planktischer Massenkultur untersucht. Wie Abb. 24 zeigt, steigt die Zellenzahl zunächst in geometrischer Progression an; dann aber biegt die Wachstumskurve um, die Zellvermehrung/Zeit (Generationsdauer) wird immer kleiner. Schließlich wird nach etwa 7 Tagen ein stationärer Zustand erreicht, in dem überhaupt keine Zellvermehrung mehr stattfindet. Beim Übergang zum stationären Zustand beginnen die Zellen zu verfetten. Die Fetteinlagerung geht vom 4. bis zum 12. Tage so weit, daß „praktisch das ganze Zellvolumen von einem einzigen riesigen Fetttropfen" angefüllt erscheint und 42% des Zelltrockengewichtes auf ätherlösliche Lipoidbestandteile entfallen. Die Einstellung des Teilungswachstums und die rasche Zellverfettung kann v. DENFFER auch dadurch erreichen, daß er den Zellen „bestimmte, zum Wachstum unbedingt notwendige Elemente — wie z.B. den assimilierbaren Stickstoff — entzieht", etwa durch Übertragung in destilliertes Wasser bei Beibehaltung der Belichtung. In alten *Nitzschia*-Kulturen entwickelt sich außerdem ein Teilungshemmstoff, der die Photosynthese unbeeinflußt läßt: das Teilungswachstum wird sistiert, die Zellen verfetten (D. v. DENFFER 1948b). Damit wird durch v. DENFFER in exaktquantitativer Weise bestätigt, was schon W. BEIJERINCK (1904) auf Grund mikroskopischer Untersuchungen behauptet hatte: „So lange die Diatomeen kräftig genug wachsen, um das gebildete Öl zu assimilieren und in protoplasmatische und andere Körpersubstanzen zu verwandeln, bemerkt man keine Anhäufung desselben und die Chromatophoren enthalten davon nur Spuren. Jede Ursache aber, welche das Wachstum beeinträchtigt, bei im übrigen ungestörter

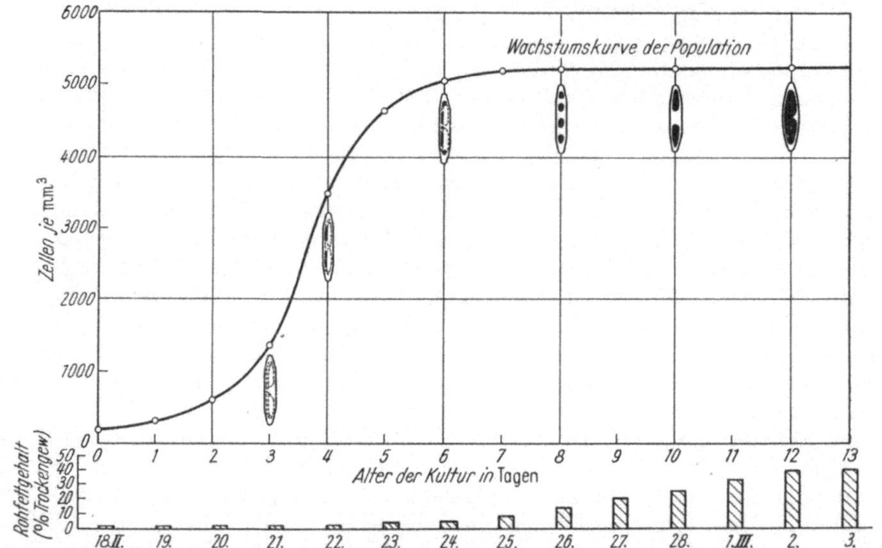

Abb. 24. *Nitzschia palea*, Planktonkultur bei 7000 Lux. Populationsentwicklung und Verfettungsgrad. Erste Anzeichen der Verfettung bei Überschreitung des Wendepunktes der Populationskurve. Maximale Verfettung zwei bis drei Tage nach Erreichung des stationären Endzustandes. (Nach D. v. DENFFER 1948a.)

Kohlensäureassimilation, gibt zu einer kräftigen und leicht sichtbaren Anhäufung von Öltropfen Veranlassung."

In sehr gründlichen Untersuchungen stellte auch H. G. AACH (1952) bei *Chlorella* fest, daß mit Erschöpfung des Nitrates der Nährlösung nicht nur (selbstverständlich) die N-Assimilation, sondern auch die Zellvermehrung ihr Ende findet und starke Fettspeicherung (von 22 auf 70% vom Trockengewicht) einsetzt. Über diese Ergebnisse berichtet auch der Beitrag von H. v. WITSCH (S. 50ff. dieses Bandes). Auch H. A. SPOEHR und H. W. MILNER (1949) fanden bei *Chlorella*-kulturen mit abnehmender N-Gabe im Medium einen sinkenden Protein- und steigenden Fettgehalt. Der von ihnen für letzteren angegebene Maximalwert von 85% beruht allerdings nicht auf direkter Fettbestimmung, sondern auf der (zweifellos etwas unsicheren) Berechnung aus dem „*R*-Wert (s. S. 210).

Die Untersuchungen von D. M. COLLYER und G. E. FOGG (1955) sind dadurch besonders wichtig, weil die Verfasser 12 Algenarten aus sehr verschiedenen Stämmen auf ihr Fettbildungsvermögen geprüft haben. Sie berechnen den Gehalt an Fettsäuren als Funktion des Trockengewichtes, welches wiederum einen Ausdruck des Alters der Kultur darstellt.

$$\log f = a + k \log w$$

wobei f Gewicht der gesamten Fettsäuren, w Trockengewicht der Kultur, a Konstante, k Wachstumsfaktor ist.

Es folgt also

wenn $k > 0$: *Zunahme* des Fettgehaltes *mit dem Alter*

$k < 0$: *Abnahme* des Fettgehaltes *mit dem Alter*.

Folgende Werte für k wurden gefunden:

Chlorophyceae

Chlorella vulgaris: $k = 1,400 + 0,123$

Chlorella pyrenoidosa: $k = 1,743 \pm 0,031$

Scotiella sp.: $k = 1,583 \pm 0,047$

Xanthophyceae

Monodus subterraneus: $k = 1,727 \pm 0,017$

Tribonema aequale: $k = 1,763 \pm 0,428$

Bacillariophyceae

Navicula pelliculosa: $k = 1,448 \pm 0,107$

Euglenales

Euglena gracilis v. bacillaris: $k = 0,801 \pm 0,047$

Rhodophyceae

Porphyridium cruentum: $k = 0,735 \pm 0,192$

Myxophyceae (= Cyanophyceae)

Microleus vaginalis: $k = 0,844 \pm 0,084$

Anabaena cylindrica: $k = 0,878 \pm 0,065$

Bei den *Chlorophyceen*, *Xanthophyceen* und *Diatomeen* nimmt also der Fettgehalt mit steigendem Alter zu, der prozentuale Proteingehalt ab. Die Verfasser selbst führen das auf den zunehmenden N-Mangel des Mediums von alternden Kulturen zurück. Bei *Euglena*, *Porphyridium* und den beiden Blaualgen haben die Fette offenbar eine andere Rolle im Verwendungsstoffwechsel. *Anabaena* ist überdies eine N-bindende Form, so daß es möglicherweise auch in den alternden Kulturen zu gar keinem wirksamen N-Mangel kommt.

Daß auch der Hydraturfaktor die Fettbildung steuert, ist in der Tatsache angedeutet, daß die Verfettung auf Agarmedium weiter geht als auf flüssigen Nährböden.

In den Grundzügen stimmen also die Bedingungen für die Fettbildung bei den Algen und bei den Pilzen gut überein. Vor allem wird auch bei vielen Algen die gleiche Gegenläufigkeit von Eiweiß- und Fettbildung gefunden, die bei den heterotrophen Mikroorganismen durch viele Beispiele belegt ist. Eine besondere Variante bedeutet es bei den Algenversuchen, daß sich hier durch Variation der Lichtstärke das Angebot des Ausgangsmaterials für alle synthetischen Vorgänge in der Zelle verändern läßt (vgl. insbesondere H. G. Aach 1952).

3. Die Reservestoffnatur der Algenfette.

Alle Autoren, welche die Frage überhaupt an Hand quantitativer Analysen geprüft haben, stimmen darin überein, daß das Algenfett als echter Reservestoff fungiert. Die Fettdepots können unter entsprechenden Kulturbedingungen wieder verbraucht werden, wie H. v. Witsch (1948) und H. G. Aach (1952) für *Chlorella*, D. v. Denffer (1948a) für *Nitzschia palea* ausdrücklich angegeben haben. Das gilt besonders, wenn Kulturbedingungen gegeben werden, die eine neuerliche Aufnahme der Teilungstätigkeit ermöglichen. Bei Kultur im Schwachlicht zeigen dann stark verfettete Kulturen sogar einen Vorsprung gegenüber weniger verfetteten Kontrollen. Während *Nitzschia*zellen im Dunkeln normalerweise keine Teilungen durchführen, kommt es bei starker Fettspeicherung auch ohne Belichtung auf frischer Nährlösung noch zu einem Teilungsschritt, wobei das Fett der Mutterzelle fast völlig verbraucht wird (D. v. DENFFER 1948a).

In diesen Ergebnissen ist auch eingeschlossen, daß die Verfettung der Zellen von *Chlorella* und *Nitzschia keinerlei degenerativen Effekt* hinsichtlich der Teilungsfähigkeit der Zellen einschließt. Das wird sowohl durch H. v. WITSCH wie durch D. v. DENFFER ausdrücklich auf Grund von einschlägigen Experimenten hervorgehoben.

4. Die Fette der Algen als Produkte der Photosynthese.

Bis vor kurzem wäre die Frage nach der Rolle der Fette als Produkt der Photosynthese wohl nur in einem indirekten Sinne aufgefaßt worden: wie die als unmittelbares Produkt der Photosynthese gebildeten Kohlenhydrate sekundär in Fette transformiert werden. Im Rahmen der von M. CALVIN und Mitarbeiter entwickelten und begründeten Vorstellungen vom „Photosynthesecyclus“ ist die Frage in einem unmittelbaren, wörtlichen Sinne erlaubt. Es braucht nur auf das von M. CALVIN (1956) entworfene Schema (Abb. 25) über die Verknüpfung der Fett-(und Kohlenhydrat- und Protein-)synthese mit dem Photosynthesecyclus verwiesen zu werden. Wegen aller Einzelheiten muß auf das eben zitierte Sammelreferat CALVINS sowie auf Band V dieses Handbuches verwiesen werden.

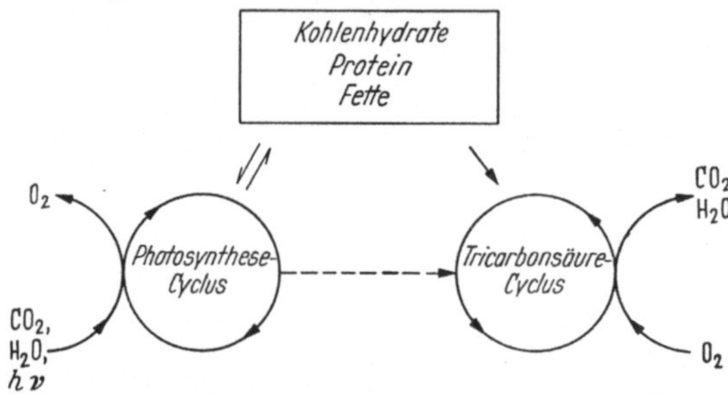

Abb. 25. Schema der Beziehungen zwischen dem Photosynthesecyclus, dem Tricarbonsäurecyclus und den Speicherprodukten in der Pflanze. (Nach M. CALVIN 1956.)

Tabelle 33. *Spezifische Aktivität der Lipoidbestandteile von Scenedesmus obliquus nach kurzer Photosynthese mit* $C^{14}O_2$. (Nach K. A. CLENDENNING 1950.)

	Menge mg	Spezifische Aktivität Impulse/min/mg
a)		
Gesamtlipoidfraktion vor Verseifung	137	*205*
Nach Verseifung		
Unverseifbares	25,0	*137*
Fettsäuren		
nicht flüchtig	75,2	*234*
flüchtig	4,3	57
Wasserlösliches		
nicht flüchtig	22,5	*258*
flüchtig	3,9	62,5
Unlöslicher Rückstand	6,1	27
b)		
Rohfettsäuren	69	*220*
Gesättigte Fettsäuren	23	*256*
Ungesättigte Fettsäuren	39,5	*133*

In der Tat liegen aus den letzten Jahren einige Arbeiten vor, welche zeigen, daß isotopen markiertes CO_2 unter Umständen bei der Photosynthese sehr rasch in die *Lipoid*fraktion der Zelle eingebaut wird.

A. H. BROWN, E. W. FAGER und H. GAFFRON (1948) arbeiteten mit *Scenedesmus*. $C^{13}O_2$ ging im Licht und im Dunkeln bevorzugt in die in Wasser und in Äthanol löslichen Zellbestandteile. Nach 10 min wurden aber 10% des gesamten in der Zelle enthaltenen C^{13} in der (in Äthanol und Benzol löslichen) *Lipoid*fraktion aufgefunden. K. A. CLENDENNING (1950) setzte diese Untersuchungen mit C^{14} fort. *Scenedesmus obliquus* wurde unter Zugabe von $Na_2C^{14}O_3$ 40 sec dem Licht exponiert und dann sofort in kochendes Wasser gegossen. Aus dem Wasserunlöslichen wurde durch

Extraktion (Methanol, Aceton, Benzol) der Lipoidanteil (20% des Algentrockengewichtes) isoliert und nach Verseifung in üblicher Weise fraktioniert und auf Radioaktivität geprüft. Tabelle 33a und b bringt die Ergebnisse.

Unverseifbares, Fettsäuren und Glycerin weisen also eine hohe Isotopenkonzentration auf, die gesättigten Fettsäuren enthielten deutlich mehr C^{14} als die ungesättigten. Die durch Chromatographie sorgfältig gereinigte Chlorophyllfraktion war übrigens völlig isotopenfrei. Es ergibt sich also, daß bei der Photosynthese das C-Isotop aus $C^{14}O_2$ sehr rasch in die Lipoidfraktion einverleibt wird.

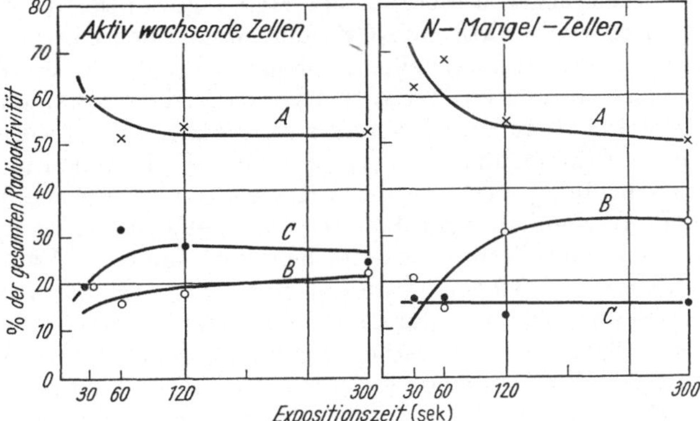

Abb. 26. Zeitliche Veränderung der C^{14}-Verteilung zwischen drei Zellfraktionen von *Navicula pelliculosa* bei aktiv wachsenden und N-Mangelzellen aus C^{14}-Bicarbonat im Licht. A: Äthanollösliche, benzolunlösliche Fraktion, B: Äthanol- und benzollösliche Fraktion, C: Äthanolunlösliche Fraktion. (Nach G. E. FOGG 1956.)

Grundsätzlich ähnliche Ergebnisse erhielt G. E. FOGG (1956) mit der Diatomee *Navicula pelliculosa*. Auch er verwendete $Na_2C^{14}O_3$, aber mit verschiedenen Expositionszeiten zwischen 30 und 300 sec. Von besonderem Interesse dürften hier die Ergebnisse von zwei Parallelversuchen sein, in denen mit aktiv wachsenden, gut mit N versorgten und mit N-verarmten Zellen, also unter für Fettbildung ungünstigen und günstigen Bedingungen experimentiert wurde. Das Zellmaterial wurde in drei Fraktionen zerlegt:

A Äthanollöslich, benzolunlöslich: Kohlenhydrat.
B Äthanollöslich, benzollöslich: Lipoide.
C Äthanolunlöslich: Proteine und Membranstoffe.

Wie Abb. 26 zeigt, liegt bei den aktiv wachsenden Zellen die Radioaktivität der „Proteinfraktion“ C stets über der „Lipoidfraktion“ B, während bei N-armen Zellen B schon nach 60 sec C deutlich überflügelt. Die je nach den Ernährungsbedingungen verschiedene Neigung der Zellen zur Synthese von Eiweiß oder Fett (s. S. 227ff. und S. 239ff.) findet also hier in kurzfristigen Isotopenversuchen eine schöne Bestätigung.

IV. Zur Frage der technischen Fettsynthese mit Hilfe von Mikroorganismen.

Es kann kaum übersehen werden, daß zahlreiche, wertvolle Untersuchungen über die Physiologie der Fettbildung durch Mikroorganismen im Hinblick auf eine mögliche technische Verwertung durchgeführt wurden. In der Tat hätte eine solche praktische *Verwertung* des Fettbildungsvermögens von Mikroorganismen eine gründliche *Kenntnis* der Bedingungen für eine maximale, ökonomische und rasche Fettsynthese durch die lebende Zelle zur Voraussetzung.

Ohne auf Einzelheiten einzugehen, die außerhalb des Rahmens dieser Darstellung liegen würden, sei nur auf einige zusammenfassende Referate über diesen Gegenstand verwiesen: z.B. B. HEINZE (1918), H. FINK, H. HAEHN, W. HOERBURGER (1937), W. SCHWARTZ (1937), M. SCHMIDT (1938), G. HAESELER und H. FINK (1940), L. REICHEL (1940), E. SCHMIDT (1943), H. DAMM (1943), K.

Bernhauer (1950). Im Anschluß an die zuletzt zitierte Veröffentlichung seien einige wenige Probleme angedeutet, die bei der Übertragung der im physiologischen Laboratorium gewonnenen Erfahrungen in den großtechnischen Betrieb auftauchen.

Die im technischen Verfahren notwendige rasche *Umsatzgeschwindigkeit* scheint bei Mycelpilzen besser als bei Hefen gewährleistet zu sein. Da Oberflächenkulturen aus praktischen Gründen, wegen der Kosten und des Raumaufwandes, von vornherein ausscheiden, spielt die Frage der *Belüftung* eine große Rolle: sie läßt sich in dichten Hefesuspensionen leichter lösen als bei dichtem Mycelwuchs. Die im Laboratoriumsversuch verwandten hochwertigen *Ausgangsmaterialien* scheiden für technische Verfahren aus. Es kommen hier nur Rohstoffe in Frage, die sehr billig und für die Ernährung von Mensch und Tier ungeeignet sind (z. B. Buchenholzsulfitablaugen; die Verwertbarkeit der Pentosen spielt im Zusammenhang damit eine entscheidende Rolle. — Vgl. S. 221). Die *Gefahr einer Infektion* ist geringer, wenn mit stark sauren Substraten gearbeitet werden kann. Die Schwierigkeiten der *Abtrennung der Pilzernte* vom Substrat sind sich bei Mycelpilzen geringer als bei Hefen. Umgekehrt ist die verlustfreie *Isolierung der Fette* aus Mycelien meist leichter als bei Hefen, usw.

Die *Wirtschaftlichkeit* aller bisher in Betracht gezogenen und versuchsmäßig angewandten Verfahren ist ausgesprochen schlecht, so daß sie höchstens für Notzeiten in Frage kommen, in denen die Kostenfrage eine geringere Rolle spielt. Selbst dann darf von einer Fetterzeugung mit Hilfe heterotropher Mikroorganismen nicht sehr viel erhofft werden. Bernhauer berechnet z. B., daß aus dem gesamten Zucker der in Deutschland anfallenden Buchenholzsulfitablaugen (130000 t) maximal 20000 t/Jahr Fett erzeugt werden können. Der Fettbedarf Deutschlands von 1939 war 1,7 Millionen Tonnen im Jahr, wovon 1,0 Millionen Tonnen durch Einfuhren gedeckt werden mußten. Diese Fettlücke würde sich durch die vorher genannte „Veredlung" der Buchenholzablaugen nur um 2% vermindern.

Auch die Möglichkeit der Gewinnung von Fettbegleitern der Pilzfette ändert daran nicht viel: weder die Phosphatide, die viel billiger auf Sojabasis gewonnen werden können, noch die Sterine (Ergosterin, Vitamin D), die allerdings als Nebenprodukte der Citronensäure- oder (vor allem!) der Penicillinfabrikation in Betracht kommen. Auch die in manchen Pilzfetten enthaltenen Carotinoide (Provitamin A!, Butterfarbstoff!) können billiger in beliebiger Menge aus höheren Pflanzen gewonnen werden.

Es ist verständlich, daß man sich neuerdings bei der Diskussion der Möglichkeiten einer mikrobiologischen Synthese von Fetten und Nahrungsstoffen überhaupt von den *heterotrophen* Mikroorganismen ab- und den *autotrophen*, niederen Algen zuwendet. Da H. v. Witsch in seinem Beitrag über die Lipoide der Algen auch darauf kurz eingeht, sei hier lediglich auf die von J. S. Burlew (1953) redigierte zusammenfassende Darstellung über dieses Forschungsgebiet verwiesen.

Literatur.

Aach, H. G.: Über Wachstum und Zusammensetzung von *Chlorella pyrenoidosa* bei unterschiedlichen Lichtstärken und Nitratmengen. Arch. Mikrobiol. **17**, 213—246 (1952). — Asselineau, J.: Sur la variation de la teneur en lipides du bacille tuberculeux en fonction de l'age de la culture. Ann. Inst. Pasteur **81**, 306—309 (1951). — Armstrong, E. F., and J. Allen: A neglected chapter in chemistry: the fats. J. Soc. Chem. Industr. (Lond.) **43**. 207T—218T (1924).

Barber, H. H.: The production of fat from carbohydrates and similar media by a species of *Penicillium*. Biochemic. J. **23**, 1158—1164 (1929). — Barker, H. A., and A. M. Taha: *Clostridium Kluyveri*, an organism concerned in the formation of caproic acid from ethyl alcohol. J. Bacter. **43**, 347—363 (1942). — BASS, A., u. J. HOSPODKA: Biosynthesis of fats by yeast. II. Composition of fats by various temperatures. Chem. Listy **46**, 243—245 (1952). Ref. Chem. Abstr. **47**, 168 (1953). — BAUCH, R.: Chemogenetische Untersuchungen an der Hefe. Ber. dtsch. bot. Ges. **60**, (42)—(63) (1943). — BAUTZ, E.: Untersuchungen über die Mitochondrien von Hefen. Ber. dtsch. bot. Ges. **67**, 281—288 (1954a). — Beeinflussung der Indophenolblaubildung (Nadireaktion) in Hefezellen durch Röntgenstrahlen. Naturwiss. **41**, 375—376 (1954b). — Mitochondrienfärbung mit Janusgrün bei Hefen. Naturwiss. **42**, 49—50 (1955). — BAUTZ, E., u. U. HAGEN: Spektroskopische Untersuchungen der Cytochrome bei Hefekultur mit verschiedener Anzahl Nadi-positiver Zellen. Naturwiss. **41**, 458—459 (1954). — BAUTZ, E., u. H. MARQUARDT: Die Grana mit Mitochondrienfunktion in Hefezellen. Naturwiss. **40**, 531 (1953a). — Das Verhalten oxydierender Fermente in den Grana mit Mitochondrienfunktion der Hefezellen. Naturwiss. **40**, 531—532 (1953b). — Sprunghafte Änderungen des Verhaltens der Mitochondrien von Hefezellen gegenüber dem Nadireagens. Naturwiss. **41**, 121—122 (1954). — BEADLE, G. W., and E. L. TATUM: *Neurospora.* II. Methods of producing and detecting mutations concerned with nutritional requirements. Amer. J. Bot. **32**, 678—686 (1945). — BEERSTECHER, E.: Petroleum microbiology. HOUSTON, N. Y.: Elsevier Press 1954. — BEIJERINCK, W.: Das Assimilationsprodukt der Kohlensäure in den Chromatophoren der Diatomeen. Rec. Trav. bot. néerl. **1**, 28—32 (1904). — BÉLIN, P.: Généralité de la distinction entre deux catégories de matières grasses: élément constant et élément variable. Bull. Soc. Chim. biol. Paris 8, 1081—1102 (1926a). — Influence de l'alimentation sur la constitution des grasses de réserve (élément variable). Bull. Soc. Chim. biol. Paris 8, 1121—1150 (1926b). — BENECKE, W.: Über farblose Diatomeen der Kieler Föhrde. Jb. wiss. Bot. **35**, 535—569 (1900). — BENHAM, R. W.: Cultural characteristics of *Pityrosporum* ovale—a lipophilic fungus. J. Invest. Dermat. **2**, 187—203 (1939). Ref. Chem. Abstr. **33**, 9350 (1939). — Cultural characteristics of *Pityrosporum* ovale, a lipophilic fungus. Nutrient and growth requirements. Proc. Soc. Exper. Biol. a. Med. **46**, 176—178 (1940). Ref. Chem. Abstr. **35**, 2550 (1941). — BERNHARD, K., u. H. ALBRECHT: Die Lipoide von *Phycomyces Blakesleanus.* Helvet. chim. Acta **31**, 977—988 (1948). — BERNHAUER, K.: Der derzeitige Stand der mikrobiologischen Fettsynthese durch heterotrophe Organismen. Fette u. Seifen **52**, 203—206 (1950). — BERNHAUER, K., A. NIETHAMMER u. J. RAUCH: Beiträge zur mikrobiologischen Eiweiß- und Fettsynthese. II. Mitt. Vergleichende Untersuchungen über die Eiweiß- und Fettbildung durch verschiedene Mycelpilze in der Submerskultur. Biochem. Z. **319**, 94—101 (1948). — BERNHAUER, K., u. G. PATZELT: Über Schimmelpilz-Sterine. I. Mitt. Die Sterinbildung bei *Aspergillus niger.* Biochem. Z. **280**, 388—393 (1935). — BERNHAUER, K., u. J. RAUCH: Beiträge zur mikrobiologischen Eiweiß- und Fettsynthese. I. Mitt. Die grundlegenden Bedingungen für die Eiweiß- und Fettproduktion durch Mycelpilze in der Submerskultur. Biochem. Z. **319**, 77—93 (1948a). — Beiträge zur mikrobiologischen Eiweiß- und Fettsynthese. III. Mitt. Zur Methodik der submersen Mycelzüchtung in der „Rührkultur" und deren Anwendung zur Erzeugung von Fettmycel. Biochem. Z. **319**, 102—119 (1948b). — Berichtigungen zu den Veröffentlichungen „Beiträge zur mikrobiologischen Eiweiß- und Fettsynthese". 1.—3. Mitt. Biochem. Z. **319**, 77, 94, 102 (1948) und Erwiderung auf die voranstehenden Bemerkungen von F. THOMAS. Biochem. Z. **320**, 352—354 (1950a). — Beiträge zur mikrobiologischen Eiweiß- und Fettsynthese. 6. Mitt. Zur Methodik der Pilzzüchtung in der Submerskultur. Biochem. Z. **320**, 368—371 (1950b). — BICHKOVSKAJA, A. L.: The production of fat by the fungus *Endomyces vernalis.* [Russisch mit engl. Zus.fassung] Mikrobiologija 8, 1170—1181 (1939). — BIEDERMANN, W.: Der Lipoidgehalt des Plasmas bei *Monotropa hypopitys* und *Orobanche (speciosa).* Flora (Jena) **113**, 132—154 (1920). — BILGER, F., W. HALDEN, E. MAYER-PITSCH u. M. PESTEMER: Zur Kenntnis des Fett- und Lipoidstoffwechsels der Hefen. 5. Mitt. Quantitative Beziehungen bei der biologischen Ergosterinbildung. Mh. Chem. **70**, 259—272 (1937). — BILLS, C. E., C. N. MASSENGALE and P. S. PRICKETT: Factors determining the ergosterol content of yeast. 1. Species. J. of Biol. Chem. **87**, 259—264 (1930). — BLOCH, K., and D. RITTENBERG: Sources of acetic acid in the animal body. J. of Biol. Chem. **155**, 243—254 (1944). Ref. Chem. Abstr. **38**, 6346 (1944). — BOHN, P. M.: Le mécanisme de la synthese des graisses aux dépens des glucides. C. r. Acad. Sci. Paris **193**, 441—442 (1931). — BOUDIER, E.: De l'importance que l'on doit attacher aux gouttelettes oléagineus^s contenues dans les spores chez les discomycetes. Rev. gén. Bot. **25**, 51—54 (1914). — BOUGHTON, B. W., and M. R. POLLOCK: Long-chain unsaturated fatty acids as essential bacterial growth factors. Further studies with *Corynebacterium „Q"*. Biochemic. J. **50**, XXXI (1952). — BROWN, A. H., E. W. FAGER and H. GAFFRON: Assimilation of tracer carbon in the alga *Scenedesmus.* Arch. of Biochem. **19**, 407—428 (1948). — BURLEW, J. S.: Algae culture from laboratory to pilot plant. Carnegie Inst. Wash. Publ. 600, **1953**.

CALVIN, M.: Der Photosynthese-Cyclus. Angew. Chem. **68**, 253—264 (1956). — CAMUS, L.: Formation de lipase par le „*Penicillium glaucum*". C. r. Soc. Biol. Paris, VI. s. **4**, 192—193 (1897a). — De la lipase dans les cultures de *l'Aspergillus niger.* C. r. Soc.

Biol. Paris VI. s. 4, 230 (1897b). — Oxidase positive bacteria in dairy products. Canad. Dairy a. Ice Cream. J. 19, 26—28 (1940). Ref. Chem. Abstr. 35, 2225 (1941). — CASTELL, C. H.: p-Aminodimethylanilinemonohydrochlorid as an indicator of microbial action on fats. Stain Technol. 16, 33—36 (1941). — CASTELL, C. H. and L. R. BRYANT: Action of microorganisms on fat. I. The significance of color changes in dyes, when used for the detection of microbial action on fat. Iowa State Coll. J. Sci. 13, 313 (1939). Ref. Stain Technol. 16, 33 (1941). — CASTELL, C. H., and E. H. GARRARD: Action of microorganisms on fat. II. Observations of uninoculated globules of triglycerides and fat acids and of butterfat in an agar medium. Canad. J. Res., Sect. C 18, 158—168 (1940a). Ref. Chem. Abstr. 34, 597 (1940). — Oxidase reactions of bacteria in relation to dairy products. Food. Res. 5, 215—222 (1940). Ref. Chem. Abstr. 34, 5109—5110 (1940b). — The actions of microorganisms on fat. III. Oxidation and hydrolysis of triolein by pure cultures of bacteria. Canad. J. Res., Sect. C 19, 106—110 (1941a). — The actions of microorganisms on fat. IV. Observations on the changes produced in globules of triolein by pure cultures of bacteria. Canad. J. Res., Sect. C 19, 111—117 (1941b). — CLENDENNING, K. A.: Distribution of tracer carbon among the lipids of the alga *Scenedesmus* during brief photosynthetic exposures. Arch. Bioch. a. Biophys. 27, 75—88 (1950). — COLEMAN, R. J.: Utilization of mixed substrates in fatty acid synthesis by *Fusarium lini* BOLLEY. Arch. of Biochem. a. Biophysics 45, 183—189 (1953). — COLEMAN, R. J., M. CEFOLA and F. F. NORD: On the mechanism of enzyme action. LII. The influence of 1-C^{14}-Acetate for the study of carbohydrate→fat synthesis in *Fusarium lini* BOLLEY. Arch. of Biochem. a. Biophysics 40, 102—110 (1952). — COLEMAN, R. J., and F. F. NORD: On the mechanism of enzyme action. XLIX. The effect of acetate on fat formation in *Fusarium lini* BOLLEY and the influence of added naphthquinones. Arch. of Biochem. a. Biophysics 38, 385—396 (1952). — COLLYER, D. M., and G. E. FOGG: Studies on fat accumulation by algae. J. of Exper. Bot. 6, 256—275 (1955). — CRAMER, E.: Zusammensetzung der Bakterien und ihre Abhängigkeit von dem Nährmaterial. Habil.-Schr. Heidelberg u. München: R. Oldenburg 1892. 47 S. — CRAMER, M., and J. MYERS: Effects of starvation on the metabolism of *Chlorella*. Plant Physiol. 24, 255—264 (1949).

DAMM, H.: Eine neue biochemische Fettsynthese. Chemiker-Ztg 67, 47—49 (1943). — DEIBEL, R. H., and C. F. NIVEN jr.: Reciprocal replacement of oleic acid and CO_2 in the nutrition of the „minute" *Streptococci* and *Lactobacillus Leichmannii*. J. Bact. 70, 135—140 (1955). — DENFFER, D. v.: Die planktische Massenkultur pennater Grunddiatomeen. Arch. Mikrobiol. 14, 159—202 (1948a). — Über einen Wachstumshemmstoff in alternden Diatomeenkulturen. Biol. Zbl. 67, 7—13 (1948b). — DIEMAIR, W., u. C. BORESCH: Ein Beitrag zur biologischen Fettsynthese. Lebensmittel-Unters. u. Forsch. 90, 14—26 (1950). — DREWS, B., u. H. SPECHT: Myceleiweißgewinnung durch Züchtung von *Oospora lactis* in Melasseschlempe. Biochem. Z. 321, 52—69 (1950). — DUCLEAUX, E.: Sur la nutrition intracellulaire. Ann. Inst. Pasteur 3, 97—125 (1889). — Sur la nutrition intracellulaire. (2. mémoire). Ann. Inst. Pasteur 3, 413—428 (1899).

EIBL, K.: Lipophanerose der Plastiden von *Spirogyra* durch K-Oleat und andere Seifen. Protoplasma (Berl.) 34, 343—346 (1940). — ENEBO, L., L. G. ANDERSON and H. LUNDIN: Microbiological fat synthesis by means of *Rhodotorula* yeast. Arch. of Biochem. 11, 383—395 (1946). — ENEBO, L., M. ELANDER, F. BERG, H. LUNDIN, R. NILSSON u. K. MYRBÄCK: Mikrobielle Fettsynthese in Lüftungskultur von *Rhodotorula*-Arten. Iva (Stockh.) 1, 252—267 (1944). — EULER, H.: Grundlagen und Ergebnisse der Pflanzenchemie, Bd. 2. Braunschweig: F. Vieweg & Sohn 1909.

FINK, H., H. HAEHN u. W. HOERBURGER: Über die Versuche zur Fettgewinnung mittels Mikroorganismen mit besonderer Berücksichtigung der Arbeiten des Instituts für Gärungsgewerbe. Chemiker-Ztg 61, 689—693, 723—726, 744—747 (1937). — FINK, H., G. HAESELER u. M. SCHMIDT: Zur Frage der Fettgewinnung mit Hilfe von Mikroorganismen. Über das Fettbildungsvermögen verschiedener Stämme von *Oidium lactis (Oospora lactis)*. Z. Spiritusindustr. 60, 74—77, 82—83 (1937). — FINK, H., u. F. JUST: Zur Chemie der *Torula utilis*. Über die Zusammensetzung der Hefesubstanz, insbesondere des Hefeeiweißes. Biochem. Z. 300, 84—88 (1939). — FINK, H., J. KREBS u. R. LECHNER: Nachtrag zu unseren Arbeiten über die Fetthefegewinnung in Holzzuckerlösungen. Ein Beitrag zur biologischen Eiweißsynthese. Biochem. Z. 290, 135—136 (1937). — FIORE, J. V.: On the mechanism of enzyme action. XXXII. Fat and sterol in *Fusarium lini* BOLLEY, *F. lycopersici* and *F. solani* D_2 purple. Arch. of Biochem. 16, 160—168 (1948). — FISCHER, E.: Die Chemie der Kohlenhydrate und ihre Bedeutung für die Physiologie. (Rede gehalten zur Feier des Stiftungstages der militärärztlichen Bildungsanstalten am 2. August 1894.) In E. FISCHER, Untersuchungen über Kohlenhydrate und Fermente, S. 96—115. Berlin: Springer 1909. — FLIEG, O.: Fette und Fettsäuren als Material für Bau- und Betriebsstoffwechsel bei *Aspergillus niger*. Jb. wiss. Bot. 61, 24—64 (1922). — FOGG, G. E.: Photosynthesis and formation of fats in a diatom. Ann. of Bot., N. S. 20, 264—285 (1956). — FOSTER, J. W.: Chemical activities of fungi. New York: Academic Press 1949. — FRANKE, W., L. LEE u. O. KIBAT:

Zum Stoffwechsel der säurefesten Bakterien. II. Mitt. Über das Verhalten der ungesättigten Fettsäuren. Biochem. Z. **319**, 262—282 (1948). — FRANKE, W., u. W. PERIS: Zum Abbau organischer Säuren durch Bakterien. I. Mitt. Aerobe Versuche mit „ruhenden" Bakterien. Biochem. Z. **295**, 61—90 (1938). — FRANKE, W., u. A. SCHILLINGER: Zum Fettstoffwechsel der säurefesten Bakterien. I. Mitt. Orientierende aerobe Reihenversuche. Biochem. Z. **316**, 313—334 (1944). — FRIES, N., u. H. M. YUSEF: Two fatty acid requiring mutants of *Ophiostoma multiannulatum*. Physiol. Plantarum (Copenh.) **8**, 852—858 (1955). — FÜRTH, O. v., u. E. H. MAJER: Über die Verwertung aliphatischer Säuren durch niedere pflanzliche Organismen. Biochem. Z. **263**, 332—339 (1933). — FÜRTH, O. v., e C. SCHWARZ: Arch. Fisiol. **7**, 442 (1909). Zit. nach O. v. FÜRTH u. E. H. MAJER, 1933.

GARRIDO, J. M., y TH. K. WALKER: Fettgewinnung durch Mikroorganismen. Über Faktoren, die die Fetterzeugung durch *Aspergillus nidulans* in Oberflächenkultur beeinflussen. An. Real. Soc. españ. Fisica y Quim., Ser. B **49**, 81—90 (1953a). — Fettbildung durch Mikroorganismen. Bestimmung der Optimalkonzentration von Magnesiumsulfat, Natriumphosphat und Kaliumsulfat für eine maximale Umwandlung von Glucose in Fett bei *Aspergillus nidulans* EIDAM, *Penicillium javanicum* v. BEYMA und *Penicillium spinulosum*. An. Real. Soc. españ. Fisic y Quim., Ser. B **49**, 839—846 (1953b). — GARRIDO, J. M., M. WOODBINE y TH. K. WALKER: Fettbildung durch Mikroorganismen. Untersuchung der Wirkung verschiedener Konzentrationen von Magnesium, Phosphor und Kalium auf die Fettbildung von *Aspergillus nidulans* EIDAM, *Penicillium javanicum* v. BEYMA und *Penicillium spinulosum*. An. Real. Soc. españ. Fisic y Quim., Ser. B **49**, 829—838 (1953). — GEFFERS, H.: Untersuchungen über das Fettbildungsvermögen bei Pilzen der Gattung *Oospora* WALLROTH (em. SACC.). Arch. Mikrobiol. **8**, 66—98 (1937). — GEITLER, L.: Studien über das Hämatochrom und die Chromatophoren von *Trentepohlia*. Österr. bot. Z. **72**, 76—83 (1923). — GÉRARD, E.: Sur une lipase végétale extrait du *Penicillium glaucum*. C. r. Acad. Sci. Paris **124**,370—371 (1897). — GORBACH, G., u. A. SABLATNÖG: Über die Bildung von Lipoiden durch Bakterien. I. Mitt. Die Gesamtfettbildung von *Bacillus prodigiosus* auf festen Nährmedien. Arch. Mikrobiol. **5**, 311—317 (1934a). — Über die Bildung von Lipoiden durch Bakterien. II. Mitt. Die Gesamtfettbildung *Bac. prodigiosus* von in Nährlösängen. Arch. Mikrobiol. **5**, 318—327 (1934b). — GOUPIL, R.: Recheres sur les composés phosphorés formés par *l'Amylomyces rouxii*. C. r. Acad. Sci. Paris **156**, 959—962 (1913). — Recheres sur les matières grasses, formées par *l'Amylomyces rouxii*. C. r. Acad. Sci. Paris **158**, 522—525 (1914). — GRAY, W. D.: Further studies on the alcohol tolerance of yeast: its relationship to cell storage products. J. Bacter. **55**, 53—59 (1948). — GREENE, H. C., and E. B. FRED: Maintenance of vigorous mold stock cultures. Industr. Engin. Chem. **26**, 1279—1299 (1934). — GREGORY, M. E., and M. WOODBINE: Microbiological synthesis of fat. The effect of varying p_H, temperature, nitrogen, incubation period, and sugar concentration on fat producing moulds. J. of Exper. Bot. **4**, 314—318 (1953). — GUILLIERMOND, A.: Étude sur le développement et la structure de *l'Oidium lactis*. Rev. gén. Bot. **12**, 464—479 (1900). — Nouvelles observations sur le chondriome des champignons. C. r. Acad. Sci. Paris **156**, 1781—1784 (1913a). — Sur le rôle du chondriome dans l'élaboration des produits de réserve des champignons. C. r. Acad. Sci. Paris **157**, 63—65 (1913b). — Nouvelles observations sur le chondriome de l'asque de *Pustularia vesiculosa*. C. r. Soc. Biol. Paris **75**, 646—649 (1913c). — The cytoplasm of the plant cell. Waltham Mass. U.S.A. 1941. — GUILLIERMOND, A., G. MANGENOT et L. PLANTEFOL: Traité de cytologie végétale. Paris: Le François 1933. — GYLLENBERG, H., u. A. RAITIO: Studies on the decomposition of cellular lipids of molds. Physiol. Plantarum (Copenh.) **5**, 367—371 (1952).

HAAG, F. E.: Über die Bedeutung der Doppelbindungen im Paraffin des Handels für das Wachstum von Bakterien. Arch. f. Hyg. **97**, 28—46 (1926). — Die Zersetzung der Fette durch Bakterien. Arch. f. Hyg. **100**, 270—308 (1928). — HAEHN, H., u. W. KINTTOF: Beitrag über den chemischen Mechanismus der Fettbildung aus Zucker. Chem. Zelle u. Gewebe **12**, 115—156 (1925). — HAESELER, G., u. H. FINK: Der gegenwärtige Stand der Fettgewinnung aus Mikroorganismen. Z. Spiritusindust. **63**, 89—90, 94—96 (1940). — HAIDASCH, J.: Der Einfluß von synthetischen Wuchsstoffen auf die Fettbildung von Hefen. Arch. Mikrobiol. **23**, 98—104 (1955). — HALDEN, W.: Zur Kenntnis des Fett- und Lipoidstoffwechsels der Hefen. III. Mitt. Sterin- und Fettanreicherung in untergäriger Brauereihefe. Hoppe-Seylers Z. **225**, 249—274 (1934). — HALDEN, W., F. BILGER u. R. KUNZE: Zur Kenntnis des Fett- und Lipoidstoffwechsels der Hefe. Naturwiss. **21**, 660—661 (1933). — HANAHAN, D. J., and S. J. AL-WAKIL: The biosynthesis of ergosterol from isotopic acetate. Arch. of Biochem. a. Biophysics **37**, 167—171 (1952). — HANUŠ, J.: Einige Beiträge zur Frage des Ranzigwerdens der Butter. Z. Unters. Nahrungs- u. Genußmittel **3**, 324—328 (1900). — HANUŠ, J., u. A. STOCKÝ: Über die chemische Einwirkung der Schimmelpilze auf die Butter. Z. Unters. Nahrungs- u. Genußmittel **3**, 606—614 (1900). — HASELHOFF, E., u. F. MACH: Über die Zersetzung der Futtermittel durch Schimmelpilze. Landwirtsch. Jb. **35**, 445—467 (1906). — HEBEBRAND, A.: Über die Veränderungen des Brotes beim Schimmeln. Landwirtsch. Versuchsstat. **42**, 421—427 (1893). — Heide, S.: Zur Physiologie und Cytologie der Fettbildung bei *Endomyces vernalis*. Arch. Mikrobiol. **10**, 135—188 (1939). — Heinemann, H.: Untersuchungen über die Physiologie und Cytologie der Fettbildung bei Pilzen. Diss. Univ. Bonn 1956. — Heinze, B.: Die Fettbildung durch niedere pflanzliche Organismen und ihre Verwertung. Jber. Ver.igg. angew. Bot. **15**, 1—8 (1918). — Henneberg, W.: Studium über das Verhalten einiger Kulturheferassen bei verschiedenen Temperaturen. Ein Beitrag zur Enzymtätigkeit, zur Lebensdauer, Haltbarkeit und zum Absterben der Hefen. Z. Spiritusindustr. **27**, 96—97, 105—106, 116—117, 126—127, 135—137, 146—147, 160—161, 173, 182—183, 195—196, 205—206, 213—216, 226, 239—240 (1904). — Handbuch der Gärungsbakteriologie, Bd. I. Berlin: P. Parey 1926. — Hepburn, J. S.: Eine kritische Studie über die natürlichen Veränderungen, denen Fette und Öle unterliegen. J. Franklin Inst. **168**, 365—384, 421—456 (1909); **169**, 23—54 (1910). Ref. Chem. Zbl. **1910** I, 1041—1042. — Hilditch, T. P.: Biosynthesis of unsaturated fatty acids in ripening seeds. Nature (Lond.) **167**, 298—301 (1951). — Die Bildung der Fettsäuren in den Ölsamen. Oléagineux **10**, 83 (1955). Ref. Fette u. Seifen **57**, 377 (1955). — Höfler, K.: Aus der Protoplasmatik der Diatomeen. Ber. dtsch. bot. Ges. **58**, 97—120 (1940). — Hoppe-Seyler, F.: Über Gährungsprozesse. Hoppe-Seylers Z. **2**, 1—28 (1878). — Über Gährungsprozesse. Synthese bei Gährungen. Hoppe-Seylers Z. **3**, 351—356 (1879). — Horowitz-Wlassowa, L. M., u. M. J. Livschitz: Zur Frage der Wirkung der Mikroben auf Fette. Zbl. Bakter. II, **92**, 425—435 (1935). — Hutchings, B. L., and E. Boggiano: Oleic acid as a growth factor for various lactobacilli. J. of Biol. Chem. **169**, 229—230 (1947).

Imai, Y.: The lipides of microbes. I. The effect of temperature on the lipid formation by fungi. J. Jap. Biochem. Soc. **22**, 192—196 (1950). Ref. Chem. Abstr. **45**, 9121 (1951). — Ivanow, S.: Die Klimaten des Erdballes und die chemische Tätigkeit der Pflanzen. Fortschr. naturwiss. Forsch., N. F. **1929**, 1—39.

Jensen, L. B.: The action of microorganisms on fats. J. Bacter. **33**, 98—99 (1937). Ref. Chem. Abstr. **31**, 3094 (1937). — Jensen, L. B., and D. P. Gretter: Actions of microorganisms on fats. Food. Res. **2**, 97—120 (1937). Ref. Chem. Abstr. **32**, 6285 (1938). — Jensen, O.: Studien über das Ranzigwerden der Butter. Zbl. Bakter. II **8**, 11—16, 42—46, 74—80, 107—114, 171—174, 211—216, 248—252, 278—281, 309—312, 342—346, 367—369, 406—412 (1902). — Jessen, W.: Über die Wirkung steigender Stickstoff- und Phosphorsäuregaben auf das Wachstum sowie die Eiweiß- und Fetterträge der Ölrauke. Z. Pflanzenern., Düng. u. Bodenkde **92**, 164 (1949). Ref. Fette u. Seifen **53**, 52 (1951). — Jong, L. E. D. de: Beiträge zur Kenntnis des Mineralisierungsvorganges. Diss. Delft 1926. 199 S. — Eine neue fettbildende Hefe. Nederl. Tijdschr. Hyg. **1**, 136—148 (1927). — Just, F., u. W. Schnabel: Submerse Massenzüchtung von Bakterien auf nichtkohlenhydrathaltigen Nährstoffen. I. Branntweinwirtsch. **2**, 113—115 (1948). Ref. Fette u. Seifen **52**, 52 (1950).

Kating, H.: Die Nutzung verschiedener N-Quellen durch *Endomycopsis vernalis*. Arch. Mikrobiol. **22**, 235—247 (1955a). — Über die Aktivität der Zelloberfläche bei der Assimilation von Amino- und Amidstickstoff durch *Endomycopsis vernalis*. Arch. Mikrobiol. **22**, 368—395 (1955b). — Kaufmann, W.: Untersuchungen über den Energiehaushalt der Hefezelle und die Ökonomie einiger Energiestoffwechseltypen anderer Mikroorganismen. Arch. Mikrobiol. **17**, 319—352 (1952). — Klein, O.: Über die Veränderung der Olivenpresslinge bei verschiedener Aufbewahrung. Z. angew. Chem. **1900**, 635—637. — Klein, H. P., and F. Lipmann: The relationship of Coenzyme A to lipid synthesis. I. Experiments with yeast. J. of Biol. Chem. **203**, 95—99 (1953a). — The relationship of Coenzyme A to lipid synthesis. II. Experiments with the rat liver. J. of Biol. Chem. **203**, 101—108 (1953b). — Kleinzeller, A.: Fat formation in *Torulopsis lipofera*. Biochemic. J. **38**, 480—482 (1944). — Fat production from yeast. Rep. Proc. Internat. Congr. Microbiol. **1947**, S. 544—546. 1949. — Synthesis of lipides. Adr. Enzymol. 8, 299—341 (1948). — Kleinzeller, A., u. A. Bass: Biosynthesis of fats from saccharides in yeasts. I. Composition of fats in various phases of formation. Chem. Listy **45**, 303—308 (1951). Ref. Chem. Abstr. **46**, 1615 (1952). — Kleinzeller, A., u. J. Málek: Biosynthesis of fats from saccharides in yaest. IV. Metabolism of the yeast *Rhodotorula gracilis* on glucose, D-xylose and sodium acetate. Chem. Listy **46** 674—678 (1952). Ref. Chem. Abstr. **47**, 2257 (1953). — Kleinzeller, A., J. Málek, R. Praus u. J. Škoda: Biosynthesis of fats from saccharides in yeasts. III. Formation of fat from D-xylose and glucose in yeast, *Rhodotorula gracilis*. Chem. Listy **46**, 470—474 (1952). Ref. Chem. Abstr. **46**, 11315 (1952). — Kleinzeller, A., u. J. Škoda: Comparison of the fat production by nine strains of *Fusarium*. Chem. Listy **44**, 184—185 (1950). Ref. Chem. Abstr. **45**, 6688 (1951). — Knösel, D., u. W. Schwartz: Untersuchungen zur Erdölbakteriologie. II. Über Vorkommen und Verhalten von Mikroorganismen aus Erdöl. Arch. Mikrobiol. **20**, 362—390 (1954). — Kodicek, E., u. A. N. Worden: Effect of unsaturated fat acids on *Lactobacillus helveticus* and other grampositive microorganisms. Biochem. J. **39** 78—85 (1945). Ref. Chem. Abstr. **39**, 5280—5281 (1945). — König, J., A. Spieckermann u. W. Bremer: Beiträge zur Zersetzung der Futter- und Nahrungsmittel durch Kleinwesen

I. Die fettverzehrenden Kleinwesen. Z. Unters. Nahrungs- u. Genußmittel 4, 721—744, 769—780 (1901). — KONDRATIEWA, T. M.: The effect of calcium salts on the structure and activity of yeast. Mikrobiologija 9, 114—128 (1940) [Russisch mit engl. Zus.fass.]. — KOPETZKY-RECHTPERG, O.: Über die Öltropfen in den Zellen der Conjugaten, besonders der Desmidiaceen. Beih. bot. Zbl., Abt. A 53, 595—605 (1935). — KORDES, H.: Biologische Untersuchungen über das in Dauerzellen und Hyphen verschiedener Pilze auftretende Fett. Bot. Archiv 3, 282—311 (1923). — KRUYFF, E. DE: Les bactéries hydrolysant et oxydant les graisses. Bull. Dépt. Agricult. Indes néerl. (Buitenzorg) (Microbiologie 2) 9, 1—12 (1906). — KÜSSNER, K.: Physiologische Untersuchungen über die Ernährung von *Penicillium glaucum* durch Fette. Bot. Archiv 23, 197—237 (1928). — KÜSTER, E.: Die Pflanzenzelle. Jena: Gustav Fischer 1935.

LAFAR, F.: Bakteriologische Studien über Butter. Arch. f. Hyg. 13, 1—39 (1891). — LABORDE, J.: Recherches physiologiques sur une moisissure nouvelle *l'Eurotiopsis Gayoni*. Ann. Inst. Pasteur 11, 1—43 (1897). — LARSON, L. W., and W. P. LARSON: Factors governing the fat content of bacteria and the influence of fat on pellicle formation. J. Inf. Dis. 31, 406—415 (1922). — LAXA, O.: Über die Spaltung des Butterfettes durch Mikroorganismen. Arch. f. Hyg. 41, 119—151 (1902). — LEIN, J., and P. S. LEIN: Studies on a *Neurospora* mutant requiring unsaturated fatty acids for growth. J. Bacter. 58, 595—599 (1949). Ref. Chem. Abstr. 44, 1562 (1950). — The production of acetate from fatty acids by *Neurospora*. J. Bacter. 60, 185—190 (1950). Ref. Chem. Abstr. 44, 10043 (1950). — LEIN, J., T. A. PUGLISI and P. S. LEIN: Unsaturated fatty acid metabolism in *Neurospora*. Arch. of Biochem. a. Biophysics 45, 434—442 (1953). — LEPESCHKIN, W. W.: Der thermische Effekt des Todes. Ber. dtsch. bot. Ges. 46, 591—593 (1928). — LINDEGREN, C. C.: The induction of dormancy in yeast cells by fat and carbohydrate storage and the conditions for reactivation. Arch. of Biochem. 8, 119—134 (1945). — LINDNER, P.: Zur Verflüchtigung des Bios-Begriffes. Z. techn. Biol. 7, 79—87 (1919). — Das Biosproblem und die Deutung negativer Ergebnisse bei Assimilationsversuchen. Z. techn. Biol. 8, 56—57 (1921a). — Mikrobenverfettung, die Biosfrage und die Bekämpfung des Tuberkelbazillus in seiner Eigenschaft als Fettpilz. Z. techn. Biol. 9, 100—107 (1921b). — Das Problem der biologischen Fettbildung und Fettgewinnung. Z. angew. Chem. 35, 110—114 (1922a). — Über Mikrobenverfettung, mit besonderer Berücksichtigung des *Endomyces vernalis*. Angew. Chem. 35, 591—592 (1922b). — Hefe- und Pilzwachstum bei Darbietung von Ammoniak und Alkoholdämpfen. Wschr. Brauerei 1929, 283—284. — LINDNER, P., u. ST. CZIFER: Der Alkohol, ein mehr oder weniger ausgezeichneter Nährstoff für verschiedene Pilze. Wschr. Brauerei 1912, 1—6. — LINDNER, P., u. T. UNGER: Die Fettbildung in Hefen auf festem Nährboden. Wschr. Brauerei 1919, 188. — LOCKWOOD, L. B., G. E. WARD, O. E. MAY, H. T. HERRICK and H. T. O'NEILL: The production of fat by *Penicillium javanicum* VAN BEIJMA. Zbl. Bakter., II 90, 411—425 (1934).— LUNDIN, H.: Fat synthesis by microorganisms and its possible applications in industry. J. Inst. Brewing 56, 17—28 (1950). — LYONS, R. E.: Über den Einfluß eines wechselnden Traubenzuckergehaltes im Nährmaterial auf die Zusammensetzung der Bakterien. Arch. f. Hyg. 28, 30—41 (1897). — LYNEN, F.: Lipide metabolism. Annual Rev. Biochem. 24, 653—688 (1955).

MAAS-FÖRSTER, M.: Der Fett- und Eiweißstoffwechsel von *Endomycopsis vernalis* unter dem Einfluß von Phosphor- und Kaliummangel. Arch. Mikrobiol. 22, 115—144 (1955). — MACLEOD, L. D., and I. SMEDLEY-MACLEAN: The carbohydrate and fat metabolism of yeast. V. The synthesis of fat form acetic acid: The influence of metallic ions on carbohydrate and fat storage. Biochemic. J. 32, 1571—1582 (1938). — MAGNUS-LEVY, A.: Über den Aufbau der hohen Fettsäuren aus Zucker. Arch. f. Physiol. 1902, 365—369. — MAHLER, H. R.: Role of coenzyme A in fatty acid metabolism. Federat. Proc. 12, 694—702 (1953). — MARQUARDT, H., u. E. BAUTZ: Die Wirkung einiger Atmungsgifte auf das Verhalten von Hefe-Mitochondrien gegenüber der Nadireaktion. Naturwiss. 41, 361—362 (1954). — MARTIN, H. H.: Beitrag zur Kenntnis der Morphologie und Physiologie der Nektarhefe *Candida Reukaufii* (GRÜSS) DIDDENS et LODDER. Arch. Mikrobiol. 20, 141—162 (1954). — MARTIUS, C.: Der Mechanismus des Aufbaus und Abbaus der Fettsäuren. Fette u. Seifen 55, 1—2 (1953). — MASSENGALE, O. N., C. E. BILLS and P. S. PRICKETT: Factors determining the ergosterol content of yeast. J. of Biol. Chem. 94, 213—219 (1931). — MCHARGUE, J. S., and R. K. CALFEE: Effect of manganese, copper, and zinc on growth and metabolism of *Aspergillus flavus* and *Rhizopus nigricans*. Bot. Gaz. 91, 183—193 (1931). — MERESCHKOWSKY, C.: Über farblose Pyrenoide und gefärbte Elaeoplasten der Diatomeen. Flora (Jena) 92, 77—83 (1903). — MEYER, A.: Das Assimilationssekret von Vaucheria terrestris. Ber. dtsch. bot. Ges. 36, 235—241 (1918). — MÜLLER, A., u. W. SCHWARTZ: Untersuchungen zur Erdölbakteriologie. I. Arch. Mikrobiol. 14, 291—308 (1948). — MULL, R. P., and F. F. NORD: On the mechanism of enzyme action. Part 23. Structure and action of rubrofusarin from *Fusarium graminis* SCHWABE (FGRA.) *(Gibberella saubinettii)*. Arch. of Biochem. 4, 419—433 (1944a). — On the mechanism of enzyme action. Part 24. Fat formation in *Fusaria*. Arch.

of Biochem. **5**, 283—290 (1944b). — MUNDT, J. O., and F. W. FABIAN: The bacterial oxidation of corn oil. J. Bacter. **48**, 1—11 (1944).

NAEGELI, C. v.: Über die Fettbildung bei den niederen Pilzen. Sitzgsber. bayer. Akad. Wiss., Math.-physik. Kl. **9**, 288—316 (1879). — NAEGELI, C. v., u. O. LOEW: Über die chemische Zusammensetzung der Hefe. Liebigs Ann. **193**, 322—348 (1878). — Über die Fettbildung bei niederen Pilzen. J. prakt. Chem. (2) **21**, 97—114 (1880). — NENCKI, M.: Über den chemischen Mechanismus der Fäulnis. J. prakt. Chem. **17**, 105—124 (1878). — NIELSEN, N., and N. G. NILSSON: Investigations on respiration, growth, and fat production of *Rhodotorula gracilis* when cultivated in media containing different carbohydrates. Arch. of Biochem. **25**, 316—322 (1949). — NIETHAMMER, A.: Hefen, sowie mikroskopische Pilze aus Blüten, ferner von Samen und Früchten. Arch. Mikrobiol. **13**, 42—59 (1942a). — Ernährungsphysiologische Untersuchungen an *Fusarien*, unter Hervorhebung der Ölspeicherung. Arch. Mikrobiol. **13**, 140—149 (1942b). — NILSSON, R., L. ENEBO, H. LUNDIN u. K. MYRBÄCK: Mikrobielle Fettsynthese unter Verwendung von *Rhodotorula glutinis* nach dem Lufthefeverfahren. Svensk kem. Tidskr. **25**, 41—51 (1943). — NOEL, R.: Sur des phénomènes de condensation de corps gras à la surface de mitochondries. C. r. Acad. Sci. Paris **174**, 572—573 (1922). — NORD, F. F., J. V. FIORE, G. KREITMAN and S. WEISS: On the mechanism of enzyme action. XL. The interaction of solanione, riboflavon, and nicotinic acid in the carbohydrate-fat conversion by certain fungi. Arch. of Biochem. **23**, 480—494 (1949). — NORD, F. F., J. V. FIORE and S. WEISS: On the mechanism of enzyme action. XXXIII. Fat formation in *Fusaria* in the presence of a pigment obtained from *Fusarium solani* D_2 purple. Arch. of Biochem. **17**, 345—358 (1948). — NOVELLI, G. D., and F. LIPMANN: Catalytic function of coenzyme A in citric acid synthesis. J. of Biol. Chem. **182**, 213—228 (1950). Ref. Chem. Abstr. **44**, 7355—7356 (1950). — NYMANN, M. A., and E. CHARGAFF: On the lipoprotein particles of yeast cells. J. of Biol. Chem. **180**, 741—745 (1949).

OTTKE, R. C., S. SIMMONDS and E. L. TATUM: Deuteroacetate in the biosynthesis of ergosterol by *Neurospora*. J. of Biol. Chem. **186**, 581—589 (1950). — OTTKE, R. C., S. SIMMONDS, E. L. TATUM, I. ZABIN and K. BLOCH: Isotopic acetate andi sovalerate in the synthesis of ergosterol by *Neurospora*. Biochem. J. **189**, 428—433 (1952).

PAN, S. C., A. A. ANDREASEN and P. KOLACHOV: Factors influencing fat synthesis by *Rhodotorula gracilis*. Arch. of Biochem. **23**, 419—433 (1949). — PEARSON, L. K., and H. ST. RAPER: The influence of temperature on the nature of the fat formed by living organisms. Biochem. J. **21**, 875—879 (1927). — PEPPLER, H. J.: Bacterial utilization of pure fats and their components. J. Bacter. **42**, 288—289 (1941). — PERRIER, A.: Sur la formation et le rôle des matières grasses chez les champignons. C. r. Acad. Sci. Paris **140**, 1052—1054 (1905).— PFEFFER, W.: Pflanzenphysiologie. Bd. 1: Stoffwechsel. Leipzig: Wilhelm Engelmann 1897.— POLLOCK, M. R., G. H. HOWARD and B. W. BOUGHTON: Studies on a bacterium needing long chain unsaturated fatty acids for growth. Biochemic. J. **44**, LII (1949a). — Long chain unsaturated fatty acids as essential bacterial growth factors. Substances able to replace oleic acid for the growth of *Corynebacterium* "Q" with a note on a possible method for their microbiological assay. Biochemic. J. **45**, 417—422 (1949b). — PONTILLON, CH.: Variation des acides gras du *Stérigmatocystis nigra* en fonction de la composition minerale du liquide de culture. C. r. Acad. Sci. Paris **191**, 1148—1150 (1930a). — Variations des insaponifiables et du phosphore lipidique du *Stérigmatocystis nigra* en fonction de la composition minerale du liquide de culture. C. r. Acad. Sci. Paris **191**, 1367—1369 (1930b). — Contributions à l'étude physiologique des lipides du *Stérigmatocystis nigra*. Rév. gén. Bot. **44**, 417—449, 465—483, 526—560 (1932); **45**, 21—52 (1933). — PORGES, N.: Chemical composition of *Aspergillus niger* as modified by zinc sulphate. Bot. Gaz. **94**, 197—205 (1932/33). — PRILL, E. A., P. R. WENCK and W. H. PETERSON: The chemistry of mould tissue. VI. Factors influencing the amount and nature of the fat produced by *Aspergillus Fischeri*. Biochemic. J. **29**, 21—33 (1935). — PRUESS, L. M., E. O. EICHINGER and W. H. PETERSON: The chemistry of mold tissue. III. The composition of certain molds with special reference to the lipid content. Zbl. Bakter. II **89**, 370—377 (1934). — PRUESS, L. M., H. J. GORCICA, H. C. GREENE u. W. H. PETERSON: Wachstum und Steringehalt gewisser Schimmelpilze. Biochem. Z. **246**, 401—413 (1932). — PRUESS, L. M., W. H. PETERSON, H. STEENBOCK and E. B. FRED: Sterol content and antirachitic activatibility of mold mycelia . J. of Biol. Chem. **90**, 369—384 (1931).

RAAF, H.: Beiträge zur Kenntnis der Fett- und Eiweißsynthese bei *Endomyces vernalis* und einigen anderen Mikroorganismen. Arch. Mikrobiol. **12**, 131—181 (1941). — RAHN, O.: Die Zersetzung der Fette. Zbl. Bakter. II **15**, 53—61, 422—429 (1905). — Ein Paraffin zersetzender Schimmelpilz. Zbl. Bakter. II **16**, 382—384 (1906). — REICHEL, L.: Biologische Fettsynthese. Angew. Chem. **53**, 577—579 (1940). — REICHEL, L., u. O. SCHMID: Über den Mechanismus der Synthese von Fettsäuren und Fett durch den Hefepilz *Endomyces vernalis*. Biochem. Z. **300**, 274—283 (1939). — REINARTZ, F., u. H. LAFOS: Über die Lipoide der Preßhefe. Angew. Chem. **62**, 572—575 (1950). — REINMANN, R.: Untersuchungen über die Ursachen des Ranzigwerdens der Butter. Zbl. Bakt. II **6**, 131, 139 166, 176, 209—214

(1900). — REYNOLDS, H., and E. W. HOPKINS: Action of microorganisms on fats and oils: oleic acid. Oil a. Soap **15**, 310—312 (1938). Ref. Chem. Abstr. **33**, 1165 (1939). — RICHTER, O.: Zur Physiologie der Diatomeen. II. Mitt. Die Biologie der *Nitschia putrida* BENECKE. Denkschr. Akad. Wiss. Wien, Math.-natwiss. Kl. **94**, 660—772 (1909). — RIPPEL, A.: Energetische Betrachtungen zur Ökonomie der Fettbildung bei Mikroorganismen. Arch. Mikrobiol. **11**, 271 (1940). — RIPPEL, A., u. G. BEHR: Über die Bedeutung des Kaliums im Stoffwechsel von *Aspergillus niger*. Arch. Mikrobiol. **5**, 561—577 (1934). — Über den Energieumsatz bei *Aspergillus niger* unter dem Einfluß der Kaliumversorgung. Arch. Mikrobiol. **7**, 315—322 (1936). — RIPPEL, A., u. K. NABEL: Über Eiweißbildung durch Bakterien. IV. Mitt. Kohlenstoffökonomie von *Bacillus glycinophilus* bei Glykokoll als Stickstoffquelle. Arch. Mikrobiol. **10**, 359—375 (1939). — RIPPEL-BALDES, A.: Die Energieausnützung durch Mikroorganismen in quantitativer Hinsicht. Arch. Mikrobiol. **17**, 166—188 (1952). — RIPPEL-BALDES, A., K. PIETSCHMANN-MEYER u. W. KÖHLER: *Candida (Nectaromyces) Reukaufii* als Fettbildner. Arch. Mikrobiol. **14**, 113—127 (1950). — RITTENBERG, D., and K. BLOCH: The utilization of AcOH for fatty acid synthesis. J. of Biol. Chem. **154**, 311—312 (1944). Ref. Chem. Abstr. **38**, 4985 (1944). — The utilization of acetic acid for the synthesis of fatty acids. J. of Biol. Chem. **160**, 417—424 (1945). — RITTHAUSEN, H., u. BAUMANN: Über Zerstörung von Fett durch Schimmelpilze. Landwirtsch. Versuchsstat. **47**, 389—390 (1896). — ROSENFELD, W. D.: Lipolytic activities of anaerobic bacteria. Arch. of Biochem. **11**, 145—154 (1946). — ROUSSY, A.: Sur la vie des champignons en milieux gras. C. r. Acad. Sci. Paris **149**, 482—484 (1909). — Sur la vie des champignons dans les acides gras. C. r. Acad. Sci. Paris **153**, 884—886 (1911). — RUBNER, M.: Über Spaltung und Zersetzung von Fetten und Fettsäuren im Boden und in Nährflüssigkeiten. Arch. f. Hyg. **38**, 67—92 (1900).

SARETT, H. P., and V. H. CHELDELIN: The utilization of β-alanine and pantothenic acid by yeasts. J. Bacter. **49**, 31—39 (1945). — SCHANDERL, H.: Untersuchungen über die systematische Stellung und die Physiologie des Kellerschimmels *Racodium cellare* PERSOON. Z. Bakter. II **94**, 112—127 (1936). — SCHMIDT, E.: Die technische Erzeugung von Buchenzellstoff und die Hefegewinnung aus Buchensulfitablauge. Chemie **56**, 93—95 (1943). — SCHMIDT, M.: Fettgewinnung aus Mikroorganismen. Vorratspfl. u. Lebensmittelforsch. **1**, 150—165 (1938). — SCHMIDT, R. H.: Über die Aufnahme und Verarbeitung von fetten Ölen durch Pflanzen. Flora (Jena) **49**, 300—370 (1891). — SCHMITZ, F.: Die Chromatophoren der Algen, Verh. naturhist. Ver. preuß. Rheinlande u. Westfalens **40**, 1—80 (1883). — SCHÖNBRUNNER, J.: Über die bakterielle Hydrierung von Ölsäure und Sorbinsäure und über ihre Beeinflussung durch Gallensäuren. Biochem. Z. **304**, 26—36 (1940). — SCHOENHEIMER, R., and D. RITTENBERG: Deuterium as an indicator in the study of intermediary metabolism. IX. The conversion of stearic acid into palmitic acid in the organism. J. of Biol. Chem. **120**, 155—165 (1937). Ref. Chem. Abstr. **31**, 8572 (1937). — SCHREIBER, K.: Fettzersetzung durch Mikroorganismen. Arch. f. Hyg. **41**, 328—347 (1902). — SCHULZ, G.: Der Einfluß einiger Schwermetallsalze (Zn, Cd, Mn, Fe) auf die chemische Zusammensetzung von *Aspergillus niger*. Planta (Berl.) **27**, 196—218 (1938). — SCHULZE, K. L.: Beiträge zur Physiologie und Technologie der Fettbildung bei Mikroorganismen. Arch. Mikrobiol. **15**, 315—351 (1950). — SCHWARTZ, W.: Fettsynthese durch Pilze und Bakterien. Angew. Chem. **50**, 924—926 (1937). — SHULL, G. M., R. W. THOMA and W. H. PETERSON: Amino acid and unsaturated fatty acid requirements. Arch. of Biochem. **20**, 227—241 (1949). — SMEDLEY-MACLEAN, I., and D. HOFFERT: Carbohydrate and fat metabolism in yeast. Biochemic. J. **17**, 720—741 (1923). — The carbohydrate and fat metabolism of yeast. Part II. The influence of phosphate on the storage of fat and carbohydrate in the cell. Biochemic. J. **18**, 1273—1278 (1924). — The carbohydrate and fat metabolism of yeast. Part III. The nature of the intermediate stages. Biochemic. J. **20**, 343—357 (1926). — SOBOTKA, M., W. HALDEN u. F. BILGER: Zur Kenntnis des Fett- und Lipoidstoffwechsels der Hefen. 4. Mitt. Der Vorgang der Sterin- und Fettanreicherung in untergäriger Brauereihefe. Hoppe-Seylers Z. **234**, 1—20 (1935). — SÖHNGEN, N. L: Benzin, Petroleum, Paraffinöl und Paraffin als Kohlenstoff- und Energiequelle für Mikroben. Zbl. Bakter. II **37**, 595—608 (1913). — SONDERHOFF, R., u. H. THOMAS: Die enzymatische Dehydrierung von Trideuteroessigsäure. Liebigs Ann. **530**, 195—213 (1937). — SPIECKERMANN, A.: Die Zersetzung der Fette durch höhere Pilze. I. Der Abbau des Glycerins und die Aufnahme des Fettes in die Pilzzelle. Z. Unters. Nahrungs- u. Genußmittel **23**, 305—331 (1912). — Die Zersetzung der Fette durch höhere Pilze. II. Der Abbau der Fettsäuren. Z. Unters. Nahrungs- u. Genußmittel **27**, 83—113 (1914). — SPOEHR, H. A., u. H. W. MILNER: The chemical composition of *Chlorella*, effect of environmental conditions. Plant Physiol. **24**, 120—149 (1949). — STADTMAN, E. R., and A. H. BARKER: Two new reactions producing acetyl phosphate. J. of Biol. Chem. **174**, 1039—1040 (1948). — Fatty acid synthesis by enzymic preparations of *Clostridium Kluyveri*. I. Preparation of cell-free extracts that catalyze the conversion of ethanol and acetats to butyrate and caproate. J. of Biol. Chem. **180**, 1085—1093 (1949a). — Fatty acid synthesis by enzymic preparations of *Clostridium Kluyveri*. II. The aerobic oxidation of ethanol and butyrate with the formation of acetyl phosphate. J. of Biol. Chem. **180**, 1095—1115 (1949b). — Fatty acid synthesis by enzymic preparations of *Clostridium Kluyveri*. III. The activation of molecular hydrogen and the conversion of acetyl phosphate and acetate to butyrate. J. of Biol. Chem. **180**, 1117—1124 (1949c). — Fatty acid synthesis by enzymic preparations of *Clostridium Kluyveri*. IV. The phosphoroclastic decomposition of acetoacetate to acetyl phosphate and acetate. J. of Biol. Chem. **180**, 1169—1186 (1949d). — Fatty acid synthesis by enzymic preparations of *Clostridium Kluyveri*. V. A consideration of postulated 4-carbon intermediates in butyrate synthesis. J. of Biol. Chem. **181**, 221—235 (1949e). — Fatty acid synthesis by enzymic preparations of *Clostridium Kluyveri*. VI. Reactions of acyl phosphates. J. of Biol. Chem. **184**, 769—793 (1950). — STADTMAN, E. R., T. E. STADTMAN and A. H. BARKER: Tracer experiments on the mechanism of synthesis of valeric and caproic acids by *Clostridium Kluyveri*. J. of. Biol. Chem. **178**, 677—682 (1949). — STARKEY, R. L.: Lipoid production by a soil yeast. J. Bacter. **51**, 33—50 (1946). — STEINBERG, M. P., and Z. J. ORDAL: Effect of fermentation variables on rate of fat formation by *Rhodotorula gracilis*. J. Agricult. a. Food Chem. **2**, 873—877 (1954a). — Theoretical rate of fat formation by yeasts. Science (Lancaster, Pa.) **120**, 609—610 (1954b). — STEINECKE, F.: Über Beziehung zwischen Färbung und Assimilation bei einigen Süßwasseralgen. Bot. Archiv **4**, 317—327 (1923). — STEINER, M.: Ernährung und Fettbildung bei *Endomyces vernalis*. Ber. dtsch. bot. Ges. **56**, (73)—(83) (1938). — Zur Verwendung von Permutit bei Stoffwechselversuchen mit Mikroorganismen. Biochem. Z. **307**, 330—322 (1941). — STEINER, M., u. H. HEINEMANN: Grana mit positiver Nadireaktion als Ort der primären Fettbildung in Pilzzellen. Naturwiss. **41**, 40—41 (1954a). — Über die Beziehungen zwischen den fettbildenden Grana und typischen Mitochondrien in den Zellen von *Oospora lactis*. Naturwiss. **41**, 90 (1954b). — STEINER, M., u. U. JENDRALSKI: 1956, unveröffentlicht.— STEINER, M., u. H. KATING: Über extracelluläre Umsetzungen von Aminosäuren und Amiden in Kulturen von *Endomycopsis vernalis*. Naturwiss. **40**, 487 (1953). — STEPHENSON, M., and M. D. WHETHAM: Studies in the fat metabolism of the timothy grass bacillus. II. The carbon balance sheet and respiratory quotient. Proc. Roy Soc. Lond., Ser. B **95**, 200—206 (1923). — STERN, A. M., Z. J. ORDAL and H. O. HALVORSON: Utilization of fatty acids by and lipolytic activities of *Mucor mucedo*. J. Bacter. **68**, 24—27 (1954). Ref. Chem. Abstr. **48**, 11547 (1954). — STOCKHAUSEN, F.: Alkoholassimilation durch Hefe (Demonstration der Versuche von Professor P. LINDNER). Jb. Versuchs- u. Lehranstalt Brauerei Berlin **14**, 551—556 (1911).

TARWIDOWA, H.: Über die Entstehung der Lipoidtröpfchen bei *Basidiobolus ranarum*. Cellule **47**, 205—216 (1938). — TATUM, E. L., R. W. BARRATT, N. FRIES and D. BONNER: Biochemical mutant strains of *Neurospora* produced by physical and chemical treatment. Amer. J. Bot. **37**, 38—46 (1950). — TAUSSON, W. O.: Zur Frage über die Assimilation des Paraffins durch Mikroorganismen. Biochem. Z. **155**, 357—368 (1925). — Naphthalin als Kohlenstoffquelle für Bakterien. Planta (Berl.) **4**, 214—256 (1927). — Die Oxydation des Phenanthrens durch Bakterien. Planta (Berl.) **5**, 238—273 (1928a). — Über die Oxydation der Wachse durch Mikroorganismen. Biochem. Z. **193**, 84—93 (1928b). — Über die Oxydation der Benzolkohlenwasserstoffe durch Bakterien. Planta (Berl.) **7**, 735—758 (1929). — TERROINE, E. F.: Contribution à la connaissance de la physiologie des substances grasses et lipoidiques. Ann. des natur. Zool., Sér. X **4**, 1—397 (1920). — TERROINE, E. F., et P. BÉLIN: L'élément constant des lipides: ses caractères. Bull. Soc. Chim. biol. Paris **9**, 12—48 (1927). — TERROINE, E. F., et R. BONNET: L'énergie de croissance. X. Formation des matières grasses aux dépens des gludices chez les microorganismes. Bull. Soc. Chim. Biol. Paris **9**, 588—596 (1927). — TERROINE, E. F., R. BONNET, et P. DUQUÉNOIS: L'énergie de croissance. XI. Formation des glucides aux dépens des acides gras par les moisissures. Bull. Soc. Chim. biol. Paris **9**, 597—604 (1927). — TERROINE, E. F., R. BONNET, G. KOPP et J. VÉCHOT: Sur la significance des liaisons éthyleniques des acides gras. Bull. Soc. Chim. biol. Paris. **9**, 605—620 (1927). — TERROINE, E. F., et J. E. LOBSTEIN: La formation des substances grasses et lipoidiques. I. Influence de la nature des aliments hydrocarbonées sur le teneur en substances grasses du bacille tuberculeux et les caractères de ces substances. Bull. Soc. Chim. biol. Paris **5**, 182—199 (1923). — THOMAS, F.: Bemerkungen zu K. BERNHAUER und J. RAUCH, Beiträge zur mikrobiologischen Eiweiß- und Fettsynthese. I. Mitt. Biochem. Z. **319**, 77 (1948); **320**, 350—351 (1950). — THOMAS, R. C.: Composition of the fungus hyphae. The Fusaria. Amer. J. Bot. **15**, 537—547 (1928.) — The composition of fungus hyphae. II. Amer. J. Bot. **17**, 779—788 (1930). — TIEGHEM, PH. VAN: Sur la végétation dans l'huile. Bull. Soc. bot. France **27**, 353 (1880). — Sur la végétation dans l'huile (deuxième note). Bull. Soc. bot. France **28**, 70—71 (1881a). — Recherches sur la vie dans l'huile. Bull. Soc. bot. France **28**, 137 (1881b). — TRAININA, F. L.: Die Anreicherung von Ergosterin in einigen Heferassen unter verschiedenen Kultivierungsbedingungen. Proc. Sci. Inst. Vitamin Res. People Commissar. Food. Ind. USSR. **2**, 53—62 (1937). Ref. Chem. Zbl. **1939** I, 458. — TURPEINEN, O.: Determination of lipides in microorganisms. Suom. Kemist. B 8, 17—18 (1935). — Studies on microbial lipides. The formation and chemical composition of lipides in *Geotrichoides* sp.

LANGERON et TALICE. Ann. Acad. Sci. fenn. Ser. A 46, 1—110 (1936). Ref. Chem. Abstr. 31, 3097—3098 (1937).

WAKSMAN, S. A.: Decomposition of the various chemical constituents etc. of complex plant materials by pure cultures of fungi and bacteria. Arch. Mikrobiol. 2, 137—154 (1931). — WALTER, H.: Beiträge zur vergleichenden Physiologie der Verdauung. VII. Das Verhalten der Hefezellen gegen Proteasen. Pflügers Arch. 181, 270—284 (1920). — Ein Beitrag zur Frage der chemischen Konstitution des Protoplasmas. Biochem. Z. 122, 86—99 (1921). — WANG, C. H., V. H. CHELDELIN, T. E. KING and B. E. CHRISTENSEN: Incorporation of carbon[14] from methionine into fatty acids of *Escherichia coli*. J. of Biol. Chem. 188, 759—761 (1951). Ref. Chem. Abstr. 45, 4783 (1951). — WARD, G. E., L. B. LOCKWOOD, O. E. MAY and H. T. HERRICK: Production of fat from Glucose by molds. Cultivation of *Penicillium javanicum* VAN BEIJMA in large scale laboraty apparatus. Industr. Engin. Chem. 27, 318—322 (1935). — WEINHOUSE, S., and R. S. MILLINGTON: Acetate metabolism in yeast, studies with isotopic carbon. J. Amer. Chem. Soc. 69, 3089—3093 (1947). — WEINHOUSE, S., R. S. MILLINGTON and M. STRASSMANN: Incorporation of glycine carbon into yeast lipides. J. Amer. Chem. Soc. 73, 1421—1423 (1951). — WEISS, S., J. V. FIORE and F. F. NORD: On the structure and possible functions of a pigment of *Fusarium solani* D_2 purple. Arch. of Biochem. 15, 326—328 (1947). — WENCK, P. R., W. H. PETERSON u. E. B. FRED: The chemistry of mold tissue. IX. Cultural factors influencing growth and sterol production of *Aspergillus Fischeri*. Zbl. Bakter. II 92, 330—338 (1935). — WENCK, P. R., W. H. PETERSON u. H. C. GREENE: The chemistry of mold tissue. VIII. Innate factors influencing growth and sterol production of *Aspergillus Fischeri*. Zbl. Bakter. II 92, 324—330 (1935). — WHITE, A. G. C., and C. H. WERKMAN: Assimilation of acetate by yeast. Arch. of Biochem. 13, 27—32 (1947). — Fat synthesis in yeast. Arch. of Biochem. 17, 474—482 (1948). — WHITEHILL, A. R., J. J. OLESON and Y. SUBBA ROW: A lactobacillus of cecal origin requiring oleic acid. Arch. of Biochem. 15, 31—37 (1947). — WIELAND, H., u. F. WILLE: Weitere Versuche über die Dehydrierung von Alkohol durch Hefe. Über den Mechanismus der Oxydationsvorgänge. XLI. Liebigs Ann. 515, 261—272 (1935). — WILL, H.: Vergleichende Untersuchungen an vier untergärigen Arten von Bierhefe. Zbl. Bakter. II 2, 752—763 (1896). — WILLIAMS, V. R., H. P. BROQUIST and E. E. SNELL: Oleic acid and related compounds as growth factors for lactic acid bacteria. J. of Biol. Chem. 170, 619—630 (1949). — WILLIAMS, V. R., and E. A. FIEGER: Further studies on lipide stimulation of *Lactobacillus casei*. J. of Biol. Chem. 170, 399—410 (1949). — WINDISCH, S.: Einiges über technische Wuchshefen. Brauwelt 1948, 203—206. — WITSCH, H. v.: Physiologischer Zustand und Wachstumsintensität bei *Chlorella*. Arch. Mikrobiol. 14, 128—141 (1948). — WOODBINE, M., M. E. GREGORY and T. K. WALKER: Microbiological synthesis of fat. Preliminary survey of the fat producing moulds. J. of Exper. Bot. 2, 204—211 (1951).

YUSEF, H. M.: The requirements of some Hymenomycetes for essential metabolites. Bull. Torrey Bot. Club. 80, 43—64 (1953). Ref. Chem. Abstr. 47, 4948 (1953).

ZIKES, H.: Über den Einfluß der Temperatur auf verschiedene Funktionen der Hefe. Zbl. Bakter. II 49, 353—373 (1919).

Nachtrag während der Korrektur.

Zu S. 242 und S. 244. Über die Veränderung der Menge und der Zusammensetzung des Fettes von *Aspergillus nidulans* mit dem Alter der Kultur auf Saccharose-Ammoniumnitrat-Medium haben vor kurzem H. SINGH und T. K. WALKER [Changes in the composition of the fat of *Aspergillus nidulans* with age of the culture. Biochemic. J. 62, 268—289 (1956)] berichtet. Aus ihren Analysenzahlen geht hervor, daß vom 9. Tage an der absolute und der prozentuale Fettgehalt des Mycels deutlich abnimmt (32,5 bzw. 24,8 g, 6,3 bzw. 4,1% am 9. bzw. 21. Tag): ein neuerlicher Beweis für die Reservestoffnatur des Mycelfettes.

Zum ersten Male wurden hier die einzelnen Fettsäuren in verschiedenen Phasen der Kultur (5., 7., 9., 14., 21. Tag) quantitativ erfaßt. Ihr Prozentanteil an den gesamten Fettsäuren verschiebt sich während der Kultur: Maximum der gesättigten Säuren (35,5%) am 9., der Ölsäure (57,4%) am 7., der Linolsäure (20,1 bzw. 21,0%) am 5. bzw. 21. Tag der Kultur. Da sich aber die Absolutmenge aller Säuren symbat verändert, ergibt sich keine Bestätigung der mehrfach diskutierten Annahme, daß die ungesättigten Fettsäuren sekundär durch Dehydrierung gesättigter entstehen. Der recht geringe Gehalt an *freien* Säuren hat ein Minimum zur Zeit des höchsten Fettgehaltes, also um den 9. Tag der Kultur.

Physiologie der Fettbildung und Fettspeicherung bei höheren Pflanzen.

Von

A. Zeller.

Mit 1 Abbildung.

I. In Früchten und Samen[1].

1. Einleitung.

Die zweifache Rolle der Fette im Lebensgeschehen — Betriebs- und Reservestoff — wurde für die höheren Pflanzen besonders von J. H. PRIESTLEY (1924) klar herausgearbeitet. E. F. TERROINE (1930) hat ähnliche Gedankengänge einer Übersicht über den tierischen Fettstoffwechsel zugrunde gelegt. PRIESTLEYS Forderung, nicht nur den Reservestoffwechsel der Pflanzenfette, sondern auch ihren Betriebsstoffwechsel zu untersuchen, ist bis heute kaum irgendwie erfüllt worden. Der Betriebsstoffwechsel geht ja in jeder Zelle der Pflanzen vor sich; jede Grenzschicht enthält Fett- oder Fettsäuremoleküle, die wahrscheinlich wesentlich an den Permeabilitätserscheinungen beteiligt sind, die vielleicht bei Stoffleitungen eine Rolle spielen, die bei den Umsetzungen anderer Stoffe beteiligt sein können und die wahrscheinlich selbst einem dauernden Ab- und Aufbau unterworfen sind (vgl. z. B. E. RHODES und R. M. WOODMAN 1925, R. C. JORDAN und A. C. CHIBNALL 1933). Von diesen Betriebsfetten funktionell wohl weitgehend verschieden sind die Reservefette, die wir als Betriebsstoff- und Energiespeicher in den Pflanzen vorfinden und die als Nahrungsmittel und als Industrierohstoffe eine so bedeutende Rolle spielen. Ob es tatsächlich gerechtfertigt ist, diese beiden Fettkategorien als voneinander anders als in ihren Funktionen verschieden anzunehmen, kann mindestens als fraglich hingestellt werden. Untersuchungen mit radioaktivem Kohlenstoff haben für Tiere gezeigt, daß dem Organismus einverleibter markierter Kohlenstoff in sehr kurzer Zeit in fast allen Substanzen, auch im Reservefett, gefunden werden kann (J. W. CORNFORTH 1953), und auf der Atomenergietagung in Genf 1955 wurden verschiedentlich ähnliche Beobachtungen berichtet, deren auffälligste vielleicht die ist, daß sogar die Kohlenstoffatome des im Milchsaft von *Hevea* enthaltenen Kautschuk selbst in blattlosen Bäumen in meßbarem Umfang am Stoffwechselgeschehen teilnehmen und ausgetauscht werden (N. J. SCULLY und Mitarbeiter 1956, besonders S. 384, sowie andere Beiträge in dieser Veröffentlichung). Andererseits haben nunmehr W. M. CROMBIE (1956) und W. M. CROMBIE und R. COMBER (1956) Hinweise dafür erhalten, daß sogar ein chemischer Unterschied zwischen den beiden Fettarten bestehen dürfte, die ihre Unterscheidung trotz hundertfachen Überschusses der Reservefette möglich machen kann (vgl. hierzu S. 302 und S. 314).

[1] Über die Zusammensetzung der meisten hier erwähnten Frucht- und Samenfette unterrichtet der Beitrag von M. L. MEARA, S. 10ff. dieses Handbuchbandes.

Eine Unterscheidung von Molekülen, die als Reservestoffe unverändert abgelagert bleiben und solchen, die als Betriebsstoff einem dauernden Ab- und Aufbau unterworfen sind, ist also oft nicht leicht möglich. Freilich wird man hier zwischen lebenden und toten Reservestoffbehältern (J. WEISSFLOG 1924) unterscheiden müssen, da Fette ja immer in lebenden Zellen gespeichert werden, aber daß in den vegetativen Organen der Pflanzen ähnliche unerwartete Molekülumbauten in großem Umfang vor sich gehen, haben R. L. WEINTRAUB und Mitarbeiter (1952), WEINTRAUB (1952) gezeigt. Sie trafen wenige Tage nach der Aufbringung von Dichlorphenoxyessigsäure, welche in der Seitenkette mit radioaktivem Kohlenstoff markiert war, den radioaktiven Kohlenstoff in praktisch allen Inhalts- und Baustoffen der untersuchten Pflanzen an. F. SHAFIZADEH und M. L. WOLFROM (1956) haben gezeigt, daß in Form von d-Glucose zugeführter radioaktiver Kohlenstoff (Markierung am Kohlenstoffatom 6) von Baumwollfrüchten in die Fettsäuren und in alle 3 Kohlenstoffatome des Glycerins des Baumwollsamenöles eingebaut wird. Zieht man die beträchtlichen und auch heutzutage meist noch unüberwindlichen Schwierigkeiten in Betracht, die einer experimentellen Bearbeitung der am sog. Betriebsstoffwechsel beteiligten Fette entgegenstehen, dann ist es nicht verwunderlich, daß praktisch alle auf den folgenden Seiten erwähnten Ergebnisse bei Untersuchungen über die Physiologie der Reservefette der höheren Pflanzen gewonnen wurden.

2. Allgemeines.

Früchte und Samen sind bekanntlich die bevorzugten Bildungs- und Speicherorte der Fette. Die qualitative und quantitative Zusammensetzung sowie die Menge der in den höheren Pflanzen gefundenen Fette scheint vor allem von 3 Faktorengruppen bestimmt zu sein, die etwa als *phylogenetische*, *genetische* und *geographische* unterschieden werden können.

a) Phylogenie[1].

J. B. McNAIR (1934) hat wohl die Frage des Zusammenhanges zwischen *der phylogenetischen Stellung* einer Pflanzenfamilie und den chemischen Eigenschaften der von den Pflanzen dieser Familie gebildeten Fette am eingehendsten erörtert. Er stellte fest, daß man bei Betrachtung von Pflanzen ein und derselben Klimazone findet, daß eine höhere Stellung im phylogenetischen Pflanzensystem mit der Tendenz gekoppelt ist, stärker ungesättigte Fette, also Fette mit höheren Jodzahlen, zu bilden. Für tropische Pflanzen und für die Numerierung der Familien nach C. G. DALLA TORRE und H. HARMS (1900—1907) konnte sogar eine Regressionsgleichung berechnet werden, mit Hilfe der es möglich ist, aus der Familiennummer x die durchschnittliche Jodzahl y der von dieser Familie stammenden Samenfette zu berechnen ($y = 78{,}98834 + 0{,}00013287\,x$).

Die Annahme, daß höher entwickelte Pflanzen kompliziertere Inhaltsstoffe bilden, wurde dann auch mit Hilfe von Untersuchungen an Alkaloiden und aromatischen Ölen zu beweisen versucht (J. B. McNAIR 1934, 1935). Auch Hinweise dafür wurden gefunden (J. B. McNAIR 1941), daß phylogenetisch höherstehende Familien höhermolekulare und zahlreichere Fettsäuren enthalten als phylogenetisch niedriger stehende Pflanzen. (Über ähnliche Verhältnisse bei den Kryptogamen s. McNAIR 1943.) In konsequenter Verfolgung dieser Gedankengänge wurden dann die chemischen Eigenschaften der Pflanzenfette dazu benützt,

[1] Die hier angedeuteten Fragen, insbesondere die Beziehungen zwischen Fettzusammensetzung und Pflanzenverwandtschaft, werden von M. L. MEARA in seinem Beitrage "The fats of higher plants", S. 10ff. dieses Handbuchbandes, ausführlich behandelt.

um für Pflanzengruppen fraglicher entwicklungsgeschichtlicher systematischer Stellung Anhaltspunkte für ihre „richtige“ Einordnung in das phylogenetische System zu erhalten (MCNAIR 1945a). Die Magnoliaceen z. B. sind demnach als primitiver als die Berberidaceen und Ranunculaceen anzusehen und krautige Angiospermen sind im allgemeinen von holzigen abzuleiten und nicht etwa umgekehrt.

Auch S. IWANOW und seine Schule haben die phylogenetische Bedeutung der Zusammensetzung der Pflanzenfette erörtert (z. B. S. IWANOW und Mitarbeiter 1930). IWANOW hat auf Grund seiner Untersuchungen sogar ein sog. „biochemisches Grundgesetz“ aufgestellt, das er in 3 Absätze zusammenfaßte: „1. Jede Art bewahrt bei Beständigkeit der äußeren Bedingungen die konstante Fähigkeit, die ihr eigentümlichen Stoffe hervorzubringen, welche ihre physiologisch-chemischen Merkmale sind. — 2. Jede Art teilt ihre physiologisch-chemischen Merkmale mit den Arten, die mit ihr in genetischer Verbindung stehen. Je näher die Verwandtschaft ist, desto größer ist die Zahl der den Arten gemeinsamen Merkmale. — 3. Bei weiterem Abstand in der Verwandtschaft erscheinen neue Stoffe, welche in einfachen chemischen Verhältnissen zu den Grundmerkmalen stehen, aus denen sie entstanden sind. Die physiologisch-chemischen Merkmale der Pflanze unterliegen einer Evolution“ (S. IWANOW 1926).

Dieser ganze Fragenkomplex ist von MCNAIR (1945b) unter Anführung der einschlägigen Literatur ausführlich dargestellt worden.

Von etwas anderer Seite und mit etwas abweichenden Ergebnissen wurde das Problem eines Zusammenhanges zwischen entwicklungsgeschichtlicher Stellung und Chemismus der Organismenfette von HILDITCH und seiner Schule studiert. Eine zusammenfassende Veröffentlichung in Buchform liegt vor (T. P. HILDITCH 1956). Verfeinerung der fettchemischen Analysemethoden ermöglichte hier eine viel weiter ins einzelne gehende Untersuchung der Fette vieler Organismen. Hierbei kamen Gesetzmäßigkeiten zutage, die für eine allgemein biologische Betrachtung der Entwicklungsgeschichte vielleicht bedeutsamer sind, als MCNAIRs Ergebnisse. HILDITCH fand, daß die Fette im Meere lebender Tiere einen komplizierteren Bau haben als die der Landtiere: sie enthalten mehr ungesättigte Fettsäuren mit 16 bis 20 oder bis 24 C-Atomen.

Ähnliches fand sich nun auch bei der Untersuchung der Fette von Wasser- und Landpflanzen. Abnahme der Zahl der verschiedenen Fettsäuren in den Fetten und Auftreten von Linol- und Linolensäure ist charakteristisch für die Fette der Landpflanzen. Einzelne Familien, ja sogar einzelne Arten können dann wieder ganz spezifische Fettsäuren aufweisen, aber im allgemeinen scheint mit zunehmender Höhe der phylogenetischen Stellung eine relative „Vereinfachung“ der Fette einherzugehen (T. P. HILDITCH, I. A. LOVERN 1936, HILDITCH 1952), wobei die Anzahl der Fettsäurekomponenten verringert wird, die aber doch, wie MCNAIR zeigte, bei Betrachtung der zugehörigen physikalischen oder chemischen Konstanten den Eindruck einer Höherentwicklung macht.

b) Genetik.

Die *genetische Bedingtheit* der Menge und Zusammensetzung der Fette in Pflanzen war ebenfalls Gegenstand mannigfacher Untersuchungen. *Mais* dürfte trotz seines verhältnismäßig geringen Fettgehaltes hier das erste Versuchsobjekt gewesen sein. R. PEARL und J. M. BARTLETT (1911) schlossen aus ihren Vererbungsversuchen mit einer weißen und einer gelben Maissorte, daß der Fettgehalt (gemessen als Rohfett) wahrscheinlich als ein eigenes, von anderen unabhängiges

Merkmal vererbt wird. W. E. LINDSTROM und G. FISK (1926) stellten andererseits bei ähnlichen Versuchen fest, daß der Fettgehalt der Maiskörner mit dem Gehalt an Kohlenhydraten derart zusammenhängt, daß hoher Fettgehalt und hoher Gehalt an freien Zuckern parallel gehen. Ihr Zahnmais enthielt 74% des gesamten Fettes im Embryo, ihr süßer Mais aber nur 64%. Etwas später haben W. E. LINDSTROM und G. FISK (1927) Zahlen über den Zusammenhang der genetischen Struktur von Maishybriden und dem Fettgehalt ihrer Samen mitgeteilt, die nicht ganz mit den früheren Befunden übereinstimmten. Die Unstimmigkeiten finden wahrscheinlich ihre Erklärung in der ziemlich geringen Vitalität der einen Elternpflanze. F. A. ABEGG (1929) hat dann einen Einfluß des Genes für „wachsiges Endosperm" auf den Fettgehalt gefunden. Dieses Gen scheint nicht nur einen höheren Fettgehalt, sondern auch ein Fett mit mehr freien Fettsäuren (höherer Säurezahl) zu bedingen.

Besonders *Soja*-Sorten wurden nicht nur auf hohen Ölertrag, sondern auch auf „gute Qualität", d. h. hohe Jodzahl des Öles gezüchtet. C. R. FELLERS (1921) berichtete, daß der Ölgehalt von 26 untersuchten Sojasorten zwischen 14,6 und 25,6% schwankte und S. WOODRUFF und H. KLAAS (1938) geben für 14 Sorten in verschiedenen Jahren gefundene Fettgehalte an. Ältere amerikanische Ergebnisse finden sich bei W. F. WASHBURN (1916) erwähnt. Die Vererbung der Eigenschaften „hoher" bzw. niederer Ölertrag wurde auch in Kanada festgestellt (F. SHUTT 1930) und auch aus Rußland liegen ähnliche Berichte vor (N. N. IWANOFF 1948), in denen z. B. für wilde Sojabohnen Fettgehalte von nur 8—9% angegeben werden. Noch niedrigere Werte (5%) wurden von F. G. DOLLEAR und Mitarbeitern (1940) in einer wilden Soja gefunden und F. LANGELD (1930) fand in frühreifen Sojasorten mehr Öl als in Spätsorten. L. J. COLE (1927) und Mitarbeiter fanden bei ihrer Züchtungsarbeit, daß die auf hohe Jodzahl (134) gezüchtete Soja aus spät und purpurfarbig blühenden Pflanzen bestand, während die Sorte mit niedriger Jodzahl (125) sich als eine kleine, frühe, weißblühende Sorte erwies. Die Vererbung des Ölertrages war hierbei unabhängig von der Vererbung der Höhe der Jodzahl. Der Ölgehalt der Ausgangssorten schwankte von 14,6—23,2% und die Jodzahl von 126—135. Ähnliche sortenbedingte Unterschiede im Ölgehalt und in der Ölqualität wurden auch von G. S. JAMIESON und Mitarbeitern (1933) sowie von R. C. BURRELL und A. C. WOLFE (1940) berichtet und W. J. MORSE (1950) hat Material über den Einfluß von Sorten sowie Reifezustand auf die Zusammensetzung des Sojaöls zusammengetragen.

Auch andere Leguminosen zeigten je nach Sorte und Art verschiedene Zusammensetzung (N. N. IWANOFF 1927, M. A. GUILLAUME 1923, A. J. YERMAKOW 1935) und viele Untersuchungen beschäftigen sich mit den entsprechenden Verhältnissen beim *Lein*, worüber z. B. W. F. WASHBURN (1916), A. L. BUSHEY und Mitarbeiter (1927) sowie F. RABAK (1918) berichteten. I. J. JOHNSON (1932) studierte die sortenbedingten Unterschiede in der Zusammensetzung von 46 Leinsorten und fand, daß der Ölgehalt in beträchtlichem Ausmaße von denselben Faktoren abhängt, die Samengröße, Reifedatum und Zahl der Tage zwischen Blüte und Reife bestimmen. Die Jodzahl des Öles hingegen wird unabhängig von diesen Faktoren vererbt, wie der gefundene niedrige, entsprechende multiple Korrelationskoeffizient zeigte. Auch J. V. EYRE und E. A. FISHER (1915), D. A. COLEMAN und H. C. FELLOWS (1927) sowie A. C. DILLMAN und H. HOPPER (1943) hatten erwähnt, daß größere Leinsamen einen höheren Fettgehalt haben als kleinere und P. NEUMANN (1941) hat das gleiche bei Lupinen festgestellt. Kein Zusammenhang zwischen Größe der Samen und Fettgehalt wurde von W. W. GARNER, H. A. ALLARD und C. L. FOUBERT (1914) bei Baumwolle gefunden. Die Einflüsse von Sorte, Reifezustand und Umweltbedingungen auf die Zusammensetzung der Baumwollsamen wurde von W. H. THARP (1948) besprochen.

A. J. YERMAKOW (1937) berichtete Ölgehalte von 31—47% in Lein und solche von 44—62% in Sesam. R. A. GROSS und C. H. BAILEY (1937) fanden beträchtliche Unterschiede in der Jodzahl der Öle der Leinsorten „Abessynischer Gelber“ und „Bison“, die ähnlich auch von W. G. ROSE und G. S. JAMIESON (1941) gefunden wurden. Auch E. P. PAINTER und L. L. NESBITT (1943a, b) fanden Jodzahlunterschiede von 18—20 Einheiten zwischen Ölen verschiedener Leinsorten und E. P. PAINTER (1944a) konnte ein von der Jodzahl unabhängiges Variieren der Linolsäure in den Leinölen feststellen. C. R. SCHOLFIELD und W. C. BULL (1944) haben sogar Gleichungen berechnet, mit Hilfe derer es möglich sein soll, aus der Jodzahl von Sojaöl die Menge der enthaltenen gesättigten Säuren, der Ölsäure, der Linolsäure und der Linolensäure zu finden. Auch DILLMAN und HOPPER (1943) haben ausführlich über die Unterschiede in der chemischen Zusammensetzung von 4 Leinsorten berichtet. Eine ausführliche Besprechung dieser Fragen findet sich bei J. B. MCNAIR (1945b, besonders S. 28ff.).

H. KUMMER (1932) fand in *Holcus lanatus* große sortenbedingte Unterschiede in der Menge der freien Fettsäuren in den Samenfetten.

Daß auch *Polyploidie* den Fettgehalt beeinflußt, wurde verschiedentlich festgestellt. Y. NOGUTI und Mitarbeiter (1940) finden eine um 18—26% höhere Gesamtätherextraktmenge in tetraploidem Tabak *(N. rustica* und *N. tabacum)* und C. SHAO-LIN und P. S. TANG (1945) berichteten über eine etwa 20%ige Zunahme des Ätherextraktes bei colchicininduzierter autotetraploider Gerste. Über diese und andere Änderungen des Pflanzenchemismus infolge von Polyploidie informiert ein Sammelreferat von G. R. NOGGLE (1946).

All diese genetisch bedingten Unterschiede in Ölgehalt und Ölqualität dürfen aber nicht darüber hinwegtäuschen, daß doch jede Pflanzenart oder -gattung, ja selbst -familie spezifisches Fett bildet, das vor allem durch die Art und Menge der darin enthaltenen ungesättigten Fettsäuren charakterisiert ist (T. P. HILDITCH 1928, 1951).

c) Klima.

Geographische Lage und damit untrennbar verbundene Einflüsse des *Klimas* sind bei vielen Pflanzen für Unterschiede in Fettgehalt und Fettqualität verantwortlich.

Schon J. W. LEATHER (1907) hat bei der Untersuchung von 52 Proben indischen *Leins* festgestellt, daß der durchschnittliche prozentuelle Ölgehalt der Proben aus verschiedenen Provinzen verschieden war und zwischen 38,3 und 42,6% schwankte und W. W. GARNER, H. A. ALLARD und C. L. FOUBERT (1914) berichteten, daß das Klima den Ölgehalt von Baumwollsamen beeinflußt. P. PIGULEWSKI (1915) hat wohl eine der ersten ausführlicheren Arbeiten hierüber veröffentlicht und unter ungünstigen klimatischen Bedingungen eine Zunahme der ungesättigten Fettsäuren festgestellt. S. L. IWANOW (1922/23) hat gemeinsam mit K. KORDASCHEV in Samen von *Camelina sativa* aus südlichen Gegenden Rußlands fast keine Linolsäure gefunden, während in nördlicheren Gebieten gewachsenes Saatgut diese Säure in beträchtlicher Menge enthielt. Dies veranlaßte ihn, „latente physiologisch-chemische“ Eigenschaften der Pflanzen zu postulieren, die nur unter bestimmten klimatischen Bedingungen manifest würden. In groß angelegten Versuchsreihen, die sich über fast die ganze Nord-Süd-Ausdehnung Rußlands erstreckten, wies S. L. IWANOW (1926, 1929a—d, besonders auch e, 1930a, b, 1931a, b, 1932a) dann nach, daß die Qualität des Öles ein und derselben reinen Linie von Lein weitgehend von der geographischen

Breite, d. h. den Klimabedingungen abhängt, in denen die Pflanzen wuchsen. In der Gegend von Moskau und weiter nördlich (58—65° n.Br.) gezogener Lein liefert ein trocknendes Öl mit Jodzahlen zwischen 180 und 200, während in etwa 40° n.Br. nicht trocknende Öle mit Jodzahlen zwischen 150 und 160 gefunden wurden. Öl aus *Helianthus annuus* zeigte Jodzahlen von 118 (38° n.Br.) bis 142 (55° n.Br.), *Camelina sativa* solche von 140 (49° n.Br.) bis 154 (56° n.Br.)[1].

Unterschiede in der geographischen Länge der Anbauorte hatten einen viel geringeren Einfluß, wie am Beispiel des Leins gezeigt wurde, dessen Öl auf 36° östl. Länge eine Jodzahl von 175 hatte, auf 85° östl. Länge aber eine Jodzahl von etwa 190 zeigte, ein Unterschied, der wohl durch das verschiedene Klima der beiden Anbauorte und nicht durch den Unterschied in der geographischen Länge bedingt sein dürfte. Auch Unterschiede in der Höhenlage der Anbauorte beeinflussen die Jodzahl wenigstens mancher Pflanzen. In etwa 0 m Seehöhe gezogene Leinpflanzen lieferten ein Öl mit einer Jodzahl von rund 157, während Pflanzen aus 1670 m Meereshöhe ein Öl mit einer Jodzahl von 179 enthielten. In *Paeonia anomala* aus 800 m Seehöhe fand sich ein Öl mit einer Jodzahl von 142, während aus Pflanzen aus 1860 m Höhe ein Öl mit einer Jodzahl von 158 gewonnen wurde. Eine Andeutung für einen Einfluß der Seehöhe, d. h. wohl der mittleren Temperatur, auf die Jodzahl des Öles von *Cedrus deodora* berichtet R. C. MALHOTRA (1931, 1933) aus dem Himalaja. Daß diese Unterschiede wenigstens weitgehend durch die verschiedenen mittleren Temperaturen der einzelnen Standorte bedingt sind, zeigte sehr schön der in Moskau ungewöhnlich kalte Sommer 1928. Das 10jährige Mittel (1907—1916) der Jodzahl von Leinöl aus der Moskauer Gegend ist 180 ± 4; Leinöl der Ernte 1928 hatte aber Jodzahlen von 188—190 und in dem ebenfalls in diesem Jahre ungewöhnlich kalten Dnjepropetrovsk hatte das Sojaöl eine Jodzahl von 130 anstatt wie sonst etwa 122. Leider scheinen keine Berechnungen der Korrelation zwischen der Temperatur und den Jodzahlen der verschiedenen Öle vorzuliegen. Daß Temperaturextreme in Verbindung mit anderen Faktoren künstlichen Klimas einen tiefgreifenden Einfluß auf die Zusammensetzung des Leinöles haben, ergab ein Versuch (S. L. IWANOW 1932b), der darin bestand, daß in Rußland gewachsenes Saatgut von Lein in Nolinsk (Rußland), in Liebefeld bei Bern (55 m Seehöhe), in Davos (1550 m Höhe) und im Tropenhaus des Berliner Botanischen Gartens kultiviert wurde. Die geernteten Öle zeigten folgende Jodzahlen: Nolinsk 185, Liebefeld 188, Davos 190, Tropenhaus 93. Die Klimaunterschiede reichten also im allgemeinen nicht hin, um eine starke Änderung der Jodzahl zu erreichen, wohl aber hatte das „Tropenklima" (25—30° C, möglichst dampfgesättigte Luft) des Berliner Glashauses die Jodzahl des Öles der dort herangewachsenen Pflanzen auf etwa die Hälfte erniedrigt.

Pflanzen, die Öle mit nur einfach ungesättigten Fettsäuren enthalten, sind viel unempfindlicher gegen Klimaeinflüsse (S. L. IWANOW besonders 1929e). So lieferten Oliven aus 29° n.Br. ein Öl mit einer Jodzahl von etwa 86, während russisches Öl aus 43° n.Br. die Jodzahl 85 und italienische Öle aus gleicher geographischer Breite eine Jodzahl von ungefähr 86 zeigten[1]. Ähnliches fand sich für *Brassica*- und *Ricinus*-Öle. IWANOW ging auch der Frage nach, ob sich die von ihm bei einzelnen Pflanzenarten beobachtete Förderung der Bildung besonders stärker ungesättigter Fettsäuren in höheren geographischen Breiten auch in der durchschnittlichen Zusammensetzung der Öle verschiedener Pflanzenarten der tropischen und gemäßigten Klimate wiederfinde. In den nördlichen *Pinus*-Arten, *P. silvestris* und *P. cembra*, fand sich viel Linolensäure, in der italienischen *P. pinea* nur wenig, in den tropischen *P. canariensis* und *P. longifolia* dagegen überhaupt

[1] Siehe auch den Beitrag M. L. MEARA, S. 29f. dieses Handbuchbandes.

keine. Dies alles führte ihn zur Aufstellung seines „Biochemischen Grundgesetzes" (S. L. IWANOW 1926), das schon erwähnt wurde (S. 282).

Auf wesentlich breiterer Basis hat J. B. MCNAIR die Frage nach den Zusammenhängen zwischen phylogenetischer Stellung, Klima und Fettqualität in Pflanzen behandelt (J. B. MCNAIR 1929, 1930a, b, 1934, 1945a). Seine wichtigsten Ergebnisse über den Zusammenhang zwischen phylogenetischer Stellung und Fett der Pflanzen sind bereits erwähnt worden. Aus den Resultaten seiner Arbeiten über den Klimaeinfluß auf die Pflanzenfette sei erwähnt, daß die Untersuchung von 318 Pflanzenfetten aus 83 Familien zeigte, daß 59% aller trocknenden und 50% aller halbtrocknenden Öle von Pflanzen der gemäßigten Klimate gebildet werden, und daß weniger als 1% der 102 untersuchten festen Pflanzenfette und etwas über 40% der nichttrocknenden Öle von Pflanzen höherer geographischer Breiten stammten. Die tropischen Pflanzenfette haben höhere Schmelzpunkte und niedrigere Jodzahlen als die meist öligen Pflanzenfette der gemäßigten Zonen. Ähnlich fand T. P. HILDITCH (1928) für Fettsäuren tropischer Pflanzenfamilien ein mittleres Molekulargewicht von 214, während sich für die Fettsäuren aus den Fetten von Pflanzen der kühleren Klimate der Wert 310 ergab. Die entsprechenden mittleren Schmelzpunkte waren 53° C (tropische Fette) und 23° C.

Tabelle 1. *Klimaeinfluß auf Jodzahl und Linolensäuregehalt von Leinöl.* (GROSS und BAILEY 1937.)

Anbau-ort	Leinsorte			
	Abessynischer Gelber		Bison	
	Jodzahl	Linolen-säure (%)	Jodzahl	Linolen-säure (%)
I	196	25	175	12
II	195	34	177	24
III	186	34	166	9
IV	196	18	166,5	11
V	187	19	166,5	18

Außer IWANOW haben auch andere Untersucher den Klimaeinfluß auf die Ölqualität festgestellt. Schon 1918 hatte F. RABACK darauf hingewiesen, daß die Jodzahlen von Leinöl nicht nur von der Leinsorte, aus der das Öl gewonnen wird, abhängen, sondern daß auch die geographische Lage des Anbauortes darauf Einfluß hat. R. A. GROSS und C. H. BAILEY (1937) berichteten über Jodzahl- und Linolensäuregehaltsschwankungen von zwei Leinsorten, die an fünf verschiedenen Stellen (alle in Minnesota USA.) angebaut waren. Allerdings verliefen die Schwankungen der beiden Kennzahlen nicht, wie erwartet werden könnte, gleichsinnig, sondern es fanden sich auch bei gleichbleibender Jodzahl Zunahmen der Linolensäuremengen (Tabelle 1).

Ähnliche Ergebnisse erhielten bei mehr ins einzelne gehende Untersuchungen W. G. ROSE und G. S. JAMIESON (1941). K. SCHMALFUSS (1941b) berichtete über Gefäßversuche mit Lein, die ergaben, daß die Jodzahl des gebildeten Öles von der Temperatur während der Reifung der Samen derart abhängt, daß bei höherer Temperatur mehr gesättigte und bei niedrigerer Temperatur mehr ungesättigte Fettsäuren gebildet werden. Dies gilt auch für das Nachreifen von unreif geernteten Samen im Exsiccator über Calciumchlorid, wobei bei 2° C sich in 4 Wochen die Jodzahl von 179 auf 184 erhöhte, während bei 31° C nur eine Erhöhung auf 180,6 eintrat. SCHMALFUSS fand die Jodzahl von der Dauer der Reifezeit der Samen weitgehend unabhängig. Auch Feldversuche (F. GIESECKE, K. SCHMALFUSS, E. GERDUN 1937) haben gezeigt, daß das Standortklima zur Zeit der Reife weitgehend für die Zusammensetzung des geernteten Öles bestimmend ist und daß die Temperatur mehr Einfluß als andere Faktoren hat. D. A. COLEMAN und H. C. FELLOWS (1927) fanden bei der Analyse von vielen hundert Leinproben, daß mit Zunahme der geographischen Breite der Herkunftsorte von 44° auf 49° n.Br. eine schwache Zunahme des durchschnittlichen Ölgehaltes der Samen einherging.

Große Feldversuche mit 4 Leinsorten, die an 54 Stellen ausgeführt wurden und die teilweise bis zu 10 Jahre liefen, veröffentlichten A. C. DILLMAN und T. H. HOPPER (1943). Die Versuchsorte repräsentierten Gegenden von 19—65° n.Br., von 65—123° westl. Länge, von 15—2500 m Seehöhe und Niederschlagsmengen von 75—1300 mm je Jahr sowie mittlere Julitemperaturen von 11—31° C. Der Flächenertrag war hauptsächlich vom verfügbaren Wasser bedingt und zeigte eine starke negative Korrelation zur mittleren Julitemperatur. Durch Wassermangel bedingte Ertragsabnahme und hohe Julitemperatur führte immer auch zu einer Abnahme des Ölgehaltes und der Jodzahl (vgl. auch E. P. PAINTER, L. L. NESBITT, T. E. STOA 1944). Die kleinsamigen Leinsorten Bison und Rio lieferten Öle mit höherer Jodzahl als die großsamigen Sorten Linota und Redwing. Im Gegensatz zu IWANOWs Hypothese, derzufolge das kältere Klima die Bildung ungesättigter Säuren begünstigt, deuten A. C. DILLMAN und T. H. HOPPER ihre Ergebnisse dahin, daß in wärmeren Klimaten Wassermangel und hohe Julitemperaturen eine gewisse Verkürzung der Vegetationsperiode des Leins bewirken, wodurch eine Art Notreife verursacht werde und wodurch diejenigen enzymatischen Vorgänge gehemmt würden, die die Bildung der stärker ungesättigten Fettsäuren verursachen. Diese Hypothese wird durch die Tatsache gestützt, daß die Einflüsse der Trockenheit und der hohen Temperaturen, soweit feststellbar, identisch sind. Die Länge der Vegetationsperiode ist aber ausschlaggebend für die Zahl der in einem Öl vorhandenen Doppelbindungen, je länger sie ist, desto mehr werden gebildet. Dies haben auch F. H. LEHBERG, W. G. MCGREGOR und W. F. GEDDES (1939) in Kanada festgestellt. Auch von Pilzen *(Aspergillus, Rhizopus)* ist bekannt, daß die Temperatur Einfluß auf die Jodzahl der gebildeten Fette hat, vgl. L. K. PEARSON und H. S. RAPER (1927)[1].

Tabelle 2. *Vegetationsdauer und Chemismus von Sonnenblumenöl.* (BARKER und HILDITCH.)

Vegetationsdauer Monate	Sorte 1			Sorte 2		
	Jodzahl	Ölsäure %	Linolsäure %	Jodzahl	Ölsäure %	Linolsäure %
3,5	135	24	64	126	37	52
3,3	120	44	45	112	58	34
2,2	107	65	27	98	72	19

Die Versuche von G. BARKER und T. P. HILDITCH (1950a, b) mit Sonnenblumen bestätigen die amerikanischen Erfahrungen aufs beste. 2 *Sonnenblumen*sorten wurden in Zanzibar an 3 Stellen mit verschiedener Dauer der Vegetationsperiode angepflanzt. Die Ergebnisse der Analyse der geernteten Öle sind in der Tabelle 2 zusammengefaßt.

Es ist offensichtlich, daß es die Länge der Vegetationsdauer ist, von der die chemische Zusammensetzung des Öles weitgehend abhängig ist. Wie groß dieser Einfluß ist, zeigen nicht nur obige Zahlen, sondern auch die Werte, die gefunden wurden, als in England Sonnenblumensamen angebaut wurden, die in sehr verschiedenen Gegenden Afrikas gewachsen waren und deren Öle sehr verschiedene Jodzahlen hatten (Tabelle 3).

Die Unterschiede im Erntegut sind viel geringer als die Unterschiede im Saatgut und die Tendenz aller Erntegutwerte, einem Mittelwert zuzustreben, ist deutlich sichtbar.

Auch aus Australien wurde berichtet, daß die Jodzahlen und die Linolsäure- sowie Ölsäurewerte von Sonnenblumenöl aus 14—36° südl. Breite einen deutlichen Gang mit der geographischen Breite zeigen (R. E. BRIDGE, A. CROSSLEY,

[1] Siehe auch den Beitrag M. STEINER, S. 209ff., insbesondere S. 240f. dieses Handbuchbandes.

T. P. HILDITCH 1951) und es ist sehr wahrscheinlich, daß die Unterschiede, die in verschiedenen Herkünften von Saflor- und anderen Ölen gefunden wurden, wenigstens zum Teil auf ähnliche Ursachen zurückgehen (C. BARKER, A. CROSSLEY, T. P. HILDITCH 1950). Auch E. P. PAINTER und L. L. NESBITT (1934b) berichten Analysen der Öle von 4 Leinsorten, die an 14 verschiedenen Orten angebaut waren und wobei Sorten- und Witterungseinflüsse auf die chemische Zusammensetzung der Öle festgestellt wurden.

An *Soja* wurde der Einfluß einer künstlichen Verkürzung der Vegetationsperiode auf den Fettgehalt von C. R. FELLERS (1921) untersucht. Bei 2 Sorten fand sich bei Anbaudaten zwischen 3. Juni und 1. August eine graduelle Abnahme des Ölgehaltes in den Samen um 5—10% des bei der frühesten Saat erreichten Höchstwertes von 19,6% bzw. 18,9%. R. W. STARK (1924) berichtet über Standortseinflüsse auf den Ölgehalt von Sojasorten und W. F. WASHBURN (1916) untersuchte in 3 aufeinanderfolgenden Jahren das Öl von 45 Sojasorten, die unter verschiedenen klimatischen Bedingungen gewachsen waren. Ähnliche Versuche führte M. ROMAGNOLI (1949) aus. F. G. DOLLEAR, P. KRAUCZUNAS und K. S. MARKLEY (1938 und 1940) führten eine gelegentlich aufgetretene abnorme Zusammensetzung einer Sojaölprobe auf eine besonders ungünstige Konstellation klimatischer und pedologischer Faktoren zurück. K. SCHMALFUSS (1941) fand in Gewächshausversuchen bei höheren Temperaturen immer niedrigere Jodzahlen des Sojaöles als bei niedrigeren Temperaturen und bei N. N. IWANOFF (1948) finden sich Angaben über die Abhängigkeit der Jodzahl von Sojaöl von der Niederschlagsmenge und der Temperatursumme. Andererseits fand derselbe Autor (N. N. IWANOFF 1927) eine bemerkenswerte Unabhängigkeit der Zusammensetzung anderer Leguminosen (Erbse, Wicke, Linse) vom Klima, was er auf die Unabhängigkeit dieser Pflanzen vom Bodenstickstoff zurückführt.

Tabelle 3. *Herkunft und Chemismus von Sonnenblumenöl.* (BARKER und HILDITCH.)

Herkunft (Afrika)	Jodzahl		Linolsäuregehalt in %	
	Saatgut (Afrika)	Erntegut (England)	Saatgut (Afrika)	Erntegut (England)
I	143	137	72	70
II	118	138	44	70
III	130	131	57	65
IV	136	136	65	60
V	132	138	58	68

Über Klimaeinflüsse auf die Qualität von Olivenöl berichtet S. A. KALOYEREAS (1948) und M. KONDO, Y. TERASAKE und U. MOTOTARO (1942) erwähnen eine Fettzunahme in dürregeschädigtem Reis. A. J. YERMAKOW (1935) fand klima- und jahrbedingte Unterschiede im Ölgehalt von Lupinusarten.

Der Einfluß von *Bodenverhältnissen* auf den Fettgehalt der Samen von *Lupinus albus* wurde von W. HEUSER, K. BOEKHOLT und G. ULICH (1934) untersucht. Auf leichten Sandböden war der Fettgehalt bei 2 Sorten höher als auf mittelschwerem Boden, bei 1 Sorte war er niedriger und bei einer weiteren Sorte war kein Einfluß des Bodens feststellbar. W. W. GARNER, H. A. ALLARD und C. L. FOUBERT (1914) haben bei Baumwolle keinen spezifischen Einfluß des Bodens auf den Fettgehalt der Samen gefunden, während F. LANGELD (1930) über einen Einfluß des Bodens auf den Ölgehalt von Sojabohnen berichtet. Ähnliches fand auch R. W. STARK (1924). Die im Boden für das Wachstum zur Verfügung stehende *Wassermenge* hat nach L. TOMBESI und M. TARANTOLA (1950) ebenfalls Einfluß auf den Fettgehalt verschiedener Samen. *Phaseolus vulgaris, Brassica campestris var. oleifera, Arachis hypogaea, Linum usitatissimum* und *Vicia faba* wurden untersucht. Erhöhung der Wassermenge bewirkte eine Zunahme des Fettgehaltes der Sprosse, aber eine Abnahme des Fettgehaltes der Samen.

3. Der Verlauf der Fettbildung und Fettspeicherung.

Ältere zusammenfassende Darstellungen des Verlaufes der Fettbildung in Früchten und Samen finden sich bei F. CZAPEK (1913) und E. F. TERROINE (1920), der besonders um die Herausarbeitung allgemeiner Gesichtspunkte bemüht war. Er unterscheidet bei der Fettbildung in Früchten und Samen 4 Abschnitte: I. Die Phase des Wachstums, während der Früchte und Samen ohne wesentliche Fetteinlagerung nahezu ihre endgültige Größe erreichen; II. die Phase der rapiden Fetteinlagerung, in der nahezu der endgültige Fettgehalt erreicht wird; III. die Phase stationären Fettgehaltes vor der Reife und IV. die Phase des Fettabbaues knapp vor der Vollreife. Diese letzte Phase eines schwachen Fettabbaues soll sich aber nur bei Früchten, nicht aber bei Samen finden. TERROINEs 4 Phasen wurden, wie die folgenden Angaben zeigen werden, auch von vielen späteren Untersuchern da und dort mehr oder minder deutlich ausgeprägt gefunden, doch wird es niemand wundernehmen, wenn man in ihnen keineswegs eine genaue Wiedergabe des Verlaufes der Fettbildung in allen Samen und Früchten sehen darf. Moderne, mehr oder minder aphoristische Zusammenfassungen stammen z. B. von K. S. MARKLEY (1947) und L. P. MILLER (1954).

Tabelle 4. *Entwicklung des Fettgehaltes in reifenden Oliven.* (HARTWICH und ULLMANN.)

Datum	Durchschnitts-Fruchtgewicht (relativ)	Öl-%	Ölmenge je Frucht (relativ)
18. VII.	19	0,55	0,3
10. VIII.	42	5,0	7
12. IX.	54	16,3	29
20. X.	83	21,3	59
15. XII.	96	28,3	91
16. I.	100	30,1	100
18. II. (reif)	92	25,7	88
100 =	2,4 g		720 mg

Wohl die ältesten wissenschaftlichen Beobachtungen über den Verlauf der Fettbildung in Früchten stammen von G. PRESTA, der 1785 berichtete, in der Zeit vom 2. September bis 15. April 13mal den Ölgehalt von *Oliven* bestimmt zu haben und Ölwerte von 0,09—25,3% in der Frischsubstanz gefunden zu haben (zit. nach C. GERBER 1901). S. DE LUCCA (1861, 1862) sowie C. HARZ (1870), A. ROUSILLE (1878), A. FUNARO (1880) und F. SCURTI und G. TOMMASI (1910) teilten ähnliche Zahlen mit. C. HARTWICH und W. ULLMANN (1902) fanden folgende Ölmengen in reifenden Oliven[1] (Tabelle 4).

Es darf nicht unerwähnt bleiben, daß bei keiner dieser Untersuchungen eine fehlerfreie Probenentnahme gewährleistet war und daß auch Veränderungen während des Transportes der Proben nicht ausgeschlossen waren. Obige Zahlenreihe zeigt aber, aus welchen Unterlagen E. F. TERROINE (1920) die erwähnte 4-Phasentheorie ableitete. Wie die Relativzahlen klar erkennen lassen, geht zunächst die Bildung von Fruchtsubstanz in ziemlichem Umfang vor sich, ohne daß entsprechende Fettmengen gebildet würden. Erst später, wenn schon etwa 83% der Fruchtsubstanz gebildet sind, erreicht und überschreitet die Bildungsgeschwindigkeit der Fette die der anderen Inhaltsstoffe. Auch Fettbildung aus anderen Inhaltsstoffen kann natürlich hierbei in größerem Ausmaß erfolgen.

Über die Entwicklung des Ölgehaltes von *Lein* liegen viel verläßlichere und neuere Angaben vor. Vielfach hat bei diesen Untersuchungen die Frage nach dem richtigen Zeitpunkt der Leinernte (zur Erzielung höchster Ölerträge) eine Rolle gespielt (B. B. ROBINSON 1931), da man einen etwaigen Ölverlust durch zu späte Ernte natürlich vermeiden will. Im allgemeinen fand man, daß die Leinsamen

[1] Zur Erhöhung der Übersichtlichkeit würden in allen folgenden Tabellen soweit als möglich die Fruchtgewichte und die Fettmengen bzw. Fettprozente in Prozent der erreichten Höchstwerte angegeben. Diese absoluten Höchstwerte sind am Ende der Tabellen angeführt, so daß dem interessierten Leser ein Berechnen der Originalwerte leicht möglich ist.

bereits 9—14 Tage nach der Blüte ihr größtes Volumen erreichen (A. C. DILLMAN 1928, I. J. JOHNSON 1932a), daß ihr Trockengewicht aber weiterhin ziemlich gleichmäßig bis zu etwa 24—28 Tagen nach der Blüte zunimmt, worauf erst kurz vor der Reife eine weitere Zunahme und später eventuell wieder Abnahme erfolgt. Der Ölgehalt in Prozent der Trockensubstanz beginnt am 7.—9. Tag nach der Blüte rapid anzusteigen und erreicht bereits um den 18. Tag nahezu seinen Höchstwert (s. auch A. L. BUSHEY und Mitarbeiter 1927, E. R. THEIS, J. S. LONG und G. F. BEAL 1929, M. F. BARKER 1932, F. H. LEHBERG, W. G. MCGREGOR und W. F. GEDDES 1939, E. P. PAINTER 1944b, L. L. NESBITT, A. J. PICKNEY, T. E. STOA, E. P. PAINTER 1943). Bei so manchen dieser Untersuchungen wurden die Flachsblüten beim Aufblühen markiert und nur wirklich gleich alte Samen wurden dann für die Analysen verwendet. E. P. PAINTER (1944b) hat hierzu im Verlaufe von 2 Jahren über 100000 Blüten markiert und für die Analysen der jungen Entwicklungsstadien bis zu 14000 Fruchtstände gleichen Alters je Analyse verwendet. Seine Zahlen können wohl als die genauesten verfügbaren angesehen werden und seien in der folgenden Tabelle im Vergleich mit denen einiger anderer Untersucher aufgeführt.

Tabelle 5. *Entwicklung des Fettgehaltes in reifenden Leinsamen.*

Autor	E. P. PAINTER (1944b)			A. C. DILLMAN (1928)			J. V. EYRE (1931)			J. V. EYRE (1931) (*Linum cribrosum*)		
Samenalter Tage nach Blüte	1000 Korngewicht (relativ)	Öl %	Öl relativ in 1000 Samen	1000 Korngewicht (relativ)	Öl %	Öl relativ in 1000 Samen	1000 Korngewicht (relativ)	Öl %	Öl relativ in 1000 Samen	1000 Korngewicht (relativ)	Öl %	Öl relativ in 1000 Samen
8	22	9,3	5	15	1,8	0,6	—	—	—	—	—	—
9	—	—	—	20	2,4	1,2	—	—	—	—	—	—
10	30	16,3	11	—	—	—	—	—	—	16	0,7	0,3
12	35	25,0	20	32	8,2	6	23	2,2	1,3	—	—	—
14	42	33,2	31	—	—	—	29	2,9	2,3	27	3,1	2,5
15	—	—	—	32	26,0	20	—	—	—	32	4,1	4
16	51	38,6	34	—	—	—	33	5,5	4,9	—	—	—
17	—	—	—	—	—	—	—	—	—	37	6,5	7
18	60	42,5	58	49	34,7	42	37	9,0	8,9	—	—	—
19	—	—	—	—	—	—	40	11,6	12	—	—	—
20	67	44,5	67	—	—	—	42	16,0	18	40	11,1	11
21	—	—	—	64	35,6	57	—	—	—	—	—	—
22	74	45,9	77	—	—	—	—	—	—	—	—	—
23	—	—	—	—	—	—	50	23,4	31	47	16,6	22
24	80	45,8	82	73	37,5	67	—	—	—	—	—	—
25	—	—	—	—	—	—	59	29,2	46	—	—	—
27	90	46,7	95	75	40,3	75	64	32,6	55	—	—	—
28	—	—	—	—	—	—	—	—	—	57	27,8	44
30	96	46,2	100	97	40,8	97	69	34,8	63	—	—	—
32	—	—	—	—	—	—	75	36,0	73	66	33,2	61
33	96	44,6	97	98	40,8	99	—	—	—	—	—	—
36	—	—	—	100	40,6	100	—	—	—	75	35,9	75
37	98	44,5	98	—	—	—	—	—	—	—	—	—
38	—	—	—	—	—	—	87	38,9	91	—	—	—
39	—	—	—	93	40,7	93	—	—	—	—	—	—
40	100	44,0	99	—	—	—	—	—	—	—	—	—
44	—	—	—	—	—	—	—	—	—	99	36,1	100
46	—	—	—	—	—	—	99	38,8	99	—	—	—
52	—	—	—	—	—	—	—	—	—	99	35,9	99
66	—	—	—	—	—	—	100	37,4	100	100	35,1	98
73	—	—	—	—	—	—	—	—	—	99	35,3	97
76	—	—	—	—	—	—	100	37,4	100	—	—	—
84	—	—	—	—	—	—	—	—	—	99	35,2	97
100 =	7,26 g		3,22 g	7,22 g		2,94 g	4,79 g		1,79 g	4,77 g		1,70 g

PAINTERs Zahlen zeigen, daß eine Abnahme des prozentischen Fettgehaltes natürlich nicht auch eine Abnahme der Gesamtfettmenge bedeutet. Überdies aber scheinen seine so exakten Versuche eine Abnahme auch der tatsächlichen Fettmenge nach dem 30.—32. Tag anzudeuten. EYREs Zahlen zeigen im großen und ganzen einen Verlauf der Fettbildung an, wie er auch aus den Versuchen von PAINTER und DILLMAN hervorgeht, lassen aber auf eine wesentlich längere Entwicklungsdauer seiner Versuchspflanzen schließen.

H. HALVORSEN (1955) hat die Atmungsquotienten von Leinsamen und Leinkapseln verschiedenen Reifezustandes untersucht und gefunden, daß der R.Q. allein keine Anhaltspunkte über die Intensität der Fettbildung liefert. Bei Temperaturen um 0° C fanden sich R.Q. der Samen von etwa 1, bei höheren Temperaturen ergaben sich höhere R.Q.s mit einem Maximum zwischen 10 und 20° C, was vielleicht auf ein Temperaturoptimum der Fettbildung in diesem Bereich schließen läßt. Die verschiedenen Entwicklungsstadien der Samen waren weniger durch verschiedene Höhe der Atmungsquotienten als durch verschiedene Atmungsintensität gekennzeichnet. Ein Höhepunkt der Fettbildung dürfte etwa 25 Tage nach der Blüte liegen (vgl. auch S. 325).

Die Entwicklung des Fettgehaltes von *Soja*bohnen wurde ebenfalls verschiedentlich verfolgt. N. N. IWANOFF (1948) erwähnt 1934 veröffentlichte Untersuchungen von M. I. SMIRNOWA und M. M. LAWROWA, die bei Soja 8 Reifungsphasen unterscheiden. Für die ersten drei geben sie Fettgehalte von 5,5%, 11,4% und 17,3% an; von der 4. Phase an ist der Fettgehalt konstant etwa 23,5%. Auch B. REWALD und W. RIEDE (1933) fanden während des Reifungsprozesses der Sojabohnen keine nennenswerten Änderungen des prozentualen Fettgehaltes. Neuere Zahlen stammen von R. O. SIMMONS und F. W. QUACKENBUSH (1954b), A. C. WOLFE, J. B. PARK und R. C. BURELL (1942) sowie von E. WOODRUFF und H. KLAAS (1938), die mehrere Sojasorten untersuchten und während der Zeit von etwa 25. August bis Ende September nur verhältnismäßig geringfügige Schwankungen der Fettprozente fanden (Zunahmen um 10—20% des Anfangswertes). Anders wird das Bild freilich, wenn man z. B. mit WOLFE und Mitarbeitern den durchschnittlichen Fettgehalt je Bohne berechnet. Die Relativwerte der folgenden Tabelle lassen sich aus WOLFEs Zahlen ableiten.

Die Entwicklung des Ölgehaltes geht auch hier im allgemeinen etwas langsamer vor sich als das allgemeine Wachstum der Samen, wie es in der Zunahme des durch-

Tabelle 6. *Entwicklung des Fettgehaltes in 3 Sojasorten.* (Nach WOLFE und Mitarbeitern 1942.)

Sorte	Bansei			Illini			Kura		
Samenalter Tage nach Saat	Fett %	Fett je Samen (relativ)	relatives Samengewicht	Fett %	Fett je Samen (relativ)	relatives Samengewicht	Fett %	Fett je Samen (relativ)	relatives Samengewicht
79	14,5	12	19	—	—	—	—	—	—
86	18,4	37	50	—	—	—	—	—	—
88	—	—	—	—	—	—	15,5	17	33
91	19,1	72	74	18,2	30	45	—	—	—
93	—	—	—	—	—	—	18,0	33	46
97	20,6	92	90	20,7	61	69	—	—	—
98	—	—	—	—	—	—	18,1	61	72
103	20,3	95	100	21,0	80	91	—	—	—
104	—	—	—	—	—	—	19,8	78	81
109	19,3	100	100	21,1	100	100	—	—	—
110	—	—	—	—	—	—	20,7	93	90
116	—	—	—	—	—	—	19,0	100	100
100 =		38,1 mg	588 mg		25,4 mg	323 mg		52,8 mg	769 mg

schnittlichen Samentrockengewichtes sichtbar ist. Die Veränderungen der Fettprozente sind natürlich nur von zweitrangigem Interesse und ihre relative Konstanz während der Samenentwicklung von Soja wurde auch von K. SCHMALFUSS (1941b) beobachtet. Es ist bedauerlich, daß nicht alle über Veränderungen des Fettgehaltes während der Reifung veröffentlichten Zahlen ein so klares Herausarbeiten der relativen Bedeutungslosigkeit der bloßen Angabe von Fettprozenten gestatten. Man darf ja schließlich nicht vergessen, daß die Fettprozente durch alle übrigen Bestandteile der Trockensubstanz genauso beeinflußt werden wie durch das Fett selbst. So spiegeln z. B. wahrscheinlich A. STROBELs (1926) Standraumversuche mit Lein, die Zunahme des Fettgehaltes mit Abnahme des Standraumes ergaben (39,7 statt 35,9%), eher Unterschiede der Assimilationstätigkeit und der Kohlenhydrat-(Membran-)bildung wider, als Unterschiede in der Fettbildung.

Tabelle 7. *Verlauf der Fettbildung in Sonnenblumenfrüchten.* (Nach BÜRKLE.)

Samenalter Tage	Relatives Trockengewicht von 1 Samen	Öl-%	Ölmenge je Samen (relativ)
5	22	30	13
10	22	29	12
15	31	39	25
20	32	35	22
25	58	48	55
30	51	51	52
45	83	60	100
50	100	49	98
55	85	49	84
60	73	54	78
100 =	42,5 mg		21,2 mg

Den Verlauf der Fettbildung in reifenden *Erdnüssen (Arachis hypogaea)* haben J. S. PATEL und C. R. SESHADRI (1935) sowie T. A. PICKETT (1950) untersucht. Vom etwa 35.—70. Tag nach der Blüte ist die Fettzunahme je Nuß von einer nahezu gleichstarken Eiweißzunahme begleitet, die Kohlenhydrate (Stärke, Saccharose, reduzierende Zucker, vor allem die Stärke) nehmen auf etwa $^1/_3$ bis $^1/_4$ ab, machen mengenmäßig bei der Reife aber nur etwa $^1/_{10}$ des Fettgewichtes aus.

Über die Entstehung der Fette in reifenden *Sonnenblumen*früchten hat B. BÜRKLE (1929) eingehend berichtet. Das unregelmäßige Reifen der Fruchtstände macht die Gewinnung verläßlich vergleichbarer Proben sehr schwer und dies mag der Grund für manche Unregelmäßigkeiten in BÜRKLEs Zahlen sein. Er berechnet das Samenalter in Tagen nach dem Aufblühen der letzten Blüten des betreffenden Blütenstandes (Tabelle 7).

Tabelle 8. *Verlauf der Fettbildung in Walnüssen.* (LECLERC DU SABLON.)

Datum	Relatives Trockengewicht je Nuß	Öl-%	Ölmenge je Nuß (relativ)
6. VII.	19	3	1
1. VIII.	20	16	5
15. VIII.	44	42	30
1. IX.	79	59	75
4. X.	100	62	100
100 =	10,51 g		6,62 g

In den Sonnenblumen finden sich also höchste Ölprozente und höchste tatsächliche Fettmengen zur selben Zeit. Auffallend ist die starke Trockengewichts- und Ölabnahme, die etwa 10 Tage vor der Vollreife beginnt. Zur Erzielung höchster Ölerträge müssen Sonnenblumen also schon *vor* der Vollreife geerntet werden. Dies hat auch M. IWANOW (1930) festgestellt. Bemerkenswert ist außerdem die außerordentlich rapide Entwicklung des Samens und der Gesamtfettmenge zwischen 20. und 25. Tage nach der Blüte.

Für *Walnüsse* hat LECLERC DU SABLON (1897) Werte über den Verlauf der Fettbildung veröffentlicht, aus denen sich die Zahlen der folgenden Tabelle ableiten lassen.

Ähnliche Zahlen fand auch F. M. McCLANAHAN (1909, 1913).

Für *Mandeln* hat ebenfalls LECLERC DU SABLON (1896) Zahlen über die Zunahme der Ölprozente angegeben. Der Ölgehalt betrug in Prozent der Trockensubstanz am 9. Juni 2%, am 6. Juli 10%, am 1. August 14,5%, am 1. September 44% und am 4. Oktober 46%.

Die Fettzunahme in reifenden *Raps*samen hat A. MÜNTZ (1886) verfolgt. Aus seinen Angaben lassen sich die Relativwerte der Tabelle 9 berechnen.

Tabelle 9. *Verlauf der Fettbildung in Rapssamen.* (MÜNTZ.)

Datum	Relatives Trockengewicht von 100 Samen	Fett-%	Fettmenge in 100 Samen (relativ)
1. VI.	22	14,2	8
7. VI.	28	21,0	14
16. VI.	35	31,5	27
27. VI.	69	44,5	74
2. VII.	90	43,6	95
7. VII.	100	41,5	100
13. VII.	91	41,7	91
100 =	549 mg		228 mg

Tabelle 10. *Verlauf der Fettbildung in Baumwollsamen.* (GRINDLEY.)

Samenalter Tage nach der Blüte	Relatives Gewicht der Samen aus 100 Kapseln	Fett-%	Fettmenge in den Samen aus 100 Kapseln (relativ)
21	14	2,2	1,2
25	21	2,5	2,2
31	35	2,4	3,5
34	43	3,6	6,4
41	62	10,3	27,2
51	100	21,7	91,4
60	94	25,3	100,0
100 =	239,1 g		56,7 g

Für *Ölrettich* und *Ölmadie* hat K. WEISSENBÖCK (1948) einige Prozentangaben gemacht, die aber keinen näheren Einblick in den Verlauf der Fettbildung vermitteln.

Die Fettbildung in *Baumwollsamen* wurde verschiedentlich untersucht, doch scheint der Sorten- und Wettereinfluß hier besonders groß zu sein. Alle Untersucher finden übereinstimmend, daß die Fettbildung in den Baumwollsamen erst sehr spät, d. h. etwa 20—30 Tage nach der Blüte, in größerem Umfang einsetzt, dann aber sehr intensiv vor sich geht. D. N. GRINDLEY (1950) gibt Zahlen an, aus denen sich die Relativwerte der Tabelle 10 ergeben.

Schon etwa 20 Jahre früher haben W. D. GALLUP (1927, 1928) und C. CASKEY und W. D. GALLUP (1931) die Baumwollölbildung verfolgt und zunächst gefunden (1927, 1928), daß nach dem Öffnen der Kapseln keine Änderung des Fettgehaltes mehr eintritt. Genauere Versuche mit der Baumwollsorte „Oklahoma Triumph 44" unter Markierung der Blüten lieferten dann Zahlen (1931), die zur Tabelle 11 verarbeitet wurden.

Tabelle 11. *Verlauf der Fettbildung in den Samen der Baumwollsorte „Oklahoma Triumph 44".* (CASKEY und GALLUP.)

Samenalter in Tagen nach der Blüte	Relatives Gewicht der Samen aus 100 Kapseln	Fett-%	Fettmenge in den Samen aus 100 Kapseln (relativ)
21	45	8,3	18
24	46	11,7	26
27	57	16,7	45
30	64	20,6	64
34	77	20,3	78
37	82	20,8	83
41	100	(20,4)	(100)
44	85	20,2	83
50	71	20,9	72
100 =	386 g		(79 g)

(Die eingeklammerten Zahlen sind leider nicht sicher, da die Fettprozente für die 41 Tage alten Samen fehlen. Das Gesamtbild des Verlaufes der Fettbildung wird dadurch aber wohl nicht beeinflußt.)

Auffällig ist an diesen Zahlen die beträchtliche Abnahme der Fettmenge vom 40. zum 50. Tag nach der Blüte unter gleichzeitiger Abnahme der übrigen Inhaltsstoffe (gleichbleibende Fettprozente, Abnahme des Trockengewichtes je Sameneinheit), ganz ähnlich wie es oben für die Sonnenblumen berichtet wurde.

In den Früchten von *Hicoria pecan (Carya olivaeformis)*, den *Pecannüssen*, haben C. J. B. THOR und C. L. SMITH (1935) den Verlauf der Fettbildung verfolgt, nachdem schon J. G. und N. C. WOODROOF (1927) über ähnliche Untersuchungen an 2 Pecanarten berichtet hatten. In den Pecannüssen wird zuerst die Samen- und Fruchthülle entwickelt, während der Embryo in 12 Wochen nicht über das 64-Zellenstadium hinauskommt. In den folgenden 5 Wochen entwickelt sich der Embryo weiter und die Speichergewebe füllen die Samenschale aus. Hierbei werden praktisch alle Inhaltsstoffe frisch von außen zugeleitet. Die Tabelle 12 zeigt den Verlauf der Ölbildung in diesen Nüssen.

Tabelle 12. *Verlauf der Fettbildung in Pecannüssen.* (THOR und SMITH.)

Fruchtalter in Tagen nach der Blüte	Fett-%	Fettmenge je Frucht (relativ)
92	0	0
144	20	2
155	58	20
169	70	53
177	72	69
184	72	89
191	72	98
193	74	100
219	75	89
100 =		4,5 g

Auch hier finden wir in der Vollreife eine Abnahme der Fettmenge und der Gesamttrockensubstanz in etwa gleichem Maße (Konstanz der Fettprozente). Außerdem scheint hier die sehr späte Entwicklung des Fettgehaltes im Gegensatz zu anderen Pflanzen derart vor sich zu gehen, daß die Fettbildung zunächst rascher erfolgt als die Bildung der übrigen Trockensubstanz. A. H. FINCH und C. W. VAN HORN (1936) haben vor allem morphologische Beobachtungen an reifenden Pecannüssen angestellt.

In der *Avocadobirne* haben D. APPLEMAN und L. NODA (1941) bei der Vollreife eine Fettabnahme ähnlich der in Pecannüssen gefunden. Auch die Entwicklung des Fettgehaltes geht in ähnlicher Form wie bei der Pecannuß vor sich, wobei bis November etwa ein Drittel des Fettes eingelagert wird, das sich schließlich in der im folgenden Mai/Juni vollreifen Frucht findet. Späte Fettbildung findet sich nach W. W. JONES (1937, 1939) und W. W. JONES und L. SHAW (1943) auch in den Samen von *Macadamia ternifolia var. integrifolia*, wobei ebenfalls ein Vorauseilen der Fettbildung vor Bildung der übrigen Trockensubstanz beobachtet werden kann. Erst 90 Tage nach der Blüte setzt die Fettbildung in nennenswertem Umfang ein und 85% des Fettes des reifen Samens werden in 30% der Entwicklungszeit gebildet. Eine Fettabnahme bei der Vollreife wurde hier nicht gefunden (Tabelle 13).

Tabelle 13. *Verlauf der Fettbildung in Macadamianüssen.* (W. W. JONES 1939.)

Samenalter in Tagen nach der Blüte	Öl-%	Ölmenge je Nuß (relativ)
90	5	0,7
100	20	8
125	50	32
150	57	60
175	63	79
200	68	92
215	70	100
100 =		1390 mg

In den Früchten von *Guizotia abyssinica*, der *Nigersaat*, haben D. L. SAHASHRABUDDHE und N. P. KALE (1933) den Verlauf der Fettbildung und die gleichzeitig vor sich gehenden Änderungen des Gehaltes an verschiedenen Kohlenhydraten verfolgt. Die Abb. 1 zeigt, was sie über den Verlauf dieser Umwandlungen fanden. Veränderungen in den Zuckermengen können natürlich nur sehr beschränkt zur Erklärung der Veränderungen der Fettmengen herangezogen werden, da ja ein dauernder Zustrom von Assimilaten in die Samen erfolgt und da daher die Zuckerbestimmungen höchstens Anhaltspunkte über etwaige Gleichgewichtskonzentrationen, nicht aber über umgesetzte Stoffmengen liefern können. Dies trifft natürlich nicht nur für *Guizotia*, sondern für alle an der Mutterpflanze reifenden Ölsamen und Ölfrüchte zu.

W. W. JONES und L. SHAW (1943) haben Veränderungen des Fettchemismus bei der Reife von *Macadamia*-Nüssen untersucht. Dieser Baum trägt in Honolulu zu Anfang Juni Früchte aller Reifestadien und stellt aus diesem und anderen Gründen (z. B. vorübergehende Kohlenhydratspeicherung in den Zweigen vor

Einlagerung der Fettreserven in den Früchten) ein sehr günstiges Versuchsobjekt dar. Die Früchte reifen etwa 215 Tage nach der Blüte, größere chemische Unterschiede im Fett der Embryonen sind aber nur in den zwischen dem 90. und dem 110. Tage nach der Blüte geernteten Früchten gefunden worden.

Andere, viel Öl enthaltende Früchte, bei denen die Entwicklung des Fettes während der Reifung verfolgt wurde, sind z. B. die *Ölpalme* (H. N. BLOOMENDAAL 1925, W. M. CROMBIE 1956, vgl. auch unten S. 301), *Okra (Hibiscus esculentus)* (C. L. WORLEY, T. J. RATCLIFFE und R. G. GROGAN 1942), der *Tungbaum (Aleurites fordii)* (H. M. SELL, F. A. JOHNSTON, F. S. LAGASSE 1946), *Gymnocladus dioica* (G. J. RALEIGH 1930) und *Ligustrum japonicum* (F. SCURTI und G. TOMMASI 1911).

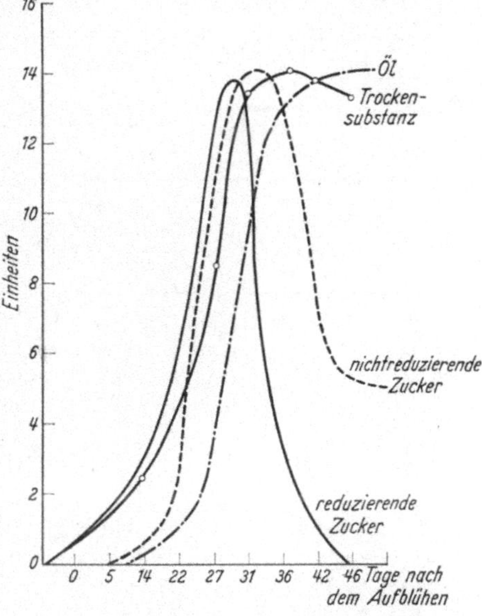

Abb. 1. Öl und Zucker in reifender Nigersaat. (Aus SAHASHRABUDDHE und KALE.) 1 Einheit Öl = 20 mg/100 Samen. 1 Einheit Trockensubstanz = 25 mg/100 Samen. 1 Einheit reduzierende Zucker = 2 mg/100 Samen. 1 Einheit nicht reduzierende Zucker = 3 mg/100 Samen.

Von den Untersuchungen über die Veränderungen im Fettgehalt reifender *Mehlfrüchte* seien die folgenden kurz erwähnt. N. B. GUERRANT (1927) hat 7 *Hirse*sorten in 3 Entwicklungsstadien (Milchreife, Knetreife, Vollreife) auf ihren Rohfettgehalt untersucht und Durchschnittswerte von 1,95; 2,58 und 3,10% des Trockengewichtes gefunden. Nach J. W. EVANS (1941) nehmen die ätherlöslichen Substanzen in reifenden *Mais*körnern vom 15. zum 57. Tag nach der Blüte von 3,8% auf 5,2% des Trockengewichtes zu, aber J. W. EVANS und D. R. BRIGGS (1941) fanden auch, daß eingehendere Analysen der ätherlöslichen Stoffe nur Fettwerte von 0,2—0,5% der Stärkewerte ergaben. Nicht unerwähnt sei, daß L. H. PULKKI und K. PUUTULA (1941) einen fast linearen Zusammenhang zwischen Fett- und Thiamingehalt von *Weizen*körnern fanden.

Tabelle 14. *Trockengewicht und Fett in reifenden Samen von Lupinus albus.* (NEUMANN.)

Relatives Trockengewicht von 1 Samen	Fett-%	Fettmenge je Samen (relativ)
11,4	2,5	2,6
11,8	2,7	2,9
17,3	3,1	4,7
19,5	3,4	5,9
22,7	3,5	7,0
27,8	4,4	10,9
30,4	5,1	13,9
30,0	5,4	14,5
49,5	7,6	33,5
76,9	10,2	70,0
78,2	10,5	74,5
100,0	11,2	100,0
100 = 478 mg		53,4 mg

In reifenden Samen von *Vicia faba* fand A. BLAGOWESTSCHENSKI (1925) kaum Veränderungen des prozentuellen Fettgehaltes trotz einer Zunahme der in 100 Samen enthaltenen Fettmenge von 11 auf 622 mg. P. NEUMANN (1941) hat den Fettgehalt reifender Samen von *Lupinus albus* verfolgt. Er verwendet als Bezugsgröße nur das durchschnittliche Trockengewicht der untersuchten Samen und nicht auch das Samenalter, das freilich viel schwieriger festzustellen ist. Die Tabelle 14 wurde aus NEUMANNS Zahlen berechnet und zeigt deutlich die unterschiedliche Geschwindigkeit der Trockengewichts- und Fettzunahme, die sich in einem wesentlichen Nachhinken der Fettbildung bis zur Erreichung von etwa 75% des Endtrockengewichtes der Samen äußert.

In den Samen von *Eriobotrya japonica*, der japanischen Mispel, fand A. L. KURSSANOW (1932) Werte für Trockensubstanz und Öl, aus denen die Zahlen von Tabelle 15 abgeleitet werden können.

Samen mit derart kleinen Fettmengen können natürlich schwer mit den fettspeichernden echten Ölsamen verglichen werden, aber die vorübergehende Abnahme der absoluten Fettmengen vor der Reife ist jedenfalls auffallend. Wieweit das Fett hier überhaupt als Reservestoff angesprochen werden kann, ist eine Frage, auf die nicht eingegangen sei.

Tabelle 15. *Trockengewicht und Fettgehalt der Samen der japanischen Mispel.* (KURSSANOW.)

Datum	Relatives Samentrockengewicht je Frucht	Öl-%	Ölmenge in den Samen aus 1 Frucht (relativ)
23. V.	50	0,99	79
27. V.	88	0,49	68
29. V.	84	0,48	63
1. VI.	100	0,58	90
3. VI.	86	0,75	100
100 =	2,25 g		14,5 mg

Den Verlauf der Fettspeicherung im Integument und Embryo von Alberta-*Pfirsichen* haben F. A. LEE und H. B. TUCKEY (1942) untersucht. Aus dem Integument verschwindet während der Reifung ein Großteil (nahezu 80%) des zunächst angesammelten Fettes. Im Embryo nahm vom 21. Juli zum 3. August die Fettmenge auf das 2,4fache zu, stieg aber bis zur Reife am 27. September nur mehr um etwa 10% an. — Einige Zahlen über die Fettzunahme (bei Abnahme der Fettprozente) in reifenden Samen der *Kornrade* finden sich bei M. KORSAKOW (1912); R. ULRICH (1939) veröffentlichte Zahlen über die Zunahme der Fettsäuren in *Efeu*früchten während der Reifung. Die Tabelle 16 ist aus seinen Werten zusammengestellt.

Hier finden wir also zunächst, nämlich solange die Samen erst weniger als 10% ihres Trockengewichtes gebildet haben, ein sehr beträchtliches Zurückbleiben der Bildung der Fette hinter der Bildung anderer Substanzen. Erst wenn mehr

Tabelle 16. *Der Fettgehalt von Efeufrüchten während der Reife.* (ULRICH.)

Trockengewicht (relativ)	Fett-%	Fettmenge je Frucht (relativ)
2,3	3,3	0,31
3,2	2,5	0,31
6,2	3,7	0,87
22	8,8	9,1
43	18,3	30,0
55	—	—
100	26,4	100,0
100 = 136 mg		100 = 35,6 mg

Tabelle 17. *Reifestadium und Fettgehalt.* (Nach KLEEBERGER.)

Samen bzw. Frucht	Fettgehaltsprozente in der			
	Milchreife (relativ)	Gelbreife (relativ)	Vollreife (relativ)	Vollreife (absolut)
Raps . . .	17	25	100	42,6
Rübsen . .	4	25	100	41,2
Mohn . . .	9	43	100	40,5
Lein. . . .	12	43	100	34,8
Leindotter .	16	60	100	26,5
Hanf . . .	50	82	100	29,5

als 50% der Trockensubstanz gebildet sind, holen die Fette auf. KLEEBERGER (1921) hat über Bestimmungen der Fettprozente in 3 Entwicklungsstadien (Milchreife, Gelbreife und Vollreife) der Samen bzw. Früchte von *Raps, Rübsen, Mohn, Lein, Leindotter* und *Hanf* berichtet. In relativen Zahlen ausgedrückt, fand er die Werte der Tabelle 17.

Hanf scheint eine gewisse Sonderstellung einzunehmen und dürfte sein Fett relativ früher bilden als die anderen Pflanzen. Andererseits aber sind die Entwicklungsstadien Milchreife und Gelbreife reichlich willkürlich; sie sind bei verschiedenen Pflanzen natürlich nicht notwendigerweise stoffwechselphysiologisch vergleichbar.

Versucht man in etwa, alle vorhandenen, recht dürftigen Untersuchungen über den Verlauf der Entwicklung des Fettgehaltes in verschiedenen Samen und Früchten zu überblicken, so kann man zusammenfassend vielleicht feststellen, daß TERROINES (l. c.) 4 Stadien den *Typus des Verlaufes* der Fettbildung im großen und ganzen richtig wiedergeben, daß jedoch der knapp vor der Reife eintretende Fettabbau nicht nur in einzelnen Früchten (Oliven, Avocado), sondern auch bei manchen Samen gefunden werden kann (Baumwolle, Raps z. B.).

Es sei hier darauf hingewiesen, daß anscheinend verschiedene Pflanzen — wenigstens in manchen Stadien der Fettbildung — ganz verschiedene Wege und Ausgangsstoffe zur Fettbildung zu verwenden scheinen. So liegen Untersuchungen vor (F. M. McCLENAHAN 1909, 1913) aus denen hervorgeht, daß in keinem Entwicklungsstadium der *Walnuß* Kohlenhydrate in nennenswerten Mengen gefunden werden können. Ganz ähnliche Angaben stammen schon von LECLERC DU SABLON (1897). Merkwürdigerweise liegen über das Verhalten der *Kohlenhydrate* in reifenden Fettsamen nur wenige exakte Untersuchungen vor. Ältere Literatur findet sich bei S. IWANOW (1912) besprochen, und die Ergebnisse der Untersuchungen W. M. CROMBIES (1956) an reifenden Früchten der Ölpalme sind weiter unten erwähnt (S. 301). B. BÜRKLE (1929) erwähnt eine „gallertartige Stärkevorstufe“ in jungen *Helianthus*früchten und J. B. THOMAS (1941) fand bei unreifen *Dammar*samen *(Shorea Wiesneri)* bei der Reife eine Umwandlung von Kohlenhydraten in Fett. V. ARASIMOVIČ (1933) hat Angaben über die Verhältnisse bei der Reifung von *Kürbis*samen gemacht und F. A. LEE und H. B. TUKEY (1942) haben in reifenden *Pfirsich*kernen nur geringe Zuckermengen gefunden, die bei der Reife nicht verschwanden. Wohl die ausführlichste Untersuchung über Zucker- und Fettgehalt bei der Samenreifung ist die von D. L. SAHASHRABUDDHE und N. P. KALE (1933) an Samen von *Guizotia abyssinica* („Niger seed“), deren Ergebnisse in obiger Abb. 1 wiedergegeben sind. Es sei hier ausdrücklich darauf hingewiesen, daß die Kurven dieser Abbildung durch passende Maßstabwahl auf gleiche Größe gebracht wurden und daß daher gleiche Höhe der Kurven zweier Bestandteile keineswegs gleiche Mengen dieser Stoffe andeutet. W. W. JONES und L. SHAW (1943) haben in reifenden *Macadamia*-Nüssen die Veränderungen des Zuckergehaltes (reduzierende Zucker und Saccharose) untersucht, T. A. PICKETT (1950) hat an reifenden *Erdnüssen* ähnliche Untersuchungen angestellt (vgl. oben S. 292) und D. N. GRINDLEY (1950) hat den Versuch gemacht, bei der Reifung von *Baumwollsamen* die Gesamtkohlenhydrate als Differenz zwischen dem Gesamttrockengewicht, der Rohfaser, dem Eiweiß und dem Rohfett zu erfassen. Wie die Tabelle 18 zeigt, lassen diese Zahlen keine Möglichkeit, die Zunahme des Rohfettes aus einer Abnahme der Kohlenhydrate zu erklären, da z. B. vom 41. zum 51. Tage nach der Blüte in den Samen aus 100 Kapseln das Rohfett um 37 g, das Eiweiß um 2 g und die Rohfaser um 36 g zunehmen, während die „Kohlenhydrate“ nur um 7 g abnehmen.

Tabelle 18. *Zusammensetzung verschieden alter Baumwollsamen.* (GRINDLEY.)

Tage nach der Blüte	In den Samen aus 100 Baumwollkapseln			
	Rohfett g	Eiweiß g	Rohfaser g	„Kohlenhydrate“ (Rest) g
21	0,7	5,3	0,8	25
25	1,2	9,8	1,3	36
31	2,0	12	3,4	63
34	3,6	15	6,9	72
41	15	27	10	91
51	52	49	46	84
60	57	46	45	68

Diese Zahlen zeigen klar, daß die Fettbildung hier wie in der Nigersaat SAHASHRABUDDHES, den Sojabohnen (WOLFE, PARK, BURELL 1942) und den *Macadamia*-Nüssen (W. W. JONES und L. SHAW 1943, M. E. HARTUNG und W. B. STOREY 1939) praktisch ausschließlich aus frisch zugeführtem Baumaterial und nicht aus vorher bereits im Samen abgelagerten Kohlenhydraten erfolgt, wie dies z. B. in den Samen von *Paeonia* (W. PFEFFER 1872) der Fall zu sein scheint. G. O. BURR und E. S. MILLER (1938) schließen aus den zeitweise unter 1 liegenden Atmungsquotienten reifender *Ricinus*samen, daß hier möglicherweise eine Zuleitung von Fettstoffen stattfinden könnte. Schon R. GODLEWSKI (1882) hatte an reifenden *Ricinus*- und *Mohn*früchten Atmungsquotienten bestimmt, aber nur Perioden mit Werten über 1 erfaßt. Zeitweise über und zeitweise unter 1 liegende Atmungsquotienten hatte C. GERBER (1897) bei reifenden Oliven gemessen.

Aus all diesen Angaben kann wohl geschlossen werden, daß eingehendere Untersuchung des Fettbildungsvorganges in verschiedenen Samen und Früchten wohl auch verschiedene chemische Wege der Fettbildung aufdecken würde.

4. Qualitätsänderungen des Fettes bei der Samenreifung.

Die ungeheure Kompliziertheit und Wandlungsfähigkeit der chemischen Umsetzungen in den Pflanzen lassen es nicht verwunderlich erscheinen, daß eine chemische Untersuchung der Pflanzenfette zu verschiedenen Zeitpunkten des Reifungsvorganges der Früchte und Samen beträchtliche Unterschiede und Veränderungen in ihrem Chemismus aufzeigt. Unsere Kenntnis dieser Veränderungen ist noch außerordentlich mangelhaft. Dies ist verständlich, da ja die Umständlichkeit der meisten Methoden der Fettchemie, zusammen mit unserem geringen Wissen um die die Fettsäurebildung steuernden Fermentsysteme, es bis jetzt fast unmöglich machte, tieferen Einblick in die Fettveränderungen während der Reife zu gewinnen. Es ist außerdem wahrscheinlich, daß viele, besonders ältere Untersuchungen durch das Eintreten von sekundären Veränderungen während der Extraktion und Aufarbeitung der Pflanzenteile wenigstens teilweise verfälscht wurden. Man wird wohl nicht fehlgehen, wenn man annimmt, daß der jeweilige Chemismus der Fette in reifenden Pflanzenteilen vor allem durch gerade gegebene Gleichgewichtslagen verschiedener Stoffwechselvorgänge bedingt ist. Mit dem fortschreitenden Alter der Samen und Früchte verschieben sich diese, und der nachlassende Zustrom von Assimilaten aus den Blättern, der abnehmende Wassergehalt reifender Samen, sowie einsetzende Aktivierungen und Inaktivierungen von Enzymsystemen tun ein Übriges zum Eintreten chemischer Veränderungen der Fette. Da die Biochemie der Fettbildung in den Pflanzen in einem anderen Abschnitt des vorliegenden Handbuches behandelt wird, seien im folgenden nur einige der auffälligsten Fettveränderungen während der Reifung von Früchten und Samen besprochen.

Am häufigsten wurden natürlich die Veränderungen des *Sättigungsgrades* der Fette untersucht, da der Gehalt der Fette an ungesättigten Fettsäuren ja vielfach den Gebrauchs- und Handelswert der Öle bestimmt (trocknende Öle). Wie T. P. HILDITCH (1951) betonte, deutet vieles darauf hin, daß die gesättigten und ungesättigten Fettsäuren auf verschiedenen Wegen entstehen und nicht wie IWANOW annahm, ineinander übergehen. Rein methodisch ist zu sagen, daß es auch bei Untersuchung dieser Frage nicht hinreicht, den Sättigungsgrad der Öle etwa aus einer Jodzahlbestimmung festzustellen, es ist vielmehr notwendig, die Anzahl der je 1 Samen vorhandenen Doppelbindungen zu bestimmen und während der Reife zu verfolgen. Raschere Neubildung ungesättigter als gesättigter Säuren kann ja sonst (wenn keine andere Bezugsgröße als die Gewichtseinheit Fett verwendet wird) vortäuschen, daß die ungesättigten Säuren aus den gesättigten entstehen oder umgekehrt. Anwendung radioaktiven Kohlenstoffes hat in den letzten Jahren gezeigt, daß frisch zugeführte Kohlenstoffatome in folgender

Reihenfolge in Fettsäuren der Sojabohnen auftreten: Ölsäure, gesättigte Säuren, Linolsäure, Linolensäure (R. O. SIMMONS und F. W. QUACKENBUSH 1954a). Da die Ölsäure konstant mehr der neuen Kohlenstoffatome enthält als die anderen Fettsäuren, ist es nicht unmöglich, daß Umwandlung von Ölsäure in andere Säuren in gewissem Umfang stattfindet.

Meist findet man, daß mit zunehmender Reife von Ölfrüchten und Ölsamen die *ungesättigten Fettsäuren* relativ *zunehmen*, daß das Öl also ungesättigter und — wenn überhaupt — trocknender wird (vgl. z. B. für Oliven S. A. KALOYEREAS 1948, für Leinöl E. D. PAINTER 1944, E. R. THEIS, J. S. LONG und G. F. BEAL 1929, EYRE 1931, M. F. BARKER 1932, für Sonnenblumen B. BÜRKLE 1929, für Nigersaat D. L. SAHASHRABUDDHE und N. P. KALE 1933, für Lupinen P. NEUMANN 1941, für Baumwolle D. N. GRINDLEY 1950 und für Soja K. SCHMALFUSS 1941b). Der Grad der Zunahme der Ungesättigtheit, den man findet, hängt natürlich weitgehend davon ab, wie früh oder wie spät nach der Blüte man die Untersuchungen beginnt. Für manche Pflanzen, vielleicht auch nur für manche Sorten oder spezielle Kulturbedingungen, scheint es charakteristisch zu sein, daß die relativen Mengen gesättigter und ungesättigter Fettsäuren oder, vorsichtiger ausgedrückt, die relative Anzahl von Doppelbindungen sich während recht großer Abschnitte der Samenentwicklung und Samenreifung kaum ändern (Baumwolle zwischen 25. und 60. Tag nach der Blüte, D. N. GRINDLEY 1950, Sonnenblume, K. H. BAUER 1934, keine Änderung zwischen 27. September und 15. November; Avocadobirne, D. APPLEMAN und L. NODA 1941, keine Änderung zwischen 18. Dezember und 4. Juni; Leinsamen, I. J. JOHNSON 1932a, keine Änderung zwischen 16. und 35. Tag nach der Blüte, E. P. PAINTER 1944, keine Änderung zwischen 12. und 32. Tag im Jahre 1941 und zwischen 20. und 40. Tag im Jahre 1942). Bei anderen Pflanzen (Nigersaat, D. L. SAHASHRABUDDHE und N. P. KALE 1933) bzw. bei Betrachtung anderer Zeitspannen oder Anwendung anderer Versuchsmethodik konnte man beträchtliche Zunahmen der Jodzahlen während der Samenreifung feststellen (vgl. auch F. H. LEHBERG, W. G. MCGREGOR und W. F. GEDDES 1939). In der folgenden Tabelle 19 sind einige Zahlenreihen aus oben erwähnten Arbeiten zusammengefaßt.

Tabelle 19. *Jodzahlen der Öle aus reifenden Samen (Früchten), Relativwerte.*

Tage nach der Blüte (etwa)	Nigersaat (SAHASHRABUDDHE 1933)	Sonnenblumen (BÜRKLE 1929)	Lein						Baumwolle (GRINDLEY 1950)
			(EYRE 1931)	(BARKER 1932)	(THEIS u. Mitarb. 1929)	(JOHNSON 1932)	Bison-Lein 1941 (PAINTER 1944)	Bison-Golden-Lein 1942 (PAINTER 1944)	
5	—	72,9	—	—	—	65,7	—	68,2	—
10	—	75,0	50,7	59,4	78,0	75,0	92,6	71,3	—
15	70,6	80,0	—	65,2	92,6	95,6	100,0	92,0	—
20	—	75,0	—	67,7	95,6	100,0	96,3	97,5	88
25	74,6	94,4	—	86,0	98,9	98,8	95,6	98,5	96,4
30	—	92,3	—	89,6	91,5	98,8	—	100,0	95,0
35	81,0	83,6	94,0	93,8	100,0	98,8	96,3	98,5	98,3
40	—	91,5	—	95,9	92,6	—	—	97,5	98,3
45	89,0	93,0	—	98,0	—	—	—	—	—
50	—	100,0	—	99,0	—	—	—	—	100,0
55	—	93,6	100,0	100,0	—	—	—	—	—
60	—	95,2	—	—	—	—	—	—	98,3
80	100,0	—	—	—	—	—	—	—	—
100 =	126	140	192	192	177	160	162	198	117

Es ist höchst wahrscheinlich, daß ein Großteil der kleineren Schwankungen obiger Tabelle entweder auf Unvollkommenheiten der angewandten Untersuchungs- oder Probenentnahmemethoden zurückzuführen ist, oder daß sich darin eine besondere Abhängigkeit der Jodzahl, besonders des Leinöles, von Außenbedingungen widerspiegelt. Daß niedere Temperaturen während der Samenreife und Verlängerung der Vegetationszeit Leinöle mit höherer Jodzahl entstehen lassen, wurde schon oben (S. 284f.) erwähnt.

Aus der obigen Tabelle ist auch zu ersehen, daß besonders beim Lein in den meisten Fällen die höchsten Jodzahlen einige Zeit *vor* der Vollreife auftreten. Auch A. L. BUSHEY, L. PURE und A. N. HUME (1927) sowie E. N. TARAN (1937) haben dies gefunden und darauf hingewiesen, daß die beste Ölqualität *vor* der Reife zu finden ist.

Auch ein *Abnehmen der Jodzahl* im Laufe des Reifungsvorganges wurde gelegentlich beobachtet, so von A. L. KURSSANOW (1932) bei der allerdings nur sehr wenig Fettsäuren enthaltenden japanischen Mispel (Jodzahlabnahme von 103 auf 79 zwischen 23. Mai und 3. Juni) und bei Avocadobirnen von D. APPLEMAN und L. NODA (1941), die zwischen 24. November und 18. Dezember eine schwache Abnahme der Jodzahl fanden.

Besonders in neuerer Zeit wurden außer der Jodzahl auch *andere Kennzahlen* der Fette während der Reifung untersucht. Sehr viele Zahlen finden sich bei E. R. THEIS, J. S. LONG und G. F. BEAL (1929) für Öl aus reifenden Leinsamen. Kaum eine der Zahlenreihen zeigt aber in überzeugender Weise irgendeinen Gang und viele der gefundenen Unterschiede dürften eher Unvollkommenheiten der chemischen Methoden als biologisch bedeutungsvolle Veränderungen widerspiegeln. Viel klarer sind die Zahlen von E. PAINTER (1944), der während der Leinreifung ein relatives Konstantbleiben (1941) oder anfängliches Abnehmen und dann Konstantbleiben des Ölsäuregehaltes des Leinöles feststellte. Linol- und Linolensäuregehalte schwanken stärker und zeigten wenigstens 1942 regelmäßiges, gegenläufiges Verhalten (Linolsäure: Abnahme von 20 auf 7%, dann Zunahme auf 8%; Linolensäure: Zunahme von 27% auf 65%, dann Abnahme auf 62%). All diese Schwankungen stellen aber in keinem Falle Abnahmen der tatsächlichen Mengen der betreffenden Fettsäure dar. Es handelt sich nur um Änderungen der Bildungsgeschwindigkeit der einzelnen Säuren, was zu Verschiebungen der gegenseitigen Mengenverhältnisse, aber nicht zu Ab- oder Umbau bereits gebildeter Fettsäuremoleküle führt. Auf diese stärkere Veränderlichkeit der relativen Mengen der stärker ungesättigten Fettsäuren hat auch T. P. HILDITCH (1951) wieder aufmerksam gemacht.

Außer in den Jodzahlen spiegeln sich die Änderungen des Chemismus der Fette reifender Samen und Früchte besonders in den *Säurezahlen* wieder, die bekanntlich den Gehalt der Fette an freien, nicht mit Glycerin zu Fetten veresterten Fettsäuren anzeigen. Nahezu ausnahmslos wird eine beträchtliche relative Abnahme der freien Fettsäuren während des Reifungsvorganges berichtet, wie die Tabelle 20 zeigt (vgl. auch E. F. TERROINE 1920).

Am interessantesten sind wohl die Zahlen von B. BÜRKLE (1929), der für die *Helianthus*früchtchen auch die „Säurezahl je Samen“ berechnete und nur ziemlich unregelmäßige Schwankungen (12, 11, 14, 17, 11, 10, 14, 12, 11, 14, 12, 11) statt der nahezu gleichmäßigen Abnahme von 4,4 auf 0,65 fand. Es ist klar, daß derartige Zahlen viel wertvoller wären als die üblichen Angaben der relativen Säurezahlen. Aus allen vorliegenden Zahlen geht nicht eindeutig hervor, in welchem Umfang freie Fettsäuren, die vielleicht bei ihrer Bildung sozusagen den Anschluß an ein Glycerinmolekül verpaßt haben, später noch verestert werden.

Für Ölpalmen wird sogar angegeben (H. N. BLOOMENDAAL 1935), daß die Säurezahlen nach der Vollreife etwas zunehmen.

All diese Angaben müssen aber mit beträchtlicher Vorsicht, ja Skepsis betrachtet werden, besonders seit W. M. CROMBIE (1956) mit neu entwickelter chemischer, chromatographischer und ultraviolettspektroskopischer Methodik begonnen hat, den Chemismus der Fette in reifenden Früchten der *Ölpalme (Elaeis Guineensis)* zu untersuchen. Die Ölpalme stellt wohl eines der interessantesten Objekte für die Erforschung des Fettstoffwechsels dar, da sie nicht nur in den Früchten zwei chemisch weitgehend verschiedene Fette von wirtschaftlicher Bedeutung bildet (das Palmöl des Pericarps und das Palmkernöl des Endosperms), sondern außerdem in der Testa, in Wurzeln, Sproß und Blättern weitere unterschiedliche Fette enthält. W. M. CROMBIEs Analysenverfahren erlaubt Identifizierung und Bestimmung individueller Fettsäuren mit mehr als

Tabelle 20. *Säurezahlen (freie Fettsäuren) in Fetten reifender Samen und Früchte.*

Pflanze	Autor	Anfangswert	Endwert
Arachis hypogaea . .	PATEL und SESHRADI 1935	52	5,1
Linum usitatissimum	PAINTER 1944	2,8	0,4
Linum usitatissimum	EYRE 1931	8,6	0,5
Linum cribrosum . .	EYRE 1931	83	0,1
Gossypium	GRINDLEY 1950	19	1
Helianthus annuus .	BÜRKLE 1929	4,4	0,65
Guizotia nigra . . .	SAHASHRABUDDHE 1933	110	5
Lupinus albus . . .	NEUMANN 1941	27	2
Lupinus albus . . .	NEUMANN 1941	50	0,5
Zea mays	EVANS 1941	34	7

6—8 C-Atomen noch im Milligrammbereich und stellt damit einen gewaltigen methodischen Fortschritt dar, der sicher noch größer ist als der, den R. O. SIMMONS und F. W. QUACKENBUSH (1954a) unter Zuhilfenahme radioaktiven Kohlenstoffs erzielten. Die große Variabilität der Ölpalmenfrüchte, die in ihrer Zusammensetzung nicht nur je nach Sorte, sondern selbst von Baum zu Baum und innerhalb des einzelnen Fruchtstandes stark schwanken [auch H. N. BLOOMENDAALs Untersuchungen (1925) waren durch dieses Verhalten erschwert worden], machte Beschränkung der Untersuchungen auf die Früchte von 3 Bäumen (Zuchtmaterial des Westafrikanischen Instituts für Ölpalmenforschung in Nigeria) notwendig. Die Blüten dieser Bäume wurden künstlich befruchtet, so daß das Alter der untersuchten Früchte genau bekannt war. Die Proben wurden 10 bis 16 Wochen nach der gleichzeitig erfolgten Bestäubung aller Fruchtstände der 3 Bäume in wöchentlichen Abständen sowie nach 19 und 20 Wochen genommen, wobei jedesmal 20 Nüsse aus der Mitte eines Fruchtstandes geerntet wurden. Perikarp und Früchte wurden getrennt in Aceton konserviert, vorläufig liegen aber nur die Ergebnisse der Untersuchung der Fruchtkerne (inklusive der Testa) vor. 11—14 Wochen nach der Befruchtung (je nach Baum verschieden) setzt ein steiler Anstieg des Trockengewichtes und der Fettmenge in den Früchten ein und hält bis zur Reife nach 20 Wochen unvermindert an. Das Fett der jungen Entwicklungsstadien ist chemisch sehr verschieden von dem der älteren und reifen Früchte und enthält z. B. nur Spuren der Laurinsäure, die etwa 50% der Fettsäuren der reifen Nüsse ausmacht. Das Innere der Früchte ist zuerst flüssig, wird dann gelatinös und schließlich fest. Bei dieser Reifung, die während etwa 10 Wochen vor sich geht, treten beträchtliche Änderungen in der Fettzusammensetzung ein. Die Ölsäure nimmt zunächst zu, dann ab und bildet von etwa der 15. Woche ab konstant ungefähr 15—16% aller Fettsäuren. Laurin- und Myristinsäure nehmen beträchtlich zu, Linol- und Stearinsäure nehmen ab. Nach der 15. Woche bleibt die Fettzusammensetzung recht konstant und W. M. CROMBIE spricht die Vermutung aus, daß die geringen Mengen ungesättigter Fette der jungen Früchte eher als Plasmafett denn als Reserfevett anzusprechen seien. Das stark gesättigte Reservefett wird offenbar zusätzlich zum Plasmafett eingelagert. Eine ähnliche Interpretation erfordern wohl auch die bereits erwähnten von W. W. JONES und L. SHAW (1943) bei *Macadamia* gefundenen Veränderungen im Chemismus der Fette der reifenden Früchte. In keinem Entwicklungsstadium der Ölpalmenfrüchte treten freie Fettsäuren auf (mit gewöhnlicher Methodik erhaltene Hinweise hierauf erwiesen sich als auf störende Nicht-Fettsäuren zurückzuführen!), und auch Umwandlungen gesättigter Fettsäuren in ungesättigte oder umgekehrt kommen offenbar nicht vor. Stärke, Rohrzucker und reduzierende Zucker zeigten in den Früchten der 3 Ölpalmen ein unterschiedliches und unregelmäßiges Verhalten, wobei Stärke von etwa 0,4% auf etwa 0,5—2% des Trockengewichtes zunimmt, Rohrzucker von 0,0—0,1% auf 0,8—1,4% ansteigt und die reduzierenden Zucker je nach Baum ab- (von 16% auf 0,2% und weniger) oder zunehmen (von 0% über 2% auf etwa 0,4%). Diese 3 Kohlenhydrate (andere wurden nicht bestimmt) machen nur 0,4—0,5% des fettfreien Trockengewichtes der Ölpalmenkerne aus, und über etwaige Kohlenhydratvorstufen der Fette liegt keine Information vor.

Die *unverseifbaren Bestandteile* der Pflanzenfette machen bei der Samenreifung ebenfalls wenigstens relative Mengenveränderungen durch. D. N. GRINDLEY (1950) hat berichtet, daß sie in Baumwollsamen von 28% auf 46% zunehmen (18 Tage nach der Blüte) und dann bis auf 1,3% des Rohfettes absinken. Umrechnung dieser Zahlen auf die Samen aus je 100 Kapseln zeigt aber, daß der relative Höchstwert nur etwa 1% des Samengewichtes oder 0,3 g entspricht. Dieser Wert nimmt bis zum 34. Tag nach der Blüte nur auf etwa 0,5 g zu, fällt am 41. Tag auf 0,4 g, steigt am 51. Tag auf 0,9 g und beträgt am 60. Tag 0,7 g. GRINDLEY macht diese geringen Mengen unverseifbarer Stoffe für die halbfeste Konsistenz des Baumwollfettes bis zum 51. Tag nach der Blüte verantwortlich. — In Leinsamen nimmt ab 16. bis 20. Tag nach der Blüte das Unverseifbare nach PAINTER (1944) etwa parallel mit dem Ölgehalt zu.

GRINDLEY (1950) hat auch das mittlere *Molekulargewicht* der Fettsäuren des Baumwollsamenfettes in verschiedenen Reifestadien bestimmt und fand eine langsame Abnahme von 300 (31 Tage nach der Blüte) auf 275 (60 Tage nach der Blüte).

5. Die Fette bei Nachreife und Stratifizierung.

Von besonderem pflanzenphysiologischen und chemischen Interesse sind aber jene Fälle, in denen in Samen und Früchten auch nach Abtrennung von der Mutterpflanze der Vorgang der Fettbildung weiterläuft. Leider sind nur ganz wenige dieser Fälle bisher mit moderner Methodik untersucht worden, obgleich hier ein vielleicht relativ leichter Zugang zum Chemismus der Fettbildung und der Stärke-Fettumwandlung offen zu sein scheint. W. PFEFFER erwähnte schon 1872, daß *Pfingstrosen*samen, wenn sie zur Zeit des höchsten Stärkegehaltes geerntet werden und der Nachreife überlassen bleiben, die Umwandlung von Stärke in Öl und die Bildung von Aleuronkörnern durchführen, so, wie wenn sie an der Mutterpflanze ausreiften. Auch in *Oliven* wurde Ölzunahme (wenigstens relativ) bei der Nachreife beobachtet (vgl. C. GERBER 1901). Bei *Lein* hat A. SCHISCHKIN (1872) die Erscheinung beschrieben, und J. V. EYRE (1931) hat das Verhalten der Ölprozente und der Jodzahl bei 15tägigem Nachreifen 15 bis 20 Tage nach der Blüte geernteter Leinsamen untersucht. In den 15-Tagesamen stiegen die Ölprozente bei 5tägigem Nachreifen um 20% und erreichten am 15. Tag der Nachreife 124% des Ausgangswertes; die Jodzahlen betrugen zu Beginn der Nachreife 125, nach 5tägiger Nachreife 129 und nach 15tägiger Nachreife 146. Samen, die 20 Tage nach der Blüte geerntet wurden, zeigten bei der Nachreife keine Zunahme des relativen Fettgehaltes, wohl aber nahm die Jodzahl des Öles von 130 auf 155 zu (15 Tage Nachreife). M. F. BARKER (1932) berichtete ebenfalls über eine Erhöhung der Qualität von Leinöl während einer 2monatigen Nachreife von *Flachs*, der eigentlich zur Fasergewinnung geerntet worden war, und K. SCHMALFUSS (1944) fand, daß Nachreife bei niederer Temperatur eine stärkere Erhöhung der Jodzahl als Nachreife bei höherer Temperatur bewirkt. KLEEBERGER (1921) hat milchreife, gelbreife und vollreife Samen von *Raps*, *Rübsen Mohn*, *Lein*, *Leindotter* und *Hanf* einer zweimonatigen Nachreife überlassen. Die milchreifen Samen zeigten hierbei kaum eine Zunahme ihres geringen Fettgehaltes, bei gelbreifen Samen trat eine bedeutende Zunahme der Fettprozente ein, ohne bis zum Fettgehalt der vollreifen Samen zu führen. Die vollreif geernteten Samen zeigten keine wesentlichen Veränderungen während der Lagerung. Eine gewisse Ausnahmestellung scheint der *Hanf* auch hier innezuhaben, da seine Körner schon zur Zeit der Gelbreife praktisch ihren vollen Fettgehalt erreicht haben und daher bei der Nachreife keine weitere Zunahme der Fettprozente erkennen lassen. Ähnliches fanden C. G. CHURCH und E. M. CHACE (1922) bei der Nachreife von *Avocadobirnen* verschiedenen Reifegrades. R. TIETZ (1953) berichtet über Fettabnahme und Zuckerzunahme während der Nachreife von *Oenothera*, *Nigella* und *Amaranthus*. Die ausführlichsten Untersuchungen zur Frage der Fettbildung und Fettveränderung bei der Nachreife hat wohl P. NEUMANN (1941) an *Lupinen* ausgeführt. Er findet bei der Nachreife sehr junger und fast ausgereifter Samen viel geringere Veränderungen als bei der Nachreife von Samen, die etwa die Hälfte ihres endgültigen Fettgehaltes gebildet haben. Beim Nachreifen wird der höchste Fettgehalt (auch absolut) etwa nach 3—4 Tagen erreicht und sinkt dann wieder etwas ab. Reifen die Samen in den Früchten nach, dann werden nicht nur Baumaterial für Fette, sondern auch andere Inhaltsstoffe aus den Fruchtschalen in die Samen zugeleitet (Zunahme der absoluten, aber geringe Zunahme der relativen Fettmenge). Die Jodzahl steigt bei der Nachreife an, die freien Säuren nehmen ab, doch werden nirgends die bei natürlicher Endreife beobachteten Werte erreicht. Temperaturen von 30° C sowie eintägige Lagerung der Samen in Wasser hemmte die Nachreife mehr als gleichlange Lagerung in n/10 KCl- oder n/10 K_2SO_4-Lösung; Sauerstoff fördert die Nachreife nur geringfügig, Stickstoff hemmt sie beträchtlich, und Blausäure bringt sie fast ganz zum Stillstand.

Auch die *Stratifizierung* von Saatgut, das zunächst nicht oder schlecht keimt, stellt eine Art Nachreifung dar, und S. ECKERSON (1913) hat bei *Crataegus gloriosa* festgestellt, daß bei der Nachreife der Samen bei etwa 5° C — bevor nach etwa 90 Tagen die Keimung einsetzt — eine Zunahme der freien Säuren, eine Abnahme der Fette und das Auftreten von Zuckern zu beobachten ist. Behandlung mit verdünnten Säuren verkürzte die Dauer der Nachreife auf etwa die Hälfte und zwar auch bei Samen, deren Testa entfernt worden war und die dadurch schon eine auf etwa $^1/_3$ verkürzte Nachreifeperiode hatten. Eine ähnliche Fettabnahme und Zucker- und Stärkezunahme hat in nachreifenden *Wacholder*samen D. A. PACK (1921a) beobachtet, wobei in vielen Fällen das Fett in eine leicht verlagerbare und wieder in Fett rückverwandelbare Substanz überzugehen schien, die kein Kohlenhydrat, sondern eher freie Säure zu sein scheint (D. A. PACK 1921b). Während der etwa 100tägigen Nachreife nahm das Rohfett von 54% auf 44% der Trockensubstanz ab, die Säurezahl (als % Rohfett) nahm von 1,9 auf 5,5 zu, die Jodzahl blieb unverändert.

6. Mineralstoffernährung und Fettgehalt.

Zahlreiche Untersuchungen beschäftigen sich mit dem Einfluß der Mineralstoffernährung der Pflanzen auf den Fettgehalt ihrer Samen. Schon A. SCHISCHKIN (1872) hat erste tastende Versuche mit *Lein* angestellt, und weder er noch irgendeiner der späteren Untersucher konnte einen größeren spezifischen Einfluß der Mineralnährstoffe auf den Fettgehalt der Leinsamen feststellen. J. V. EYRE und E. A. FISCHER (1915) sowie H. FABIAN (1928) haben sehr genaue Zahlen mitgeteilt. Der Einfluß der Mineralnährstoffe auf den Samenertrag und damit auf den Ölertrag ist immer viel größer als ein etwaiger Einfluß auf die Ölprozente, die in diesem Zusammenhang ja überhaupt ziemlich bedeutungslos sind, da sie, wie schon erwähnt, selbstverständlich von allen anderen Bestandteilen der Samen genau so abhängen wie vom Fettgehalt. Auch die Jodzahl des Leinöles wird von der Mineralernährung kaum beeinflußt, geringe, unregelmäßige Schwankungen (meist weit unter 5% der Jodzahl) wurden immer wieder festgestellt, so z. B. von K. SCHMALFUSS und H. MICHEEL (1935). Hohe Wassergaben ermöglichen die Bildung von Leinöl mit höherer Jodzahl, Sulfat ergibt, besonders im Vergleich zu Chlorid, eine deutliche Senkung der Jodzahl um 6—8 Einheiten, aber größere Mengen Kalium erhöhen die Jodzahlen (K. SCHMALFUSS 1936a, 1939). M. GROSS (1925) berichtete, daß physiologisch saure Stickstoffdünger den relativen Fettgehalt der Leinsamen erhöhen, daß schwach alkalische ihn nur in größeren Gaben erhöhen und daß alkalische Stickstoffdünger keinen Einfluß auf den Fettgehalt haben. Versuche mit steigenden Stickstoffgaben zeigten bei 80 kg N/ha ein Absinken der Fettprozente und der Jodzahl ohne Einbuße an Fettertrag. CaO in Mengen von 40 q/ha ergab höhere Fettprozente und etwas geringere Jodzahlen als geringere CaO-Gaben (K. SCHMALFUSS 1937).

Auch bei *Soja* übertrifft der ertragsteigernde Einfluß einer Mineraldüngung meist bei weitem jeden Einfluß auf den relativen Fettgehalt, wenn auch Ca eine Senkung (C. R. FELLERS 1918, R. W. STARK 1924) und Sulfate im Vergleich zu Chloriden eine Erhöhung der Fettprozente zu bewirken scheinen (M. STERZ 1940). Kalium soll nach H. WILFARTH und G. WIMMER (1902) und nach F. GIESECKE und YI LUNG LIU (1940) den Fettgehalt und Fettertrag erhöhen, andere Untersucher aber betonen (H. LÜDECKE, K. SAMMET und W. LESCH 1941, K. SCHMALFUSS 1941a), daß Sortenunterschiede praktisch weit wichtiger sind als Kalium- oder Phosphateinflüsse auf den Fettgehalt und die Jodzahl. Manche der erhaltenen Ergebnisse hängen sicher auch davon ab, ob man mit Kalimangelpflanzen oder mit halbwegs gut mit Kali bzw. Phosphor versorgten Pflanzen beginnt und den Einfluß zusätzlicher Mineralgaben untersucht. Auch mit *Winterraps*, *Ölrauke*, *Raps*, *Rübsen*, *Hanf* und *Ölrettich* wurden einige Versuche ausgeführt, ohne daß etwas anderes gefunden wurde, als der Einfluß der Düngung auf den Ertrag und allenfalls ein Einfluß der Stickstoffdüngung auf den Eiweißgehalt sowie damit verbundene geringe Verschiebungen der Fettprozente und unbedeutende Änderungen der Jodzahlen (R. KAYSER 1925, K. SCHMALFUSS 1938, 1941a, K. SCHARRER und R. SCHREIBER 1941a, b). Genauere Zahlen für Mohn finden sich bei K. SCHMALFUSS (1936b); russische Untersuchungen an Lupinen wurden von M. J. SMIRNOWA (1938) berichtet und H. J. SNIDER (1952) fand in KCl-gedüngtem *Mais* eine Erhöhung des Fettgehaltes. An *Baumwolle* führten W. W. GARNER, H. A. ALLARD und C. L. FOUBERT (1914) sowie M. GIEGER (1941) Untersuchungen aus, über Versuche an *Aleurites*bäumen berichten S. R. GREER, T. E. ASHLEY, G. F. POTTER und E. ANGELO (1944) und *Hibiscus esculentus* wurde von C. L. WORLEY, T. J. RATCLIFFE und R. G. GROGAN (1942) mit verschiedenen Phosphatgaben kultiviert und genau analysiert. An *Oliven* erzielte

S. A. KALOYEREAS (1948) Ölertragszunahmen durch Injektion der Bäume mit Mineralsalzlösungen.

Auch der *Zeitpunkt* der Mineraldüngung (W. SCHROPP und B. ARENZ 1940, K. SCHMALFUSS 1941a) sowie andere Kulturmaßnahmen (M. GRIEGER 1941 an Baumwolle) sowie insbesondere der Einfluß verschiedener *Saatzeiten* (A. L. BUSHEY, L. PUHR und A. L. HUME 1927, A. ERMAKOV 1933, I. J. JOHNSON 1932a, F. H. LEHBERG, W. G. MCGREGOR, W. F. GEDDES 1939, M. ROMAGNOLI 1949) wurden in Zusammenhang mit dem Fettgehalt der Samen studiert und W. BRAUN (1939) berichtete, daß *Bestrahlung* von Leinpflanzen mit Ultraviolett von ungefähr 365 mμ keinen Einfluß auf den Fettgehalt hatte.

7. Die Fette bei der Lagerung.

Soweit Untersuchungen darüber vorliegen, fand man verständlicherweise, daß in lebenden, trocken lagernden Samen kaum kurzfristige chemische Veränderungen vor sich gehen. Freilich spielt hierbei der Wassergehalt die Hauptrolle, aber auch die Lagerungstemperatur hat beträchtlichen Einfluß. Die Veränderungen der Fette in zu feucht lagernden Samen gehen vor allem auf die Tätigkeit der in allen fetthaltigen Samen vorhandenen Lipasen zurück und bestehen in einer beträchtlichen Zunahme der freien Fettsäuren infolge lipatischer Fettspaltung. Ausführliche Untersuchungen hierüber liegen z. B. an *Baumwollsamen* vor, in denen auch Eiweißdenaturierung und Keimfähigkeitsverlust eintreten. Bei Wassergehalten von mehr als 13% tritt eine rasche Zunahme der freien Fettsäuren ein (F.R. ROBERTSON und J. G. CAMPBELL 1933, M. D. SIMPSON 1942, M. L. KARON und A. M. ALTSCHUL 1944, L. KYAME und A. M. ALTSCHUL 1946, A. M. ALTSCHUL 1948); in wärmeren Gegenden und bei anderen Baumwollsorten kann dies schon bei Wassergehalten von 8—10% eintreten (D. M. SIMPSON 1935). In neuester Zeit wurde gefunden (M. G. LAMBOU, N. S. PARKER und H. R. CARUS 1956), daß Spritzen entlaubter Baumwollstauden mit Maleinsäurehydrazid das Auftreten freier Fettsäuren auch in feucht gelagertem Saatgut durch fast $^1/_2$ Jahr verhindern kann. Durch Behandlung mit Ammoniakdämpfen vor der Lagerung (etwa 1 Std, bis das p_H eines Samenbreies etwa 8 beträgt) läßt sich dieses Verderben der Baumwollsamen ziemlich weitgehend verhindern (A. M. ALTSCHUL, M. L. KARON, L. KYANE und C. M. HALL 1946). Auch künstliche Trocknung kann natürlich durchgeführt werden (R. A. RUSCA und F. L. GERDES 1942). Samen mit mehr als 1—2% freien Fettsäuren im Öl sind kaum mehr keimfähig (R. A. RUSCA und F. L. GERDES 1942, besonders aber C. L. HOFFPAUIR, D. H. PETTY und J. D. GUTHRIE 1947 und 1948).

An vier indischen *Raps*sorten (*Brassica campestris* L. *var. toria* D. und F., *var. dichotoma* Watt, *B. trilocularis* H. G. und T. und *B. juncea* H. f. und T.) haben A. ZAFER und S. AHMAD (1946) den Einfluß 2jähriger Lagerung auf Fettgehalt, Säurezahl, Verseifungszahl und Jodzahl untersucht und überall nur unwesentliche Veränderungen gefunden. Über das Lagerverhalten der Früchte von *Carya pecan*, *Juglans regia*, *Juglans nigra*, *Prunus amygdalus*, *Corylus avellana* und *Macadamia ternifolia* wurde von R. C. WRIGHT (1941) berichtet. Chemische Methoden erwiesen sich ihm weniger brauchbar als Geschmacksproben, und er fand, daß Pecannüsse am raschesten verderben. Bei 21° C verderben sie innerhalb von 3 Monaten, Walnüsse *(J. nigra)* und Mandeln halten etwa 8 Monate, *Juglans regia* und Haselnüsse verderben noch später, und die Haltbarkeit der *Macadamia*-nüsse betrug etwa 10 Monate. Am längsten halten sich alle Nüsse bei Lagerung im Vakuum; nicht alle Nüsse zeigen das Auftreten von echter Ranzigkeit, viele bekamen nur einen dumpfen Geschmack, für den keine chemischen Anhaltspunkte gefunden werden konnten. J. B. McGLAMEREY und M. P. HOOD (1951) fanden, daß Wärmebehandlung das Ranzigwerden der Pecannüsse beträchtlich verzögert. An Walnüssen haben D. D. MUSCO und W. V. CRUESS (1954) das Auftreten von freien Fettsäuren bei der Lagerung verfolgt und es wie bei den Baumwollsamen weitgehend vom Wassergehalt abhängig gefunden. Ein schwaches Zunehmen des Säuregrades von Samenextrakten von Leguminosen (*Vicia villosa, Trifolium pratense, Phaseolus, Melilotus albus, Medicago sativa*) nach längerer (bis zu 8jähriger) Lagerung fand J. K. WILSON (1929), und es ist nicht ausgeschlossen, daß es sich auch hier wenigstens teilweise um das Auftreten freier Fettsäuren handelt. Daß verschiedene *Erdnuß*sorten in verschiedenem Ausmaß zu Peroxydranzigkeit neigen, haben T. A. PICKETT und K. T. HOLLEY (1951) gefunden. In stratifizierten Samen von *Juniperus scopulorum* haben M. AFANASIEV und M. CRESS (1942) in qualitativer Weise Veränderungen des Fettgehaltes beobachtet.

Auch in lagerndem *Getreide* ist die Zunahme freier Fettsäuren ein guter Gradmesser für seinen Qualitätszustand (D. W. ROBERTSON, C. C. FIFIELD, L. ZELENY 1939, V. L. KRETOVICH 1945, L. ZELENY 1948, 1949), wobei allerdings bei der sehr fettarmen Gerste kaum verläßliche Werte zu erhalten sind (P. H. BLUM 1954). Daß z. B. beim Mais hier auch genetisch bedingt Sortenunterschiede auftreten, hat F. A. ABEGG (1939) gefunden.

Das rasche Verderben lagernder *Oliven* ist allgemein bekannt (vgl. z. B. J. WIESNER 1927) und F. E. HUELIN und R. A. GALLOP (1951) haben gezeigt, daß während der Lagerung von *Äpfeln* (Sorte "Granny Smith") auf der Schale eine Zunahme des ausgeschiedenen Fettes und eine Erhöhung seiner Jodzahl eintreten.

8. Zur physiologischen Bedeutung der Samenfette.

Die Fette stellen bekanntlich in vielen Samen die Hauptmasse der Reservestoffe dar — immer zusammen mit mindestens verhältnismäßig beträchtlichen Eiweißmengen. Es ist völlig müßig, zu fragen, warum einzelne Pflanzen fetthaltige Samen, andere aber stärkereiche Samen ausbilden; beide Reservestoffe ermöglichen offenbar die Existenz und Verbreitung der betreffenden Arten; wieweit darüber hinaus die Reservestoffe etwa spezielle physiologische Funktionen zu erfüllen haben, ist schwer zu sagen. Einzelne Untersuchungen zu dieser Frage liegen vor, lassen aber natürlich keinerlei zwingende Kausalzusammenhänge erkennen. H. KUMMER (1932) fand, daß bei Gramineen und insbesondere bei verschiedenen Sorten von *Holcus lanatus* die Lichtabhängigkeit der Keimung und die Säurezahl das Samenfettes derart gleichsinnig variieren, daß einer höheren Säurezahl ein höherer Prozentsatz im Dunkeln keimender Samen entspricht. Auch andere stark lichtbedürftige Samen wie die von *Nicotiana tabacum, Lythrum salicaria* und *Epilobium roseum* enthalten Fette mit sehr kleiner Säurezahl. H. TIETZ (1953) hat Anhaltspunkte dafür erbracht, daß dies mit lichtabhängiger Aktivierung der Samenlipasen zusammenhängen könnte, wodurch bei manchen Samen nur im Licht die offenbar notwendige Bildung freier Fettsäuren in ausreichendem Maße vor sich ginge. P. PARIJA und P. MALLIK (1940) haben gefunden, daß fetthaltige Samen resistenter gegen hohe Temperaturen sind als stärkehaltige.

Daß in Ausnahmefällen Fette auch physiologische Funktionen haben können, die von ihren normalen Aufgaben völlig verschieden sind, zeigt der von E. HEINRICHER (1915) berichtete Fall der Loranthaceae *Arceutholobium Oxycedri*, der Wacholdermistel, bei der die winzig kleinen weiblichen Blüten einen Tropfen fetten Öles ausscheiden, der offenbar die Aufgabe hat, Pollen aufzufangen und der später wieder in die Blüte eingesogen wird.

II. Die Fette in Stammgeweben von Holzpflanzen und in sonstigen Speicherorganen[1].

Daß das Holz der Bäume neben all seinen anderen Funktionen wenigstens zeitweise auch als Speicherorgan fungiert, wurde schon von N. J. C. MÜLLER (1875) beobachtet, der in der Fichte das Verhalten der sog. transitorischen Stärke untersuchte, ihr periodisches Auftreten und Verschwinden im Laufe des Jahres verfolgte und feststellte, daß gegen Ende November die Rinde aller untersuchten Bäume frei von Stärke war. E. MER (1879) hat kurz darauf in der Fichte und in anderen Bäumen beobachtet, daß ihr Stärkegehalt im Winter abnimmt, daß aber zu dieser Zeit ein vermehrtes Auftreten von Ölkügelchen festzustellen ist. Wenige Jahre später hat E. RUSSOW (1882) mehrjährige Beobachtungen an 92 Baum- und Straucharten mitgeteilt, bei denen er fand, daß im Phloem ein spätherbstliches Stärkemaximum mit darauffolgender wenigstens teilweise temperaturabhängiger Stärkelösung auftritt. Er nahm eine Umwandlung der Stärke in Fett an. Auf der Odessaer Tagung der russischen Naturforscher und Ärzte wurde dann 1884 über A. GREBNITZKYs Untersuchungen über die jährliche Periode der Stärkespeicherung in den Zweigen der Bäume berichtet, wobei ein Maximum im Herbst und ein zweites vor Knospenausbruch angegeben wurde, und wobei die Linde als ein Baum erwähnt wird, in dem die Stärke im Winter völlig verschwindet. In der Diskussion bemerkte J. BARANETZKY, daß nach seinen Beobachtungen im Winter in vielen Bäumen, so z. B. auch in der Linde, statt Stärke Fett als Reservestoff auftritt, daß andere Bäume im Winter sowohl Stärke als auch Fett, wieder andere dagegen nur Stärke enthalten. J. SUROŽ (1890) beobachtete, daß in *Tilia*, *Caragana* und *Populus* im Herbst ein Zerfall der Stärkekörner eintritt und daß dann zwischen den entstandenen kleinen Bruchstücken Fetttröpfchen verschiedener Größe auftreten. Bei *Betula* und *Prunus* aber wandeln sich die Stärkekörner in kleisterähnliche Tropfen um, die bald keine Jodreaktion mehr geben, sich aber nun mit Osmiumsäure schwärzen und bei *Betula* bald durch Tropfen echten fetten Öles ersetzt werden.

Im Jahre 1891 hat dann A. FISCHER seine bekannte Arbeit über die Physiologie der Holzgewächse veröffentlicht und darin die Typen der „Fettbäume“ und der „Stärkebäume“ aufgestellt. Die meist weichholzigen Fettbäume wie Linde, Birke, Föhre, Erle, Pappel und Robinie enthalten während des winterlichen Stärkeminimums keine oder fast keine Stärke, weder im Holz noch in der Rinde, sondern vorwiegend Fett im Holz. Im Winter ins Zimmer gebrachte Zweige dieser Bäume regenerieren Stärke bei Temperaturen von 15—30° C, nicht aber unter 5° C. In den meist hartholzigen Stärkebäumen verschwindet die Stärke im Winter nur in der Rinde ganz, in Holz und Mark bleibt sie vom Herbst bis Mai nahezu unverändert, wird in der Rinde aber im Frühjahr neu gebildet. Eiche, Haselnuß, Ulme, Platane, *Acer*arten, Esche und Flieder gehören in diese Gruppe. FISCHER unterschied 8 Phasen der Stärkeumwandlung im Laufe des Jahres, die sich um die 2 Stärkemaxima im April und Herbst und um die 2 Stärkeminima im Winter und Mai gruppieren. Als winterliches Umwandlungsprodukt der Stärke hat FISCHER neben Glucose und Fett auch andere unbekannte Substanzen angenommen. E. MER (1891) veröffentlichte ähnliche Beobachtungen über winterliche Stärkelösung in Bäumen und unterschied (E. MER 1898) 5 jahreszeitliche Phasen der Stärkeumwandlung in den Bäumen, wobei allerdings nicht etwa Glucose, sondern ein unbekannter Stoff aus der Stärke zu entstehen schien. K. G. LUTZ (1895a, b) hat in Buchen im Herbst die Umwandlung von Stärke in Fett gesehen,

[1] Wegen der Zusammensetzung der Fette von Rinden und Wurzeln vgl. den Beitrag von M. L. MEARA, insbesondere S. 22f. dieses Handbuchbandes.

und auch in Dänemark wurden ähnliche Beobachtungen über das Verhalten der Stärke in Bäumen gemacht (O. G. PETERSEN 1896, 1898) und A. J. J. VANDEVELDE (1897) fand — abgesehen von einem Minimum als Folge des Frühjahrsaustriebes — keine wesentlichen jahreszeitlichen Veränderungen von Fett und Eiweiß und beobachtete eine deutliche Temperaturabhängigkeit des Stärke-Glucose-Gleichgewichtes *(Fagus silvatica, Populus italica, Salix babylonica, Taxus baccata, Ilex aquifolium, Aucuba japonica).*

D. G. IONESCU (1894) hat den wechselnden Fettgehalt von Zweigstücken von Bäumen für die von ihm gefundenen wassergehaltsunabhängigen jahreszeitlichen Schwankungen elektrischer Eigenschaften von Zweigstücken verantwortlich gemacht, die ein verschiedenartiges und verschieden leichtes Durchschlagenwerden der Zweigstücke durch elektrische Funken bewirken.

J. D'ARBAUMONT (1901) untersuchte dann nahezu 100 Baum- und Straucharten und fand, daß FISCHERS Verallgemeinerung, wonach Härte des Holzes und winterlicher Stärkegehalt mehr oder minder Hand in Hand gehen, kaum zutrifft. An Proben aus frisch gefällten, etwa 90jährigen Fichtenstämmen hat L. FABRICIUS (1905) ausführliche Untersuchungen über den Fett- und Stärkegehalt im Laufe des Jahres ausgeführt. Seine mikroskopischen Beobachtungen zeigten ziemliche Unterschiede zwischen zwei am selben Tage gefällten gleichalten Bäumen und machten ein praktisch völliges Verschwinden der Stärke aus dem Holz bei gleichzeitigem starkem Zunehmen des Fettgehaltes von etwa April-Mai bis August wahrscheinlich. Ende September setzte im Holz wieder Stärkebildung ein, und im Winter war nur wenig Fett vorhanden. Das Stärkemaximum fand sich etwa vor dem Austrieb.

LECLERC DU SABLON (1904) scheint der erste gewesen zu sein, der die fast ausschließlich auf mikroskopischen Farb- oder unspezifischen Reduktionsreaktionen (Osmiumsäure) gegründeten Angaben früherer Untersucher über Stärke- und Fettmengen in Bäumen durch chemische Analysen zu überprüfen versuchte. Hierbei fand er in Zweigen von Kastanien, Weiden, Birnen, Pfirsich usw. zwar Schwankungen des Rohfettgehaltes im Laufe des Jahres, doch waren sie viel zu klein, als daß sie das Fett als Umwandlungsprodukt oder Gleichgewichtskörper zu Stärke hätten erscheinen lassen können. Mengenmäßig hinreichende Veränderungen fand er hingegen bei den Reservecellulosen, die er daher als Hauptreservestoffe der Bäume bezeichnet und denen gegenüber im Holz weder Stärke noch Fett eine nennenswerte Rolle spielen. In einer späteren Arbeit (LECLERC DU SABLON 1906) wird darauf hingewiesen, daß der Zeitpunkt des Maximums der Reservecellulosen für sommergrüne Bäume zur Zeit des Laubfalles, für wintergrüne aber beim Frühjahrsaustrieb zu finden sei. H. C. SCHELLENBERG (1905) hat auf mikroskopischem Wege Ablagerung und Wiederauflösung der Reservecellulosen in Bäumen beobachtet und beschrieben.

Ebenfalls auf dem Wege der chemischen Analyse hat B. NIKLEWSKI (1906) in Linden-, Vogelkirschen-, Flieder- und Birkenholz das jahreszeitliche Verhalten von Fett und Stärke untersucht. Er konnte die winterliche Fettzunahme bestätigen, allerdings nur in bescheidenem Umfang und unter Verwendung des Trockengewichtes als Bezugsgröße. Die Stärkeumwandlung fand er in anderer Weise von der Temperatur abhängig als Fettbildung und Fettlösung. Temperaturerhöhung beschleunigte die Fettbildung, eine Temperaturerniedrigung hatte keinen Einfluß auf die Fettbildung, bewirkte aber Verzuckerung der Stärke. Die bei höherer Temperatur gesteigerte Atmung ist schwer zu schätzen, beeinflußte durch den eintretenden Zuckerverlust aber die Ergebnisse, so daß NIKLEWSKI mit aller Vorsicht den Schluß ziehen konnte, daß außer der Stärke noch andere Körper an der Bildung veratembarer Kohlenhydrate im Stoffwechsel der Bäume beteiligt sein müssen.

Auch fortgesetzte mikroskopische Untersuchungen zeigten immer deutlicher, daß FISCHERs Typen der Fett- und Stärkebäume eher Abstraktionen als Wirklichkeiten sind. C. SCHMIDT (1909) vergrößerte FISCHERs Liste der Fettbäume z. B. um zahlreiche *Salix*arten, fand bei anderen *Salix*formen und in anderen Bäumen nur Stärke oder sogar weder Stärke noch Fett oder nur Fettspuren (*Salix glauca* z. B.). *Robinia* schien ihm eher zu den Stärkebäumen als (wie FISCHER gefunden hatte) zu den Fettbäumen zu gehören. Die Untersuchungen von F. WEBER (1909) haben dann klar gezeigt, daß die in den Bäumen beobachteten Veränderungen der Reservestoffe mannigfaltiger sind als die FISCHERsche Klassifizierung in Fett- und Stärkebäume vermuten ließ. *Picea* und *Abies* zeigten das Stärkeminimum nicht im Winter, sondern im Sommer, zugleich mit einem Fettmaximum. Bei *Aesculus* fand sich das ganze Jahr über der gleiche Fettgehalt und bei *Gingko* war in der Rinde im Winter viel Stärke und wenig Fett, im Holz aber viel Fett und wenig Stärke zu finden. Die Buche zeigte ein Fettmaximum im März, wenn die Linde ihr Fettminimum hatte. Zu gleichartigen Ergebnissen kamen auch J. F. PRESTON und F. J. PHILLIPS (1911) in Amerika, die in einer leider schwer zugänglichen zusammenfassenden Arbeit auf Grund der Literatur und eigener Untersuchungen für die Bäume des gemäßigten Klimas zur Folgerung kommen, daß die Verminderung des Stärkegehaltes der Baumstämme im November und Dezember eine allgemeine Erscheinung ist, daß aber das Ausmaß dieses Stärkeschwundes recht verschieden sein kann. In wenigen Fällen kann gleichzeitig eine Fettzunahme beobachtet werden, meist fehlt sie jedoch ganz oder fast ganz. Nichts deutet wirklich darauf hin, daß die verschwindende Stärke in Fett umgewandelt wird. Es liegt kein Grund dafür vor, Hartholzbäume als Stärkebäume und Weichholzbäume als Fettbäume zu bezeichnen. Da der spätherbstliche Stärkeschwund nicht von parallelen Zunahmen an löslichen Kohlenhydraten begleitet ist, steht nichts der Annahme im Wege, daß die verschwindende Stärke wenigstens zum Teil in Reservecellulosen umgewandelt wird. In den Wurzeln treten wesentlich weniger ausgeprägte Stoffumwandlungen ein als in den Stämmen, und der stärkste Stärkeschwund ist zur Zeit des Austreibens zu beobachten. Alte und junge Stämme und Zweige zeigen vielfach ein wenigstens quantitativ verschiedenes Verhalten, wobei die Stoffumsetzungen in den älteren Pflanzenteilen geringer zu sein scheinen als in den jüngeren.

Auch O. ISHIBE (1935) hat an 5 laubabwerfenden (*Castanea pubinervis*, *Alnus japonica*, *Robinia pseudacacia*, *Tilia miqueliana* und *Populus nigra*) und an 2 immergrünen Bäumen Japans (*Quercus glauca* und *Pinus densiflora*) beobachtet, daß die Änderungen im Fettgehalt der Zweige und Stämme unabhängig von den Änderungen des Stärkegehaltes sind und daß überall ein Fettmaximum im Hochwinter und ein Minimum im Sommer zu finden ist. Die Fetteinlagerung beginnt aber schon im Sommer, und zu keiner Zeit des Jahres fehlt Fett völlig. In den Wurzeln finden sich kaum irgendwelche jahreszeitliche Schwankungen des geringen Fettgehaltes. Der Stärkegehalt zeigte im allgemeinen ein Maximum vor dem Laubfall und ein kleineres vor dem Austreiben der Knospen, nur *Pinus* hatte das stärkere Maximum zur Zeit des intensiven Frühjahrswachstums und ein Minimum Ende Juli. Auch traten nur in den *Pinus*-Wurzeln die Stärkegehaltsänderungen zur gleichen Zeit wie in den oberirdischen Teilen ein, in allen anderen Wurzeln waren sie um mehrere Wochen verzögert.

Daß vielleicht der Wassergehalt der Bäume mit dem Verschwinden der Stärke und dem Auftreten von Fett in Zusammenhang stehen könnte, wurde von K. PURIEWITSCH (1898), J. DMOCHOWSKI (1934) und E. W. SINNOT (1918) vermutet. Auch Erörterungen eines möglichen Zusammenhanges des Stärkeschwundes mit der Bildung von ,,Gerbstoffen" wurden angestellt (A. RENVALL 1912). E. W. SINNOT (1918) hat über 300 Baum- und Straucharten aus mehr als 100 Gattungen untersucht und hierbei außer Fett und Stärke auch andere unbekannte Reservestoffe gesehen. Zerstreutporige Hölzer fanden sich fast ausschließlich bei fetthaltigen Bäumen, während ringporige fast nur von stärkereichen Bäumen stammten. Reservestoffe, die weder Stärke noch Fett zu sein schienen, hat auch E. ANTEVS (1916) in allen Laubbäumen im Winter oder Frühjahr gesehen. C. COSTER (1925) hat in Java das Vorkommen von Fett in tropischen und in eingeführten Bäumen studiert und in 19% der untersuchten 63 tropischen Bäume Fett gefunden, während von den 23 eingeführten Baumarten 74% Fett enthielten. Aus keinem anderen tropischen Gebiet scheinen ähnliche Untersuchungen vorzuliegen, es ist daher nicht möglich, zu beurteilen, ob und wieweit eine Verallgemeinerung dieser Beobachtungen COSTERs zulässig ist. Weitere Mitteilungen über das Verhalten von Stärke und Fett im Laufe des Jahres in verschiedenen Bäumen, Sträuchern und ausdauernden Pflanzen liegen vor von L. F. C. NOTTER (1903), A. LARKUM (1914), E. ANTEVS (1916), O. R. BUTLER, T. O. SMITH, B. E. CURRY (1917), G. M. TUTTLE (1921), A. POJARKOVA (1924), M. J. PRIKHOD'KO (1928), E. V. LAING (1932) und L. A. STODDART (1941) und zum Teil ausführlichere Besprechungen mancher der hier kurz erwähnten Arbeiten finden sich in F. CZAPEKs Biochemie der Pflanzen und besonders bei J. F. PRESTON und F. J. PHILIPPS (1911) sowie bei M. BÜSGEN und E. MÜNCH (1927).

Tabelle 21. *Rohfett in Prozent des Frischgewichtes im Holz zu verschiedener Zeit gefällter, etwa gleichalter Fichten und Tannen.* (Relativwerte, nach GÄUMANN 1928.)

Fällungstag	Fichte		Tanne	
	Kernholz	Splintholz	Kernholz	Splintholz
13. IX.	69	32	36	24
18. X.	64	55	43	35
15. XI.	55	56	50	39
13. XII.	69	69	18	13
17. I.	82	30	28	10
14. II.	94	91	80	56
14. III.	100	99	98	80
18. IV.	97	100	100	100
16. V.	80	82	66	65
17. VI.	84	79	39	54
18. VII.	76	66	38	37
15. VIII.	71	59	49	28
100 =	1,53	1,76	1,29	0,71

Daß in Baumzweigen Verlagerungen von Fetten oder Fettvorstufen in beträchtlichem Umfang vor sich gehen, hat durch Ringelungsversuche H. D. HOOKER (1927) gezeigt, nachdem schon K. PURIEWITSCH (1898) gefunden hatte, daß Fett und Zucker aus Lindenzweigen in angesetzte Gipsblöcke und das umgebende Wasser geleitet werden. Auch J. SUROŽ (1890) hatte berichtet, daß in Bäumen wiederholt Verlagerungen und Wanderungen von Fett stattfinden.

Die großzügigsten Untersuchungen über Reservestoffveränderungen in Bäumen stammen aber wohl von E. GÄUMANN, der frisch gefällte Fichten und Tannen (1928) sowie Buchen (1935a, b) eingehend analysierte. Etwa um die Mitte jedes Monats wurde 1 Jahr hindurch bei Morgengrauen je 1 Fichte und Tanne im Alter von etwa 120—130 Jahren gefällt, und Holzproben aus vergleichbaren Stammgegenden wurden chemisch analysiert (Rohfett). Als Bezugsgröße wurde das Frischgewicht verwendet. Die Zahlen der Tabelle 21 wurden erhalten.

Die Fichten waren also deutlich fettreicher als die Tannen, und der Fichtensplint enthielt etwas mehr Fett als das Kernholz, das geringere Schwankungen des Fettgehaltes zeigte als der Splint. Die Tannen enthielten im Kern bedeutend mehr Fett als im Splint und zeigten in beiden Stammteilen etwa gleiche Schwankungen des Fettgehaltes. Das Maximum des Fettgehaltes fand sich bei beiden Bäumen zwischen Februar und April, also vor dem Einsetzen der vollen Vegetationstätigkeit. Irgendein Hinweis darauf, daß das Fett aus der im Herbst verschwindenden Stärke stammen würde, findet sich nicht. Bei allem Interpretieren solch mühevoll erarbeiteter Zahlenreihen darf natürlich nicht übersehen werden, daß sie weitgehend von der gewählten Bezugsgröße und Analysenmethode bestimmt sind. GÄUMANN hat nur das Rohfett bestimmt und betont, daß darin natürlich vielleicht wechselnde Anteile Harz enthalten sein können, die eventuell die Ergebnisse beeinflußt haben könnten. — Jedenfalls liefern GÄUMANNs Zahlen keinen Hinweis auf das Auftreten von 2 Fettmaxima im Laufe des Jahres.

Für die Untersuchungen über das Fett in Buchenstämmen standen 33 frisch gefällte etwa 110jährige Buchen zur Verfügung. Rinde, Jungholz, Reifholz, Astrinde, Astholz, Zweige und Wurzeln wurden auf ihren Gehalt an Fett und den übrigen Reservestoffen untersucht. Als Bezugsgröße diente diesmal die Trockensubstanz. Es zeigte sich überraschenderweise, daß im ganzen Baum nur die Fette (und die Kohlenhydrate der Rinde) bei der Wiederaufnahme der Lebenstätigkeit im Frühjahr veratmet werden. Während etwa 2 Monaten werden da rund 5,5 kg Fett pro Baum veratmet, d. h. etwa 90% des Gesamtfettes verschwinden; eine „Rückumwandlung" der Fette in Kohlenhydrate ist nicht feststellbar. Aus der Rinde verschwinden die Fette nie ganz, wohl aber aus dem Holz von Stamm, Ästen und Wurzeln. Die folgende Tabelle 22 wurde aus GÄUMANNs Zahlen berechnet.

Tabelle 22. *Fettgehalte in Prozent des Trockengewichtes in Buchen zu verschiedenen Zeiten des Jahres.* (Relativwerte, nach GÄUMANN 1935a.)

Fällungstag	Stamm			Äste		Zweige	Wurzel	
	Rinde	Jungholz	Reifholz	Rinde	Holz		Rinde	Holz
6. I.	94	100	87	—	—	86	—	100
13. I.	83	81	48	88	75	90	81	88
15. II.	83	90	48	85	56	73	83	48
13. III.	70	50	39	71	69	50	100	74
3. IV.	42	19	39	48	56	40	7	24
18. IV.	28	28	23	42	56	29	23	12
4. V.	25	28	16	54	69	29	15	7
15. V.	26	0	0	52	0	20	13	0
1. VI.	18	12	0	22	0	25	1	0
15. VI.	18	28	3	60	0	24	1	5
17. VII.	21	41	0	57	31	38	26	0
14. VIII.	22	44	35	56	75	42	30	31
13. IX.	28	60	29	54	87	41	64	24
12. X.	31	60	68	50	38	43	20	21
26. X.	17	16	29	40	50	44	25	0
16. XI.	42	69	65	100	100	55	28	12
18. XII.	100	94	100	87	87	100	66	64
100 =	2,93	0,32	0,31	1,83	0,16	1,63	0,96	0,42

Stämme aufgetaut: 13. III.; Laubausbruch: 15. V.; Laubfall beendet: 16. XI.

Der starke Fettschwund nach dem Laubausbruch ist auf eine Veratmung der Fette und nicht etwa auf eine Umwandlung in Stärke zurückzuführen. Auffällig ist das relative Fettminimum in Stamm und Wurzel des am 26. Oktober gefällten Baumes, und es wäre vielleicht denkbar, daß es mit zu dieser Zeit im Baum vor sich gehenden Stoffverschiebungen im Zusammenhang steht. GÄUMANN hat berechnet, daß eine rund 1775 kg schwere Buche im Durchschnitt etwa 6,1 kg Fett enthält, die ungefähr so auf die einzelnen Teile des Baumes verteilt sind, wie die folgende Tabelle 23 zeigt.

Fruchtende Buchen enthielten etwa 40% mehr Fett als nicht fruchtende und veratmen während Blüte und Fruchtbildung etwa 2mal soviel Kohlenhydrate als neu gebildet werden. Beim Laubfall verliert jeder Baum etwa 0,4 kg Fett.

Überblickt man all die Untersuchungen, die sich mit dem Fettstoffwechsel der Bäume befassen, dann kann man sich wohl kaum des Eindrucks erwehren, daß eine wirkliche Vertiefung unserer Kentnisse dieses Gebietes wohl sicher nicht von mikroskopischen Untersuchungen kleiner Stamm- oder Aststückchen kommen kann, sondern daß vielmehr nur großzügige chemische Untersuchungen mit Aufstellung von Reservestoffbilanzen wirklich der Mühe wert wären. Ohne sozusagen

Tabelle 23. *Durchschnittliche Fettverteilung in einer Buche.* (Nach GÄUMANN 1935b.)

Organ	Trockengewicht (relativ)	Fett-%	Fettmenge (relativ)
Zweige	0,6	1,5	3,3
Äste			
Rinde	0,6	1,5	3,3
Holz	7,3	0,1	1,6
Stamm			
Rinde	4,2	2,5	31,0
Jungholz	16,9	0,3	15,0
Reifholz	62,0	0,2	36,0
Stock und Wurzeln			
Rinde	1,1	0,8	3,3
Holz	7,3	0,3	6,5
	100 = 1775 kg	gewogenes Mittel = 0,35%	100 = 6,1 kg

ganze Wälder zu analysieren (wenn vielleicht auch nur in Form zahlreicher Bohrproben), wird die Frage, was nun wirklich in den Bäumen während des Winters oder etwa in Tropengebieten vor sich geht, kaum gelöst werden können.

Über den Fettstoffwechsel von Rhizomen, unterirdischen Stammteilen, Knollen usw. ist praktisch kaum etwas bekannt. O. ROSENBERG (1896) hat einige Rhizome und Knollen zu verschiedenen Zeitpunkten während des Jahres auf ihren Gehalt an Stärke untersucht, macht jedoch keine Erwähnung von Fett, und H. J. DE CORDEMOY (1893) hat fettes Öl im Rhizom der Dracaenacee *Colmi flabelliformis* gefunden. U. GERLOFF (1936) hat Kletten- und Pfingstrosen-Wurzelöle untersucht, ohne auf die Physiologie ihrer Bildung näher einzugehen.

III. Die Fette in Blättern[1].

Die ältesten der wenigen über die Fette in Blättern vorliegenden Untersuchungen beschäftigen sich mit Olivenblättern; zwischen Mai und November gepflückte Blätter ergaben hierbei Fettgehalte, die von 5,4 auf etwa 3,7% abnahmen (A. ROUSILLE 1878) (Schwefelkohlenstoffextrakte), während Ätherextrakte (A. FUNARO 1880) zwischen September und Februar geernteter Blätter unregelmäßig zwischen 5,5 und 10% schwankende Werte ergaben. Ähnliche Zahlen fanden F. SCURTI und G. TOMMASI (1910) bei der Analyse der Blätter von 2 Olivensorten. Für Kastanienblätter hat LECLERC DU SABLON (1904) die folgenden Werte für Äther- und Petrolätherextrakte angegeben (Tabelle 24).

In den immergrünen Nadeln von *Pinus austriaca* fanden sich 6,2—8,5% Rohfett, das Minimum lag im August, das Maximum im Oktober und Jänner.

[1] Wegen der Zusammensetzung der Fette von Blättern vgl. den Beitrag von M. L. MEARA, insbesondere S. 23ff. dieses Handbuchbandes.

Ein- und dreijährige Nadeln zeigten von November bis Jänner Abnahmen der Rohfettgehalte von 8,2 auf 6,7% bzw. von 10,1 auf 9,1%. Die etwa 18 Monate lebenden Blätter von *Evonymus japonicus* zeigten hingegen zwischen März und Oktober Fettgehaltsabnahmen von 4,3 auf 3,5% in jungen und von 3,8 auf 2,6% in alten Blättern. Bei *Quercus ilex* hingegen enthalten die alten Blätter mehr Rohfett als die jungen und die Fettgehalte nehmen von Ende bis Anfang Oktober zu (von 1,6 auf 2,2 bzw. von 2,4 auf 3,7%, LECLERC DU SABLON 1906). Unregelmäßige Schwankungen des Rohfettgehaltes von 5,7—7,4% fanden F. SCURTI und M. FORNAINI (1911) in den Blättern des *Liguster* zwischen Ende September und Ende Dezember.

Tabelle 24. *Fettgehalt von Kastanienblättern, bestimmt mit Äther bzw. Petroläther.* (Relativwerte, nach LECLERC DU SABLON 1904.)

Erntedatum	Ätherextrakt	Petrolätherextrakt
	% der Trockensubstanz (relativ)	
20. V.	50	24
22. VI.	47	18
26. VII.	58	40
12. IX.	100	100
19. X.	94	80
100 =	9,3	7,4

Aus neuerer Zeit liegen Untersuchungen von R. C. JORDAN und A. C. CHIBNALL (1933) vor, die für die Fettsäuren aus den in 1000 g frischen Bohnenblättern enthaltenen Fette die Werte der Tabelle 25 berichten.

Tabelle 25. *Fettsäuren aus dem Fett von 1000 g frischen Bohnenblättern.* (JORDAN und CHIBNALL 1933.)

Alter der Blätter in Wochen	Primärblätter g	Sekundärblätter g
1	5,6	—
3	3,2	3,3
7	2,7	4,0
10	2,0	3,5
12	—	3,5
20 (abgestorben)	—	6,1

Hier findet sich also eine ziemlich beträchtliche Anreicherung von Fetten in den alternden Blättern und auch in anderen Blättern liegen vielleicht ähnliche Verhältnisse vor. So fand O. B. STANLEY (1931) Fetttropfen nur in alten Blättern von *Arctostaphylos uva ursi*, und auch das Auftreten der Ölkörper der Oenotheraceen, Melastomaceen und Hippuridaceen, die F. STEIN (1915) beschrieb, dürfte wenigstens zum Teil vom Alter der Blätter abhängen. In ähnlicher Richtung könnten die Befunde von R. M. ADAMS und F. C. MOUNCE (1932) weisen, die fanden, daß in Preiselbeerblättern Fettgehalt und Wachstumsintensität etwa gegenläufig variieren. Stärke scheint in diesen Blättern nur transitorisch aufzutreten, und das eigentliche Reservematerial scheint Fett zu sein. In Buchenknospen fand E. GÄUMANN (1935b) eine von den Kohlenhydraten unabhängige beträchtliche Zunahme der Fette von 0,03 mg auf 1,3 mg zwischen Mitte Juli und Mitte Oktober. — Angaben über Fett in wachsenden Tabakblättern finden sich bei S. IGAKI (1929). Eine Umwandlung von Fetten in Stärke unter dem Einfluß von Lichtmangel hat J. GRÜSS (1927) von den Fettzellen der Luftblätter der Nymphaeaceen berichtet, und W. E. TOTTINGHAM (1932) fand höheren Fettgehalt in Blättern von Tomaten, die mehr Ultraviolett und Infrarot erhalten hatten als die Kontrollen. — F. M. SCOTT (1955) hat mit Hilfe von Sudanschwarz-Vitalfärbung ein lokalisiertes Vorkommen von 0,05—1 μ großen Fetttröpfchen in den Plasmafibrillen, die alle Plastiden umgeben, im parietalen Plasma und in Plasmodesmen festgestellt. Außer 35 Samenpflanzen wurden auch Farne, Moose und Algen untersucht.

Periodisches winterliches Verschwinden der Stärke in den Blättern immergrüner Pflanzen wurde schon von E. MER (1876) beobachtet, aber A. MEYER (1918) hat darauf hingewiesen, daß weder die von MER beobachteten stark lichtbrechenden Kügelchen noch E. SCHULZs (1888) Fettfunde in vielen Blättern, noch irgendwelche andere der von F. CZAPEK (1913) erwähnten Arbeiten wirklich darauf schließen lassen, daß die Stärke z. B. der Koniferenblätter im Winter in Fett umgewandelt wird. Auch K. MIYAKE (1902) hat nur auf Stärke geprüft und in Alkohol konservierte Blätter untersucht, und B. LIDFORSS (1896, 1907) hat ebenfalls vor allem den Stärkegehalt der Blätter untersucht und nur gelegentlich Ölvorkommen in immergrünen Blättern festgestellt. G. M. TUTTLE (1919) hingegen berichtet über Wiederumwandlung von Stärke in Fett bei Temperaturen zwischen $+2^0$ und -2^0 C in Blättern immergrüner Pflanzen, die vorübergehend bei höheren Temperaturen gehalten worden waren und Stärke gebildet hatten. Fett im Mesophyll von *Picea* wurde von F. J. LEWIS und G. M. TUTTLE (1920) beobachtet und G. M. TUTTLE (1921) fand zur Zeit des Laubfalles in keinem der abgeworfenen Blätter Stärke, vielfach aber viel Fett und beobachtete jahreszeitliche Fettspiegelschwankungen in den Blättern und in Knospen. Eine Nachuntersuchung der Inhaltsstoffe von Koniferenblättern im schottischen Winter durch J. DOYLE und P. E. CLICH (1927) sowie eine Zusammenfassung der bisher veröffentlichten Ergebnisse durch J. DOYLE (1938) führte nur zur Feststellung, daß bisher nicht genügend Unterlagen dafür vorhanden sind, anzunehmen, daß Fette in winterlichen Koniferenblättern in nennenswertem Ausmaß umgesetzt werden oder eine Rolle als Reservestoffe spielen. Makrochemische Untersuchung der Blätter verschiedener Koniferen in verschiedenen Klimaten im Laufe eines Jahres würde wohl helfen, die hier aufgeworfene Frage nach der Rolle der Fette in den winterlichen Koniferenblättern zu klären. Sorgfältigste Auswahl der Bezugsgröße wäre wohl eine der Vorbedingungen für den Erfolg derartiger Untersuchungen.

Auf experimentellem Wege haben A. C. CHIBNALL und P. N. SAHAI (1931) versucht, tiefer in die Physiologie der Blattfette einzudringen. Reife Blätter von Kohlsprossen (*Brassica oleracea*) wurden mit den Stielen in Wasser gestellt und dunkel gehalten. Nach etwa 8 Tagen war das Chlorophyl völlig zersetzt, die Blätter waren im Absterben, ein beträchtlicher Eiweißabbau war erfolgt, Kohlenhydrate waren noch reichlich vorhanden, aber keinerlei signifikante Veränderungen der Fette oder Phosphatide waren feststellbar. Später ausgeführte Versuche mit Bohnenblättern (R. C. JORDAN und A. C. CHIBNALL 1933) zeigten, daß hier im Dunkeln die echten Fette verbraucht werden, daß die Phosphatide aber viel langsamer und nur im selben Ausmaß wie die Eiweißkörper verschwinden. Hier scheinen also die Fette der Blätter echte Reservestoffe zu sein.

Das Verschwinden des Samenfettes in ergrünenden Kotyledonen und das Auftreten von Blattfett wurde in jüngster Zeit von W. M. CROMBIE und R. COMBER (1956) an Lichtkeimlingen von *Citrullus vulgaris* mit moderner Methodik näher verfolgt. Das Fett der assimilierenden Kotyledonen (etwa 14—16 Tage nach Keimungsbeginn) enthält zum Unterschied vom Samenfett der Melonen einen beträchtlichen Prozentsatz (35—45% der Fettsäuren) 3fach ungesättigter C_{18}-Säuren („Linolensäure“) und ist wohl als ein Assimilationsprodukt und nicht als ein Umwandlungsprodukt des abgebauten Samenfettes aufzufassen. Seine Menge beträgt nur etwa 1% des ursprünglich vorhandenen Reservefettes.

Literatur.

Bei manchen schwer zugänglichen Veröffentlichungen werden auch Referate der betreffenden Arbeit zitiert. Zitate, die nicht im Original eingesehen wurden, sind mit * gekennzeichnet.

ABEGG, F. A.: Some effects of the waxy gene in maize on fat metabolism. J. Agric. Res. 38, 183—193 (1929). — ADAMS, R. M., and F. C. MOUNCE: Further notes on the nutrient requirements and the histology of the cranberry with special reference to the sources of nitrogen. Plant Physiol. 7, 643—656 (1932). — AFANASIEV, M., and M. CRESS: Changes within the seeds of *Juniperus scopulorum* during the process of after-ripening and germination. J. Forestry **40**, 798—801 (1942). Ref.: Biol. Abstr. **17**, 6927 (1943). — ALTSCHUL, A. M.: Biological processes of the cotton seed. Fats and Oils. Cottonseed and Cottonseedproducts, herausgeg. von A. E. BAILEY, S. 157—212. New York 1948. — ALTSCHUL, A. M., M. L. KARON, L. KYANE and C. M. HALL: Effect of inhibitors on the respiration and storage of cottonseed. Plant Physiol. **21**, 573—587 (1946). — ANTEVS, E.: Zur Kenntnis der jährlichen Wandlungen der stickstofffreien Reservestoffe der Holzpflanzen. Ark. Bot. (Stockh.) **14** (16), 1—25 (1916). Ref.: Bot. Zbl. **131**, 313—314 (1916). — APPLEMAN, D., and L. NODA: Biochemical studies of the Fuerte Avocado fruit. A preliminary report. Calif. Avocado Assoc. Yearbook 1941, 60—63. Rev. Plant Physiol. **1**, 190 (1950). — ARASIMOVIČ, V.: Biochemische Studien an Kürbissamen (Russisch). *Trudy prikl Bot. i pr. (III. Physiol., Biochem. a. Anat. of Plants) **1**, 73—99 (1933). Ref.: Ber. ges. Physiol. **78**, 232 (1934). — D'ARBAUMONT, J.: Sur l'evolution de la chlorophylle et de l'amidon dans le tige. Ann. des Sci. natur., VIII. s., Bot. **13**, 319—423; **14**, 125—212 (1901).

BARKER, C., A. CROSSLEY and T. P. HILDITCH: African drying oils. II., III., IV. J. Soc. Chem. Industr. **69**, 13—15, 15—16, 16—20 (1950). — BARKER, C., and T. P. HILDITCH: The influence of environment upon the composition of sunflower seed oils. I. Individual varieties of sunflowers grown in different parts of Africa. J. Sci. Food a. Agric. **1**, 118—121 (1950a). — The influence of environment upon the composition of sunflower seed oils. II. Composition of the seed oils of sunflowers grown in English gardens from five specimens of different African sunflower seed. J. Sci. Food a. Agric. **1**, 140—146 (1950b). — BARKER, M. F.: Development of oil in the seed of the flax fibre crop and the variations in its character with maturity. J. Soc. Chem. Industr. **51**, 218T—222T (1932). — BAUER, K. H.: Über die Veränderung in der Zusammensetzung des Sonnenblumenöles während der Reifung der Samen. Fettchem. Umschau **41**, 1—2 (1934). — BLAGOWESTSCHENSKI, A.: Untersuchungen über die Samenreifung. Biochem. Z. **157**, 201—219 (1925). — BLOOMENDAAL, H. N.: Preliminary paper on the ripening of oil palm fruits. Meded. Allgemeine Proefstation der A.V. R.O.S. (Allgemeene Vereenigung Rubberplanters Oostkust Sumatra), (a) Allg. Ser. 20, 43 S., 1925. Ref.: Exper. Stat. Rec. **57**, 141 (1928). — BLUM, P. H.: Grain spoilage: A review. Wallerstein Lab. Communic. **17**, (56) 35—45 (1954). Ref.: Biol. Abstr. **28**, 28300 (1954). — BRAUN, W.: Untersuchungen über den Einfluß langwelliger UV-Strahlen auf die Inhaltsstoffe verschiedener Arzneipflanzen. Beitr. Biol. Pflanz. **26**, 331—400 (1939). — BRIDGE, R. E., A. CROSSLEY and T. P. HILDITCH: The influence of environment upon the composition of sunflower seed oils. II. Oils from sunflower seeds grown in different regions of Australia. J. Sci. Food a. Agric. **2**, 472 (1951). — BÜRKLE, B.: Physiologische Untersuchungen über Umwandlungen des Öles in reifenden Sonnenblumensamen. Bot. Archiv **26**, 385—436 (1929). — BÜSGEN, M., u. E. MÜNCH: Bau und Leben unserer Waldbäume, 3. Aufl. Jena: Gustav Fischer 1927. — The structure and life of forest trees, translated by T. THOMAS. London: Chapman & Hall 1929. 436 S. — BURR, G. O., and E. S. MILLER: Synthesis of fat by green plants. Bot. Gaz. **99**, 773—785 (1938). — BURRELL, R. C., and A. C. WOLFE: A comparative study of the chemical composition of five varieties of soybeans. Food Res. **5**, 109—113 (1940). — BUSHEY, A. L., L. PUHR and A. N. HUME: A study of certain physical and chemical characteristics of flaxseed and of linseed oil. South Dakota Agric. Exper. Stat. Bull. **228**, 1—11 (1927). Ref.: Biol. Abstr. **3**, 21369 (1929). — BUTLER, O. R., T. O. SMITH and B. E. CURRY: Physiology of the apple. *New Hampshire Agric. Exper. Stat. Techn. Bull. **13**, 21 S. (1917). Ref.: Exper. Stat. Rec. **43**, 28 (1920).

CASKEY, C., and W. D. GALLUP: Changes in the sugar, oil and gossypol content of the developing cotton boll. J. Agric. Res. **42**, 671—673 (1931). — CHIBNALL, A. C., and P. N. SAHAI: Observations on the fat metabolism of leaves. I. Detached and starved mature leaves of brussels sprout *(Brassica oleracea)*. Ann. of Bot. **45**, 489—502 (1931). — CHURCH, C. G., and E. M. CHACE: Some changes in the composition of California Avocados during growth. U.S. Dept. Agric. Bull. 1073, **1922**, 22 S. — COLE, L. J., E. W. LINDSTROM and C. M. WOODWORTH: Selection for quality of oil in soy beans. J. Agric. Res. **35**, 75—95 (1927). — COLEMAN, D. A., and H. C. FELLOWS: Oil content of flax seed, with comparison of tests for determining oil content. U.S. Dept. Agric. Bull. 1471, **1927**. — CORDEMOY, H. J. DE: Sur les tissues secondaires de réserve des monocotylédones arborescentes. C. r. Acad. Sci. Paris **117**, 132—134 (1893). — CORNFORTH, J. W.: Die Indikatormethode in der Biosynthese. Endeavour **12**, 61—67 (1953). — COSTER, C.: Die Fettumwandlung im Baumkörper in den Tropen. Ann. Jard. bot. Buitenzorg **35**, 71—104 (1925). — CROMBIE, W. M.: Fat metabolism in the West African oil palm *(Elaeis Guineensis)*. I. Fatty acid formation in the maturing kernel. J. of Exper. Bot. **7** (20), 181—193 (1956). — CROMBIE, W. M., and R. COMBER: Fat metabolism in germinating *Citrullus vulgaris*. J. of Exper. Bot. **7** (20), 166—180 (1956). — CZAPEK, F.: Biochemie der Pflanzen, 2. Aufl. Jena: Gustav Fischer 1913.

DALLA TORRE, C. G. DE, u. H. HARMS: Genera siphonogamarum ad systema Englerianum conscripta. Leipzig 1900—1907. — DILLMAN, A. C.: Daily growth and oil content of flaxseeds. J. Agric. Res. 37, 357—377 (1928). — DILLMAN, A. C., and T. H. HOPPER: Effect of climate on the yield and oil content of flaxseed and on the iodine number of linseed oil. U.S. Dept. Agric. Techn. Bull. 844, 1943, 69 S. — DMOCHOWSKI, J.: Recherches sur la quantité d'huile dans les graines du lin et du chanvre pendant les différent périodes de la maturation et de la germination ainsi que dans le bouleau et dans le tilleul pendant les différentes phases et l'hiver. Rocnik Nauk Rolniczych i Lesnych (Poln. Jb. Land- u. Forstwirtsch.) 32, 35—77 (1934), Polnisch mit franz. Zus.fass. S. 76—77. — DOLLEAR, F. G., P. KRAUCZUNAS and K. S. MARKLEY: Composition of a soybean oil of abnormally low iodine number. Oil a. Soap 15, 263—264 (1938). — The chemical composition of some high iodine number soybean oils. Oil a. Soap 17, 120—121 (1940). — DOYLE, J.: Notes on the supposed accumulation of fat in conifer leaves in winter. New Phytologist 37, 375—376 (1938). Ref.: Biol. Abstr. 12, 15778 (1938) u. Ber. wiss. Biol. 49, 467. — DOYLE, J., and P. E. CLINCH: Seasonal changes in conifer leaves, with special reference to enzymes and starch formation. Proc. Roy. Irish Acad. B 37, 373—414 (1927).

ECKERSON, S.: A physiological and chemical study of after-ripening. Bot. Gaz. 55, 286—299 (1913). — ERMAKOV, A.: Einfluß der Saatzeit auf Menge und Qualität des Öles in den Samen der Ölfrüchte [Russisch mit engl. Zus.fass.]. *Trudy prikl. Bot. i. pr. (III. Physiol., Biochem. a. Anat. of Plants) 1, 31—69—71 (1933). Ref.: Ber. wiss. Biol. 30, 89 u. Ber. ges. Physiol. 78, 232. — EVANS, J. W.: Changes in the biochemical composition of the corn kernel during development. Cereal Chem. 18, 468—473 (1941). — EVANS, J. W., and D. R. BRIGGS: The fatty acid composition of the lipids of corn starch at various stages during the development of the corn kernel. Cereal Chem. 18, 465—468 (1941). Ref.: Biol. Abstr. 18, 9144 (1944). — EYRE, J. V.: Notes on oil development in the seed of a growing plant. Biochemic. J. 25, 1902—1908 (1931). — EYRE, J. V., and E. A. FISHER: Some considerations affecting the growing of linseed as a farm crop in England. I. Variations in the oil content. J. Agric. Sci. 7, 120—134 (1915).

FABIAN, H.: Der Einfluß der Ernährung auf die wertbestimmenden Eigenschaften von Bastfaserpflanzen (Flachs und Nessel) unter besonderer Berücksichtigung der Ausbildung ihrer Fasern. Faserforsch. 7, 1—56, 69—115 (1928). — FABRICIUS, L.: Untersuchungen über den Stärke- und Fettgehalt der Fichte auf der oberbayrischen Hochebene. Naturwiss. Z. Land- u. Forstwirtsch. 3, 137—176 (1905). — FELLERS, C. R.: The effect of inoculation, fertilizer treatment and certain minerals on the yield, composition and nodule formation of soybean. Soil Sci. 6, 81—119 (1918). — Soy bean oil; factors which influence its production and composition. J. Industr. a. Engng. Chem. 13, 689—691 (1921). — FINCH, A. H., and C. W. VAN HORN: The physiology and control of pecan nut filling and maturity. Arizona Agricult. Exper. Stat. Techn. Bull. 62, 420—472 (1936). Ref.: Exper. Stat. Rec. 76, 190 (1937). — FISCHER, A.: Beiträge zur Physiologie der Holzgewächse. Jb. Bot. 22, 73—160 (1891). — FUNARO, A.: Studien über die Bildung der fetten Öle und über die Reifung der Oliven. Landwirtsch. Versuchsstat. 25, 52—56 (1880).

GÄUMANN, E.: Die chemische Zusammensetzung des Fichten- und Tannenholzes in den verschiedenen Jahreszeiten. Flora (Jena) 123, 344—385 (1928). — Der Stoffhaushalt der Buche im Laufe eines Jahres. Ber. schweiz. bot. Ges. 44, 157—334 (1935a). — Über den Stoffhaushalt der Buche *(Fagus silvatica)*. Ber. dtsch. bot. Ges. 53, 366—377 (1935b). — GALLUP, W. D.: The gossypol content and chemical composition of cottonseed during certain periods of development. J. Agric. Res. 34, 987—992 (1927). — A chemical study of the development of cotton bolls and the rate of formation of gossypol in the cotton seed. J. Agric. Res. 36, 471—480 (1928). — GARNER, W. W., H. A. ALLARD and C. L. FOUBERT: Oil content of seeds as affected by the nutrition of the plant. J. Agric. Res. 3, 227—249 (1914). — GERBER, C.: Recherches sur la formation des réserves oléagineuses des graines et des fruits. C. r. Acad. Sci. Paris 125, 658—661 (1897). — Recherches sur la respiration des olives et sur les relations existant entres les valeurs du quotient respiratoire observé et la formation de l'huile. J. de Bot. 15, 9—22, 88—94, 121—136 (1901). — GERLOFF, U.: Beiträge zur Kenntnis der Zusammensetzung fetter Öle in verschiedenen Teilen einer Pflanze. Planta (Berl.) 25, 667—688 (1936). — GIEGER, M.: The effect of fertilization and cultural practices on the oil and ammonia content of cottonseed grown in Yazoo-Mississippi Delta soils. J. Agric. Res. 63, 49—54 (1941). — GIESECKE, F., K. SCHMALFUSS u. E. GERDUM: Experimentelle Studien zur Physiologie und Ernährung des Leins im Hinblick auf die Ausbildung von Faser und Öl. II. Feldversuche. Bodenkde. u. Pflanzenernähr. 4, 340—357 (1937). — GIESECKE, F., u. YI LUNG LIU: Steigende K_2SO_4-Gaben in ihrem Einfluß auf Ertrag und Zusammensetzung der Sojabohne. Ernähr. Pflanze 36, 73—76 (1940). — GODLEWSKI, E.: Beiträge zur Kenntnis der Pflanzenatmung. Jb. wiss. Bot. 13, 491—543 (1882). — GREBNITZKY, A.: Über die jährliche Periode der Stärkespeicherung in den Zweigen unserer Bäume [Russisch]. *Sitzgsber. Bot. Sect. der 7. Versammlung russ. Naturforscher u. Ärzte, Odessa 1883, Sitzg v. 24. Aug. Ref.: Bot. Zbl. 18, 157—158. — GREER, S. R., T. E. ASHLEY, G. F. POTTER and E. ANGELO: Tung culture in Southern Mississippi.

*Mississippi Agric. Exper. Stat. Bull. 409, **1944**, 26 S. Ref.: Biol. Abstr. **21**, 1742 (1947). — Grindley, D. N.: Changes in composition of cottonseed during development and ripening. J. Sci. Food a. Agric. **1**, 147—151 (1950). — Gross, M.: N-Düngung und Flachs. Faserforsch. **5**, 37—51 (1925). — Gross, R. A., and C. H. Bailey: Chemical constitution of the oils from superior and inferior flaxseed. Oil a. Soap **14**, 260—263 (1937). — Grüss, J.: Die Luftblätter der Nymphaeaceen. Ber. dtsch. bot. Ges. **45**, 454—458 (1927). — Guerrant, N. B.: Some relations of the phospholipins in seeds to other constituents. J. Agric. Res. **35**, 1001—1019 (1927). — Guillaume, M. A.: Sur les huiles retirées des graines de lupin. C. r. Soc. Biol. Paris **89**, 887—889 (1923).

Halvorsen, H.: The gas exchange of flax seeds in relation to temperature. I. Experiments with immature seeds and capsules. Physiol. Plantarum (Copenh.) **8**, 501—511 (1955). — Hartung, M. E., and W. B. Storey: The development of the fruit of *Macadamia ternifolia*. J. Agricult. Res. **59**, 397—406 (1939). — Hartwich, C., u. W. Uhlmann: Betrachtungen über den Nachweis des fetten Öles und seine Bildung, besonders in der Olive. Arch. Pharmaz. **240**, 471—480 (1902). — Harz, C.: Über die Entstehung des fetten Öles in der Olive. Sitzgsber. Akad. Wiss. Wien, Math.-naturwiss. Kl. **61**, 930—946 (1870). — Heinricher, E.: Über Bau und Biologie der Blüten von *Arceutholobium Oxycedri* (DC.) MB. Sitzgsber. Akad. Wiss. Wien, Math.-naturwiss. Kl., Abt. I **124**, 481—504 (1915). — Heuser, W., K. Boekholt u. G. Ulich: Der Gehalt der Samen von *Lupinus albus* an Eiweiß, Fett und Alkaloiden im Vergleich zu anderen Lupinenarten und unter dem Einfluß äußerer Bedingungen. Pflanzenbau **11**, 129—138 (1934). — Hilditch, T. P.: The chemical composition of vegetable seed fats in relation to the natural orders of plants. Proc. Roy. Soc. Lond., Ser. B **103**, 111—117 (1928). — The chemical constitution of natural fats, 3. Aufl. London: Chapman & Hall 1956. 664 S. — Biosynthesis of unsaturated fatty acids in ripening seeds. Nature (Lond.) **167**, 298—301 (1951). — Pflanzliche Samen- und Fruchtfette. Endeavour **11**, 173—182 (1952). — Hilditch, T. P., and J. B. Lovern: The evolution of natural fats. A general survey. Nature (Lond.) **137**, 478—481 (1936). — Hooker, H. D.: Movement of fat in apple shoots. Proc. Amer. Soc. Horticult. Sci. **24**, 185—187 (1927). — Huelin, F. E., and R. A. Gallop: Studies in the natural coating of apples. I. Preparations and properties of fractions. II. Changes in the fractions during storage. Austral. J. Sci. Res. B **4**, 526—532, 533—543 (1951).

Igaki, S.: Quantitative changes in the constituents of each part of the tobacco plant during growth. *Bulteno Sci. Fakult. Terkult., Kjusu Imp. Univ. Fukuoka, Japan **3**, 317—326 (1929). Ref.: Biol. Abstr. **6**, 6519 (1932). — Ionescu, D. G.: Weitere Untersuchungen über die Blitzschläge in Bäume. Ber. dtsch. bot. Ges. **12**, 129—136 (1894). — Ishibe, O.: The seasonal changes in starch and fat reserves in some woody plants. Mem. Coll. Sci. Kyoto Imp. Univ., Ser. B **11** (1), 1—53 (1935). — Iwanoff, N. N.: Über die Stabilität der chemischen Zusammensetzung der Pflanzen. Biochem. Z. **182**, 88—98 (1927). — Biochemie der Leguminosen und Fouragepflanzen. Amsterdam: W. Junk 1948. 115 S. — Iwanow, M.: Accumulation of nutritive substances by the sunflower under field conditions. *J. Agric. Sci. S.E.U.S.S.R. 8, 161—213 (1930). Ref.: Biol. Abstr. **6**, 24542 (1932). — Iwanow, S. L.: Über den Stoffwechsel beim Reifen ölhaltiger Samen mit besonderer Berücksichtigung der Ölbildungsprozesse. Beih. bot. Zbl. **28** (I), 159—191 (1912). — Der Einfluß klimatischer Faktoren auf die physiologisch-chemischen Eigenschaften der Pflanzen [Russisch]. *Bull. Appl. Bot. a. Plantbreed. **13**, 483—491 (1922/23). Ref.: Bot. Abstr. **14**, 7504 (1925). — Die Evolution des Stoffes in der Pflanzenwelt und das Grundgesetz der Biochemie. Ber. dtsch. bot. Ges. **44**, 31—39 (1926). — Zur Kenntnis der Pflanzenöle in der U.S.S.R. I.—IX. Chem. Umschau, Gebiete Fette, Öle, Wachse, Harze **36**, 305—308 (1929a); **36**, 308—310 (1929b); **36**, 322—324 (1929c); **36**, 401—403 (1929d). **37**, 83—84 (1930a); **37**, 349—350 (1930b). **38**, 53—55 (1931a); **38**, 301—305 (1931b). **39**, 33—36 (1932a). — Zur Biochemie der Fette in den Pflanzen. Biol. generalis (Wien) **5**, 579—586 (1929e). — Über die klimatische Beeinflussung der Qualität des Öles von *Linum usitatissimum* beim Reifen. Allg. Öl- u. Fettztg. **29**, 149—150 (1932b).

Jamieson, G. S., W. F. Baugham and R. S. McKinney: Oil content of 9 varieties of soybean and characteristics of the extracted oils. J. Agric. Res. **46**, 57—58 (1933). — Johnson, I. J.: The relation of agronomic practice to the quantity and quality of the oil in flax seed. J. Agric. Res. **45**, 239—255 (1932a). — Correlation studies with strains of flax with particular reference to the quantity and quality of the oil. J. Amer. Soc. Agron. **24**, 537—544 (1932b). — Jones, W. W.: The physiology of oil production in the macadamia (*Macadamia integrifolia*, Maiden et Betche). Proc. Amer. Soc. Horticult. Sci. **35**, 239—245 (1937). — A study of developmental changes in composition of the macadamia. Plant Physiol. **14**, 755—768 (1939). — Jones, W. W., and L. Shaw: The process of oil formation and accumulation in the Macadamia. Plant Physiol. **18**, 1—7 (1943). — Jordan, R. C., and A. C. Chibnall: Observations on the fat metabolism of leaves. II. Fats and phosphatides of the runner bean *(Phaseolus multiflorus)*. Ann. of Bot. **47**, 163—186 (1933).

KALOYEREAS, S. A.: Studies on olive oil standardization. J. Amer. Oil Chem. Soc. **25** (1), 22—24 (1948). Ref.: Biol. Abstr. **23**, 615 (1949). — KARON, M. L., and A. M. ALTSCHUL: Effect of moisture and of treatment with acid and alkali on rate of formation of free fatty acids in stored cotton seed. I. Plant Physiol. **19**, 310—325 (1944). — KAYSER, R.: Hat eine Mineraldüngung Einfluß auf die wertbestimmenden Eigenschaften von Ölpflanzen und ändert sich durch die Düngung das Öl in seiner Zusammensetzung? Bot. Archiv **10**, 349—386 (1925). — KLEEBERGER: Die Entwicklung des Eiweiß- und Ölgehaltes in den Samen von Öl- und Gespinstpflanzen. Chem. Umschau, Gebiete Fette, Öle, Wachse, Harze **28**, 2—5 (1921). — KONDO, M., Y. TERASAKA and U. MOTOTARO: The influence of drought upon rice. II. *Rep. Ohara Agric. Inst. (Japan) **33**, 25—49 (1942). Ref.: Biol. Abstr. **23**, 14623 (1949). — KORSAKOW, M.: Recherches sur la variation des matières grasses, des sucres et de la saponine au cours de la maturation des graines de *Lychnis Githago*. C. r. Acad. Sci. Paris **155**, 1162—1164 (1912). — KRETOVICH, V. L.: Physiologico-biochemical bases of the storage of grain. *Izdatel'stvo Akademii Nauk USSR., Moskau 1945, 135 S. Ref.: Biol. Abstr. **20**, 5708 (1946). — KUMMER, H.: Fett und Fettsäuregehalt bei Gramineensamen in Beziehung zur Lichtbedürftigkeit bei der Keimung. Ber. dtsch. bot. Ges. **50**, 300—303 (1932). — KURSSANOW, A. L.: Biochemische Untersuchungen über die Fruchtreife der japanischen Mispel *(Eriobotrya japonica)*. Planta (Berl.) **15**, 752—766 (1932). — KYANE, L., and A. M. ALTSCHUL: Comparison of respiration, free fatty acid formation and changes in the spectrum of the seedling oil during the storage of cotton seed. Plant Physiol. **21**, 550—561 (1946).

LAING, E. V.: Studies in tree roots. Forest Commission (Gt. Britain) Bull. **13**, 72 S., 1932. — LAMBOU, M. G., N. S. PARKER and H. R. CARNS: Effect of maleic hydrazide applied to the cotton plant on the development of free fatty acids in the seed. J. Amer. Oil Chem. Soc. **33**, 199—202 (1956). Ref.: Chem. Abstr. **50**, 9528h (1956). — LANGELD, F.: Oil content in the seeds of truck and garden crops and soybeans. *J. Agric. Sci. S.E.U.S.S.R. **8** (2), 355—365 (1930). Ref.: Biol. Abstr. **6**, 25432 (1932). — LEATHER, J. W.: The composition of oil seed of India. Mem. Dep. Agric. India, Chem. Ser. **1**, 13—38 (1907). — LECLERC DU SABLON: Sur la formation des réserves non azotèes dans la Noix et dans l'Amande. C. r. Acad. Sci. Paris **123**, 1084—1086 (1896). — Sur les réserves oléagineuses de la Noix. Rev. gén. Bot. **9**, 313—317 (1897). — Recherches physiologiques sur les matières de réserves des arbres. Rev. gén. Bot. **16**, 341—368, 386—401 (1904). — Recherches physiologiques sur les matières des réserves des arbres. II. Rev. gén. Bot. **18**, 5—25, 82—96 (1906). — LEE, F. A., and H. B. TUCKEY: Chemical changes accompanying growth and development of seed and fruit of the Elberta peach. Bot. Gaz. **104**, 348—355 (1942). — LEHBERG, F. H., W. G. MCGREGOR and F. W. GEDDES: Flax studies. IV. The physical and chemical characteristics of flaxseed at progressive stages of maturity. Canad. J. Res., Sect. C **17**, 181—194 (1939). — LARKUM, A.: * Beiträge zur Kenntnis der Jahresperiode unserer Holzgewächse. Diss. Göttingen 1914 Zit. nach E. ANTEVS 1916. — LEWIS, F. J., and G. M. TUTTLE: Osmotic properties of some plant cells at low temperatures. Ann. of Bot. **34**, 405—416 (1920). — LIDFORSS, B.: Zur Physiologie und Biologie der wintergrünen Flora. Bot. Zbl. **68**, 33—44 (1896). — Die wintergrüne Flora. Lunds Univ. Arsskr., N. F. **2** (13), 76 S. (1907). — LINDSTROM, E. W., and G. FISK: Inheritance of carbohydrates and fat in crosses of dent and sweet corn. Iowa Agric. Exper. Stat. Res. Bull. **98**, 257—277 (1926). — Inheritance of chemical characters in maize. Iowa State Coll. J. Sci. **2**, 9—18 (1927). Ref.: Biol. Abstr. **3**, 4564 (1929). — LUCCA, S. DE: Recherches sur la formation de la matière grasses dans les olives. C. r. Acad. Sci. Paris **53**, 380—384 (1861); **55**, 470—473, 506—510 (1862); **57**, 520—522 (1863). — LÜDECKE, H., K. SAMMET u. W. LESCH: Untersuchungen über den Nährstoffbedarf und den Einfluß steigender Gaben Phosphorsäure und Kali auf Ertrag und Beschaffenheit der Sojabohne. Bodenkde u. Pflanzenernähr. **25** (**70**), 1—31 (1941). — LUTZ, K. G.: Beiträge zur Physiologie der Holzgewächse. Ber. dtsch. bot. Ges. **13**, 185—188 (1895). — Fünfstücks Beitr. z. wiss. Bot. **1**, 1—80 (1897).

MALHOTRA, R. C.: Changes in plants during low temperatures. I. Oil synthesis at various elevations of the Himalayas and its physiological influence on the protoplasm of *Cedrus deodora*. J. Indian Bot. Soc. **10**, 293—310 (1931). — The effect of elevation on the synthesis and some properties of the oil of *Cedrus deodora*. Biol. generalis (Wien) **9**, 249—256 (1933). — MARKLEY, K. S.: Synthesis of fatty acids by plants. Fatty acids, their chemistry and physical properties, S. 563—569. New York 1947. 668 S. — MCCLENAHAN, F.-M.: The development of fat in the black walnut *(Juglans regia)*. I. J. Amer. Chem. Soc. **31**, 1093—1098 (1909). — II. J. Amer. Chem. Soc. **35**, 485—493 (1913). — MCGLAMERY, J. B., and M. P. HOOD: Effect of two heat treatments on rancidity development in unshelled pecans. Food Res. **16**, 80–84 (1951). – MCNAIR, J. B.: The taxonomic and climatic distribution of oils, fats and waxes in plants. Amer. J. Bot. **16**, 832—841 (1929). — The taxonomic and climatic distribution of oil and starch in the seeds in relation to the physiological and chemical properties of both substances. Amer. J. Bot. **17**, 662—668 (1930a). — A study of some characteristics of vegetable oils.

Field Museum Nat. History, Bot. Ser. **9** (2), 47—68 (1930b). — The evolutionary status of plant families in relation to some chemical properties. Amer. J. Bot. **21**, 427—453 (1934). — Angiosperm phylogeny on a chemical basis. Bull. Torrey Bot. Club **62**, 515—532 (1935). — Indications of an increase in number of C-atoms in acids and number of acids in seed fats with advance in evolutionary position. Science (Lancaster, Pa.) **94**, 422 (1941). — Advance in phylogenetic position in the cryptogames as indicated by their fats. Lloydia **6** (2), 115—156 (1943). — Some comparison of chemical ontogeny with chemical phylogeny in vascular plants. Lloydia 8 (3), 145—169 (1945a). — Plant fats in relation to environment and evolution. Bot. Review **11**, 1—59 (1945b). — MER, E.: De la constitution et des fonctions des feuilles hivernales. Bull. Soc. bot. France **23**, 231—238 (1876). — De la repartition de l'amidon dans les rameaux des plantes ligneuses. Des causes qui y président. — De son influence sur la ramification. Bull. Soc. bot. France **26**, XLIV—LIII (1879). — Repartition hivernale de l'amidon dans les plantes ligneuses. C. r. Acad. Sci. Paris **112**, 964—966 (1891). — Des variations qu'éprouve la réserve amylacee des arbres aux diverses epoques de l'année. Bull. Soc. bot. France **45**, 299—309 (1898). — MEYER, A.: Die angebliche Fettspeicherung immergrüner Laubblätter. Ber. dtsch. bot. Ges. **36**, 5—10 (1918). — MILLER, L. P.: Genesis of fats and oils in plants. Monograph series Amer. Chem. Soc. **123** (Vegetable fats and oils, 836 p. New York 1954), S. 208—216. — MIYAKE, K.: On the starch of evergreen leaves and its relation to photosynthesis during the winter. Bot. Gaz. **33**, 321—340 (1902). — MORSE, W. J.: Chemical composition of soybean seed. Soybeans and Soybean products, herausgeg. von K. S. MARKLEY, 2 Bde. Bd. 1, S. 135—154. New York 1950. — MÜLLER, N. J. C.: Botanische Untersuchungen. IV. Untersuchungen über die Molekularkräfte im Baum. Heft V. Die einjährige Periode. *Heidelberg: C. Winter 1875. 75 S. Ref.: Justs Bot. Jber. **3**, 1034 (1875). — MÜNTZ, A.: Recherches chimiques sur la maturation des graines. Ann. Sci. Nat., VII. s., Bot. **3**, 45—74 (1886). — MUSCO, D. D., and W. V. CRUESS: Food rancidity. Studies on deterioration of walnut meats. J. Agric. a. Food Chem. **2**, 520—523 (1954).

NESBITT, L. L., A. J. PICKNEY, T. E. STOA and E. P. PAINTER: Oil formation in flax seed (Changes in oil constituents of seed from flax plants harvested at regular intervals after full bloom). North Dakota Agric. Exper. Stat. Techn. Bull. **323**, 19 S. (1943). — Exper. Stat. Rec. **90**, 147 (1944). — NEUMANN, P.: Über den Fettstoffwechsel reifender und keimender Leguminosensamen. Biochem. Z. **308**, 141—174 (1941). — NIKLEWSKI, B.: Untersuchungen über die Umwandlung einiger N-freier Reservestoffe während der Winterperiode der Bäume. Beih. bot. Zbl. **19**, 68—117 (1906). — NOGGLE, G. R.: The physiology of polyploidy in plants. Review of the literature. Lloydia **9** (3), 153—173 (1946). — NOGUTI, Y., H. OKA and T. ÔTUKA: Growth rate and chemical analysis of diploid and its autotetraploid in *Nicotiana rustica* and *N. tabacum*. Japan. J. Bot. **10**, 343—364 (1940). — NOTTER, L. F. C.: *Beitrag zur Physiologie der Holzgewächse. — Die jährlichen Wandlungen der stickstoffreien Reservestoffe. Diss. Heidelberg 1903. Zit. nach E. ANTEVS 1916.

PACK, D. A.: After-ripening and germination of *Juniperus virginiana* seeds. Bot. Gaz. **71**, 32—60 (1921a). — Chemistry of after-ripening, germination and seedling development of Juniper seeds. Bot. Gaz. **72**, 139—150 (1921b). — PAINTER, E. P.: Some relationship between fat acid composition and the iodine number of linseed oil. Oil a. Soap **21**, 343—346 (1944a). — Fat acid formation during oil deposition in flaxseed. Arch. of Biochem. **5**, 337—348 (1944b). — PAINTER, E. P., and L. L. NESBITT: Thiocyanogen absorption of linseed oils. Industr. Engng. Chem., Anal. Ed. **15**, 123 (1943a). — Fat acid composition of linseed oil from different varieties of flaxseed. Oil a. Soap **20**, 208—211 (1943b). — PAINTER, E. P., L. L. NESBITT and T. E. STOA: The influence of seasonal conditions on oil formation and changes in the iodine number during growth of flaxseed. J. Amer. Soc. Agron. **36**, 204—213 (1944). — PARIJA, P., and P. MALLIK: Nature of reserve food in seeds and their resistance to high temperature. J. Indian Bot. Soc. **19**, 223—230 (1940). Ref.: Biol. Abstr. **15**, 19773 (1941). — PATEL, J. S., and C. R. SESHADRI: Oil formation in groundnut with reference to quality. Indian J. Agricult. Sci. **5**, 165—175 (1935). — PEARL, R., u. J. M. BARTLETT: The Mendelian inheritance of certain chemical characters in maize. Z. Abstammgslehre **6**, 1—28 (1911). — PEARSON, L. K,. and H. S. RAPER: The influence of temperature on the nature of the fat formed by living organisms. Biochemic. J. **21**, 875—879 (1927). — PETERSEN, O. G.: État de l'amidon dans nos arbres à feuilles caduques durant le repos hivernal. Études sur les phénomènes vitaux des racines des arbres (Dänisch). Oversight kong. Danske Vidensk. Selsk. Forh. **1896**, 50—66; **1898**, 1—57; franz. Zusammenfassung beider Arbeiten: Oversight kong. Danske Vidensk. Selsk. Forh. **1898**, 58—68. — PFEFFER, W.: Untersuchungen über die Proteinkörner und die Bedeutung des Asparagins beim Keimen der Samen. Jb. Bot. 8, 429—574 (1872). — PICKETT, T. A.: Composition of developing peanut seed. Plant Physiol. **25**, 210—224 (1950). — PICKETT, T. A., and K. T. HOLLEY: Susceptibility of types of peanuts to rancidity development. J. Amer. Oil Chem. Soc. **28**, 478—479 (1951). — *PIGULEWSKI, G.: J. russ. phys.-chem. Ges. **1915**. Zit. nach CZAPEK, Biochemie, Bd. III, S. 791. — POJARKOVA, A.: Winterruhe, Reservestoffe und Kälteresistenz bei Holzpflanzen. Ber. dtsch.

bot. Ges. **42**, 420—429 (1924). — PRESTON, J. F., and F. J. PHILLIPS: Seasonal variation in the food reserve of trees. Forest Quarterly **9**, 232—243 (1911). — PRIESTLEY, J. H.: The fundamental fat metabolism of the plant. New Phytologist **23**, 1—19 (1924). Ref.: Bot. Zbl. **5** (147), 404. — PRIKHOD'KO, M. I.: Periodic transformation of reserve materials in certain trees [Russisch]. *Bull. Polytechn. Inst. Tiflis **3**, 61—68 (1928). Ref.: Biol. Abstr. **5**, 10751 (1931). — PULKKI, L. H., u. K. PUUTULA: Über das Thiamin im Weizenkorn. Biochem. Z. **308**, 122—127 (1941). — PURIEWITSCH, K.: Physiologische Untersuchungen über die Entleerung der Reservestoffbehälter. Jb. Bot. **31**, 1—76 (1898).

RABAK, F.: Influence on linseed oil of the geographical source and variety of flax. U.S. Dept. Agric. Bull. 655, **1918**, 1—16. — RALEIGH, G. J.: Chemical condition in maturation, dormancy and g'rmination of seeds of *Gymnocladus dioica*. Bot. Gaz. **89**, 273—294 (1930). — RENVALL, A.: Über die Beziehungen zwischen der Stärketransformation der Holzgewächse in der Winterperiode in ihrem Gehalt an sogenanntem Gerbstoff. Beih. bot. Zbl., Abt. I **28**, 282—306 (1912). — REWALD, B., u. W. RIEDE: Das Verhalten von Fett, Phosphatiden und Eiweiß während der Samenreife. Biochem. Z. **260**, 147—152 (1933). — RHODES, E., and R. M. WOODMAN: The fatty substances of the plant growing point. Proc. Leeds Philos. Soc. **1** (1), 27—36 (1925). — ROBERTSON, D. W., C. C. FIFIELD and L. ZELENY: Milling, baking and chemical properties of Colorado-grown Marquis and Kanred wheat stored 9 to 17 years. J. Amer. Soc. Agron. **31**, 851—856 (1939). — ROBERTSON, F. R., and J. G. CAMPBELL: Some observations on the increase of free fatty acid in cotton seed. Oil a. Soap **10**, 146—147 (1933). — ROBINSON, B. B.: The time to harvest fibre flax. U. S. Dept. Agric. Techn. Bull. 236, **1931**, 23 S. — ROMAGNOLI, M.: Note relative a ricerche compiute sulla coltivazione della soia in Italia. Riv. Agric. Subtrop. e Trop. **43** (7/9), 154—160 (1949). Ref.: Biol. Abstr. **25**, 2716 (1951). — ROSE, W. G., and G. S. JAMIESON: The comparison of 7 American linseed oils. Oil a. Soap **18**, 173 (1941). — ROSENBERG, O.: Die Stärke der Pflanzen im Winter. Bot. Zbl. **66**, 337—340 (1896). — ROUSILLE, A.: Recherches relatives à la maturation des olives. C. r. Acad. Sci. Paris **86**, 610—613 (1878). — RUSCA, R. A., and F. L. GERDES: Effects of artificially drying seed cotton on certain quality elements of cotton seed in storage. U.S. Dept. Agric. Circular 651, **1942**, 19 S. — RUSSOW, E.: Über den Inhalt der parenchymatischen Elemente der Rinde usw. *Sitzgsber. Dorpater naturforsch. Ges. **6**, 350—389 (1882). Ref.: Bot. Zbl. **14**, 271—275 (1883).

SAHASHRABUDDHE, D. L., and N. P. KALE: A biochemical study of the formation of the oil in niger seed *(Guizotia abyssinica)*. Indian J. Agric. Sci. **3**, 57—88 (1933). — SCHARRER, K., u. R. SCHREIBER: Über die Wirkung magnesiumarmer und magnesiumreicher Kalidüngemittel auf den Eiweiß- und Fettertrag der Ölrauke *(Eruca sativa)*. Bodenkde. u. Pflanzenernähr. **24** (**69**), 55—64 (1941a). — Über die Wirkung magnesiumarmer und magnesiumreicher Kalidüngemittel auf Eiweiß- und Fettertrag von Sommerrübsen. Bodenkde. u. Pflanzenernähr. **25** (**70**), 228—239 (1941b). — SCHELLENBERG, H. C.: Über Hemicellulosen als Reservestoffe bei unseren Waldbäumen. Ber. dtsch. bot. Ges. **23**, 36—45 (1905). — SCHISCHKIN, A.: Einfluß einiger Düngemittel auf das Wachstum des Leins, seine Entwicklungsdauer, seine Zusammensetzung und den Ölgehalt seiner Samen. Landwirtsch. Versuchsstat. **15**, 161—168 (1872). — SCHMALFUSS, K.: Experimentelle Studien zur Physiologie und Ernährung des Leins im Hinblick auf die Ausbildung von Faser und Öl. Teil 1. Gefäßversuche. Bodenkde. u. Pflanzenernähr. **1** (**46**), 1—39 (1936a). — Forschungsdienst (Sonderh.) **6**, 173—180 (1937). — Über den Einfluß der Ernährung der Pflanzen auf die Beschaffenheit des Mohnöls. Angew. Bot. **18**, 345—347 (1936b). — Weitere Studien zur Fettbildung im Leinsamen unter dem Einfluß von Umwelt und Ernährung der Pflanze. Bodenkde. u. Pflanzenernähr. **5** (**50**), 37—46 (1937). — Über die Wirkung von Stickstoff- und Kalidüngung zu Winterraps. Bodenkde. u. Pflanzenernähr. **6** (**51**), 254—258 (1938). — Die Wirkung des Kalium und der Kalisalzionen auf die Ausbildung des Leinöls im Felddüngungsversuch. Ernähr. Pflanze **35**, 65—66 (1939). — Düngung und Qualität bei Öl- und Fettpflanzen. *Forschungsdienst (Sonderh.) **16**, 169—171 (1941a). Ref.: Biol. Abstr. **25**, 5985 (1951). — Zur Kenntnis der Fettbildung in Pflanzensamen. Bodenkde. u. Pflanzenernähr. **24** (**69**), 321—341 (1941b). — SCHMALFUSS, K., u. H. MICHEEL: Über die Abhängigkeit der Beschaffenheit des Leinöls von der Mineralernährung der Pflanzen. Angew. Bot. **17**, 199—206 (1935). — SCHMIDT, C.: Über Stärke- und Fettbäume. Bot. Ztg. Abt. II **67**, 130—131 (1909). — SCHOLFIELD, C. R., and W. C. BULL: Relation between the fatty acid composition and the iodine number of soybean oil. Oil a. Soap **21**, 87—89 (1944). — SCHROPP, W., u. B. ARENZ: Düngungsversuche zu Winterölfrüchten. Bodenkde. u. Pflanzenernähr. **18** (**63**), 315—330 (1940). — SCHULZ, E.: Über Reservestoffe in immergrünen Blättern. Flora (Jena) **71**, 223—241, 248—257 (1888). — SCOTT, F. M.: The distribution and physical appearance of fats in living cells. Introductory survey. Amer. J. Bot. **42**, 475—480 (1955). — SCULLY, N. J., W. CHORNEY, G. KOSTAL, R. WATANABE, J. SKOK and J. W. GLATTFIELD: Biosynthesis in C^{14}-labelled plants, their use in agricultural and biological research. Proceedings of the international conference on the peaceful uses of atomic energy, Genf, Aug. 1955. **12**,

376—385 (1956). — Scurti, F., e M. Fornaini: Sulla formazione del grasso nei frutti oleaginosi. III. Richerchi sperimentali eseguite sulle foglie di ligustro. Ann. R. Staz. Chimica Sper. di Roma, Ser. II 5, 223—239 (1911). — Scurti, F., e G. Tommasi: Sulla formazione del grasso nei frutti oleaginosi. I. Richerche sperimentali eseguite sull'olivo. Ann. R. Staz. Chimica Agrar. Sper. di Roma, Ser. II 4, 253—286 (1910). — II. Richerche sperimentali eseguite sul Ligustro *(Ligustrum japonicum)*. Ann. R. Staz. Chimica Agrar. Sper. di Roma, Ser. II 5, 103—121 (1911). — Sell, H. M., F. A. Johnston and F. S. Lagasse: Changes in the chemical composition of the tung fruit and its component parts. J. Agricult. Res. **73**, 319—334 (1946). — Shafizadeh, F., and M. L. Wolfrom: Biosynthesis of C^{14}-labelled cotton seed oil from d-glucose-6-C^{14}. J. Amer. Chem. Soc. **78** (11), 2498—2499 (1956). — Shao-Liu, C., and P. S. Tang: Studies on colchicine induced autotetraploid barley. III. Physiological studies. Amer. J. Bot. **32**, 177—181 (1945). — Shutt, F. T.: Influence of heredity and environment on the protein and oils contents of soybeans, as grown at Harrow and Ottawa; crops of 1928—1929. Dominion of Canada, Department of Agriculture, Dominion Exper. Farms. Division of Chemistry, Report of the Dominion Chemist for the year ending March 31. 1930, p. 54—57. Ottawa 1931. — Simmons, R. O., and F. W. Quackenbush: The sequence of formation of fatty acids in developing soybean seeds. J. Amer. Oil Chem. Soc. **31** (11), 441—443 (1954a). — Comparative rates of formation of fatty acids in the soybean seed during its development. J. Amer. Oil Chem. Soc. **31** (12), 601—603 (1954b). — Simpson, D. M.: Relation of moisture content and methods of storage to deterioration of stored cotton seed. J. Agric. Res. **50**, 449—456 (1935). — Simpson, M. D.: Factors affecting the longevity of cotton seed. J. Agric. Res. **64**, 407—419 (1942). — Sinnott, E. W.: Factors determining character and distribution of food reserve in woody plants. Bot. Gaz. **66**, 162—175 (1918). — Smirnova, M. I.: Biochemistry of lupine. *Biokhim. Kul'turnych Rastenii **2**, 270—327 (1938). Ref.: Herb. Abstr. **10**, 136 (1940) u. Biol. Abstr. **16**, 21103 (1942). — *Smirnowa, M. I., and M. N. Lawrowa: Bull. Appl. Bot. Genetic. a. Plant Breeding, Ser. 3 (5) **1934**, 73—97 (Russisch). Zit. in N. N. Iwanoff, 1948, S. 27. — Snider, H. J.: Fertilizing corn for yield, protein and oil. Better Crops With Plant Food **36**, 26, 44—45 (1952). Ref.: Biol. Abstr. **28**, 4298 (1954). — Stanley, O. B.: Fat deposits in certain *Ericaceae*. *Butler Univ. Bot. Studies **2**, 33—41 (1931). Ref.: Biol. Abstr. **7**, 16939 (1931). — Stark, R. W.: Environmental factor affecting the protein and the oil content of soybeans and the iodine number of soybean oil. J. Amer. Soc. Agron. **16**, 636—645 (1924). — Stein, F.: Über Ölkörper bei Oenotheraceen. Österr. bot. Z. **65**, 43—49 (1915). — Sterz, M.: Über den Einfluß verschiedenartiger Mineraldüngung auf die Fettbildung in den Samen von *Glycine Soja*. Bodenkde. u. Pflanzenernähr. **24** (**69**), 34—63 (1940). — Stoddard, L. A.: Chemical composition of *Symphoricarpus rotundifolius* as influenced by soil, site and date of collection. J. Agric. Res. **63**, 727—739 (1941). — Strobel, A.: Ein Standraumversuch mit Lein. Faserforsch. **5**, 227—228 (1926). — Surož, J.: Öl als Reservestoff der Bäume. *VIII. Congr. Russ. Naturforsch. u. Ärzte, Botanik, S. 24—28, St. Petersburg 1890 [Russisch]. Ref.: Beih. bot. Zbl. **1**, 342—343 (1891).

Taran, E. N.: Einige Eigenschaften des Ölbildungsvorganges während der letzten Stufen der Samenreife der Ölpflanzen. [Russisch mit engl. Zus.fass.] *Biochimija **2**, 741—744 (1937). Ref.: Ber. wiss. Biol. **46**, 204. — Terroine, E. F.: Fat metabolism. Annual Rev. Biochem. **5**, 227—247 (1936). — Tharp, W. H.: Cottonseed composition. Relation to variety, naturity and environment of the plant. Cottonseed and Cottonseed products. Their chemistry and chemical technology, herausgeg. von A. C. Bailey, S. 117—154. New York 1948, 936S. — Theis, E. R., J. S. Long and G. F. Beal: Drying oils. XIII. Changes in linseed oil, lipase and other constituents of the flaxseed as it matures. Industr. a. Engng. Chem. **22**, 768—771 (1929). — Thomas, J. B.: On the metabolism of Damar seeds. *Ann. Bot. Gard. Buitenzorg **51**, 94—114 (1941). Ref.: Biol. Abstr. **16**, 13402 (1942). — Thor, C. J. B., and C. L. Smith: A physiological study of seasonal changes in the composition of the pecan during fruit development. J. Agric. Res. **50**, 97—121 (1935). — Tietz, R.: Über mögliche Beziehungen des Fettstoffwechsels zum Lichtkeimproblem. Biochem. Z. **324**, 517—529 (1953). — Tombesi, L., e M. Tarantola: La composizione del semi di alcune specia vegetali in relazione alla disponibilità idriche del suolo. *Ann. Staz. Chimico-Agraria Sper. di Roma, Ser. III **30**, 1—10 (1950). Ref.: Biol. Abstr. **28**, 6804 (1954). — Tottingham, W. E.: Are leaf lipides responsive to solar irradiation ? Science (Lancaster, Pa.) **75**, 223—224 (1932). — Tuttle, G. M.: Induced changes in reserve-materials in evergreen herbaceous leaves. Ann. of Bot. **33**, 201—210 (1919). — Reserve food materials in vegetative tissues. Bot. Gaz. **71**, 146—151 (1921).

Ulrich, R.: Variations quantitatives des acides gras, des sterols et des phospholipides au cours de la maturation des fruits de Lierre. C. r. Acad. Sci. Paris **208**, 664—666 (1939).

Vandevelde, A. J. J.: Beiträge zur chemischen Physiologie der Baumstämme. [Flämisch]. Bot. Jaarboek (Gent) **9**, 94—120 (1897). Ref.: Chem. Zbl. **69**, 466 (1898 I).

Washburn, W. F.: I. Soy bean oil; II. Flax studies. *North Dakota Agric. Exper Stat. Bull. **118**, 35—48 (1916). Ref.: Exper. Stat. Rec. **36**, 206 (1917). — Weber, F.: Untersuchungen über die Wandlungen des Stärke- und Fettgehaltes der Pflanze, insbesondere der Bäume. Sitzgsber. Akad. Wiss. Wien, Math.-naturwiss. Kl., Abt. I **118**, 967—103 (1909). — Weintraub, R. L.: *Symposium Report. Amer. Chem. Soc. 122. Meeting p. 266—273, 1952. Zit. in Rev. Plant Physiol. **4**, 263 (1953). — Weintraub, R. L., J. N Yeatman, J. A. Lockhart, J. H. Reinhart and M. Fields: Metabolism of 2,4-dichlorophenoxyacetic acid. II. Metabolism of the side chain by bean plants. Arch. of Biochem a. Biophysics **40**, 277—285 (1952). — Weissenböck, K.: Fett- und Phosphatidstoffwechsel keimender und reifender Ölsamen. 2. Mitt. Lipoidstoffwechsel von Ölsamen während des Reifungsvorganges. Österr. bot. Z. **94**, 314—329 (1948). — Weissflog, J.: Leben und Lebensdauer in den Reservestoffbehältern keimender Samen. Bot. Archiv **5**, 283—312 (1924). — Wiesner, J.: Die Rohstoffe des Pflanzenreiches, 4. Aufl. Leipzig: Wilhelm Engelmann 1927. — Wilfarth, H., u. G. Wimmer: Die Wirkung des Kaliums auf das Pflanzenleben Arb. dtsch. landwirtsch. Ges. **68**, 1—106 (1902). — Wilson, J. K.: Acidity changes in stored legume seeds. J. Amer. Soc. Agron. **21**, 815—817 (1929). — Wolfe, A. C., J. B. Park and R. C. Burrell: A study of the chemical composition of soybeans during maturation Plant Physiol. **17**, 289—295 (1942). — Woodroof, J. G., and N. C. Woodroof: The development of the pecan nut *(Hicoria pecan)* from flower to maturity. J. Agric. Res. **34**, 1049—106 (1927). — Woodruff, S., and H. Klaas: A study of soybean varieties with reference to their use as food. Univ. Illinois, Agric. Exper. Stat. Bull. **443** **1938**, 421—467. — Worley C. L., T. J. Ratcliffe and R. G. Grogan: The changes of crude lipids in the okra when grown on three levels of phosphorus. *J. Tennessee Acad. Sci. **17**, 210—223 (1942). Ref.: Biol Abstr. **16**, 21055 (1943). — Wright, R. C.: Investigation on the storage of nuts. U. S. Dept Agric. Techn. Bull. 770, **1941**, 1—35. Ref.: Biol. Abstr. **15**, 15469 (1941).

Yermakov, A. J.: Intraspezifische und individuelle Variabilität des Ölgehaltes von Samen. [Russisch mit engl. Zus.fass.] *Bull. Acad. Sci. USSR., Cl. Sci. math. et natur. Ser. biol. **6**, 1853—1870 (1937). Ref.: Ber. wiss. Biol. **47**, 584. — Yermakov, A. J. u. Mitarb. Örtliche und Speziesdifferenzen im Ölgehalt verschiedener Lupinensamen. Trudy prikl Bot. i pr. (III. Physiol., Biochem. a. Anat. of Plants) **10**, 6—23 (1935). [Russisch mit engl Zus.fass. S. 23—24.] Ref.: Ber. wiss. Biol. **40**, 49.

Zafar, A., and S. Ahmad: The effect of storage on rape-seed and rape-oil. Indian J Agric. Sci. **16**, 279—283 (1946). Ref.: Biol. Abstr. **23**, 10511 (1949). — Zeleny, L.: Fat acidity of grain as an index of soundness and risk of storage. *Milling production **14**, 1 (1949). Ref. Biol. Abstr. **23**, 23392 (1949). — Chemical and nutritive changes in stored grain. Trans Amer. Assoc. Cereal Chem. **6**, 112—125 (1948). Ref.: Biol. Abstr. **23**, 6484 (1949).

Die Mobilisierung der Fette bei der Keimung.

Von

A. Zeller.

Mit 2 Abbildungen.

A. Einleitung.

Schon im Kapitel über die Bildung der Fette in Samen und Früchten von Pflanzen wurde darauf hingewiesen, daß es unmöglich ist, einen Grund dafür anzugeben, warum manche Pflanzen ausschließlich oder überwiegend Fette als Reservestoffe speichern, während viele andere ausschließlich oder überwiegend Kohlenhydrate enthalten. Bei der Keimung von fetthaltigen Samen nimmt nun verständlicherweise der Gehalt der Samen an Fett in anderer Weise ab als bei kohlenhydrathaltigen Samen die Kohlenhydrate bei der Keimung verschwinden, wenn sie als Atmungsmaterial verbraucht werden. Im Falle der Fette ist es aus chemischen Gründen offensichtlich, daß sie nicht, wie etwa die Reservekohlenhydrate, unmittelbar oder nach verhältnismäßig geringen Umwandlungen als Baustoffe für den jungen Keimling verwendet werden können. Ist es doch klar, daß aus Fetten das Baumaterial für die Zellwände und andere Bestandteile des jungen Keimlings nur durch weitgehenden chemischen Umbau der Moleküle erhalten werden kann und außerdem ist es wohl wahrscheinlich, daß die Fette vor allem über den Weg der Umwandlung in Kohlenhydrate oder deren Vorstufen in den Atmungsstoffwechsel einbezogen werden. Ein Einfluß möglicher Abbauprodukte höherer Fettsäuren (nämlich von Sorbinsäure, Buttersäure, β-Oxybuttersäure und β-Ketobuttersäure) auf den biologischen und fermentativen Kohlenhydrat-(Stärke-, Rohrzucker-)aufbau wurde von A. Zeller (1935, vgl. auch A. Zeller 1942) und K. Enser (1935) in Kürbiskeimlingen und ihren Autolysaten festgestellt. Damit schien auch experimentell der Übergang von Fettsäuren in Verbindungen, die als Kohlenhydratvorstufen oder Kohlenhydrate dem Atmungsstoffwechsel und Baustoffumsatz zugänglich sind, chemisch vorstellbar gemacht zu sein. Freilich sind ja jetzt alle diese Vorstellungen durch die Aufklärung der Fettveratmung und der Fett-Kohlenhydratumwandlung im tierischen Organismus längst überholt (vgl. z. B. Lehninger 1953), aber es ist doch sehr bedauerlich, daß von botanischer Seite nur vereinzelte Ansätze dazu vorliegen, mit moderner physiologisch-chemischer Methodik den Fettstoffwechsel der Pflanzen zu erforschen (vgl. z. B. W. Franke, H. Frehse 1954, T. E. Humphreys und P. K. Stumpf 1955, P. K. Stumpf und G. A. Barber 1956, N. Coppens 1956).

B. Der Atmungsquotient keimender Fettsamen.

Messungen des im Atmungs-Gasstoffwechsel aufgenommenen Sauerstoffes und des dabei abgegebenen Kohlendioxyds erlauben bis zu einem gewissen Grade einen Rückschluß auf die Natur des zur Energiegewinnung verwendeten Reservematerials. Da aber dieser Teil des Keimungsstoffwechsels an anderer Stelle des

21*

vorliegenden Handbuches eingehend behandelt wird, seien hier nur einige Hinweise auf die speziellen Verhältnisse bei der Keimung der Fettsamen gebracht.

Wenn ausschließlich Kohlenhydrate veratmet werden, dann ist ein Respirationsquotient (R. Q.) von genau 1 zu erwarten, da das Verhältnis des entstehenden Kohlendioxyds zum verbrauchten Sauerstoff, CO_2/O_2, gleich 1 ist gemäß folgender Umsetzung

$$C_6H_{12}O_6 + 6\,O_2 \rightarrow 6\,CO_2 + 6\,H_2O.$$

Bei der Veratmung sauerstoffreicherer Verbindungen (z. B. organischer Säuren) sind höhere R. Q.s zu erwarten, wie etwa die folgenden Schemen der Veratmung von Apfelsäure und Oxalsäure mit R. Q. von 1,33 nnd 4,0 zeigen.

$$C_4H_6O_5 + 3\,O_2 \rightarrow 4\,CO_2 + 3\,H_2O$$

$$2\,C_2H_2O_4 + O_2 \rightarrow 4\,CO_2 + 2\,H_2O.$$

Sauerstoffarme Verbindungen dagegen, wie die Fette, lasse R. Q.s, die kleiner als 1 sind, erwarten. Für Triolein z. B. ergäbe sich folgende Umsetzung

$$C_{57}H_{104}O_6 + 80\,O_2 \rightarrow 57\,CO_2 + 29\,H_2O$$

entsprechend einem R. Q. von ziemlich genau gleich 0,7. Umwandlung der Fettsäuren in Kohlenhydrate hingegen ist ohne oder fast ohne CO_2-Abspaltung denkbar und die entsprechenden R. Q.s können daher sogar 0 sein. Die folgende Umwandlung von Ricinolsäure in Saccharose veranschaulicht dies:

$$2\,C_{18}H_{34}O_3 + 14\,O_2 \rightarrow 3\,C_{12}H_{22}O_{11} + H_2O.$$

Da nun in keimenden Fettsamen sowohl Veratmung als auch andere Umwandlungen der Fette (meist wohl Kohlenhydratbildung) vor sich gehen, ist es klar, daß R. Q.s von weniger als 0,7 auftreten können. Je tiefer der R. Q. unter 0,7 liegt, desto mehr sauerstoffverbrauchende Vorgänge, die keine Kohlensäure liefern, müssen neben der Fettveratmung (deren R. Q. ja nicht unter 0,7 sinken kann) ablaufen.

Die erste Erwähnung einer Bestimmung eines Atmungsquotienten von Pflanzenteilen findet sich in einem Vortrag, den P. BERT am Naturhistorischen Museum in Paris hielt (P. BERT 1870, S. 32), und E. GODLEWSKI (1882) war wohl der erste, der ausführliche Untersuchungen über Atmungsquotienten von Fett- und Kohlenhydratsamen und -keimlingen anstellte. (Es ist klar, daß alle Untersuchungen über den Keimungsstoffwechsel im Dunkeln ausgeführt werden, um die Vorgänge unabhängig vom Assimilationsstoffwechsel erfassen zu können.) Während der Quellung fand er sowohl bei Fett- als auch bei Stärkesamen den Atmungsquotienten nicht von 1 verschieden. Dies stellten auch A. J. ERMAKOFF (1931), A. ERMAKOFF und N. N. IWANOFF (1931) fest. Beim Sichtbarwerden der Keimwurzel fand E. GODLEWSKI jedoch in Fettsamen (Rettich, Hanf, Lein, Luzerne) Atmungsquotienten von 0,55—0,65, während sich bei Stärkesamen (Erbsen und Weizen) kein wesentliches Absinken unter 1 zeigte. Wenige Jahre nach E. GODLEWSKI berichteten G. BONNIER und L. MANGIN (1884), beim keimenden Leinsamen Atmungsquotienten von 0,3 gefunden zu haben. Ähnliche, nur wenig höhere Werte fanden sich auch bei Keimlingen von *Lepidium sativum*, *Lupinus luteus* und *Vicia faba*, während bei Gerste nur vorübergehend ein Absinken des R. Q. unter etwa 0,9 gefunden wurde. Am Ersten Internationalen Botanikerkongreß in Paris berichtete C. GERBER (1900) ausführlich über den Atmungsquotienten und über relative Atmungsintensitäten reifender und keimender (auch unreif geernteter) Samen von Lein, Erbsen und *Ricinus*. Die Fettsamen zeigten in manchen Keimungsstadien immer Werte von 0,6—0,5, während der R. Q. bei Erbsen nie unter 0,8 sank (für Eiweiß beträgt der R. Q. etwa 0,8). Ein Jahr später hat C. GERBER (1901) auch über Bestimmungen des Atmungsquotienten von reifenden Oliven berichtet und H. HALVORSEN (1955) hat an reifenden Leinsamen und Leinkapseln Atmungsquotienten und ihre Temperaturabhängigkeit bestimmt (vgl. auch oben S. 291).

In neuerer Zeit haben W. STILES und W. LEACH (1933) sowie W. LEACH und K. W. DENT (1934) über die Atmungsquotienten keimender Kohlenhydrat- und Fettsamen berichtet. Sie haben bei *Zea*, *Pisum*, *Vicia faba* und *Lathyrus odoratus* ein vorübergehendes Fallen des R. Q., das auf Fettveratmung hindeutet, gefunden. *Lupinus luteus* und *Tropaeolum majus* zeigten stärker schwankende Quotienten unter 1, bei *Ricinus* wurde ein Fallen von 0,8 auf 0,5 gefunden, *Helianthus* zeigte zuerst ein Ansteigen, später ein Fallen und *Cucurbita* zeigte überraschenderweise einen nur zwischen 0,8 und 0,9 uncharakteristisch schwankenden Atmungsquotienten. Bei *Ricinus*, *Helianthus* und *Cucurbita* fiel in einer Stickstoffatmosphäre die Kohlensäurebildung reversibel stark ab; beim Zurückbringen in sauerstoffhaltige Umgebung traten dann zunächst wesentlich niedrigere R. Q.s als bei den Kontrollen auf. Sie normalisierten sich aber bald. In keimenden Damarsamen (*Shorea Wiesneri*) fand J. B. THOMAS (1941) Atmungsquotienten von 0,4—0,6, während er in reifenden Samen Quotienten über 1 und in den reifen Samen solche von 0,7—1 antraf. Auch an Kürbissamen und -kotyledonen wurde gefunden (R. BROWN 1942), daß während des größten Teiles der ersten 48 Std der Keimung der ursprünglich hohe Atmungsquotient nur etwa 0,5 beträgt; in diesen ersten Keimungsstadien scheint also teilweise anaerobe Atmung vorzuliegen.

J. KNABE (1943) hat bei Senfsamen am 1. Keimungstag einen R. Q. von 0,7, am 3. Keimungstag einen solchen von 0,5 und am 8. Keimungstag einen Wert von 0,8 festgestellt. G. R. BARTLETT (1945) fand bei Tomatenkeimlingen einen R. Q. von etwa 0,3 (ähnlich F. N. CRAIG 1936 bei Lupinen), und H. G. ALBAUM und B. EICHEL (1943) fanden sogar Hinweise dafür, daß selbst Haferkeimlinge in den ersten 72 Keimungsstunden mehr Fett als Kohlenhydrate veratmen. Nebenbei sei erwähnt, daß G. O. BURR und E. S. MILLER (1938) in reifenden *Ricinus*samen einen engen Zusammenhang zwischen Atmungsquotientem und Ausmaß der Fettbildung in den Samen gefunden haben. Zur Zeit der intensivsten Fettbildung stieg der R. Q. bis auf nahezu 1,2, lag aber etwa 7 Tage nach der Befruchtung sowie im reifen Samen bei 0,7.

Die Frage nach der anatomischen Lokalisierung der Umwandlung von Fett in Kohlenhydrate bei der Keimung wurde von J. R. MURLIN und Mitarbeitern ausführlich bearbeitet (J. R. MURLIN, W. R. MURLIN, J. D. WATKEYS 1933, H. B. PIERCE, D. E. SHELDON, J. R. MURLIN 1933, R. G. DAGGS, H. C. H. WARDLAW 1932, R. G. DAGGS, H. S. HALCRO-WARDLAW 1933). Bei *Ricinus* erwies sich eindeutig das Endosperm als Sitz der Umwandlung Fett-Kohlenhydrat. Der niedrigste Gesamt-R. Q. (etwa 0,3) fand sich zur Zeit einer Hypocotyllänge von 20—35 mm; der Atmungsquotient des jungen Keimlings selbst (ohne Endosperm) liegt bei 0,8—1,0 und das Endosperm allein weist zu dieser Zeit einen R. Q. von 0,4—0,5 auf. Daß der Gesamt-R. Q. immer noch tiefer als der des Endosperms allein lag, ist merkwürdig. Hier scheint also — wenigstens im isolierten Keimling — die Atmung größtenteils auf dem Wege über die Kohlenhydrate zu erfolgen.

Untersuchungen über den Wirkungsgrad des Keimungsstoffwechsels (ökonomischer Koeffizient nach PFEFFER 1895, S. 257 ff.) wurden unter Anwendung calorimetrischer Methoden von E. F. TERROINE, R. BONNET und P. H. JOESSEL (1924) ausgeführt. Reis- und Hirsekeimlinge enthielten etwa 73% der in den Samen enthalten gewesenen Energie; für Keimlinge aus eiweißhaltigen Samen

(Linse und Erbse) war die entsprechende Zahl 63% und für Keimlinge aus fetthaltigen Samen (Lein,Erdnuß) nur 53%. Die Energieverluste scheinen also in den Fettsamen höher als in anderen zu sein, jedoch berechnen TERROINE und seine Mitarbeiter, daß auf Grund der chemischen Zusammensetzung der verschiedenen Samen und der entsprechenden relativen Energiegehalte der Reservestoffe geschlossen werden muß, daß die Energieverluste in Wirklichkeit in den eiweißhaltigen Linsen und Erbsen größer als in den fetthaltigen Samen sind.

Z. KRASINSKA (1929) hat ähnliche Versuche mit Sonnenblumen durchgeführt und fand die Zahlen der Tabelle 1.

Etwa 40% des Energiegehaltes der während 6 Keimungstagen verschwundenen Fette wurden verbraucht, und der durchschnittliche Atmungsquotient für alle 6 Tage ist 0,52, also deutlich kleiner als einer reinen Fettveratmung entspräche. Z. KRASINSKA stellt fest, daß ihre Zahlen mit der Annahme übereinstimmen, daß etwa 94% der in den ersten Keimungstagen frei gewordenen Energie aus Fettumsetzungen stammen und daß nur 6% bei anderen Reaktionen, wahrscheinlich Eiweißspaltungen, frei werden. Etwas weniger als 56% der verschwindenden Fette werden in Kohlenhydrate umgewandelt und etwa 44% werden veratmet; etwa 77% der frei gewordenen Energie dürfte aus völliger Veratmung von Fetten und etwa 23% aus ihrer Umwandlung in Kohlenhydrate stammen (Wärmetönung Fett → Kohlenhydrate 2150 cal), und zur Zeit des intensivsten Stoffumsatzes dürften etwa 66% der verbrauchten Fette in Kohlenhydrate umgewandelt werden. U. WEBER (1936) hat für Kürbissamen aus Atmungsversuchen eine Umwandlung von 11—22% der umgesetzten Fette in Kohlenhydrate erschlossen, während L. WIENKE (1938) fand, daß keimende Sonnenblumensamen manchmal mehr Kohlendioxyd ausscheiden, als aus dem verschwindenden Fett (und Eiweiß) entstehen könnte. An keimenden Samen von *Sinapis* hat J. KNABE (1943) Umwandlung von etwa 14% der abgebauten Fettsäuren in Kohlenhydrate und spätere Veratmung dieser Kohlenhydrate aus Messungen des Atmungsgaswechsels erschlossen. W. HEUMANN (1944) gibt an, daß zwischen 3. und 7. Tag der Keimung von Kürbissamen 63% der Fette in Kohlenhydrate umgewandelt werden.

Tabelle 1. *Verbrennungswärme von Helianthussamen und -keimlingen.* (Z. KRASINSKA 1929.)

Keimlingsalter	Verbrennungswärme in cal		Atmungsquotient	Fettgehalt
Tage	1 g Trockensubstanz	1 Keimling		%
0	7330	383	etwa 0,93	58,2
2	7210	368	—	—
4	6160	325	0,49	—
6	5100	278	etwa 0,60	22

Wirklich verläßliche Messungen des Atmungsquotienten sind sehr schwierig auszuführen, und es ist fraglich, ob die während einer gewissen Zeitspanne aufgenommenen und abgegebenen Gasmengen wirklich ein genaues Abbild der gleichzeitig ablaufenden Stoffwechselvorgänge sind. Ist doch bekannt, daß etwa CO_2 in beträchtlichem Ausmaße in Pflanzenzellen gespeichert und bei Lichteinfall dann assimiliert werden kann. So ist es nicht verwunderlich, wenn die überwiegende Mehrzahl aller Untersuchungen des Chemismus keimender Fettsamen unter Vermeidung gasanalytischer Methoden mit Hilfe anderer chemischer Verfahren ausgeführt wurde.

C. Der Fettgehalt keimender Samen.

Nachdem L. C. TREVIRANUS, J. F. MEYER und TH. DE SAUSSURE in den 30er Jahren des 19. Jahrhunderts die Fette der Pflanzen als Reservestoffe erkannt hatten (vgl. F. CZAPEK 1913, S. 713 und 733), hat auf Grund von Elementaranalysen H. HELLRIEGEL (1855, S. 103) als erster postuliert, daß bei der

Keimung von Fettsamen eine Umwandlung des Fettes in Zucker stattfinden müsse, und J. SACHS (1859) war der erste, der mit Hilfe des Mikroskopes in Keimlingen den Übergang von Fett in Stärke beobachtete. In der Folgezeit haben viele Untersucher sich bemüht, diesen vom chemischen Standpunkt aus so überraschenden und undurchsichtigen Vorgang der Umwandlung von Fetten in Kohlenhydrate, wenn schon nicht zu erklären, so doch möglichst eingehend und richtig zu beobachten und zu beschreiben. Die ältere Literatur über diesen Gegenstand findet sich in den einschlägigen Monographien der 70er und 80er Jahre des vorigen Jahrhunderts ausführlich dargestellt; erwähnt seien etwa W. DETMERs Habilitationsschrift (1875) und sein Buch über die Keimungsphysiologie (1880) sowie F. NOBBEs Handbuch der Samenkunde (1876). Später haben F. CZAPEK (1913) und E. F. TERROINE (1920) den Fettstoffwechsel keimender Samen eingehend besprochen. Ausführliche historische Abschnitte finden sich auch in den Arbeiten von E. C. MILLER (1910), N.-F. DELEANO (1909), S. IVANOW (1912), E. H. TOOLE (1924) und E. MATTHES (1927).

Mikroskopische Untersuchungen über die Umwandlung von Fett in Stärke bei der Samenkeimung wurden in größerem Umfang von E. MESNARD (1893, 1894) ausgeführt, und E. H. TOOLE (1924) hat den Fettschwund im Embryo keimenden Maises mikroskopisch verfolgt. Daß aber die Fette in keimenden Samen kaum als solche, sondern wohl nur als Kohlenhydrate geleitet werden, hat J. B. RHINE (1926) klargestellt.

Für eingehendere Studien des Fett- und Kohlenhydratstoffwechsels keimender Samen wurde wohl am häufigsten *Ricinus* verwendet. Die älteren Untersuchungen (G. FLEURY 1865, R. H. SCHMIDT 1891, LECLERC DU SABLON 1893, 1895, L. MAQUENNE 1898) sind infolge unvermeidlicher experimenteller Schwierigkeiten und methodischer Unvollkommenheiten natürlich nicht so aufschlußreich wie modernere, von denen zunächst die Arbeit von N.-F. DELEANO (1909) hervorgehoben sei, da er wohl als erster klar den Gedanken aussprach, daß die Umwandlung von Fetten in Kohlenhydrate als ein bewegliches Gleichgewicht zwischen mehreren Fermenten und ihren Reaktionsprodukten aufzufassen sei. Die Tatsache, daß *Ricinus*samen, und besonders keimende, reich an Lipase sind, ohne daß in den keimenden Samen eine nennenswerte Menge freier Fettsäuren auftritt, war wohl mitbestimmend für die Aufstellung dieser Hypothese. N.-F. DELEANO war auch einer der ersten, der die Fettgehalte nicht in Prozenten der Trockensubstanz ausdrückte. Er verwendet als Bezugsgröße 100 Durchschnittskeimlinge, die er aus einer größeren Anzahl gleichzeitig zur Keimung gebrachter auswählte. Seine Zahlen geben daher ein sehr klares Bild der Vorgänge bei der Keimung, wenn natürlich auch aus verständlichen Gründen keine Bestimmungen höherer Kohlenhydrate ausgeführt werden konnten. N.-F. DELEANO keimte seine *Ricinus*samen bei 20—22° C, also bei einer Temperatur, die für *Ricinus* eigentlich etwas niedrig ist. Als Folge davon zeigen seine Zahlen die chemischen Vorgänge bei der Keimung sozusagen im Zeitlupentempo und verzögert. Was er während etwa 15 Tagen ablaufen sah, fand J. HOUGET (1942, 1943), der bei höheren Temperaturen arbeitete, in etwa 5 Tagen vor sich gehend, wie gleich erwähnt werden soll. Die folgende Tabelle 2 wurde aus N.-F. DELEANOs Zahlen zusammengestellt bzw. berechnet[1].

Charakteristisch scheint der rapide Fettabbau zu sein, der nach einer anfänglichen Periode ohne Fettabnahme (oder vielleicht sogar mit geringer

[1] In dieser und manchen weiteren Tabellen wurden vielfach aus Gründen der Übersichtlichkeit manche Analysenergebnisse in relative Werte umgerechnet. Die entsprechenden absoluten Zahlen können mit Hilfe der am Ende der Tabellen jeweils erwähnten Umrechnungsschlüssel im Bedarfsfalle leicht gefunden werden.

Fettneubildung) innerhalb von 2—3 Tagen den Fettgehalt auf etwa ein Drittel sinken läßt, sowie die Tatsache, daß dieser energische Fettabbau ohne wesentlichen Trockensubstanzverlust vor sich geht, während gleichzeitig eine gewisse Vermehrung von Glucose und Saccharose eintritt, ohne daß sie mengenmäßig auch nur annähernd den großen verschwundenen Fettmengen entspräche. H. B. PIERCE, D. E. SHELDON und J. R. MURLIN (1933) haben ähnliche Zahlen mitgeteilt und vermuten eine teilweise Umwandlung der Fette in Oxydationsprodukte von Pentosen. Stärke- und Pentosenbestimmungen wurden aber nicht ausgeführt.

Tabelle 2. *Fette und Zucker in keimenden Ricinussamen (T = 20—22° C).* (N.-F. DELEANO 1909.)

Keimlingsalter	Wurzellänge	Frischgewicht je 100 Keimlinge	Trockengewicht je 100 Keimlinge	Fett		Glucose + Saccharose	
				in % der Trockensubstanz	relativ je 100 Keimlinge	in % der Trockensubstanz	relativ je 100 Keimlinge
Tage	cm	relativ	relativ				
0	—	1,0	100	68,6	100	Spur	
4	0,2	2,0	96	73,3	102	0,9	1,0
5	0,4	2,4	95	71,4	99	0,4	0,5
6	0,6	3,0	93	73,4	100	1,1	1,2
7	1,5	3,6	93	73,4	100	2,4	2,6
8	2,5	4,6	92	73,4	98	4,3	4,6
9	3,2	5,8	90	50,4	67	5,2	5,5
10	4,7	6,6	90	33,3	44	7,3	7,6
11	5,5	7,1	89	23,2	30	—	—
12	6,0	7,7	88	13,5	18	11,5	11,9
13	7,0	8,3	86	10,1	13	13,7	13,7
14	8,5	6,1	56	11,3	9	12,2	7,9
15	10,0	6,3	49	2,3	1,7	17,3	10,0
		1 = 26,7 g	100 = 24,5 g		100 = 16,8 g		1 = 0,21 g

Tabelle 3. *Fette und Zucker in keimenden Ricinussamen (T = 28,5° C).* (J. HOUGET 1942/43.)

Keimlingsalter	Fett in 100 Samen	Reduzierende Zucker in 100 Samen	Nicht-reduzierende Zucker in 100 Samen	Gesamtzucker in 100 Samen
Tage	relativ	relativ	relativ	relativ
0	100	1,0	1,0	1,0
1	99,5	0,74	1,0	0,97
2	97,6	0,12	1,0	0,98
3	92,2	4,4	1,9	2,2
4	39,1	24,5	8,6	10,2
5	1,9	34,7	15,6	17,5
	100 = 24,6 g	1,0 = 0,117 g	1,0 = 1,0 g	1,0 = 1,12 g

Auf den folgenden Seiten wird immer wieder berichtet werden müssen, daß sorgfältige Versuchsanstellung bei den verschiedensten keimenden Fettsamen

Tabelle 4. *Fett und Kohlenhydrate in keimendem Ricinus. Alle Werte bezogen*

Alter Tage	Frischgewicht			Trockengewicht						Fette					
	Endosperm	Keim	Gesamt	Endosperm		Keim		Gesamt		Endosperm		Keim		Gesamt	
	g	g	g	g	rel.	g	rel.	g	rel.	g	rel.	g	rel.	g	rel.
0	41,0	—	41,0	37,6	100	—	—	37,6	100	26,24	100	—	—	26,24	100
4	103,4	20,0	123,4	37,0	98	1,95	1	39,0	104	24,9	95	0,07	1	25,0	95
6	146,3	131,1	277,4	25,6	68	19,44	10	45,1	120	10,0	38	0,77	11	10,77	41
8	131,5	217,0	348,5	18,4	49	25,52	13	43,9	117	3,95	15	1,45	21	5,40	21
11	44,4	439,2	483,5	4,0	11	34,42	18	38,4	102	0,66	2,5	1,12	16	1,78	6,8

immer wieder diesen gerade erwähnten, verzögert einsetzenden, aber dann sehr intensiv vor sich gehenden Fettabbau fand. Der auch gegenwärtig vielleicht noch nicht ganz geklärten Frage nach dem tatsächlichen Verhalten der Gesamttrockensubstanz bei der Keimung von Fettsamen werden auch noch einige Sätze zu widmen sein.

Etwa 10 Jahre nach H. B. PIERCE und Mitarbeitern wurden wiederum ähnliche Untersuchungen an *Ricinus* ausgeführt (J. HOUGET 1942, 1943), wobei allerdings als Temperatur bei der steril durchgeführten Keimung 28,5° C gewählt wurde. Daher zeigt die Tabelle 3 den rapiden Fettabbau, den N.-F. DELEANO zwischen dem 8. und dem 11.—12. Keimungstag fand, schon zwischen dem 3. und dem 4.—5. Tag der Keimung und zwar in verstärktem Ausmaß, da in diesen 48 Std der Fettgehalt auf etwa ein Fünfzigstel fällt. (Auch Unterschiede im Saatgut mögen natürlich mitspielen.) Bei den Zuckern wurde — wie schon von N.-F. DELEANO — nach anfänglicher vorübergehender Abnahme eine starke, aber doch nicht mit der Fettabnahme vergleichbare Zunahme gefunden.

J. HOUGET ließ auch von den Keimlingen abgetrennte Endosperme sich weiterentwickeln und fand darin innerhalb von 24 Std eine Fettabnahme von 4,35 g auf 1,78 g Fett je 100 Endosperme sowie Zunahmen der reduzierenden Zucker von 1,05 auf 2,2 g und der nichtreduzierenden Zucker von 5,6 auf 9,8 g je 100 Endosperme. Hiermit scheint also auch auf Grund chemischer Analysen bewiesen zu sein, daß wenigstens bei *Ricinus* das Endosperm der Hauptsitz der Umwandlung der Fette in Kohlenhydrate ist. Auch scheinen damit die *Ricinus*-endosperme als ein Versuchsobjekt wahrscheinlich gemacht, an dem die Fettmobilisierung und -umwandlung in Kohlenhydrate ungestört durch den Aufbaustoffwechsel des jungen Keimes untersucht werden könnte.

In neuester Zeit haben R. DESVEAUX und M. KOGANE-CHARLES (1952) sorgfältige Untersuchungen an keimendem *Ricinus* ausgeführt und haben bei einer Temperatur von 25° C gearbeitet. Ihre Ergebnisse sind in der Tabelle 4 zusammengestellt.

Auch hier finden wir die rapide Fettabnahme nach dem 4. Keimungstag, die in diesem Falle aber außerdem mit einer beträchtlichen Zunahme des Gesamttrockengewichtes um etwa 20% verbunden ist. Eine Trockengewichtszunahme keimender Fettsamen ist nicht allzu verwunderlich, wenn man bedenkt, daß bei der Umwandlung von Fett in Kohlenhydrate eine beträchtliche Sauerstoffeinlagerung stattfinden muß. Nimmt man die auf S. 324 erwähnte Umlagerung von Ricinolsäure in Saccharose als Grundlage, dann findet man leicht, daß je Gramm verschwundene Ricinolsäure 1,74 g Saccharose auftreten. Tritt aber Bildung organischer Säuren ein, dann kann dieser Umwandlungsfaktor noch

auf 100 Durchschnittskeimlinge. (R. DESVEAUX und M. KOGANE-CHARLES 1952.)

Reduzierende Zucker			Nichtreduzierende Zucker			Stärke und ähnliche Kohlenhydrate			Gesamtkohlenhydrate			Rohfaser		
Endosperm g	Keim g	Gesamt g	Endosperm g	Keim g	Gesamt g	Endosperm g	Keim g	Gesamt g	Endosperm g	Keim g	Gesamt g	Endosperm g	Keim g	Gesamt g
0,0	—	0,0	1,02	—	1,02	0,49	—	0,49	1,51	—	1,51	0,49	—	0,49
0,23	0,34	0,57	3,07	0,14	3,21	0,93	0,39	1,32	4,23	0,87	5,10	0,45	0,11	0,56
1,74	1,96	3,70	8,80	2,10	10,90	1,72	1,89	3,61	12,26	5,95	18,21	(0,27)	0,69	0,96
1,40	3,27	4,67	7,27	4,99	12,26	0,86	5,54	6,40	9,53	13,80	23,33	0,46	0,83	1,29
0,24	7,70	7,94	0,50	4,85	5,35	0,44	3,96	4,40	1,18	16,51	17,69	0,37	2,32	2,69

beträchtlich größer werden. Oxalsäure etwa — um einen Extremfall herauszugreifen — könnte nach der folgenden Bruttogleichung in etwa der 2,7fachen Menge der verschwindenden Ricinolsäure entstehen:

$$2\,C_{18}H_{34}O_3 + 41\,O_2 \rightarrow 18\,C_2H_2O_4 + 16\,H_2O.$$

Die von R. Desveaux und M. Kogane-Charles gefundene Fettabnahme vom 4. zum 6. Keimungstag betrug etwa 14,2 g je 100 Keimlinge und die gleichzeitig gefundene Kohlenhydrat- und Rohfaserzunahme betrug etwa 13,5 g, wobei eine Zunahme des Gesamttrockengewichtes von 6,1 g auftrat. Nehmen wir an, daß die Differenz zwischen der Gewichtszunahme der Kohlenhydrate und der Gesamttrockengewichtszunahme (13,5—6,1 g) durch völlige Veratmung von 7,4 g Fettsäuren bedingt ist, dann wurden zur Bildung der 13,5 g Kohlenhydrate insgesamt 14,2—7,4 = 6,8 g Fette verbraucht, was einem Umwandlungskoeffizienten von ziemlich genau 2 entspricht und bedeutet, daß nahezu 50% der zwischen 4. und 6. Keimungstag abgebauten Fette in Kohlenhydrate umgewandelt wurden. Wenn diese „Bilanz“ natürlich auch nicht sehr gut stimmt (zumal wenn man bedenkt, daß gleichzeitig je 100 Keimlinge auch eine etwa 0,45 g Oxalsäure entsprechende Menge freier Nichtfettsäuren gebildet wurde, was den Fett-Kohlenhydrat-Umwandlungskoeffizienten eigentlich auf 2,2 erhöhen müßte), so läßt sich doch erkennen, daß die Zunahme des Gesamttrockengewichts um etwa 20% durchaus der Größenordnung der vor sich gehenden Fettumwandlungen entsprechen kann.

Tabelle 5. *Trockengewicht keimender Kürbissamen.* (A. Zeller 1935.)

Keimlingsalter Tage	Trockengewicht von 20 Keimlingen g	Durchschnittstrockengewicht eines Keimlings und mittlerer Fehler (20 Einzelbestimmungen)	
		g	relativ
2	5,346	0,267 ± 0,014	100 ± 5,2
3	5,242	0,267 ± 0,013	100 ± 4,9
4	5,284	0,264 ± 0,013	99 ± 4,9
5	5,401	0,270 ± 0,014	101 ± 5,2
6	5,423	0,271 ± 0,017	101,5 ± 6,4
7	5,305	0,265 ± 0,017	99,5 ± 6,4

Eine Zunahme des Trockengewichts keimender Fettsamen wurde verschiedentlich berichtet. A. Zeller (1935, S. 124ff.) und W. Heumann (1944, S. 8) haben die älteren Angaben von H. Hellriegel (1855), von P. Mazè (1902), der über 15% Trockengewichtszunahme isolierter *Arachis*-Kotyledonen bei 17tägiger „Keimung“ berichtete, von U. Jegorow (1904) sowie neuere von E. Matthes (1927) und A. J. Ermakoff und N. N. Iwanoff (1931) besprochen. A. Zeller hat an sorgfältigst ausgelesenen Kürbiskeimlingen an den ersten 7 Keimungstagen je 20 Einzeltrockengewichtsbestimmungen ausgeführt und aus den erhaltenen Werten Mittelwerte samt mittleren Fehlern berechnet. Es fand sich nur eine unregelmäßige Streuung der Werte für die einzelnen Keimungstage, die weit innerhalb der zufällig zu erwartenden Schwankungen lag (Tabelle 5). Die Differenz zwischen größtem und kleinstem Wert ist viel geringer als der mittlere Fehler jedes dieser Einzelwerte, alle gefundenen Werte müssen daher als nur zufällige Abweichungen von einem konstanten Wert angesehen werden.

A. Zeller schloß aus dem Gleichbleiben des Trockengewichtes und aus einem in der Literatur (W. Stiles und W. Leach 1933) angegebenen R. Q. von 0,8—0,9, daß vermutet werden könne, daß die Kürbiskeimlinge auch zur Zeit der stärksten Umwandlung von Fetten in Kohlenhydrate ihren großen Sauerstoffbedarf nicht aus der Luft, sondern aus anderen sauerstoffreichen Verbindungen decken. Es ist wohl zum mindesten ziemlich fraglich, ob W. Heumann (1944) diese Vermutung dadurch wirklich widerlegt hat, daß er zeigte, daß Kürbiskeimlinge in einer Stickstoffatmosphäre ersticken. In A. Zellers Veröffentlichung finden sich übrigens keine Anhaltspunkte dafür, daß er den Kürbiskeimlingen, wie W. HEUMANN sagt, einen intramolekularen Atmungsstoffwechsel zuschreibt. Bei seinen eigenen Versuchen fand W. HEUMANN eine fehlerstatistisch nicht bearbeitete Trockengewichtszunahme, die am 6.—7. Keimungstag mit etwa 8% ihren Höchstwert erreichte.

Die Schwierigkeiten, die einer exakten Beantwortung der anscheinend so einfachen Frage, ob und wie sich das Trockengewicht der Fettsamen bei Dunkelkeimung ändert, entgegenstehen, sind teils methodischer Art (Schwierigkeiten des verlust- und oxydationsfreien Trocknens unter Ausschluß von Fermentwirkungen), sind aber auch durch die große Veränderlichkeit der Samen und Keimlinge bedingt, die ein Erfassen kleiner Trockengewichtsunterschiede (weniger als etwa 10—13%) in einer jeder Kritik standhaltenden Weise bisher unmöglich gemacht haben. Anwendung von Gefriertrocknung und dgl. könnte hier vielleicht weiter führen. Eine allgemeine Antwort auf obige Frage scheint also noch nicht möglich zu sein, aber es kann als sicher angenommen werden, daß in keimenden Fettsamen im Gegensatz zu keimenden Kohlenhydratsamen keine stärkere Abnahme des Trockengewichts bei der Dunkelkeimung stattfindet, bevor der normale Stoffwechsel in den prämortalen Erschöpfungsstoffwechsel übergeht.

Tabelle 6. *Relative Fettsäuremengen in quellenden Kürbissamen.* (A. ZELLER 1935.)

Quellungsdauer Stunden	Fettsäuregehalt von 100 Samen (relativ)
0	100
3	85
6	(94)
12	91
18	95
24	102

Die Versuche von R. DESVEAUX und M. KOGANE-CHARLES haben gezeigt, daß in den Samen und in den ersten Keimungsstadien von *Ricinus* keine reduzierenden Kohlenhydrate vorkommen und daß auch später diese Kohlenhydratform im allgemeinen in geringerer Menge als die nichtreduzierenden Zucker vorkommt. Nach etwa 10—11 Keimungstagen dürften bei den *Ricinus*keimlingen, die bei 25° C gehalten werden, alle Reserven mobilisiert und umgewandelt sein, und es scheint bereits ein Abbau neugebildeter Stoffe angefangen zu haben.

Viele der Untersuchungen über die Mobilisierung der Fette in keimenden Samen wurden mit *Kürbis*samen ausgeführt. E. PETERS (1861) hat an Kürbiskeimlingen, die im Freien in Sägespänen heranwuchsen, das Verschwinden der Fette und die Zunahme der Kohlenhydrate bis zur Entwicklung des ersten Laubblattes verfolgt und hat — im Gegensatz zu allen Untersuchern, die mit etiolierten Keimlingen arbeiteten — keine Saccharose gefunden. N. LASKOVSKYs (1874) Untersuchungen litten sehr unter ungleichmäßigen Außenbedingungen, und U. JEGOROW (1904) untersuchte seine Kürbiskeimlinge nur in großen zeitlichen Abständen (6, 10, 20, 28 Tage alt). Er fand in 6 Tage alten Keimlingen um etwa 5% mehr Trockensubstanz und ungefähr 3% mehr Fett als in den Samen. Wieweit methodische Unzulänglichkeiten diese Resultate erklären können, ist aus dem Zentralblattreferat von U. JEGEROWS Arbeit nicht zu ersehen. H. LUNDEGÅRDH (1914) berichtete, daß in Kürbissamen schon durch Einquellen in Wasser Stärkebildung hervorgerufen wird und daß die Stärke beim Rücktrocknen wieder verschwindet. A. ZELLER (1935) konnte diese Angaben bestätigen und bringt Zahlen über das Schwanken der Gesamtfettsäuren in quellenden Kürbissamen während der ersten 24 Std der Keimung. Hierbei scheint eine Quellungsdauer von 6 Std irgendwie einen kritischen Zeitpunkt darzustellen, da nur hier die mit verschiedenen Samenmustern angestellten Parallelversuche schlecht übereinstimmten (Tabelle 6).

Über den weiteren Verlauf des Fettsäureabbaues in keimenden Kürbissamen wurden von A. ZELLER und F. MASCHEK (1942) von J. FUCHS (1933) erhaltene

Zahlen mitgeteilt; W. HEUMANN (1944) hat unter vergleichbaren Versuchsbedingungen, aber natürlich an ganz anderem Saatgut und mit anderer Methode arbeitend (Fett- statt Gesamtfettsäurebestimmungen), überraschend gut mit J. FUCHS' Zahlen übereinstimmende Werte erhalten (Tabelle 7).

Tabelle 7. *Fett- bzw. Fettsäuren in keimenden Kürbissamen.* (A. ZELLER und F. MASCHEK 1942 nach J. FUCHS 1933, W. HEUMANN 1944.)

Keimlingsalter Tage	Fettsäurengehalt (relativ) (J. FUCHS 1933)	Fettgehalt (relativ) (W. HEUMANN 1944)
0	100	100
2	94	—
3	—	92
4	—	76
5	67	61
6	—	45
7	43	39
8	30	27
10	30	—
11	25	23
		100 = 14,79 g je 100 Keimlinge

Zwischen 4. und 6. Tag der Keimung erfolgt also der intensivste Fettabbau und ist, wie A. ZELLER (1935) und K. ENSER (1935) gezeigt haben, von ähnlich intensiven Kohlenhydratumsetzungen begleitet (Tabelle 8).

Auffällig ist hier das gegenläufige Verhalten von Glucose und Saccharose sowie die Tatsache, daß Glucose und Stärke (neben etwas Pentosen) von den hier untersuchten Kohlenhydraten die wichtigsten Umwandlungsprodukte der Fette zu sein scheinen.

Mit moderner Methodik (Säulenchromatographie, Ultraviolettspektroskopie) wurde der Fettstoffwechsel im Lichte keimender Samen von *Citrullus vulgaris* von W. M. CROMBIE und R. COMBER (1956) untersucht. Auch hier findet sich eine geradezu explosionsartige Umwandlung der Fettsäuren zwischen dem 6. und 7. Keimungstag (mittlere Keimungstemperatur 31° C mit Extremwerten von 24 und 45° C),

Tabelle 8. *Die Kohlenhydrate in keimenden Kürbissamen, Mittelwerte.* (K. ENSER 1935.)

Keimlingsalter Tage	Vergärbare reduzierende Zucker als Glucose	Saccharose als Glucose	Pentosen	Reduzierende Nichtzucker als Glucose	Stärke als Glucose
0	1,0	100	1,0	1,0	1,0
4	3,5	62	1,9	1,5	9,8
6	14,9	19,5	5,0	1,5	81
	1,0 = 11,9	100 = 64,4	1,0 = 4,7	1,0 = 23,2	1,0 = 1,25
	Milligramm je 10 Keimlinge				

wobei in 24 Std etwa 80% der noch vorhandenen 82% der Samenfette verschwinden. Je Samen werden in dieser Zeit 9 mg Fettsäuren (die 25% des Trockengewichtes entsprechen) umgesetzt, und dabei nimmt das Frischgewicht der Kotyledonen um 86% zu, ihr Trockengewicht aber um 33% ab. Daß die Hautpmenge der aus den Kotyledonen verschwindenden Stoffe in Wurzel und Sproß wandert, ergibt sich aus der nur etwa 6% betragenden Abnahme des Gesamttrockengewichtes der Keimlinge (bei 100% Zunahme des Frichsgewichtes).

Tabelle 9. *Fettgehalt keimender Rapssamen.* (LECLERC DU SABLON 1895.)

Keimlingsalter Tage	Fett-% (relativ)	Keimlingsalter Tage	Fett-% (relativ
0	100	4	43
2	96	5	37
2,5	72	6	35
3	68	7	29
	100 = 41,0%		

Eines der ersten Objekte, an denen die Mobilisierung der Samenfette untersucht wurde, war der *Raps* (H. HELLRIEGEL 1855, G. FLEURY 1865, A. MÜNTZ 1871, A. NOBBE 1876, W. DETMER 1880). Die ersten (und praktisch einzigen) genauen Zahlen stammen von LECLERC DU SABLON (1895), der vom 2.—7. Keimungstag die in Tabelle 9 angegebenen Fettprozente feststellte.

Hier scheint der Fettabbau am 3. und 4. Keimungstag mit besonderer Intensität abzulaufen. An den keimenden Rapssamen ausgeführte Glucosebestimmungen ergaben eine Zunahme von 0 auf 3,5% der Trockensubstanz, während

Tabelle 10. *Verlauf des Fettabbaues in keimenden Erdnüssen.*

Keimlingsalter	LECLERC DU SABLON (1895)				L. MAQUENNE (1898)			
	Fett		Glucose	Saccharose	Fett		Verzuckerbare Stoffe als Saccharose	Cellulose
Tage	%	relativ	%	%	%	relativ	%	%
0	49,5	100	0	3,1	51,4	100	11,6	2,5
2	49,3	100	Spuren	3,9	—	—	—	—
4	40,6	82	0,8	5,8	—	—	—	—
6	30,0	61	2,0	11,0	49,8	97	8,4	3,5
8	19,6	39	4,3	10,6	—	—	—	—
10	12,0	24	5,9	8,2	36,2	71	11,1	5,0
12	9,7	19	10,2	5,8	29,0	56	12,5	5,2
18	—	—	—	—	20,5	40	12,3	7,3
28	—	—	—	—	12,3	24	9,5	9,5

die Saccharose gleichzeitig von 2,3% auf 1,6% abnahm. Nach H. LUNDEGÅRDH (1914) bildet sich in quellenden Rapssamen keine Stärke oder doch nur in wenig reproduzierbarer Weise. K. WEISSENBÖCK (1948) hat in unregelmäßig (bei 10—13° C) keimendem Raps in 9 Tagen eine Abnahme des Gesamtlipoidgehaltes von 48,8% auf 40,0% der Trockensubstanz festgestellt.

An *Arachis* haben vor allem R. H. SCHMIDT (1891), LECLERC DU SABLON (1895) und L. MAQUENNE (1898) den Verlauf der Fettmobilisierung verfolgt. R. H. SCHMIDT hat zu wenige Keimungsstadien untersucht, und LECLERC und L. MAQUENNE haben offenbar verschiedene Keimungstemperaturen angewandt, so daß ihre Ergebnisse nur schwer vergleichbar sind (Tabelle 10).

Tabelle 11. *Fettgehalt keimender Baumwollsamen.* (H. S. OLCOTT und T. D. FONTAINE 1941.)

Keimlingsalter	Trockengewicht	Fettgehalt von 100 Keimlingen
Tage	relativ	relativ
0	100	100
1,75	94	95
2,8	89	93
3,8	86	66
4,7	86	50
5,8	84	45
	100 = 6,40 g je 100 Samen	100 = 2,75 g je 100 Samen

LECLERC DE SABLON fand also hier einen gleichmäßigen, ziemlich steilen Fettabfall zwischen 2. und 10. Keimungstag, während L. MAQUENNE eine wesentlich langsamere und nicht vor dem 6. Tag beginnende Fettabnahme findet. In *Arachis* scheint also die Dynamik des Fettabbaues etwas anders als in *Ricinus* und *Cucurbita* zu sein. Ein Vergleich der für die Kohlenhydrate angegebenen Werte mit den von N.-F. DELEANO bzw. J. HOUGET (Tabelle 2 und 3) bei *Ricinus* und den von K. ENSER (Tabelle 8) bei *Cucurbita* mit modernerer Methodik gefundenen Zahlen ist natürlich kaum möglich, doch scheint die starke, bei *Cucurbita* nachgewiesene Saccharoseabnahme hier nicht vorzukommen.

Über das Verhalten des Fettgehaltes in keimenden *Baumwoll*samen scheinen nur die Angaben von H. S. OLCOTT und T. D. FONTAINE (1941) vorzuliegen, die in Tabelle 11 zusammengefaßt sind.

Zwischen 3. und 6. Tag tritt offenbar auch hier ein ähnlich rapides Fallen des Fettgehaltes ein wie bei *Ricinus* und *Cucurbita*. Das Trockengewicht ändert sich während dieser Zeit kaum.

Bei *Lein* wurde der Fettabbau während der Keimung von LECLERC DU SABLON (1895), A. J. ERMAKOFF und N. IWANOFF (1931) sowie von R. DESVEAUX und M. KOGANE-CHARLES (1952) untersucht. LECLERC DU SABLON gibt nur Ölprozente an, bei R. DESVEAUX und Mitarbeiterin finden sich keine Zahlen für die Zeit bis zum 5. Keimungstag. Die Tabelle 12 gibt eine Übersicht über beide Versuchsreihen.

Tabelle 12. *Fettabbau in keimendem Lein.* (LECLERC DU SABLON, R. DESVEAUX.)

Keimlingsalter Tage	LECLERC (1895) Öl-% (relativ)	R. DESVEAUX und M. KOGANE-CHARLES (1952) Öl je 1000 Keimlinge (relativ)	Trockengewicht je 1000 Keimlinge (relativ)	Reduzierende und leicht hydrolysierbare Kohlenhydrate (relativ) je 1000 Keimlinge
0	100	100	100	100
2	93	—	—	—
3	80	—	—	—
4	73	—	—	—
5	58	57	80	105
6	46	—	—	—
7	37	—	—	—
8	—	52	88	123
9	29	—	—	—
10	—	22	67	45
11	25	—	—	—
12	—	18	58	32
	100=37,9% der Trockensubstanz	100=1,19 g	100=3,97 g	100=0,65 g

LECLERC DU SABLONS Zahlen legen die Annahme nahe, daß die Fettprozente vom 2. bis etwa 7. Tag der Keimung linear abnehmen, während R. DESVEAUXS Ziffern, wenigstens ab dem 5. Tag, einen unregelmäßigen Verlauf andeuten. Reduzierende Zucker (in Tabelle 12 nicht separat angeführt) sind in den Samen nach R. DESVEAUX in minimalen Mengen vorhanden, nach LECLERC fehlen sie. Sie nehmen bei R. DESVEAUX nur vorübergehend zu, bis sie etwa 50% des Gewichtes des gleichzeitig vorhandenen Fettes erreichen (5. und 8. Tag); am 12. Keimungstag sind sie aber auf etwa 10% des Fettgewichtes abgesunken. Nichtreduzierende Zucker machen nach R. DESVEAUX in den Samen etwa 5% des Fettgewichtes aus und verschwinden während der Keimung vollständig. Auffällig sind in R. DESVEAUXS Ergebnissen die hohen Werte für „leicht hydrolysierbare Kohlenhydrate", die schon in den Samen, die nach R. H. SCHMIDT (1891) stärkefrei sind, etwa 50% des Fettgewichtes betragen. Methodische Gründe (20 min Hydrolyse mit 3% HCl im Autoklaven) dürften für diese Zahlen verantwortlich sein. LECLERC fand während der Keimung Zunahme der Glucose von 0% auf etwa 4% und Zunahme der Saccharose von 1,6% auf ebenfalls etwa 4%. — Weitere Zahlen über den Fettgehalt keimender Leinsamen finden sich auch bei A. J. ERMAKOFF und N. IWANOFF (1931) sowie bei J. DMOCHOWSKI (1934).

Für *Hanf* wurden Analysen ebenfalls im wesentlichen nur von LECLERC DU SABLON (1894, 1895) und von R. DESVEAUX und M. KOGANE-CHARLES (1952) mitgeteilt. Sie sind in Tabelle 13 zusammengefaßt.

Der Verlauf des Fettabbaues ist hier praktisch identisch mit dem für Lein gefundenen; sogar eine vorübergehende schwache Trockengewichtszunahme findet sich in beiden Fällen. Bei LECLERC nimmt die Glucose von 0% auf 3% und die Saccharose von 1,7% auf 2,7% zu, bei R. DESVEAUX nehmen die reduzierenden Zucker von 0 auf 150 mg je 100 Pflanzen zu und wieder auf 0 ab; die nichtreduzierenden Zucker schwanken von 30—80 mg je 100 Pflanzen, und die leicht hydrolysierbaren Kohlenhydrate nehmen von 160 auf 280 mg je 100 Keimlinge zu und dann auf 60 ab. Nach R. H. SCHMIDT (1891) ist auch der Same des Hanf stärkefrei. Vereinzelte Angaben über den Fettgehalt keimender Hanfsamen findet man noch bei W. DETMER (1875, 1880) und bei J. DMOCHOWSKI (1934).

Tabelle 13. *Fettabbau in keimendem Hanf.* (LECLERC DU SABLON, R. DESVEAUX.)

Keimlingsalter	LECLERC	R. DESVEAUX	
	Fett-%	Fettmenge in 100 Keimlingen	Trockengewicht von 100 Keimlingen
Tage	(relativ)	(relatv)	(relativ)
0	100	100	100
1	100	—	—
2	80	—	—
4	57	—	—
5	—	57	76
6	47	—	—
8	—	29	81
14	—	14	62
18	—	9	40
	100 = 30%	100=0,56 g	100=1,58 g

Die stärkefreien (H. LUNDEGÅRDH 1914) Samen von *Sinapis* wurden von J. KNABE (1943) während der Keimung (22° C) auf das Verhalten des Fettes untersucht. J. KNABE gibt die Werte in Prozenten des Ausgangsgewichtes an und teilt auch die Ergebnisse von Parallelserien mit. Die Tabelle 14 ist aus seinen Zahlen abgeleitet.

Am zweiten Keimungstag setzt ein starker Fettabbau ein, der bis zum 4. Tag anhält und dann nur sehr verlangsamt weitergeht. Im Gegensatz zu diesen Untersuchungen steht die Mitteilung von R. DUPERON (1949), der bei Wurzellängen von etwa 5—12 mm deutlich (etwa 12%) mehr Fett je Senfkeimling fand, als die Samen enthalten hatten.

Ein sehr rapider Fettabbau scheint auch in keimenden Samen von *Raphanus sativus* vor sich zu gehen, bei dem A. MÜNTZ (1871) zwischen 3. und 4. Tag der Keimung einen Rückgang des Fettgehaltes von 5 g Samen von etwa 89% auf etwa 45% des Ausgangswertes beobachtete.

Tabelle 14. *Fettgehalt keimender Samen von Sinapis.* (J. KNABE 1943.)

Keimlingsalter Tage	Fettgehalt je 1 g Saatgut (relativ)
0	100
2	91,5
3	65,3
4	42,3
6	37
8	27,7
	100=300 mg

Die recht langsam keimenden Samen des *Tungbaumes (Aleurites fordii)* wurden von F. A. JOHNSTON und H. M. SELL (1944) untersucht. Von 1000 in Kistchen im Freien in Florida zum Keimen ausgelegten Samen wurden nach 16, 23 und 32 Tagen (Hauptwurzeln etwa 7 cm, Seitenwurzeln etwa 1—2 cm) Proben von je 25 Samen gleichen Entwicklungszustandes entnommen und analysiert. Die Ergebnisse sind in der folgenden Tabelle 15 zusammengestellt.

Auffällig ist neben der rapiden Fettabnahme zwischen dem 16. und 32. Tag die Abnahme der Saccharosewerte, denen eine starke Zunahme der reduzierenden Zucker und der Stärke gegenübersteht. Ein ganz ähnliches gegenläufiges Verhalten von reduzierenden Zuckern und Saccharose findet sich beim Kürbis (A. ZELLER 1935, K. ENSER 1935).

In keimenden Samen von *Pisum maritima* hat S. BOUDON (1953) ebenfalls ein anfängliches starkes Absinken des Fettgehaltes (auf etwa $^1/_8$ in 25—30 mm langen Keimlingen) beobachtet, das später nur in vermindertem Umfang weitergeht.

Die rapide Fettabnahme, die bei allen bisher erwähnten Pflanzen einige Zeit nach Keimungsbeginn zu beobachten ist, scheint bei manchen anderen in noch früheren Keimungsstadien zu erfolgen. So wurde die Keimung der Fett und

Tabelle 15. *Keimung der Samen von Aleurites fordii.* (F. A. JOHNSTON und H. M. SELL 1944.)

Keimlingsalter Tage	Trockengewicht je Same (relativ)	Fett je Same (relativ)	Reduzierende Zucker je Same mg	Saccharose je Same (relativ)	Stärke je Same mg	Andere Polysaccharide je Same (relativ)
0	100	100	0	100	0	100
16	104,5	95,4	27	59,6	0	110
23	102,2	73,5	86,5	44,2	18,5	95
32	106,2	29,8	199	14,6	160,5	101
	100 = 2,01 g	100 = 1,31 g		100 = 0,102 g		100 = 0,115 g

Inulin enthaltenden Samen von *Cichorium intybus* von V. GRAFE und V. VOUK (1912) näher untersucht. Der Fettabbau ging in den ersten 2 Keimungstagen viel rascher vor sich als später (Abnahme von 100 auf 56; vom 2. zum 12. Tag dann von 56 auf 33). Reduzierende Zucker hatten am 2. Keimungstag auf die Hälfte abgenommen, stiegen dann zunächst wieder an (auf etwa das 4fache der Ausgangsmenge am 6. Keimungstag), nahmen später aber wieder ab auf das etwa 2,5fache der Ausgangsmenge. Das Inulin nahm insgesamt auf etwa das 3fache des Ausgangswertes zu und zeigte dabei ein sprunghaftes Ansteigen zwischen 4. und 6. Keimungstag.

E. REUHL (1936) hat in den ersten 48 Std der Keimung von *schwarzem Senf* ebenfalls schon eine so starke Abnahme des Fettes gefunden (um über 55%), daß anzunehmen ist, daß der Fettabbau in den (nicht untersuchten) späteren Keimungsstadien nur langsamer vor sich gehen dürfte.

Für alle bisher besprochenen Fälle der Fettmobilisierung bei der Keimung scheint es charakteristisch zu sein, daß entweder sofort (*Cichorium*, schwarzer Senf) oder nach einer verschieden langen „Induktionsperiode" geringen Fettabbaues eine sehr intensive Umwandlung der Fette einsetzt, die nach verhältnismäßig kurzer Zeit (etwa 2—4 Tage) nur geringe Fettmengen im Keimling übrig läßt, die dann langsam weiter abgebaut werden; hierbei wird vielleicht der Übergang vom Reservestoffwechsel zu einem Hunger- oder Erschöpfungsstoffwechsel vollzogen.

In Sonnenblumen, Mohn, Nüssen, Mandeln und wohl noch anderen Früchten und Samen scheinen nun etwas andere Verhältnisse vorzuliegen.

*Sonnenblumen*kerne wurden schon von Beginn der Untersuchungen über Fette in Pflanzen vielfach als Beobachtungsobjekt herangezogen (J. SACHS 1859, R. H. SCHMIDT 1891, S. FRANKFURT 1894, R. v. FÜRTH 1904, W. BIALOSUKINA 1908, H. LUNDEGÅRDH 1914), jedoch haben sich nur wenige Forscher wirklich eingehend mit der Mobilisierung der Fette in keimenden Sonnenblumen befaßt. E. C. MILLER (1910, 1912) hat seine Versuchspflanzen in kohlensäurefreier Umgebung aufgezogen und Kotyledonen sowie Hypocotyle und Wurzeln der Pflanzen getrennt auf ihren Fett- und Kohlenhydratgehalt untersucht. Heute haben wohl nur mehr seine Fettbestimmungen ein gewisses Interesse, zumal auch E. MATTHES (1927) und G. TILENIUS (1938) den Verlauf des Fettabbaues in etiolierten *Helianthus*keimlingen verfolgt haben. In der folgenden Tabelle 16 sind die Ergebnisse aller dieser Untersuchungen zusammengefaßt.

Leider fehlen bei allen 3 Autoren die so wichtigen Untersuchungen ganz junger (1 und 2 Tage alter) Keimlinge, so daß über den wirklichen Verlauf des Fettabbaues (trotz der verhältnismäßig befriedigenden Übereinstimmung der 3 Kurven), besonders im Vergleich zum so viel genauer untersuchten Kürbis, kaum etwas Sicheres gesagt werden kann. Es scheint aber doch wahrscheinlich

Tabelle 16. *Fettgehalt keimender Sonnenblumen.* (E. C. MILLER 1910, E. MATTHES 1927, G. TILENIUS 1938.)

Keimlingsalter etwa Tage	E. C. MILLER Kotyledonen, Fettmenge je 100 Keimlinge (relativ)	E. MATTHES Fettgehalt je 100 Keimlinge (relativ)	E. MATTHES Trockengewicht je 100 Keimlinge (relativ)	G. TILENIUS Fettgehalt je 100 Keimlinge (relativ)
0	100	100	100	100
3	78	84	85	72
4	—	83	85	—
5	67	76	84	—
6	—	72	91	68
7	34	63	87	—
8	—	43	84	—
9	—	32	85	23
10	13,2	—	—	—
11	—	31	87	—
12	—	18	91	—
13	8,4	—	—	—
15	—	—	—	7,1
21	—	—	—	2,1
28	—	—	—	1,9
	100=3,79 g	100=2,77 g	100=5,50 g	100=2,38 g

zu sein, daß *Helianthus* einen Typus des Verlaufes des Fettabbaues zeigt, der verschieden vom *Cucurbita*typus ist. Ob den in allen 3 Versuchsreihen bei graphischer Darstellung aufscheinenden vorübergehenden Änderungen des Fettabbaues etwa am 5. Tag irgendeine Realität zukommt, kann auf Grund des vorhandenen Versuchsmaterials wohl kaum entschieden werden.

Beim *Mohn* fand A. MÜNTZ (1871) eine gleichmäßige Fettabnahme bei der Keimung; seine Werte betrugen (für je 20 g Samen) nach 2 Tagen 71% und nach 4 Tagen 44% des Ausgangsfettgehaltes, und S. IVANOW (1912) fand nach 0, 4 und 8 Keimungstagen 47%, 39% und 31% Fett in der Trockensubstanz keimenden Mohnes. Ausführlichere Untersuchungen stellte LECLERC DU SABLON (1895) an, doch sind seine Zahlen so wie die S. IVANOWs leider nur auf das Trockengewicht bezogen (Tabelle 17); sie deuten eine ziemlich gleichmäßige Fettabnahme an.

Tabelle 17. *Fettgehalt keimender Mohnsamen.* (LECLERC DU SABLON 1895.)

Keimlingsalter Tage	Fett-% in der Trockensubstanz (relativ)	Keimlingsalter Tage	Fett-% in der Trockensubstanz (relativ)
0	100	4	62,0
1	95,4	5	55,8
2	92,5	6	42,4
3	82,0	7	37,8
			100 = 43,9%

Über den Fettgehalt keimender *Walnüsse* liegen alte Angaben von LECLERC DU SABLON (1897b) sowie neuere von U. WEBER (1936) und von S. SKRAUP und U. WEBER (1936) vor. Der langsamen Keimung entsprechend findet eine sehr langsame Fettabnahme statt, die in 56 Tagen den Fettgehalt von etwa 31% auf etwa 5% abnehmen läßt.

In süßen und bitteren *Mandeln* hat nach G. FLEURY (1865) LECLERC DU SABLON (1897a) das Verhalten des Fettes bei der Keimung untersucht. Er bezog diesmal seine Zahlen nicht auf das Keimlingsalter, sondern auf die Wurzellänge, und er fand ein ziemlich gleichmäßiges Abnehmen der etwa 50 Fettprozente des Samens mit einer wahrscheinlich etwas intensiveren Fettzersetzung bei Wurzellängen von 13—15 cm. Glucose nahm auf etwa 10% der Trockensubstanz zu, Saccharose sank in den süßen Mandeln auf etwa die Hälfte des Anfangswertes von 4%, während sie in den bitteren Mandeln von 5 auf 8% des Trockengewichtes zunahm.

Ein weiterer Typus des Verlaufes der Fettmobilisierung in keimenden Samen scheint in den im folgenden besprochenen Pflanzen vorzuliegen.

*Soja*bohnen sind nicht Fettsamen in demselben Sinne, in dem etwa die Samen von Kürbis oder *Ricinus* Fettsamen sind. Der relative Fettgehalt der Sojasamen beträgt nur etwa die Hälfte oder ein Drittel dessen, was wir in den ganz typischen Fettsamen finden (20—25% gegenüber 40—70% der Trockensubstanz). Der hohe Gehalt an besonders hochwertigem Eiweiß macht die Soja bekanntlich zu einem der ganz wenigen pflanzlichen Nahrungsmittel, die imstande sind, tierisches Eiweiß in der menschlichen Nahrung weitgehend oder vollständig zu ersetzen. Gerade der Fettgehalt der Sojabohnen schwankt aber je nach Sorte sehr beträchtlich und kann bei Wildformen bis auf etwa 5% heruntergehen, so daß die Vermutung berechtigt erscheint, daß Soja eigentlich nicht eine primäre Fettpflanze ist, sondern daß der höhere Fettgehalt der Kultursorten mehr das Ergebnis züchterischer Auslese als natürlicher Anlage ist. Es liegen Anzeichen dafür vor, daß sich bei der Keimung der Sojabohnen das Fett anders verhält als das der Keimlinge der meisten anderen Fettpflanzen. Zunächst fand LECLERC DU SABLON (1895) ein gleichmäßiges Abnehmen der Fette in keimenden Sojabohnen, und W. ZIMMERMANN und A. BAUMANN (1935) haben bei Kultur der Soja in Sand und Erde einen bis zum 4. Tag langsam, dann aber stark beschleunigten Fettabbau beobachtet. S. SHÛIKU (1936) hat in einer sehr sorgfältigen Arbeit nicht nur über den Verlauf des Fettabbaues, sondern auch über das Verhalten verschiedener Kohlenhydratfraktionen berichtet. Seine Ergebnisse sind in Prozent des Gewichtes der ausgesäten Samen angegeben, sind also frei von den täuschenden Einflüssen einer Berechnung der Versuchsergebnisse in Prozenten der Trockensubstanz. Die Abb. 1 gibt seine Resultate wieder.

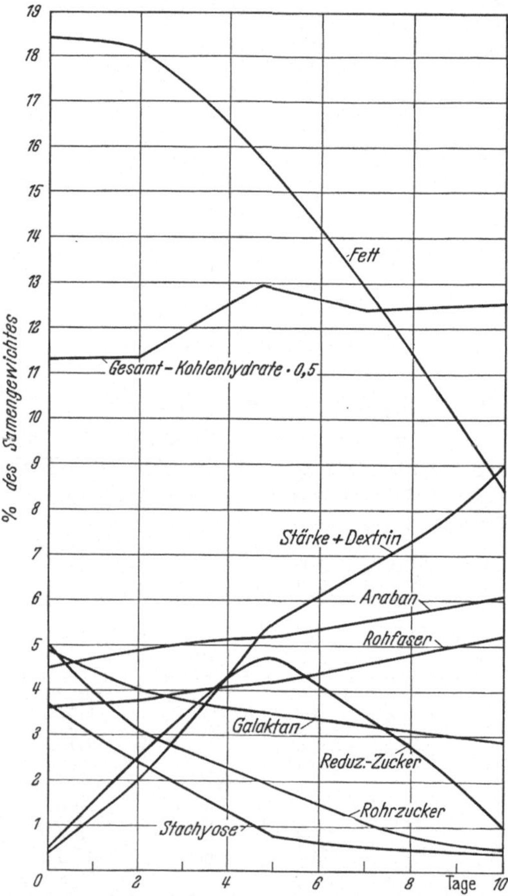

Abb. 1. Fett und Kohlenhydrate in keimenden Sojabohnen. (SHÛIKU 1936.)

Einem in den ersten 2 Tagen der Keimung nur langsam fortschreitenden Fettabbau folgt eine ziemlich regelmäßige Abnahme der Fette und mancher Zucker; demgegenüber steht eine regelmäßige starke Zunahme der Stärke. Die reduzierenden Zucker nehmen bis zum 5. Tag beträchtlich zu (auf nahezu das

10fache des Ausgangswertes), nehmen aber dann bis zum 10. Keimungstag wieder auf nur den doppelten Ausgangswert ab. In der Abbildung ist auch eine Kurve für die Gesamtkohlenhydrate eingezeichnet, jedoch aus Gründen der Raumersparnis und Übersichtlichkeit nur im halben Maßstab. Sie zeigt, wie unbedeutend die mengenmäßigen Veränderungen des Gesamtkohlenhydratgehaltes bei der Keimung der Soja sind und stellt diese Pflanze auch damit in starken Gegensatz zu den übrigen Fettpflanzen, bei denen einer Abnahme der Fette eine Zunahme der Kohlenhydrate entspricht.

V. ZAMBOTTI (1940) untersuchte den Fettgehalt von Sojakeimlingen von einer Länge bis zu 19 cm und fand eine zunächst langsame, dann raschere Abnahme der Fette je Samen, ganz entsprechend den Befunden von S. SHÛIKU.

Tabelle 18. *Fettgehalt keimender Sojabohnen.* (P. NEUMANN 1941.)

Keimlingsalter	Trockengewicht von 1 Keimling		Fettgehalt von 1 Keimling	
	in Luft	in O_2	in Luft	in O_2
Tage	(relativ)		(relativ)	
0	100	100	100	100
0,5	—	103	—	104
1,5	—	99	—	103
2	90	—	120	—
2,5	—	94	—	96
3	88	—	123	—
3,5	—	94	—	89
4	88	—	114	—
4,5	—	93	—	76
5	103	—	102	—
6	100	—	94	—
6,5	—	91	—	68
	100 = 117,3 mg	100 = 137,5 mg	100 = 20,63 mg	100 = 27,45 mg

Überraschende Ergebnisse erhielt dagegen P. NEUMANN (1941), der in keimenden Sojabohnen eine deutliche Fettzunahme fand, die erst nach dem 5. Keimungstag den Fettgehalt unter den Ausgangswert sinken ließ. Die Fettzunahme während der Keimung war von P. NEUMANN zunächst bei 20° C gefunden worden, in seinen Hauptversuchen hat er aber die bei 30° C rascher ablaufende Keimung untersucht (Tabelle 18). Außerdem hat P. NEUMANN Sojabohnen in reinem Sauerstoff bei 30° C keimen lassen und das Verhalten des Fettes verfolgt. Wie aus der Tabelle 18 zu ersehen ist, bewirkt das Verbringen in Sauerstoff zwar eine Beschleunigung des Fettabbaues, die aber nicht imstande ist, das andeutungsweise Zunehmen der Fette kurz nach Keimungsbeginn zu unterdrücken. Zu berücksichtigen ist auch, daß der Sauerstoffversuch erst mehr als 1 Jahr nach dem gewöhnlichen Keimungsversuch ausgeführt wurde und daß dabei anderes Saatgut verwendet wurde.

P. NEUMANNs Ergebnisse bieten also starke Hinweise dafür, daß in den Sojabohnen wahrscheinlich je nach Sorte oder durch andere Gründe bedingt, verschiedene Verläufe des Fettabbaues vorkommen können. Es liegen keinerlei Untersuchungen darüber vor, auf Kosten welcher Stoffe die Fettzunahme bei der Keimung der Soja etwa erfolgt, jedoch sind P. NEUMANNs Zahlen soweit außerhalb der Schwankungsbreite seiner Methode, daß es wohl kaum möglich ist, an der Realität seiner Beobachtung zu zweifeln. Eine Deutung ist freilich schwer möglich, aber vielleicht wäre es nicht ganz abwegig, daran zu denken, daß in rasch oder notgereiften Sojabohnen bei der Keimung noch eine Art Nachreifevorgang eintreten kann, bevor der eigentliche Keimungsstoffwechsel einsetzt. Undenkbar ist es ja nicht, daß der Stoffwechsel zu Beginn der Keimung etwa dort wieder einsetzt, wo er bei der Reifung (etwa durch Wassermangel) unterbrochen wurde und daß er somit in gewissem Ausmaß von der Art, wie die Reifung vor sich ging, abhängig ist. Rasches Austrocknen eines reifenden Samens mag vielleicht Stoffwechselvorgänge sozusagen „eintrocknen" und fixieren, die bei langsamer Ausreifung noch zu Ende, d. h. bis zu einem stabilen Gleichgewicht abgelaufen wären. Spätere Wasserzufuhr mag dann zunächst bewirken, daß dieses Gleichgewicht angestrebt wird — bis die einsetzenden Keimungsvorgänge eine Umschaltung des Stoffwechsels in Richtung Mobilisierung der Reserven herbeiführen. Auch mag berücksichtigt werden müssen, daß in Soja das Fett vielleicht nicht in dem Sinne Reservestoff ist wie in den eigentlichen Fettsamen und daß daher sein Stoffwechsel hier mehr flexibel als in anderen Samen ist.

Bis zu einem gewissen Grade denkbar wäre auch ein Einfluß methodischer Schwierigkeiten auf P. NEUMANNs Zahlen, obwohl keineswegs anzunehmen ist, daß er so groß sein könnte, daß er die Ergebnisse wesentlich beeinflussen würde. S. SHŬIKO (1936), der keine Fettzunahme in keimenden Sojabohnen fand, hat seine Fettbestimmungen in Kohlensäureatmosphäre ausgeführt und das Fett über Calciumchlorid im Dunkeln in CO_2 aufbewahrt. R. T. HOLMAN (1948) hat Sojakeimlinge in flüssiger Luft eingefroren, zerkleinert und gefriergetrocknet, bevor er das Fett mit Pentan extrahierte. Auch er fand in den ersten Keimungstagen eine geringe Fettzunahme, die etwa ab dem 4. Keimungstag in eine bis zum 12. Tag gleichmäßig fortschreitende Fettabnahme überging. Allerdings sind diese Ergebnisse nur in Prozent des Trockengewichtes der Keimlinge angegeben, lassen also kaum einen Schluß auf die absoluten Änderungen der Fettmengen zu. Außerdem hat R. T. HOLMAN Keimlinge verwendet, die im Freien im Licht herangewachsen waren und seine Zahlen können daher kaum zur Stützung oder Widerlegung der NEUMANNschen Befunde herangezogen werden. LEE WEI-YOUNG (1938) scheint ebenfalls keine Zunahme der Fette in keimenden Sojabohnen gefunden zu haben. K. WEISSENBÖCK (1948) bestimmte die Gesamtlipoide in Sojabohnen, die bei verhältnismäßig niederer Temperatur keimten. Er fand eine deutliche (bis zu 20% betragende) Lipoidzunahme, die selbst nach 12 Keimungstagen, als die Keimlinge schon abstarben, noch nicht abgeklungen war. Der Tod der Keimlinge dürfte aber hier eher auf die niedrigen Keimungstemperaturen als auf ein Erschöpfen der Reserven zurückzuführen gewesen sein. Aus Unterschieden in der Atmung sowie aus Wachstumskurven schloß A. L. HAFENRICHTER (1928), daß in keimenden Sojabohnen abwechselnd verschiedene Reservestoffe veratmet werden, wobei die Übergänge recht unvermittelt erfolgen sollen. Wenn sich diese Befunde bei sorgfältigster Überprüfung bewahrheiten sollten, könnten sie vielleicht dazu beitragen, das Rätsel der Fettmobilisierung in keimenden Sojabohnen zu lösen.

Bei mikroskopischen Untersuchungen von Sojakeimlingen hat F. W. v. OHLEN (1931) eine starke Fettabnahme und Stärkezunahme während der Keimung beobachtet. Er hat seine Untersuchungen bis zu dem etwa nach 30 Tagen erfolgenden Absterben der etiolierten Keimlinge ausgedehnt und gibt eingehende Beschreibungen der vorgefundenen histochemischen Verhältnisse.

Keimende *Lupinen*samen wurden zuerst von M. MERLIS (1897) auf ihren Fettgehalt untersucht (7,5% in den Samen, 1,6% in etwa 2 Wochen alten etiolierten Keimlingen). A. HÉE und L. BAYLE (1932) berichteten über Lupinenkeimlinge, bei denen in sterilem destilliertem Wasser die Fettprozente auf etwa ein Viertel ihres Ausgangswertes absanken, wobei zwischen 5. und 10. Keimungstag die Abnahme rascher als vor- und nachher erfolgte. P. NEUMANN (1941) hat am 2. Keimungstag (T = 20° C) von *Lupinus albus* Anhaltspunkte für ein Hinausgehen des Fettgehaltes je Keimling über den Ausgangswert gefunden (Tabelle 19).

Bei Fettzunahmen um derart geringe Beträge wie 3% ist es natürlich sehr schwer, die Bedeutung der gefundenen Zahlen abzuschätzen, wenn Angaben über Unterschiede zwischen Parallelbestimmungen und Reproduzierbarkeit der Werte bei Aufarbeitung verschiedener Samenmuster fehlen.

Tabelle 19. *Fettgehalt keimender Samen von Lupinus albus.* (P. NEUMANN 1941.)

Keimlingsalter Tage	Trockengewicht je Keimling (relativ)	Fettgehalt je Keimling (relativ)
0	100	100
1	93,5	98
2	95,5	103
3	92	93,5
4	98	96
	100 = 260 mg	100 = 24,6 mg

Von manchen anderen Pflanzen liegen nur bruchstückartige oder schwer zugängliche Angaben über den Verlauf der Fettmobilisierung bei der Keimung vor, so daß darauf verzichtet wird, sie im einzelnen wiederzugeben, würden sie doch nicht erlauben, einen Zusammenhang mit den bisher besprochenen Typen des Verlaufes des Fettabbaues herzustellen. Dies gilt z. B. für *Euphorbia lathyrus* (G. FLEURY 1865), *Feigen* (G. SANI 1904), *Oliven* (G. SANI 1900), *Gymnocladus dioica* (G. J. RALEIGH 1930), *Trifolium repens* (S. G. BROOKER 1932), *Juniperus scopulorum* (M. AFANASIEV und M. CRESS 1942), *Eicheln* (E. R. GIRTON und E. R. PARK 1942), *Tomaten* (G. R. BARTLETT 1945), *Oenothera biennis* (N. TIETZ 1953).

Ausführlichere mikroskopische Untersuchungen liegen vor für keimende *Zwiebel*samen (J. SACHS 1863) sowie für keimende *Cyclamen*samen (H. GRESSNER 1874); A. E. GRIFFITS (1938) hat mit qualitativen mikrochemischen Methoden keimende *Salat*samen untersucht, und A. POLJAKOFF-MAYBER und A. M. MAYER (1955) haben den Einfluß von Cumarin und Thioharnstoff auf die Fettumwandlung bei der Keimung dieser Samen studiert und Hydrolysehemmung durch Cumarin gefunden, während Thioharnstoff die Ansammlung freier Fettsäuren begünstigte. Da aber, wie W. M. CROMBIE und R. COMBER (1956) gezeigt haben, die üblichen Verfahren des Nachweises und der Bestimmung der sog. freien Fettsäuren in Samen und Keimlingen irreführende Werte ergeben, wird man eine Überprüfung der Cumarin- und Harnstoffwirkung mit verläßlicherer Methodik abwarten müssen.

Auch in Samen, die nicht als Fettsamen gelten können, wurde verschiedentlich das Verhalten der geringen, darin vorkommenden Fettmengen bei der Keimung untersucht. Wieweit es sich hierbei mehr um Untersuchungen des Betriebsstoffwechsels als um solche des Reservestoffwechsels handelt, sei nicht näher erörtert. R. C. MALHOTRA (1932) fand in *Erbsen* während 8 Keimungstagen eine Abnahme des Fettes um knapp 30%, wobei eine vorübergehende geringe Zunahme am 4. Tag wohl kaum außerhalb der Fehlerbreite der Bestimmungen liegt. An im Licht heranwachsenden Keimlingen von *Phaseolus multiflorus* haben R. C. JORDAN und A. C. CHIBNALL (1933) den Fettsäuregehalt der Kotyledonen von je 1000 Pflanzen bestimmt und folgende Werte gefunden: Samen 18,4 g, Keimlinge, 8tägig: 16,5; 15tägig: 6,4; 30tägig: 2,4 g. *Phaseolus aureus* (Mungbohne) wurde von H.-C. KAO (1936) untersucht; er fand in den etiolierten Keimlingen zuerst eine Abnahme (7%) und in den 10—15 Tage alten eine Zunahme (etwa 15%) der Fettsäuren, deren Gesamtmenge 3% des Samengewichtes kaum je überstieg.

In *Reis* fand W.-S. TAO (1930) keine Veränderung des Fettgehaltes während 7 Keimungstagen. — In *Weizen* fand R. C. MALHOTRA (1932) keine Änderung des Fettgehaltes (1,87%) bis zum 6. Keimungstag; am 8. Tag waren aber etwa 30% des Fettes abgebaut. — In keimender *Gerste* wurde das Fett nur im Zusammenhang mit der Mälzerei (Biererzeugung) untersucht (M. WALLENSTEIN 1897 K. TÄUFEL und M. RUSCH 1929a, b) und eine etwa 12%ige Fettabnahme festgestellt. Gleichzeitig findet eine beträchtliche Zunahme des Unverseifbaren statt. An isolierten Gerstenembryonen haben W. O. und A. L. JAMES (1940) Atmungsquotienten von weniger als 0,3 gemessen und auf intensive Fettumwandlung in Abwesenheit des Endosperms geschlossen. — Der *Mais*embryo enthält etwa 30% fettes Öl, das Endosperm etwa 1%. Während der Keimung nimmt nach E. H. TOOLE (1924) der Fettgehalt im Embryo nur langsam ab (10,5; 10,2 und 8 mg je Embryo bei Wurzellängen von 0,2; 1—2 und 4—8 cm); im Endosperm bleibt das Fett praktisch unverändert, scheint also nicht als Reservestoff zu dienen. R. C. MALHOTRA (1935) hat nur die Abnahme der Gesamtfettprozente in keimendem Mais untersucht und seine Angaben stimmen mit den Befunden E. H. TOOLEs praktisch völlig überein. — Der Fettgehalt in keimender *Hirse* wurde von N. B. GUERRANT (1927) an 2 Sorten untersucht, die 3,2 und 4,1% Fettgehalt hatten; die relativen Änderungen des prozentischen Fettgehaltes waren aber während der Keimung beider Sorten praktisch gleich groß, und der Fettgehalt betrug nach 2, 4, 6, 8 und 10 Keimungstagen 93%, 87%, 81%, 76% und 69 (74)% des Ausgangswertes, zeigte also eine langsame, genau lineare Abnahme.

Anatomische Untersuchungen über die Fettkörper in Samen und keimenden Samen von Gramineen liegen vor von E. RALSKI (1924).

Über den *Einfluß von Außenfaktoren* auf den Fettabbau bei der Keimung liegen nur ganz vereinzelte Angaben vor. Dies ist verständlich, da es kaum möglich ist, die Wirkung solcher Einflüsse von der gleichzeitig stattfindenden Beeinflussung der Keimung zu trennen und da ja andererseits der unbeeinflußte Ablauf des Fettabbaues schwierig genug zu erfassen ist. Licht beeinflußt gerade bei der Keimung nicht nur in vielen Fällen den Vorgang selbst, sondern bringt durch Ermöglichung des Ergrünens und der Assimilation auch einen ganz anderen Typus des Stoffwechsels hervor. Um so überraschender ist es, daß alle vorhandenen Daten darauf hindeuten, daß im Verlaufe der Lichtkeimung der Fettabbau nur wenig anders als in der Dunkelheit vor sich geht, so daß man den Eindruck bekommt, daß Reservestoffmobilisierung und Aufbaustoffwechsel getrennt nebeneinander ablaufen können. Über echte Fettsamen liegen allerdings nur ganz wenige Angaben vor. H. L. WHITE (1919) hat den Einfluß von diffusem Licht und von direktem Sonnenlicht auf den Fettabbau in Sojakeimlingen verfolgt und hat gefunden, daß im diffusen Licht der Fettabbau etwas langsamer vor sich ging als im Dunkeln, daß er aber im vollen Sonnenlicht viel rascher ablief. In etwa 6—12 cm großen Pflanzen waren im Dunkeln 55% des Ausgangsfettgehaltes verschwunden, im diffusen Licht waren es 45% und im Sonnenlicht 80%. Freilich handelt es sich hier nur um Abgaben des Fettgehaltes in Prozent der Trockensubstanz, so daß über den tatsächlichen Fettabbau je Keimling kaum eine sichere Aussage möglich ist. Derselbe Einwand ist auch gegen P. L. MACLACHLANS (1936b) Werte zu erheben, der in den Kotyledonen 2 Wochen alter Sojakeimlinge im Licht 61% des Gehaltes an Fettsäuren verschwunden fand, während im Dunkeln nur 39% fehlten. R. TIETZ (1953) fand, daß in Samen von *Oenothera biennis* während 24 Std Lichtkeimung 17% der Fette abgebaut wurden, während bei gleich langer Dunkelkeimung nur 7% verschwunden waren. V. GRAFE und V. VOUK (1912) fanden bei Licht- und Dunkelkeimung (12 Tage) der Samen von *Cichorium intybus* keinen nennenswerten Unterschied im Verlauf des Fettabbaues. Eine schwache Lichthemmung des Fettabbaues stellte N. B. GUERRANT (1927) bei *Sorghum* fest. Leider sind diese Zahlen so wie die für *Cichorium* nur als Prozent der Trockensubstanz angegeben.

Abschließend seien hier die Versuche von R. DUPÉRON (1949) erwähnt, der den Einfluß einer *Vernalisierung* des Saatgutes von *Sinapis alba* auf den Fettabbau während der Keimung untersuchte. Er fand, daß der Fettabbau — wenn die Keimwurzeln etwa 15 mm lang geworden waren — in vernalisierten und unbehandelten Samen ganz gleich verläuft, daß der Fettsäurengehalt des jarowisierten Saatgutes bei gleicher Wurzellänge der Keimlinge aber etwas größer ist. Der Grund dafür liegt darin, daß die bereits erwähnte (S. 335) von R. DUPÉRON in frühen Keimungsstadien beobachtete Zunahme der Fettsäuren in den vernalisierten Keimlingen viel geringer als in den Kontrollen war (etwa von 480 auf 500 mg je 500 Keimlinge im Vergleich zu 480 auf etwa 570 in den Kontrollen), ohne daß eine in den Kontrollen beobachtete kurze Periode rapider Fettabnahme auftrat, bevor der kontinuierliche Fettabbau einsetzte. Die löslichen Zucker nahmen in den vernalisierten Samen und in den unbehandelten zunächst um etwa 20% zu, worauf in den normalen Keimlingen eine zuerst etwas raschere, aber sehr bald sehr gering werdende Abnahme eintrat, die den Gehalt an löslichen Zuckern bei Erreichung einer Wurzellänge von 50 mm etwa auf den Ausgangswert zurückbrachte. In den vernalisierten Keimlingen hingegen trat nach der anfänglichen Zunahme zuerst ebenfalls eine etwas raschere Abnahme ein, die aber sehr bald in eine sehr langsame Zunahme der löslichen Zucker überging, so daß sie bei einer Wurzellänge von 50 mm auf etwa 130% des Ausgangswertes gestiegen waren.

Versuchen wir nun, *zusammenfassend* den Verlauf der Fettmobilisierung in keimenden Samen zu betrachten, so finden wir vor allem, daß bei den meisten Samen eine Periode außerordentlich raschen Fettabbaues zu finden ist, die bei einigen Pflanzen sofort bei Keimungsbeginn einzusetzen scheint (*Cichorium*, schwarzer Senf), der aber meist eine kürzere oder etwas längere Periode geringen oder nahezu unmerklichen Fettabbaues vorangeht (*Ricinus*, Kürbis, *Aleurites*, weißer Senf, Lein, Hanf, Rettich, Erdnuß und Baumwolle). Von diesem Typus der Fettmobilisierung verschieden scheint die annähernd gleichmäßige Abnahme der Fette zu sein, wie sie während der Keimung von Sonnenblumenkernen, von Walnüssen, Mandeln und Mohn beobachtet wurde. — Ein 3. Typus der Fettmobilisierung liegt vielleicht in jenen Samen vor, die wie *Soja*, *Lupinus* und vielleicht weißer Senf und *Cucurbita maxima*, wenigstens manchmal (vielleicht auch sorten- oder saatgutbedingt) früher oder später während der Keimung eine vorübergehende Zunahme der Fette erkennen lassen. Ob diese 3 Typen der Fettmobilisierung wirklich real sind und wirklich verschiedenen Typen oder gar verschiedenen Wegen des chemischen Stoffwechsels entsprechen, ist auf Grund der sehr dürftigen exakten Unterlagen heute wohl noch nicht feststellbar. Nur sorgfältigste Versuche, die das Schicksal des Fettes nicht relativ zum Trockengewicht sondern für den Einzelkeimling verfolgen lassen, können dies entscheiden. Soweit wenigstens halbwegs verläßliche Angaben darüber vorliegen, scheinen bei der Mobilisierung der Samenfette auch mehrere Typen des Verhaltens der *Kohlenhydrate* in den Keimlingen aufzutreten. Die Glucose z. B. dürfte während der Keimung von *Ricinus* zuerst ab- und später wieder zunehmen, bei Kürbis und *Aleurites* aber nimmt sie wohl sicher ziemlich gleichmäßig zu. Ebenso auffällig verschieden scheint das Verhalten der Saccharose zu sein. Bei *Ricinus* bleibt sie zunächst unverändert und nimmt dann zu, beim Kürbis und dem Tungbaum dagegen nimmt sie nur ab und verschwindet schließlich ganz oder fast ganz. In jüngster Zeit hat R. DUPÉRON (1954) auf papierchromatographischem Weg gefunden, daß in verschiedenen Fettsamen (Cruciferen, *Arachis*, *Papaver*, *Linum*) beträchtliche Kohlenhydratmengen (bis zu 12%) vorliegen, die neben Saccharose vor allem aus Raffinose und Stachyose bestehen. Bei der Keimung verschwinden diese Kohlenhydrate bevor noch eine Umwandlung von Fetten in Kohlenhydrate (nicht in Raffinose oder Stachyose) stattfindet. — Über ein etwaiges unterschiedliches Verhalten der Stärke lassen sich wohl noch keine verallgemeinernden Aussagen machen, liegt doch viel zu wenig exaktes Beobachtungsmaterial vor und scheint doch, nach den mikroskopischen Untersuchungen H. LUNDEGÅRDHS (1914) zu schließen, Vorhandensein oder Fehlen von Stärke in den Samen selbst keineswegs ein absolutes Charakteristikum einer Art zu sein. — Auch die Frage, ob während der Keimung der Fettsamen infolge stärkerer Sauerstoffixierung (eventuell aus Wasser) bei der Umwandlung von Fetten in Kohlenhydrate nicht manchmal eine wenigstens vorübergehende Zunahme des Trockengewichtes eintritt, scheint noch nicht mit einer jeden Zweifel ausschließenden Sicherheit beantwortet werden zu können. — Schließlich sei noch darauf hingewiesen, daß wir in der Sojabohne vielleicht eine (oder eine der) Pflanze(n) vor uns haben, bei der die Mobilisierung der Fette bei der Keimung nicht zu einer wesentlichen Zunahme der Gesamtkohlenhydrate führt. Es ist keineswegs unmöglich oder unwahrscheinlich, daß sorgfältige Untersuchung der Keimung bisher wenig genau erforschter Fettsamen noch manche neue Typen der Fettmobilisierung ans Tageslicht bringen würde.

D. Die chemischen Veränderungen der Fette bei der Keimung.

Wenn im folgenden versucht wird, einen Überblick über die Arbeiten zu geben, die sich mit den chemischen Veränderungen der Fette bei der Samenkeimung befassen, so muß man sich darüber klar sein, daß diesem ganzen Problem im Lichte der in den letzten Jahren erarbeiteten Vorstellungen über den Verlauf der Fettumsetzungen im tierischen Organismus (und damit wohl prinzipiell bei allen Lebewesen) nicht mehr jene Bedeutung zukommt, die man früher geneigt war, ihnen beizulegen. Seit wir durch Auffindung des Krebscyclus und ähnlicher enzymatischer Kreisprozesse ganz neue Vorstellungen vom Ablauf der Stoffwechselvorgänge auch in den Pflanzen gewonnen haben, wissen wir ja, daß chemische Analysen verschiedener Entwicklungsstadien von Pflanzen prinzipiell nicht imstande sind, wirkliche Einsicht in Art und Weg von Stoffwechselvorgängen zu geben. Was wir vielleicht mit solchen Analysen erfassen können, sind nur vorübergehende End- oder Gleichgewichtszustände der uns eigentlich interessierenden Stoffwechselvorgänge. Wir können auf diese Weise immer nur feststellen, welche Produkte die chemische Fabrik, mit der ein Keimling verglichen werden könnte, gerade verlassen oder in ihr gerade auf Lager gelegt werden, wir lernen aber praktisch gar nichts darüber, wie sie hergestellt werden. Solche Einsicht scheint nur mit Hilfe der Enzymchemie zugänglich zu sein[1].

Die folgende Übersicht kann also kaum etwas anderes sein, als eine kurze Zusammenfassung eines nunmehr weitgehend historisch gewordenen Kapitels pflanzenphysiologischer Forschung.

Daß wenigstens in vielen Pflanzen das Fett bei der Keimung trotz mengenmäßiger Abnahme und wenigstens scheinbarer Änderung mancher Kennzahlen

[1] Wegen der Enzymatik des Fettabbaus sei auf die Beiträge von E. BAMANN und E. ULLMANN (Lipasen; S. 109ff.) und von W. FRANKE und H. FREHSE (Fettsäure-Oxydasen, -Dehydrasen und Reduktasen; α- und β-Oxydation; S. 137ff.) in diesem Handbuchbande verwiesen. In diesen Beiträgen wird eingehend auf die in Samen und Keimlingen gefundenen Fermente und die von ihnen kontrollierten Umsetzungen eingegangen.

chemisch weitgehend unverändert bleibt, hat schon R. H. SCHMIDT (1891) daraus geschlossen, daß der Schmelzpunkt der aus verschieden weit entwickelten Keimlingen gewonnenen Fette praktisch gleich war. Wenn trotzdem in vielen Veröffentlichungen über während der Keimung vor sich gehende Änderungen der chemischen Zusammensetzung der Keimlingsfette berichtet wird, so kann man sich oft des Eindruckes nicht erwehren, daß methodische Unzulänglichkeiten das Auftreten von chemischen Veränderungen während der Aufarbeitung der Proben nicht ausschlossen oder sogar geradezu begünstigten. So erwähnt z. B. K. PIRSCHLE (1926), daß A. MÜNTZ (1871) bewiesen habe, daß in keimenden Fettsamen nach 3—4 Keimungstagen 96—98% des Fettes in freie Fettsäuren übergegangen sei. A. MÜNTZ schloß dies aber nur aus der Löslichkeit der Fettfraktion in absolutem Alkohol, und keiner der später mit exakterer Methodik arbeitenden Forscher hat ähnlich hohe Gehalte an freien Fettsäuren gefunden.

Ricinus dürfte ein typisches Beispiel dafür darstellen, wie die spezielle chemische Zusammensetzung eines Untersuchungsobjektes es sehr schwierig macht, den Chemismus der Keimlingsfette unbeeinflußt von sekundären Veränderungen zu erfassen. *Ricinus*samen und *Ricinus*keimlinge enthalten bekanntlich eine sehr aktive Lipase, und es ist daher nicht verwunderlich, wenn J. R. GREEN (1890, 1906) bei der Analyse seiner *Ricinus*keimlinge beträchtliche Mengen freier Fettsäuren fand. Daraus darf aber nicht ohne weiters geschlossen werden, daß sie auch im lebenden Keimling in freier Form vorlagen. LECLERC DU SABLON (1893) fand in *Ricinus*keimlingen nur minimale Mengen freier Fettsäuren (Säurezahlen 0,1—0,5), O. v. FÜRTH (1904) hingegen fand höhere Werte (etwa 42 in 9 Tage alten Keimlingen), und N.-F. DELEANO (1909) fand überhaupt keine Zunahme der freien Fettsäuren in 4—15 Tage alten *Ricinus*keimlingen. R. DESVEAUX und M. KOGANE-CHARLES (1952) haben in *Ricinus*-Samen nur geringe Mengen (1,6%) freier Fettsäuren gefunden. Bei der Keimung nahm ihre Gesamtmenge vorübergehend auf etwa das Doppelte zu und hatte am 11. Keimungstag wieder den Ausgangswert erreicht. Die gleichzeitig vor sich gehende starke Abnahme des Gesamtfettes auf knapp 7% des Ausgangswertes täuscht bei prozentualer Berechnung eine gewaltige (völlig irreführende) Zunahme der freien Fettsäuren vor. Es sei hier aber darauf hingewiesen, daß im Lichte neuester methodischer Erfahrungen (W. M. CROMBIE 1956, besonders S. 185f.) höchst zweifelhaft ist, ob die bisher angewandten Verfahren zur Bestimmung freier Fettsäuren im Pflanzenmaterial wirklich freie höhere Fettsäuren erfassen.

J. HOUGET (1943) hat ein annäherndes Konstantbleiben der Jodzahl der Fette keimender *Ricinus*samen festgestellt, doch fanden sich, besonders am 4. und 5. Keimungstag, höhere Werte für Hydroxyl- und Verseifungszahl; Isolierung und nähere Untersuchung einer Fettfraktion des 4. Keimungstages lieferte Anhaltspunkte für das Vorhandensein einer ungesättigten Fettsäure mit einer Alkoholgruppe. R. DESVEAUX und M. KOGANE-CHARLES (1952) fanden besonders im Endosperm von *Ricinus* eine ziemlich beträchtliche Zunahme (etwa 28%) der Jodzahl während der Keimung. Die Verseifungszahl nahm bis zum 8. Keimungstag auf etwa 115% des Ausgangswertes zu und fiel bis zum 11. Tag dann auf 111% ab, die Hydroxylzahl war bis zum 6. Keimungstag konstant und sank dann bis zum 11. Keimungstag auf etwa 80% des Ausgangswertes.

In keimenden Samen von *Cucurbita maxima* fand U. JEGOROW (1904) bis zum 20. Keimungstag eine Abnahme der Jodzahl von 114 auf 105, eine Zunahme der flüchtigen Säuren von 4 auf 16 und einen Anstieg der Säurezahl von 0,7 auf 59, nachdem am 6. Tag 2,9 und am 10. Tag 3,1 gemessen worden war.

Ein unregelmäßiger Verlauf der Jodzahlkurve keimender *Kürbisse* wurde von U. WEBER (1936) gefunden, der Maxima am 6. und 12. Keimungstag (je etwa 108% des Ausgangswertes), getrennt durch ein Minimum am 8. Keimungstag (etwa 101% des Ausgangswertes) angibt. In 0,05% Kaliumacetatlösung fand sich ein zusätzliches, etwas tieferes Minimum etwa am 20. Tag. A. ZELLER und F. MASCHEK (1942) haben das Jodbindungsvermögen je Durchschnittskeimling an sorgfältigst ausgewählten Kürbiskeimlingen untersucht und den in der Abb. 2 dargestellten Verlauf während der Keimung gefunden. In dieser Abb. 2 sind auch die in der gleichen Veröffentlichung erwähnten, von J. FUCHS (1933) im gleichen Laboratorium unter vergleichbaren Bedingungen gefundenen, relativen Fettsäurewerte keimender Kürbissamen eingezeichnet. Der vollständige Gleichlauf, ja die weitestgehende Identität dieser beiden Kurven kann wohl nur so gedeutet werden, daß daraus das Ausbleiben von Änderungen des Sättigungsgrades der noch nicht abgebauten Fette während der Keimung entnommen wird. Zu diesem Schluß kam auch W. HEUMANN (1944), der aus seinen Bestimmungen von Kali-, Jod- und Rhodanzahl in Kürbiskeimlingen schloß, daß jedes Fettsäuremolekül, das abgebaut wird, so schnell zerlegt wird, daß Zwischenprodukte von vielleicht anderem Sättigungsgrad nicht erfaßt werden können und daß ungesättigte und gesättigte Fettsäuren wahllos abgebaut werden. Darauf, daß auf Grund aller vorliegenden Erfahrungen nicht erwartet werden kann, daß sich irgendwelche Zwischenprodukte des Fettabbaues in den Keimlingen zu irgendeinem Zeitpunkt anreichern, hatte schon A. ZELLER (1935, S. 142 und 147) ausdrücklich aufmerksam gemacht. W. HEUMANN fand in den älteren Kürbiskeimlingen mehr freie Fettsäuren als in den Samen. Gegen seine Versuchsmethodik können aber ähnliche Einwände wie gegen die oben erwähnten Versuche von R. DESVEAUX und M. KOGANE-CHARLES erhoben werden: die Möglichkeit einer lipatischen oder hydrolytischen Fettspaltung war keineswegs ausgeschaltet. Aus den Ergebnissen seiner Bestimmungen von Verseifungs- und Esterzahlen der Öle der Kürbiskeimlinge schließt W. HEUMANN auf einen etwas bevorzugten Abbau von Triglyceriden im Vergleich zu Diglyceriden.

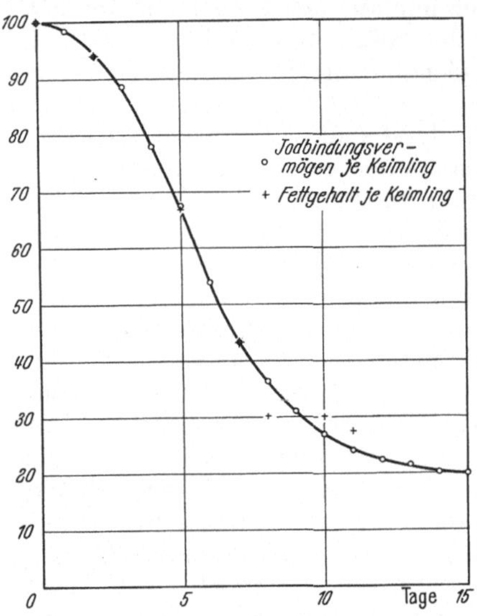

Abb. 2. Verlauf von Jodbindungsvermögen und Fettgehalt in keimenden Kürbissamen (Relativwerte, bezogen auf einen Durchschnittskeimling). (Nach A. ZELLER und F. MASCHEK 1942.)

Trotz allem scheint aber der Kürbis auf Grund der exakten Versuche von A. ZELLER, A. ZELLER und F. MASCHEK sowie W. HEUMANN die erste und wohl einzige Pflanze zu sein, bei der schon *vor* Erforschung des fermentchemischen Mechanismus der Fett-Kohlenhydratumwandlung die Voraussetzungen dafür geschaffen worden waren, daß jetzt geradezu auf experimenteller Grundlage die Hypothese vertreten werden kann, daß der Fett-Kohlenhydratumbau in Tier und Pflanze im Prinzip gleich oder analog verläuft.

Daß der Fettabbau in keimenden Samen von *Citrullus vulgaris* ohne Auftreten freier oder kurzkettiger Fettsäuren vor sich geht und daß alle wichtigeren Fettsäuren gleichmäßig abgebaut werden, haben W. M. CROMBIE und R. COMBER

(1956) gezeigt. Nur die Ölsäure, die etwa 7% der Fettsäuren des Melonensamenfettes ausmacht, scheint etwas rascher umgesetzt zu werden. „Säurezahl" und „Verseifungszahl" der Fette geben infolge Auftretens störender Stoffe keine richtige Auskunft über das Vorliegen freier Fettsäuren.

Über chemische Veränderungen des Fettes bei der Keimung von *Lein*samen liegen nur ganz wenige Angaben vor. A. J. Ermakoff und N. Iwanoff (1931) fanden keine Änderung der Jodzahl des Öles verschieden alter Leinkeimlinge. R. Desveaux und M. Kogane-Charles (1952) hingegen fanden eine Abnahme der Jodzahl (um etwa 23%), sowie geringe (etwa 5%) Schwankungen der Verseifungszahl neben vorübergehenden Veränderungen der Menge freier Fettsäuren bei gleichzeitiger Zunahme anderer freier Säuren auf nahezu das 20fache. Für keimenden *Hanf* werden angegeben: eine Zunahme der Jodzahl um etwa 5% am 5. Keimungstag, dann Abnahme auf etwa 66% des Ausgangswertes am 18. Keimungstag, geringfügige Veränderungen der Gesamtmenge freier Fettsäuren und Zunahme der sonstigen freien Säuren auf etwa das 40fache des Ausgangswertes. Bezüglich der freien Fettsäuren sind natürlich hier ebenfalls die oben (S. 345) erwähnten neueren methodischen Erfahrungen zu berücksichtigen.

In keimenden *Erdnüssen* hat R. H. Schmidt (1891) nur geringe Veränderungen der Säure- und Jodzahlen gefunden, und F. A. Johnston und H. M. Sell (1944) haben in am Licht keimenden Samen von *Aleurites fordii* praktisches Gleichbleiben von Verseifungs- und Jodzahl, aber starkes Ansteigen der Säurezahl (von 0 auf 77) gefunden.

Darauf, daß praktisch alle *Baumwoll*samen mit mehr als etwa 2% freier Fettsäure im Öl nicht mehr keimfähig sind, wurde oft hingewiesen (vgl. Literatur S. 305). Andererseits fanden H. S. Olcott und T. D. Fontaine (1941) in keimender Baumwolle während 6 Keimungstagen den Prozentsatz freier Fettsäuren im Öl von 1,8% auf über 20% ansteigend. Daß künstliche Trocknung das Auftreten freier Fettsäuren in Baumwollsamen verzögert und damit die Keimfähigkeitsdauer und Ölqualität erhöht, haben R. A. Rusca und F. L. Gerdes (1942) betont, und D. M. Simpson (1942) hat den Einfluß von Lagerungstemperatur und Wassergehalt auf das Auftreten der freien Fettsäuren in Baumwollsamen und die dadurch gegebene Beeinflussung der Keimfähigkeit untersucht. Eine zusammenfassende Besprechung der biologischen Prozesse in Baumwollsamen stammt von A. M. Altschul (1948).

Ganz anders als Baumwolle dürften sich Samen verhalten, deren Keimung vom Licht beeinflußt wird und bei denen der Keimungsprozentsatz im Licht um so größer ist, je höher die Säurezahl der Samenfette ist. Bei *Nigella damascena* z. B. stiegen bei Zunahme der Säurezahl von 26 auf 33 die Lichtkeimungsprozente von 0 auf 96 (R. Tietz 1953), und nach W. A. Gardner (1921) scheint auch bei anderen lichtempfindlichen Samen das Licht eine Aktivierung des lipolytischen Systems mit Zunahme der freien Fettsäuren zu bewirken.

Über den Chemismus der Fette keimender *Sonnenblumen* liegen ausführliche neuere Untersuchungen von E. C. Miller (1910, 1912), E. Matthes (1927) und G. Tilenius (1938) vor, so daß es nicht nötig ist, andere Arbeiten (R. H. Schmidt 1891, O. v. Fürth 1904, J. Lemarchands 1929) im einzelnen zu erwähnen. E. C. Miller und auch E. Matthes finden, daß die Jodzahl des Sonnenblumenfettes während der Keimung konstant bleibt, wenigstens bis etwa 80% des Fettes verbraucht sind und bis ein deutlich veränderter Hunger- oder Absterbestoffwechsel der im Dunkeln oder CO_2-frei im Licht (Miller) gezogenen Keimlinge einsetzt. Dann findet A. Matthes eine Zunahme der Jodzahl, E. C. Miller dagegen eine Abnahme, und G. Tilenius berichtet, daß seine Keimlinge zuerst (4—5 Keimungstage) eine schwache (2%) Zunahme der Jodzahl zeigten, auf die bis zum 17. Keimungstag eine ziemlich regelmäßige Abnahme bis auf etwa 80% des Ausgangswertes folgte, worauf eine neuerliche Zunahme (20. Keimungstag 88% des Anfangswertes) zu beobachten war. Die Säurezahl wird entweder unregelmäßig schwankend gefunden (E. MATTHES) oder als längere Zeit unverändert niedrig und dann stark ansteigend berichtet (E. C. MILLER), wobei allerdings auch die freien Nichtfettsäuren stark zunehmen (auf etwa das 1000fache!) so daß wieder die Frage offen ist, ob nicht am Ende ein Teil der Zunahme der freien Fettsäuren als sekundär und als durch die anderen freien Säuren bei der Aufarbeitung gebildet anzusehen ist. Außerdem findet E. C. MILLER kurz vor dem Absterben seiner Keimlinge eine starke Zunahme der Hydroxylgruppen und spricht vom Auftreten von Glyceriden niederen Molekulargewichtes. E. MATTHES hat noch eine ganze Anzahl weiterer Fettkonstanten festgestellt und versucht, für die aufgefundenen Veränderungen Erklärungen zu finden doch ist es nicht wahrscheinlich, daß diese jetzt noch zu unserer Kenntnis des Chemismus des Fettstoffwechsels wesentlich beitragen können. Auch die Versuche, die G. TILENIUS über den Einfluß von den Wurzeln gebotenen 0,05% Kaliumacetat auf den Fettstoffwechsel ausführte, würden jetzt, da „aktiviertes Acetat" (Acetat mit der Sulfhydrylgruppe von Coenzym A verestert) als das zentrale Zwischenprodukt des Fettsäureabbaues erkannt ist, anders geplant werden müssen.

In keimenden *Walnüssen* stellte U. WEBER (1936) eine langsame Zunahme der Jodzahl bis zum 29. Keimungstag (+16%) mit folgender Abnahme auf den Ausgangswert (56. Keimungstag) fest.

Oben wurde berichtet (S. 338ff.), daß die Mobilisierung der Fette bei der Keimung von Sojabohnen vermutlich etwas anders verläuft als bei anderen fetthaltigen Samen. Chemische Veränderungen des Sojaöles während der Keimung scheinen aber kaum wesentlich verschieden von den bei anderen Pflanzen beobachteten zu sein. Sie wurden von H. K. WHITE (1919) zu erfassen versucht, der hoffte, während der Keimung ein stärker ungesättigtes Öl zu finden, was industrielle Bedeutung gehabt hätte. Er fand aber keine Änderung des Sättigungsgrades des Öles während der Keimung, auch nicht bei Änderung der Lichtverhältnisse. Die starken und unregelmäßigen Schwankungen der Säurezahl, die WHITE fand, sind wohl kaum zu deuten. W. ZIMMERMANN und A. BAUMANN (1935) fanden bei der Untersuchung der Keimung verschiedener Sojasorten in Sand und Erde ebenfalls keine wesentliche Änderungen der Jodzahl, bevor nach dem 4. Keimungstag mit rapider Fettabnahme eine meist deutliche Abnahme der Jodzahl eintrat. Zu praktisch den gleichen Ergebnissen kaum auch P. L. MACLACHLAN (1936b) sowie V. ZAMBOTTI (1940), der außerdem geringe Schwankungen der Verseifungszahl, der Acetylzahl, der REICHERT-MEISSL-Zahl und der POLENSKE-Zahl feststellte. Auch P. NEUMANN (1941) fand in keimenden Sojabohnen (sowohl in Luft als auch in O_2) in den ersten Keimungstagen entweder unregelmäßige Jodzahlen (in Luft) oder (in O_2) etwas unregelmäßig ein wenig abnehmende Werte. Die Verseifungszahlen zeigten keine wesentlichen Veränderungen, während die Säurezahlen in Luft auf das 3- (5 Keimungstage) bzw. 5- (6 Keimungstage) und 13fache (6 Tage in O_2) zunahmen, wobei freilich erwähnt werden muß, daß beim Keimungsversuch in O_2 die Anfangssäurezahlen nur etwa 20% der des Keimungsversuches in Luft betrugen. Zunahmen der Säurezahl und besonders der Acetylzahl waren schon von S. SHÛIKU (1936) neben geringen und späten Jodzahl- und Esterzahländerungen gefunden worden. Eine auffällige Sterinveresterung bei der Sojakeimung stellte P. L. MACLACHLAN (1936a) fest, und auch R. T. HOLMAN (1948) fand erst nach dem 5. Keimungstag ein Absinken der Jodzahl der Fette keimender Sojabohnen. Er stellte außerdem fest, daß vom 2. Keimungstag an Linol- und Linolensäure vielleicht etwas rascher als das Gesamtfett verschwinden und daß von den beiden Säuren die Linolensäure zeitweise rascher abgebaut wird als die Linolsäure. Leider sind R. T. HOLMANS Zahlen nur als Prozente der Trockensubstanz angegeben, sagen also nur sehr wenig über allfällige tatsächliche Änderungen der Fettwerte aus.

Bei keimenden *Lupinen* hat P. NEUMANN (1941) das Verhalten der Jodzahl, der Säurezahl und der Verseifungszahl festgestellt. Die Jodzahl stieg bis zum 4. Keimungstag etwas (2%) an, die Säurezahl nahm innerhalb der ersten 24 Keimungsstunden auf etwa das 5fache zu, stieg dann bis zum 4. Keimungstag aber nur mehr um 50% des 24-Std-Wertes an. Die Verseifungszahl zeigte nur unbedeutende und unregelmäßige Schwankungen, schien aber nach Quellung der Samen in n/10 KCl zwischen 2. und 4. Keimungstag deutlich abzusinken (8%).

In *Mungbohnen (Phaseolus aureus)*, die nur etwa 1% Fett enthalten, hat H.-CH. KAO (1936) festgestellt, daß in den ersten 5 Keimungstagen eine beträchtliche Zunahme (60%) der Jodzahl der flüssigen Fettsäuren der Kotyledonen eintritt, die dann bis zum 15. Keimungstag annähernd gleich bleibt. Die flüssigen Fettsäuren der anderen Teile des Mungbohnenkeimlings zeigen eine etwas geringere 5-Tagezunahme der Jodzahl (etwa 33%), die dann bis zum 15. Keimungstag wieder etwas zurückgeht (bis auf etwa 115% des Ausgangswertes). Die festen Fettsäuren aller Keimlingsteile zeigen bis zum 5. Keimungstag eine Jodzahlzunahme von etwa 40—50% ohne später weitere Veränderungen zu zeigen.

Zusammenfassend kann vielleicht festgestellt werden, daß keine Untersuchungsergebnisse in Widerspruch mit der Hypothese stehen, daß in allen bisher sorgfältig untersuchten Pflanzen die Mobilisierung der Reservefette chemisch in einer Weise vor sich geht, die keine wesentlichen Änderungen an dem noch nicht umgesetzten Teil der Fette bedingt. In späteren Stadien der Keimung (Hunger- und Erschöpfungszustände bei Dunkelkeimung, Ende des Keimungsstoffwechsels bei Keimung in Licht) treten dann natürlich Änderungen des Stoffwechsels ein (die im Falle der Dunkelkeimung wohl richtiger als Störungen des Stoffwechsels zu bezeichnen wären), die den schon sehr verlangsamten Fettabbau in andere Bahnen lenken bzw. die ihn — im Falle der Lichtkeimung — in den normalen Assimilations- und Wachstumsstoffwechsel überleiten. Und wenn nicht alles trügt, hat die moderne Ferment- und Stoffwechselphysiologie und -biochemie nunmehr Methoden und Hilfsmittel zur Verfügung, die es ermöglichen werden, eines der schwierigsten Probleme der chemischen Pflanzenphysiologie zu lösen: die Aufklärung des Weges der pflanzlichen Fettumsetzungen.

Literatur.

Bei manchen schwerer zugänglichen Veröffentlichungen werden auch Referate der betreffenden Arbeit zitiert. Zitate, die nicht im Original eingesehen wurden, sind mit * gekennzeichnet.

AFANASIEV, M., and M. CRESS: Changes within the seeds of *Juniperus scopulorum* during the processes of afterripening and germination. J. Forestry **40**, 798—801 (1942). — ALBAUM, H. G., and B. EICHEL: The relationship between growth and metabolism in the oat seedling. Amer. J. Bot. **30**, 18—22 (1943). — ALTSCHUL, A. M.: Biological processes of the cotton seed. Fats and Oils. Cottonseed and Cottonseedproducts, herausgeg. von A. E. BAILEY, S. 157—212. New York 1948.

BARTLETT, G. R.: Notes on the biochemistry of the tomato seedling. *Diss. Univ. Chicago, 1945. Ref.: Biol. Abstr. **21**, 4443 (1947). — BERT, P.: Leçons sur la physiologie comparée de la respiration. Paris: J.-B. Bailliere et Fils 1870. 588 S. — BIALOSUKINA, W.: Produkte der intramolekularen Atmung bei sistiertem Leben der Fettsamen. Jb. wiss. Bot. **45**, 644—660 (1908). — BONNIER, G., et L. MANGIN: Recherches sur la respiration des tissus. Ann. des Sci. natur., Ser. VI **18**, 293—382 (1884). — BOUDON, S.: Étude de la lipolyse au cours de la germination des graines de Pin maritime. Rev. gén. Bot. **60**, 284—313 (1953). — BROOKER, S. G.: Chemical changes during the germination of seeds of white clover

(Trifolium repens). *Thesis Univ. of New Zealand, Wellington 1932, 29 p. Ref.: Biol. Abstr. **10**, 361 (1936). — Brown, R.: The gaseous exchange of seeds and isolated cotyledons of *Cucurbita Pepo.* Ann. of Bot., N. S. **6**, 293—321 (1942). — Bürkle, B.: Physiologische Untersuchungen über Umwandlungen des Öles in reifenden Sonnenblumensamen. Bot. Archiv **26**, 385—436 (1929). — Burr, G. O., and E. S. Miller: Synthesis of fat by green plants. Bot. Gaz. **99**, 773—785 (1938).

Coppens, N.: Biosynthesis of fatty acids by seeds of *Ricinus communis.* Nature (Lond.) **177**, 279 (1956). — Craig, F. N.: A fat oxidation system in *Lupinus albus.* J. of Biol. Chem. **114**, 727—746 (1936). — Crombie, W. M.: Fat metabolism in the West-African oil palm *(Elaeis Guineensis).* I. Fatty acid formation in the maturing kernel. J. of Exper. Bot. **7** (20), 181—193 (1956). — Crombie, W. M., and R. Comber: Fat metabolism in germinating *Citrullus vulgaris.* J. of Exper. Bot. **7** (20), 166—180 (1956). — Czapek, F.: Biochemie der Pflanzen, 2. Aufl., Bd. 1. Jena 1913.

Daggs, R. G., and H. S. Halcro-Wardlaw: The conversion of fat to carbohydrate in the germinating castor bean. II. The combustion respiratory quotient as determined by a modified oxycalorimeter. J. Gen. Physiol. **17**, 303—309 (1933). — Daggs, R. G., and H. C. H. Wardlaw: A study of the conversion of fat to carbohydrate in the germinated castor bean *(Ricinus)* by means of a modified oxycalorimeter. Amer. J. Physiol. **101**, 27 (1932). Ref.: Ber. Physiol. **69**, 294. — Deleano, N.-F.: Recherches chimiques sur la germination. Zbl. Bakter. II **24**, 130—146 (1909). — *Arch. Sci. biol. St. Petersburg **25**, 1—24 (1910). — Desveaux, R., et M. Kogane-Charles: Étude sur la germination des quelques graines oléagineuses. Ann. Inst. Nat. Rech. agronom. **3**, 385—416 (1952). — Detmer, W.: Physiologisch-chemische Untersuchungen über die Keimung ölhaltiger Samen. Habil-.Schr., Fromann, Jena 1875. — Vergleichende Physiologie des Keimungsprozesses der Samen. Jena: Gustav Fischer 1880. — Dmochowski, J.: Recherches sur la quantité d'huile dans les graines du lin et du chanvre pendant les différent périodes de la maturation et de la germination ainsi que dans le bouleau et dans le tilleul pendant les différentes phases et l'hiver. Rocznik Nauk Rolniczych i Lesnych (Poln. Jb. Land- u. Forstwirtsch.) **32**, 35—77 (1934), Polnisch mit französischer Zus.fass. S. 76—77. — Dupéron, R.: Influence de la printanisation sur l'évolution des lipides et des glucides chez *Sinapis alba.* C. r. Acad. Sci. Paris **228**, 192—194 (1949). — Les glucides des graines oleagineuses. Leur rôle au cours de la germination. Rev. gén. Bot. **61**, 261—284 (1954).

Enser, K.: Untersuchungen über die Umwandlung höherer Fettsäuren in Kohlehydrate bei der Keimung von Kürbissamen. Jb. wiss. Bot. **82**, 158—169 (1935). — Ermakoff, A. J., u. N. N. Iwanoff: Über die Atmung der Samen von Ölpflanzen. Biochem. Z. **231**, 79—91 (1931). — Ermakov, A.: Die ersten Keimungsstadien der Ölpflanzensamen. Ber. wiss. Biol. **20**, 205. — *Trudy prikl. Bot. i pr. **25** (1), 135—162 (1931).

Fleury, G.: Recherches chimiques sur la germination. Ann. de Chim., Ser. IV **4**, 38—65 (1865). — Franke, W., u. H. Frehse: Über Oxydationsfermente aus höheren Pflanzen. II. Zur Kenntnis von Lipoxydase und „Lipodehydrase" und ihrer Beziehungen zueinander. Z. physiol. Chem. **298**, 1—26 (1954). — Frankfurt, S.: Über die Zusammensetzung der etiolierten Keimpflanzen von *Cannabis sativa* und *Helianthus annua.* Landw. Versuchsstat. **43**, 143—182 (1894). — Fuchs, J.: Dissertation der Philosophischen Fakultät der Universität Wien, 1933 (unveröffentlicht). — Fürth, O. v.: Über das Verhalten des Fettes bei der Keimung ölhaltiger Samen. Hofmeisters Beitr. **4**, 430—437 (1904).

Gardner, W. A.: Effect of light on germination of light sensitive seeds. Bot. Gaz. **71**, 249—288 (1921). — Gerber, C.: Étude comparée de la respiration des graines oléagineuses pendant leur développement et pendant leur germination. Actes du 1. Congr. Internat. de Botanique, Paris 1900, p. 59—101. — Recherches sur la respiration des olives et sur les relations existant entres les valeurs du quotient respiratoire observé et la formation de l'huile. J. de Bot. **15**, 9—22, 88—94, 121—136 (1901). — Girton, R. E., and E. R. Park: Respiration studies on germinating white oak acorns. *Proc. Indian Acad. Sci. **51**, 83—86 (1942). Ref.: Biol. Abstr. **17**, 2950 (1943). — Godlewski, E.: Beiträge zur Kenntnis der Pflanzenatmung. Jb. wiss. Bot. **13**, 491—543 (1882). — Grafe, V., u. V. Vouk: Untersuchungen über den Inulinstoffwechsel bei *Cichorium Intybus* L. I. Keimungsstoffwechsel. Biochem. Z. **43**, 424—433 (1912). — Green, J. R.: On the germination of the seed of the castor oil plant *(Ricinus communis).* Proc. Roy. Soc. Lond. **47**, 146—147; **48**, 370—392 (1890). — On the changes in the endosperm of *Ricinus communis* during germination. Ann. of Bot. **4**, 383—385 (1890). — Green, R. J., and J. Jackson: Further observation on the germination of the seeds of the castor oil plant *(Ricinus communis).* Proc. Roy. Soc. Lond., Ser. B **77**, 69—85 (1906). — Gressner, H.: Zur Keimungsgeschichte von *Cyclamen.* Bot. Ztg. **1874**, 801—814, 816—825, 830—840. — Griffiths, A. E.: Observations on the germination of lettuce seed. Contrib. Boyce Thompson Inst. **9**, 329—337 (1938). — Guerrant,

N. B.: Some relations of the phospholipids in seed to other constituents. J. Agricult. Res. **35**, 1001—1019 (1927).

Hafenrichter, A. L.: The respiration of the soy bean. Bot. Gaz. **85**, 271—298 (1928). — Hée, A., et L. Bayle: Recherches chimiques sur le germination. I. Evolution des substances grasses et du phosphore lipoidique chez le *Lupinus albus* au cours des premiers stades du développement. Bull. Soc. Chim. biol. Paris **14**, 758—782 (1932). — Halvorsen, H.: The gas exchange of flax seeds in relation to temperature. I. Experiments with immature seeds and capsules. Physiol. Plantarum (Copenh.) **8**, 501—511 (1955). — Hellriegel, H.: Beiträge zur Keimungsgeschichte der ölgebenden Samen. J. prakt. Chem. **64**, 94—107 (1855). — Heumann, W.: Über den Fettstoffwechsel keimender Kürbissamen. Planta (Berl.) **34**, 1—16 (1944). — Hoffpauir, C. L., D. H. Petty and J. D. Guthrie: Germination and free fatty acid in individual cotton seeds. Science (Lancaster, Pa.) **106**, 344—345 (1947). Ref.: Biol. Abstr. **22**, 10005 (1948). — J. Amer. Oil Chem. Soc. **25**, 127—128 (1948). Ref.: Biol. Abstr. **23**, 614 (1949). — Hoffpauir, C. L., S. E. Poe, L. U. Wiles and M. Hicks: Germination and free fatty acids in seed stock lots of cottonseed. J. Amer. Oil. Chem. Soc. **27**, 347—348 (1950). Ref.: Biol. Abstr. **26**, 26042 (1952). — Holman, R. T.: Lipoxidase activity and fat composition of germinating soy beans. Arch. of Biochem. **17**, 459—466 (1948). — Houget, J.: Sur la formation des glucides au cours de la germination du Ricin. C. r. Acad. Sci. Paris **215**, 387—388 (1942). — Sur la mécanisme de la transformation des lipides au glucides au cours de la germination de Ricin. C. r. Acad. Sci. Paris **216**, 821—822 (1943). — Humphreys, T. E., and P. K. Stumpf: Fat metabolism in higher plants. IV. Preparation of soluble fatty acid oxidases from peanut microsomes. J. of Biol. Chem. **213**, 941—949 (1955).

Ivanow, S.: Über die Umwandlung des Öles in der Pflanze. Jb. wiss. Bot. **50**, 375—386 (1912).

James, W. O., and A. L. James: The respiration of barley germinating in the dark. New Phytologist **39**, 145—176 (1940). Ref.: Biol. Abstr. **15**, 2938 (1941). — Jegorow, U.: Über Stoffmetamorphose bei der Samenkeimung von *Cucurbita maxima*. *Ann. Inst. agron. Moskau **10** (1904). Ref.: Bot. Zbl. **101**, 597—599. — Johnston, F. A., and H. M. Sell: Changes in the chemical composition of Tung *(Aleurites fordii* Hemsl.) kernel during germination. Plant Physiol. **19**, 694—698 (1944). — Jordan, R. C., and A. C. Chibnall: Observations on the fat metabolism of leaves. II. Fats and phosphatides in the runner bean *(Phaseolus multiflorus)*. Ann. of Bot. **47**, 163—186 (1933).

Kao, H.-Ch.: Lipoid metabolism in the mung bean during germination. Biochemic. J. **30**, 202—207 (1936). — Knabe, J.: Über die gegenseitigen Beziehungen des Gas- und Fettstoffwechsels etiolierter Senfkeimlinge in Wasser, Kaliumphthalat- und Kaliumlaktatlösungen. Planta (Berl.) **33**, 388—423 (1943). — Krasinska, Z.: Contribution à l'étude du metabolisme énergétique de la germination *(Helianthus annuus)*. Acta Biol. exper. (Warschau) **3**, 101—141 (1929), Polnisch mit franz. Zus.fass. S. 101—103. Ref.: Chem. Abstr. **24**, 1882 (1930). — Kummer, H.: Fett und Fettsäuregehalt bei Gramineensamen in Beziehung zur Lichtbedürftigkeit bei der Keimung. Ber. dtsch. bot. Ges. **50**, 300—303 (1932).

Laskovsky, N.: Über einige chemische Vorgänge bei der Keimung der Kürbissamen. Landwirtsch. Versuchsstat. **17**, 219—244 (1874). — Leach, W., and K. W. Dent: Recherches on plant respiration. III. The relationship between the respiration in air and in nitrogen of certain seeds during germination. (a) Seeds in which fat constitutes the chief food reserve. Proc. Roy. Soc. Lond., Ser. B **116**, 150—169 (1934). — Leclerc du Sablon: Sur la germination du Ricin. C. r. Acad. Sci. Paris **117**, 524—527 (1893). — Sur la germination des graines oléagineuses. C. r. Acad. Sci. Paris **119**, 610—612 (1894). — Recherches sur la germination des graines oléagineuses. Rev. gén. Bot. **7**, 145—165, 205—215, 258—269 (1895). — Sur la germination des amandes. Rev. gén. Bot. **9**, 5—16 (1897a). — Sur les réserves oléagineuses de la noix. Rev. gén. Bot. **9**, 313—317 (1897b). — Lee, W.-Y.: Metabolic studies in germinating soy beans. I. General metabolic changes in the germinating soybean. *J. Chin. Chem. Soc. **6**, 15—22 (1938) (erschienen 1940). Ref.: Chem. Abstr. **35**, 5151 (1941). — Lehninger, A. L.: From fats to energy. J. Agricult. a. Food Chemistry **1**, 1195—1198 (1953). — Lemarchands, J.: Recherches sur les transformations et plus spécialment sur la saponification des réserves grasses dans la graines au cours de la germination. C. r. Acad. Sci. Paris **189**, 375—377 (1929). — Lundegårdh, H.: Einige Bedingungen der Bildung und Auflösung der Stärke. Jb. wiss. Bot. **53**, 421—462 (1914).

MacLachlan, P. L.: Fat metabolism in plants with special reference to sterols. J. of Biol. Chem. **113**, 197—204 (1936a). — Fat metabolism in plants with special reference to sterols. II. Differential changes in the cotyledons and with roots, stems and leaves. J. of Biol. Chem. **114**, 185—191 (1936b). — Malhotra, R. C.: Biochemical studies of seeds during germination. Beih. bot. Zbl., Abt. I **50**, 1—7, 8—14, 15—19 (1932); **51**, 524—530, 531—540 (1935). — MAQUENNE, L.: Sur les changements de composition qu'éprouvent les graines oléagineuses au cours de la germination. C. r. Acad. Sci. Paris **127**, 625—628 (1898). — MATTHES, E.: Physiologische Untersuchungen über Umwandlungen des Öles in keimenden Sonnenblumensamen. Bot. Archiv **19**, 79—135 (1927). — MAZÉ, P.: Recherches sur la digestion des reserves dans les graines en voie des germination et leur assimilation par les plantules. C. r. Acad. Sci. Paris **130**, 424—427 (1900). — Sur la transformation des matières grasses en sucres dans les graines oléagineuses en voie de germination. C. r. Acad. Sci. Paris **134**, 309—311 (1902). — MERLIS, M.: Über die Zusammensetzung der Samen und der etiolierten Keimpflanzen von *Lupinus angustifolius* L. Landwirtsch. Versuchsstat. **48**, 419—454 (1897). — MESNARD, E.: Sur les transformations que subissent les substances de réserve pendant le germination des graines. Bull. Soc. bot. France **40**, 35—42 (1893). — Recherches sur la formation des huiles grasses (et des huiles essentielles) dans les vegetaux. Ann. des Sci. natur., Ser. VII **18**, 257—317 (1894). — MILLER, E. C.: A physiological study of the germination of *Helianthus annuus*. I. Ann. of Bot. **24**, 693—726 (1910). — A physiological study of the germination of *Helianthus annuus*. II. The oily reserve. Ann. of Bot. **26**, 889—901 (1912). — MÜNTZ, A.: Sur la germination des graines oléagineuses. Ann. de Chim., Ser. IV **22**, 472—485 (1871). — MURLIN, J. R., W. R. MURLIN and J. D. WATKEYS: The conversion of fat to carbohydrate in the germinating castor bean. I. The respiratory metabolism. J. Gen. Physiol. **17**, 283—302 (1933).

NEUMANN, P.: Über den Fettstoffwechsel reifender und keimender Leguminosensamen. Biochem. Z. **308**, 141—174 (1941). — NOBBE, F.: Handbuch der Samenkunde. Berlin 1876.

OHLEN, F. W. v.: A microchemical study of soybeans during germination. Amer. J. Bot. **18**, 30—49 (1931). — OLCOTT, H. S., and T. D. FONTAINE: Composition of cottonseeds. IV. Lipase of germinated seeds. J. Amer. Chem. Soc. **63**, 825—827 (1941).

PETERS, E.: Zur Keimungsgeschichte des Kürbissamens. Landwirtsch. Versuchsstat. **3**, 1—18 (1861). — PFEFFER, W.: Über Elektion organischer Nährstoffe. Jb. wiss. Bot. **28**, 205—268 (1895, S. 257). — PIERCE, H. B., D. E. SHELDON and J. R. MURLIN: The conversion of fat to carbohydrate in the germinating castor bean. III. The chemical analysis and correlation with respiratory exchange. J. Gen. Physiol. **17**, 311—325 (1933). — PIRSCHLE, K.: Acetaldehyd als Zwischenprodukt bei der Keimung fetthaltiger Samen. Biochem. Z. **169**, 482—489 (1926). — POLJAKOFF-MAYBER, A., and A. M. MAYER: Some quantitative changes in the fat metabolism of germinating lettuce seed and their inhibition by coumarin. J. of Exper. Bot. **6** (17), 287—292 (1955).

RALEIGH, G. J.: Chemical condition in maturation, dormancy and germination of seeds of *Gymnocladus dioica*. Bot. Gaz. **89**, 273—294 (1930). — RALSKI, E.: Die Fettkörper der Gramineensamen. *Kosmos (Leopol) **49**, 62—99 (1924) [Polnisch]. Ref.: Bot. Abstr. **14**, 5834 (1935). — REUHL, E.: Oxygen-intake of oily and starchy seeds. Proc. Acad. Amsterdam **38**, 879—886 (1935). — Notes on the metabolic changes in the germination of seeds. Rec. Trav. bot. néerl. **33**, 1—76 (1936). — RHINE, J. B.: Translocation of fats as such in germinating fatty seeds . Bot. Gaz. **82**, 154—169 (1926). — RUSCA, R. A., and F. L. GERDES: Effects of artificially drying seed cotton on certain quality elements of cotton seed in storage. U.S. Dpt. Agricult. Circular 651, 19 S., 1942. Ref.: Exper. Stat. Rec. **88**, 43 (1943).

SACHS, J.: Über das Auftreten von Stärke bei der Keimung ölhaltiger Samen. Bot. Ztg. **1859**, 177—183, 185—188. — Über die Keimung des Samens von *Allium cepa*. Bot. Ztg. **1863**, 57—62, 65—70. — SANI, G.: Intorno alla germinazione dell'olivo. Atti R. Accad. Lincei, Roma, V. ser. **9** (1), 47—51 (1900). — Richerche intorno alla germinazione del faggio. Atti R. Accad. Lincei, Roma, V. ser. **13/2**, 382—385 (1904). — SCHMIDT, R. H.: Über Aufnahme und Verarbeitung von fetten Ölen durch Pflanzen. Flora (Jena) **74** (49), 300—370 (1891). — SHÛIKU, S.: Studies on some constituents of soybean seeds and their transformation during germination. J. Dept. Agric. Kyushu Imp. Univ. **5**, 51—116 (1936). Ref.: Chem. Abstr. **31**, 4697 (1937). — SIMPSON, D. M.: Factors affecting the longevity of cotton seed. J. Agricult. Res. **64**, 407—419 (1942). — SKRAUP, S., u. U. WEBER: Untersuchungen zum Fettstoffwechsel der Pflanzen. Ber. physik.-med. Ges. Würzburg **60**, 4—8 (1936). — STILES, W., and W. LEACH: Researches on plant respiration. II. Variation in the respiration quotient during germination of seeds with different food reserves. Proc. Roy. Soc. Lond., Ser. B **113**, 405—428 (1933). — STUMPF, P. K., and G. A. BARBER: Fat metabolism in higher plants. VI. β-Oxidation of fatty acids by peanut mitochondria. Plant Physiol. **31** (4), 304—308 (1956).

TÄUFEL, K., u. M. RUSCH: Über den Einfluß des Mälzungsprozesses auf das Fett der Gerste. Biochem. Z. **209**, 55—61 (1929a). — Zur Kenntnis des Fettes der Gerste und ihrer Mälzungsprodukte. Z. Unters. Lebensmitt. **57**, 422—431 (1929b). — TAO, W.-S.: Biochemical studies in rice starch. Bull. Chem. Soc. Japan **5** (2), 64—73 (1930). Ref.: Biol.

Abstr. **7**, 7798 (1933). — Terroine, E. F.: État actuel de nos connaissances sur la formation des graisses au cours de la maturation des graines et fruits oléagineuses et sur l'utilisation des graisses au cours de la germination. Ann. des Sci. natur. Bot., X. s. **2**, I—LXIII (1920). — Terroine, E. F., R. Bonnet et P. H. Joessel: L'energie de croissance. II. La germination. Bull. Soc. Chim. biol. Paris **6**, 357—393 (1924). — Thomas, J. B.: On the metabolism of Damar seeds. *Ann. Bot. Gard. Buitenzorg **51** (1), 94—114 (1941). Ref.: Biol. Abstr. **16**, 13402 (1942). — Tietz, R.: Über mögliche Beziehungen des Fettstoffwechsels zum Lichtkeimproblem. Biochem. Z. **324**, 517—529 (1953). — Tilenius, G.: Untersuchungen an keimenden Sonnenblumen, über die Veränderung des Öles und seine Beeinflussung durch Azetate. Planta (Berl.) **28**, 429—452 (1938). — Toole, E. H.: The transformation and course of development of germinating maize. Amer. J. Bot. **11**, 325—350 (1924).

Wallenstein, M.: Veränderungen des Fettes während der Keimung und deren Bedeutung für die chemisch-physiologischen Vorgänge der Keimung. *Forschungsber. Lebensmitt. **3**, 372 (1896). Ref.: Beih. bot. Zbl. **6**, 493—494 (1896). — Weber, U.: Untersuchungen über den Fettstoffwechsel keimender Kürbisse. Ber. dtsch. bot. Ges. **54**, (70)—(75) (1936). — Weissenböck, K.: Fett- und Phosphatidstoffwechsel keimender und reifender Ölsamen. I. Lipoidstoffwechsel bei der Keimung von Raps und Soja. Österr. bot. Z. **94**, 301—313 (1948). — White, H. L.: The modification of the composition of vegetable oils, with special reference to increasing unsaturation. J. Industr. a. Engng. Chem. **11**, 648—651 (1919). — Wienke, L.: Über die Kohlendioxydabgabe etiolierter Keimlinge von *Cucurbita* und *Helianthus* in Wasser und Salzlösungen. Planta (Berl.) **28**, 205—226 (1938).

Zambotti, V.: Metabolismo lipidico durante la germinazione dei semi oleosi. Nota I. Variazione della constante chimiche. Annali Bot. **22**, 81—112 (1940). — Zeller, A.: Untersuchungen über die Umwandlung höherer Fettsäuren in Kohlehydrate bei der Keimung von Kürbissamen. I. Jb. wiss. Bot. **82**, 123—157 (1935). — Zeller, A., u. F. Maschek: Über eine Saturase höherer Fettsäuren aus Kürbiskeimlingen. Biochem. Z. **312**, 354—369 (1942). — Zimmermann, W., u. A. Baumann: Fett- und Phosphatidabbau in der keimenden Sojabohne. Fettchem. Umschau **42**, 65 (1935).

Die wirtschaftliche Bedeutung der Pflanzenfette und Fettpflanzen.

Von

K. Schmalfuß.

Von den wirtschaftlich genutzten Fetten sind etwa 40% tierischer und 60% pflanzlicher Herkunft. Diese Tatsache kennzeichnet eindrucksvoll die Bedeutung der pflanzlichen Fette für die menschliche Wirtschaft.

Fette oder fette Öle kommen weit verbreitet in höheren und niederen Pflanzen vor. Jede pflanzliche Zelle oder jedes Gewebe der Pflanze kann Fette enthalten. Allerdings sind die Gehalte der Pflanzensubstanz an Fetten sehr unterschiedlich und fallen wegen ihrer Geringfügigkeit oft nicht weiter auf. Wir kennen Fette aus Wurzeln, aus Blättern, aus Sproßorganen, besonders häufig aber aus Samen und Früchten. In diesen kommen Fette und Öle gewöhnlich angereichert in Speichergeweben vor. Sie haben hier als Reservestoffe für die Entwicklung des Keimlings zu gelten. Ähnliches trifft auch für vegetative Speicherorgane der Pflanzen zu. Schätzungsweise enthalten etwa 80% der höheren Pflanzen Samenfette in größeren Mengen.

Von den Pflanzen mit fetthaltigen Organen kann aber nur ein kleiner Teil zur Gewinnung der Fette herangezogen werden. Das liegt einfach daran, daß eine Verarbeitung von Pflanzensubstanzen mit geringeren Mengen Fett mit den bis heute zur Verfügung stehenden Mitteln vielfach unwirtschaftlich ist. Erst wenn die Samen oder Früchte oder die fetthaltigen Pflanzenorgane allgemein eine gewisse Mindestmenge an Fett führen, wird die betreffende Pflanze zur Öl- oder Fettpflanze im wirtschaftlichen Sprachgebrauch.

Die Fettgehalte der in dieser Hinsicht genutzten Pflanzen schwanken in einem weiten Bereich. So enthält die Sojabohne mit etwa 17% Öl nur verhältnismäßig wenig Fett, das Fruchtfleisch der Kokosnuß hingegen hat einen Fettgehalt von meist über 60%. Zwischen diesen beiden Extremen liegen ungefähr die mittleren Fettgehalte der anderen genutzten Fettpflanzen der Weltwirtschaft.

Aber nicht nur die Fett*gehalte* von Samen oder Früchten bestimmen den Nutzungswert gewisser Pflanzen zur Fettgewinnung. Eine weitere Voraussetzung ist die, daß die fetthaltigen Pflanzenorgane in größeren einheitlichen Mengen, sei es durch Anbau, sei es durch sonst ein Gewinnungsverfahren, für eine großtechnische Verarbeitung geeignet und verfügbar sind. Da eine technisch hochentwickelte Ölindustrie das ganze Jahr über arbeitet, müssen sich die ölhaltigen Samen auch längere Zeit lagern und auf weitere Entfernungen transportieren lassen, damit sie kontinuierlich der Verwertung zu Gebote stehen.

Ferner kommt bei unseren landwirtschaftlichen Kulturpflanzen, die Fette liefern, noch hinzu, daß der Fettertrag je Hektar mit diesen Pflanzen bestandener Fläche auch eine Mindesthöhe erreichen muß, die wirtschaftlich vertretbar ist. So liefert im gemäßigten Klima Mitteleuropas der Winterraps mit einer Samenernte von etwa 1500—2500 kg je Hektar den höchsten Ölertrag von allen hier anbaufähigen Ölpflanzen. Der Sommerraps hingegen erbringt bei ungefähr gleichem prozentischen Fettgehalt der Samen durchschnittliche Ertragsleistungen von etwa 1000—1500 kg Samenernte je Hektar. Aus diesen Gründen beschränkt sich der Anbau nur auf verhältnismäßig wenige ertragssichere Ölpflanzen, die ihre Fette fast ausnahmslos in Samen oder Früchten speichern, die wiederum durch technische Verfahren leicht zu gewinnen sind.

Aus dem gleichen Grunde haben andere auch Öl führende Pflanzen bis heute keine dauernde Bedeutung als Ölpflanzen gewinnen können *(Cucurbita, Lallemantia, Xanthium* u. a.*)*. Das gilt insbesondere auch für niedere Pflanzen, wie Pilze und Algen.

A. Die Gewinnung der Pflanzenfette.

Für die Gewinnung der Fette aus dem pflanzlichen Rohmaterial der Ölsamen und Ölfrüchte gibt es nur zwei großtechnisch brauchbare Verfahren. Bei dem einen werden die Fette aus ihren natürlichen Trägern durch *Pressung* gewonnen, das andere Verfahren besteht in der *Extraktion* des Fettes aus dem fetthaltigen Rohstoff mit flüchtigen Fettlösungsmitteln. Beide Verfahren können auch kombiniert angewendet werden.

Die Gewinnung der Pflanzenfette durch Auspressen des Rohmaterials wurde früher fast ausschließlich angewandt und hat heute noch große Bedeutung. Die gereinigten ölhaltigen Samen oder Früchte oder sonstiges fetthaltiges Material werden auf einen bestimmten Feuchtigkeitsgrad eingestellt, vorgewärmt und durch Walzenmühlen zerkleinert. Hierauf wird das Öl mit großen hydraulischen Pressen bei Drucken von etwa 150—300 atü abgepreßt. Das Vorwärmen verfolgt den Zweck, das Öl dünnflüssiger zu machen und die Strömungswiderstände in dem Material zu verringern.

Die Pressung kann in der Kälte oder bei höheren Wärmegraden erfolgen. Die kalt gepreßten Öle sind stets reiner als die warm gepreßten, die in größerer Menge anfallen und meistens noch verschiedene andere Stoffe neben den Fetten enthalten. Feine Speiseöle, wie Olivenöl, werden kalt gepreßt, technische Öle hingegen zur Erhöhung der Ausbeute bei höheren Temperaturen. Die Pressung kann einmal angewendet, sie kann auch wiederholt werden. Die reinsten Öle ergibt die erste Pressung, zumal bei niederen Temperaturen. Die Ölgewinnung durch Pressung hat den Nachteil, daß etwa 5—10% des Öles vom Rückstand festgehalten und durch mechanische Kräfte nicht herausgebracht werden können.

Will man das Fett möglichst vollständig gewinnen, so schließt man an die Pressung noch eine Extraktion an, die sich bestimmter flüchtiger Extraktionsmittel (Benzin, Trichloräthylen, Tetrachlorkohlenstoff u. a.) bedient. Die Extraktion kann auch von vornherein ohne Preßverfahren allein angewandt werden.

Die Extraktion des Rohmaterials mit Fettlösungsmitteln gewinnt in dem Maße an Bedeutung, wie die Verarbeitung ölarmer Saaten oder noch ölhaltiger Preßkuchen wirtschaftlich oder aus irgendeinem anderen Grunde notwendig wird. Die Extraktion hat sich erst in der Zeit nach dem ersten Weltkriege als ein der Pressung gleichwertiges Entölungsverfahren durchsetzen können.

Die Rückstände der Ölgewinnung sind im Falle des Preßverfahrens die sog. Ölkuchen, während die Extraktionsrückstände in Form eiweißreicher Mehle oder Schrote anfallen. In beiden Fällen können diese Rückstände, falls sie keine schädlichen Stoffe enthalten, als Futtermittel oder auf jeden Fall als Düngemittel angewendet werden. Die Preßkuchen enthalten noch gewisse Mengen (4—6%) Fett, während der Ölgehalt der Extraktionsmehle oder -schrote 1% meist nicht überschreitet.

Im Falle der Anwendung von Extraktionsverfahren ist es meist notwendig, sowohl die Rückstände wie auch die gewonnenen Öle oder Fette von Lösungsmittelresten sorgfältig zu reinigen, wenn die einen als Futtermittel, die anderen zu Speisezwecken verwendet werden sollen. Auch sonst enthalten die gewonnenen Öle und Fette oft noch andere, vielfach auch giftige Stoffe, die gleichfalls vor Gebrauch entfernt werden müssen. In diesen Fällen werden die Fette einer Raffination unterzogen und gereinigt. Dabei wird das Öl mit Natronlauge zur Entfernung unangenehm riechender freier Fettsäuren behandelt und die entstehende Seife mit Wasser ausgewaschen. Andere Riechstoffe werden durch überhitzten Wasserdampf im Vakuum entfernt. Ferner werden die Öle zur Entfärbung und zum Bleichen noch mit Kalkmilch, Natronlauge, Superoxyden, Bleicherden und anderen Stoffen behandelt. Auf diese Weise können unerwünschte Bestandteile aus den Ölen und Fetten so entfernt werden, daß diese völlig geruch- und geschmacklos erscheinen.

B. Die wirtschaftliche Bedeutung der Pflanzenfette und ihre Nutzung.

Die Pflanzenfette werden überwiegend als Speisefette genutzt, zum geringeren Teil in mannigfaltiger Weise für technische Zwecke verwendet. Für die Art ihrer Nutzung sind vielfach ihre spezifischen Eigenschaften maßgebend. Dazu ist einmal zu erwähnen, daß manche Pflanzenfette bei Raumtemperatur feste Konsistenz besitzen, während andere bei gewöhnlicher Temperatur flüssig sind und deswegen meist als Öle bezeichnet werden.

Da die in den Glyceriden der Pflanzenfette hauptsächlich vorkommenden Fettsäuren in der Regel eine größere Anzahl von Kohlenstoffatomen besitzen, wird die Höhe des Schmelzpunktes der Pflanzenfette mehr vom Sättigungsgrad der vorwiegend vorhandenen Fettsäuren als von deren Molekülgröße bestimmt. Es gibt feste, aber noch mehr flüssige Pflanzenfette; jene stammen in der Regel von Pflanzen warmer, tropischer Gebiete, während die Fette subtropischer Pflanzen oder solcher aus gemäßigten Klimagebieten meist flüssig sind.

Bei den fetten Ölen der Pflanzenwelt können Fettsäuren mit verschiedenem Sättigungsgrad in den Glyceriden vorhanden sein (Palmitinsäure, Stearinsäure, Ölsäure, Linolsäure, Linolensäure), es können aber auch einzelne Fettsäuren weitaus dominieren (Ölsäure im Olivenöl, Erucasäure in Cruciferenölen). Von dem Fettsäurenbestand hängen aber die Eigenschaften der Fette ab und nach diesen Eigenschaften richtet sich weitgehend die vielfache wirtschaftliche Nutzung der Pflanzenfette.

Dabei ist vor allem noch ein Punkt zu berücksichtigen. Die pflanzlichen Fette unterliegen nach ihrer Gewinnung oder noch in ihren Trägern leicht dem Verderben. Dazu gehört vor allem das Ranzigwerden der Fette, das auf einer Veränderung beruht, die unter dem Einfluß von Licht, Luft sowie Mikroorganismen beim Lagern eintreten kann. Dabei entstehen neben freien Fettsäuren namentlich übelriechende Aldehyde (Heptylaldehyd) und Ketone, die den Gebrauchswert der Fette sehr beeinträchtigen können. Deshalb müssen nicht nur die gewonnenen Produkte, sondern vor allem die Ölsaaten in besonderen Speichern oder Silos trocken bei niedriger Temperatur und unter Luftabschluß gelagert werden, wenn eine Lagerung aus betrieblichen Gründen in den Ölfabriken notwendig ist.

Dem Verderben der Fette wirken schon einige in ihnen gelöst enthaltene natürliche Antioxydantien entgegen, so Sesamol (3,4-Methylendioxyphenol), eine phenolische Verbindung aus Sesamöl, und Tokopherole. Als ein künstliches Antioxydans wird tert. Butyl-Hydroxyanisol, für Pflanzenfette die einfache Zugabe von geringen Mengen Schwefel empfohlen.

a) Fette als Nahrungsmittel.

Die Hauptnutzung der gewonnenen pflanzlichen Fette liegt zweifellos in ihrer Verwendung als Nahrungsmittel. Diesem Zweck können sie unmittelbar ohne Vorbehandlung oder nach einer gewissen Reinigung und Raffination oder besonderen Aufbereitung dienen. Das gilt besonders für das Kokosfett, das Olivenöl und noch für einige andere pflanzliche Fette.

Für die Herstellung besonderer Speisefette oder von Margarine werden die festen oder wenigstens halbfesten Fette gewöhnlich bevorzugt. Zu diesem Zweck werden aber auch viele natürliche Pflanzenöle gehärtet und dadurch in feste Fette übergeführt.

Die Härtung besteht in der Absättigung der Doppelbindungen ungesättigter Fettsäuren in Gegenwart geeigneter Katalysatoren (Nickel) durch Wasserstoff. Dieser im Jahre 1902 von W. NORMANN entdeckte Vorgang ist heute technisch durchgearbeitet. Durch die Addition von Wasserstoff an eine oder mehrere Äthylenbindungen werden die in den Glyceriden vorhandenen ungesättigten Fettsäuren (Öl-, Linol-, Linolensäure) in gesättigte (Stearinsäure) übergeführt. Der zur Härtung benötigte Wasserstoff kann aus Wassergas oder auch in reinerer Form elektrolytisch gewonnen werden.

Zur Erzeugung von Kunstspeisefetten haben die natürlichen Hartfette oder die gehärteten Fette von Kokos, Soja, Erdnuß- oder Baumwollsaat die größte Bedeutung. Pflanzenbutter (Palmin) ist nichts anderes als ein hochraffiniertes Kokosfett. Auch für die Gewinnung von Margarine werden vielfach oder hauptsächlich pflanzliche Fette verwendet. Sie wird aus dem Fett mit gesäuerter Milch oder Wasser durch Emulgieren hergestellt. Als pflanzliche Fettbasis dienen vielfach Kokosfett, Palmkernfett oder auch andere oft gehärtete Öle.

Der besonders große Wert der Fette für die menschliche Ernährung beruht auf ihrem hohen Energiegehalt im Vergleich zu anderen Stoffen, wie etwa den Kohlenhydraten. Die Verbrennungswärme der Fette liegt im Durchschnitt bei etwa 9,3 Calorien je Gramm gegenüber 4,2 Calorien bei Kohlenhydraten und Eiweißstoffen. Die hohe Verbrennungswärme der Fette ist auf ihren hohen Kohlenstoffgehalt (74—78%) im Vergleich zu den Kohlenhydraten (etwa 40% C) und ihre Sauerstoffarmut zurückzuführen. Neben diesen bedeutungsvollen Eigenschaften als Energiematerial wäre schließlich noch auf ihren Gehalt an wichtigen Begleitstoffen (Phosphatide, Sterine, Carotinoide, Vitamine) hinzuweisen, wie sie natürliche Pflanzenfette gelöst enthalten.

Eine besondere Bedeutung bei der Verwendung der Pflanzenöle als Nahrungsfette haben noch die in ihnen vorkommenden sog. essentiellen Fettsäuren. Das sind, soweit bekannt ist, Linol-, Arachidon- und Linolensäure. Es hat sich gezeigt, daß der tierische Organismus diese Fettsäuren unbedingt benötigt, aber offenbar nicht selbst aufzubauen vermag und daß sie ihm deshalb mit den pflanzlichen Nahrungsfetten zugeführt werden müssen, wenn nicht schwere Schädigungen auftreten sollen. Es kann angenommen werden, daß auch der menschliche Organismus auf die Zufuhr dieser essentiellen Fettsäuren in pflanzlichen Fetten angewiesen ist.

b) Verwendung der Fette für technische Zwecke.

Neben der Verwendung als Speisefett, das die Hauptnutzung der Pflanzenfette ausmacht, wird ein großer Teil der pflanzlichen Öle aber auch für wichtige technische Zwecke gebraucht.

Allbekannt ist die Verwendung der Fette für die Herstellung von *Seifen*. Darunter versteht man die Alkalisalze höherer Fettsäuren, die mehr als

6 Kohlenstoffatome im Molekül besitzen. Die bekannten Seifeneigenschaften sind insbesondere bei den Alkalisalzen der Fettsäuren mit 12—18 Kohlenstoffatomen am stärksten ausgeprägt. Zur Seifenbereitung sind die meisten Pflanzenöle oder -fette gut brauchbar.

Heute werden durch Oxydation von Paraffinen hergestellte synthetische Fettsäuren auch vielfach mit zur Seifenfabrikation verwendet. Dabei ist aber wichtig, daß von dem technischen Paraffin-Carbonsäuregemisch diejenigen synthetischen Fettsäuren für Seifen Verwendung finden, deren Kohlenstoffzahl etwa zwischen C_{10} und C_{20} liegt.

Dabei ist ferner von Bedeutung, daß vor allem Feinseifen zur Erzielung einer guten Waschwirkung auch noch Ölsäure enthalten müssen. Für die Kernseifenfabrikation eignen sich in der Hauptsache Palmitin-, Stearin- und Ölsäure, während Schmierseifen überwiegend aus den ungesättigten Säuren hergestellt werden.

Größere Mengen gesättigter Fettsäuren (Stearinsäure) werden auch noch zur Herstellung von Kerzen verwendet.

Bei der Verseifung oder sonstigen Fettspaltungen mit dem Ziel der Gewinnung der Fettsäuren fallen auch noch bestimmte Mengen an Glycerin an, das ein wichtiges Nebenprodukt der entsprechenden Industrien bildet. Es wird nach einer Reinigung durch Konzentrierung und Destillation gewonnen und für die verschiedensten Zwecke gebraucht.

Gelegentlich werden pflanzliche Fette auch als Schmiermittel oder als Zusatz zu solchen benutzt; namentlich kommen hierzu Olivenöl, Rüböl, Palmöl und Ricinusöl in Frage, jedenfalls solche Fette, die keine trocknenden Eigenschaften besitzen. So gilt Ricinusöl wegen seiner großen Zähigkeit und wegen seines wenig veränderten Verhaltens bei tiefen und hohen Temperaturen als ein gutes Schmieröl, besonders für Flugzeugmotoren. Auch sog. geblasene Öle werden als Schmiermittel verwendet.

Gelegentlich werden pflanzliche Öle auch auf Faktisse verarbeitet, das sind knetbare elastische Stoffe, die dem Naturkautschuk zugesetzt werden oder als Ersatzmittel für diesen dienen. Weißer Faktis entsteht durch Behandlung des Öles mit Chlorschwefel (S_2Cl_2), braune Faktisse werden durch Erwärmen der Öle mit Schwefel gewonnen.

Eines der wichtigsten technischen Anwendungsgebiete pflanzlicher Öle ist der Gebrauch als Firnis oder, allgemeiner gesagt, für irgendwelche technischen Zwecke, bei denen die Trockeneigenschaften bestimmter pflanzlicher Öle maßgebend sind.

Als Ölfirnisse werden Produkte bezeichnet, die aus einem trocknenden Öl mit gewissen, das Trocknen beschleunigenden Zusätzen (Sikkativen) bestehen. Das Trocknen des Öles beruht in einer Aufnahme von Sauerstoff aus der Luft und auf Polymerisationsvorgängen, die noch nicht ganz durchsichtig sind. Dabei bildet sich, wenn das Öl an der Luft ausgestrichen wird, in kurzer Zeit ein fester, hautartiger Film.

Die Firnisse werden in der Regel aus Leinöl, aber auch aus anderen trocknenden Ölen hergestellt. Sie werden vielfach gekocht und zum Teil durch Einblasen von Luft zubereitet. Die Sikkative bestehen gewöhnlich aus Bleiglätte (Bleioxyd PbO) oder Mangan- oder Kobaltverbindungen.

Auch sog. Standöle werden durch Erhitzen trocknender Öle, hauptsächlich des Leinöls, aber unter Luftabschluß, erhalten. Wie bei den Ölfirnissen liegt der Wert der Standöle vornehmlich in ihrem Gebrauch als Anstrichbindemittel, namentlich für Rostschutz, Farben und für wetterfeste Anstriche. Ferner ist noch ihre Verwendung für Öllacke, zur Bereitung von Druckfarben und für andere Zwecke hervorzuheben.

Zur Herstellung von Lacken werden den Firnissen Harze und Verdünnungsmittel (Terpentinöl) zugefügt. Dadurch wird erreicht, daß nach Verdunstung des Verdünnungsmittels das ausgestrichene Öl in dünner Schicht trocknet und hart wird. Was die Druckfarben anlangt, so werden in größter Menge schwarze Farben (Druckerschwärze) benötigt, die durch Anreiben von Ruß und anderen Stoffen mit Leinölfirnis (Standöl) zubereitet werden. Schließlich werden Firnisse auch zum Wasserdichtmachen von Geweben, zur Herstellung von Linoleum, zu Glaserkitt mit Schlämmkreide als Füllstoff und zu vielen anderen Dingen verwendet, wo es auf die eigentümliche Fähigkeit des Leinöls und weniger anderer Pflanzenöle, zu trocknen, ankommt.

Vielfach wurde auch nach Ersatzmitteln für trocknende Öle oder aus ihnen hergestellte Firnisse gesucht. Namentlich in den letzten Jahren ist in dieser Hinsicht manches geschehen. Es ist aber bisher kein Stoff bekanntgeworden, der einen Leinölfirnis in seiner Beschaffenheit und in seiner Haltbarkeit, namentlich für wetterfeste Anstriche, vollständig zu ersetzen vermag.

C. Die wirtschaftlich wichtigsten Fettpflanzen und ihre Fette.

Aus der riesigen Zahl fetthaltiger und wirtschaftlich in dieser Hinsicht verwertbarer Pflanzen sollen nachstehend nur die wichtigsten und ihre hauptsächlichste Verwendung erwähnt werden. Es ist dabei zweckmäßig, die Pflanzen nach ihrer systematischen Stellung und nach der Art der von ihnen gebildeten Fette einzuteilen. Natürlich sind die Übergänge, wie überall in der Natur, nicht scharf, so daß es oft schwierig ist, ein Fett zu dieser oder zu jener Gruppe zu zählen.

Betreffend nähere Einzelheiten über die Verbreitung und die Beschaffenheit der nachstehend beschriebenen Pflanzenfette sei auf den Beitrag von M. L. Meara "The fats of higher plants" im gleichen Band dieses Handbuches S. 10ff. nachdrücklich hingewiesen.

a) Nutzpflanzen mit vorwiegend festen Fetten.

Die hierher zu rechnenden Pflanzen sind durch Fette charakterisiert, die bei normaler Raumtemperatur gewöhnlich feste Konsistenz haben. Es handelt sich meistens um tropische Pflanzen.

Das wichtigste hier zu nennende Fett ist das Kokosnußfett der **Kokospalme** *(Cocos nucifera* L.*)* aus der Familie der *Palmae*. Die Heimat der Kokospalme sind vermutlich die Inseln der Südsee, sie wird aber heute in den tropischen Küstengebieten der verschiedenen Erdteile kultiviert (Ceylon, Sundainseln, Malaya, Philippinen, Zentralamerika, Küsten Afrikas). Der Baum beginnt etwa im 5. Lebensjahr zu fruchten, liefert die höchsten Erträge zwischen dem 25. und 65. Lebensjahr mit mittleren Ernten von etwa 50 reifen Früchten je Baum und Jahr und stirbt etwa mit 100—120 Lebensjahren ab. Die Frucht ist die bekannte, etwa 30 cm lange Kokosnuß. Diese trägt unter der Oberhaut, dem Exokarp, eine sehr faserige Bastschicht, das Mesokarp, das die Kokosfaser liefert. Die innerste Schicht, das Endokarp, bildet eine 5—10 mm dicke, fast kugelige, steinharte Schale, die den Steinsamen umschließt, dessen Nährgewebe aus einer weißen, ölreichen Masse besteht. Zur Gewinnung des Öles wird nach Aufbrechen der Nuß das feste Nährgewebe in Stücke geschnitten und getrocknet und bildet die sog. Kopra. Diese ist das Rohmaterial für Öl und Ölkuchen. Der Ölgehalt der trockenen Kopra beträgt etwa 55—70%. Das Gewicht des frischen Samenfleisches der Kokosnuß macht mit etwa 300 g ungefähr 25% des Nußgewichtes aus (= etwa 165 g Kopragewicht je Nuß = etwa 17% des Nußgewichtes). Der mittlere Ertrag je Hektar und Jahr an den kontinuierlich reifenden Nüssen kann mit etwa 6000 kg angegeben werden. Der mittlere Ertrag an Kopra je Hektar liegt demnach bei etwa 1000 kg im Jahre.

Die nachfolgende Tabelle 1 gibt einen Überblick über die Welterzeugung der wichtigsten Produktionsländer an Kopra.

Tabelle 1. *Welterzeugung an Kopra* (in 1000 metrischen Tonnen).
[Aus Yearbook, VI/1 (1953), S. 86, Tabelle 40.]

	1933/38	1949	1950	1951
Nord-, Mittel- und Südamerika zusammen	—	*100*	*120*	*80*
Ceylon	215	215	193	249
Indonesien	715	502	448	507
Malaya	188	125	152	163
Philippinen	583	860	994	1056
Asien insgesamt	*1800*	*1980*	*2070*	*2250*
Afrika insgesamt	—	*100*	*120*	*100*
Ozeanien insgesamt	—	*210*	*220*	*240*
Weltproduktion gesamt	—	*2390*	*2530*	*2710*

Aus dieser Tabelle ist die große Bedeutung der asiatischen Küsten und der Südseeinseln als Produktionsstätten für Kopra und die steigende Erzeugung in den letzten Jahren ersichtlich.

Das Fett der Kopra wird meistens durch Pressung gewonnen und nach einer Raffination teils für Speisezwecke (Palmin), teils zur Seifen- und Kerzenfabrikation benutzt. Der Schmelzpunkt liegt bei etwa 23—26° C. Das Fett zeigt daher in den Tropen gewöhnlich flüssige, im gemäßigten Klima feste Konsistenz. Charakteristisch ist sein Gehalt an niedriger molekularen Fettsäuren. Diese bestehen zu etwa je 10% aus Caprylsäure, Caprinsäure, Ölsäure und Palmitinsäure, zu etwa 20% aus Myristinsäure und etwa 40% aus Laurinsäure. Dementsprechend ist die Verseifungszahl mit etwa 230—260 sehr hoch, die Jodzahl infolge nur geringen Anteils ungesättigter Säuren mit etwa 7—9 sehr niedrig. Damit hängt ferner auch die Eigenschaft der Kokosseife zusammen, als einzige Seife mit salzhaltigem Meerwasser Schaum zu geben. Die Preßrückstände bilden als Kokosnußkuchen ein gutes Viehfutter.

Eine weitere wichtige Ölpflanze der Weltwirtschaft ist die aus dem feuchten tropischen Afrika stammende, heute aber auch namentlich auf den Sundainseln kultivierte **Ölpalme** *(Elaeis guineensis* JACQUIN*)*, die gleichfalls zur Familie der Palmen gehört. Aus ihrem Fruchtfleisch wird das Palmöl, aus ihren Samen das Palmkernfett gewonnen. Die Pflanze wird etwa 80—120 Jahre alt, beginnt im 4. Jahr zu fruchten und liefert durchschnittlich etwa 60 Jahre lang Erträge. Ein Fruchtstand, der etwa 800—4000 Früchte von Oliven- oder Pflaumengröße mit einem Durchschnittsgewicht von 5—10 g enthält, hat ein Gewicht bis zu 50 kg. Eine Hektarernte an Früchten beträgt etwa zwischen 3000—8000 kg.

Das Mesokarp der Frucht bildet eine 4—5 mm dicke, fleischige, stark ölhaltige Schicht um den von einem steinharten Endokarp umschlossenen Samen, der ein gleichfalls ölhaltiges Nährgewebe enthält. Der Fettgehalt im Fruchtfleisch liegt, bezogen auf das Gewicht der ganzen Frucht, bei etwa 20—30%, bezogen auf das Fruchtfleisch kann er bis zu 70% betragen. Das Palmöl ist meistens gelb bis dunkel gefärbt, hat einen veilchenartigen Geruch und einen Schmelzpunkt zwischen 25—40° C. Es wird nahezu ausschließlich in den Produktionsländern an Ort und Stelle (s. Tabelle 11) gewonnen, da die Früchte mit ihrem Fruchtfleisch wegen leichter Verderblichkeit schlecht transportfähig sind. Das Palmöl hat eine häufig vorkommende Verseifungszahl von etwa 200 und eine Jodzahl von etwa 50—60. In den Glyceriden befinden sich an Fettsäuren etwa 40% Palmitinsäure und 5% Stearinsäure neben 40—45% Ölsäure und etwas Linolsäure. Ein großer Teil der Fettsäuren ist im Palmöl meist in freier Form enthalten. Das Öl wird auf Speisefett sowie zu Seife, Kerzen und auf andere Erzeugnisse verarbeitet.

Aus den Samen der Ölpalme wird nach Aufbrechen der Steinschale das Palmkernfett (s. Tabelle 11) gewonnen. Da die Palmkerne nach Entfernung des Fruchtfleisches leicht transportabel sind, kann deren Verarbeitung durch die Ölindustrie der gemäßigten Klimagebiete vorgenommen werden. Die Palmkerne enthalten etwa 50% Fett, das sind 5—12% des gesamten Fruchtgewichtes. Das Palmkernfett hat einen Schmelzpunkt bei 23—28° C. In seinem Fettsäurenbestand überwiegen Laurinsäure mit etwa 50% und Myristinsäure mit etwa 15%; daneben kommen noch Palmitinsäure mit etwa 10%, Ölsäure mit etwa 15% sowie Caprinsäure und Caprylsäure vor. Die Jodzahl liegt etwa zwischen 10 und 20. Das Palmkernfett ist ganz ähnlich zusammengesetzt wie das Kokosfett (Verseifungszahl = 240—250) und verhält sich auch wie dieses. Es wird in großen Mengen auf Speisefett und Seife verarbeitet, aber auch für andere technische Zwecke genutzt. Die Preß- oder Extraktionsrückstände werden als Futtermittel verwendet.

In neuerer Zeit, seit dem ersten Weltkriege, hat das Samenfett der **Babassupalme** *(Attalea speciosa* MART.*)* Brasiliens größere Bedeutung gewonnen (s. Tabelle 11). Der Fettgehalt der Babassukerne liegt bei 65%. Sie werden in Europa ähnlich wie die Kopra verarbeitet. Auch das Babassufett hat eine ähnliche Beschaffenheit und Zusammensetzung wie Kokosfett und dementsprechend ähnliche Kennzahlen (Verseifungszahl = 250, Jodzahl = 15). Es wird auch für die gleichen Zwecke verarbeitet und genutzt.

Von den festen Pflanzenfetten verdient auch noch das Fett des Fruchtfleisches von *Sapium sebiferum* ROXB. *(Euphorbiaceae)* Erwähnung, das den **Chinesischen Talg** (Stillingiatalg) liefert. Auch die Samen enthalten ein Öl, das allerdings eine andere Zusammensetzung aufweist als das des Fruchtfleisches. Der Chinesische Talg hat einen Schmelzpunkt von etwa 40—45° C, eine Jodzahl von 30—40. In China dient er hauptsächlich zur Kerzenfabrikation, in Europa zu verschiedenen technischen Zwecken.

Ein wichtiges Pflanzenfett von fester Konsistenz ist die Kakaobutter. Der **Kakaobaum** *(Theobroma cacao* L.*)* aus der Familie der *Sterculiaceae* ist eine Pflanze des tropischen Südamerikas. Der Baum wird nach 4—5 Jahren ertragsfähig und bleibt es etwa 30 Jahre. Seine Kultur ist von Amerika aus auf alle tropischen Gebiete ausgedehnt worden. Die aus den Früchten gewonnenen Samen, die Kakaobohnen, werden geröstet und nach Entfernung der Schale gemahlen. Die dabei entstehende feine, rohe Kakaomasse wird entweder direkt weiter auf Schokolade verarbeitet oder es wird das Fett zur Gewinnung von Kakaopulver abgepreßt. Die Bohnen enthalten etwa 50% Fett. Die Kakaobutter schmilzt bei etwa 33° C und hat eine Jodzahl von etwa 35. Sie besteht hauptsächlich aus

Glyceriden der Palmitin-, Stearin- und Ölsäure (Verseifungszahl = 195). Kakaobutter dient in Gestalt der Schokolade als Speisefett, sie wird aber vielfach auch für pharmazeutische und kosmetische Zwecke verwendet. Die Weltproduktion an Kakao bewegte sich in den letzten Jahren zwischen 0,7—0,8 Millionen Tonnen.

Gleichfalls ein Fett von fester Konsistenz wird aus den Samen des im tropischen Westafrika und Indien heimischen Baumes *Bassia Parkii* D. C. aus der Familie der *Sapotaceae* gewonnen, die sog. **Shea-Butter.** Von der beerenartigen Frucht muß zur Gewinnung des Samens zunächst das Fruchtfleisch entfernt werden. Der Samen enthält etwa bis zu 50% Fett und wird zur Gewinnung desselben in der Ölindustrie verarbeitet. Der Schmelzpunkt dieses Fettes liegt bei 30—40° C. Die Fettsäuren sind hauptsächlich Stearinsäure, Palmitinsäure und Ölsäure. Die Jodzahl schwankt etwa zwischen 50—60. Das Öl ist ferner durch einen verhältnismäßig hohen Gehalt an unverseifbaren Bestandteilen gekennzeichnet. Die Shea-Butter wird hauptsächlich als Speisefett und zur Seifenfabrikation verwendet.

Eine andere *Bassia*-Art, *B. longifolia* L., liefert die sog. **Mowrah-Butter,** die gleichfalls aus den Samen gewonnen wird. Diese enthalten etwa 50% Fett, das nach seiner Zusammensetzung und seinen Eigenschaften dem Palmöl ähnelt. Es wird auch in Europa viel eingeführt und dient zur Herstellung von Speisefetten, Seifen und Kerzen.

b) Nutzpflanzen mit nicht trocknenden Ölen.

Die hier zu nennenden Pflanzenfette sind bei Raumtemperatur flüssig (Öle), trocknen nicht und sind durch Jodzahlen unter 100 gekennzeichnet.

Zu dieser Gruppe von Ölpflanzen gehört die **Erdmandel** *(Cyperus esculentus* L.*)* aus der Familie der *Cyperaceae,* ein Gewächs der Mittelmeergebiete und der Küsten Westafrikas. Die grasartige Pflanze treibt eine große Zahl unterirdischer Ausläufer, an deren Enden rundliche bis eiförmige Knollen gebildet werden. Die Erträge je Hektar und Jahr werden auf etwa 12000 kg angegeben. Die Knollen (Erdmandel) enthalten neben Stärke auch größere Mengen (etwa 16%) Öl, das in den Anbaugebieten der Pflanze als Speiseöl oder für industrielle Zwecke benutzt wird. Die Knöllchen können aber auch als Ganzes als Nahrungsmittel dienen.

Gleichfalls hauptsächlich als Speiseöl genutzt wird das Öl der bekannten **Haselnuß** *Corylus avellana* L. aus der Familie der *Betulaceae.* Die Haselnußkerne enthalten etwa 60% Öl mit dem Hauptbestandteil Ölsäure neben gesättigten Fettsäuren. Es ähnelt dem Olivenöl und dient kosmetischen und medizinischen Zwecken.

Der aus Zentralasien und Indien stammende Baum *Moringa oleifera* LAM. *(Moringaceae)* liefert das **Behenöl.** Dieses ist durch einen höheren Gehalt seines charakteristischen Bestandteiles Behensäure $CH_3 \cdot (CH_2)_{20} \cdot COOH$ gekennzeichnet und dient als feines Schmieröl und kosmetischen Zwecken.

Eine gewisse wirtschaftliche Bedeutung hat auch das Mandelöl aus den Samen von *Amygdalus communis* L. *(Rosaceae).* Der **Mandelbaum** ist eine Pflanze der Mittelmeerländer. Die süßen Mandeln enthalten etwa 55% Öl, die bitteren weniger. Das Mandelöl besteht zum größten Teil aus Glyceriden der Ölsäure und wird für pharmazeutische, kosmetische Zwecke, aber auch als Speiseöl verwendet.

Eine der wichtigsten Ölpflanzen der Weltwirtschaft ist die **Erdnuß** *(Arachis hypogaea* L.*)* aus der Familie der *Leguminosae.* Die Pflanze stammt aus Brasilien und wird jetzt allgemein in den tropischen und subtropischen Ländern, aber auch in Südeuropa in verschiedenen Sorten kultiviert. Sie ist durch die Eigentümlichkeit der Reifung ihrer Früchte gekennzeichnet. Nach der Befruchtung verlängert sich die Blütenachse zu einem langen Stiel und bohrt die sich entwickelnde Frucht dadurch in die Erde ein, wo sie reift (Geokarpie).

Die Frucht zeigt eine Hülse mit einem Netz von Längs- und Querrippen, die mehrere bohnenähnliche Samen umschließt. Der Anteil der Schale an der gesamten Frucht liegt etwa bei 25%. Der Hektarertrag an Nüssen schwankt zwischen 1000 und 4000 kg.

Einen Überblick über die Welterzeugung an Erdnüssen gibt Tabelle 2.

Tabelle 2. *Welterzeugung an Erdnüssen* (in 1000 metrischen Tonnen). [Aus Yearbook, VI/1 (1953), S. 76, Tabelle 34.]

	1934/38	1949	1950	1951
Europa gesamt	*25*	*20*	*25*	*25*
Mexiko	10	38	64	69
USA	540	851	917	760
Argentinien	79	61	93	148
Brasilien	—	136	118	151
China	2739	2400	2500	2250
Indien	3196	3433	3492	3086
Indonesien	—	363	384	292
Belgisch Kongo	128	146	162	160
Französisch Westafrika	713	850	704	878
Nigeria	356	463	430	615
Südafrikanische Union	14	86	92	97
Weltproduktion (außer USSR)	*8900*	*9800*	*10000*	*9600*

Die von der Hülse befreiten Samen aus tropischen Kulturen enthalten etwa 45—50% Öl. In den weniger warmen Gebieten ist der Fettgehalt geringer. Das Erdnußöl wird nach dem Enthülsen der Früchte aus den auch von ihrer roten Samenhaut befreiten Samen durch Pressung gewonnen. Es wird meistens raffiniert und gebleicht, gegebenenfalls auch gehärtet. Das Erdnußöl enthält hauptsächlich Ölsäure (50—70%) und Linolsäure (15—25%) als ungesättigte, ferner Arachinsäure ($CH_3 \cdot (CH_2)_{18} \cdot COOH$) sowie Stearin- und Palmitinsäure (zusammen etwa 15%) als gesättigte Fettsäuren. Dementsprechend beträgt die Verseifungszahl etwa 190, die Jodzahl 85—95. Es wird namentlich zu Speisezwecken genutzt, aber auch als Rohstoff für die Seifenindustrie. Die Erdnußkuchen werden als Viehfutter verwendet.

Wohl das am längsten und am weitesten bekannte und genutzte Pflanzenöl ist zweifellos das Olivenöl aus der Frucht des **Ölbaumes** *(Olea europaea* L.*)* aus der Familie der *Oleaceae.* Die Heimat des Ölbaumes liegt im Mittelmeergebiet, wo er seit Jahrtausenden und auch heute noch in vielen Formen und Spielarten kultiviert wird. Er ist jedoch schon vor Jahrhunderten nach Mittelamerika, nach Südafrika und nach Australien verpflanzt worden, wo das Klima seinen Ansprüchen genügt.

Die Tragfähigkeit des Ölbaumes beginnt etwa vom 8.—10. Jahre und bleibt dann 100 Jahre und mehr erhalten. Der Ertrag bewegt sich zwischen 30—200 kg Oliven je Baum. Das Gewicht der Olive schwankt etwa zwischen 2—6 g, wovon die Samenkerne 10—30% ausmachen. Das Olivenöl bildet sich im Fruchtfleisch (Mesokarpium) der Frucht. Der Ölgehalt in Prozenten des Fruchtfleisches beträgt bei reifen Früchten 50—60%, in Prozenten der ganzen Frucht ausgedrückt macht er bis zu 40% aus. Das Olivenöl ist also ein typisches Fruchtfleischöl. Es ist dabei aber zu erwähnen, daß auch der von einer Steinschale umgebene Samen, der den Kern der Olive bildet, ein ungesättigtes Öl führt, das bis zu 30% des reinen Samengewichtes ausmacht. Hauptsächliche Erzeugungsländer für Olivenöl sind die Länder um das Mittelmeer, aber auch Kalifornien und Südamerika. Einen Überblick über die Erzeugung von Oliven und Olivenöl gibt die Tabelle 3.

Tabelle 3. *Welterzeugung an Oliven und Olivenöl* (in 1000 metrischen Tonnen).
[Aus Yearbook, VI/1 (1953), S. 74, Tabelle 32.]

	Oliven				Olivenöl			
	1934/38	1949	1950	1951	1934/38	1949	1950	1951
Frankreich	39	32	21	40	4	9	5	10
Griechenland	40	1112	188	706	115	249	42	160
Italien	1267	1101	1004	2051	213	181	178	353
Portugal	—	694	258	754	46	98	40	106
Spanien	1824	1924	902	—	353	388	172	700
Europa insgesamt	*4100*	*4900*	*2400*	*7000*	*740*	*940*	*440*	*1330*
Nord- und Südamerika zusammen	*32*	*70*	*60*	*90*	*2*	*5*	*5*	*5*
Syrien	54	120	26	91	8	10	4	9
Türkei	229	241	265	227	37	43	52	39
Asien gesamt	*374*	*440*	*330*	*390*	*59*	*70*	*60*	*60*
Algerien	104	125	135	180	12	27	17	24
Französisch Marokko	68	78	62	—	10	12	10	27
Afrika gesamt	*540*	*950*	*550*	*700*	*69*	*160*	*80*	*100*
Weltproduktion gesamt	*5000*	*6400*	*3400*	*8200*	*870*	*1200*	*600*	*1500*

Die Gewinnung des Olivenöles erfolgt gewöhnlich nach dem Preßverfahren, und zwar in der nächsten Umgebung der Erzeugungsstätten, da die Oliven wegen der Empfindlichkeit des Fruchtfleisches langes Lagern und vor allem einen Transport schlecht vertragen.

Das Olivenöl ist ein nicht trocknendes; hellgelb bis grüngelb gefärbtes, klares und dünnflüssiges Öl. Es beginnt bei Temperaturen von 10° C abwärts zu erstarren. Sein Fettsäurenbestand setzt sich zusammen aus etwa 80% Ölsäure, 5—10% Linolsäure und 10—15% Palmitinsäure + Stearinsäure. Die Verseifungszahl bewegt sich zwischen 180—200 und die Jodzahl etwa zwischen 80—90. Die Hauptnutzung des Olivenöles ist wohl die als Speiseöl am Ort seiner Erzeugung; das gilt namentlich für das Öl erster Pressung bei geringen Drucken. Es wird aber auch im großen Ausmaß zur Seifenerzeugung oder auch für andere technische und ähnliche Zwecke genutzt.

c) Nutzpflanzen mit schwach trocknenden Ölen.

In die Gruppe der schwach trocknenden Öle fallen solche vegetabilischen Fette, deren Jodzahlen etwa 100—130 betragen. Es handelt sich dabei um Pflanzen der gemäßigten bis wärmeren Klimagebiete der Erde.

Hierher gehören zunächst die Getreideöle, die aus den Keimen von Weizen, Roggen oder auch anderen Getreidearten stammen, in denen sie zu etwa 10—12% enthalten sind.

Das wichtigste dieser Öle ist das aus den Körnern des **Maises** *(Zea mays* L.*)* aus der Familie der *Gramineae*. Die Pflanze stammt aus Amerika und ist als landwirtschaftliche Kulturpflanze allgemein bekannt und angebaut. Das Maiskorn enthält etwa 5—10% Öl, das vor allem im Keim enthalten ist (Ölgehalt bis zu 50%). Je größer der Keimling im Verhältnis zum Gesamtkorn (Keimlingsanteil etwa 12—15%) ist, um so fettreicher ist der Mais. Das Material für die Ölgewinnung bilden die Maiskeime, die in bestimmten Industrien, wo der Mais auf Stärkeerzeugnisse (Maizena) verarbeitet wird, abfallen und daher zur Ölgewinnung genutzt werden können.

Das Maisöl wird den Keimen meistens durch Extraktion entzogen und eignet sich besonders für die Seifenfabrikation, raffiniertes Öl kann auch für Speisezwecke dienen. Der Fettsäurenbestand ist charakterisiert durch etwa je 40% Ölsäure und Linolsäure, den Rest bilden gesättigte Säuren, namentlich Palmitin- und Stearinsäure. Die Verseifungszahl liegt bei etwa 190, die Jodzahl bei 120.

Von den schwierig auseinanderzuhaltenden Cruciferenölen sei zunächst das fette Senföl, das entweder von **weißem Senf** *(Sinapis alba* L.*)* oder von **schwarzem Senf** *(Brassica nigra* [L.] Koch*)* stammt. Dieses fette Senföl ist wohl zu unterscheiden von dem ätherischen, schwefelhaltigen Senföl, das gleichfalls für die Familie der *Cruciferae* charakteristisch ist, aber in den Zellen der Pflanzen glucosidisch gebunden vorliegt. Sowohl der weiße wie auch der schwarze Senf werden in West-, Mittel- und Südeuropa, aber auch in anderen Kontinenten als landwirtschaftliche Kulturen angebaut. Die Samenerträge bei den angebauten Senfarten liegen im großen Durchschnitt bei 1000 kg je Hektar. Der Gehalt der Samen an fettem Öl beträgt 25—35%.

Von den wichtigsten Fettsäuren des Senföles seien genannt: etwa 20% Ölsäure, rund 20% Linolsäure und gegen 50% Erucasäure ($CH_3 \cdot (CH_2)_7 \cdot CH{:}CH \cdot (CH_2)_{11} \cdot COOH$), die für alle Cruciferenöle charakteristisch ist. Die Jodzahlen bewegen sich etwa zwischen 90—105. Das fette Senföl wird für verschiedene technische Zwecke, aber auch als Speiseöl verwendet. Aus den Preßrückständen bei der Ölgewinnung wird Tafelsenf hergestellt.

In diesem Zusammenhang sei noch auf die fetten Samenöle verschiedener Varietäten von *Brassica oleracea* L. hingewiesen, die als Kohlgemüse bekannt sind und deren Öle demzufolge als Kohlsaatöle bezeichnet werden.

Unter dem Namen Rüböl sind neben den Samenölen anderer Cruciferen hauptsächlich solche des **Rapses** *(Brassica napus* L.*)* und des **Rübsens** *(Brassica rapa* L.*)* zusammengefaßt. Der Raps wie der Rübsen werden namentlich in Mitteleuropa und besonders auch in Deutschland in je einer Winterform und einer Sommerform weithin angebaut. Sowohl der Winterraps wie auch der Winterrübsen geben im Vergleich zu den Sommerformen größere Erträge, weshalb man den Anbau in der Winterform bevorzugt. Die durchschnittlichen Samenernten beim Winterraps liegen etwa bei 2000 kg je Hektar, während der Winterrübsen diese Erträge meistens nicht erreicht.

Der Gehalt der Samen an Öl schwankt zwischen 35—45% in weitem Bereich, der Ölgehalt des Rübsens liegt aber gewöhnlich um einige Prozente niedriger als der des Rapses. Die Gewinnung des Öles geschieht nach Zerkleinerung der Samen gewöhnlich durch Pressung.

Die Fettsäuren in den Glyceriden der Rüböle bestehen neben einem geringeren Anteil anderer Säuren zu etwa 25% aus Ölsäure, 15% Linolsäure und zu 50% aus der höher molekularen Erucasäure. Wegen der Anwesenheit dieser Säure ist die Verseifungszahl der Rüböle niedrig (165—175). Die Jodzahl liegt um 100—105.

Das Rüböl wird für manche technischen Zwecke, als Zusatz zu Schmierölen und Faktis, benutzt, die Hauptmenge wird nach der Raffination zu Speisefett und Margarine verarbeitet. Zur Seifenfabrikation ist das Rüböl weniger geeignet. Einige Produktionszahlen von Rapssaat enthält die Tabelle 4.

Tabelle 4. *Welterzeugung an Rapssaat* (in 1000 metrischen Tonnen).
[Aus Yearbook, VI/1 (1953), S. 82, Tabelle 37.]

	1934/38	1949	1950	1951
Deutschland	73	257	—	—
Schweden	—	114	172	209
Europa gesamt	*210*	*710*	*690*	*760*
USSR	88	—	—	—
Nord- und Südamerika zusammen	*51*	*23*	*13*	*10*
Indien	745	806	760	914
Pakistan	232	240	282	307
Weltproduktion gesamt (außer USSR)	*3870*	*4900*	*5100*	*5500*

Vom landwirtschaftlichen Standpunkt wäre noch zu erwähnen, daß namentlich der Raps eine sehr starke Düngung, besonders mit Stickstoff, benötigt, wenn er höchste Erträge liefern soll. Eine hohe Stickstoffdüngung senkt aber in gewissem Ausmaß gewöhnlich den prozentischen Fettgehalt der Samen, während zugleich deren Eiweißgehalt ansteigt. Das wirkt sich aber praktisch nicht ungünstig aus, da durch den infolge Stickstoffdüngung gewöhnlich erzielten großen Mehrertrag auch der Fettertrag je Hektar wesentlich gesteigert wird. Diese Beobachtung gilt nicht nur für die *Brassica*-Ölsaaten, sondern mehr oder weniger auch für alle Ölpflanzen.

Eine weitere Crucifere, die brauchbares Samenöl liefert, ist der **Leindotter** *(Camelina sativa* CRTZ.*)*, der auch gelegentlich angebaut wird. Die Samenerträge liegen bei etwa 1000 kg je Hektar. Der Fettgehalt der Samen erreicht bis zu 30%. Wie die Jodzahl (etwa 135) anzeigt, bildet dieses Öl bereits den Übergang zu den stark trocknenden Ölen. Leindotteröl wird hauptsächlich für Seife, aber auch als Speiseöl genutzt.

Eine der wichtigsten Weltwirtschaftspflanzen, die Öl liefert, ist die **Sojabohne** *(Glycine hispida* MAXIM.*)* aus der Familie der *Leguminosae*. Sie stammt aus Ostasien, wo sie in China, der Mandschurei, Japan und an anderen Orten in zahllosen Spielarten seit Jahrtausenden angebaut wird. Sie ist in neuerer Zeit auch in Amerika, in geeigneten Gebieten Afrikas sowie in Europa (Balkanländer) eingebürgert worden. In Deutschland hat sie sich aber aus verschiedenen Gründen noch nicht recht durchsetzen können. Die Samenernten guter Kulturen erreichen bis zu 2000 kg je Hektar.

Einen Eindruck von der Weltproduktion an Sojabohnen und eine Zusammenstellung der Haupterzeugungsländer vermittelt die Tabelle 5. Aus dieser Tabelle geht besonders hervor, daß die ostasiatischen Gebiete und die Vereinigten Staaten von Amerika enorme Mengen von Sojabohnen erzeugen. Sojabohnen werden aus den Erzeugungsländern auch in großem Maßstab ausgeführt. Die Weltanbaufläche betrug in den Jahren 1935—1939 11,6 Millionen Hektar und belief sich 1953 auf 16,25 Millionen Hektar; die entsprechende Produktion an Sojabohnen betrug 12,6 Millionen Tonnen und im Jahre 1953 18,0 Millionen Tonnen. Auch für die kommenden Jahre wird eine Erweiterung des Sojabohnenanbaues angekündigt.

Die Samen enthalten etwa 15—20% fettes Öl und, was bei dieser Pflanze von nicht geringerem Wert ist, einen hohen Gehalt von etwa 35—45% Protein. Ein höherer Proteingehalt der Samen ist, wie auch bei anderen Ölpflanzen, stets mit niedrigerem Ölgehalt gekoppelt und umgekehrt. Das Öl wird teils durch Pressung mit einer Ausbeute bis zu 75% oder durch Extraktion mit einer solchen bis zu 95% des Gesamtgehaltes gewonnen.

Das Öl ist von gelblicher bis gelbbrauner Farbe und ähnelt in seinen Eigenschaften dem Olivenöl. Als Nebenprodukt bei der Ölgewinnung fällt Lecithin an. Die im Sojaöl vorkommenden Fettsäuren sind ungefähr 50% Linolsäure neben

Tabelle 5. *Welterzeugung an Sojabohnen* (in 1000 metrischen Tonnen). [Aus Yearbook, VI/1 (1953), S. 75, Tabelle 33.]

	1934/38	1949	1950	1951
Europa gesamt	*27*	—	—	—
USA	1164	6284	8145	7688
China	6093	4880	—	—
Mandschurei	3851	1800	3375	3025
Indonesien	—	264	255	270
Japan	321	218	447	474
Asien gesamt	*11060*	*7570*	*9700*	*9400*
Afrika gesamt	—	*15*	*10*	*20*
Weltproduktion gesamt (außer USSR)	*12260*	*14000*	*18000*	*17300*

etwas (5%) Linolensäure, 25% Ölsäure, etwa 15% gesättigte Fettsäuren, namentlich Palmitinsäure, und einige in geringeren Mengen vorkommende andere Komponenten. Die Jodzahl liegt etwa zwischen 120—140.

Das Öl kann nach einer Raffination entweder unmittelbar als Speiseöl dienen, es läßt sich auch durch Wasserstoffanlagerung, hauptsächlich an die Linolsäure, gut härten (Margarine). In geringerem Ausmaß wird das Sojaöl auch zur Herstellung von Seife, in den letzten Jahren jedoch stark zunehmend in der Farben- und Firnisindustrie sowie zur Herstellung von Linoleum verwendet. Die Preß- oder Extraktionsrückstände (Sojakuchen und Sojamehl) geben ein vorzügliches Viehfutter. Sie sind dazu ganz besonders geeignet, da das Eiweiß der Sojabohne zu den wenigen Pflanzeneiweißen gehört, die für den tierischen Organismus als biologisch vollwertig gelten.

Ein wegen seiner Eigenschaften und seiner spezifischen Wirkung auch allgemein bekanntes Öl ist das **Ricinusöl**. Es wird aus den Samen von *Ricinus communis* L. *(Euphorbiaceae)* gewonnen. Die Ricinuspflanze stammt aus Ostindien oder nach anderen Annahmen aus Nordostafrika. Die Pflanze ist schon aus dem Altertum (Ägypten) bekannt und wird heute in Indien, Japan, den Mittelmeerländern und Mittelamerika, ja fast in allen tropischen und subtropischen Ländern angebaut. Sie hat in den Tropen und subtropischen Gebieten Baumgröße, während sie, in kühlerem Klima angebaut, nur einjährig ist und strauchartig bleibt. Die Samenerträge werden mit 1000—1800 kg je Hektar angegeben.

Das Öl ist in den Samen zu etwa 45—50% enthalten und wird durch kaltes oder warmes Pressen oder durch Extraktion gewonnen. Die jährliche Weltproduktion beträgt etwa 200000 Tonnen. Es ist dickflüssig, grünlichgelb gefärbt und wird gewöhnlich einer Raffination unterzogen, um unerwünschte Bestandteile abzuscheiden.

In bezug auf sein Verhalten zu Fettlösungsmitteln ist hervorzuheben, daß es sich, abweichend von anderen Ölen, in Petroläther nicht, aber in reinem

Alkohol löst. Von den in den Glyceriden des Ricinusöles enthaltenen Fettsäuren ist als eine spezifische Eigentümlichkeit dieses Öles die Ricinolsäure zu erwähnen, eine ungesättigte Oxyfettsäure mit 18 C-Atomen. Die Ricinolsäure macht etwa 80% des gesamten Fettsäurenbestandes aus. Dazu kommen noch 10% Ölsäure + Linolsäure und ein Rest gesättigter Fettsäuren. Die Jodzahl des Ricinusöles liegt zwischen 80 und 90.

Verwendet wird die reinste Qualität des Ricinusöles für medizinische Zwecke. Ein Teil wird in der Textilindustrie verbraucht, und schließlich wird es meist in Mischungen mit Mineralöl in größeren Mengen als Schmiermittel für Verbrennungsmotoren, namentlich für Flugzeuge, verwendet. Aber auch die Lederindustrie benutzt es, um das Leder geschmeidig und wasserdicht zu machen. Schließlich kann es, da es sich leicht verseift, auch zur Herstellung von Seifen genutzt werden.

Eine weltwirtschaftlich höchst bedeutsame Pflanze, die auch erhebliche Mengen Öl liefert, ist die **Baumwolle** *(Gossypium*arten aus der Familie der *Malvaceae)*. Das Baumwollsamenöl ist namentlich unter der Bezeichnung Cottonöl im Welthandel bekannt. Die Baumwollpflanze ist eine uralte Kulturpflanze und stammt in verschiedenen Arten aus den tropischen und subtropischen Gebieten zum Teil der alten, zum Teil der neuen Welt. Heute wird sie in den ihr klimatisch zusagenden Gebieten zunächst für Zwecke der Textilindustrie überall angebaut. Aber auch für die Ölgewinnung hat sie einen großen Wert.

Der Samen dieser Pflanze ist mit einem dichten Pelz von Samenhaaren besetzt, die aus reiner Zellulose bestehen und das Ausgangsmaterial für die Baumwolle darstellen. Nach Entfernung der Samenhaare, was heute meistens maschinell geschieht, kann der rundliche, harte, dunkelgefärbte Samen nach entsprechender Vorbereitung und Zerkleinerung auf Öl verarbeitet werden. Meistens wird heute die Saat auch vorher geschält, um keine zu dunklen Öle zu erhalten, und das Öl anschließend raffiniert. Der Fettgehalt der Samen beträgt 15—25% des Samengewichtes.

Die Welterzeugung an Baumwollsaat läßt die Tabelle 6 erkennen. Diese Tabelle macht ersichtlich, daß die Baumwolle eine wichtige Kulturpflanze der

Tabelle 6. *Welterzeugung an Baumwollsaat* (in 1000 metrischen Tonnen). [Aus Yearbook, VI/1 (1953), S. 78, Tabelle 35.]

	1934/38	1949	1950	1951
Europa	*60*	*75*	*100*	*130*
USSR	1290	—	—	—
Mexiko	127	346	443	485
USA	4927	5950	3724	5738
Argentinien	146	299	198	261
Brasilien	778	780	774	620
Peru	139	118	127	149
China (mit Mandschurei)	1520	820	1150	1450
Indien	—	1036	1186	1114
Pakistan	—	451	459	478
Syrien	11	25	65	113
Türkei	104	185	192	309
Asien gesamt	*4140*	*2650*	*3200*	*3600*
Sudan	101	117	176	102
Belgisch Kongo	66	94	91	88
Ägypten	771	697	707	676
Uganda	117	142	126	134
Afrika insgesamt	*1160*	*1240*	*1330*	*1260*
Weltproduktion gesamt (außer USSR)	*11500*	*11500*	*10000*	*12400*

Tropen und Subtropen ist. Haupterzeugungsländer für Baumwollsaat sind die Vereinigten Staaten, die Sowjetunion, China, Ägypten und Indien.

Das Cottonöl besteht hauptsächlich aus Glyceriden der Linolsäure (45%), der Ölsäure (30%), der Palmitinsäure (20%), der Rest sind Stearinsäure und andere Fettsäuren. Die Jodzahl liegt etwa zwischen 105—115, die Verseifungszahl bei 195. Baumwollsaatöl wird zu einem Teil in der Seifenindustrie verbraucht, weitaus größer ist der Anteil raffinierter Öle bei der Nutzung zu Speisezwecken. Außerdem dient es technischen Zwecken, so z. B. zur Herstellung von Faktis. Die Preßrückstände werden in Gestalt von Baumwollsaatkuchen und -mehlen in den Handel gebracht und als Viehfutter benutzt.

Im Zusammenhang mit der Baumwolle ist auch der **Wollbaum** *(Eriodendron anfractuosum* D. C.*)* aus der Familie der *Bombacaceae* als Ölpflanze zu nennen. Die Pflanze liefert zunächst den Kapok, eine Art weißliche Wolle aus Haaren, die aber nicht aus den Samen, sondern aus der Innenseite der Kapselwand hervorgehen. Er wird auch technisch genutzt (Polstermaterial). Das Kapoköl ist in den Samen zu etwa 25% enthalten und zeigt ähnliche Eigenschaften wie das Cottonöl. Es wird auch wie dieses verwertet.

Eine größere Bedeutung als vegetabilisches Fett hat noch das Samenöl des **Sesams** *(Sesamum indicum* L.*)* aus der Familie der *Pedaliaceae*. Es ist eine einjährige strauchartige Pflanze, die aus Indien stammt, deren Kultur sich aber über die meisten tropischen und subtropischen Länder erstreckt.

Einen Überblick über die Weltproduktion an Sesamsaat gibt die Tabelle 7. Danach sind die Hauptproduktionsländer China und Indien.

Tabelle 7. *Welterzeugung an Sesamsaat* (in 1000 metrischen Tonnen).
[Aus Yearbook, VI/1 (1953), S. 83, Tabelle 38.]

	1934/38	1949	1950	1951
Europa	*12*	*11*	*14*	*13*
Nord-, Mittel- und Südamerika	*25*	*110*	*110*	*130*
China	851	830	—	—
Indien	397	438	460	448
Pakistan	43	25	34	34
Türkei	27	35	30	28
Sudan	30	79	168	38
Äthiopien	—	30	26	35
Weltproduktion (ohne USSR)	*1600*	*1740*	*2000*	*1830*

Der Ölgehalt der Samen liegt bei etwa 45—55%. Zur Gewinnung des Öles wird die Saat zerquetscht und anschließend hauptsächlich durch Pressung entölt. Das Öl wird meistens anschließend raffiniert, da es für pharmazeutische oder Speisezwecke verbraucht wird. Außer der Nutzung für Genußzwecke dient es auch zur Herstellung von Seifen. Es enthält hauptsächlich Öl- und Linolsäure (je etwa 40%), aber auch Stearin- und Palmitinsäure (15%). Die Jodzahl liegt zwischen 110—120.

Ein gewisses Interesse als Ölpflanze beansprucht auch der **Kürbis**, und zwar die gewöhnliche *Cucurbita Pepo* L. sowie insbesondere der **Ölkürbis** *(C. maxima* Duch.*)*, beide aus der Familie der *Cucurbitaceae*. Der Kürbis stammt aus Amerika und hat sich als Kulturpflanze namentlich im südlichen Mitteleuropa (z. B. Steiermark) eingebürgert.

Die Frucht trägt im Fruchtfleisch eingebettet zahlreiche flache, große Samen, bei denen $^1/_4$ ihres Gewichtes etwa auf die Schale entfällt. Die schalenhaltigen

Formen enthalten bis zu 40%, schalenlose Sorten hingegen über 50% Fett. Im feldmäßigen Anbau kann eine Kürbisernte von 100000 kg je Hektar bei einem Gehalt an 35% Öl einen Ertrag von über 500 kg Öl liefern. Ein Problem bildet noch die Erleichterung der technischen Gewinnung der Samen aus der Frucht.

Die Fettsäuren des Kürbissamenöles bestehen zu etwa 45% aus Linolsäure, 30% Ölsäure, der Rest entfällt auf Palmitin- und Stearinsäure. Die Jodzahl bewegt sich im Bereich von 115—130. Das Öl wird in den Anbaugebieten der Pflanze hauptsächlich als Speiseöl genutzt.

d) Nutzpflanzen mit stark trocknenden Ölen.

Die hierher gehörenden Pflanzen sind durch fette Öle gekennzeichnet, deren Jodzahlen etwa zwischen 130—200 liegen. Diese Öle haben die besondere Eigenschaft, in dünner Schicht an der Luft ausgebreitet nach kurzer Zeit zu trocknen und einen harten Film zu bilden, wobei Sauerstoff aufgenommen wird. Die trocknenden Öle werden gegebenenfalls nach vorheriger Zubereitung meistens für spezifische technische Zwecke genutzt, die mit der Eigenschaft zu trocknen zusammenhängen.

Eine hier zu nennende Pflanze, allerdings nur von begrenzter wirtschaftlicher Bedeutung, ist der aus dem Orient stammende **Walnußbaum** (*Juglans regia* L.) aus der Familie der *Juglandaceae*; dessen Samenkerne enthalten etwa 50—60% Fett, das auch durch Pressung gewonnen werden kann. Es ist ein trocknendes Öl, dessen Fettsäurengemisch aus etwa 75% Linolsäure, etwa 15% Ölsäure und einem Rest Linolensäure sowie festen Fettsäuren besteht und dessen Jodzahl etwa zwischen 140—150 liegt. Es kann als Speiseöl oder zur Firnisbereitung benutzt werden.

Als Ölpflanze von beschränkterer Wichtigkeit ist auch der **Hanf** (*Cannabis sativa* L., Familie der *Cannabinaceae*). Die aus Westasien stammende Pflanze wird schon jahrtausendelang ganz besonders auch zur Fasergewinnung genutzt. In Deutschland wird der Hanf besonders gern auf Niederungsmoor angebaut. Die Kornerträge erreichen in unseren mitteleuropäischen Anbaugebieten bis zu 1000 kg je Hektar. Die Hanffrüchte enthalten etwa 30—35% fettes Öl. Das Öl wird meistens durch Pressen gewonnen, ist etwas grünlich gefärbt und enthält an Fettsäuren etwa 50% Linolsäure, 20% Linolensäure, 10% Ölsäure sowie noch gesättigte Säuren. Die Jodzahlen bewegen sich zwischen 140—160. Hanföl wird teils zur Herstellung geringwertiger Firnisse, grüner Schmierseifen sowie auch als Speiseöl verwendet. Die Weltproduktion an Hanfsaat wird für den Zeitraum 1934—1938 mit 400000 metrischen Tonnen im Jahresdurchschnitt angegeben, wovon der weitaus größte Teil in der Sowjetunion erzeugt wurde.

Von größerer Bedeutung als der Hanf ist der **Mohn** (*Papaver somniferum* L., Familie der *Papaveraceae*) als Ölpflanze unserer gemäßigten Klimagebiete. Die Kultur des Mohns ist schon uralt und auch heute in vielen Gegenden Europas und Asiens, aber auch in den anderen Erdteilen verbreitet. Die Samenerträge in den mitteleuropäischen Anbaugebieten erreichen etwa bis 1500 kg je Hektar.

Der Ölgehalt der Samen beträgt etwa 45%. Von diesem Öl können bis zu 40% durch Pressen gewonnen werden. Die Fettsäuren der Glyceride bestehen zu etwa 60% aus Linolsäure, 30% Ölsäure und einem Rest aus Linolen- und gesättigten Säuren. Die Jodzahl erreicht Werte von etwa 130—150. Das Mohnöl wird sowohl als Speiseöl wie auch als gut trocknendes in der Farbenindustrie sowie für pharmazeutische Zwecke benutzt. Die Preßrückstände sind ein gutes Futtermittel. Als Nebennutzung können bei Mohnanbau die entleerten Samenkapseln auf Morphin verarbeitet werden.

Ein gewisses wirtschaftliches Interesse haben die sog. **Oiticica-Öle**, die von dem zur Familie der *Rosaceae* gehörenden, in Brasilien beheimateten Baum *Licania rigida* BENTH. oder von ihm verwandten Arten stammen. Die Samen enthalten etwa 60% Öl, das durch hohe Jodzahlen (140—180) gekennzeichnet ist. Dementsprechend ist das gut trocknende Öl zur Herstellung von Anstrichfarben, Linoleum, Faktis und ähnlichen Produkten geeignet.

Das wohl wichtigste trocknende Öl ist das Leinöl. Es wird aus den Samen der **Lein-** oder **Flachspflanze** *(Linum usitatissimum* L., Familie der *Linaceae)* gewonnen. Der Flachs ist eine der ältesten und bekanntesten Kulturpflanzen überhaupt. Er wird auch in den temperierten Gebieten der ganzen Welt, die für seinen Anbau geeignet sind, kultiviert; in Mitteleuropa sind es namentlich die feuchteren Landschaften in den Mittelgebirgen oder an den Küsten.

Es werden je nach dem Hauptnutzungszweck Faserleine oder Ölleine angebaut. Die Faserleine sind meist wenig verzweigt und haben Samen von einem geringeren Korngewicht als die Ölleine, die größere Samen und einen reicher verzweigten Habitus aufweisen. Die Samenerträge erreichen beim Faserlein etwa 750, beim Öllein jedoch bis zu 2000 kg je Hektar. Der Ölgehalt der Samen schwankt beim Faserlein um etwa 35%, bei Ölleinsorten um etwa 40%.

Zur Ölgewinnung wird die Saat gereinigt, auf Walzenmühlen zerkleinert und vorwiegend durch Pressen entölt. Das warm gepreßte Leinöl zeigt oft ziemliche Verunreinigungen und muß von diesen durch Raffination befreit werden. Von diesen Verunreinigungen sind besonders die aus den Samen stammenden Schleimstoffe zu nennen.

Die Glyceride des Leinöles enthalten an Fettsäuren etwa 40—60% Linolensäure, 10—30% Linolsäure, 10—20% Ölsäure und einen Rest von 5—10%, der aus gesättigten Fettsäuren besteht. Diese Verhältnisse können je nach Anbaubedingungen, Klima und Düngung starke Abweichungen erfahren. Dementsprechend schwanken denn auch die Jodzahlen für einzelne Leinöle in einem Bereich von etwa 150 bis über 200, die Verseifungszahl schwankt nur geringfügig etwa zwischen 190—195.

Es ist bekannt, daß sich an zur Reifezeit kühlen Standorten Öle mit hohen Jodzahlen und einem hohen Gehalt an Linolensäure und mit vorzüglichen Trockeneigenschaften ausbilden. An warmen Standorten dagegen entstehen mehr gesättigte Fettsäuren, wodurch die Jodzahl auf viel geringere Werte (bis unter 150) absinken kann; diese Öle zeigen nur schwache Trockenfähigkeit. Bedingt durch Veränderungen im Wärmehaushalt der Pflanzen, die primär durch solche des Wasserhaushaltes und der Düngung hervorgerufen werden, treten auch entsprechende Veränderungen in der chemischen Beschaffenheit des Leinöles, aber auch anderer trocknender Öle, soweit sie untersucht sind, auf. Diese Eigenschaften sind für die Verwendung der Öle in der Technik von größter Bedeutung.

Was die Erzeugung an Leinsaat anlangt, so gibt darüber die nachfolgende Tabelle 8 Auskunft. Die Hauptproduktionsgebiete der Welt an Leinsaat waren bis vor kurzer Zeit neben der Sowjetunion Argentinien und Indien. Neuerdings sind auch die Vereinigten Staaten und Kanada sowie einige andere Gebiete stärker in den Vordergrund getreten.

Der größte Teil des Leinöles wird wohl auf Firnis verarbeitet und in der Farben- und Lackindustrie gebraucht. Wie eingangs schon erwähnt, ist Leinölfirnis ein Leinöl, dessen natürliche Trockenfähigkeit durch Zusatz von Sikkativen erhöht ist. Außerdem kann das Leinöl durch Erhitzen auf Standöl oder Dicköl verarbeitet werden. Bei diesem Kochprozeß wird die Viscosität gesteigert, während die Jodzahl abnimmt. Leinöl dient ferner insbesondere noch zur Herstellung von Linoleum und von Wachstuch, von Faktis und schließlich auch in der Seifenindustrie zu Schmierseife. Wegen seiner trocknenden Eigenschaften hat das Leinöl eine einzigartige Stellung unter den technisch nutzbaren Pflanzenfetten. Einen vollwertigen Ersatz für dieses Öl gibt es bis jetzt noch nicht. Es wäre noch zu erwähnen, daß auch die Preßrückstände in Gestalt von Ölkuchen ein ausgezeichnetes Viehfutter abgeben.

Tabelle 8. *Welterzeugung an Leinsaat* (in 1000 metrischen Tonnen). [Aus Yearbook, VI/1 (1953), S. 80, Tabelle 36.]

	1934/38	1949	1950	1951
Europa insgesamt	*120*	*280*	*250*	*260*
USSR	845	—	—	—
Kanada	32	58	119	251
USA	209	1116	1022	881
Argentinien	1702	676	559	313
Uruguay	89	75	90	142
Indien	433	418	367	314
Afrika gesamt	*11*	*120*	*40*	*60*
Weltproduktion (ohne USSR)	*2710*	*2960*	*2630*	*2400*

Ein sehr gut trocknendes Öl ist auch das sog. Holzöl oder Tungöl, das in Ostasien, namentlich in China, in den letzten Jahren aber zunehmend auch in den USA aus besonderen Pflanzungen gewonnen wird, aber auch bei uns in Europa und in Amerika an Stelle von Leinöl in der Lack- und Firnisindustrie Verwendung findet. Es wird aus den Samen des **Holzölbaumes** *(Aleurites Fordii* HEMSL., *A. montana* WILS.*)* und anderer *Aleurites*arten (Familie *Euphorbiaceae*) gewonnen. Der Fettgehalt der Samen liegt etwa bei 50%. Das Öl besteht zum überwiegenden Teile (75%) aus dem Glycerid der Elaeostearinsäure, einer Fettsäure mit 18 C-Atomen und 3 Doppelbindungen wie bei der Linolensäure; der Rest ist Olein-, Linol- und Palmitinsäure, Linolensäure fehlt. Die Jodzahl schwankt zwischen 160—175. Beim Erhitzen bildet das Holzöl eine gelierende Masse. Es trocknet, in dünner Schicht ausgebreitet, schneller als Leinöl und wird hauptsächlich in der Lackindustrie (Möbel, Fußböden, Boote) zu wasserdichten Anstrichen verwendet. Einige Angaben über die Produktion an Tungöl enthält die Tabelle 11.

Anhangsweise sei noch das **Bankulnußöl** erwähnt, das auch aus den Samen anderer *Aleurites*arten der Südsee gewonnen wird.

Eine häufig erwähnte, gleichfalls ein trocknendes Samenöl liefernde Pflanze ist *Lallemantia iberica* FISCH. und MEY. *(Labiatae)*. Der Fettgehalt der Samen liegt bei etwa 30%. Beim Anbau und bei der Samengewinnung von dieser Pflanze treten aber verschiedene Schwierigkeiten auf, weshalb ihre Kultur bisher wohl kaum Anklang gefunden hat.

Bekannter als Ölpflanze ist die in Ostasien und Nordindien heimische Labiate *Perilla ocymoides* L. Die Früchte enthalten bis zu 45% fettes Öl. Seine Jodzahl ist sehr hoch (180—205). Die hauptsächlichsten Fettsäuren des Perillaöles sind Linolensäure (45%), Linolsäure (40%), der Rest verteilt sich auf Ölsäure und gesättigte Säuren. Es trocknet ähnlich wie das Leinöl. Es wird vielfach zur Lack- und Firnisbereitung (Amerika) und zum Imprägnieren von Kleidern und Papier, aber auch gelegentlich für Speisezwecke (Indien) benutzt.

Aus der großen Familie der *Compositae* wären schließlich noch zwei Pflanzenarten zu nennen, die wenigstens zum Teil anbauwürdig sind und ein gut trocknendes Öl liefern, die Sonnenblume und der Saflor.

Die **Sonnenblume** *(Helianthus annuus* L.*)* ist von beiden wohl die wichtigere. Die Pflanze stammt aus Mittelamerika, wird aber heute in den verschiedenen Teilen der Welt, namentlich in Südost-Europa, Südrußland und den anderen wärmeren Gegenden, neuerdings auch in Deutschland, im großen angebaut. Die Körnererträge schwanken je nach Sorte und Anbauort im weiten Bereich und bewegen sich etwa zwischen 1000—4000 kg je Hektar.

Tabelle 9. *Welterzeugung an Sonnenblumensaat* (in 1000 metrischen Tonnen). [Aus Yearbook, VI/1 (1953), S. 85, Tabelle 39.]

	1934/38	1949	1950	1951
Europa gesamt	*347*	—	—	—
USSR	1949	—	—	—
Argentinien	154	712	1021	648
Welterzeugung (ohne USSR)	*515*	*1770*	*2030*	*1760*

Das Hauptterzeugungsland an Sonnenblumensaat dürfte die Sowjetunion sein, an zweiter Stelle steht Argentinien (Tabelle 9).

Die Körnerfrüchte der Sonnenblume bestehen zu etwa 40% aus Schalen und 60% aus den ölhaltigen Samen. Der Fettgehalt liegt, auf die ganze Frucht bezogen, bei etwa 30%, auf die schalenfreien Samen bezogen, bei etwa 50%. Zur Ölgewinnung wird die entschälte Saat auf Walzenstühlen zerkleinert und gewöhnlich einmal kalt und hinterher warm abgepreßt. Auch die Extraktion kann zur Ölgewinnung angewendet werden.

Das Sonnenblumenöl enthält etwa 50% Linolsäure, 40% Ölsäure, der Rest sind gesättigte Fettsäuren. Die Jodzahl liegt zwischen 120—140. Da dem Öl die Linolensäure fehlt, ist sein Trockenvermögen weniger gut als das des Leinöles. Es findet Verwendung als Speiseöl oder zur Margarinebereitung, ferner zu einem gewissen Teil als Ersatz für Leinöl sowie zur Seifenherstellung. Die Preßrückstände bilden ein gutes Viehfutter, wobei ein großer Unterschied im Wert der Kuchen besteht, je nachdem, ob sie von geschälter oder ungeschälter Saat stammen.

Der **Saflor** *(Carthamus tinctorius* L.*)* tritt gegenüber der Sonnenblume als Ölpflanze weit zurück. Die aus dem Mittelmeergebiet stammende Pflanze wurde früher zur Farbstoffgewinnung (Färberdistel) angebaut. Die Pflanze ist gegen Dürre weitgehend unempfindlich und bringt Erträge zwischen 1000—3000 kg Körner je Hektar.

Die Früchte, die ähnlich denen der Sonnenblume zu 45% aus Schalen und 55% aus ölhaltigen Samenkernen bestehen, enthalten, auf die ganze Frucht bezogen, etwa 30% Öl oder 50% im schalenlosen Kern. Das Safloröl ähnelt dem Öl der Sonnenblume. Sein Fettsäurenbestand setzt sich wesentlich aus etwa 70% Linolsäure und 20% Ölsäure zusammen, daneben sind gesättigte Säuren, Palmitinsäure und Stearinsäure, vertreten. Die Jodzahl beträgt zwischen 130—150. Neben einer Nutzung als Speiseöl und zur Seifenbereitung kann das Safloröl wegen seiner Trockeneigenschaften teilweise als Ersatz für Leinöl benutzt werden.

D. Die Erzeugung an Pflanzenfetten.

Die Weltproduktion an Pflanzenfetten zeigt für die wichtigsten von ihnen nach den vorliegenden statistischen Angaben in der ersten Jahrhunderthälfte folgendes Bild (Tabelle 10).

Tabelle 10. *Weltproduktion an Ölsaaten* (in 1000 metrischen Tonnen, Öläquivalent). [Aus Fats and Oils, Bull. 13 (1949), S. 46, Tabelle 4.]

	1909—1913	1924—1928	1929—1933	1934	1935	1936
Erdnüsse	1150	1975	2480	2230	2260	2615
Sojabohnen . .	620	1645	1865	1635	1680	1815
Baumwollsaat .	1555	1845	1835	1715	1910	2295
Raps- u. Senfsaat	1625	1275	1295	1335	1330	1360
Leinsaat	940	1265	1185	1175	1135	1210
Kopra	385	800	915	1020	1005	1030
Palmöl	120	215	280	330	415	480
Tungöl	30	55	60	65	75	85
Olivenöl	590	750	850	825	955	740

Aus diesen Angaben ist vor allem ein starker Anstieg in der Erzeugung von Erdnüssen und Sojabohnen von der Zeit vor dem ersten Weltkriege bis zum Jahre 1936 ersichtlich. Auch die Baumwollsaat- und Leinsaaterzeugung sowie die Erzeugung von Kopra und Palmöl haben während dieser Zeit erheblich zugenommen.

Die heutige Welterzeugung an Ölen, Fetten und fetthaltigen Rohstoffen wird nach den zuletzt zugänglich gewesenen Informationen auf insgesamt 24,2 Millionen Tonnen für das Jahr 1953 veranschlagt; davon entfallen 14,2 Millionen Tonnen auf Fette und Öle pflanzlicher Herkunft (Tabelle 11).

Tabelle 11. *Weltproduktion an pflanzlichen Fetten, Ölen und Ölsaaten* (in 1000 t). (Aus Auslandsinformationen für Ernährung und Landwirtschaft, Teil A: Agrar- und Ernährungswirtschaft, Jahrgang VII, Nr. 5/6. 2. 1954, 1. Februarausgabe, S. 10.)

	ø 1935—1939	ø 1945—1949	1952	1953
Baumwollsaat	1560	1220	1725	1740
Erdnüsse	1515	1700	1605	1765
Sojabohnen	1230	1500	1785	1760
Sonnenblumenkerne	560	750	850	860
Oliven	885	835	700	1050
Sesam	655	640	660	680
Speiseöle insgesamt	6405	6645	7325	7855
Kokosnüsse	1930	1515	1940	1805
Palmkerne	370	320	390	445
Palm	960	900	1105	1155
Babassukerne	25	40	35	35
Palmöle insgesamt	3285	2775	3470	3440
Leinsaat	1040	1000	1015	995
Ricinus	180	190	195	215
Raps	1205	1420	1630	1540
Oiticica	9	10	7	9
Tung	135	105	115	130
Perilla	60	5	5	5
Industrieöle insgesamt . . .	2629	2730	2967	2894

Die vorliegende Tabelle läßt bezüglich der Gewinnung der wichtigsten Pflanzenfette in der Welt während der letzten Jahre eine steigende Tendenz erkennen, obzwar in vielen Fällen die Vorkriegsproduktion noch nicht wieder erreicht ist. Im einzelnen betrifft dies besonders das Palmöl und die Palmkerne, Ricinus, Sonnenblumenkerne, Sojabohnen und auch teilweise Erdnüsse. Andere pflanzliche Fettträger hingegen zeigen kaum eine Veränderung bezüglich der

Produktionsergebnisse in den letzten Jahren; das gilt insbesondere für Oliven, für Sesam, für Kokosnüsse, für Leinsaat sowie für Rüböle.

Diese Erscheinungen sind teils durch technische Besonderheiten beim Anbau oder bei der Gewinnung der Ölsaaten bedingt, teils spielen auch wirtschaftliche Erwägungen, wie etwa die Preisgestaltung, dabei eine Rolle. Insgesamt lassen die Produktionsziffern an Pflanzenölen deren gewaltige Bedeutung in der Weltwirtschaft eindrucksvoll erkennen.

Literatur.

Auslandsinformationen für Ernährung und Landwirtschaft, Teil A: Agrar- und Ernährungswirtschaft, Jahrgang VII, Nr 5/6. 2. 1954, 1. Februarausgabe.

Bauer, K. H.: Die trocknenden Öle. Stuttgart: Wissenschaftliche Verlagsgesellschaft 1928. — Braun, K.: Die Fette und Öle. Berlin: W. de Gruyter & Co. 1945.

Davidsohn, J., u. H. Stadlinger: Hilfsbuch für das Gebiet der Fette und Fettprodukte. Leipzig: S. Hirzel 1930.

Eckey, E. W.: Vegetable Fats and Oils. New York: Reinhold Publishing Corporation 1954.

Fats and Oils: Commodity Series, Bull. 13, Aug. 1949, Washington 1949.

Hackbarth, J.: Die Ölpflanzen Mitteleuropas. Stuttgart: Wissenschaftliche Verlagsgesellschaft 1944. — Hefter, G.: Technologie der Fette und Öle. Berlin: Springer 1910. — Hefter, G., u. H. Schönfeld: Chemie und Technologie der Fette und Fettprodukte. Bd. 1, Chemie und Gewinnung der Fette. 1936. Bd. 2, Verarbeitung und Anwendung der Fette. 1937. Bd. 4, Seifen und seifenartige Stoffe. 1939. Wien: Springer. — Honcamp, F.: Handbuch der Pflanzenernährung und Düngerlehre, Bd. 1. Berlin: Springer 1931.

Kaufmann, H. P.: Studien auf dem Fettgebiet. Berlin: Verlag Chemie 1935.

Lüde, R.: Die Gewinnung von Fetten und fetten Ölen. Dresden u. Leipzig: Theodor Steinkopff 1948.

Paech, K.: Biochemie und Physiologie der sekundären Pflanzenstoffe. In Lehrbuch der Pflanzenphysiologie, Bd. 1/2. Berlin-Göttingen-Heidelberg: Springer 1950.

Schmalfuss, K.: Über Fettbildung in Pflanzensamen. Sitzgsber. Dtsch. Akad. Landwirtschaftswiss. Berlin **3**, 15 (1954). — Sprecher v. Bernegg, A.: Tropische und subtropische Weltwirtschaftspflanzen. Bd. 1, Stärke- und Zuckerpflanzen. 1929. Bd. 2, Ölpflanzen. 1929. Bd. 3/1, Kakao und Kola. 1934. Stuttgart: Ferdinand Enke.

Tobler, F., u. H. Ulbricht: Koloniale Nutzpflanzen. Leipzig: S. Hirzel 1945.

Ubbelohdes Handbuch der Chemie und Technologie der Öle und Fette. Bd. 1, Chemie und Technologie der Öle, Fette und Wachse. Allgemeiner Teil. 1929. Bd. 2/1, Chemie und Technologie der pflanzlichen Öle und Fette. 1932. Bd. 3/1, Chemie, Analyse, Technologie der Fettsäuren, des Glyzerins und der Türkischrotöle. 1929. Bd. 3/2, Chemie und Technologie der Seifen und Waschmittel. 1930. Bd. 4, Chemie, Technologie und Analyse der oxydierten, polymerisierten und reduzierten fetten Öle und der Wachse. 1926. Leipzig: S. Hirzel.

Wittka, F.: Gewinnung der höheren Fettsäuren durch Oxydation der Kohlenwasserstoffe. Leipzig: Johann Ambrosius Barth 1940.

Yearbook of Food and Agricultural Statistics—Production—1952, Bd. VI, S. 1. Rom 1953.

The phosphatides and glycolipids.

By

J. A. Lovern.

I. Introduction.

In general far less is known about the chemistry and, in particular, the metabolism of the phosphatides and glycolipids of plants than of similar substances in animals. The relatively greater role of carbohydrates in the composition of plant than of animal tissues is reflected in their lipid chemistry; glycolipids of many types are of widespread occurrence in plant tissues. The typical glycolipids of animals, cerebrosides, have, however, never been found in plants. Nor has the related animal phosphatide sphingomyelin. A compound known as "cerebrin", which has long been known as a constituent of fungal lipids (*cf.* the review by Celmer and Carter), is the amide of a fatty acid and a nitrogenous base closely related to sphingosine[1]. The same base has recently been isolated from the inositol phosphatide fractions of soyabean and maize and shown to be 1,3,4-trihydroxy-2-amino-octadecane (Carter, Celmer, Lands, Mueller and Tomizawa), whereas sphingosine is 1,3-dihydroxy-2-amino-octadecene-4. It has been named "phytosphingosine". It is not yet known whether cerebrin or the phytosphingosine of these inositol lipid fractions is incorporated into more complex molecules, *e.g.* into glycolipids analogous to the cerebrosides.

The classical lecithin and cephalin of animal tissues, now often named phosphatidylcholine and phosphatidylethanolamine respectively, have been reported as occurring in many plant tissues, but have often been accompanied by carbohydrate which may or may not have been chemically linked with the lecithin and, particularly, with the cephalin. This point will be amplified below. The "acetal phosphatides", or "plasmalogens" have generally been found only in animal tissues, but two reports have appeared of their occurrence in plants; in the crude phosphatides of soyabean and groundnut (Lovern [2]) and in olive oil (Thiele). It should, however, be noted that both workers only really detected the presence of aldehydes in the lipids and did not isolate a plasmalogen-like lipid.

It seems preferable to consider separately the lipid chemistry of microorganisms, of seeds and other tissues of higher plants. This is done under appropriate subheadings. Lastly a brief account is given of the little that is known of the metabolism of these substances by plants. No account is included of sterol glycosides, since this subject is really an aspect of the much greater field of sterol chemistry. The section on bacterial lipids inevitably overlaps with the review given by Asselineau, this volume p. 90—108, since phosphatides and, in particular, glycolipids form so large a part of the total lipids of bacteria. The account given here is a condensed one, designed to fit into a general account of the phosphatides and glycolipids of plants. The approach has been somewhat different in the two chapters and the reader is recommended to consult them both.

---

[1] For details on fungal cerebrins see also Steiner, this volume, p. 81.

II. Microorganisms.

Among the microorganisms bacteria often possess a lipid pattern very different from that of higher plants, whereas fungi generally resemble higher organisms in their lipid composition. Bacterial lipids are, in general, remarkable for the complete absence of sterols, the presence of unusual and often highly complex fatty acids, for a lack of the classical lecithins and cephalins but the presence of glycolipids or glycophosphatides, and for the presence of large proportions of free fatty acids. Ordinary triglycerides are often completely lacking. The contribution of ASSELINEAU, p. 90—108 of this volume, should be consulted for details.

In considering the literature on bacterial lipids it is essential to note that the organisms investigated have been grown in quantity in pure culture and it is well established (*e.g.* CREIGHTON, CHANG and ANDERSON, DE SÜTÖ-NAGY and ANDERSON [1], PUDLES) that both the content and the composition of bacterial lipids vary widely according to the precise strain of organism, its age, and the ingredients of the culture medium. This is the probable reason for the considerable discrepancy often to be found between the results of different workers. The lipids of the organisms growing under natural conditions may well be different again.

1. Acid-fast bacteria.

The bacteria to be discussed now are certain *Mycobacteria*, namely the human, bovine and avian varieties of the tubercle bacillus *(Mycobacterium tuberculosis)*, the non-pathogenic timothy grass bacillus *(M. phlei)* and the so-called leprosy bacillus *(M. leprae)*. The lipids of this group of organisms have been studied in much more detail than those of other acid-fast bacteria. To neutral solvents they may yield from 8% *(M. phlei)* to 24% (human tubercle bacillus) of lipids, on a dry weight basis, and an additional 12–19% of "firmly bound" lipids may be obtained with acidified solvents.

The lipids are a singularly complex mixture, including three types of compound appropriate for discussion in this chapter, namely phosphatides, esters of the disaccharide trehalose, and esters of complex polysaccharides. The fatty acids of all classes of lipids from this group of microorganisms generally include both normal straight chain acids, *e.g.* palmitic and oleic acids (although normal acids seem to be lacking in certain "wax" fractions, *e.g.* from the timothy and avian tubercle bacilli), and a great range of branched chain acids, both saturated and unsaturated (CASON and SUMRELL, CASON, FREEMAN and SUMRELL, POLGAR [1, 2, 3]). In the following account only the main constituents will be described, largely as they have been discovered by the extensive researches of ANDERSON.

a) Phosphatides.

The phosphatides are acidic in reaction and have usually been reported as almost nitrogen-free, but it seems undesirable to refer to them as phosphatidic acids in view of the growing modern use of the term "phosphatidyl" to denote a radical derived from a difatty acid ester of glycerophosphoric acid. MACHEBOEUF and FAURE used the term "complex phosphatidic acids" but also reported the occurrence in tubercle bacillus phosphatides of simple phosphatidic acids. Traces of nitrogen in these acid-fast bacterial phosphatides are not derived from choline or ethanolamine[1] and until recently have not been characterized, apart from ammonia in certain cases. Particular interest attaches, therefore, to the

[1] Ethanolamine occurs in the phosphatides of *M. marianum* (MICHEL and LEDERER).

report by BARBIER and LEDERER [1] that the phosphatide of *M. phlei* contains the amino acid hydroxylysine and that other amino acids are present in the phosphatides of the tubercle bacillus (see also GENDRE and LEDERER, VILKAS and LEDERER, MICHEL and LEDERER). The hydroxylysine is linked to the phosphatide through its hydroxyl group, with the amino group free, *i.e.* in the normal manner for the bases of glycerophosphatides, but this phosphatide is far more complex than, *e.g.* lecithin, with a minimum molecular weight of about 16,000 and containing only about 1% hydroxylysine.

After hydrolysis the phosphorus is found in the water-soluble fraction, which amounts to some 40% of the original lipid, in two forms: as a neutral phosphorus-containing polysaccharide and as a mannose-glycerol-diphosphoric acid. This latter substance may, under certain conditions, be replaced by inositol-glycerol-diphosphoric acid (DE SÜTÖ-NAGY and ANDERSON [2]). The polysaccharide, on further hydrolysis, yields mannose and inositol in the molecular ration of 2:1. Glucose has been reported to accompany mannose and inositol in the phosphatides of *M. marianum* (MICHEL and LEDERER).

The phosphatides are, therefore, fatty acid esters of these phosphorylated carbohydrates, although whether or not the two classes of phosphate ester are combined into some larger structure is not known. Chromatography of the phosphatides of certain *Mycobacteria* on silicic acid/celite columns has shown that more than one compound is present (VILKAS and LEDERER, MICHEL and LEDERER). The phosphatide of one strain of *M. tuberculosis* seems to contain about 60% of a mixture of (mainly) phosphatidyl D-mannose and phosphatidyl inositol, the mannose and inositol being additionally esterified with one molecule of fatty acid (VILKAS and LEDERER). The phosphatides of the human tubercle bacillus are notable for containing the optically active "phthioic" acid (not found in the other organisms), which on injection into animals produces typical tubercular lesions. The "phthioic" acid of ANDERSON is now known to be a complex mixture, one important component of which has been found to be 2,4,6-trimethyltetracos-2-enoic acid ("mycolipenic" acid: POLGAR [2]). This is apparently the same substance as the "C_{27}-phthienoic" acid of CASON and SUMRELL.

b) Trehalose esters.

Trehalose esters of fatty acids include derivatives of the interesting "tuberculostearic" acid (10-methylstearic acid) and, in human tubercle bacillus lipids, phthioic acid. The leprosy bacillus lipid contains a particularly complex range of both normal and branched chain acids. The acetone-soluble "fat" fraction of *Mycobacteria* is composed mainly of free fatty acids and trehalose esters, with complete absence of triglycerides, even although glycerol is present in the culture medium and is incorporated into lipids, *e.g.* the phosphatides. Complex "glycerides" are found in the wax fraction of some of these bacteria, *e.g.* the timothy bacillus. Trehalose has only been actually isolated from the acetone-soluble "fat" of the human tubercle and leprosy bacilli, but is probably present in the same fractions from the other organisms, especially as it has been found in the wax fractions from the avian tubercle and timothy bacilli, where it is presumably esterified with acids of the mycolic type (see below).

c) Polysaccharide esters.

Fatty acid esters of complex polysaccharides make up a large part of the "wax" of *Myobacteria*. Only in the human tubercle bacillus do they seem to occur in the "free" chloroform-soluble wax fraction, but they occur in the "firmly bound" lipids of the avian tubercle and leprosy bacilli, as well as of the human tubercle bacillus. The firmly bound lipids of the bovine tubercle and timothy bacilli do not seem to have been studied in detail. The polysaccharides in question give a precipitin reaction with immune serum and are largely esterified with hydroxy-acids of very high molecular weight, the "mycolic" acids.

The mycolic acids differ in the various organisms and are undoubtedly not the individual substances originally described by ANDERSON (*cf.* STÄLLBERG-STENHAGEN and STENHAGEN, DEMARTEAU and LEDERER, ASSELINEAU [2]). Some 37 individual acids have been described (PUDLES). Mycolic acids decompose on heating under reduced pressure to give a normal higher fatty acid and an unidentified, non-volatile, neutral residue. Thus one mycolic acid from human and bovine tubercle bacilli, which has a probable empirical formula of $C_{88}H_{176}O_4$, yields hexacosanoic acid while avian β-mycolic acid yields tetracosanoic acid. This behaviour on pyrolysis suggests that mycolic acids possess a long unbranched paraffin chain attached to the α-carbon atom (with reference to the carboxyl group) and a hydroxyl group on the β-carbon atom (ASSELINEAU and LEDERER [1]). Considerable progress has been made in elucidating the structure of some of the mycolic acids (ASSELINEAU and LEDERER [2], BARBIER and LEDERER [2]). The timothy and leprosy bacilli waxes contain similar complex hydroxy-acids, "phleimycolic" and "leprosinic" acids. These mycolic acids are apparently responsible for the property of "acid fastness" in the staining behaviour of the organisms.

The polysaccharides are of unknown structure and only some of the hydrolysis products have been identified. Thus the polysaccharide of the wax of the human tubercle bacillus yields D-arabinose, D-galactose, D-mannose, small amounts of glucosamine and inositol, an unidentified, phosphorus-containing acid, together with much unidentified material. All reducing sugar after hydrolysis is, however, accountable for as arabinose, galactose and mannose. The polysaccharide of each organism yields a different range of hydrolysis products. ASSELINEAU [1] found that the polysaccharide from human strains, in contrast to bovine strains, contains glutamic acid, alanine and the unusual 2,6-diamino pimelic acid, probably in the form of a tripeptide. The so called "cord factor" of *M. tuberculosis* appears to be an ester of mycolic acids and a glucoside, the latter as yet only partially characterized (NOLL and BLOCH).

2. Other bacteria.

a) Diphtheria bacillus.

Second only to those of the *Mycobacteria*, the lipids of the diphtheria bacillus, *Corynebacterium diphtheriae* have been more studied than those of other bacteria. CHARGAFF found that the cells contain about 4% of "fat" and only about 0.4% of acetone-insoluble phosphatide. The latter contained 1.4% P and 0.8% N and on hydrolysis yielded, besides normal fatty acids, others of relatively high molecular weight including one called "corynin", with a probable formula of $C_{50}H_{100}O_4$. No organic base was identified and the aqueous phase after hydrolysis apparently contained an aldohexose. The "fat" fraction was mainly free acids, including a whole series of unsaturated derivatives of high molecular weight, but apparently glycerol and carbohydrate were absent. In general CHARGAFF's findings were confirmed by TAKAHASHI who concluded that corynin, now named "corynolic acid", has the formula:

$$CH_3 \cdot CH(OH) \cdot (CH_2)_7 \cdot CH(CH_3) \cdot CH(OH) \cdot CH(CH_3) \cdot CH(CH_3) \cdot CH(COOH) \cdot$$
$$(CH_2)_{17} \cdot CH(CH_3) \cdot (CH_2)_{14} \cdot CH_3$$

This Japanese work, however, revealed dihydroxyacetone as the sole water-soluble hydrolytic fragment of both "fat" and phosphatide fractions.

In complete contrast to this picture, Russian workers (GUBAREV and LUBENETS, GUBAREV, LUBENETS, KANCHUKH and GALAEV, GUBAREV, LUBENETS and GALAEV) have reported a whole range of fatty acids, combined with various carbohydrates, including galactose, trehalose and a galactose-mannose-phosphoric acid polysaccharide.

Yet a third picture of the diphtheria bacillus lipids is given by the work of PUDLES, who isolated "coryno-mycolic" and "coryno-mycolenic" acids as the main constituents of the higher molecular weight fatty acid fraction. These acids, as the names imply, show some structural resemblance to the mycolic acids of the *Mycobacteria* in that they are of high molecular weight, have a hydroxyl group on the β-carbon atom and a long aliphatic chain on the α-carbon atom (with reference to the carboxyl group). They are, however, simpler than the true mycolic acids, having only 32 carbon atoms. Corynomycolic acid has been shown (PUDLES) to be 2-*n*-tetradecyl-3-hydroxy-stearic acid, while coryno-mycolenic acid is 2-*n*-tetradecyl-3-hydroxy-octadec-11-enoic acoid.

The aqueous phase after hydrolysis of the phosphatide was shown by PUDLES to contain choline and inositol, but not ethanolamine or amino acids. HARA claims to have isolated a phosphatide resembling beef heart cardiolipin from *C. diphtheriae*, while still another type of phosphatide has been found by ČMELIK [4] which lacks corynomycolic acid, polysaccharide and choline, although containing nitrogen.

b) Agrobacterium (phytomonas) tumefaciens.

If this organism is grown on a glycerol-containing medium it contains about 2% of lipids, of which 44% is phosphatide, whereas on a sucrose-containing medium it has 6% of lipids, of which 64% is phosphatide. The composition of the fatty acids also varies markedly with the medium. The phosphatides are apparently a mixture of about equal parts of the classical lecithin and cephalin, giving glycerophosphoric acid, choline and ethanolamine on hydrolysis, together with an amino acid. Along with normal fatty acids there is about 18% of branched chain acids, predominantly "phytomonic" acid (GEIGER and ANDERSON, VELICK). Phytomonic acid has recently been found by HOFMANN and TAUSIG to be identical with "lactobacillic" acid (see below).

c) Salmonella bacteria.

ČMELIK [1, 2] has studied the lipids of various species of *Salmonella*. The phosphatides of *S. ballerup* and *S. paratyphi* C are notable for containing a large range of known amino acids and a particularly high concentration of an unknown amino acid. They do not contain choline, ethanolamine, inositol or carbohydrate. They contain about 4% P and 55—65% fatty acids, which appear to be of normal type, *e.g.* palmitic. The phosphatides are acidic. On the other hand, the phosphatides of certain strains of *S. typhosa* gave ethanolamine (apparently derived from phosphatidyl ethanolamine) on hydrolysis, with lesser proportions of amino acids (ČMELIK [3, 5]). Fractionation shows a mixture of phosphatides to be present, some containing carbohydrate (galactose). The amino acids are considered to be present in a lipid-peptide complex.

d) Lactobacillus acidophilus.

CROWDER and ANDERSON found that this organism contained about 7% of lipids, of which about 32% were phosphatides. The latter contained 1.4% P and 1.2% N and on acid hydrolysis yielded 55% of fatty acids, glycerophosphoric acid, choline, and about 21% of reducing sugar (as glucose). The choline only accounted for a small fraction of the total nitrogen, the rest being unidentified. Ethanolamine was absent. The reducing sugars were a mixture of galactose, glucose and fructose. Progressive alkaline hydrolysis showed that the sugars were present as a phosphorylated non-reducing polysaccharide, apparently an isomer of raffinose. The fatty acids, both saturated and unsaturated, were mainly of the normal series and no indication was found of the presence of the "lactobacillic" acid which HOFMANN, LUCAS and SAX reported to be a major constituent of the total fatty acids of *L. arabinosus* and *L. casei*. The different culture conditions employed may be important (HOFMANN *et al*). Lactobacillic acid is remarkable in possessing a propane ring, being a saturated methylene-octadecanoic acid. The position of the ring is not yet known with certainty, but recent work (MARCO and HOFMANN) suggests it bridges the 11th and 12th carbon atoms.

e) Pseudomonas aeruginosa.

A most unusual type of glycolipid has been isolated in crystalline form from this organism when grown on a particular medium (JARVIS and JOHNSON). It is acidic and on acid hydrolysis yields two moles of L-rhamnose and two moles of an acid tentatively identified as 3-hydroxy-*n*-decanoic acid. HAUSER and KARNOVSKY have studied the effect of various factors on the production of this lipid.

f) Other bacterial phosphatides and glycolipids.

WITTCOFF has summarized information on a variety of bacterial phosphatides additional to those discussed above, while extensive information on bacterial phosphatides and glycolipids is also given by ASSELINEAU, p. 90—108 of this volume. WESTPHAL and LÜDERITZ have reviewed our knowledge of the glycolipids of Gram-negative bacteria. Reference should also be made to the phospholipid-protein complex found in *Micrococcus pyogenes* (MITCHELL and MOYLE [1]) which, it is suggested, might form a continuous lipid membrane immediately beneath the cell wall. The phosphatide of the complex contains 30% of the total lipid P of the organism and on hydrolysis gives, besides glycerophosphoric acid, smaller proportions of some unidentified polyolphosphoric acid. The cell wall itself appears to contain the same two organic phosphoric acids combined with protein, but relatively free from lipid. The phosphorus of the phospholipid, but not that of the phosphoprotein of the cell, exhibits rapid "turnover" during respiration and growth of the cells (MITCHELL and MOYLE [2]).

3. Fungi[1].

a) Yeasts.

Ordinary yeast, *Saccharomyces cerevisiae*, has received more attention from lipid investigators than has any other fungus. The dry organism contains about 1.3% of phosphatides, which are a mixture of about four parts of lecithin and one of cephalin, the former being the classical type (phosphatidyl choline). The cephalin was until very recently considered to be solely phosphatidyl ethanolamine but chromatographic studies (HANAHAN and RHODES) have now shown that it contains three major components. The fatty acids are all of the normal type, being a mixture of palmitic, stearic, hexadecenoic and oleic for both phosphatides (SALISBURY and ANDERSON). It is of interest that an individual, completely unsaturated lecithin, namely dipalmitoleoyl-L-α-glycerophosphorylcholine, has been isolated from yeast (HANAHAN and JAYKO).

[1] See also STEINER, this volume p. 59—89.

Saccharomyces cerevisiae cells contain discrete lipoprotein particles which can be isolated by differential centrifugation of the crushed cells (NYMAN and CHARGAFF). These particles, obtained in a yield of 0.4—0.7% of the fresh yeast, contained 22—26% of lipids, of which about one eighth was phosphatide.

"Food yeast", *Torula utilis*, has been studied by a number of workers (*e.g.* DIRR and RUPPERT, DIEMAIR and POETSCH). The lipid, some 4% of the dry cells, is largely firmly-bound and after extraction is mixed with lipoprotein material and with degradation products. From 50—80% of the total lipid is phosphatide, containing both lecithin and cephalin. Of the total nitrogen some 77% can be found as choline and ethanolamine, the remaining 23% being still unidentified. The fatty acids, both saturated and unsaturated, are of normal type. A purified lecithin containing 16% of choline was, however, deficient in fatty acids (55% instead of about 70%) and, to a slight extent, in glycerol (DIEMAIR and KOCH).

b) Other fungi.

Oidium lactis has a lipid and phosphatide pattern very similar to that of *Torula utilis* (DIEMAIR and POETSCH). The pathogenic fungi *Blastomyces dermatitidis* and *Monilia albicans* are very rich in lipid, much of which is bound and requires drastic extraction procedures (PECK and HAUSER [1, 2, 3]). The total lipid content is about 14% of the dry weight. The readily-extractable lipids of *B. dermatitidis* (8—10% of the fungus) are about one-third phosphatide, which after hydrolysis yields 65% of fatty acids of normal type, glycerophosphate, choline and ethanolamine, thus presumably consisting largely of typical lecithins and cephalins. There is, however, a little carbohydrate present. The corresponding lipid extract of *M. albicans* (5.3% of the fungus) contains only 3% of acetone-insoluble phosphatide which, so far as identified, appears very similar to that of *B. dermatitidis*. In neither case was the carbohydrate identified. Among the lipids extractable from both fungi with boiling ethanol are a phosphatide with a 2:1 N/P ratio and a glycolipid containing only traces of phosphorus but 7% of nitrogen and yielding fatty acids and unidentified polysaccharides on hydrolysis.

Aspergillus sydowi contains 7—8% of total lipids (dry weight) but only about 0.5% of phosphatide (WOOLLEY, STRONG, PETERSON and PRILL). This phosphatide is a mixture of lecithins and cephalins, yielding glycerophosphoric acid, choline, ethanolamine and normal fatty acids on hydrolysis.

A most interesting study (IMAI) with *Penicillium chrysogenum* grown at temperatures ranging from 4° to 26° C revealed that at higher temperatures the mould is richer in both total lipid and phosphatide, but that the fatty acids of the phosphatides are more unsaturated at lower temperatures, this, of course, being in line with many observations on the fatty acids of seed fats. The cephalin fraction of the phosphatides probably contains inositol phosphatides (*cf.* seed phosphatides), since inositol, inositol phosphate, glycerol and glycerophosphoric acid were isolated after hydrolysis.

Further details on mould phosphatides are given by WITTCOFF.

III. Seeds (and similar tissues).

1. Phosphatides.

Determination of the phosphatide content of seeds is particularly difficult owing to the large proportions of other phosphorus- and nitrogen-containing substandes present in the crude lipid extracts and to the difficulty of securing complete extraction of the phosphatides, which occur largely as complexes with carbohydrates and proteins (*cf.* WITTCOFF). Some of these complexes may even involve the linkage of a phosphatide molecule to both carbohydrate and protein, one or the other linkage being split according to the solvent used (ANTENER and HÖGL [1]). Thus boiling ethanol will liberate a phosphatide-sucrose complex from wheat germ, containing 49% phosphatides (mainly lecithin, with some cephalin), 7.7% magnesium phosphatidate and 43.4% sucrose. In spite of its high sugar content, the complex is readily soluble in ether (ANTENER and HÖGL [2]). Similarly phosphatide-protein complexes, soluble in light petroleum, can be extracted from wheat flour (BALLS, HALE and HARRIS). The method of treatment of the material before extraction may markedly alter the degree of lipid binding, *e.g.* complex formation may be encouraged (OLCOTT and MECHAM, MECHAM and WEINSTEIN). Hence it is not surprising to find gross discrepancies in the reported phosphatide content of various seeds. A similar lack of agreement exists among the results of numerous studies of the variation in phosphatide content of seeds during ripening and germination (MACLEAN and MACLEAN, WITTCOFF). In general it appears that seeds may contain from 0.1% to as much as nearly 4%, depending on species and portion of seed examined. The embryo is particularly rich. Thus the whole wheat berry contains about 0.5% phosphatides and wheat germ about 1.5%. The gluten fraction of wheat flour contains about 10% phosphatides. Soyabean, cottonseed and groundnut are particularly rich in phosphatides (average about 2%).

These phosphatide values have usually been determined by applying a conventional conversion factor of about 25 to a lipid phosphorus value, the assumption being that the seed phosphatides closely resemble animal phosphatides. Actually, as will be seen from the examples quoted later, this assumption is unjustified. In particular the "cephalin" fraction of seed phosphatides almost invariably contains large proportions of carbohydrate material, some of which may well be an integral component of a lipid molecule.

Seed phosphatides contain the units found in animal phosphatides and choline, ethanolamine and glycerophosphoric acid have long been known as constituents of hydrolysates. Serine has recently been added to the list; in groundnut cephalin (HUTT, MALKIN, POOLE and WATT) and wheat and rye germ phosphatides (SCHULTE and KRAUSE). It is established that true lecithin, *i.e.* phosphatidyl choline, occurs in seed phosphatides, *e.g.* in soyabeans (THORNTON, JOHNSON and EWAN). The "cephalin" fraction of seed phosphatides is usually a complex mixture which, although it yields ethanolamine and glycerophosphoric acid on hydrolysis, cannot thereby be proved to contain the classical cephalin, phosphatidyl ethanolamine. Thus the ethanolamine of WOOLLEY's "lipositol" from soyabeans and of MALKIN and POOLE's inositol lipid from groundnuts (see below) cannot be attributed to such a simple phosphatide. Nevertheless, much of the ethanolamine-containing lipid can be separated from the inositol-containing lipid (*e.g.* MALKIN and POOLE; SCHOLFIELD and DUTTON [1]) and recently phosphatidyl ethanolamine has been isolated in satisfactory purity from soyabean phosphatides (SCHOLFIELD and DUTTON [2]).

Inositol-containing phosphatides are of widespread, probably universal, occurrence in seeds. The first report of a definite seed lipid of this type was by WOOLLEY, but his "lipositol" was probably a mixture since subsequent work has proved that soyabeans contain more than one type of inositol phosphatide. These inositol lipids are probably more complex than, for instance, the "diphospho-inositide" of brain lipids, since they are usually found to contain carbohydrate components, *e.g.* lipositol contained galactose. The inositol lipids of the soyabean can be separated by counter-current distribution into a more hexane-soluble and a more methanol-soluble lipid, both probably still far from pure.

The greater part of the "bound" sugar (as distinct from the "free" carbohydrate found in the crude lipid extracts) accompanies the more hexane-soluble inositol phosphatide, which on hydrolysis yields galactose, mannose and arabinose (SCHOLFIELD, DUTTON and DIMLER).

These sugars are probably linked chemically into a complex phosphatide, *e.g.* HAWTHORNE and CHARGAFF obtained evidence of a galactoside and an arabinoside of inositol monophosphate as components of soyabean inositol phosphatides. NOMURA found glycerophosphate as an additional component. A nitrogen- and sugar-free inositol lipid has been found in the more methanol-soluble fraction of soyabean "cephalin" (SCHOLFIELD and DUTTON). This may, perhaps, be the simple phosphatidyl inositol which OKUHARA and NAKAYAMA claim to have isolated from soyabeans by an entirely different fractionation technique. From groundnut phosphatides MALKIN and POOLE isolated a purified inositol phosphatide which is possibly an N-glycosyl derivative of the ethanolamine ester of phosphatidyl inositol phosphate. The carbohydrate components are two molecules of L-arabinose and one of D-galactose per molecule of phosphatide, and a disaccharide of arabinose and galactose is obtained on mild hydrolysis. A similar compound containing serine instead of ethanolamine is probably also present (HUTT, MALKIN, POOLE and WATT).

The counter-current separation of inositol phosphatides has been mentioned. Two main groups of such compounds have been sharply separated whenever the procedure has been applied, *e.g.* to soyabean (SCHOLFIELD, DUTTON, TANNER and COWAN), maize (SCHOLFIELD, MCGUIRE and DUTTON) and linseed (MCGUIRE and EARLE). There is still much to learn about the complexity of these lipid fractions, *e.g.* CARTER, CELMER, LANDS, MUELLER and TOMIZAWA found that "phytosphingosine" (see introduction) is a component of both types of inositol lipid preparations from soyabean and maize. A simple phosphatidyl inositol was isolated from wheat germ by FAURE and MORELEC-COULON, although they mention the possibility of it being a product of enzymic degradation of a more complex lipid.

Other types of phosphatide have also been reported to occur in seeds, *e.g.* "sitolipin" (UROMA and LOUHIVUORI), an acidic, nitrogen-free phosphatide obtained from wheat germ and apparently very similar to cardiolipin (PANGBORN), even to having similar syphilitic-serological properties (see HARA's finding with the lipids of *C. diphtheriae*). Phosphatidic acids, which are probably artifacts of autolytic origin (ACKER and ERNST), have earlier been reported as important constituents of wheat germ lipids, *e.g.* CHANNON and FOSTER found at least 42% of the total phosphatide to be in that form. They have also been reported in soyabeans (NIELSEN). An acidic phosphoamino lipid (N/P = 2), obtained from soyabean phosphatides, is considered by VAN HANDEL to be a possible vegetable counterpart of sphingomyelin.

2. Fatty acids of seed phosphatides.

In view of the multiple nature of total phosphatide preparations from seeds, it is unfortunate that detailed fatty acid analyses (*e.g.* by fractional distillation of the methyl esters) are, in general, only available for total or partially fractionated, *e.g.* total "cephalin" (alcohol-insoluble) preparations. An exception is the analysis by THORNTON, JOHNSON and EWAN of the fatty acids of an almost pure soyabean lecithin. They found palmitic 16, stearic 6, oleic 13, linoleic 63 and linolenic acids 2%. This is in agreement with LEVENE and ROLF's qualitative results on a similar lecithin. HILDITCH and his school have examined the phosphatides of soyabean, rapeseed, cottonseed, sunflower seed, linseed and groundnut and the results may, perhaps, be summed up by saying that there is a broad general resemblance between the fatty acids of the seed glycerides and the seed phosphatides. When a seed glyceride is notable for the presence of an unusual fatty acid, or for appreciable amounts of an acid normally found in minor amounts, the phosphatide fatty acids show a similar trend, *e.g.* erucic acid in rapeseed, higher (above C_{18}) saturated acids in groundnut, linolenic acid in linseed. However, the total saturated acid percentage is usually higher in the phosphatides than the glycerides. Palmitic, oleic and linoleic acids are, in general, the most plentiful constituents of the phosphatide acids. The fatty acids of seed phosphatides are sharply distinguished from those of animal glycerophosphatides by the virtual absence from the former of highly-unsaturated derivatives of 20 and 22 carbon atoms (KLENK and DEBUCH).

3. Alleged glycolipids.

In conclusion of this section brief reference may be made to a type of lipid once considered (*e.g.* TAYLOR and SHERMAN) to exist in cereal grains, namely esters of fatty acids and α-amylose (β-amylose is lipid-free). SCHOCH's finding that the fatty acid could be liberated by such solvents as methanol, and that the lipid-free starch could recombine with fatty acids, disproved the ester concept and led to the view that the linkage was of an adsorption type. The work of MIKUS, HIXON and RUNDLE, however, on the X-ray and optical properties of the complex suggests that it is an inclusion compound (analogous to the urea inclusion compounds of fatty acids), in which the fatty acid is enclosed within a helical chain configuration of the amylose. Amylose having such a helical configuration will bind fatty acid, amylose with an extended chain configuration will not.

IV. Vegetative tissues.

A review of the phosphatide content of leaves, roots, flowers, etc., is given by WITTCOFF. There is the difficulty of extractability and association with carbohydrate common to all studies of plant phosphatides. There is also evidence of complex formation between phosphatides, proteins and chlorophyll in the leaf chloroplasts (BOT, COMAR). This section will be restricted to those tissues the phosphatides of which have been examined in some detail.

1. Grass.

According to REWALD, dried grass is quite rich in total lipids (over 7%) and contains at least 0.5% of phosphatides. SMITH and CHIBNALL had earlier examined the phosphatides of cocksfoot grass, *Dactylis glomerata*, using a procedure which REWALD showed would give incomplete extraction from the grass. They found the product (about 0.13% of the dry weight) to be a mixture of lecithin, cephalin (N/P = 1) and salts of phosphatidic acid, although since considerable amounts of decomposition products were found they did not discuss the quantitative aspects of their study. The fatty acids of the cocksfoot phosphatides were remarkable for the apparent absence of oleic acid, the principal fatty acids being apparently linoleic (with some linolenic), palmitic and stearic. SHORLAND, in agreement with SMITH and CHIBNALL, found from 0.1–0.3% of phosphatides in air-dried or alcohol-preserved cocksfoot, and also in rye grass *(Lolium perenne)*, but obtained 1.5–1.7% from freshly-cut grass. The increased proportion of free fatty acids in the extract from dried grass indicates extensive autolysis during drying or preservation. The crude phosphatide contained only about 2% P, but a product was obtained from it, in about 15% yield, having 3.4% P and 1.4% N, *i.e.* a relatively pure lecithin-cephalin fraction. Phosphatidic acid (or its salts) was also present, but may well have been an artifact (see below). The fatty acids of cocksfoot phosphatides were found to contain mainly palmitic (11%), hexadecenoic (6%) and polyunsaturated C_{18} acids (76%). In agreement with SMITH and CHIBNALL, only traces of oleic acid could be detected in the C_{18} fraction.

2. Cabbage.

CHANNON and CHIBNALL, using an extraction procedure of doubtful efficiency, obtained about 0.03% of crude calcium phosphatidate from fresh cabbage leaves. They found very little lecithin or cephalin and considered this phosphatidate to be the chief phosphatide of cabbage leaf. Hydrolysis yielded glycerophosphoric acid and fatty acids (palmitic, stearic, linoleic and linolenic). This work is almost certainly invalidated by the finding (HANAHAN and CHAIKOFF [3]) of an enzyme in cabbage leaf which rapidly liberates choline from lecithin. These workers obtained about 0.3% of total phosphatide from either raw or steamed cabbage (*i.e.* about 3% on the dry weight). The material from the raw leaves was essentially similar to that obtained by CHANNON and CHIBNALL, but the product from steamed leaves had an N/P ratio of 1.3, a choline/P ratio of 0.4 (4% of choline) and was evidently largely a mixture of lecithin and cephalin.

3. Carrot.

Carrot root phosphatides are subject to the same rapid enzymic attack as those of the cabbage leaf. HANAHAN and CHAIKOFF [1, 2] found that the phosphatide (0.2–0.6% of the dry weight) contained about 3% phosphorus, 14% glycerophosphoric acid and 60% fatty acids in all cases, but from raw carrots contained only 0.5% nitrogen and no choline, against 1.2% nitrogen (N/P = 0.9) and 4.3% choline in the product from steamed carrots. Hence the phosphatidic acid extracted from raw carrots is an artifact and the original phosphatide must have been a mixture of lecithin and cephalin.

4. Rubber latex.

TRISTRAM found about 0.09% of phosphatides in the latex of *Hevea brasiliensis,* of which about one third was lecithin and another third phosphatidic acid. Choline was isolated but ethanolamine was virtually absent. However, the same story of enzymic attack, which makes all findings of phosphatidic acids suspect, has been told again. SMITH [1], using a more thorough extraction technique, found about 0.02% of lipid P (say 0.5% of phosphatide) in the fresh latex. When extraction involved boiling ethanol, the phosphatide had a choline/P ratio of about 0.7 and hence was essentially lecithin, but at extraction temperatures below 40° C choline was progressively liberated by enzyme action and phosphatidic acid produced. This enzyme was even active during storage of the frozen latex. SMITH [2, 3] subsequently showed that the undegraded phosphatides were a mixture of a "lecithin" which contained reducing sugar in roughly equimolecular ratio to phosphorus, metal phosphatidates containing reducing sugar and *meso*-inositol, again in approximately equimolecular amount to phosphorus, and phosphatidyl ethanolamine. By making a number of assumptions it was possible to calculate that there was about 79% "lecithin" (2.8% P, 16% reducing sugar), 16% metal (mainly K with a little Mg) phosphatidate (2.8% P, 16% each of sugar and inositol) and 5% phosphatidyl ethanolamine. Evidence was obtained that the reducing sugar and the inositol were chemically incorporated into the phosphatides, since compounds of them with glycerol were isolated after partial hydrolysis. The reducing sugar was mainly galactose, together with some glucose and an unidentified sugar which was probably a ketose.

V. Algae.

In examining the lipids of a number of marine algae, SIRAHAMA studied in some detail the phosphatide from the blades of *Alaria crassifolia*. He obtained about 0.7% of phosphatide from the fresh alga, having the solubility properties of cephalin. It contained 1.3% N (all amino-N), 2.8% P (N/P = 1) and 50% fatty acids. The latter contained about 50% of saturated components, mainly palmitic with some myristic, the unsaturated acids being of 16 and 18 carbon atoms, the latter polyethylenic. This is a higher proportion of saturated acids than is found in algal lipids as a whole (LOVERN, MILNER), recalling the position with the seed phosphatides.

VI. Metabolism.

It has been mentioned earlier that numerous studies of the changes in phosphatide content of seeds during ripening and germination fail to show any consistent pattern, probably because of inadequate and variable experimental technique. It is well known that during germination the seed triglycerides, which greatly exceed the phosphatides in amount, are converted into sugars. HOUGET, studying the germination of *Ricinus sanguineus* at 28.5° C (about the optimum) found that production of sugars from fatty acids began on the third day and was completed by the fifth day. Phosphatides (lipid P) doubled during the first three days, when total fatty acids and sugars showed virtually no change, and remained at a high level until much of the fatty acid had disappeared. HOUGET considered that phosphatides probably play a role in fat: carbohydrate transformation by presenting the fatty acids in a reactive form. More evidence would be needed that the fatty acids of phosphatides are, in fact, more readily available to enzymes than are those of glycerides before this hypothesis could be accepted. DUCET likewise found a synthesis of lecithin, involving the use of free choline, during germination of the soyabean, with a reverse trend during ripening.

Plant tissues of many types contain enzymes which rapidly attack phosphatides and the rate of phosphatide "turnover" is presumably high during actual growth, although tracer studies, such as have been made with animal tissues, are sadly lacking for plants, the first such study having just been described (MAZELIS and STUMPF). The mitochondria of the groundnut incorporate added phosphate *in vitro*, through the intermediate stage of adenosine triphosphate, into a phosphatide of unknown structure, the other phosphatides, *e.g.* lecithin, remaining unlabelled. An equilibrium level of labelling is attained in about 30 minutes with labelled inorganic phosphate or about 10 minutes with labelled ATP. There is as yet no clue to the stages between ATP and phosphatide, all tests with possible intermediates or precursors, such as glycerol, glycerophosphate, inositol and choline giving negative results.

The best known enzyme is phospholipase (lecithinase) C (or D)[1], which liberates only the organic base of lecithin, and of phosphatidyl ethanolamine, to yield a phosphatidic acid. Reference has already been made to this enzyme, which has been found, for instance, in carrot roots, cabbage leaves, rubber latex, cereal grains and cottonseed (HANAHAN and CHAIKOFF [2, 3], SMITH, ACKER and

[1] The terminology of phospholipases (lecithinases) C and D is confused.

ERNST, TOOKEY and BALLS). In the leaf the enzyme is apparently concentrated entirely in the chloroplasts (KATES [1, 2]). In the carrot root it is also exclusively in the plastid fraction (KATES [2]), and in cereal grains is largely concentrated in the embryo, where its activity increases greatly on germination (ACKER and ERNST).

KATES [1, 2] working with pure synthetic substrates, found a rapid liberation of choline and production of phosphatidic acid, followed by a much slower liberation of inorganic phosphate and water-soluble organic phosphate. The chloroplast enzymes could liberate inorganic phosphate from L-α-glycerophosphoric acid, but could not attack phosphorylcholine or glycerophosphorylcholine. From this work it appears that phosphatidic acid is an inevitable intermediate in enzymic breakdown of lecithin in the leaves studied (cabbage, spinach and sugar beet). The results of HOLDEN, suggesting a different series of reactions in the tobacco leaf, are not conclusive since her "lipid P" fractions almost certainly must have contained other forms of phosphorus and it is difficult to envisage lipid phosphorus becoming water-soluble in advance of lipid nitrogen.

The position with some bacteria appears to be different. HAYAISHI and KORNBERG, also working with pure synthetic substrates, studied enzyme preparations and suspensions of resting cells from *Serratia plymuthicum* and obtained evidence that both lecithin and cephalin are attacked firstly by a lecithinase A, with loss of one fatty acid molecule and production of a lysophosphatide. This enzyme is equally effective on fully-saturated or fully-unsaturated lecithins, and it is the most potent of the phosphatide-splitting enzymes detected. The lysophosphatide is probably next attacked by a lecithinase B, with loss of the remaining fatty acid and production of glycerophosphorylcholine (or glycerophosphorylethanolamine from cephalin). It is possible that this enzyme also attacks, to a subsidiary extent, the intact phosphatide, but the probability is that lysophosphatides are an invariable intermediate. Finally a third enzyme attacks the glycerophosphorylcholine to yield the free base and glycerophosphoric acid. The organism could not split phosphorylcholine, which was hence an unlikely intermediate.

The organism *Clostridium welchii* produces a lecithinase D (or C) which appears to be, in fact, the specific α-toxin of the organism (MACFARLANE and KNIGHT). This splits lecithin into phosphorylcholine and a diglyceride. HANAHAN and VERCAMER have shown that the enzyme (from *Cl. perfringens*) is an effective agent for the laboratory preparation of D-1,2 diglycerides. This same enzymic degradation of lecithin may by produced by the plastid enzymes of higher plants, side by side with liberation of choline and production of phosphatidic acid (KATES [3]).

There is, of course, no evidence that these various degradation routes are merely reversed during phosphatide synthesis in the plant, or indeed that they are representative of the catabolism of normal "turnover". The much more extensive studies with animal tissues suggest caution in making any such assumptions. Nevertheless, the enzyme studies are suggestive, as is the finding of certain intermediates as normal components of plant cells, *e.g.* glycerophosphorylcholine (DUCET) and glycerophosphorylethanolamine (CAMPBELL, SIMMONDS and WORK).

Literature.

ACKER, L., u. G. ERNST: Über das Vorkommen eines phosphatidspaltenden Ferments in Cerealien. Biochem. Z. **325**, 253—257 (1954). — ANDERSON, R. J.: [1] The chemistry of the lipids of the tubercle bacillus and certain other microorganisms. Fortschr. Chem. organ. Naturstoffe **3**, 145—202 (1939). — [2] Structural peculiarities of acid-fast bacterial lipids. Chem. Reviews **29**, 225—243 (1941). — ANTENER, I., u. O. HÖGL: [1] Vergleichende Untersuchungen über die Fettbestimmungsmethoden in Mehlprodukten. Mitt. Geb. Lebensmittelunters. u. Hyg. **38**, 207—225 (1947). — [2] Studien über die Phosphatide in Weizenkeimen. Mitt. Geb. Lebensmittelunters. u. Hyg. **38**, 226—244 (1947). — ASSELINEAU, J.: [1] Lipides du bacille tuberculeux. Constitution chimique et activité biologique. Prog. Explor. Tbc. **5**, 1—55 (1952). — [2] Isolement de nouveaux acides mycoliques des souches humaines Test, R et L-25 de *Mycobacterium tuberculosis*. Présence d'un acide mycolonique dans une souche humaine. Biochim. et Biophysica Acta **10**, 453—461 (1953). — ASSELINEAU, J., and E. LEDERER [1]: Sur la constitution chimique des acides mycoliques de deux souches humaines virulentes de *Mycobacterium tuberculosis*. Biochim. et Biophysica Acta **7**, 126—145 (1951). — [2] Sur la structure chimique de l'acide α-mycolique de la souche humaine Test de *Mycobacterium tuberculosis*. Bull. Soc. chim. France **1953**, 335—339.

BALLS, A. K., W. S. HALE and T. H. HARRIS: A crystalline protein obtained from a lipoprotein of wheat flour. Cereal Chem. **19**, 279—288 (1942). — BARBIER, M., and E. LEDERER: [1] Sur un acide aminé du phosphatide de *Mycobacterium phlei*. Biochim. et Biophysica Acta **8**, 590—591 (1952). — [2] Sur l'isolement et la constitution chimique des acides mycoliques de *Mycobacterium phlei* et de *Mycobacterium smegmatis*. Biochim. et Biophysica Acta **14**, 246—258 (1954). — BOT, G. M.: The chemical composition of chloroplast granules (grana) in relation to their structure. Chron. Bot. **7**, 66—67 (1942).

CAMPBELL, P. N., D. H. SIMMONDS, and T. S. WORK: The occurrence of glycerylphosphorylethanolamine in extracts of liver and yeast. Biochemic. J. **49**, XVI. Proc. (1951). — CARTER, H. E., W. D. CELMER, W. E. M. LANDS, K. L. MUELLER, and H. H. TOMIZAWA: Biochemistry of the sphingolipides. VIII. Occurrence of a long chain base in plant phosphatides. J. of Biol. Chem. **206**, 613—623 (1954). — CASON, J., N. K. FREEMAN, and G. SUMRELL: The principal structural features of C_{27}-phthienoic acid. J. of Biol. Chem. **192**, 415—424 (1951). — CASON, J., and G. SUMRELL: Investigation of a fraction of acids of the phthioic type from the tubercle bacillus. J. of Biol. Chem. **192**, 405—413 (1951). — CELMER, W. D., and H. E. CARTER: Chemistry of phosphatides and cerebrosides. Physiologic. Rev. **32**, 167—196 (1952). — CHANNON, H. J., and A. C. CHIBNALL: The phosphatides of forage grasses. I. Cocksfoot. Biochemic. J. **26**, 1345—1357 (1932). — CHANNON, H. J., and C. A. M. FOSTER: The phosphatides of wheat germ. Biochemic. J. **28**, 853—864 (1934). — CHARGAFF, E.: Über das Fett und das Phosphatid der Diphtheriebakterien. Hoppe-Seylers Z. **218**, 223—240 (1933). — ČMELIK, S.: [1] Über Bakterienlipoide. 2. Die Lipoide von *Salmonella ballerup (Paracolobacterium ballerup)*. Hoppe-Seylers Z. **293**, 222—229 (1953). — [2] Über Bakterienlipoide. 3. Untersuchung verschiedener Lipoidfraktionen von *Salmonella paratyphi* C. Hoppe-Seylers Z. **296**, 67—73 (1954). — [3] Zur Kenntnis der Lipoide von Typhusbakterien. Hoppe-Seylers. Z. **299**, 227—234 (1955). — [4] Die komplex gebundenen Lipoide von Diphtheriebakterien aus verschiedenen Nährlösungen. Hoppe-Seylers Z. **300**, 167—173 (1955). — [5] Über Phosphatide von Typhusbakterien. Hoppe-Seylers Z. **302**, 20—28 (1955). — COMAR, C. L.: Chloroplast substance of spinach leaves. Bot. Gaz. **104**, 122—127 (1942). — CREIGHTON, M. M., L. H. CHANG, and R. J. ANDERSON: The chemistry of the lipides of tubercle bacilli. LXVII. The lipides of the human tubercle bacillus H-37 cultivated on a dextrose-containing medium. J. of Biol. Chem. **154**, 569—579 (1944). — CROWDER, J. A., and R. J. ANDERSON: A contribution to the chemistry of *Lactobacillus acidophilus*. III. The composition of the phosphatide fraction. J. of Biol. Chem. **104**, 487—495 (1934).

DEMARTEAU, H., et E. LEDERER: Sur un acide mycolonique (nouveau type d'acide mycolique) isolé d'une souche bovine de *Mycobacterium tuberculosis*. C. r. Acad. Sci. Paris **235**, 265—267 (1952). — DE SÜTÖ-NAGY, G. I., and R. J. ANDERSON: [1] The chemistry of the lipides of tubercle bacilli. LXXV. Further studies on the polysaccharides of the phosphatides obtained from cell residues in the preparation of tuberculin. J. of Biol. Chem. **171**, 749—760 (1947). — [2] The chemistry of the lipides of tubercle bacilli. LXXVI. Concerning inositol glycerol diphosphoric acid, a component of the phosphatide of human tubercle bacilli. J. of Biol. Chem. **171**, 761—765 (1947). — DIEMAIR, W., u. J. KOCH: Beitrag zur Kenntnis der Hefephosphatide I. Angew. Chem. A **60**, 155—157 (1948). — DIEMAIR, W., u. W. POETSCH: Phosphatide als Fettbegleitstoffe der Nährhefen. Biochem. Z. **319**, 571—591 (1949). — DIRR, K., u. A. RUPPERT: Über den Wert der Wuchshefen für die menschliche Ernährung. VI. Die Phosphatide der Wuchshefen. Biochem. Z. **319**, 163—173 (1948). — DUCET, G.: La choline des vegetaux. Intern. Congr. Biochem. Abstr. of Communications. lst Congr. Cambridge, Engl., p. 495—496, 1949.

FAURE, M., et M. J. MORELEC-COULON: Isolement d'un acide glyceroinosito-phosphatidique contenu dans le germe de blé. C. r. Acad. Sci. Paris **236**, 1104—1106 (1953).

GEIGER jr., W. B., and R. J. ANDERSON: The chemistry of *Phytomonas tumefaciens*. I. The lipids of *Phytomonas tumefaciens*. The composition of the phosphatide. J. of Biol. Chem. **129**, 519—529 (1939). — GENDRE, T., et E. LEDERER: Sur les substances azotées des phosphatides de quelques mycobactéries. Ann. Acad. Sci. fenn., Ser. A, II **60**, 313—320 (1955). —

GUBAREV, E. M., i E. K. LUBENETS: Some structural peculiarities of the lipids of the diphtheria microbe. Dokl. Akad. Nauk SSSR. **60**, 413—416 (1948). — GUBAREV, E. M., E. K. LUBENETS i Y. V. GALAEV: Chemical composition of some fractions of lipids of diphtheria bacteria. Biokhimija **18**, 37—46 (1953). — GUBAREV, E. M., E. K. LUBENETS, A. A. KANCHUKH i Y. V. GALAEV: Fractionation and composition of some lipid fractions of diphtheria bacteria. Biokhimija **16**, 139—145 (1951).

HANAHAN, D. J., and I. L. CHAIKOFF: [1] The phosphorus-containing lipides of the carrot. J. of Biol. Chem. **168**, 233—240 (1947). — [2] A new phospholipide-splitting enzyme specific for the ester linkage between the nitrogenous base and the phosphoric acid grouping. J. of Biol. Chem. **169**, 699—705 (1947). — [3] On the nature of the phosphorus-containing lipides of cabbage leaves and their relation to a phospholipide-splitting enzyme contained in these leaves. J. of Biol. Chem. **172**, 191—198 (1948). — HANAHAN, D. J., and M. E. JAYKO: The isolation of dipalmitoleoyl-L-α-glycerophosphorylcholine from yeast. A new route to (dipalmitoyl) L-α-lecithin. J. Amer. Chem. Soc. **74**, 5070—5073 (1952). — HANAHAN, D. J., and D. N. RHODES: Nature of yeast cephalins. Federat. Proc. **15**, 267 (1956). — HANAHAN, D. J., and R. VERCAMER: The action of lecithinase D on lecithin. The enzymatic preparation of D-1, 2-dipalmitolein and D-1, 2-dipalmitin. J. Amer. Chem. Soc. **76**, 1804—1806 (1954). — HANDEL, E. VAN: The Chemistry of Phosphoaminolipids. Amsterdam: D. B. Centen 1954. — HARA, I.: Chemical and biochemical studies on cardiolipin. 1. Isolation of cardiolipin from bacteria and other sources. J. Chem. Soc. Jap., Pure Chem. Sect. **76**, 910—914 (1955). — HAUSER, G., and M. L. KARNOVSKY: Studies on the production of glycolipide by *Pseudomonas aeruginosa*. J. Bacter. **68**, 645—654 (1954). — HAWTHORNE, J. N., and E. CHARGAFF: A study of inositol-containing lipides. J. of Biol. Chem. **206**, 27—37 (1954). — HAYAISHI, O., and A. KORNBERG: Metabolism of phospholipides by bacterial enzymes. J. of Biol. Chem. **206**, 647—663 (1954). — HILDITCH, T. P.: The chemical constitution of natural fats. 2nd. edn., p. 205—209. London: Chapman & Hall 1947. — HOFMANN, K., R. A. LUCAS and S. M. SAX: The chemical nature of the fatty acids of *Lactobacillus arabinosus*. J. of Biol. Chem. **195**, 473—485 (1952). — HOFMANN, K., F. and TAUSIG: On the identity of phytomonic and lactobacillic acids. A reinvestigation of the fatty acid spectrum of *Agrobacterium (Phytomonas) tumefaciens*. J. of Biol. Chem. **213**, 425—432 (1955). — HOLDEN, M.: The fractionation and enzymic breakdown of some phosphorus compounds in leaf tissue. Biochemic. J. **51**, 433—442 (1952). — HOUGET, J.: Sur le mécanisme de la transformation des lipides en glucides au cours de la germination du Ricin. C. r. Acad. Sci. Paris **216**, 821—822 (1943). — HUTT, H. H., T. MALKIN, A. G. POOLE, and P. R. WATT: Structure of kephalin: presence of serine and a disaccharide in groundnut kephalin. Nature (Lond.) **165**, 314—315 (1950).

IMAI, Y.: The lipides of microbes. 1. The effect of temperature on the lipide formation by fungi. J. Jap. Biochem. Soc. **22**, 192—196 (1950.)

JARVIS, F. G., and M. J. JOHNSON: A glycolipide produced by *Pseudomonas aeruginosa*. J. Amer. Chem. Soc. **71**, 4124—4126 (1949).

KATES, M.: [1] Lecithinase activity of chloroplasts. Nature (Lond.) **172**, 814—815 (1953). — [2] Lecithinase systems in sugar beet, spinach, cabbage and carrot. Canad. J. Biochem. a. Physiol. **32**, 571—583 (1954). — [3] Hydrolysis of lecithin by plant plastid enzymes. Canad. J. Biochem. a. Physiol. **33**, 575—589 (1955). — KLENK, E., und H. DEBUCH: Zur Frage des Vorkommens der hochungesättigten Fettsäuren C_{20} und C_{22} in den Pflanzenphosphatiden. Hoppe-Seylers Z. **286**, 33—37 (1950).

LEVENE, P. A., and I. P. ROLF: Plant phosphatides. I. Lecithin and cephalin of the soyabean. J. of Biol. Chem. **62**, 759—766 (1925). — LOVERN, J. A.: [1] Fat metabolism in fishes. IX. The fats of some aquatic plants. Biochemic. J. **30**, 387—390 (1936). — [2] Occurrence of plasmalogens in vegetable phosphatides. Nature (Lond.) **169**, 969 (1952).

MACFARLANE, M. G. and B. C. J. G. KNIGHT: The biochemistry of bacterial toxins. I. The lecithinase activity of *Cl. welchii* toxins. Biochemic. J. **35**, 884—902 (1941). — MACHEBOEUF, M., and M. FAURE: Sur l'existence, dans les bacilles tuberculeux, d'acides phosphatidiques complexes constitués par l'acide glycérophosphorique lié par esterification, d'une part à des acides gras, et d'autre part à des polyalcools non azotés. C. r. Acad. Sci. Paris **209**, 700—702 (1939). — MACLEAN, H., and I. S. MACLEAN: Lecithin and allied substances. London: Longmans, Green & Co. **1927**. — MALKIN, T., and A. G. POOLE: The structure of the glycerinositophosphatide of groundnut. J. Chem. Soc. **1953**, 3470—3478. — MARCO, G. J., and K. HOFMANN: Structural studies on lactobacillic acid and other long chain fatty acids containing the cyclopropane ring. Federat. Proc. **15**, 308 (1956). — MAZELIS, M., and P. K. STUMPF: Fat metabolism in higher plants. VI. Incorporation of P^{32} into peanut mitochondrial phospholipids. Plant Physiol. **30**, 237—243 (1955). — MCGUIRE, T. A., and F. R. EARLE: Fractionation of linseed phosphatides. J. Amer. Oil Chem. Soc. **28**, 328—331 (1951). —

MECHAM, D. K., and N. E. WEINSTEIN: Lipide binding in doughs. Effects of dough ingredients. Cereal Chem. **29**, 448—455 (1952). — MICHEL, G., et E. LEDERER: Étude chromatographique du phosphatide de *Mycobacterium marianum*. C. r. Acad. Sci. Paris **240**, 2454—2455 (1955). — MIKUS, F. F., R. M. HIXON, and R. E. RUNDLE: The complexes of fatty acids with amylose. J. Amer. Chem. Soc. **68**, 1115—1123 (1946). — MILNER, H. W.: The fatty acids of chlorella. J. of Biol. Chem. **176**, 813—817 (1948). — MITCHELL, P., and J. MOYLE: [1] The glycerophospho-protein complex envelope of *Micrococcus pyogenes*. J. Gen. Microbiol. **5**, 981—982 (1951). — [2] Paths of phosphate transfer in *Micrococcus pyogenes:* Phosphate turnover in nucleic acids and other fractions. J. Gen. Microbiol. **9**, 257—272 (1953).

NIELSEN, K.: Composition of the nonhydratable soyabean phosphatides. Acta chem. scand. (Copenh.) **9**, 173—174 (1955). — NOLL, H., and H. BLOCH: The chemistry of the cord factor of *Mycobacterium tuberculosis*. J. of Biol. Chem. **214**, 251—265 (1955). — NOMURA, D.: Inositolmonophosphoric acid in soybean phosphatide. J. Jap. Chem. **3**, 145—147 (1949). — NYMAN, M. A., and E. CHARGAFF: On the lipoprotein particles of yeast cells. J. of Biol. Chem. **180**, 741—746 (1949).

OKUHARA, E., and T. NAKAYAMA: Studies on the conjugated lipides. VI. On the structure of inositol phospholipide. J. of Biol. Chem. **215**, 295—302 (1955). — OLCOTT, H. S., and D. K. MECHAM: Characterization of wheat gluten. I. Protein-lipide complex formation during doughing of flours. Lipoprotein nature of the glutenin fraction. Cereal Chem. **24**, 407—414 (1947).

PANGBORN, M. C.: The composition of cardiolipin. J. of Biol. Chem. **168**, 351—361 (1947). — PECK, R. L., and C. R. HAUSER: [1] Chemical studies of certain pathogenic fungi. I. The lipids of *Blastomyces dermatiditis*. J. Amer. Chem. Soc. **60**, 2599—2603 (1938). — [2] Chemical studies of certain pathogenic fungi. II. The lipids of *Monilia albicans*. J. Amer. Chem. Soc. **61**, 281—284 (1939). — [3] Chemical studies of certain pathogenic fungi. III. Further studies of *Blastomyces dermatiditis* and *Monilia albicans*. J. of Biol. Chem. **134**, 403—412 (1940). — POLGAR, N.: [1] Component acids of the lipids of human tubercle bacilli. Part I. Biochemic. J. **42**, 206—211 (1948). — [2] Constituents of the lipids of tubercle bacilli. III. Mycolipenic acid. J. Chem. Soc. (Lond.) **1954**, 1008—1010. — [3] Constitutents of the lipids of tubercle bacilli. IV. Mycoceranic acid. J. Chem. Soc. (Lond.) **1954**, 1011—1012. — PUDLES, J., Étude chimique des lipides du bacille diphtérique. Thesis, Faculty of Sciences, University of Paris, for the degree of Docteur de l'Université 1953.

REWALD, B.: The fats and phosphatides in grass. Oil a. Soap **21**, 50—52 (1944).

SALISBURY, L. F., and R. J. ANDERSON: The chemistry of the lipids of yeast. III. Lecithin and cephalin. J. of Biol. Chem. **112**, 541—550 (1936). — SCHOCH, T.: Absence of combined fatty acid in cereal starches. J. Amer. Chem. Soc. **60**, 2824—2828 (1938). — SCHOLFIELD, C. R., and H. J. DUTTON: [1] Separations of soybean inositide fractions of low partition coefficient. J. of Biol. Chem. **208**, 461—469 (1954). — [2] Preparation of phosphatidyl ethanolamine from soybean phosphatides. J. of Biol. Chem. **214**, 633—638 (1955). — SCHOLFIELD, C. R., H. J. DUTTON, and R. J. DIMLER: Carbohydrate constituents of soybean "lecithin". J. Amer. Oil Chem. Soc. **29**, 293—298 (1952). — SCHOLFIELD, C. R., H. J. DUTTON, F. W. TANNER jr., and J. C. COWAN: Components of "soybean lecithin". J. Amer. Oil Chem. Soc. **25**, 368—372 (1948). — SCHOLFIELD, C. R., T. A. MCGUIRE, and H. J. DUTTON: Comparative composition of soybean and corn phosphatides. J. Amer. Oil Chem. Soc. **27**, 352—355 (1950). — SCHULTE, K. E., and H. KRAUSE: Beiträge zur Kenntnis der Phosphatide. I. Die Stickstoffkomponenten der Phosphatide des Weizen- und Roggenkeimlings. Biochem. Z. **322**, 168—173 (1951). — SHORLAND, F. B.: Leaf lipids of forage grasses and clovers. Nature (Lond.) **153**, 168 (1944). — SIRAHAMA, K.: Unsaponifiable matter and phosphatides in marine algae fat. J. Faculty Agricult. Hokkaido Univ. **49**, 1—93 (1942). — SMITH, R. H.: [1] The phosphatides of the latex of *Hevea brasiliensis*. I. Biochemical changes. Biochemic. J. **56**, 240—250 (1954). — [2] The phosphatides of the latex of *Hevea brasiliensis*. II. Purification and analysis. Biochemic. J. **57**, 130—139 (1954). — [3] The phosphatides of the latex of *Hevea brasiliensis*. III. Carbohydrate and polyhydroxy constituents. Biochemic. J. **57**, 140—144 (1954). — SMITH, J. A. B., and A. C. CHIBNALL: The phosphatides of forage grasses. I. Cocksfoot. Biochemic. J. **26**, 1345—1357 (1932). — STÄLLBERG-STENHAGEN, S., and E. STENHAGEN: On the nature of two carboxylic acids of high molecular weight obtained from the waxes of acid fast bacteria. J. of. Biol. Chem. **165**, 599—604 (1946).

TAKAHASHI, H.: Bacterial components of *Corynebacterium diphtheriae*. V. Phospholipids. 2. Structure of Chargaff's corynin. J. Pharmaceut. Soc. Japan **68**, 292—296 (1948). — TAYLOR, T. C., and R. T. SHERMAN: Carbohydrate-fatty acid linkings in corn alpha amylose. J. Amer. Chem. Soc. **55**, 258—264 (1933). — THIELE, O. W.: Die Acetalphosphatide des Blutserums bei alimentärer Lipämie. Z. exper. Med. **121**, 246—253 (1953). — THORNTON, M. H., C. S. JOHNSON and M. A. EWAN: The component fatty acids of soybean lecithin.

Oil a. Soap **21**, 85—87 (1944). — TOOKEY, H. I., and A. K. BALLS: Plant phospholipase D. 1. Studies on cottonseed and cabbage phospholipase D. J. of Biol. Chem. **218**, 213—214 (1956). — TRISTRAM, G. R.: The phosphatides of *Hevea brasiliensis*. Biochem. J. **36**, 400—405 (1942).

UROMA, E., and A. LOUHIVUORI: Preparation of sitolipin. Ann. med. exper. et biol. fenn. **29**, 227—232 (1951).

VELICK, S. F.: The chemistry of *Phytomonas tumefaciens*. IV. Concerning the structure of phytomonic acid. J. of Biol. Chem. **156**, 101—107 (1944). — VILKAS, E., et E. LEDERER: Purification chromatographique de phosphatides de Mycobactéries. C. r. Acad. Sci. Paris **240**, 1156—1157 (1955).

WESTPHAL, O., u. O. LÜDERITZ: Chemische Erforschung von Lipopolysacchariden gram-negativer Bakterien. Angew. Chem. **66**, 407—417 (1954). — WITTCOFF, H.: The phosphatides. New York: Reinhold Publ. Corpn. 1951. — WOOLLEY, D. W.: Isolation and partial determination of structure of soybean lipositol, a new inositol-containing phospholipide. J. of Biol. Chem. **147**, 581—591 (1943). — WOOLLEY, D. W., F. M. STRONG, W. H. PETERSON and E. A. PRILL: The chemistry of mold tissue. X. The phospholipides of *Aspergillus sydowi*. J. Amer. Chem. Soc. **57**, 2589—2591 (1935).

Namenverzeichnis. — Author Index.

Die *kursiv* gesetzten Seitenzahlen beziehen sich auf die Literatur.

Page numbers in *italics* refer to the bibliography.

Sachverzeichnis.

(Deutsch-Englisch.)

Bei gleicher Schreibweise in beiden Sprachen sind die Stichworte jeweils einfach aufgeführt
Ä, Ö, Ü sind wie Ae, Oe, Ue eingereiht.
cis-, trans-, n-, D-, L-, α-, β- und ähnliche Isomere sind unter dem Anfangsbuchstaben der Verbindungen und nicht unter dem Präfix eingereiht. Alle iso-Verbindungen finden sich unter Iso-.
Salze organischer Säuren sind unter dem Namen der Säure aufgeführt, also Formiat unter Ameisensäure, Acetat unter Essigsäure usw.
Grundsätzlich wird den wissenschaftlichen (lateinischen) Pflanzennamen der Vorzug gegeben.
Bei häufig gebrachten Trivialnamen wird auf die wissenschaftlichen Namen verwiesen.

Subject Index.

(English-German.)

Ä, Ö, Ü are taken as Ae, Oe, Ue.
Where English and German spelling of a word is identical, the italiced (German) entry is omitted.
cis-, trans-, n-, D-, L-, α-, β- and similar isomers are listed according to the first letter of the following word. All iso-compounds are to be found under Iso-.
Salts of organic acids are listed under the name of the acids, *e.g.* formate under formic acid, acetate under acetic acid a. s. o.
As a rule, preference is given to the scientific (latin) names of plants. Trivial plant names commonly used are referred to the scientific names.